Springer Proceedings in Business and Economics

D1806737

Springer Proceedings in Business and Economics brings the most current research presented at conferences and workshops to a global readership. The series features volumes (in electronic and print formats) of selected contributions from conferences in all areas of economics, business, management, and finance. In addition to an overall evaluation by the publisher of the topical interest, scientific quality, and timeliness of each volume, each contribution is refereed to standards comparable to those of leading journals, resulting in authoritative contributions to the respective fields. Springer's production and distribution infrastructure ensures rapid publication and wide circulation of the latest developments in the most compelling and promising areas of research today.

The editorial development of volumes may be managed using Springer's innovative Online Conference Service (OCS), a proven online manuscript management and review system. This system is designed to ensure an efficient timeline for your publication, making Springer Proceedings in Business and Economics the premier series to publish your workshop or conference volume.

More information about this series at http://www.springer.com/series/11960

Vicky Katsoni · Ciná van Zyl
Editors

Culture and Tourism in a Smart, Globalized, and Sustainable World

7th International Conference of IACuDiT,
Hydra, Greece, 2020

 Springer

Editors
Vicky Katsoni 🆔
University of West Attica
Athens, Greece

Ciná van Zyl 🆔
University of South Africa
Pretoria, South Africa

ISSN 2198-7246   ISSN 2198-7254 (electronic)
Springer Proceedings in Business and Economics
ISBN 978-3-030-72471-9   ISBN 978-3-030-72469-6 (eBook)
https://doi.org/10.1007/978-3-030-72469-6

This Springer imprint is published by the registered company Springer Nature Switzerland AG
The registered company address is: Gewerbestrasse 11, 6330 Cham, Switzerland

Preface

We are pleased to present the proceedings of the 7th International Conference of the International Association of Cultural and Digital Tourism (IACuDiT) entitled *Culture and Tourism in a Smart, Globalized, and Sustainable World* at the Greek island of Hydra, from September 2 to 4, 2020. The proceedings present the work of academics and practitioners from across the world conducted on various aspects of digital and cultural innovation, sustainability, and progress in destination management.

The International Association of Cultural and Digital Tourism (IACuDiT) is a global network of people, projects, and events that bear on a wide range of issues of concern and interest in cultural and digital tourism, in an era of major global changes. IACuDiT is a nonprofit international association which values creative, ethical, and progressive action aimed at the improvement of global hospitality and tourism research on cultural and digital issues.

IACuDiT brings together a wide range of academics and industry practitioners from cultural, heritage, communication, and innovational tourism backgrounds and interests. It mainly promotes and sponsors discussion, knowledge sharing, and close cooperation among scholars, researchers, policymakers, and tourism professionals. It is based on the notion that: "Technological changes do not influence the missions of cultural tourism actors in the areas of promotion and product development, but rather the manner of carrying them out." It provides its members with a timely, interactive, and international platform to meet, discuss, and debate cultural, heritage, and other tourism issues that will affect the future direction of hospitality and tourism research and practice in a digital and innovational era.

Bearing in mind the global situation with COVID-19, the conference organizing committee was animated toward providing more convenient access to video conferencing. For that purpose, a virtual presentation support team was set up to help delegates present remotely at the conference. Judging from the warm welcome, the lively discussions, the friendly, unofficial, and warm atmosphere, both inside and outside conference rooms, we believe we succeeded in our goals.

It is real pleasure to express our sincere gratitude to the people and organizations for their contributions, help, and support for this conference. We would like to express our deep gratitude to the members of our international scientific committee for their valuable and vitalizing ideas, comments, suggestions, and criticism on the scientific program of the conference. We express our sincere appreciation to all our keynote speakers, that is to Ciná van Zyl from the University of South Africa and Co-Editor of these proceedings; Alexandros Vassilikos, President of the Hellenic Chamber of Hotels; George Petrakos from Panteion University and General Director of the Research Institute of Tourism, and last but not least, Dr. George Koukoudakis, Mayor of Hydra Island.

It would not be possible to organize this symposium without the support of the Greek Ministry of Tourism and the Municipality of Hydra; their full support, understanding, and encouragement made the life easy for us. Many thanks to our virtual presentation support team, Ms. Artemis Giourgali and Mr. Adrián Martínez.

Quoting Confucius: "Wherever you go, go with all your heart"; this is exactly what we at IACuDiT try to do every time, and we are really blessed we share our journey with all of you!

Athens, Greece

Dr. Vicky Katsoni
President of IACuDiT

Pretoria, South Africa
September 2020

Dr. Ciná van Zyl

Introduction

On behalf of the University of South Africa (UNISA) and the University of West Attica, Greece's collaboration, we are honored and delighted to express our warmest welcome and gratitude that you have joined our 7th International Conference of the International Association of Cultural and Digital Tourism (IACuDiT). Though we could not all be there together physically, the show did go on and most of us attended live from our living rooms. The COVID-19 pandemic has devastated the travel and tourism sector. The world has experienced a 20–30% decrease in global international tourist arrivals, and this setback has led to a potential loss of US$ 30–50 billion (UNWTO, 2020a). In many of the world's cities, planned travel has decreased by 80–90%. The cancelation of flights has kept us locked up in our own countries. For the first time in history, almost 90% of the world's population has lived in countries with travel restrictions (IATA, 2020). An estimated 25 million aviation jobs and 100 million travel and tourism jobs are at risk (UNWTO, 2020b). The year 2020 called for new ways of thinking about our next travel experience.

The World Travel and Tourism Council (WTTC, 2020) reiterated the benefits that come from planning your next holiday or travel experience which indicates the importance new research paradigms. With sustainable travel here to stay, travelers will want their next holiday to be more meaningful. "Philantourism" is the act of choosing a holiday or safari experience in Africa in order to consciously support a destination that has suffered the total absence of tourists during lockdown. The importance of keeping up to date with the latest travel trends is of such importance in academic literature. For example, future travelers will want to "travel better." Industry experts in the lead have already planned a multi-month and multi-destination family sabbatical to Africa in 2021. The family will take their own tutor along to keep the children up to date with the school curriculum while they are away. The sabbatical will involve a key element of "giving back," where the family will be volunteering to help out in conservation activities and wildlife monitoring.

Nature-based wellness experiences will help us realize that we do not need to be connected to the Internet at all hours of the day, and that the connection we have with our immediate surroundings and the people close to us are more meaningful and important than those we have with people behind a digital screen. Of the top ten rated luxury safaris in Africa, all offer luxury spa experiences in the bush, but little in terms of a digital detox and a way needs to be paved for a new tourism construct—the "healing tourism safari" or "nurture in nature tourism safari." The new generation of safaris as a digital detox will allow us to nurture ourselves in nature. A post-COVID-19 approach to marketing safari tourism calls for rethinking the existing safari offers. Personalized individual e-mails that document an individual experience and include more emotion and humanity are far more effective marketing methods. A call for the development of ecotourism through a new lens is necessary. A tentative research agenda for the healing or nurturing tourism safari that incorporates a digital detox safari is paved. These safaris need to be more people-centered by building up a resilience and by contributing to the health and wellness of their visitors so as to remind people that they could thrive without technology in the past and can thrive again.

IACuDiT's mission is to contribute to the paradigm changes needed, in order to achieve cultural tourism development on all levels which closely links to this year's conference's theme of Culture and Tourism in a Smart, Globalized, and Sustainable World. This fully reflects the importance of tourism as a modern-day engine for cross-cultural growth and development in a digital era.

Now, more than ever, the understanding of diversity on how different cultures respond to the coronavirus and the changed travel patterns on a global level is experienced. In Greece, tourism directly employs more employees than any other sector and remains a vital contributor to the Grecian economy. Hydra Island, Greece, is the first Mediterranean Island to successfully host the conference in a hybrid format. The possibilities of the digital era are emphasized. Stretching the imagination by creating new ways of networking and conversation with members, colleagues, and persons from a wide variety of national and international settings. The aim of the conference to promote constructive, critical, and interdisciplinary conversations on the challenges emerging in tourism from the digital transformation of the industry by bringing together researchers, communities, industry, and government stakeholders, is hereby fully achieved.

An array of more than 31 conference topics was included in the call for papers. Papers submitted through virtual and poster presentations were considered for publication. All papers for the conference were double-blind peer reviewed in two phases. All the abstracts submitted for the conference were firstly peer reviewed by two experts in the field for relevance to the conference theme and contribution to the academic debate. A total of 132 abstracts were reviewed, and 78 were found acceptable. All accepted papers were then given the opportunity to submit to full conference proceeding papers utilizing the comments from the first phase of double-blind peer review for the improvement and submission of a full conference paper. Forty-nine full conference proceedings were submitted for inclusion in the Springer proceedings in Business and Economics. These full paper submissions were again double-blind peer reviewed by two experts in the field as well as the editor. During the second review process, each submission was reviewed for (a) relevance to the conference theme (b) quality of the paper in terms of theoretical relevance and significance of the topic, and (c) contribution to the academic debate.

The oral presentations covered a wide range of interdisciplinary themes. The contributions were thematically selected for each group and are arranged in order of presentation in the proceedings. The sub-themes to be covered by the conference were further classified into six distinctive parts, which are as follows:

- Part One—Sports Tourism
- Part Two—Sustainable Tourism
- Part Three—Trending World Changes
- Part Four—Factors Affecting Destinations
- Part Five—Digital Innovation
- Part Six—Cultural Innovation

The editors and IACUDiT anticipate that readers of this volume will find the papers informative, thought provoking, and of value to their niche research areas.

The papers as classified into each of the four parts are to be discussed next.

Part One—Sports Tourism

The first part consists of *nine chapters*. Its focus is on introducing a fast-growing sector in tourism, namely sport tourism which contributes $7.68 billion to the global travel industry (UNWTO, 2020b). Tourists all over the world are interested in sport activities, either as participants or as spectators. The value of various kinds and sizes of sports events to a destination cannot be underestimated. Sports events have the potential to provide long-term benefits for local communities by increasing revenues, supporting local economies, and improving existing infrastructure (Mouratidis et al., 2020). A growing interest in sports tourism research is acknowledged by scholars, which recognized that sports events consist a strategy for local, regional, and national tourism development (Mouratidis, 2018). Papers incorporated in this part reflect different forms of special interest tourism, i.e., cycling, eSports, fishing, and scuba diving tourism, conceptualizing motivational and negotiation factors in the field of sport as well as the economic impact of sports tourism events have on a destination.

In First Chapter "Sport Tourism: An Analysis of Possible Developmental Factors in Sport and Recreation Centers," the authors *Georgia Yfantidou, Charalampos Spiliakos, Ourania Vrondou, Dimitris Gargalianos, Antonia Kalafatzi*, and *Eleni Mami* introduce sport tourism by analyzing the possible developmental factors at sport and recreation centers. The authors noted that the development of sport tourism is determined by several factors, while various models' and factors' typologies have been identified and reported in literature. The primary aim of their research was threefold: firstly, the identification and classification of factors affecting the development of sport tourism, then investigating the level to which these factors are met by a sport and recreation center, and lastly, comparing opinions of the users of the center regarding these factors. The authors followed a structured questionnaire in their methodological approach to investigate sport and recreation center under study. The results revealed differences in the opinions of the users of the center. From the research, it is clear that the most important factors to users are the changing bathrooms, security, healthcare and injury rehabilitation, and the accommodation facilities; while the less important factors were the proximity to a port and marinas, as well as the access through the railway network. Noteworthy is that these factors were met by the sport and recreation center under study at a very high to adequate level. In this respect, clear direction is given for the development of future sport and recreation centers.

In Second Chapter "Cycling Tourism: Characteristics and Challenges for the Developments and Promotions of a Special Interest Product," the authors *Elina Tsitoura, Paris Tsartas, Efthymia Sarantakou*, and *Alexios-Patapios Kontis* explore the challenges and possibilities of developing cycling tourism in a highly competitive marketplace. Cycling tourism has the potential to offer multiple economic, social, and environmental benefits to a country. The authors examined important aspects associated with cycling tourism in Greece as a new international trend, focusing on conditions for its development. Qualitative opinion research was used to understand the conditions for development of cycling tourism in Greece for both local and international cyclists. The survey highlighted the pertinent considerations on developing cycling tourism in Greece and established guidelines that can be used by stakeholders who would like to develop cycling tourism.

In Third Chapter "Evaluating the Economic Impact of Active Sports Tourism Events: Lessons Learned from Cyprus," the work of *Achilleas Achilleos, Christos Markides, Michalis Makrominas, Andreas Konstantinides, Rafael Alexandrou, Effie Zikouli, Elena Papacosta, Panos Constantinides*, and *Leondios Tselepos* aims at evaluating the economic impact of active sports tourism events and reports on valuable lessons learned from Cyprus in their case study research. The economic impact of major and mega sport events to the host destination is widely recognized in tourism literature. Furthermore, the authors argued that most research examine and demonstrate the economic impact of major (passive) sport events, while the importance of active sports events in promoting sports tourism is under researched. In this case study approach, the economic impact of active sports events is explored through a major international swimming event held in Cyprus. The research on which the study is based made use of a local research in start-ups project, which developed a Web platform and recommender system dedicated to a niche market (i.e., active sports events). The paper mainly focused on the analysis of the survey results, in which the authors demonstrate the economic impact of active sports events and the potential that such events offer to the growth of a niche market, i.e., active sports tourism.

Fourth Chapter "The Effects of PUSH and PULL Factors on Spectators' Satisfaction Attitudes. A Mediation Analysis of Perceived Satisfaction from a Small-Scale Sport's Event," written by *Konstantinos Mouratidis* and *Maria Doumi,* focuses on the effects of push and pull factors on spectators' satisfaction attitudes. The authors performed a mediation analysis of perceived satisfaction from a small-scale sport event, the Half-Marathon, in an island destination of Skyros. The research once again confirmed the importance of the islands as a major attraction to Greece as a country. According to SETE Statistics, the total international arrivals coming to Greece reached 31.3 million foreign visitors, in 2019, and the total revenue of the

Greek tourism industry was nearly €17.7 billion. The Greek Ministry of Tourism is of the opinion to reduce the seasonality of demand and the negative impacts of mass tourism. This is achieved by attempts to enrich and diversify the national tourism product through special and alternative forms of tourism. The results of the survey offer basic information of travel motives of spectators', who attended the small-scale sport event. The paper indicates that modern-day tourists are seeking for travel experiences related to the consumption of special tourism products and unique services. A particular traveling behavior of spectators'– visitors' and the direct or indirect relationship between their expressed motives and their satisfaction from the small-scale sport event and the hosting destination is explored in this research. The results revealed that by understanding the complexity of spectators'– visitors' motives, behaviors, and satisfaction attitudes, the organizing committee, policymakers, and other co-organizers of such sport events proactively market the event strategically to the specific needs of the market.

Fifth Chapter "Strategic Negotiation Factors in Participating at Recreational Sport Activities Aiming at the Well-being and the Presentation of Perma Scale for the Greek Population," prepared by *Georgia Yfantidou, Alexia Noutsou, Panagiota Balaska, Evangelos Bebetsos, Alkistis Papaioannou, and Eleni Spryridopoulou*, discusses how modern societies' physical, spiritual, and mental health and community quality of life can be determined by people's engagement with physical activity. This paper looks at how participation in recreational sports and creation of green recreational areas increase individual viability and mental health through PERMA-Profiler questionnaires in urban areas of Athens, Greece. The PERMA-Profiler scales served as a useful well-being tool to measure relevant psychometric properties and help individuals understand how to create a more fulfilling life through exercise.

In Sixth Chapter "Exploring Scuba Diving Tourism Sector in Malta and Its Sustainable Impact on the Island," *Simon Caruana* and *Tiffany Sultana* explore scuba diving as a sport and niche sector contributing to Malta's tourism product. The study notices that all marine resources have to be used with caution to ensure sustainability of the marine experience. The research followed a qualitative approach with interviews held among the key stakeholders. Various key legislative documents such as the 2011 Master Plan related to scuba diving were examined. The paper finds that more laws are needed to protect the marine environment upon which the industry depends. Most of the stakeholders interviewed alluded to the insufficient marine law enforcement as well as a lack of coordination between the different bodies entrusted with different aspects of marine law enforcement. The study provided a solution to have only one body with representatives from all the agencies to ensure better coordination of law enforcement efforts.

Seventh Chapter "eSports Tourism: Sports Tourism in a Modern Tourism Environment," written by *Ioannis A. Nikas* and *Ioulia Poulaki* , covered the topic of eSports tourism which introduces sports tourism in a modern tourism environment. eSports are a modern Web-oriented, technological trend that has the potential to create new tourism products and destinations. The "eSports tourism" has many similarities to traditional sports tourism, with the exception that the main attraction, in the case of eSports tourism, is that the speculators are watching video game players instead of athletes participating in traditional individual or team sports.

The worldwide growing market of eSports creates promising opportunities for the development of eSports tourism in Greece as well. Thus, this paper aims to investigate the potentials of such a special tourism form by studying and describing the context of eSports tourism, under the following aspects: (a) the consuming behavior of a eSports tourists, (b) the economic impact, in terms of tourism-related revenue, mainly spread in the destination, and (c) the case of eSports tourism as an event or congress tourism form as well as its differences with sports tourism and other special and alternative tourism forms. The opportunities arising in Greece and especially in cities like Patras are discussed, mostly due to the congress venues and the existing technological infrastructures. The paper is conceptual, and the methodology followed focuses mainly on Internet research, related articles, social media, and scientific papers.

Eighth Chapter "Nostalgia Sport Tourism: An Examination of an Underestimated Post-event Tourism Proposal," written by *Ourania Vrondou*, covered the topic of the "nostalgia" sport tourism type which has been overlooked and, thus, is in need for further analysis. Sport events historical sites, venues' magnitude and architectural significance, and more vividly, funs emotions constitute a complex but also an appealing tourism proposal especially after the completion of large sport events. With the contemporary inclusion of technology, the sport nostalgia experience could be enhanced further leading to new tourism development paths combining different leisure industries. The study investigates the role of the sport bodies, the local cultural/historical organizations, and the tourism business that consist the policy community needed to materialize related proposals especially post-events. Paradoxically, this sport tourism approach despite its powerful role in creating cultural, societal, sport, and tourism synergies around the historical sport events areas, venues, and legacies remains underestimated by scholars, policymakers, and the industry. The present study uses cross-case analysis and theoretical paradigms to conceptually set the basis of a generalizable platform for future planners through the understanding of all critical aspects involved in the development of nostalgia sport tourism post-events.

Ninth Chapter "The Role of Sports Tourism Infrastructures and Sports Events in Destinations Competitiveness," *Dália Liberato*, *Pedro Liberato*, and *Catarina Moreira*, reports two thematic areas that have been very relevant to Oporto as a tourism destination, focusing on attraction and competitiveness: sports and tourism. In this context, their research aims to understand the relevance of sports tourism, in Oporto as a tourism destination, using the analysis of two infrastructures, relevant to the city, which are still poorly explored, the Dragão Stadium and the FC Porto Museum. This study intends to be an important contribution to the scientific literature in the field of sports tourism, namely in the relationship between the Porto Football Club and Oporto as a tourism destination, regarding the definition of future strategies for collaboration and strengthening partnerships in a competitive perspective for both. Visioning the research objectives, a quantitative methodology was applied through the application of a questionnaire to 400 tourists/visitors in the city of Oporto.

The data gathered confirm that the main reasons for the trip influence the visit to the Dragão stadium and the FC Porto Museum, and the assistance to a football game at the Dragão stadium influences the visit to the Dragão Stadium and the FC Porto Museum.

Part Two—Sustainable Tourism

The *second part* of the book consists of *seven chapters* that aim to capture sustainable tourism and the application thereof of in the field. *Sustainable tourist development* remains at the heart of global and local policy efforts in both tourist developed and developing areas, seeking to reap the economic benefits of tourist development as well as manage carrying capacity aspects of available resources in destinations and sustain cultural integrity, essential ecological processes, biological diversity, and life support systems (Stratigea & Katsoni, 2015). Topics discussed in this part include sustainable business models, Area Life Cycle (TALC), over tourism and tourism carrying capacity as well as providing sustainable tourism transport services. These papers shed some light on developments in methodologies followed, tools and approaches that are capable of dealing with sustainable tourist development perspectives and paths to be taken.

More specifically, in Tenth Chapter "Case Study Protocol for the Analysis of Sustainable Business Models," *Joaquin Sanchez-Planelles* and *Marival Segarra-Oña* are in the process to develop theory for analyzing sustainable business models. The research design made use of a case study approach. The study made use of a combination of frameworks to help researchers understand how sustainability is integrated through the company value chain. Environmental practices are incorporated through a set of questions which aims to identify the value thereof to company's activities.

In Eleventh Chapter "Examining the Relationship Between Tourism Seasonality and Saturation for the Greek Prefectures: A Combined Operational and TALC-Theoretic Approach," *Thomas Krabokoukis, Dimitrios Tsiotas*, and *Serafeim Polyzos* examined the relationship between tourism seasonality and saturation for the Greek prefectures with a combined operational and Tourism Area Life Cycle (TALC)-theoretic approach. The uneven distribution of tourism destination's demand remains a complex phenomenon. Sustainable tourism development is linked to it due to periods of resource intensive activities during the high seasons versus inactive tourism capital usage for the rest of the year. Data of overnight tourist stays for the period 1998–2018 were used. The results of the study provide insights about the relationship between tourism seasonality and saturation for the case of Greece. The authors managed to develop a quantitative tool for tourism managers to be used for regional tourism development as it allows for taking into account the temporal and spatial dimensions of the tourism seasonality phenomenon.

In Twelfth Chapter "Sustainable Tourism Development in the Ionian Islands. The Case of Corfu Island," prepared by *Konstantinos Mouratidis*, the topic of sustainable tourism development in the Ionian Islands, more specifically Corfu Island, is analyzed. The author set the scene by elaborating on the importance of tourism's economic contribution worldwide as well as the devastating effect of the COVID-19 pandemic (2020) on international tourism arrivals. The research analyzed a priori the major trends of tourism in Greece, Ionian Islands region, and Corfu Island over the last decade, as well as exploring the emergence of sustainable tourism development for the island. The methodological approach is that of a critical analysis of available literature on the topic. In the results, the author elaborates on the implementation of sustainable development programs in order to protect the environment and cultural characteristics of each region and thus to determine the long-run benefits of the sustainability agenda.

In Thirteenth Chapter "Sustainable Tourism; Vector of the Social and Solidarity Economy: Case of Region Souss Massa, South of Morocco," *Asma Edaoudi, M'bark Houssas*, and *Abdelhaq Lahfidi* stress the importance of sustainable tourism, as a path to the social and solidarity economy in the region of Souss Massa, south of Marocco. The authors noted the negative effects of tourism activities, which is indicative of a desirable and politically appropriate approach to tourism development. By conducting a comprehensive literature review on the concept of sustainable tourism and its role in social and solidarity economy for the region, the background is outlined. The authors questioned how the development of sustainable tourism can promote socioeconomic growth in the Souss Massa region in order to reach the full tourism potential of Morocco.

Fourteenth Chapter "Overtourism and Tourism Carrying Capacity: A Regional Perspective for Greece," written by *Sophia Panousi* and *George Petrakos*, investigates over tourism and tourism carrying capacity from a regional perspective for Greece, if not managed appropriately. Through quantitative analysis, the research explores the construction of two composite indicators measuring over tourism and tourism carrying capacity in a specific geographical region. By doing so, regions close to saturation and those with potential for further tourism development are identified. The value of these indicators proofs to be a valuable tool for planning tourism policy.

Fifteenth Chapter "Tourism Transportation Services Provided on the Principle of Sharing Economy," by *Radka Marčeková, L'ubica Šebová, Kristína Pompurová*, and *Ivana Šimočková*, elaborates on tourism transportation services based on the principle of the sharing economy in Slovakia. The paper analyzed the public awareness of transport services provided, the usage, and motivation of potential providers of services. The paper questioned the following: "What is the motivation for the potential providers of the transportation services in the sharing economy platforms? What needs to be changed in Slovakia in order to expand the possibilities of the offering the tourism transportation services provided on the principle of the sharing economy?" A total of $N = 727$ responses were received from a structured questionnaire on both current and potential users and providers of transportation services.

In Sixteenth Chapter "Integration of Sustainable Practices in Firms: The Specifics of the Tourism, Leisure and Hospitality Sectors," *Inés Díez Martínez* and *Ángel Peiró*

Signes argue that geographical areas, and industries approach sustainability differently, and there is a wide spectrum of patterns regarding sustainability integration tools, triggers, and the implementation of these practices. In the case of firms related to the tourism, leisure and hospitality industries, many sustainability management practices are put in place. Their paper explores the three dimensions of sustainability: environmental, social, and economic, showing best practices developed by firms belonging to the above sectors. On the environmental dimension, practices target the creation of a more eco-friendly service or product along with the focus on improving resource efficiency related to energy and water consumption and reduce waste. On the social side, practices focus on increasing the well-being of the community or the staff, frequently merging both through the promotion of local employability and training. Finally, the economic dimension is mainly evidenced by cost-saving initiatives.

Part Three—Trending World Changes

The *third part* of the book, consisting of *nine chapters*, aims at presenting recent trends that impacts the world to change at an unprecedented rate. What does the future hold for tourism, attractions, and destinations? Will destinations survive all the issues and challenges with which the sector is currently faced? A collection of different views provided by these papers offers valuable insight into decisionmakers.

More specifically, Seventeenth Chapter "Environmentally Friendly Tourists In Morocco," *by Charef Kenza* and *D. M. Bark Houssas*, explores the topic of environmentally friendly tourists in the Moroccan context. The authors reported on the threat of climate change and the impact it has on the natural, social, and economic environment. The importance of preserving natural resources and the management thereof becomes evident. Cooperation between all relevant parties is called upon. The paper investigates what tourism behavior can be considered sustainable or environmentally friendly. The results revealed that consensus about who an environmentally friendly tourists are still needs to be established. The research method followed review of secondary literature on environmentally friendly behavior.

In Eighteenth Chapter "Tourism and Contact Tracing Apps in the COVID-19 Era," *Agisilaos Konidaris, Ourania Stellatou, Spyros E. Polykalas*, and *Vicky Kats*oni explore tourism and contract tracing apps during the COVID-19 period. The COVID-19 pandemic brought major changes to humanity in a very short period of time. Things that had been established for years no longer apply, and a "new normal" has been abruptly introduced into people's lives. Tourism has also been tremendously affected. A part of the "new-normal" in tourism will probably include contact tracing smartphone applications. This paper initially aims to explain what contact tracing applications are, how they work as well as the differences between existing implementation approaches. The authors then continue to discuss the main concerns that arise by their adoption, mainly, security, data privacy issues, and interoperability between different country-specific app implementations. The authors present proposals for contact tracing application implementation in tourism and answers to questions that destinations and local health authorities could benefit from and propose possible improvements to the Apple/Google implementation of the Exposure Notification System. These improvements include a notification feedback mechanism that will potentially improve the validity of future notifications and a built-in global COVID-19 information feature. Finally, they present and analyze their findings acquired through a survey that was conducted in Greece.

In Nineteenth Chapter "Tourism Customer Attitudes During the COVID-19 Crisis," *Ivanka Vasenska* and *Blagovesta Koyundzhiyska-Davidkova* deal with tourism customer attitudes during the COVID-19 crisis. The authors questioned the status quo as the COVID-19 pandemic bringing an end to the golden era of global tourism. The paper aims to establish Bulgarian tourists' attitudes toward traveling before and after the COVID-19 pandemic. Primary research is gathered from a survey collected from $N = 121$ respondents. The results revealed the extent to which customers incline to travel before, in the post self-isolation period and after the crisis. The survey of the study provides the opportunity to be used as a model for analyzing data of tourist's attitudes during a crisis.

In Twentieth Chapter "Solo Female Travelers as a New Trend in Tourism Destinations," *Lubica Sebova, Kristina Pompurova, Radka Marcekova*, and *Alica Albertova* introduce the new emerging trend that of independant travel, or solo travel for female travelers in Slovakia. The emancipation of women, longing for rest, and discovering new adventures are introduced in this paper. An analysis of the requirements and preferences of women-independent travel as a new tourism trend is reasoned from literature. The contribution of the paper is to increase awareness for solo female travelers which may result in a positive impact on the domestic and inbound tourism market of Slovakia. The chapter provides an interesting view on the new trend of female solo travel.

In Twenty-First Chapter "Strategic Hotel Management in the "Hostile" International Environment," *Ioannis Rossidis, Dimitrios Belias*, and *Labros Vasiliadis* through thorough research present a pilot study by using two measurement scales (questionnaires), namely a) to assess intentions of use and b) quality of service (Servqual) for the use of robots in the hotel industry. The paper outlined the 28-item intentions questionnaire and the 22-item Servqual questionnaire as distributed among $N = 157$ Greece hotel staff. The factors as extracted from each questionnaire are given, three for Servqual of robot and six for intention–expectation of robot use in the hotel industry. The study's results revealed a novel application of robots in the hotel industry which makes a valuable contribution.

In Twenty-Second Chapter "Strategic Human Resource Management in the International Hospitality Industry. An Extensive Literature Review," *Ioannis Rossidis, Dimitrios Belias*, and *Labros Vasiliadis* make their review through a theoretical literature review, and the paper aims to explore human resource management in a hotel business by focusing on topics such as recruitment, training, and compensation to ensure a successful future organization. Various cases and good practice from the international area are analyzed, the evolution in the field acknowledged which provide a practical solution to hotels all over.

In Twenty-Third Chapter "From Mass Tourism and Mass Culture to Sustainable Tourism in the Post-covid 19 Era: The Case of Mykonos," *Konstantinos Skagias, Labros Vasiliadis, Dimitrios Belias,* and *Papademetriou Christos* deal with mass tourism and its impact on local culture. Indeed, during the past years, tourist development had relied on mass tourism, which meant that the destinations had to accommodate a significant number of tourists, something that had an impact on the destination's culture. Such an example is the case of Mykonos. This a case of an island which has constructed its brand name as a high-end cosmopolitan destination by including mass tourist activities on its product offering but also with mass culture activities, such as major dance music events. Nonetheless, the current situation has found Mykonos, like many other destinations, without tourist demand and with the need to reposition its tourist product. In the post-COVID-19 tourist industry, it seems that sustainable tourism can be the answer on how Mykonos shall develop so to recover from the current crisis. Therefore, the suggested strategy is the shift from mass tourism and mass culture into sustainable tourism and emphasis on high culture. This means that the destination will have to rely in a "less tourists and more income per tourist" model of development while Mykonos can focus on cultural tourism as part of its shift into sustainable tourism.

In Twenty-Fourth Chapter "Smart Analysis of Volatility Visualization as a Tool of Financial and Tourism Risk Management," *Ani Stoykova* and *Mariya Paskaleva* focus on volatility, as it is important for options traders because it affects options prices. Generally, higher volatility makes options more valuable and vice versa. During periods of high volatility, financial markets are relatively more efficient. Volatility falls when markets rise and rise as markets fall. This paper focuses on examining the dynamics of European capital markets, S&P 500, GEPU, and VIX. Also, the authors observe the performance of S&P 500, Stoxx 600 and Stoxx Europe 600 Travel, and Leisure price index. The analyzed period is 2003–2016. The results show that studied stock indices tend to move synchronously during the examined period. The authors conclude that there is a strong correlation between VIX, GEPU, and S&P500 measuring "investor risk appetite." GEPU fluctuates around consistently high levels since mid-2011 until the beginning of 2013. The dynamics of GEPU and VIX are not synchronized for the period 2013–2016. In the crisis period, travel and leisure stock

In Twenty-Fifth Chapter "The Productivity Puzzle in Cultural Tourism at Regional Level," *Eleonora Santos, Inês Lisboa, Jacinta Moreira,* and *Neuza Ribeiro* focus on the distribution of labor productivity of tourism firms across regions to address this concern, and an empirical analysis at firm level is used to calculate the labor productivity across seven Portuguese NUTS II regions. Using data from 2866 firms' financial reports obtained from SABI database, over four years, this paper analyzes the regional inequalities and its consequences in terms of economic and tourism development. The findings show that, at the national level, the mean labor productivity has increased around 1.9% on average, from 2015 to 2017 and dropped 3.9% in 2018. At the regional level, the results suggest that firm performance regarding productivity is not related to the dynamics observed in neighboring regions but to the resources and conditions available in each region separately. Such results confirm the assumption of spatial inequality in the distribution of labor productivity among cultural tourism firms. The results presented in the article can be of significant use to further understand the productivity dimension of tourism production at regional level. Finally, by identifying the areas where tourism management needs to be improved, this paper provides suggestions on measures to increase regional productivity, and thus, regional competitiveness.

Part Four—Factors Affecting Destinations

The *fourth part* of the book, consisting of *seven chapters*, introduces a variety of factors impacting destinations. Destinations form part of Leiper's (1990) tourism system and remain in many respects one of the most important parts as destinations attract tourists, motivate the visit, and therefore energize the whole tourism system. Sustainable planning and management of a destination can contribute to a destination's competitiveness, and continuous research on the topic is introduced in this part of the book.

In Twenty-Sixth Chapter "Capital Structure Determinants of Greek Hotels: The Impact of the Greek Debt Crisis," *Panagiotis E. Dimitropoulos* and *Konstantinos Koronios* discuss the impact of the Greek debt crisis on the capital structure of Greek hotels. The paper used a random effect panel regression analysis to test the efficiency of trade-off theory and pecking order theory, before (2003–2010) and after (2011–2016) the Greek sovereign debt crisis. It found that asset tangibility is positively and directly related to total leverage and long-term leverage, specifically during the debt crisis. Non-debt tax shields and tax payments negatively impact total leverage for the period. The study was among the first of its kind reported literature. Future research into the topic is hereby called upon to address the gap in research.

Twenty-Seventh Chapter "Human Resource Empowerment and Employees' Job Satisfaction in a Public Tourism Organization: The Case of Greek Ministry of Tourism," by *Alkistis Papaioannou, George Baroutas, Ioulia Poulaki, Georgia Yfantidou*, and *Alexia Noutsou*, explores human resource empowerment and employees' job satisfaction in a public tourism organization. Their paper aims to examine (a) the extent to which human resource empowerment is applied to the Greek Ministry of Tourism, (b) the extent to which the job satisfaction of employees was existed in this public tourism organization, and (c) the relationship between human resource empowerment and employees' job satisfaction. For the purpose of this study, two questionnaires were used: (a) the questionnaire of human resource empowerment and (b) the questionnaire of employees' job satisfaction. The results of the data analysis indicated that human resource empowerment is applied to a mediocre extent by the Greek Ministry of Tourism, while the overall employees' job satisfaction was above average. Furthermore, significant and positive relationships among the key factors of human resource empowerment and employees' job satisfaction were highlighted. This study is useful in extending the concept of human resource empowerment to the public tourism sector. Additionally, the outcomes may be useful in helping directors to further understand the human resource empowerment process in their respective tourism organizations.

In Twenty-Eighth Chapter "The City of Thessaloniki as a Culture Tourism Destination for Israeli Tourist," *Efstathios Velissariou* and *Iliostalakti Mitonidou* explore Thessaloniki as a culture tourism destination for Israeli tourists. The city of Thessaloniki has been a multicultural city that attracts people from all walks of life. It was also a refuge destination for many persecuted Jewish people in Europe between 1942 and 1943, which led to a large Jewish community, thereby attracting a large number of Israeli tourists each year. The aim of the paper was to determine the motivations, characteristics, and interests of tourists from Israel to Thessaloniki. The study showed that most visitors were over the age of 45 and have a very high level of education. Most visitors from Israel stayed for 3–7 nights and visited museums or monuments during their stay. 95.7% of respondents organized their own trips to Thessaloniki. The existing cultural tourism features of history, traditions, monuments, and infrastructure are present in Thessaloniki and need to be improved in order for the city to become a more popular destination for Israeli tourists. The research in the chapter is well articulated from a city tourism perspective and stimulate future research of a similar nature.

Twenty-Ninth Chapter "Place Attachment Genesis: The Case of Heritage Sites and the Role of Reenactment Performances," is introduced by *Simona Mălăescu*. Place attachment and its role as precursor of destination loyalty have mainly focused on recreational and beach tourism in the last decade. The study aimed to fill a gap in existing literature regarding the perceived instrumental role of reenactment representations in tourists' experience at heritage sites using correlation, multilinear regression, and mediation analysis. Variables such as the strength of destination's identification with a reenactment site and the perceived role of reenactment in enhancing patriotic feelings resulted in an explanation of the place identity component in place attachment genesis. The paper makes a fine contribution to place attachment and a destination's loyalty.

In Thirtieth Chapter "Conditions for Creating Business Tourism Offers and the Regional Potential in Poland," an empirical analysis of the readiness of Polish regions to develop business tourism products is produced by *Ewa Lipianin-Zontek* and *Zbigniew Zontek*. The aim was to identify and analyze business tourism in Poland and determine which internal and external conditions are important for business tourism development. The study showed that certain conditions for shaping business tourism exist in Poland and that diversity of business tourism is visible. More successful business tourism regions were those with a wide range of tourist attractions. The paper stimulated thought on the topic of business tourism products to a destination's attractiveness.

In Thirty-First Chapter "The Role of Fashion Events in Tourism Destinations: DMOs Perspective," *Dália Liberato, Benedita Barros e Mendes, Pedro Liberato,* and *Elisa Alén* identify that fashion events add value to host cities and create competitive advantage to other destinations, despite fashion tourism not being properly valued by the tourism industry. The authors aimed to provide more scientific knowledge on the topic through their study, particularly understanding the role of fashion events in the northern region of Portugal, with additional focus on the city of Porto. Interviews determined that wider promotion of fashion events would make these events more advantageous to the tourism sector, due to the image and the personality of Porto, and its active, cultured, and cosmopolitan lifestyle. Those who participate in fashion events use the event to get to know the city and often return to future events, extending their stay to travel with family and friends. City tourism being a valuable research topic for destinations is articulated well in this chapter.

In Thirty-Second Chapter "Wellness Tourism Resorts: A Case Study of an Emerging Segment of Tourism Sector in Greece," *Marilena Skoumpi, Paris Tsartas, Efthymia Sarantakou*, and *Maria Pagoni* aim to fill in the research gap that exists about wellness resorts in Greece, as well as about the specific characteristics of the visitors and the prospects of growth that wellness tourism has in Greece. Initially, it presents statistics related to health and wellness tourism, as well as to wellness resorts in Greece and worldwide. Secondly, it includes the primary research that has been conducted at two levels: the first level of research, which was conducted on the general public, resulted to conclusions on the public's viewpoints about wellness resorts. The second level of research was conducted on wellness resorts in Greece, in order to define the specific characteristics of their visitors. Their study offers conclusions as well as potential policies for the state and the entrepreneurs that can be implemented for the publicity and the promotion of this profitable form of tourism.

Part Five—Digital Innovation

The *fifth part* of the book, consisting of *ten chapters*, aims at presenting a variety of different aspects of digital transformation in the tourism industry. Changes in consumption patterns driven by the gradual increase of Internet use indicate that there is great potential in investing in companies' digital transformation (Poulaki & Katsoni, 2020).

In Thirty-Third Chapter "Policy Responses to Critical Issues for the Digital Transformation of Tourism SMEs: Evidence from Greece," *Panagiota Dionysopoulou* and *Konstantina Tsakopoulou* argue that tourism businesses present a varying degree of digital transformation, depending on the size and type of business. Tourism SMEs are struggling to remain competitive and reap the benefits of digital transformation. Empowering tourism SMEs in the digital age is increasingly becoming a strategic priority for governance. However, research on policies to enhance the digitilization of tourism SMEs is limited. The paper attempts to fill this gap. Based on a review of official policy documents and governmental and industry associations' sites, it looks into ongoing policy initiatives in Greece to support the digital transformation of Greek tourism SMEs at a national level. Desk research was coupled with in-depth interviews, conducted in autumn 2019, with key informants in Greek tourism industry associations and in the tourism departments of regional authorities. Policies to build digital skills, to support start-ups and bring innovation to the market, to enhance e-governance, and to provide financial assistance have been implemented in Greece to tackle structural barriers to digital transformation owing to the limited resources, the traditional business model, and the state of ownership (family run businesses) of Greek tourism SMEs. The aforementioned policy responses are in line with critical issues identified in the academic literature and official policy reports at EU and OECD level concerning the digital transformation of tourism SMEs.

In Thirty-Fourth Chapter "Security and Safety as a Key Factor for Smart Tourism Destinations: New Management Challenges in Relation to Health Risks," *Salvador Ruiz-Sancho, Maria José Viñals, Lola Teruel,* and *Marival Segarra* discuss the concept of competitiveness of a smart tourist destination, based on three pillars: its natural and cultural attractions, its tourism infrastructures and facilities, and the quality of the services it has to offer. These services include security and safety, both of which are key elements in the tourist's choice of destination. Their aim is, first, to analyze the elements that make up the concepts of security and safety in tourism. Then, an in-depth review of the existing literature on security and safety is carried out, and a risk classification is presented. A distinction has been made between territorial, socioeconomic, and political risks as well as the diverse reactions (action measures) to fight them. In addition, the key factors that influence the tourist's perception of security have also been addressed. Beyond the personality traits of an individual, other factors such as the visibility and tangibility of how dangerous the risk is, the information about security or the public administration's capacity to address risks have also been considered. Furthermore, the evolution of security and safety issues in the tourist context has been studied up until the current, unprecedented, situation caused by the coronavirus COVID-19 pandemic, together with the importance of knowing how to address new management challenges when faced with health risks.

In Thirty-Fifth Chapter "An Evaluation of Hotel Websites' Persuasive Characteristics: A Segmentation of Four–Star Hotels in Greece," the authors *Konstantinos Koronios, Lazaros Ntasis, Panagiotis Dimitropoulos, John Douvis, Genovefa Manousaridou,* and *Andreas Papadopoulos* made use of latent class segmentation. The study examined a total of six factors of Web site persuasiveness on a sample of $N = 404$ four-star hotels' Web sites. The results of the study demonstrated current Webpages persuasiveness and correspondence and provide valuable guidance for its advancement in a particular hotel industry.

In Thirty-Sixth Chapter "Pilot Study for Two Questionnaires Assessing Intentions of Use and Quality of Service of Robots in the Hotel Industry," *Dimitrios Belias* and *Labros Vasiliadis*, through thorough research, presents a pilot study by using two measurement scales (questionnaires), namely a) to assess intentions of use and b) quality of service (Servqual) for the use of robots in the hotel industry. The paper outlined the 28-item intentions questionnaire and the 22-item Servqual questionnaire as distributed among $N = 157$ Greece hotel staff. The factors as extracted from each questionnaire are given, three for Servqual of robot and six for intention–expectation of robot use in the hotel industry. The study's results revealed a novel application of robots in the hotel industry which makes a valuable contribution.

In Thirty-Seventh Chapter "Spatial Patterns of Tourism Activity Through the Lens of TripAdvisor's Online Restaurant Reviews: A Case Study from Corfu," *Christina Beneki* and *Thanassis Spiggos* analyzed TripAdvisor reviews to Corfu to understand patron's behavioral patterns. The study's results revealed valuable spatial clustering patterns across the island which gives a clear indication of the extent to which different areas have undergone "tourismification." The authors noted that spatial patterns may be used in local-level policymaking by various official bodies in the region. The added advantage of the research will be useful in the formulation of strategic tourism plans to improve destination management.

In Thirty-Eighth Chapter "Online Platforms for Tourist Accommodation from Economic Policy Perspective in Greece: Case for Further Digitalization," *Vesna Luković* focuses on the economic viewpoint of digitalization enabled through Internet in Greece. The paper is organized with a twofold aim. On the one hand, the aim is, by analyzing data from the Eurostat database and other sources, to explore the evolution of digitalization enabled through Internet in Greece in the last few years. On the other hand, by exploring data about financing opportunities available via European Union funds, the aim is to review the argument that, as a result of further digitalization, households' disposable income would get a boost, and hence, aggregate demand would rise. The conclusion of the in-depth look at the economic argument is that further digitalization would help Greece toward achieving its macroeconomic policy goals. That is crucial considering that the country has very high public debt and that the economy has not yet fully recovered from the financial crisis that started in 2008. This is especially relevant now considering the coronavirus outbreak in the first quarter of 2020 causing a setback to achieving some goals.

In Thirty-Ninth Chapter "The E-Tour Facilitator Platform Supporting an Innovative Health Tourism Marketing Strategy," *Constantinos Halkiopoulos, Eleni Dimou, Aristotelis Kompothrekas, Giorgos Telonis*, and *Basilis Boutsinas* argue that the hotel industry could benefit from its cooperation with thematic tourism platforms to address the seasonality, expand its target audience, and enhance the effectiveness of its marketing strategy. Nevertheless, most tourism providers in Greece, being small family businesses, lack the necessary information and communication technologies (and inherent technologies) to become globally competitive. Given the growing trend in the interest in health tourism, this paper aims to present the case of "e-Tour Facilitator Platform," an intelligent information system aiming at supporting an innovative health tourism strategy. The platform focuses on the end users, namely the patients/tourists, matching their profile to characteristics of both medical and tourism services. It exploits state-of-the-art machine learning techniques in order both to help end users to view/select the proper health tourism product (recommender system) with respect to their profile as well as to automatically handle their comments (text mining) for evaluating purposes.

In Fortieth Chapter "The Evolution of Online Travel Agencies in the Last Decade: E-Travel SA as an Exceptional Paradigm," *Dimitra Psefti, Ioulia Poulaki, Alkistis Papaioannou*, and *Vicky Katsoni* highlight the development of the online travel agencies (hereinafter OTAs) and the way that technology has contributed to this direction by investigating the case of an OTA company, namely e-Travel SA. The Internet seems to have changed dramatically and in a positive way the tourism industry. Information and Communication Technologies (ICTs), such as artificial intelligence and machine learning, virtual and augmented reality, Internet of things as well as gamification, offer valuable tools and applications both to businesses and to customers. At the same time, the increased use of mobile devices, digital tools, and social media along with the unlimited access to information have made consumers demanding and updated. Consequently, they require more and more useful services based on their needs. Considering the aforementioned, it is easy to understand that marketing strategies of many companies have changed and that these companies are now exploiting the state of the art in digital marketing tools in order to get the best possible result. E-travel is a real example of the OTAs' evolution during the last decade, and their success proves that opportunities may be generated by exploiting the technological advancements.

In Forty-First Chapter "Do Hotels Care? A Proposed Smart Framework for the Effectiveness of an Environmental Management Accounting System Based on Business Intelligence Technologies," *Christos Sarigiannidis, Constantinos Halkiopoulos, Konstantinos Giannopoulos, Fay Giannopoulou, Anastasios E. Politis, Basilis Boutsinas*, and *Konstantinos Kollias* investigate the characteristics of an environmental management accounting system (EMA). It also illustrates the innovative structure of accounting for the environmental costs which are normally hidden into overheads or are totally ignored. More specifically, this research proposes a smart framework that focuses on the Industry 4.0 innovations in order to collect the appropriate data in a hotel setting. Since a large amount of such data emerges from devices connected to the Internet of things (IoT), business intelligence techniques are proposed for big data analytics. Finally, an EMA system is used to transform data into monetary units. Thus, the proposed smart framework comprises a system of measuring environmental costs in both physical units and in monetary terms for hotels sector, implementing Industry 4.0 elements. This framework can be used as an innovative tool for waste management and the enhancement of sustainability and environmental protection.

In Forty-Second Chapter "An Innovative Recommender System for Health Tourism," *Antiopi Panteli, Aristotelis Kompothrekas, Constantinos Halkiopoulos*, and *Basilis Boutsinas* present an innovative recommender system aiming at matching health tourist preferences to health/tourism providers. It focuses on providing complete health tourism products, by matching the user profile to characteristics of both health and tourism service providers, in order that users receive the treatment they choose, in the right location, the right period, at the right cost. The proposed recommender system is implemented by applying a facility location problem which employs a parameter that controls the diversity of the recommendation list and thus the variety of the proposed results. It incorporates a database of cases, i.e., medical, wellness, and tourism service providers. A case is described by a set of attributes such as medical service category, spa category, wellness category, cost, infrastructure, accreditations, communication languages, and so on. Users that have already acquired a health tourism package provide ratings for certain categories of attributes. A new user expresses her preferences in the form of a query, and then the system tries to match this query to the cases that exist in the database. At first, the best possible cases are extracted, by applying a sorting procedure based on comparisons to the ideal one. Then the facility location method is applied to provide the final recommended cases to the user that are both similar to the provided query and diverse to each other.

Part VI—Cultural Innovation

The *sixth part* of the book, consisting of *seven chapters*, focuses on the dynamic nature of culture in the way it is adapted in a new digital era which greatly encourages the performing and recreation of various cultural expressions.

In Forty-Third Chapter "Enhancing Revisit Intention Through Emotions and Place Identity: A Case of the Local Theme Restaurant," *Alexander M. Pakhalov* and *Liliya M. Dosaykina* investigate food tourism as a unique segment of tourism that contributes to the development of tourist destinations. Recent researches show that local food experience can positively affect trip satisfaction, revisit intention, and destination's attractiveness. However, there is still a lack of research focusing on the relationship between experience of regional theme restaurants and destination's place identity. Their study aims to examine the relationships among tourist experience, emotions, and behavioral intentions based on a case of the local-specific theme restaurant Zaboi ("Coalface") located in Kemerovo, Russia's largest coal-mining region. Research design is based on a combination of a content analysis of online reviews and a visitors' survey followed by regression analysis. They focus on various aspects of visitor emotions, behavioral intentions, and relationships between visitor experiences and place identity. In line with previous studies, their results confirm a significant positive relationship between emotions and revisit intention and provide new evidence on the role of place identity in shaping visitors' emotions. The authentic coal-mining theme provokes a positive emotional response for both tourists and local residents. This response, in turn, is positively associated with intentions to revisit and to recommend.

In Forty-Fourth Chapter "Anticipated Booking on Touristic Attractions: Flamenco Show in Seville," *Fernando Toro Sánchez* investigates Flamenco, an UNESCO Universal Heritage since 1990. As a tourist attraction in most destinations in Spain, it has special significance in Andalusia and especially in high growth destinations such as Seville.The anticipation of the reservation in a "Tablao" Flamenco in Seville and in two periods distinguished by the milestone of the realization of marketing actions by own online channels versus that of external channels and online travel agencies (OTAS) before.To do this, he uses a sample of 2759 bookings to a local flamenco show in Seville, of daily representation and that are offered by various channels both online and offline and which, the anticipation on the reservation is evaluated. Hence, a logarithmic scale regression is presented in addition to an analysis of associations between different categorical variables.To determine this partnership, machine learning techniques have been used with the big ML program that leads us to conclusions about advance bookings that can be of practical use in the marketing actions of local tourist attraction operators.As a consequence, a good practice own channels drives to get a better results on booking anticipation based on branding techniques of branding: The purpose of tourist marketing is to anticipate the reservations.

In Forty-Fifth Chapter "The Fisheries Local Action Groups (Flags) and the Opportunity to Generate Synergies Between Tourism, Fisheries and Culture," *Luis Miret-Pastor, Ángel Peiro-Signes, Marival Segarra-Oña*, and *Paloma Herrera-Racionero* aim to identify, quantify, and characterize cultural projects related to fishing financed in Spain through European fishing funds. A total of 136 cultural projects in five different categories have been found. The most popular projects through the two founding programs are related to tangible and intangible heritage projects and museums. The number of projects more or less maintained throughout the two periods is analyzed. Projects related to routes have been reduced by half from the first founding program, while festival projects more than doubled in the second founding program over the first.

In Forty-Sixth Chapter "Cultural and Tourism Promotion Through Digital Marketing Approaches. A Case Study of Social Media Campaigns in Greece," *Constantinos Halkiopoulos, Maria Katsouda, Eleni Dimou*, and *Antiopi Panteli* present the case study of the European Program named "Regio-Gnosis—Information and updating on cohesion policy in Greece." The project has launched at May of 2020, and it is based mainly on the use of promoted social media campaigns and many contests with cultural and tourism material. It managed to accomplish its target to reach over 4.5 million people. By analyzing data for the Facebook campaigns of "Regio-Gnosis," it is examined how helpful are the promoted social media campaigns to the increasing of popularity of a social media page, what is the role of social media contests and how the large audiences can be reached.

In Forty-Seventh Chapter "Challenges and Opportunities for the Use of Indoor Drones in the Cultural and Creative Industries Sector," *Virginia Santamarina-Campos, María de-Miguel-Molina, Blanca de-Miguel-Molina*, and *Marival Segarra-Oña* present the AiRT project's results, achieved jointly by all partners between January 2017 and January 2019. Significant innovative progress has been made in this short period of time. The far-reaching effects on the drone manufacturing industry as a whole, as well as in the cultural and creative industry (CCI), will likely only be precisely quantified in the coming years. This is primarily due to the fact that the innovative integrated indoor positioning system not only brings advantages to the creative industries, but will certainly inspire other industries in the future that will adapt this innovative product to their specific needs. Overall, however, the authors hope that the AiRT project has drawn attention to the needs of the CCCIs, especially for SMEs, so that they will be able to strengthen their economic position within the EU through the solution they offer.

In Forty-Eighth Chapter "Intangible Cultural Heritage in Spata Greece: From Mythology to Gastronomic Folklore and from Tradition to Contemporary Culture," *Dionysia Fragkou, Loukia Martha*, and *Maria Vrasida* critically engage with the articulations and manifestations of a clear shift UNESCO made by suggesting a new definition of intangible cultural heritage. The historic development of the notion of intangible cultural heritage within UNESCO is considered in order to explore some of the implications and opportunities that may arise from this shift in approach

(Whitaker, 2017). The focus of this paper is on a folklore (intangible cultural activity) taking place in Spata Greece and involves festivities and the communal cooking of a traditional meal. The images of communal co-creation and participation present a living tradition that has been neglected and overlooked for many years. This article aims to explore the potential for revitalizing and promoting an old tradition, through the use of technology. Issues such as networking, community participation, digital community participation, physical and digital experience, and more will be discussed in detail. Using and analyzing the example of Sparta, the proposed approach aims at conceiving folklore and intangible heritage not only as the ways of a distant past but also as a symbolic and living space to be appropriated by local communities who are the bearers of a collective and active memory. Following the trend for digital experiences, the new tradition is not only revived or viewed as a representation but it becomes an active agent for creating and promoting social cohesion, sense of community, and civic pride far beyond the physical frontiers.

In Forty-Ninth Chapter "Silk Road Regionalism and Polycentric Tourism Development," *Stella Kostopoulou, Evina Sofianou,* and *Dimitrios Kyriakou* focus on designing a polycentric network of regional tourism destinations related to Silk Road cultural heritage, where interregional tourism alliances will connect regions through entrepreneurial connectivity, based on the notion of polycentricity. Polycentricity indicates the connection of neighboring centers that have common characteristics and their integration in wider spatial networks. The interest is focused on cross-border linkages among polycentric regional networks of Silk Road cultural heritage destinations, creating tourism development entrepreneurial networks that could take advantage of historically cumulative relations. These polycentric tourism networks may emerge due to a number of factors falling under the rubric of Silk Road regionalization and its attendant tourism dimensions. The potential for building an inter-regional Black Sea Silk Road tourism cooperation polycentric network focusing on active societies engaging in new types of cultural tourism experiences is being put in place.

Please enjoy and share our book.

References

Cooper, C. (2012). *Essentials of tourism.* Harlow: Financial Times Prentice Hall Publishing.

IATA. (2020). *Monthly statistics.* Montreal: International Air Transport Association (IATA).

Leiper, N. (1990). *Tourism systems.* Auckland: Massey University Department of Management Systems Occasional Paper 2.

Poulaki, I., & Katsoni, V. (2020). Current trends in air services distribution channel strategy: Evolution through digital transformation. In Katsoni, V., Spyriadis, T. (Eds.), *Cultural & tourism innovation in the digital era* (pp. 257–267). 10.1007/978-3-030-36342-0_21

Stratigea, A., & Katsoni, V. (2015). A strategic policy scenario analysis framework for the sustainable tourist development of peripheral small Island Areas—The case of Lefkada-Greece Island. *European Journal of Futures Research (EJFR), 3*(5), 1–17. 10.1007/s40309-015-0063-z

UNTWO. (2020a). *International tourist arrivals could fall by 20–30% in 2020.* Madrid: World Tourism Organization.

UNTWO. (2020b). *Sports tourism.* Madrid: World Tourism Organization.

WTTC. (2020). *Covid 19.* London: World Travel and Tourism Council.

Contents

Digital Innovation

Cultural Innovation

Sports Tourism

Sport Tourism: An Analysis of Possible Developmental Factors in Sport and Recreation Centers

Georgia Yfantidou, Charalampos Spiliakos, Ourania Vrondou, Dimitris Gargalianos, Antonia Kalafatzi, and Eleni Mami

1 Introduction

The individuals who participate in sport tourism have different characteristics from ordinary tourists (Aicher et al., 2015; Gibson, 2011). Their interest and motivation are based on various factors, such as the popularity of the sporting events, the international/national/regional sporting events (Green & Chalip, 1998), the pursuit of a change in their daily life routine (Bouchet et al., 2004), the sport itself (Gammon & Robinson, 2005), the exploration and the escape in personal or interpersonal level (Caber & Albayrak, 2016), the improvement of health and fitness (Hagger & Chatzisarantis, 2005), the relaxation and recreation (Cohen, 2003), social–psychological motives for traveling to a destination (tourism-oriented, sport tourism enthusiasts, and sport-oriented (Hungenberg et al., 2016), and the sense of belonging (Hagger & Chatzisarantis, 2005). Important role for the expression of this kind of tourist's behavior plays their gender, age, and nationality race (Bull, 2011), as well as their physical activity profile (shapers, all-around pursuers, sport performers, and decompress seekers) (Osti et al., 2018), which affect their decision to participate in sports.

C. Spiliakos · O. Vrondou
Department of Sports Organization and Management, University of Peloponnese, Sparti, Greece

G. Yfantidou (✉) · D. Gargalianos
Department of Physical Education and Sport Science, Democritus University of Thrace, Komotini, Greece
e-mail: gifantid@phyed.duth.gr

A. Kalafatzi
Faculty of Health Sciences and Sport, University of Stirling, Stirling, UK

E. Mami
Department of Business Administration, School of Business, Athens University of Economics and Business, Athens, Greece

© The Author(s), under exclusive license to Springer Nature Switzerland AG 2021
V. Katsoni and C. van Zyl (eds.), *Culture and Tourism in a Smart, Globalized, and Sustainable World*, Springer Proceedings in Business and Economics, https://doi.org/10.1007/978-3-030-72469-6_1

3

Many factors and their typologies, which influence the development of sport tourism of a specific place, are suggested in the literature, including but not restricted to self-expressiveness depending on activity's perceived difficulty, perceived effort, perceived importance, potential for self-realization (Bosnjak et al., 2016), ease of access, political/economical/cultural/environmental aspects (Bull, 2011), sport tourism's products, resources and funding, policies, design of tourist activities, availability of information, agricultural/environmental/aquatic places (Weed & Bull, 2009), tourist product, environment in which the services are offered, invisible organizations and systems, services' providers and other customers (Harrison-Hill & Chalip, 2005), accommodation facilities, sporting facilities and infrastructure, sporting event or activity (Shock & Chelladurai, 2008), level of orientation toward environmental friendliness, ecotourism, or sport activities, sport tourism (vacation, green, action-oriented, and active tourists) (Singh et al., 2016).

However, although various authors have identified several factors that affect the development of sport tourism based on different criteria, no effort has been made for their classification creating a gap in the existing literature. By identifying and classifying these factors, relevant businesses, including sport and recreation centers, might increase their possibilities to succeed.

The aims of the present study were (a) the identification and classification of the most and least important factors for the development of sport tourism, (b) the investigation of the level by which these factors are met in a sport and recreation center, and (c) the comparison of opinions of the users of the center regarding these factors.

2 Literature Review

Many definitions have been proposed in the literature for sport tourism, which tend to include mainly basic dimensions and features of tourism. Weed (2008: 7) defined sport tourism as follows: «…a social, economic and cultural phenomenon arising from the unique interaction of activity, people and place», Gibson (2013: 12) approached it from an attraction perspective as: «… sport-based travel away from the home environment for a limited time where sport is characterized by unique rule sets, competition related to physical prowess and a playful nature», while Van Rheene et al. (2017) underlined the need for further examination of the epistemological boundaries of sport tourism.

Regarding the models and typologies for sport tourism and/or sport tourists, the criteria used in each occasion are different and include the demand (Glyptis, 1982), the competitiveness rate and the level of activity (Hall, 1992), the participation motives (Kurtzman & Zauhar, 1997; Kurtzm, 2005), and the participation level (Gibson, 1998, 2003; Standeven & De Knop, 1999). Other authors suggest typologies based on the type of tourists, with the dominant element being their distinction into «active» and «passive» (Standeven & De Knop, 1999). Krsmanović and Gajić (2016) suggested that sport tourism is distinguished into (a) competition (attending or participating in an event) and (b) leisure/recreational tourism. Weed and Bull (2009) suggested a model typology, which summarizes all these points and discern sport tourism in (a) supplementary or tourism with sports content, (b) participation, (c) training, (d) event, and (e) luxury adding significant dimensions. These dimensions are inter-excluded between them and refer to multiple or single sports and active or passive or vicarious activities. Every type of sport tourism described above can potentially demonstrate these features/dimensions, even if they constitute a determinant part of a specific type of sport tourism (Weed, 2008).

As sport tourists have different characteristics than other kinds of tourists, it is important to examine their motives (Gibson, 2005, 2011). It seems that many factors, such as the popularity of particular world level sporting events (e.g., Olympic Games, Football World Cup, etc.), contribute in increasing their interest and motivation toward sports (Green & Chalip, 1998). According to Kurtzman and Zauhar (2005), the individual's decision to participate in sport tourism is distinguished in two basic options: (a) the fake option, which constitutes the result of combinational factors, such as family, friends, and media messages and (b) the intentional option, which occurs when the tourist has an internal tendency and preference for sports and for this reason they choose to arrange a trip. Additionally, Gammon and Robinson (1997, 2005) pointed out that a person travels and participates in sport tourism pushed by sports either primarily, or secondarily.

As individuals' behavior is led by their biological and social–psychological needs, desires, and motives, the typical motivation theories can contribute in understanding their behavior regarding recreation, sports, and tourism (Higham, 2011). It has been found that socio-demographic (gender, age) and socioeconomic (income, level of education) factors play an important role in the selection of a sport tourism destination (Slak-Valek et al., 2014). The natural environment and safety/security issues have been found to play a significant role in a decision for participation in sport events (Perić et al., 2018). Also, sport tourists' perception of destination image, perceived quality, and perceived value positively influence their satisfaction and revisit intention (Allameh et al., 2015).

In personal level, significant motives constitute the improvement of health and physical condition (Hagger & Chatzisarantis, 2005), as well as the relaxation and recreation (Cohen, 2003), while in interpersonal level very important motive constitutes the individual's need to belong in a team or in the society (Hagger & Chatzisarantis, 2005). The overall event image also impacts the traveling parents' subjective norms, attitudes, and behavioral intentions (Kaplanidou & Gibson, 2012). Moreover, the participation in sport tourism is also influenced by cultural factors, as the tourist, by watching or by participating in local sporting events, acquires the experience of participation in a cultural event (Choi et al., 2016; Hinch & Higham, 2005). Recent studies also indicate the significant role of sport infrastructure such as marine sport tourism attractions (Hsiao-Ching, 2017), as well as this of wellness and spa services (Hashemi et al., 2015), which directly affect the travel experience, the perceived value, and the revisit intention of tourists.

In tourism research, the idea of combining the highest individual motivation level with an activity or an environment could also be perceived as a need of combing the "push factors," which refer to the reasons why the individuals want to escape from daily routine and distance themselves from their usual place of residence and the "pull factors," which refer to the reasons why an individual desires to be at a particular destination (Caber & Albayrak, 2016; Turco, et al., 2002). Recent studies also indicate that involvement, travel motives, and decision factors significantly affect motivational factors, and motivational factors significantly affect decision factors. A combination of positive motivation factors, positive involvement, and negative travel motives can significantly influence decision factors (Fotiadis et al., 2016).

The nature of sport tourism contains experiences from both sports and tourism (Standeven & De Knop, 1999). A series of factors referring to the development of sport tourism (natural characteristics, services, entertainment, transportation, facilities, accommodation, legacy, comforts, organizations, etc.) derive by combining the different factors suggested separately for sports and tourism (Standeven & De Knop, 1999). Bull and Weed (2012) suggested the "Wheel of Policy for Sports and Tourism," which includes products of sport tourism, facilities, resources and funding, policy and planning, information and promotion, agricultural, environmental and aquatic elements, while Fatemeh et al., (2014) argued that advertisement factors highly contribute to the development of sport tourism industry. Kabanova & Vetrova (2017) pointed out that services differentiation level, quality and prices incompatibility, strategic creation of positive country image, and lack of professionals involved in sport tourism are key developmental sport tourism factors. Moreover, Bouchet et al., (2004) suggested that the total experience obtained by the consumer of sport tourism is influenced by variables, which is associate with (a) the specific individual and their self-esteem (perceived danger, extensions and applications of sports, searching for the optimum, searching for variety and innovation), (b) the space (region where the consumer can observe/consume/interact/share experiences), and (c) interpersonal variables.

The local/national/international sporting events constitute significant factors for tourism development (Mason & Duquette, 2008) and compare to the capabilities of the host city (Higham, 1999) because major sporting events demand huge investments in infrastructure and sporting facilities (Preuss & Solberg, 2006). With the intention, though, a sporting activity/event to contribute to the development of tourism at a destination, such an activity should integrate with other tourist products and services available there. Taleghani and Ghafary (2014) point out that services related to hill climbing, beach sports, and desert and hunting attractions could form an effective developmental model of sport tourism, while it is underlined that cultural barriers and infrastructure problems should be overcome.

King et al., (2018) argued that a crucial developmental factor of sport tourism, which is elated to major sport event organizers, is the involvement and satisfaction level of the local sport clubs and the connections with local places, while Shock and Chelladurai (2008) suggested that the following features of sport tourism are critical for the quality of the services offered and the development of sport tourism: (a) access to destination, sporting event or facility, accommodation, (b) lodging, which includes interactions, environment, and value, (c) sporting facilities and infrastructure, which include interactions, environment, and value, and (d) sporting event or activity, which includes the process and the product/service. Moreover, Higham (2005) suggested that the gyms and the sporting classes, the infrastructure for aquatics (swimming pools, lakes, springboards), the infrastructure for other sports (sporting routes, routes for marathons), the medical infrastructure and the training fields, and the service and the staff in sporting and recreation centers constitute important elements playing a particular role in the decision-making process of athletes to visit this kind of centers. Myburgh et al., (2018) supported that lifestyle, event attributes, and travel behavior are positively related to athletes' commitment to sport events, which indicates the role the strategic management of these factors can play in the sport tourism development.

2.1 Tourism in Greece

In 2017, tourism contributed to the country's GDP at 27.3% and to the employment at 24.8% (934.500 jobs). There were 9.783 hotels (806.045 beds) in the country with 70% concentration, and the international tourist arrivals reached 27.2 million people (cruise arrivals are not included), with an average expense of €522 per person (SETE (2017a, 2017b). It is important to mention that in this particular year Greece was in the 24th place in competitiveness of its tourism, and Greek tourism elicited 4% of the European market and 2% of the worldwide one (Ikkos, 2018). Its main weakness is the seasonality, as in 2017, 57.3% of the arrivals and 61.2% of the revenue occurred in the third quarter of the year, while only 5.8% of the arrivals and 3.5% of the revenue occurred in the first quarter (SETE (2017a, 2017b).

2.2 Sport Tourism and Sport and Recreation Centers in Greece

Sport tourism in Greece is intensively associated with ecotourism and health tourism, while the sport tourism markets in the country are the sporting events, the resorts and the recreation facilities, outdoors programs, and activities/programs in sporting centers or sport and recreation camps (Costa & Clinia, 2004). In 2015, in Greece, there were 249 physical activity facilities, 218 spas and springs, and 341 sporting and entertainment clubs (Table 1).

2.3 The Sport and Recreation Center "Sportcamp"

In terms of infrastructure and annual economic cycle, Sportcamp (www.sportc amp.gr) is the biggest private sporting and training center of the country. It is located in the city of Loutraki, 75 km away from Athens and 5 km away from the sea, near

Table 1 Sporting centers of physical activity and recreation subcategories

	Category[*]	Total establishments	Total sales ($m)	Total employment
Physical fitness facilities	Minor 1	220	39	1.88
Physical fitness facilities	Minor 1	29	6	324
Athletic club and gymnasiums, membership	Minor 2	41	13	761
Health club	Minor 2	78	28	1.58
Spas	Minor 2	218	25	1.45
Weigh reducing clubs	Minor 1	4	1	31
Reducing facility	Minor 2	0	0	2
Slenderizing salon	Minor 2	2	0	8
Exercise facilities	Minor 1	16	2	107
Aerobic dance and exercise classes	Minor 2	20	3	152
Exercise salon	Minor 2	53	5	333
Membership sports and recreation clubs	Minor 1	341	61	3.32

Source Barnes Reports: Worldwide Industry Market (2015)

the Gerania Mountains. It expands in an area of 75.000 m^2, and it has 850 beds, as well as facilities for the provision of integrated services of hospitality, boarding, sports, recreation, and tourism at its disposal. In cooperation with specialists, it offers ergometric, education, physical condition, and rehabilitation services. Every year it hosts many individuals who engage into sport tourism, so it is particularly useful to be investigated as a case study.

3 Methodology

3.1 Participants

The sample of this study constituted of 206 individuals, classified into four categories (athletes, executives, team managers, coaches), who were staying in the center, in summer 2015. They were selected through simple random sampling from the list of visitors, meaning that all had the same possibility to be selected to participate in the study (Thomas et al., 2015).

3.2 Instruments

To achieve the aims of the study, the structured questionnaire was selected (Gratton & Jones, 2010) and was compiled by the researcher. It included eight broader questions (close end and open end) and three subsections. In the first section, the respondents were asked to express their opinion regarding how important 34 items, derived from the review of literature and presented in Table 2 in random order, were for the development of sport tourism in a sport and recreation center. In the second section, the participants were asked to express the level in which they thought that the center met these factors for the development of sport tourism. On both sections, the answers were given on a Likert-type scale ($1 =$ not important, $5 =$ very important) (Joshi et al., 2015). Both sections' reliability was estimated with Cronbach's (1951) alpha index, and it was found to be satisfying ($a = 0.89$ and $a = 0.93$). In the third section, the respondents were asked to complete specific demographics (gender, age, nationality, and category), the physical activity they are engaged in, and the duration of their stay in the center. The questionnaire was anonymous and compiled in three languages (Greek, English, Russian).

3.3 Procedure

The questionnaires were distributed and collected inside the center, with the investigator being present in order to facilitate the participants in case they needed any assistance.

3.4 Pilot Study

In order to time the duration of the questionnaire's completion (so as not to exceed 10 min) and to identify possible omissions, ambiguities, misconstructions, syntactic, and spelling mistakes in all three languages, a pilot study was conducted, involving five employees and ten guests (7.3% of the total).

3.5 Statistical Analysis

The data was analyzed with the use of the SPSS 21.0 program. The frequency distribution of the sample was examined, and a t-test control and Levene's test were conducted.

Table 2 Examined items

1. Proximity to the beach and the sea	18. Accommodation facilities (camp, rooms)
2. Proximity to airports	19. Promotional posters and informative leaflets
3. Access through the road network	20. Sporting programs (flexibility and variety)
4. Access through the national bus network (KTEL)	21. Training and supervision (specialized coaches/trainers)
5. Access through the railway network	22. Swimming pools (outdoor—indoor)
6. Climate conditions	23. Electronic provisions and Internet
7. Parking spaces	24. Proximity to sights, nightlife
8. Healthcare and injuries' rehabilitation	25. Proximity to historical monuments
9. Security	26. Proximity to malls
10. Restroom, bathroom, changing room facilities	27. 24 h information desk service
11. Catering facilities	28. Awareness of the name of the exercise and recreation center
12. Image and name of the country/city/region of destination	29. Gym and sporting equipment
13. Organizing sporting event, festivals (national/regional)	30. Delivery and transfer infrastructure for travelers/visitors
14. Scenery	31. Web site, blogs, and social media
15. Facilities' hygiene conditions	32. Offers, sales, and loyal customers' reward
16. Stadiums, playing fields (outdoor/indoor)	33. Proximity to a port, marinas
17. Facilities' signage, transfer guidelines (signs, maps)	34. Customer service

3.6 Limitations of the Study

This study was conducted in a specific sport and recreation center, and the participants were guests there in a specific period of time; therefore, the findings could not be generalized. Also, only the guests' opinions of the sport and recreation center were investigated and not the employees' ones.

4 Results

The participants of the study were 206 individuals, coincidentally distributed evenly between men (103–50%) and women (103–50%). The highest percentage (49.51%) was 15–24 years old, 23.3% was < 14 years old, 15% was 25–34 years old, 6.80% was 35–44 years old, 3.4% was 45–54 years old, and only 1.94% was 55+ years old. Most of them were Greeks (51.94%), 27.18% Russian, 11.65% German, followed by Egyptians, Albanians and Danes (each 2.43%), Lebanese (0.95%), Ukrainians (0.49%), and Poles (0.49%). The total number of athletes was 157 (76.21%) and the team officials 49 (23.78%).

The majority of the participants were amateur (50.97%) and professional athletes (25.24%), while 10.68% were coaches, 8.25% were team managers, curators, or individuals in charge of teams, while 4.85% were administrators of teams. The majority of athletes (35.92%) was engaging into a team sport (soccer, basketball, volleyball, handball, rugby, beach volley/soccer/handball), followed by aquatic sports (swimming, water polo, synchronized and technical swimming) (28.64%), individual sports (cycling, archery, triathlon, artistic and rhythmic gymnastics, fencing, shooting) (15.53%), heavy sports (weightlifting, wrestling, boxing) (5.34%), combat sports (Tae Know Do, karate, judo) (5.83%), racquet sports (tennis, table tennis, squash, badminton) (1.94%), water sports (water ski, canoe-kayak, rowing, dragon boats) (1.46%), recreation activities (dance, yoga), and other activities (5.34%). The majority (41.26%) stayed at the center 6–10 days, 26.21% stayed 14+ days, 18.93% stayed 5 days, and 13.59% stayed 11–14 days.

It could be argued that the most important developmental factors of sport tourism for a sport and recreation center are the those evaluated with a score of 4–5 (positions 1–15), of average importance could be categorized the factors evaluated with a score of 3–4 (positions 16–32), while of low importance could be categorized those evaluated with a score <3 (positions 33–34) (Table 3).

The frequencies and relative frequencies' distribution analysis showed that none of the respondents characterized the healthcare and the injuries' rehabilitation as not at all important for the development of sport tourism in a sport and recreation center, while a percentage smaller than 1% of the sample characterized as not important the items of security, catering and accommodation facilities, restroom, bathroom, changing room facilities, gym and sporting equipment, customer service (0.97% each), climate conditions, and stadiums—playing fields (0.49% each). On the other hand, more than 50% of the sample characterized as very important the items of security (68.45%), healthcare and injuries' rehabilitation (65.53%), accommodation facilities (61.17%), facilities' hygiene conditions (60.19%), gym/sporting equipment (53.80%), customer service (53.40%), stadiums/playing fields/swimming pools (50.97%), and electronic provisions and Internet (50%).

More than 30% of the participants characterized neither important—nor not important the items of access through the railway network (39.32%), proximity to a port and marinas (35.92%), proximity to airports (33.01%), proximity to historical museums (31.07%), and image and name of the country/city/region of the destination

Table 3 Mean and standard deviation scores for the developmental factors conditions of sport tourism for a sport and recreation center

	M	SD
1. Restroom, bathroom, changing room facilities	4.54	0.77
2. Security	4.53	0.78
3. Healthcare and injuries' rehabilitation	4.49	0.80
4. Accommodation facilities (camp, rooms)	4.46	0.80
5. Gym and sporting equipment	4.36	0.82
6. Catering facilities	4.31	0.87
7. Customer service	4.31	0.89
8. Facilities' hygiene conditions	4.30	1.04
9. Swimming pools (outdoor—indoor)	4.25	0.90
10. Stadiums, playing fields (outdoor—indoor)	4.25	0.90
11. Electronic provisions and Internet	4.18	0.97
12. Training and supervision (specialized coaches—trainers)	4.08	1.08
13. Climate conditions	4.04	0.85
14. Sporting programs (flexibility and variety)	4.01	0.98
15. Proximity to the beach and the sea	4.00	1.00
16. Facilities' signage, transfer guidelines (signs, maps)	3.88	0.96
17. Delivery and transfer infrastructure for travelers/visitors	3.77	0.99
18. Proximity to sights, nightlife	3.76	1.00
19. Scenery	3.68	0.96
20. Organizing sporting event, festivals (national/regional)	3.65	1.04
21. Access through the road network	3.63	1.10
22. 24 h information desk service	3.61	1.19
23. Web site, blogs, and social media	3.61	1.10
24. Offers, sales, and loyal customers' reward	3.60	1.17
25. Awareness of the name of the exercise and recreation center	3.52	1.04
26. Access through the national bus network (KTEL)	3.30	1.20
27. Proximity to airports	3.29	1.14
28. Image and name of the country/city/region of destination	3.26	1.14
29. Parking spaces	3.20	1.21
30. Proximity to historical monuments	3.16	1.16
31. Promotional posters and informative leaflets	3.09	1.24
32. Proximity to malls	3.08	1.18
33. Proximity to a port, marinas	2.99	1.08
34. Access through the railway network	2.81	1.14

(30.58%). This practically means that the tourists who visit the sport and recreation centers do not pay particular attention to these factors.

Subsequently, the level to which Sportcamp meets the above developmental conditions of sport tourism was examined. The respondents indicated that the center meets:

- In a very high level ($M > 4$) the following conditions: stadiums/playing fields (outdoor–indoor), security, and gym and sporting equipment.
- Adequately ($M < 4 - > 3$) the following conditions: climate condition, swimming pools (indoor/outdoor), facilities' hygiene, accommodation facilities (camp–rooms), changing rooms/bathrooms/resting rooms facilities, health-care and injuries' rehabilitation, sporting programs (flexibility–variety), training and supervision (specialized coaches/trainers), facilities' signage, transfer guidelines (signs, maps), organizing sporting events/festivals (national and regional), catering facilities, name awareness of the center, space's beauty, access through the road network, image/name of the country/city/region of the destination, delivery and transfer infrastructure of the travelers/visitors, Web site/blogs/social media, parking spaces, 24 h information desk service, electronic provisions and Internet, offers/sales/loyal customers' reward, proximity to attractions, nightlife, promotional posters/informative leaflets, proximity to the beach/sea, and access through the national bus network (KTEL).
- Not adequately ($M < 3$) the following conditions: proximity to historical monuments, to airports, to malls, access through the railway network, and proximity to a port and marinas.

A rather high number of the sample's participants indicated that Sportcamp does not meet the condition of proximity to a port (31.55%), access through the railway network (27.67%), and proximity to airports (24.76%). On the other hand, a significant part of the sample thinks that the center meets at a great extent (more than 30%) the following conditions developing sport tourism in an exercise and recreation center: playing fields (40.29%), gym and sporting equipment (38,83%), security (37.86%), stadiums, swimming pools (33.50%), and customer service (32.52%). Moreover, approximately more than 40% of the sample thinks that climate conditions (53.88%), sporting programs (flexibility and variety) (42.72%) security (39.81%), and accommodation facilities (39.32%) are met at a great extent.

A significant part of the examined sample, more than 30%, marked moderately (neither important—nor not important) Sportcamp regarding the following factors: sales and loyal customers' reward (36.89%), image and name of the country/city/region of destination (36.41%), access through the road network (35.44%), scenery (34.95%), access through the national bus network (KTEL) (33.01%), offers, proximity to sights, nightlife (33.01%), awareness of the name of the center (32.52%), facilities' signage, transfer guidelines (signs, maps) (31.55%), and parking spaces (30.10%).

The opinions of the visitors/recipients of the sport center's tourism product were also examined with the use of the criterion t and the independent samples t-test. The independent variables were the respondents' opinions regarding the importance of each of the 34 factors for the development of sport tourism and the extent to which Sportcamp meets these conditions. The dependent variable is referred to the individual's status (professional/amateur athlete, coach, team manager/curator/team responsible, top executive) (Table 4).

In aggregate, the average scores of the athletes regarding the importance of the factors developing sport tourism were significantly different in relation to the average scores of the team executives in the following factors: F.1 (proximity to the beach and the sea), F.3 (access through the road network), F.4 (facilities' hygiene conditions), F.5 (access through the railway network), F.6 (proximity to airport), F.9 (security), F.10 (restroom, bathroom, changing room facilities), F.13 (organizing sporting

Table 4 Group statistics, factors developing sport tourism/status

Factor developing sport tourism	Status	M	SD	SEM
F.1	Athlete	3.91	1.04	0.08
	Team executive	4.30	0.82	0.11
F.3	Athlete	3.50	1.13	0.09
	Team executive	4.02	0.94	0.13
F.4	Athlete	4.19	1.11	0.08
	Team executive	4.67	0.68	0.09
F.5	Athlete	2.68	1.13	0.09
	Team executive	3.24	1.10	0.15
F.6	Athlete	3.14	1.15	0.09
	Team executive	3.79	0.95	0.13
F.9	Athlete	4.48	0.82	0.06
	Team executive	4.71	0.57	0.08
F.10	Athlete	4.47	0.83	0.06
	Team executive	4.79	0.45	0.06
F.13	Athlete	3.56	1.05	0.08
	Team executive	3.93	0.96	0.13
F.17	Athlete	3.80	0.95	0.07
	Team executive	4.14	0.97	0.13
F.19	Athlete	2.95	1.28	0.10
	Team executive	3.55	0.98	0.14
F.23	Athlete	4.11	1.01	0.08
	Team executive	4.42	0.79	0.11
F.31	Athlete	3.50	1.08	0.08
	Team executive	3.97	1.07	0.15
F.32	Athlete	3.42	1.17	0.09
	Team executive	4.18	0.97	0.13

events, festivals—national and regional), F.17 (facility's signage, transfer guidelines—signs, maps), F.19 (promotional posters and informative leaflets), F.23 (electronic provisions and Internet), F.31 (Web site, blogs, and social media), and F.32 (offers, sales, and loyal customers' reward). Also, in most of the cases, the teams' executives considered these factors as more important than athletes did (Table 5).

In aggregate, the average scores of the athletes regarding the extent to which Sportcamp meets the conditions developing sport tourism in sport and recreation centers in Greece were significantly different in relation to the average scores of the team executives in the cases of the following factors: F.1.S (proximity to the beach and the sea), F.3.S (access through the road network), F.12.S (image and name of the country/city/region of destination), F.13.S (organizing sporting events, festivals—national and regional), F.15.S (access through the national bus network—KTEL), F.17.S (facilities' signage, transfer guidelines—signs, maps), F.19.S (promotional posters and informative leaflets), F.29.S (gym and sporting equipment), F.31.S (Web site, blogs, and social media), and F.32.S (offers, sales, and loyal customers' reward) (Table 6).

The above results were confirmed by the *t*-test for equality of means analysis (Table 7).

5 Conclusion

The major conclusions of this study were that (a) the most important developmental factors of sport tourism in a sport and recreation center in Greece are the changing rooms/bathrooms/resting rooms, the security, the healthcare and the injuries' rehabilitation, and the accommodation facilities (camp–rooms), and the less important factors were the proximity to a port and marinas, as well as the access through the railway network, (b) Sportcamp meets the changing rooms/bathrooms/resting rooms, the healthcare and the injuries' rehabilitation, and the accommodation facilities (camp–rooms) factors in adequate level and the security factor in a very high level, and (c) in many cases the athletes' opinions regarding the importance of the factors developing sport tourism are significantly different in relation to the opinions of the team executives.

The present study's results could be useful to sport and recreation centers for decision-making and strategy formulation regarding the required improvements and investments. It would be useful in the future to (a) collect longitudinal data from the same center, (b) repeat this study in other sport and recreation centers, and (c) conduct qualitative studies, which would investigate the opinions of the administrators of the sport and recreation centers regarding the factors influencing the development of sport tourism.

Table 5 Independent samples test factors developing sport tourism/status

| | | Levene's test | | t-test for equality of means | | | | | 95% Conf. Interv. of Diff | |
		F	Sig	t	df	Sig. (2-tailed)	Mean Diff	Std. Err. Diff	Lower	Upper
F.1	Equal variances assumed	0.56	0.45	-2.43	204	0.01	-0.39	0.16	-0.71	-0.07
	Equal variances not assumed			-2.75	100.29	0.00	-0.39	0.14	-0.68	-0.11
F.3	Equal variances assumed	4.25	0.04	-2.86	204	0.00	-0.51	0.17	-0.86	-0.15
	Equal variances not assumed			-3.14	94.48	0.00	-0.51	0.16	-0.83	-0.18
F.4	Equal variances assumed	15.70	0.00	-2.85	204	0.00	-0.48	0.16	-0.81	-0.14
	Equal variances not assumed			-3.63	131.66	0.00	-0.48	0.13	-0.74	-0.21
F.5	Equal variances assumed	0.09	0.76	-3.05	204	0.00	-0.56	0.18	-0.92	-0.19
	Equal variances not assumed			-3.08	81.65	0.00	-0.56	0.18	-0.92	-0.20
F.6	Equal variances assumed	2.66	0.10	-3.61	204	0.00	-0.65	0.18	-1.01	-0.29
	Equal variances not assumed			-3.98	95.21	0.00	-0.65	0.16	-0.98	-0.32
F.9	Equal variances assumed	8.35	0.00	-1.81	204	0.07	-0.23	0.12	-0.48	0.02

(continued)

Table 5 (continued)

	Levene's test		t-test for equality of means						95% Conf. Interv. of Diff	
	F	Sig	t	df	Sig. (2-tailed)	Mean Diff	Std. Err. Diff		Lower	Upper
Equal variances not assumed			−2.17	114.99	0.03	−0.23	0.10		−0.43	−0.02
F.10 Equal variances assumed	22.37	0.00	−2.59	204	0.01	−0.32	0.12		−0.57	−0.07
Equal variances not assumed			−3.48	150.78	0.00	−0.32	0.09		−0.50	−0.14
F.13 Equal variances assumed	2.85	0.09	−2.20	204	0.02	−0.37	0.16		−0.70	−0.03
Equal variances not assumed			−2.30	86.42	0.02	−0.37	0.16		−0.69	−0.05
F.17 Equal variances assumed	0.09	0.76	−2.17	204	0.03	−0.34	0.15		−0.64	−0.03
Equal variances not assumed			−2.13	78.30	0.03	−0.34	0.15		−0.65	−0.02
F.19 Equal variances assumed	6.96	0.00	−2.97	204	0.00	−0.59561	0.20		−0.98	−0.20
Equal variances not assumed			−3.42	104.32	0.00	−0.59	0.17		−0.94	−0.25

(continued)

Table 5 (continued)

		Levene's test		t-test for equality of means						95% Conf. Interv. of Diff	
		F	Sig	t	df	Sig. (2-tailed)	Mean Diff	Std. Err. Diff		Lower	Upper
F.23	Equal variances assumed	3.27	0.07	−1.98	204	0.04	−0.31	0.15		−0.62	−0.00
	Equal variances not assumed			−2.26	101.54	0.02	−0.31	0.13		−0.58	−0.03
F.31	Equal variances assumed	1.32	0.25	−2.68	204	0.00	−0.47	0.17		−0.82	−0.12
	Equal variances not assumed			−2.70	81.46	0.00	−0.47	0.17		−0.82	−0.12
F.32	Equal variances assumed	4.00	0.04	−4.08	204	0.00	−0.75	0.18		−1.12	−0.39
	Equal variances not assumed			−4.51	95.88	0.00	−0.75	0.16		−1.08	−0.42

Table 6 Group statistics, factors developing sport tourism—SPORTCAMP/status

Factor developing sport tourism	Status	M	SD	SE M
F.1.S	Athlete	2.92	1.41	0.11
	Team executive	3.53	1.26	0.18
F.3.S	Athlete	3.38	1.01	0.08
	Team executive	3.83	0.85	0.12
F.12.S	Athlete	3.36	1.05	0.08
	Team executive	3.87	0.85	0.12
F.13.S	Athlete	3.40	1.17	0.09
	Team executive	4.08	0.83	0.11
F.15.S	Athlete	2.94	1.27	0.10
	Team executive	3.36	0.95	0.13
F.17.S	Athlete	3.50	1.05	0.08
	Team executive	3.83	0.87	0.12
F.19.S	Athlete	3.00	1.20	0.09
	Team executive	3.69	0.89	0.12
F.29.S	Athlete	4.12	0.92	0.07
	Team executive	3.81	0.97	0.13
F.31.S	Athlete	3.34	1.19	0.09
	Team executive	3.85	0.97	0.13
F.32.S	Athlete	3.09	1.20	0.096
	Team executive	3.53	1.00	0.14

Table 7 Independent samples test factors developing sport tourism—Sportcamp status

		Levene's test		t-test for equality of means					95% Conf. Interv. of Diff	
		F	Sig	t	df	Sig. (2-tailed)	Mean Diff	Std. Err. Diff	Lower	Upper
F.1.S	Equal variances assumed	0.80	0.37	−2.66	204	0.00	−0.60	0.22	−1.04	−0.15
	Equal variances not assumed			−2.82	88.96	0.00	−0.60	0.21	−1.02	−0.17
F.3.S	Equal variances assumed	3.48	0.06	−2.83	204	0.00	−0.45	0.16	−0.77	−0.13
	Equal variances not assumed			−3.11	94.55	0.00	−0.45	0.14	−0.74	−0.16
F.12.S	Equal variances assumed	3.03	0.08	−3.11	204	0.00	−0.51	0.16	−0.83	−0.18
	Equal variances not assumed			−3.46	97.03	0.00	−0.51	0.14	−0.80	−0.21
F.13.S	Equal variances assumed	12.29	0.00	−3.73	204	0.00	−0.67	0.18	−1.02	−0.31
	Equal variances not assumed			−4.43	111.64	0.00	−0.67	0.15	−0.97	−0.37
F.15.S	Equal variances assumed	3.84	0.05	−2.15	204	0.03	−0.42	0.19	−0.81	−0.03
	Equal variances not assumed			−2.50	106.41	0.01	−0.42	0.16	−0.76	−0.08
F.17.S	Equal variances assumed	4.66	0.03	−1.97	204	0.05	−0.32	0.16	−0.65	0.00

(continued)

Table 7 (continued)

		Levene's test		t-test for equality of means						95% Conf. Interv. of Diff	
		F	Sig	t	df	Sig. (2-tailed)	Mean Diff	Std. Err. Diff		Lower	Upper
	Equal variances not assumed			−2.17	95.35	0.03	−0.32	0.15		−0.62	−0.02
F.19.S	Equal variances assumed	4.62	0.03	−3.72	204	0.00	−0.69	0.18		−1.06	−0.32
	Equal variances not assumed			−4.34	107.05	0.00	−0.69	0.15		−1.01	−0.37
F.29.S	Equal variances assumed	1.25	0.26	1.98	204	0.04	0.30	0.15		0.00	0.60
	Equal variances not assumed			1.93	77.35	0.05	0.30	0.15		−0.00	0.61
F.31.S	Equal variances assumed	5.83	0.01	−2.73	204	0.00	−0.51	0.18		−0.88	−0.14
	Equal variances not assumed			−3.03	96.30	0.00	−0.51	0.16		−0.84	−0.17
F.32.S	Equal variances assumed	0.24	0.62	−2.29	204	0.02	−0.43	0.18		−0.80	−0.06
	Equal variances not assumed			−2.52	94.93	0.01	−0.43	0.17		−0.77	−0.09

References

Aicher, T. J., Brenner, J., & Aicher, T. J. (2015). Individuals' motivation to participate in sport tourism: A self-determination theory perspective. *International Journal of Sport Management, Recreation & Tourism, 12*, 167–190.

Allameh, S. M., J, Khazaei, Pool, Jaberi, A., Salehzadeh, R., & Asadi, H. (2015). Factors influencing sport tourists' revisit intentions: The role and effect of destination image, perceived quality, perceived value and satisfaction. *Asia Pacific Journal of Marketing & Logistics, 27*(2), 191–207.

Barnes, C. & Co. (2015). *Fitness and recreational sports centers (NAICS 71394).* 2015 World Industry & Market Outlook, Barnes Reports: Worldwide Industry Market Report.

Bosnjak, M., Brown, C. A., Lee, D. J., Yu, G. B., & Sirgy, M. J. (2016). Self-expressiveness in sport tourism: Determinants and consequences. *Journal of Travel Research, 55*(1), 125–134.

Bouchet, P., Lebrun, A., & Auvergne, S. (2004). Sport tourism consumer experiences: A comprehensive model. *Journal of Sport Tourism, 9*(2), 127–140.

Bull, C. (2011). Sport tourism destinations resource analysis. In J. Higham (Ed.), *Sport tourism destinations.* NY: Routledge.

Bull, C., & Weed, M. (2012). *Sports tourism: Participants, policy and providers.* Routledge.

Caber, M., & Albayrak, T. (2016). Push or pull? Identifying rock climbing tourists' motivations. *Tourism Management, 55*, 74–84.

Choi, D. W., Shonk, D. J., & Bravo, G. (2016). Development of a conceptual model in international sport tourism: Exploring pre-and post-consumption factors. *International Journal of Sport Management, Recreation & Tourism, 21*, 21–47.

Cohen, E. (2003). The social psychology of tourist behaviour. *Annals of Tourism Research, 15*(1), 29–46.

Costa, G., & Clinia, E. (2004). Sport tourism in Greece. *Journal of Sport Tourism, 9*(3), 283–286.

Cronbach, L. J. (1951). Coefficient alpha and internal structure of tests. *Psychometric, 16*, 297–333.

Fatemeh, E. K., Mehrdad, M., & Sohrab, G. (2014). The role of advertisement factors in development of sport tourism industry of Fars province. *Physical Education of Students, 18*(3), 61–66.

Fotiadis, A., Xie, L., Li, Y., & Huan, T. C. T. (2016). Attracting athletes to small-scale sports events using motivational decision-making factors. *Journal of Business Research, 69*(11), 5467–5472.

Gammon, S., & Robinson, T. (1997). Sport and tourism: A conceptual framework. *Journal of Sport Tourism, 4*(3), 11–18.

Gammon, S., & Robinson, T. (2005). A question of primary and secondary motives: Revisiting and applying the sport tourism framework. *Journal of Sport Tourism, 3*, 221–233.

Gibson, H. J. (1998). Sport tourism: A critical analysis of research. *Sport Management Review, 1*(1), 45–76.

Gibson, H. J. (2003). Sport tourism: An introduction to the special issue. *Journal of Sport Management, 17*, 205–213.

Gibson, H. J. (2005). Towards an understanding of why sport tourist do what they do. *Sport in Society, 8*(2), 198–217.

Gibson, H. J. (2011). Understanding sport tourism experiences. In J. Higham (Ed.), *Sport tourism destinations.* Routledge.

Gibson, H. J. (Ed.). (2013). *Sport tourism: Concepts and theories.* . NY: Routledge.

Glyptis, S. A. (1982). *Sport and tourism in Western Europe.* . London: British Travel Education Trust.

Gratton, C., & Jones, I. (2010). *Research methods for sports studies.* Routledge.

Green, C. B., & Chalip, L. (1998). Sport tourism as the celebration of subculture. *Annals of Tourism Research, 25*(2), 275–291.

Ikkos, A. (2018). The contribution of tourism to the Greek economy in 2017. SETE Intelligence. Retrieved from www.insete.gr. Accessed on August 28, 2018.

Hagger, M. S., & Chatzisarantis, N. L. D. (2005). *The social psychology of exercise and sport.* . Buckingham: Open University Press.

Hall, 1992.Hall, C. M. (1992). Adventure, sport and health tourism. *In Special Interest Tourism* (pp. 141–158). London: Weiler B. Hall CM. Belhaven Press.

Harrison-Hill, T. & Chalip, L. (2005). Marketing sport tourism: Creating synergy between sport and destination. *Sport in Society, 8*(2), 302–320.

Hashemi, S. M., Jusoh, J., Kiumarsi, S., & Mohammadi, S. (2015). Influence factors of spa and wellness tourism on revisit intention: The mediating role of international tourist motivation and tourist satisfaction. *International Journal of Research, 3*(7), 1–11.

Higham, J. (1999). Sport as an avenue of tourism development: An analysis of the positive and negative impacts of sport tourism. *Current Issues in Tourism, 2,* 82–90.

Higham, J. (2005). *Sport tourism destinations: Issues, opportunities and analysis.* . Burlington: Elsevier Butterworth-Heinemann.

Higham, J. E. S. (Ed.). (2011). *Sport tourism destinations: Issues, opportunities and analysis.* . NY: Routledge.

Hinch, T. D., & Higham, J. E. S. (2005). Sport, tourism and authenticity. *European Sport Management Quarterly, 5*(3), 243–256.

Hsiao-Ching, H. (2017). A study of sport tourists participate motivation, travel experience, perceived value, and behavioral intention in marine sport tourism. *Global Journal of Management & Business Research (online), 17*(5).

Hungenberg, E., Gray, D., Gould, J., & Stotlar, D. (2016). An examination of motives underlying active sport tourist behavior: A market segmentation approach. *Journal of Sport & Tourism, 20*(2), 81–101.

Joshi, A., Kale, S., Chandel, S., & Pal, D. K. (2015). Likert scale: Explored and explained. *British Journal of Applied Science & Technology, 7*(4), 396.

Kabanova, E. E. & Vetrova, E. A. (2017). Cluster approach as tourism development factor. *Journal of Environmental Management & Tourism, 8*(8, 24), 1587–1594.

Kaplanidou, K., & Gibson, H. J. (2012). Event image and traveling parents' intentions to attend youth sport events: A test of the reasoned action model. *European Sport Management Quarterly, 12*(1), 3–18.

King, K., Shipway, R., Lee, I. S., & Brown, G. (2018). Proximate tourists and major sport events in everyday leisure spaces. *Tourism Geographies,* 1–19.

Krsmanović, V., & Gajić, I. (2016). Sports tourism: A new significant dimension of development of tourism. In *3rd International Conference, Higher Education in Function of Sustainable Development of Tourism in Serbia and Western Balkans* (pp. 369–376). Paper Proceedings.

Kurtzam, J. R. (2005). Sports tourism categories. *Journal of Sport Tourism, 10*(1), 15–20.

Kurtzam, J. R., & Zauhar, J. (1997). A wave in time: The sports tourism phenomena. *Journal of Sports Tourism, 4*(2), 5–20.

Kurtzam, J. R., & Zauhar, J. (2005). Sports tourism consumer motivation. *Journal of Sport Tourism, 10*(1), 21–31.

Osti, L., Cicero, L., & Moreschini, M. (2018). Tourists' motivations for practicing physical activity: A home-holiday comparison. *Journal of Sport & Tourism, 22*(3), 207–226.

Mason, D. S., & Duquette, G. H. (2008). Exploring the relationship between local hockey franchises, and tourism development. *Tourism Management, 28,* 1157–1165.

Myburgh, E., Kruger, M., & Saayman, M. (2018). Aspects influencing the commitment of endurance athletes: A tourism perspective. *Journal of Sport & Tourism,* 1–27.

Perić, M., Đurkin, J., & Vitezić, V. (2018). Active event sport tourism experience: The role of the natural environment, safety and security in event business models. *International Journal of Sustainable Development & Planning, 13*(5), 758–772.

Preuss, H., & Solberg, H. A. (2006). Attracting major sporting events: The role of local residents. *European Sport Management Quarterly, 6*(4), 391–411.

SETE (2017a). Annual Report 2017. Greek Tourism Confederation. Retrieved from www.sete.gr. Accessed on August 16, 2018.

SETE (2017b). *Greek tourism: Facts & figures.* Athens: GREEK TOURISM CONFEDERATION. Retrieved from www.sete.gr. Accessed on August 16, 2018.

Shock, D. J., & Chelladurai, P. (2008). Service quality, satisfaction, and intent to return in event sport tourism. *Journal of Sport Management, 22*, 587–602.

Singh, S., Dash, T. R., & Vashko, I. (2016). Tourism, ecotourism and sport tourism: The framework for certification. *Marketing Intelligence & Planning, 34*(2), 236–255.

Slak-Valek, N., Shaw, M., & Bednarik, J. (2014). Socio-demographic characteristics affecting sport tourism choices: A structural model. *Acta Gymnica, 44*(1), 57–65.

Sport Camp Official site. https://www.sportcamp.gr/. Accessed on August 16, 2018.

Sport Camp Official site https://www.sportcampkids.gr/. Accessed on August 16, 2018.

Standeven, J., & De Knop, P. (1999). *Sport tourism.* . Champaign: Human Kinetics.

Taleghani, G. R., & Ghafary, A. (2014). Providing a management model for the development of sports tourism. *Procedia—Social & Behavioral Sciences, 120*, 289–298.

Thomas, J. R., Silverman, S., & Nelson, J. (2015). *Research methods in physical activity* (7th edn). Human Kinetics.

Turco, D. M., Riley, R., & Swart, K. (2002). *Sport tourism.* . Morgantown, USA: Fitness Information Technology Inc.

Van Rheenen, D., Cernaianu, S., & Sobry, C. (2017). Defining sport tourism: A content analysis of an evolving epistemology. *Journal of Sport & Tourism, 21*(2), 75–93.

Weed, M. E. (2008). *Olympic tourism.* . Great Britain: Butterworth-Heinemann.

Weed, M. E, & Bull, C. J. (2009). *Sports tourism: Participants, policy and providers.* (2nd ed.). Great Britain: Butterworth-Heinemann.

World Economic Forum (2017). The Travel and Tourism Competitiveness Report 2017. Accessed on August 16, 2018, from www.weforum.org.

World Tourism Organisation (UNWTO) (2017). Tourism Highlights, 2017 Edition. Accessed on August 16, 2018, from www.e-unwto.org.

Cycling Tourism: Characteristics and Challenges for the Developments and Promotions of a Special Interest Product

Elina Tsitoura, Paris Tsartas, Efthymia Sarantakou, and Alexios-Patapios Kontis

1 Introduction

Cycling tourism represents a concrete expression of sustainable tourism, offering a number of economic, social and environmental benefits to the regions and local communities where it is developed, such as the enrichment of the tourism product, extension of the tourist season, reduction of the carbon footprint, growth of entrepreneurship and job positions, involvement of local communities in planning and management of this type of tourism.

Cycling tourism shows a strong trend of growth at international level, including the market of "active" tourists, who seek through their travels and holidays unique experiences and knowledge, related to the local characteristics of each region and its inhabitants.

The European Union is increasingly supporting cycling tourism by participating in the financing of programmes and projects, in cooperation with cycling and tourism organizations. As a result, cycling tourism in Europe has been integrated into sustainable transport and development policies leading to the planning and development of regional, national and pan-European networks of cycling routes that meet specific quality standards.

E. Tsitoura
Hellenic Open University, Thessaloniki, Greece

P. Tsartas (✉)
Harokopio University, Kallithéa, Greece
e-mail: tsartas@hua.gr

E. Sarantakou
University of West Attica, Egaleo, Greece

A.-P. Kontis
University of Aegean, Mytilene, Greece

© The Author(s), under exclusive license to Springer Nature Switzerland AG 2021
V. Katsoni and C. van Zyl (eds.), *Culture and Tourism in a Smart, Globalized, and Sustainable World*, Springer Proceedings in Business and Economics, https://doi.org/10.1007/978-3-030-72469-6_2

In Greece, the use of bicycles faces a number of challenges as the country has taken few steps towards the development of cycling tourism, unlike other European countries that for many years have successfully developed this form of tourism (i.e. France, the Netherlands, Austria, Denmark). The reduced road safety observed in Greece and the absence of cycling culture are also inhibitors for the development of cycling tourism, and although the country has a fairly well-developed public transport network, the ban on the transport of bicycles by many means and the lack of the relevant infrastructure in them (special bike ramps on coaches, parking spaces on trains and ships, etc.) are a deterrent for the tourist-cyclist.

However, despite the problems a cyclist may encounter while travelling in the country, Greece is suitable for cycling tourism. The mild climate, the unique terrain, the extremely diverse natural environment, the plethora of cultural and historical attractions, the good provincial low-traffic road network and plenty of dirt paths which with the appropriate interventions can be converted to cycling parks, make Greece an attractive destination for recreational or long-distance cyclists, and sports teams.

The objective of this paper was to explore the challenges and possibilities of developing cycling tourism in Greece as a special form of tourism with multiple economic, social and environmental benefits. The main objectives—research questions of this study were:

1. Exploring and recording the conditions for the development of cycling tourism in Greece, ensuring long-term benefits at national, regional and local level.
2. Exploring the possibilities and challenges for the development of cycling tourism in Greece for both domestic and foreign tourists.
3. Establishing strategic guidelines for tourism stakeholders (cyclists, public services involved with cycling, tourism planners and managers, entrepreneurs and local communities) at national, regional and local level.

In order to answer the research questions, the following general methodological approach was followed: important aspects and international practices related to cycling tourism were initially examined, focusing mainly on the conditions for its development. For the most part, (Internet) research in scientific articles, books and studies of European centres on the subject have been used as a method of searching the literature.

The keywords used are: cycling tourism, sustainable and alternative forms of tourism, sports tourism.

At the second section, the conditions for the development of cycling tourism in Greece were explored. The investigation highlighted the lack of bibliographical data and field surveys on the issue (Tsartas et al., 2016; Tsartas & Sarantakou, 2016; Kokkosis et al., 2020). For this reason, a primary survey was carried out, by distributing two questionnaires, one in Greek and one in English, aimed, respectively, at Greek and foreigner persons involved in cycling.

The findings of the survey feed into a SWOT analysis show the potential for the cycling tourism development in Greece. Paper draws some conclusions and policy guidelines for the development of cycling tourism for stakeholders at a national, regional and local level, from both the state and the private sector.

2 Literature Review

The concepts of "cycling tourism" and "tourist cyclist" have been approached by several parameters and criteria. A general definition that is often found in the literature states that a cycle tourist can be defined as "a person of any nationality, who at some stage or other during his or her holiday uses the bicycle as a mode of transportation, and for whom cycling is an important part of this holiday".

Studies by the Institute of Transport and Tourism, & Centre for Sustainable Transport and Tourism (2012) and the British Transport Charity Sustrans (1999) distinguishes "cycling holidays" in three types: those in which cycling is the main part of the tourist experience, those that include cycling not as an exclusive activity and the "cycling day trips". Lamont's definition (2009), according to which "cycling tourism is travelling away from the place of residence of a person whose active or passive participation in cycling is considered the main purpose of the journey" broadens the scope of the concept of tourist cyclist including persons travelling for the purpose of participating in cycling competitions.

Simonsen et al. (1998) classify tourist cyclists according to their level of commitment to cycling and the purpose of cycling holidays by placing at one end of the scale the tourist cyclists, for whom cycling is the only way to travel and the main purpose of their holidays, and at the other end of the scale tourists who ride a bike during their holidays occasionally. Between these two extremes lies the main and largest part of the tourist cyclists.

The above definitions illustrate that "tourist cyclists" are a non-homogeneous group which, although it may be motivated by a common "special interest", consists of several categories of people (and, respectively, possible market segments). Therefore, we could categorize them based either on their motivations or the kind of activity (recreational cyclists, day cyclists, touring cyclists, mountain bikers and cyclists of race or tour events) (Faulks et al., 2006).

The analysis of eight market studies in five countries (ETI, 2007; Fietsplatform, 2009a, 2009b; Ickert et al., 2005; MANOVA, 2007; Mercat, 2009; Öhlschläger, 2007; Trendscope, 2010) revealed the following generalized profile of cycling tourists: average age 45–55 years, 60% men, 40% women, a significant percentage of university education and professional status, a variety in economic profile, 20% cycle alone, 50% in pairs and 20% in small groups of 3–5 people.

The main period chosen by tourists-cyclists for cycling holidays is from May to the end of August, as temperature and rainfall are a strong determinant for cyclists (Dohmen et al., 2011). The average length of stay of cycling holidays calculated from 15 studies, in 6 countries, on 18 different routes and networks, ranges from 5 to 8 days (ARGE Donau Österreich, 2011; ETI, 2007; Fietsplatform, 2004, 2009a; Ickert et al., 2005; MANOVA, 2007).

On accommodation, surveys show a dispersion in Guesthouse/B&B/farmhouse accommodation (45%) and camping (15%), but also in typical hotels (40%) Altermodal (2007), ETI (2007), Fietsplatform (2009a), Ickert et al. (2005), MANOVA (2007), Trendscope (2010)). Many of cycling tourists prefer "bike accommodation".

The bibliographical survey highlights cycling tourism as an emerging alternative sustainable form of tourism with benefits at national, regional and local level (Lumsdon, 1996, 2000; Ritchie, 1998). In Europe, cycling tourism has been integrated into sustainable development and transport policies leading to the planning and development of regional, national and pan-European networks that meet specific quality standards.

The strengthening of local economies, the support of businesses and the maintenance or creation of new jobs, result from the direct costs of tourist cyclists, particularly in rural areas (Sotolongo, 2018). The New Zealand cycling network "Great Rides of Nga Haerenga" generated $37,400,000 in revenue in 2015 for local communities (New Zealand Cycle Trail 2016).

Two regional (Rhineland-Palatinate, Southern Austria) and four national studies (Velöschvöz, Dutch LF-routes, France, Germany) give us a more detailed display of the daily expenses of cycling tourists who stay overnight for one or more nights: on average €57.08, of which about 40% (€23) is spent on accommodation, 30% (€17) on food and drink and 30% (€17) on other expenses, such as shopping (almost half of this amount), local transport and activities (ETI, 2007; Fietsplatform, 2009a; Ickert et al., 2005; MANOVA, 2007; Mercat, 2009; Trendscope, 2008a).

Although cyclists are often considered low-budget travellers, a survey in the northern Outer Banks tourist area in North Carolina, USA found that 87% of bike visitors had an annual income of more than $50,000, half of them earning more than $100,000 a year (https://www.americantrails.org/resources/economic-impact-of-inv estments-in-bicycle-facilities.)

In addition, cycling routes increase the quality of life of local communities (Schafer et al., 2000), by providing the residents open spaces for recreation and relaxation (Lumsdon et al., 2004). This fact may result in the increase of land prices in areas with popular cycling routes (Tourism France Ministry, 2007).

In order to develop cycling tourism, the members concerned need to participate continuously and over time by promoting and maintaining cycling routes. Numerous international examples also highlight the need for an integrated approach including policies, programmes, information provision, land use planning (infrastructure), etc., and demonstrate that the synergies between the members involved add value to the cycling tourism product (Zovko, 2013).

Compared to other forms of tourism and especially mass leisure tourism, cycling tourism is significantly more sustainable and has better environmental performance, mainly in terms of the environmental footprint of the means of transport (UNWTO-UNEP-WMO, 2008) and the type of accommodation preferred, but also because of the small-scale landscape interventions needed for the development of cycling networks. According to the German Bureau of Statistics (Statitisches Bundesamt, 2008) and the German study Trendscope (2008b), 93% of cycling tourists are not using airplane to transport to the holidays destination and overall cyclist tourists use more environmentally friendly modes of transport compared to other tourists and travel shorter distances by 53%. Altermodal (2007), ETI (2007), Fietsplatform (2009a), Ickert et al. (2005), MANOVA (2007) and Trendscope (2010) showed that a significant proportion (15%) of the cycling tourists choose to stay in camps that are considered a more environmentally friendly type of accommodation compared to hotels and holiday homes.

3 Parameters for the Development of Cycling Tourism

The development of cycling as a tourist product and the provision of quality tourist services for cyclists offers opportunities to achieve international competitiveness in the tourist market. Therefore, good organization, carefully planned promotion and continuous monitoring of results are required. The natural attractions of an area may act as an incentive for many tourist cyclists, but this element alone is not enough. Cycling tourism requires several conditions to be taken into account when planning (Rotar, 2012).

3.1 Appropriate Cycling Infrastructure

Appropriate infrastructures that contribute to safety and comfort form the basis for attracting tourist cyclists (Rotar, 2012). According to CROW (https://www.cro w.nl) the five basic conditions that cycling infrastructure should meet are safety, accessibility, networking, comfort, attractiveness.

3.2 Means of Transport

As with all types of holidays, in cycling tourism there is a need to interconnect the different modes of transport to and from the destination. Cyclists tourists seem to use public transport more frequently than other tourists, as they usually cycle from one point to another. Also, many tourist cyclists prefer to ride their own bike on holiday, which is associated with a greater preference for train and ferry travel (Institute of Transport and Tourism, & Centre for Sustainable Transport and Tourism, 2012).

3.3   Certified Services and Special Marketing

An area promoted as a cycling destination should meet certain conditions to ensure the provision of quality services. Therefore, a body responsible to ensure the implementation and monitoring of standardization is essential. The Slovenian Tourist Board (https://www.slovenia.si/) has developed a number of cycling standards relating to bike-friendly accommodations, "Bicycle Information Points", certified cycling routes, guided tours etc. and Cycling Marketing Standards.

4   Cycling Tourism in Greece

4.1   EuroVelo

An integrated cycling network connecting European countries is the goal of the European Federation of Cyclists through EuroVelo, which currently includes more than 45,000 km of cycling routes, while the network, when completed, is estimated to exceed 70,000 km (https://eurovelo.gr/el/eurovelo/). Responsible for the development of EuroVelo routes in Greece is the non-governmental organization "Cities for Bicycle" (https://citiesforcycling.gr/). Three Eurovelo cycling routes traversing Greece are in the process of being prepared and marked: the Mediterranean Route EV8, (5388 km), the ROUTE EV 11 (5964 km) and Eurovelo EV 13.

4.2   Cycling Infrastructure

The cycle paths network that already exists in various Greek cities as infrastructure is not currently an incentive for long-distance tourists, mainly because the existing networks cover short-distance routes limited to the boundaries of cities. In addition, there are problems concerning the safety and maintenance of cycle paths in Greece and the hostile behaviour of car drivers, especially in areas with no bicycle tradition such as Athens (Vassi & Bakogiannis, 2013). However, the existing network could act as a starting point for some of the above tourist cyclists who know a city as "bike friendly", or be an incentive for tourists who wish to cycle for a few hours in a Greek city during their holidays. For mountain biking, there are trails throughout Greece (Table 1).

Table 1 Cycle paths in Greece

Cycle paths in Greece	
Number of municipalities having cycle paths	102
Percentage of municipalities having cycle paths	31% (102/324)
Kilometres of cycle paths (in total)	403.7 km
Mixed-use paths (cars and bikes)	34.8 km
Bike lanes	54.1 km
Bike corridors	305.4 km
Bike path metres per inhabitant	0.037

Source https://www.smu.gr/greece_cycle_map/

4.3 Use of Bicycles in Means of Transport

In Greece, each mode of transport has its own independent policy on the transport of bicycles. Specific rules and restrictions on the carriage of bicycles have been in place since 2015, while the Greek railway is in line with the provisions of Regulation 1371/2007 of the European Union, according to which railway undertakings must enable passengers to carry bicycles under specific conditions. On aeroplanes, depending on the regulations of each airline, the bicycle is given as sports equipment or as baggage, while on board ships, in general, bicycles are allowed to be transported free of charge in the car park.

4.4 "Bike Friendly" Businesses and Destinations

In Greece, the development of "Bike Friendly" hotels and destinations network was launched in 2017 at the initiative of the urban, non-profit company NATTOUR, which promotes the development of alternative forms of tourism, giving priority to cycling tourism. The "Bike Friendly" badge is awarded to both hotels and municipality-tourist destinations, which meet a number of internationally recognized criteria related to comfort, information and cycling tourism services. For the time, 80 hotels, 17 companies offering bicycle services, 2 municipalities and a region (Thessalia) have been certified (https://www.bikehotels.travel/).

4.5 Cycling Races and Events

Very important for the development of cycling tourism is the contribution of cycling clubs organizing sporting events to the Greek countryside. Popular events are, for example, the Bike Odyssey taking place in five regions (Grevena, Trikala, Evritania, Fokida and Aitoloakarnania), the road cycling races in Kos and Rhodes and the Brevets, which are long-distance routes managed in accordance with the terms and rules of the French Brevet Randonneurs Mondeaux and Audax Club Parissien.

4.6   Guided Tours and Bike Rental

In Greece, there are companies that organize bike tours and hotels that offer bicycles to their guests. For instance, www.grcycling.com organizes interesting routes close to Athens, either by escort-driver or independently, Pedal Greece (https://pedalgreece. com/trip/) offers a package of holidays that includes driver escort, accommodation, food and drinks and transportation and an increasing number of hotels and companies in Athens offer bike tours around the historical and commercial centre of the city. In addition, there are plenty of bicycle rental companies in Greece offering the clients services such as online booking, while in many cases bike delivery is available in the area desired by the customer. A special category of bicycle rental is the bike-sharing services in various cities of Greece, offering bicycles for public short-term use, from one starting point to another point (https://www.easybike.gr).

5   Primary Research Methodology and Results

The primary survey was carried out through the use of two structured question-naires, one in Greek and the other in English, aimed at domestic and foreign cyclists respectively. The purpose of completing these questionnaires was to investigate the demographic, economic and social profile of tourists, their degree of satisfaction with the treatment of Greek tourism enterprises, the motivations and preferences of cycling tourism, and the extent to which various factors such as cycling tourism infrastructure, transport, accommodation affect their travel choices.

The sample used for Questionnaire I (national cyclists) was derived from domestic persons who are systematically engaged in cycling (convenience sample) and was forwarded via Facebook and Messenger to cyclists-users (available from 1/4/19 to 17/4/2018), including a total of eight sections with closed-ended answers, gradient able responses according to "scale of importance" and predefined double-selection replies ("Yes" or "No"). It was completed by 296 people.

The sample used for questionnaire II (foreign cyclists) was derived from bicycle users participating in cycling blogs (https://www.cyclingforums.com/forums/the-bike-cafe.25/ and https://www.cyclechat.net/) and was completed by 24 people. The questionnaire was available between 24/4/19 and 1/5/2018 and included a total of four modules.

5.1 Primary Research Results (Domestic Cyclists)

The following conclusions are drawn from the research on the development of cycling tourism in Greece.

With regard to the participants in the survey aimed at domestic cyclists, we found that:

- The majority are men at 77.7%.
- Most visitors belong to the 45–65 age group and a significant percentage to the 26–44 age group.
- Most respondents have monthly earnings of 451–1000 Euros, but many also have monthly earnings of more than 1001 Euros. The majority therefore have incomes of more than 451 Euros per month (see Fig. 1).
- More than half of the participants have a university education.

Regarding the holiday "cycling" profile of participants, our research shows the following results:

- More than half of participants have taken a "cycling holiday" or a "holiday that included cycling as an activity" (and almost a third of them have taken both of these types of holidays)
- The most popular holiday season is Summer and the second most popular is Spring, but almost 1 in 10 take holidays all year round, and 1 in 10 take holidays all season except Winter (see Fig. 2).

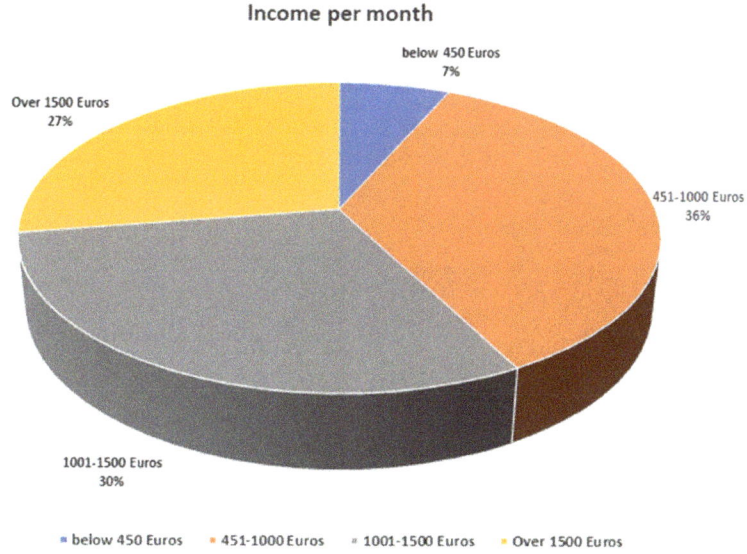

Fig. 1 Income per month

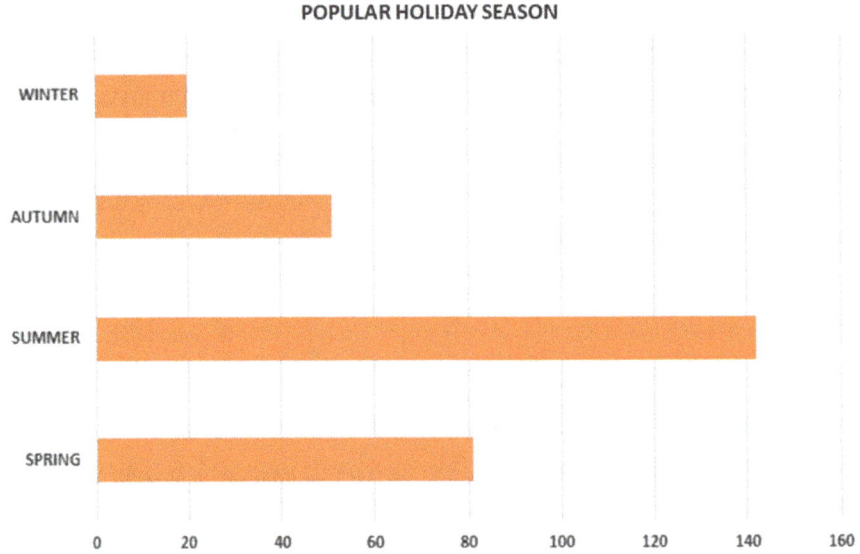

Fig. 2 Popular holiday season

- Almost half of respondents stay overnight from 2 to 4 nights, but a significant proportion (37.1%) stays overnight for more than 5 nights.
- In terms of accommodation preference, the majority of respondents prefer rented rooms and camping, and a significant proportion (25.1%) prefer accommodation in a hotel (see Fig. 3).
- The daily expenditure (for accommodation, food, entertainment, etc.) is in the majority of respondents from 26 to 50 Euros (see Fig. 4).

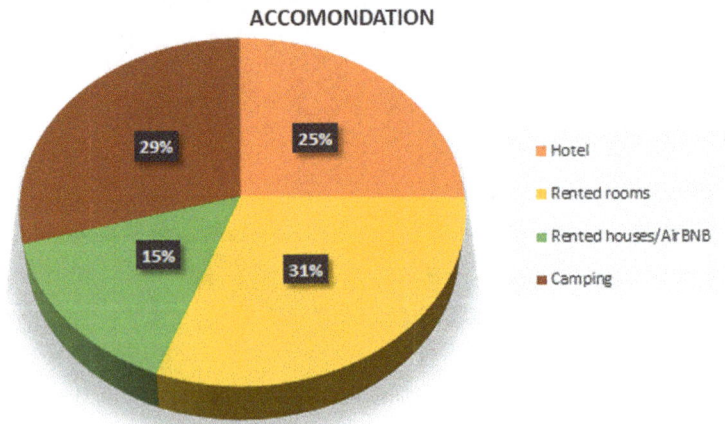

Fig. 3 Accommodation

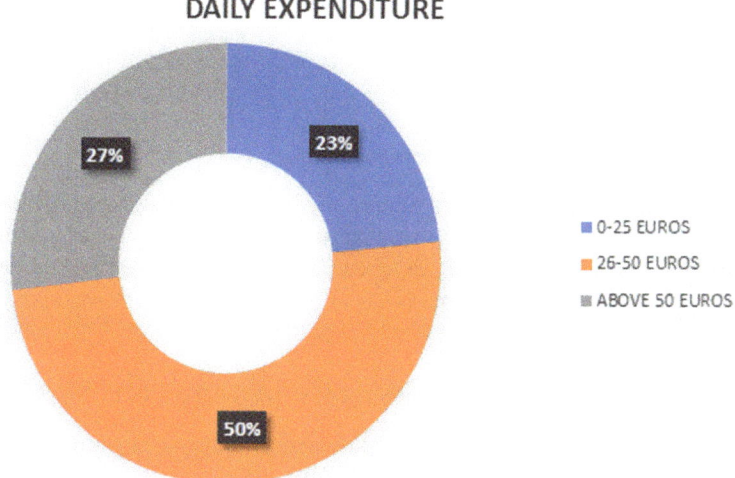

DAILY EXPENDITURE

- 0-25 EUROS
- 26-50 EUROS
- ABOVE 50 EUROS

Fig. 4 Daily expenses

- The participants show great interest in cycling races and cycling events, as most of the respondents (70.1%) have attended or taken part in cycling races or cycling events (e.g. Brevets).

  Regarding their preferences and habits, the participants:

- They are mainly engaged in street cycling, while a small percentage prefer leisure/family bike rides. Few respondents are involved in mountain biking or cycling racing.
- They are mainly using Internet for getting information on cycling routes, but also friend's recommendations, while several make use of such applications for mobiles, tablets, etc.
- They have toured by bike, beyond Attica, the Peloponnese and Central Greece, as well as other regions throughout Greece.
- The majority find businesses (accommodations, restaurants, etc.) neutral in their friendliness regarding cycling, and a significant proportion (33.7%) describes them as good.

Regarding the factors that play a role in choosing a cycling route, respondents:

- The majority of the respondents consider as very important the low traffic in vehicles and the beauty of the natural landscape on a route.
- Facilities such as accommodation, cafes, clear signage, attractions (archaeological, cultural, etc.), access by train, ship or plane and cycling on an international route (e.g. EuroVelo) seem to play a role to some extent, but are not decisive for the choice of a cycling route, as for most they are either "important" or "not very important".

Participants also appear to:

- Be very interested in travelling from their place of residence to a destination, either by means of transport (mainly by boat or train) or exclusively by bicycle.

About the facilities offered by the accommodations:

- The most important provision for almost all respondents is storage area for bicycles and secondary services such as breakfast for the needs of cyclists, information on the premises (e.g. for routes), dryer and equipment for the repair of bicycles.
- Almost all of the respondents would choose to stay overnight in an accommodation certified as "bike friendly".

More than half of participants seem interested in:

- Renting a bicycle (e.g. from a train station, from the accommodation staying on holiday, from a pick-up point of a municipality),
- Using a cycling tour available via a GPS.
- Buying a package of "cycling holidays" through an agency.
- Taking part in a "thematic" cycling tour mainly on the natural environment (and less cultural, archaeological or gastronomic interest).

Finally, a significant proportion of participants (41.6%) replied that they would be interested in renting an electric bicycle (e-bike).

It is interesting to note that the results achieved from this research show similarities and some differences with the results from the other available studies discussed earlier in this paper. In more detail, the demographic, economic and social profile of bicycle tourists is similar, although in our study there is a smaller (by 20%) percentage of women engaged in cycling. In all studies, the main period chosen by tourists-cyclists for cycling holidays is Summer and Spring. Previous research indicates that the average length of stay on cycling holidays ranges from 5 to 8 days, whereas in the present research half of respondents stay less nights (2–4 nights). Yet, a significant proportion (37.1%) stays overnight for more than 5 nights. On accommodation, all surveys show that rented rooms or houses are the main form of accommodation utilized by approximately 45% of cyclists. Hotels (40%) are the most popular second choice indicated by the foreign studies, whereas in our study the correspondents were more interested in camping (29%) and less in hotels (25%). In all surveys, almost all cycling tourists prefer "bike accommodation". The daily expenses of the domestic cyclists tourists are lower, as half of the participants spend on average 7 euros less than the cyclists tourists on foreign surveys, although 27% spend above 50 euros. Finally, overall cyclist tourists prefer using other means of transport than airplane.

Would you make a bicycling trip to Greece? (a bicycling trip is defined as any trip that includes bicycling as an activity, including trips for which bicycling may not be the primary activity)
24 απαντήσεις

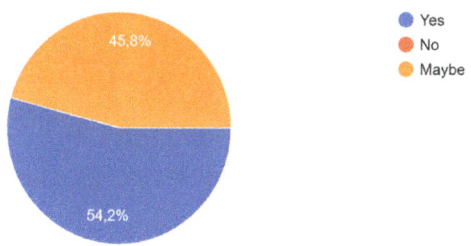

Fig. 5 "Would you make a bicycling trip in Greece?"

5.2 Primary Research Results (Foreign Cyclists)

The survey of foreign cyclists was based on a questionnaire with a limited number of questions, that was completed by 24 people, therefore the sample is quite small and does not allow safe conclusions. However, due to the experience of these cyclists, trends and characteristics that are of interest are highlighted through their responses and could be further explored in future research. The survey indicated that:

- The majority of participants (83.3%) are men.
- Most belong to the 45–65 age group (70.8%) and a significant proportion in the 26–44 age group (20.8%).
- More than half of the participants have university education (Bachelor and Academic Degree).
- All participants replied that they would (54.2%) or would maybe (45.8%) take a "cycling trip" to Greece (Fig. 5).
- The cycling "type" of interest to the majority of respondents (58%) during their holidays on a possible trip to Greece, was the "Multi-Day Tours", and second in preference the "Mountain Cycling trails" (25%).
- Half of the respondents replied that they would bring their own bike on a possible trip, while the other half would rent a bike in Greece.
- Great interest showed by foreign cyclists (83.3%) for the guided tours.

  Regarding the factors that are important in choosing a cycling route, respondents:

- The majority of them consider a low traffic in vehicles route to be very important, but many of them do not consider it important.
- All participants replied that the beautiful natural landscape of the route is important (Fig. 6).
- Facilities such as accommodation and cafes are important for most of the respondents but for many they are minor.

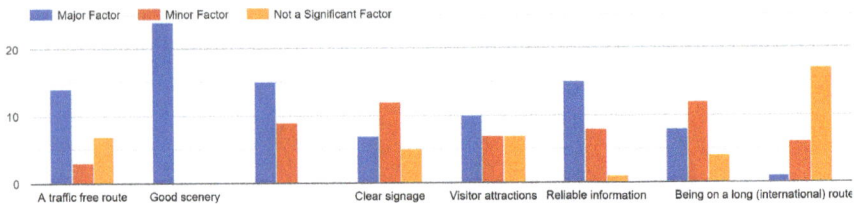

Would any of the following factors influence your decision on the route you choose while on a bicycling trip?

Fig. 6 Factors influence the decision on the chosen route on a bike trip

- Clear signage, attractions and public transport access for some of the participants are very important factors, while for others less or not at all important.
- Valid information seems to be very important for the majority of respondents.
- Cycling on an international route (e.g. EuroVelo) does not seem to play an important role in choosing a cycling route.
- Most of the respondents replied that facilities such as safe parking/bicycle storage, bicycle rental and the provision of food and drink during the journey are important, while for example the existence of bicycle repair/washing stations is of minor importance.

Finally, it is worth noting that, although in general, the responses of domestic and foreign participants are similar, the following differences were identified:

The beautiful natural landscape, clear signage and valid information seem to be very important factors in choosing a cycling route for foreign cyclists, unlike domestic cyclists for whom they are less important.

Domestic cyclists seem to show a greater interest in sightseeing, public transport access and cycling on an international cycling route (e.g. EuroVelo), as opposed to foreign cyclists who seem to be less interested.

The storage of bicycles and the provision of "breakfast for cyclists" seem to be very important for domestic cyclists, whereas mixed ratings are given for these factors by the foreign cyclists.

6   Conclusions and Recommendations

International experience from countries that have developed cycling tourism and the results from various studies and surveys converge to conclude that cycling tourism is an up-and-coming niche tourism market with economic, social and environmental benefits for the destinations where it develops.

Cycling tourism is in line with the principles of sustainability, contributing to the economic and social wellbeing of rural areas less popular to tourists.

In Greece, the development of alternative-thematic forms of tourism such as cycling could stimulate demand throughout the year and lengthen the tourist season, due to the mild climate which allows cycling tourism all year. In addition, the country has a good provincial network with asphalt routes and many paths for mountain biking, through landscapes with rich biodiversity and natural environment, thus giving the tourist-cyclist the opportunity to enjoy within a short time a continuous rotation of land, mountains, coastal lanes and islands. Many of these routes pass through areas of strong historical and cultural interest and could be combined with other thematic forms of tourism (i.e. archaeological, cultural, gastronomic). Greece, as a developed tourist country for many decades, has infrastructure such as accommodation, catering businesses and public transport.

The objective of this paper was to summarize the key parameters contributing or required for the development of cycling tourism in Greece.

The study results illustrate that the cyclists tourists of our research wish for safe and comfortable routes, natural landscapes, accommodations, cafes and other tourist businesses, attractions, etc. Key parameters are also the accessibility of the destination by public transport and the appropriate marking.

To meet the demands of current bicycle tourists, destinations interested in developing cycling tourism destination should meet certain standards in the case of housing and other services. Proper facilities (such as storage space for bicycles in accommodation, the provision of information on routes, renting bicycles services, availability of cycling routes via GPS applications, or through agencies and the development of "thematic" cycling routes), successful marketing and cooperation of stakeholders play major role in the upgrade of the tourist product.

The main prerequisite for the development of cycling tourism in Greece is the planning and development of suitable cycling routes. Yet, emphasis should be placed on the maintenance of these routes and their marking, as well as on speeding up the signalling of the three EUROVELO routes crossing Greece, to attract foreign tourists-cyclists.

The main inhibitors—disadvantages for the development of cycling tourism are the lack of coordination at a national level, the absence of investment interest, the reduced road safety and the absence of cycling culture. Although Greece has a well-developed public transport network, cycling tourism is not included in the operational and marketing strategies of the transport means. Restrictions on the use of bicycles in many means and lack of relevant infrastructure in them, such as special bike ramps on coaches, parking spaces on trains and ships, is a deterrent for the tourist-cyclist.

A major threat to the development of cycling tourism in Greece is the intense competition from European countries that for many years have successfully developed cycling tourism (such as France, the Netherlands, Austria, Denmark, etc.) but also from other countries that in recent years have invested heavily in cycling tourism (e.g. Poland, Cyprus). In addition, the major tour operators show for the time no interest in this type of tourism in Greece.

In order to successfully develop cycling tourism in Greece (**main prerequisites**) a national development strategy with clearly defined objectives within the framework of a sustainable agenda and secondly, the coordination and cooperation at regional, local and business level are required.

One of the **recommendations** proposed is to entrust coordination to a single forum body, represented by public bodies and other stakeholders, the tasks of which will include the implementation and/or coordination of marketing, monitoring and other essential actions for the development of cycling tourism in the country, in cooperation with other organizations.

In order to facilitate direct cooperation between local, regional and national stakeholders, but also between private and public bodies in different geographical areas, there should be a network linking and providing information to all stakeholders so that, for example, small businesses (e.g. accommodation providers, tour operators, catering companies) have access to cycling tourism development opportunities.

Local/regional providers of cycling tourism services should develop thematic cycling routes shaping the brand name of each destination and promote them through a successful marketing plan. Similarly, the development of a strong tourism brand related to all cycling tourism activities at national level would stimulate the demand from other countries.

Particular attention should also be paid to the monitoring of cycling tourism, with regard to its economic impact and in particular its wider impact (environmental and social) on the regions. Ideally, an organization should be responsible for the collection, analysis and dissemination of data, in order to better understand the motivations and expectations of cycling visitors and to renew the action plan for cycling tourism each time.

Finally, individual actions such as the integration of local and regional cycling tourism information into a web page and social media, the legislative strengthening of regulations concerning the transport of bicycles on public transport, support through the introduction of funding for bicycle events and races, would also help to build cycling tourism in Greece.

References

Altermodal. (2007). *2006 survey of traffic and economic impact along EuroVelo 6*. Altermodal.

ArgeDonau Österreich, 2011. Arge Donau Österreich. (2011). *Unterwegs entlang der Donau (press release)*. Oberösterreich Tourismus.

Dohmen, R., Tiffe, A., Dürhager, U., Borsbach, K., Funke, R., & Kollbach, K. (2011). *Analyse von Radverkehrsströmen zur nachhaltigen Optimierung von Radverkehrsnetzen*. TMB Tourismus-Marketing Brandenburg GmbH.

ETI. (2007). *Regionalwirtschaftliche Effekte des Radtourismus in Rheinland-Pfalz*. Europäisches Tourismus Institut an der Universität Trier GmbH.

Faulks, P., Ritchie, B., & Fluker, M. (2006). *Cycle Tourism in Australia: An investigation into its size and scope*. SUSTAINABLE TOURISM CRC.

Fietsplatform. (2004). *Informatie onderzoek LF-routes*. Stichting Landelijk Fietsplatform.

Fietsplatform. (2009a). *Informatie onderzoek LF-routes*. Stichting Landelijk Fietsplatform.

Fietsplatform. (2009b). *Zicht op Nederland Fietsland 2009*. Stichting Landelijk Fietsplatform.

Ickert, L., Rommerskirchen, S. & Weyand, E. (2005). Veloland Schweiz: Ergebnis-Band zur Gästebefragung. Zählung und Befragung 2004. ProgTrans AG.

Institute of Transport and Tourism, & Centre for Sustainable Transport and Tourism. (2012). *The European Cycle Route Network Eurovelo.* Retrieved from https://ecf.com/files/wp-content/upl oads/The-european-cycle-route-network-EuroVelo.pdf. Accessed on January 15, 2019.

Kokkosis H., Tsartas P., & Grimpa E. (2020). "Ειδικές και εναλλακτικές μορφές τουρισμού", Κριτική.

Lamont, M. (2009). Reinventing the wheel: A definitional discussion of bicycle tourism. *Journal of Sport and Tourism, 14*(1), 5–23.

Lumsdon, L., Downward, P., & Cope, A. (2004). Monitoring of cycle tourism on long distance trails: The North Sea Cycle Route. *Journal of Transport Geography, 12*(1), 13–22.

Lumsdon, L. (1996). *Cycle tourism in Britain. Insights (March).* English Tourist Board.

Lumsdon, L. (2000). Transport and tourism: A sustainable tourism development model. *Journal of Sustainable Tourism, 8*(4), 1–17.

MANOVA. (2007). *Radfahrer-Befragung 2006: Niederösterreichische Haupt-Radrouten.* MANOVA.

Mercat, N. (2009). *Spécial économie du vélo. Étude complète.* ATOUT FRANCE.

Ritchie, B.W. (1998). Bicycle Tourism in the South Island of New Zealand: planning and management issue. *Tourism Management, 19*(6), 567–582.

Rotar, J. (2012). *How to develop cycle tourism?* Maribor: CENTRAL EUROPE Programme cofinanced by the ERDF.

Schafer, C. S., Lee, B. K., & Turner, S. (2000). The tale of three greenway trails: User perceptions related to quality of life. *Journal of Landscape and Urban Planning, 49,* 163178.

Simonsen, P. S., Jørgensen, B., & Robbins, D. (1998). *Cycling tourism.* Unit of Tourism Research at Recearch Centre of Bornholm.

Sotolongo, J. (2018). The Spend Cycle: How bicycle tourism impacts small communities. *Dirtragmag.* Retrieved from https://dirtragmag.com/the-spendhttps://dirtragmag.com/the-spend-cycle-how-bicycle-tourism-impacts-smallcycle-how-bicycle-tourism-impacts-smallc ommunities/. Accessed on April 25, 2019.

Bundesamt, S. (2008). *Tourismus in Zahlen 2007.* Statistisches Bundesamt.

Sustrans. (1999). *Sustrans Information Sheets.* Retrieved from https://www.sustrans.org.uk/web files/Info%20sheets/ff28.pdf. Accessed on December 10, 2018.

Tourism France Ministry. (2007). *Mission nationale véloroutes et voies vertes,* Retrieved from https://www.tourisme.gouv.fr/fr/z2/territo/rural/veloroutes/Bilan_perspectives_2007.jsp. Accessed on April 10, 2019.

Trendscope. (2008a). *Radreisen der Deutschen 2008.* Trendscope GbR.

Trendscope. (2008b). *Radreisen der Deutschen 2008 (additional data tables).* Trendscope GbR.

Trendscope. (2010). *Radreisen der Deutschen 2010.* Cologne, Germany: Trendscope UNEP/WTO. (2005). *Making tourism more sustainable—A guide for policy makers.* United Nations Environment Programme.

Tsartas, P., & Sarantakou, E. (2016). "Tourism market trends and their effect on entrepreneurship, cultural consumption and sustainability". In Z. Andreopoulou, N. Leandros, G. Quaranta, R. Salvia, Francoangeli (Eds.), *New media, entrepreneurship and sustainable development* (pp. 25–37), Italy. ISBN 9788891751058.

Tsartas P., Sarantakou E., & Apostolopoulou E. (2016), «Κύριες αλλαγές στον κλάδο του τουρισμού: τα προϊόντα ειδικών και εναλλακτικών μορφών τουρισμού και η βιώσιμη ανάπτυξη», in the book: *"Θαλάσσια χωρικά ζητήματα"* (pp. 365–372) In S. Kivelou, Kritiki (Ed.). ISBN 9789605861124.

UNWTO-UNEP-WMO. (2008). *Climate change and tourism: Responding to global challenges.* UNWTO.

Vassi, A. & Bakogiannis, E. (2013). Progress and problems when planning for bicycle-case study Athens, In *ICHC 2013 24th International Cycling History Conference,* May 14, 2013. Lisbon.

Zovko, I. (2013). The Value of Cycle Tourism, Opportunities for the Scottish economy. *Transform Scotland*. Retrieved from https://transformscotland.org.uk/wp/wp-content/uploads/2014/12/The-Value-of-CycleTourism-full-report.pdf. Accessed on February 10, 2019.

Evaluating the Economic Impact of Active Sports Tourism Events: Lessons Learned from Cyprus

Achilleas Achilleos, Christos Markides, Michalis Makrominas,
Andreas Konstantinides, Rafael Alexandrou, Effie Zikouli, Elena Papacosta,
Panos Constantinides, and Leondios Tselepos

1 Introduction

Different forms of niche tourism are becoming increasingly important and have the potential to help a tourist destination to differentiate from the norm. Overall, development of web platform products and services into niche forms of tourism aims to and can address social and economic regeneration linked to related tourism developments. Gammon and Robinson define a framework in (Gammon and Robinson, 2003) that categorizes "hard sport tourism" as traveling to a mega or major sport event to passively watch the event. On the other hand, they refer to "soft sport tourism" when travelers are actively participating in sporting activity, e.g., hiking, canoeing, or caving. The latter is also referred to in the literature as "active sport tourism" (Greenwell et al., 2019), which is the definition adopted in this work.

In particular, in the editorial note "The Growing Recognition of Sport Tourism" (Richie and Adair, 2002), the authors define sports tourism as follows: "Sport tourism includes travel to participate in a passive (e.g., sports events and sports museums) or active sport holiday (e.g., running, cycling), and it may involve instances where either sport or tourism is the dominant activity or reason for travel." Moreover, in (Gibson

A. Achilleos (✉) · C. Markides · A. Konstantinides · R. Alexandrou
Computer Science, Frederick Research Center, Nicosia, Cyprus
e-mail: com.aa@frederick.ac.cy

E. Zikouli
SportsTraveler76 Ltd, Nicosia, Cyprus

M. Makrominas
Business, Accounting and Finance, Frederick University, Nicosia, Cyprus

E. Papacosta · P. Constantinides
Physical Education and Sports Science, Nicosia, Cyprus

L. Tselepos
Cyprus Sports Organisation, Nicosia, Cyprus

© The Author(s), under exclusive license to Springer Nature Switzerland AG 2021   43
V. Katsoni and C. van Zyl (eds.), *Culture and Tourism in a Smart, Globalized,
and Sustainable World*, Springer Proceedings in Business and Economics,
https://doi.org/10.1007/978-3-030-72469-6_3

et al., 2018), the nature and evolution of active sport tourism are portrayed, and how active sport tourism research has evolved in a short amount of time. Research on active sport tourism has evolved from the existing typologies of sport tourism and their limitations to the development of new frameworks. In addition, existing research is primarily based on the participants' perspective, while there is the need to provide new perspectives, and promote diverse methodologies and technologies to obtain insights on active sports tourism events both from the participants' and the organizers' perspectives.

This paper's key contribution is to demonstrate the economic impact of active sport events and consequently sports tourism. Most research in this area is devoted to global Mega Sport Events (MSEs) or passive sport events (Taks, 2013). Nevertheless, there is currently a huge variety and quantity of active sport events (e.g., marathons, hiking, swimming) that are taking place around the globe. The work in this paper calls for a reflection on how these active sport events contribute to tourism economies and provides evidence on that, based on the results of such an event conducted in Cyprus.

The remainder of this paper is organized as follows: Sect. 2 presents related work on active sport events and highlights the key research topics examined in these works. Section 3 provides a high-level overview of the SportsTraveler76 (ST76) web platform and recommender system implemented in this work, which facilitates Active Sport Events organization and management. The key contribution of this work is presented in Sect. 4 that refers to the analysis of the swimming event case study survey results. The survey results show the impact and the economic benefits to the local economy and showcases that such events can establish Cyprus on the map as an attractive destination for active sports tourism.

2 Related Work

This paper contributes both from a theoretical and practical perspective on the economic impact of sport events and consequently sports tourism. Most research in this area is devoted to global Mega Sport Events (MSEs), most of which refer to passive sport events (Taks, 2013). Nevertheless, there are many more sports events (i.e., non-mega sport events—NMSEs) that are organized around the globe. Many of these NMSEs refer to active sport events. The work in this paper provides evidence on the impact of active sport events on tourism and the economy, and an overview of how technology can assist in the promotion and organization of such events. This section presents some early research work on active sport events.

In (Taks, 2013), the author offers a theoretical perspective on the social impacts of sport events. This work is a comparison of the social impacts and outcomes of MSEs and NMSEs, which concludes that there is reason to believe that NMSEs can be more relevant in creating durable benefits for the local communities that host these events. The paper contrasts and compares social impacts and outcomes of both, MSEs and NMSEs, using four different perspectives: power relations, urban regeneration, socialization, and human capital. In fact, the claim in this work is that this can be explained via the concept of social capital, which indicates that since NMSEs are omnipresent, it seems that they have more lasting global benefits. This actually showcases that NMSEs appears to provide a more positive social impact and outcome opportunities for local residents compared to the passive counterparts. The paper calls for a broader research agenda focusing on the true value of small and medium-sized sport events for local communities.

The work in (Kaplanidou & Gibson, 2010) performed a research study that focused on the examination of sport event images held by active and passive sports tourists at marathon races in Germany. The study outlines some differences in the perception of event images between active and passive sports tourists, as well as in the perception of different types of destinations. In particular, for active sport tourists, the clustering was closer in terms of emotional, physical, and organizational image associations. The emotional theme is as valuable as the physical for active sports tourists. As explicitly stated in (Kaplanidou & Gibson, 2010): "for example, offering special side events which the runners experience during the race (e.g. music bands along the course or running through/past a building for which the destination is renowned)." On the other hand, the study also concluded that for passive sport tourists, social and historical image associations were clustered closer. Finally, the results of the study suggest that the type of the destination also affects and elicits different event images among active and passive sport tourists. The authors explicitly point out in the paper that their research findings have limitations mainly because a single sport event (i.e., marathon) was examined.

Another study (Hallmann et al., 2010) examined and aimed at understanding the variables that influence the behaviors of active sport tourists within the context of recurring small-scale sport events. The authors explicitly state that small-scale active sport events have not been widely observed and examined in the sport and tourism literature. Specifically, the study examined whether the participation of sport tourists in past events, the attitudes toward event participation, the participant's satisfaction with the sport event, and the destination image are accurate predictors of intentions to participate in a sport event again. The study concluded that attitudes towards event participation are important since it looks to impact behavioral intentions directly, while the sport tourist satisfaction with the event is critical as it formulates a positive attitude and most importantly it acts as a direct predictor of behavioral intentions. The authors state that the study examined one sport event with participants over 50 years old, which can limit generalizations of the results to similar events.

Moreover, the study in (Kaplanidou, 2010) attests that: "Sport events can be used as tourist attractions by destinations that seek to attract large numbers of tourists in their locale." It follows that the destination image measurement paradigm to inves-tigate sport event image perceptions of active sport participants. The study asked the active sport tourists ($N = 2,000$) that traveled abroad for participating in the event to give their opinion, by asking them to indicate three words that come to mind following the completion of the event (i.e., post-trip phase). Both a qualitative and a quantitative approach were performed. The qualitative approach classified the acquired words ($n = 1,015$) into six image themes—historical, emotional, organizational, physical, environmental, and social, while the quantitative one conducted a frequency analysis of the words associated with each dimension and revealed that the emotional theme was more frequently mentioned by the participants. In overall, the study concludes that active sport tourists' perception of the event image is associated to the six themes, but most predominantly with the emotional aspects of the event.

The above research works cover principally the social impact and the sport tourists' intentions and perceptions relevant to the themes and the image of the event. In fact, as attested in (Ritchie and Adair, 2002) there was initially little research that analyzed the links between sports and tourism, as well as the social and economic impact of this type of tourism. As can be realized from the above literature, while there is growing research work that examines the social impact of sport events on tourism, and even active sports tourism, there is basically no relevant research that examines the economic impact of active sport events on tourism. In this paper, the aim is twofold: (1) principally to perform the analysis of the survey results that showcase the economic impact of active sports events and (2) to present an overview of the developed web platform and recommender system, which enable SportsTraveller76 Ltd to organize and manage successfully active sports events.

3   SportsTraveller76

3.1   The Web Platform

The SportsTraveller76 (ST76) platform started as a unique and extensive source of reference for groups and individuals, who seek to actively participate in any sporting event taking place in Cyprus and Greece under the following categories: (1) Trending Sports: The newest sports trends available to register, (2) Bespoke Sports Events: Sporting opportunities for children and the elderly and Equality sports for people with disabilities, (3) Top Sports Destinations (Sports oriented travel attractions) and (4) Mainstream Sports Directory: A plethora of endorsed sports activities.

The innovation and originality of ST76 come from the fact that it differentiates from competitors in the following ways: the first group of competitors includes the major providers such as SportsTraveler.net, RoadTrips.com, etc., which focus on passive sport events participation. In particular, they offer the services (e.g., flights, hotel, tickets) for traveling to a country with the primary target to passively participate (i.e., watch) in a major sporting event (e.g., World Cup final, Wimbledon final). The second group of competitors includes major providers such as WorldMarathons.com, FieldsSportsTravel.com, etc. that focus on a specific category of sporting events, e.g., Marathons, and offer the services (e.g., flights, hotel, tickets) for traveling to a country with the primary target to actively participate to that sports event. ST76 is the first platform to provide the full set of services (e.g., flights, hotel, tickets) for traveling to

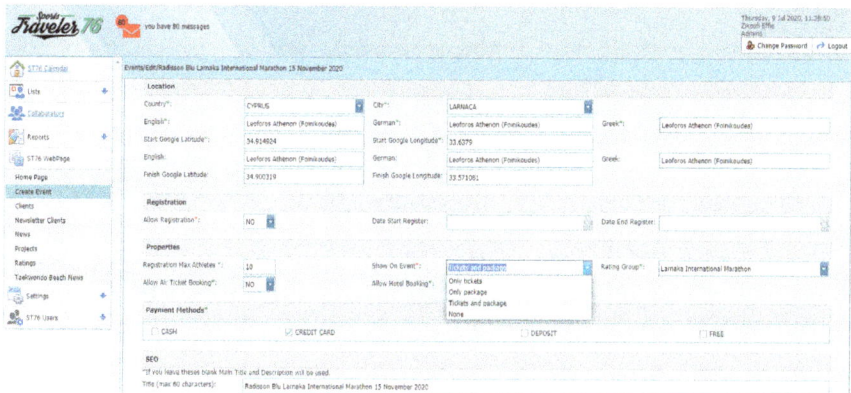

Fig. 1 ST76 Web platform—Backend

a country with the primary target to participate in any kind of sporting activity/event hosted on the platform.

The ST76 web platform is, to the best of our knowledge, the first web platform worldwide that offers a complete set of services for online booking, management, and organization of active sport events. The web platform offers to the company administrator the capability to manage sport events through the backend. Figure 1 shows on the left pane the entire set of features offered to the administrator of the platform, who apart from managing active sport events, is also able to manage users, manage newsletter clients, generate reports, generate recommendations, etc.

The administrator, when creating an event hosted on the platform, is able to select one of the following event modes: (1) Only tickets—customers are only able to purchase tickets for participating in the sport event, (2) Only package (hotel) — enables customer to book only hotels for a specific event and (3) Tickets and package (hotel) —enables a customer to book a combined ticket and hotel package price. The option to enable the iFrame for flight booking can be also enabled for an event, while additional facility services (e.g., sporting equipment) can be enabled as options for customers.

Figure 2 illustrates the end-user view when an event is published, where the customer is able to purchase a ticket or a package based on his/her requirements. For instance, in the case the customer selects a package (ticket and hotel) then the user follows a page-by-page wizard where he/she needs to select the number of rooms, the number of athletes, enter each athlete details and finalize the purchase using an external service provider, i.e., Six Payment services. Finally, the platform allows creating an event where the hotel and flight ticket can be purchased by company's external collaborators (e.g., travel agency) with the help of iFrames [W3C, 2020] that are integrated into the process flow of the customer registration and purchase wizard.

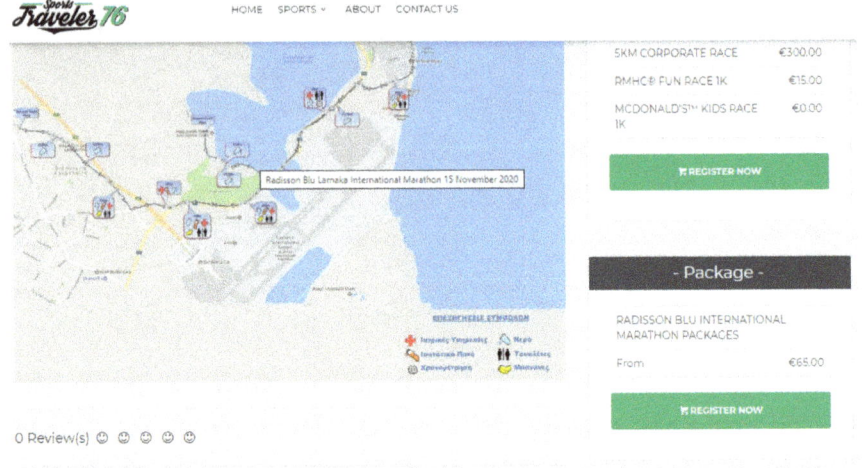

Fig. 2 ST76 Web platform—Frontend

3.2 The Recommender System

The ST76 platform new feature is the ST76 recommender system (ST76_RS). It is a domain-specific solution that aims at providing recommendations of users that are more likely to attend a specific type of event based on the similarity between users' contextual information and events' preferences. In particular, the ST76_RS is used as a Software as a Service (SaaS) to the ST76 web platform. The web service is hosted on the cloud (i.e., Windows Server) and is developed leveraging the.Net Core Framework, which ensures scalability, reliability, and reusability. Additionally, the recommendation algorithm with the K-Means machine learning algorithm, is developed on top of the scikit-learn python library, which ensures valid and efficient operation, as well as high performance; it is also hosted on the same server.

Figure 3 shows the ST76 Recommender System Model: consider a web platform wp (i.e., the SportsTraveler76 web platform) and a web service ws located on server S. Consider a dataset D that contains contextual information of several users clustered into $C_1 \ldots C_n$ clusters of users, where C_i is composed of users interested in similar events, using a K-means algorithm. Users contextual information include event type t, event intensity i, event season s, user's companion c and participation's regional information p described by the tuple CI$\{t, i, s, c, p\}$. Additionally, consider a prediction model m stored on S that is able to predict which cluster of users C_i is more suitable to attend a new event e, which is represented by the tuple CIe$\{t, i, s, c, p\}$. In this case, wp sends an HTTP Request to ws for a user recommendation based on CIe and ws responds back with a set of users Su that are most likely to attend event e based on their contextual information.

The ST76_RS algorithmic part is composed of two phases:

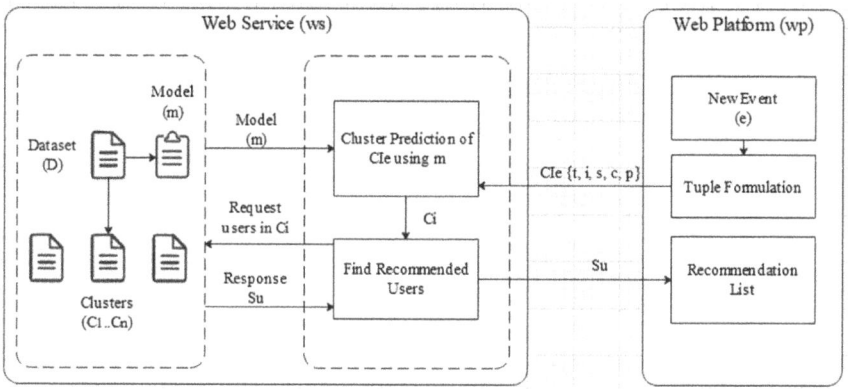

Fig. 3 ST76 Recommender system model

i. In the **offline or clustering phase**, the dataset is clustered using the K-Means machine learning algorithm (Forgy, 1965). Each cluster consists of user-related entries that include users' contextual information which refers to specific types of events. Thus, each cluster represents one or more events along with the users that are most likely to attend or have already attended the corresponding event(s) based on their contextual information. Hence, a prediction model representing the clusters of users and their contextual information is generated and stored on a central server using the Joblib python library.[1]

ii. In the **online or recommendation phase**, the prediction model of the previous phase is used to predict the cluster with users that their preferences and contextual information match the profile of a new event.

Finally, in order to provide accurate user recommendations to specific events a proper dataset is needed. In the absence of any existing dataset suitable for our needs, we generated our own ST76 dataset by collecting domain-specific data from users' contextual information. In particular, the dataset formulated is a result of 68 users answering a Google Forms questionnaire,[2] formulated in a way to obtain users' contextual information. The questionnaire was defined in such a way in order for participants to specify their answers including choices (e.g., user attends trails, marathons) and ratings (e.g., with 5/5 for trails, 3/5 for marathons) for those choices. The different answers, provided by each participant, produced a resulting dataset of approximately 2 thousand entries, with each entry providing contextual information of a user to a specific event type. Contextual information includes information regarding:

(1) the type of event (t)(Official, Leisure, Domestic, Charity),
(2) the intensity of the event (i)(Scale $1 - 5$),

---

[1] Joblib Python Library: https://joblib.readthedocs.io/.

[2] ST76_RS Questionnaire: https://forms.gle/d8Ah7VbeJLuQA3689 (EN version + 19 offline responses), https://forms.gle/6NTYjk8FXuDbBDxcA (GR version).

(3)   the season event is scheduled (*s*)(Autumn, Winter, Spring, Summer),
(4)   the user's companion (*c*)(Solo, +1, Family, Team/Friends), and
(5)   the event's locality (p)(National, European, International).

As explained above, each user may have multiple entries into the dataset and each entry describes contextual information about a specific event type, which the user has attended presenting a different user's profile perspective. More information on the ST76 platform and recommender system is out of the main context of this paper, but interested readers can refer to project's system specification deliverable.[3] The main contribution of this paper is the examination and analysis of the economic impact of active sport events, through the international active sport (swimming) event that is presented in the following case study.

4   Case Study: Experiences from Cyprus

In this section, the survey's methodology is defined, the population of the study is presented and finally the analysis of the results is demonstrated for the different dimensions considered in the study, in order to illustrate the economic impact that active sports events have on the tourism economy.

4.1   Survey Methodology

The event organization survey was conducted during the two days of the international swimming event. The survey was defined in the form of a Google Docs questionnaire[4] and the participants were invited during the registration to answer it electronically. The participants were informed that their responses will remain confidential and that responses are collected and stored in digital form are to be analyzed only by the researchers for the requirements of this study. The objective of the survey is to study and analyze the economic impact of active sport events. In fact, the goal is to identify if the participants of active sports events are also engaging in tourism during their stay, as well as the overall economic impact of such events.

4.2   Survey Population

The study presented in this paper was performed at an international swimming event that took place in Cyprus, and involved 51 athletes out of a total of 512 (i.e., 10%

---

[3]System Specification Deliverable: https://mdl.frederick.ac.cy/SportsTraveler76/Main/Results.
[4]Event Organisation Survey: https://forms.gle/SC6mxTea2ocv8daN8.

of the participants answered the survey). The participants of the event were from 29 countries across the globe, while the gender ratio was 52.9% males and 47.1% females. It is important to note that a large number of participants did not complete the evaluation questionnaire, which is a limitation of the study, since from the total of participants (i.e., 512), $N = 51$ of them are valid for analysis.

4.3 Survey Results: The Economic Impact of Active Sport Events

Figure 1 shows the most representative results/charts from the data collected by the survey, which refer to the views of the event participants. In particular, a very important point that is clearly portrayed in the results is that half (i.e., 51%) of the international travelers are actually visiting Cyprus for the first time. Moreover, nearly 70% have booked accommodation for either 2–4 nights or more than 4 nights, which is particularly important since this was a two-day event. This is confirmed by the third pie chart, since 71.7% of the participants stated that they have combined the event with holidays. This explains and confirms the reason why nearly the same percentage has booked accommodation for more than 2 nights. Another important point is that only 21.6% of the participants traveled solo, whereas nearly 80% of the participants traveled with at least another individual. The final pie chart clearly indicates that 9 out of 10 participants are planning to revisit Cyprus again, which is both an indicator of the quality of the event but also of the fact that Cyprus is indeed an attractive tourist destination. This showcases that even new visitors were satisfied and will revisit for sport and tourism purposes as a result of the swimming event experience (Fig. 4).

OceanMan was the first international active sport swimming event organized in Cyprus. It attracted 512 participants (79 from Cyprus), from which 186 were women that made a dynamic presence in the swimming event. Despite the fact that it was the first time an international active sport event of this magnitude took place in Cyprus, 433 international participants arrived in Cyprus from 29 countries. Based on the participants registered for the event approximately 1500 people are estimated to have watched the event based on combined data; event registrations and hotel bookings. In particular, from the 512 travel parties, the sample comprised of 79 residents and 433 non-residents, whereas an average of 2.17 persons per travel party was calculated from the survey results. A travel party is defined as one or more individuals traveling together to actively participate or watch the swimming event.

International visitors are the drivers of economic impact. Hence, only international visitors were considered in the calculation of the economic impact of the event. This is because without the international visitors the expenditure in the local economy would not have occurred otherwise. As aforesaid, from the 512 total participants, 51 undertook the survey, and based on their responses the economic impact analysis was performed.

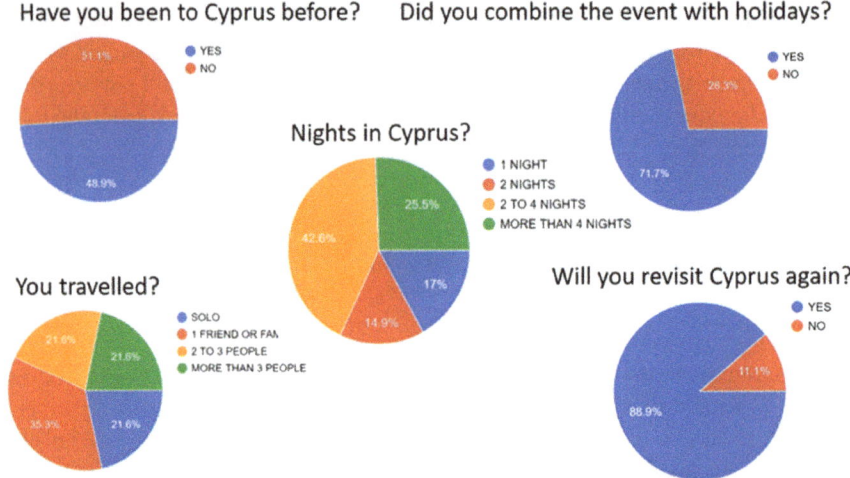

Fig. 4 High-level survey results

A sample from the visitors was asked to undertake the survey and estimate their expenditure patterns in the following three NACE (Nomenclature of Economic Activities) categories: (1) Accommodation and Food Services, (2) Transportation, and (3) Trade (Wholesale and Retail). This included principally accommodation, food and beverages, excursions, transportation, and retail shopping. Based on the data collected from the survey the IMPLAN Input–Output model was applied to calculate the economic impact of the event. More to the point, the survey data collected the average expenditure per person for the entire duration of their stay, which was then multiplied by the total number of international visitors (433 participants/travel party × 2.17 people per party = 938 visitors), yielding an overall estimate of the total visitor spending. Table 1 showcases that the international visitors estimated total direct spending is at € 365,551.92. The total amount is divided into: (1) 55.43% spending

Table 1 Regional and total economic benefits: estimated based on expenditure by international visitors

Description	Direct expenditure	Local purchasing (RPC-65%)	CY I/O multiplier (GM18)	Multiplied effects on local benefit	Multiplied effects on total benefit
Accommodation and food services	€ 202,621.79	€ 131,704.00	1.66	€ 218,628.92	€ 336,352.18
Transportation	€ 111,025.64	€ 72,167.00	1.79	€ 129,178.33	€ 198,735.90
Trade (Wholesale and Retail)	€ 51,904.49	€ 33,738.00	1.35	€ 47,907.84	€ 73,704.37
Total	€ 365,551.92	€ 237,608.75		€ 395,715.09	€ 608,792.45

on Accommodation and Food Services, (2) 30.37% spending on Transportation and (3) 14.20% spending on Trade (Wholesale and Retail), e.g., shopping.

Furthermore, the Regional Purchase Coefficient (RPC) allows measuring the true economic impact of tourist spending, e.g., when attending an event (Stynes 1997). In particular, the international visitors that have attended the event purchase goods and services from local businesses. This is in fact, money coming from outside the community that stimulates the economic activities of the region since tourism is linked with the other sectors of the local economy. It is important to note that some of the spendings leaves the community, which is the reason why the local purchasing is calculated at the RPC of 65% (Stynes 1997).

Moreover, multiplier effects for each category are considered as illustrated in Table 1. The multipliers for Cyprus economy are adopted from the study in (Giannakis and Mamuneas, 2018). For instance, the expenditure for NACE Rev.2 Classification (CY)—I—Accommodation and Food Services is € 202,621.79, while the multiplier effect for this category is 1.66 or € 218,628.92 in local earnings (MOF NACE, 2008). In terms of transportation, the multiplier effect is 1.79 or € 129,178.33 in local earnings and for trade it is 1.42 or € 47,907.84 in local earnings mainly due to shopping activities. Hence, *the economic impact at a local level is calculated at a total multiplied effect of € 395,715.09, while the total economy benefits are calculated at € 608,792.45.* It is important to note here that local benefits reveal the true economic impact of tourist spending, since it reflects money that stay within the local community.

5 Conclusions

Major sport events, commonly termed as passive sport events, are an established and proven source of sports tourism that contributes largely to the economy. Still, the economic impact of active sport events is scarcely recognized in the literature. Most research works on active sports events focus on the analysis and evaluation of the social impact of such events (see Sect. 2). In this work, the main contribution is on examining and analyzing via the case study the economic impact that active sport events have on the local economy and to what extent. At the same time, an overview of the implemented platform and recommender system is also presented in the paper. It is evident from the case study results that active sport events can contribute greatly to the local economy and have a significant impact with a positive total effect on the economy. In particular, the economic impact from the international visitors' expenditure for the swimming event is calculated at a total effect of € 608,792.45, while multiplied effects on local benefit are estimated at € 395,715.09.

Acknowledgements The work presented in this manuscript is performed based on the research funding received from the European Regional Development Fund and the Research Promotion Foundation of Cyprus as part of the Research in Startups SportsTraveler76 project (START-UPS/0618/0049–RESTART 2016–2020).

References

Forgy, E. W. (1965). Cluster analysis of multivariate data: efficiency versus interpretability of classifications. *Biometrics, 21*, 768–769.

Gammon, S., & Robinson, T. (2003). Sport and tourism: A conceptual framework. *Journal of Sport and Tourism, 8*(1), 21–26. https://doi.org/10.1080/14775080306236.

Giannakis, E., & Mamuneas, T. P. (2018). Sectoral linkages and economic crisis: An input-output analysis of the Cypriot economy. *Cyprus Economic Policy Review, 12*(1), 28–40.

Gibson, H. J., Lamont, M., Kennelly, M., & Buning, R, J. (2018). Introduction to the special issue active sport tourism. *Journal of Sport and Tourism, 22*(2), 83–91. https://doi.org/10.1080/14775085.2018.1466350

Greenwell, T. C., Danzey-Bussell, L. A., & Shonk, D. (2019). *Managing sport events* (2nd Edn, 272 p), Human Kinetics Publishers, ISBN10: 1492570958.

Hallmann, K., Kaplanidou, K., & Breuer, C. (2010). Event Image perceptions among active and passive sports tourists at marathon races. *International Journal of Sports Marketing and Sponsorship, 12*(1), 32–47. https://doi.org/10.1108/IJSMS-12-01-2010-B005.

Kaplanidou, K. (2010). Active sport tourists: sport event image considerations. *Journal of Tourism Analysis, 15*(3), 381–386(6). https://doi.org/10.3727/108354210X12801550666303.

Kaplanidou, K., & Gibson, H. J. (2010). Predicting behavioral intentions of active event sport tourists: the case of a small-scale recurring sports event. *Journal of Sport and Tourism, 15*(2), 163–179. https://doi.org/10.1080/14775085.2010.498261.

MOF (Ministry of Finance) NACE Rev. 2. (2008). Statistical Codes of Economic Activities, (Based on the Statistical Classification of Economic Activities NACE, Rev.2, of the European Union). Cyprus.

Ritchie, B., & Adair, D. (2002). The growing recognition of sport tourism. *Current Issues in Tourism, 5*(1), 1–6. https://doi.org/10.1080/13683500208667903.

Stynes, D. J. (1997). *Economic impact of tourism: A Handbook for tourism professionals*. Urbana, IL: University of Illinois, Tourism Research Laboratory.

Taks, M. (2013). Social sustainability of non-mega sport events in a global world. *European Journal of Sport and Society, 10*(2), 121–141. https://scholar.uwindsor.ca/humankineticspub/27.

World Wide Web Consortium (W3C). (2020). HTML Living Standard, The iFrame element. Retrieved from: https://html.spec.whatwg.org/#the-iframe-element. Accessed: July 22, 2020.

The Effects of PUSH and PULL Factors on Spectators' Satisfaction Attitudes. A Mediation Analysis of Perceived Satisfaction from a Small-Scale Sport's Event

Konstantinos Mouratidis and Maria Doumi

1 Introduction

TOURISM is a key sector and established and dominant industry in many advanced and emerging economies worldwide. International and domestic tourism are expanding rapidly, pointing out that local authorities should understand major tourism trends internationally to plan for sustainable tourism development that meets the special interests and expectations of modern-day tourists. Greece is one of the most attractive worldwide tourist destinations and its islands considered from the most visited destinations in Europe. According to SETE statistics, the total international arrivals coming to Greece reached 31.3 million foreign visitors, in 2019, and the total revenue of the Greek tourism industry was nearly €17.7 billion. To reduce the seasonality of demand and the negative impacts of mass tourism, the Greek Ministry of Tourism attempts to enrich and diversify the national tourism product through special and alternative forms of tourism. Taking into consideration, the above-mentioned facts and that the consumers in the global tourism industry are not homogenous in their needs and demands, and this survey provides basic information of travel motives of spectators, who attended a small-scale sport's event in an island destination of the country, the Half-Marathon in Skyros. Nowadays, sports tourism is one of the fastest-growing special forms of tourism worldwide, while the approach of sport's events as tourist products has spurred the interest of the international scientific community, which recognized that sport's events consist a strategy for local, regional, and national tourism development (Mouratidis, 2018). Sport's events could provide long-term benefits for local communities by increasing revenues, supporting local economies, and improving existing infrastructures (Mouratidis et al., 2020). Having in mind that modern-day tourists are seeking for travel experiences related to the

K. Mouratidis (✉) · M. Doumi
University of the Aegean, Chios, Greece
e-mail: mouratidis.konstantinos@outlook.com.gr

© The Author(s), under exclusive license to Springer Nature Switzerland AG 2021
V. Katsoni and C. van Zyl (eds.), *Culture and Tourism in a Smart, Globalized, and Sustainable World*, Springer Proceedings in Business and Economics, https://doi.org/10.1007/978-3-030-72469-6_4

consumption of special tourism products and unique services, this study attempts to explain a particular traveling behavior of spectators-visitors and explore the direct or indirect relationship between their expressed motives and their satisfaction from the small-scale sport's event and the hosting destination. Understanding the complexity of spectators-visitors motives, behaviors and satisfaction attitudes will guide the organizing committee, policymakers, and other co-organizers of small-scale sport's events to ensure the success of the events, by the implementation of strategic and marketing plans, which satisfies a variety of different needs and demands.

2 Literature Review

2.1 Motivation Attributes

In a tourism context, Iso-Ahola (1982) supported that the selection of a tourist destination is based on the tourists' behavioral patterns and the motivations attributed to them and distinguished two types of motives, the PUSH factor (desire to escape from the place of residence) and the PULL factor (desire to see other destinations). Siri et al. (2012), exploring Indian tourist's motivation and perception toward a visit to Bangkok, recognized PUSH (i.e., entertainment opportunities, reducing stress and escaping from the daily routine, etc.) and PULL factors (i.e., experiencing new destinations, seeing historical monuments and cultural sites, etc.). In this way, PULL-based motives refer mainly to 'place' as a tourism product that describes a destination, as Yiamjanya and Wongleede (2014) suggested. Plangmarn et al. (2012) claimed that PULL factors can be categorized into tangible and intangible elements of a specific destination that attract individuals to participate in a tourist activity and realize their needs in the light of particular travel experiences. In a sports tourism context, the literature about the spectators' motivation for the attendance of a sport's event mostly related to the investigation of internal and external spectators' motivations, game attractiveness, economic, competitive and demographic factors, stadium conditions, the value of sport to the community, and spectators' identification (Shank, 2001). Menzies and Nguyen (2012) suggested that the motives of the visitors attending a sport's event are influenced by internal (i.e., enjoyment, excitement) and external (i.e., entertainment, atmosphere) factors. Another group of factors that influence spectators' attendance related mostly to their excitement and escape, as the sports fans who are attending a sport's event expresses the need to escape from their daily routine, enjoy the excitement of the athletic competition, and feel the overall atmosphere of the sport's event (Krohn et al., 1998). To understand spectators' attendance motivations for sport's events, Zhang et al. (1995) note the values of sports to the community, such as community interaction and commitment, public's behavior, social equity, individual quality, and health awareness. Sutton et al. (1997) recognize that fans' identification refers to the personal commitment and emotional involvement of spectators' with a sports organization, and it is considered a major factor that influences spectators' behavior. Over time, an extensive literature has developed on motivational factors related to the active participation in a sport's event (Alexandris et al., 2009; Buonamano et al., 1995; Ko et al., 2008; Rohn et al., 2006; Tokuyama & Greenwell, 2011), or to understand the spectators' motivation in attending major sport's events (Bouchet et al., 2011; Prayag & Grivel, 2014), Consequently, spectators' motivation and satisfaction in attending small-scale sport's events are rarely analyzed in the literature (Wafi et al., 2017a; Yusof, et al., 2009).

2.2 Satisfaction Attributes

Satisfaction is an important mean to understand the best outcome or process of an experience (Krohn & Backman, 2011) and is considered that affects the choice of destination, consumption of products, and services (Kozak & Rimmington, 2000). In a tourism context, satisfaction is originated by the comparison of tourist's expectation before and after consumption (Aliman et al., 2016) and referred mostly to the fulfillment of customers' and tourists' needs or the consumption from their side of a product or service during and after their travel. Several authors have recognized that tourist satisfaction influenced mainly by tourist and cultural attractions, accessibility terms, facilities, activities, and basic services offered (Coban, 2012; Rajesh, 2013). Taking into consideration that prior studies have revealed several cause and effect models for measuring customer satisfaction with a destination (Kozak & Rimmington, 2000; Leong et al., 2010), the additional advantage of this survey is that attempted to measure spectators-visitors satisfaction at a small-scale sport's event, which took place in an island destination and organized mainly by local organizations and to explore the relationship between motivational (PUSH and PULL) factors and spectators' satisfaction attributes.

2.2.1 Satisfaction Attributes from Sport's Event Image

In sports tourism literature, satisfaction attributes have also been explored in prior studies which mentioned that satisfaction levels might vary depending on several factors such as service quality, emotions, experiences, and positive behavior of the visitors. Lee and Beeler (2009) emphasize that service quality and organization and visitors' involvement in the programs and activities offered by the event drive them to be satisfied and express their intention to return in the future. Previous studies on sports tourism have almost exclusively focused on the way event image and destination image can co-worked to affect behavioral intention (Jago et al., 2003; Xing & Chalip, 2006), while others found clear that satisfaction is an overall affective response to a service or product, which is positively correlated with individuals' behavioral intention (Cho et al., 2004; Yoo et al., 2003). In the sport's event context, many scholars have also demonstrated that promotional elements entertaining sport consumers have a significant impact on consumers' satisfaction during sporting events (Lee & Kang, 2015). Consequently, a more positive impression of the small-scale sport's event image, based on its organization, quality, and promotion actions, might correspond to higher levels of spectators' satisfaction and indicate that the perceived satisfaction of the event image can be considered as a strong indicator of spectators' intention to revisit the sport's event's destination.

2.2.2   Satisfaction Attributes from Destination Image

Destination image is considered as an interactive system of perceptions, feelings, visualizations, and intention toward a place (Tasci et al., 2007), which includes the beliefs, ideas, opinions, and impressions of individuals for a destination. Javier and Bign (2001) claimed that the destination image had a direct relationship with perceived quality, satisfaction attributes, and intentions of someone to revisit the destination and recommend it to others. Chen et al. (2013) complemented also that destination image influences the tourists' satisfaction levels and affects their travel decision and plans. Similarly, several scholars suggest that destination image effect consumer and tourist behavior like destination choice, decision making, and their satisfaction attitudes (Chen & Hsu, 2000; Court & Lupton, 1997) or may directly or indirectly affect their satisfaction attributes through expectations and the perceived quality and value of services (Aliman et al., 2016). However, in tourism literature, the positive relationship among destination image and satisfaction attributes is distinct for several types of destinations, including island destinations (Kozak & Rimmington, 2000; Lee et al., 2005; Prayag, 2009). Previous studies revealed that a favorable destination image positively affects on-site experiences and levels of satisfaction (Lee et al., 2005) and influence the intentions of sport tourists to revisit a destination (Kaplanidou & Vogt, 2007).

3   Methodology

3.1   Data Collection and Questionnaire Design

As a research field, Skyros Island in Greece was selected, and as a small-scale sport's event was defined the Skyros Half-Marathon. For the study, a quantitative research was conducted, and self-administered questionnaires were distributed in a random sample of 120 non-local spectators–visitors who visited Skyros and attended its Half-Marathon, which organized by local authorities. In total, 108 questionnaires were successfully collected, while 12 were canceled from the final analysis due to insufficient responses. The questionnaires were drafted in Greek and English and completed by spectators on September 15, 2018 (day of a sport's event) from 09:00 to 21:00. All of the questions used were closed-ended and rating scale to measure spectators' perceptions. Especially, spectators' motivation attributes were measured in the first section of the questionnaire, taking into consideration several statements of travel motivation that have been addressed in previous studies (Yiamjanya & Wongleedee, 2014). Spectators' satisfaction attributes were measured in the second part of the questionnaire, having in mind several satisfaction statements have been discussed in prior studies (Huang, 2009; Wafi et al., 2017b), applying 5-point scale questions, with '1' indicating 'Strongly Dissatisfied' and '5' indicating 'Strongly Satisfied.' The final section of the questionnaire covered the demographic characteristics of the respondents.

3.2 Data Analysis

The data collected were based on descriptive and inductive statistics analyzed by SPSS. Firstly, descriptive analysis was used to present the demographic characteristics of the respondents. Secondly, reliability analysis was applied to examine the internal consistency of the overall scale. Cronbach's alpha statistic presented the reliability of the motivation (0.757) and satisfaction (0.801) measurement scale. Thirdly, a principal component analysis (PCA) and Varimax with Kaiser normalization rotation method were applied to extract two fixed number of factors for motivation and satisfaction measurement scale. The Kaiser–Meyer–Olkin measure of sampling adequacy for motivation (0.664) and satisfaction (0.797) attitudes and Bartlett's Sphericity test statistic (0.000) for both measurement scales supported the implementation of the PCA method. PUSH and PULL factors were identified, which explain 42.3% of the total variance of the motivation attitudes, as the rotation sums of squared loadings revealed that the PULL factor is responsible for the 25% of variance and the PUSH factor for the 17.3% of variance. Accordingly, perceived satisfaction of sport's event organization (PSSEO) and perceived spectators' satisfaction of sport's event destination image (PSSEDI) were identified as two main components, which explain 57.4% of the total variance of the spectators' satisfaction attitudes. The rotation sums of squared loadings revealed that PSSEO is responsible for the 37.7% of variance and the PSSEDI for 19.7% of variance. Fourthly, Pearson's correlation analysis and linear regression analysis were conducted to point out the relationship among the above-mentioned components. Linear regression and path analysis were employed to examine the causal relationships among independent and dependent variables and then conclude in the research hypotheses. In the final step, a mediation analysis conducted with PROCESS macro for SPSS, developed by Hayes (2018). This software was used to test the mediating effect of perceived satisfaction of sport's event organization between PUSH or PULL factors and perceived satisfaction of sport's event destination image. Furthermore, this process was used to estimate direct and indirect effects in the single mediator model by implementing bootstrapping in SPSS, which is considered as an alternative option to prove indirect effect (Hayes, 2009).

3.3 Proposed Hypothesis Framework

This survey provides a conceptual framework (Fig. 1) based on seven hypotheses, which have been formulated to explore the direct and indirect effects among the PUSH and PULL motives and the perceived satisfaction attitudes.

H1: PUSH factor is hypothesized to positively and directly affect spectators' satisfaction of sport's event organization.

H2: PULL factor is hypothesized to positively and directly affect spectators' satisfaction of sport's event organization.

H3: Spectators' satisfaction of sport's event organization is hypothesized to positively and directly affect spectators' satisfaction of sport's event destination image.

H4: PUSH factor is hypothesized to positively and directly affect spectators' satisfaction of sport's event destination image.

H5: PULL factor is hypothesized to positively and directly affect spectators' satisfaction of sport's event destination image.

H6: PUSH factor is hypothesized to indirectly affect spectators' satisfaction of sport's event destination image through spectators' satisfaction of sport's event organization.

H7: PULL factor is hypothesized to indirectly affect spectators' satisfaction of sport's event destination image through spectators' satisfaction of sport's event organization.

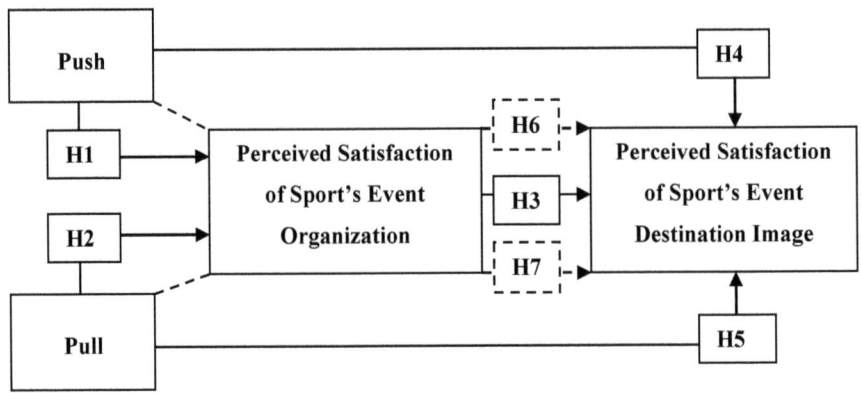

Fig. 1 Proposed hypothesis model

Table 1 Descriptive statistics of Skyros Half-Marathon spectators' ($n = 108$)

		Percentage			Percentage
Gender	Female	42.6	Nationality	Greek	87.0
	Male	57.4		Other	13.0
Age	18–35	38.0	Marital status		
	36–50	28.7		Single	60.2
	51–65	24.1		Married	39.8
	66 +	9.3			
Education	None	0.9	Employment	Private employee	30.6
	Primary	11.1		Public servant	22.2
	Secondary	31.5		Freelancer	18.5
	University	38.0		Pensioner	11.1
	Master degree	15.7		University student	10.2
	Doctoral degree	2.8		Unemployed	7.4

4 Research Findings

4.1 Profile of the Sample

The majority of the sample was male (57.4%), while the small-scale sport's event attracted mainly Greek visitors. Regarding age, the majority of spectators included in the range of 18–35 years. As for the educational level of the spectators, the results demonstrated that the majority of them have slightly a high level (university to doctoral degree), working mostly as private employees (30.6%) and public servants (22.2%) (Table 1).

4.2 Factor Analysis and Reliability

For this survey, two principal component analyses (PCA) were applied with Kaiser–Meyer–Olkin and Bartlett's Sphericity test and Varimax of 14 items of independent variables and 12 items of dependent variables. The results showed that the KMO measure of sampling adequacy for both categories of independent (0.664) and dependent variables (0.797) was greater than the minimum value (0.60) for the appropriate use of the factor analysis. Accordingly, Bartlett's Sphericity test was significant (0.000) for both measurement scales and supported the implementation of the PCA method. Table 2 presented the results of independent variables, which were classified into two components (PUSH and PULL). The factor loadings ranged from 0.525 to 0.922 for PUSH and from 0.431 to 0.719 for PULL factor. Reliability analysis of internal consistency was also conducted and the Cronbach's coefficient

Table 2 Factor analysis and reliability coefficients of independent variables

Variables	Factor loadings	Cronbach's a
PUSH factor (PUSH)		0.814
I came…		
• To interact with the new people who have similar interests	0.922	
• To support the local community	0.917	
• To support the participants-athletes	0.783	
• To interact with the locals	0.525	
PULL factor (PULL)		0.733
I came…		
• Because I have been here before and had memories of the area	0.719	
• Because the sport's event can become a notable tourist attraction	0.656	
• To experience new destinations	0.633	
• To spend time with my family	0.500	
• To feel the atmosphere of the sport's event	0.500	
• To experience the variety of the activities offered by organizers of the sport's event	0.475	
• To experience the uniqueness of the sport's event	0.466	
• To enjoy the traditional cuisine of the local community	0.466	
• To spend time with my friends and relatives	0.435	
• To enjoy the action of traveling	0.431	

alpha (0.814) for PUSH and alpha (0.733) for PULL factor indicates a high level of reliability (over 0.70).

Similarly, Table 3 displayed the results of dependent variables, which were grouped into two components (PSSEO and PSSEDI). The factor loadings ranged from 0.489 to 0.911 for PSSEO and from 0.489 to 0.797 for PSSEDI. Cronbach's alpha values are considered acceptable for PSSEO (0.901) and PSSEDI (0.676).

4.3   Factors Affecting Perceived Satisfaction of Sport's Event Organization and Perceived Satisfaction of Sport's Event Destination Image

Pearson's correlation analysis and linear regression analysis were conducted to point out the relationship among variables. Table 4 presents that there was a positive correlation between two independent variables (PUSH and PULL factor), but also illustrates a positive correlation between one independent variable (PUSH factor) and

Table 3 Factor analysis and reliability coefficients of dependent variables

Variables	Factor loadings	Cronbach's a
Perceived satisfaction of sport's event organization (PSSEO)		0.901
I am satisfied with…		
• The signs and symbols of Half-Marathon	0.911	
• Other events offered by Half-Marathon organizers	0.853	
• The support offered to the participants-athletes	0.809	
• The promotion of Half-Marathon through social media	0.800	
• Half-Marathon information provided by organizers	0.767	
• Half-Marathon promotion through media (Travelogues, TV/Radio Shows, Newspapers, etc.)	0.709	
• The directions provided to the participants-athletes and spectators by organizers	0.688	
Perceived satisfaction of sport's event destination image (PSSEDI)		0.676
I am satisfied with…		
• Skyros and its eco-cultural environment and sightseeing's	0.787	
• Information about Skyros	0.751	
• facilities provided and feel comfortable when attending Half-Marathon	0.647	
• The hospitality and locals' behavior	0.569	
• My decision to attend Half-Marathon	0.489	

Table 4 Correlations

	PSSEDI	PUSH	PULL	PSSEO
PUSH	−0.008			
PULL	0.261**	191*		
PSSEO	−0.029	0.894**	0.045	
Mean	**4.00**	**3.53**	**4.02**	**3.44**
S.D	**0.493**	**0.875**	**0.523**	**0.900**

*Correlation is significant at 0.05 level (2-tailed); **at 0.01 level (2-tailed)

the mediate variable (perceived satisfaction of sport's event organization). A positive correlation between one independent variable (PULL factor) and the dependent variable (perceived satisfaction of sport's event destination image) was also obvious in Table 4. This means that spectators with stronger PUSH travel motivations tend to feel more comfortable with the sport's event organization, and visitors with PULL travel motivations tend to feel more satisfied with the image of the sport's event destination.

The results of the survey demonstrated that there was a significant positive relationship between the mediate variable (PSSEO) and only one independent variable

(PUSH) ($r = 0.894$, $p < 0.01$). The linear regression coefficient of PUSH was $\beta = 0.92$, $p = 000$. This revealed that the PUSH factor had a positive effect on perceived satisfaction of sport's event organization at the 99% confidence level. Furthermore, the PUSH factor could explain 89.4% of the variation of perceived satisfaction of sport's event organization ($R^2 = 0.894$). In contrast, the findings illustrated that there was no significant positive relationship between the mediate variable and the other independent variable (PULL) ($r = 0.045$, p $= 0.645$). The regression coefficient of PULL was $\beta = 0.07$, $p = 645$. Consequently, H1 was accepted, and H2 was rejected. The results revealed, also, that there was not a positive correlation between the mediate variable (PSSEO) and the dependent variable (PSSEDI) ($r = -0.029$, $p = 0.769$). The coefficient of determination ($\beta = -0.01, p = 0.769$) illustrated that H3 was not accepted. The findings of the primary data revealed that there was not a positive correlation between the independent variable (PUSH) and the dependent variable (PSSEDI) ($r = 0.008, p = 0.933$). The regression coefficient ($\beta = -0.00, p = 0.933$) demonstrated that H4 was rejected. The Pearson correlation analysis results also displayed a positive correlation between the other independent variable (PULL) and the dependent variable (PSSEDI) ($r = 0.261, p < 0.01$). Spectators' satisfaction of sport's event destination image affected by PULL factor ($\beta = 0.24$, $p < 0.01$) in the positive direction and at the 99% confidence level. Furthermore, the PULL factor could explain 26.1% of the variation of perceived satisfaction of sport's event destination image ($R^2 = 0.261$). Consequently, H5 was substantiated.

4.4   Significance of the Indirect Effects

In a general context, the indirect effect of an independent variable on the dependent variable through the mediate one concludes the effect of the independent variable on the mediate variable and the effect of the mediate one on the dependent variable. Hence, for the exploration of the indirect effects, hypotheses 6 and 7 have been formulated. As previously stated, perceived satisfaction of sport's event organization was positively affected only by the PUSH factor ($\beta = 0.92$, $p = 000$) (H1) and not influenced by the PULL factor ($\beta = 0.07$, $p = 645$) (H2). Taking into consideration, the above-mentioned results that the perceived satisfaction of sport's event organization directly not caused an effect on perceived satisfaction of sport's event destination image ($\beta = -0.01$, $p = 0.769$) (H3), and it can be mentioned that through the mediate variable, PUSH and PULL factors not created indirect effects on spectators' satisfaction of sport's event destination image. Consequently, this survey determined that PUSH motivations would influence mainly the spectators' satisfaction of sport's event organization, and PULL motivations would accordingly lead to higher destination satisfaction.

To test the significance of indirect effects or mediations, the bootstrapping method was applied, and the data output displayed the bootstrapped confidence intervals (at the 95%). If the confidence interval includes ZERO (0) between the lower boundary (LL) and the upper boundary (UL), this would indicate, with 95% confidence, that

Table 5 Hypotheses testing

Hypotheses	B	Sig	Results
H1: PUSH → PSSEO	0.92	0.000	Accepted
H2: PULL → PSSEO	0.07	0.645	Rejected
H3: PSSEO → PSSEDI	−0.01	0.769	Rejected
H4: PUSH → PSSEDI	−0.00	0.933	Rejected
H5: PULL → PSSEDI	0.26	0.006	Accepted
H6: PUSH → PSSEO → PSSEDI	−0.09	0.626	Rejected
H7: PULL → PSSEO → PSSEDI	−0.00	0.754	Rejected

there is no mediation or indirect effect. In contrast, if the confidence interval NOT includes ZERO (0) between the lower boundary (LL) and the upper boundary (UL), this would indicate, with 95% confidence, that the mediation or indirect effect is significant. According to the confidence intervals for the indirect effects using bootstrap methods, it can be noticed that the indirect effects of PUSH and PULL factors on perceived satisfaction of sport's event destination image through the perceived satisfaction of sport's event organization were not significant. Focusing on the effect sizes of PUSH factor, the standardized b for the indirect effect, its value is $b = -0.095$, 95% BCa CI $[-0.639, 0.316]$. Similarly, the standardized b for the indirect effect of PULL factor is $b = -0.002$, 95% BCa CI $[-0.029, 0.018]$.

As previously stated, the indirect effect is the combined effects of paths a (PUSH → PSSEO) and b (PSSEO → PSSEDI) for the first case and the combined effects of paths a (PULL → PSSEO) and b (PSSEO → PSSEDI) for the second one. An alternative option to estimate the indirect effect and its significance can be addressed using the Sobol Test. If the Sobol Test is significant, it means that the predictor variable (PUSH or PULL factor) significantly affects the outcome variable (PSSEDI) via the mediator (PSSEO), or, in other words, there is significant mediation. According to Sobol Test results, neither PUSH factor nor PULL factor significantly affects the outcome variable (PSSEDI) via the mediator (PSSEO), as the p-values were not quite under the limit of 0.05 ($p = 0.626$ and $p = 0.754$ accordingly). Consequently, taking into consideration, the results of the Sobol Test and that all of the indirect effect size measures have confidence interval that includes ZERO, the results demonstrated that there were no significant indirect effects and not indicated significant mediation. Hence, H6 and H7 were not accepted. Table 5 summarizes the results of the research's hypotheses, and Fig. 2 presents a diagram of the mediation model and indicates it the regression coefficients, the indirect effect, and its bootstrapped confidence intervals.

5 Discussion and Conclusions

Regardless of the sport's event size, its image depends in part on the cognitive image (i.e., event organization and quality and destination features) and affective image

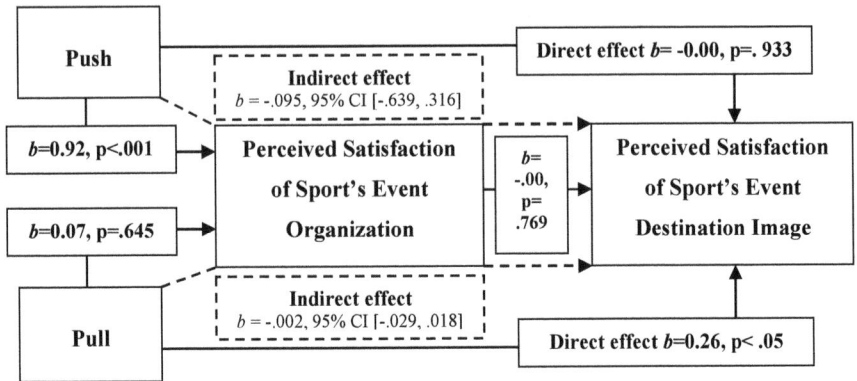

Fig. 2 Model of PUSH and PULL factor as a predictor of perceived satisfaction of sport's event destination image, mediated by perceived satisfaction of sport's event organization

(i.e., emotional and social aspects), both of which contribute to a holistic evaluation of the event (Baloglu & McCleary, 1999). Admittedly, the type of sport consumer (participant or spectator) consists of another factor that affects the evaluation of the sport's event image (Halmann et al., 2010). Hence, defining the profile of sport tourists, passive or spectators, is a prerequisite in understanding their essential travel motives. Taking into consideration the role of satisfaction, this survey measured specific contextual satisfaction (i.e., I am satisfied with my decision to attend Half-Marathon) and overall satisfaction (i.e., I am satisfied with other events offered by Half-Marathon organizers) to explore the relationships among PUSH and PULL factors and spectators' satisfaction of sport's event destination image, as well as the mediating role of perceived satisfaction of sport's event organization. The findings confirmed the positive correlation between PUSH-based motives and the perceived satisfaction of sport's event organization, and the positive correlation among PULL-based motives and perceived satisfaction of sport's event destination image. Due to the fact that no empirical work, in our knowledge, has examined the mediating role of the perceived satisfaction of sport's event organization between spectators' travel motives and spectators' satisfaction of sport's event destination image, in the context of a small-scale Half-Marathon event, the additional advantage of this survey is that demonstrated neither PUSH, nor PULL motives indirectly affect the spectators' satisfaction of sport's event destination image through spectators' satisfaction of sport's event organization. Nevertheless, even the results of our primary data do not show mediation between variables, the development of small-scale sport's event tourism should focus on active or passive sport tourists' satisfaction, and the promotional activities should be compatible with their demands. For instance, marathon event organizers should provide the participants with positive experiences that enhance the organizational, environmental, emotional, social fulfillment, and physical activity issues of the event (Kaplanidou & Vogt, 2006). Thus, the findings of this study contribute to the identification of key features influence spectators' satisfaction during a small-scale Half-Marathon event and its patterns facilitate such event organizers to establish, both a positive event and destination image or a strong reputation for the sporting events, which take place mainly in local communities. Understanding spectators' motivational and satisfaction attributes and their causal relationships in the context of small-scale sport's event tourism remains the main challenge for event organizers and could provide useful information into the development of strategic investments and marketing plans of several tourism destinations.

6 Limitations and Suggestions for Future Research

This study provides several important contributions to sports tourism literature, but its findings have several limitations. Given the small sample size and the specific sport context (i.e., spectators of a small-scale sport's event in Greece), the results should be interpreted with caution, as they expressed only a static picture of individuals' motivational and satisfaction attitudes. Consequently, the results of this study cannot be generalized to other events of the country or the world. One limitation of our implementation is that the data collected only from spectators attending the sport's event and not from the participants-athletes. Also, this study measures mainly spectators' motivational and satisfaction attributes from a specific type of small-scale sport's event, a Half-Marathon event, and consequently, the investigation of other types of sport's events such as mountain running events, cycle tours, etc., remains unexplored. This survey may consider as a starting point to explore the development of small-scale sport's events as a tourist attraction in island destinations of Greece, and its patterns are crucially important in understanding the spectators' motives and satisfaction attitudes. Future studies will benefit from the collection of experimental data to measure the direction of causality among relationships more exclusively. Future approaches should additionally focus on other groups related to small-scale sport's events, such as athletes, sponsors, and organizers. For future, researches have been suggested also apart from the quantitative technique the use of qualitative methods, to exact more qualitative data referred to sports tourists' physical and emotional engagement in sport's events activities. Moreover, the research field may be increased throughout Greece by selecting small-scale sport's events organized in several destinations of the country. Finally, it would be interesting to explore whether the patterns of this study can be replicated in small-scale sport's events, which be organized in other countries. Thus, future research is encouraged to test the conceptual model developed in this study across different small-scale sport's events in other regions or countries.

References

Alexandris, K., Kouthouris, C., Funk, D., & Giovani, K. (2009). Segmenting winter sport tourists by motivation: The case of recreational skiers. *Journal of Hospitality Marketing and Management, 18*, 480–500.

Aliman, N. K., Hashim, S. M., Wahid, S. D. M., & Harudin, S. (2016). Tourists' satisfaction with a destination: An investigation on visitors to Langkawi Island. *International Journal of Marketing Studies, 8*(3), 173–188.

Baloglu, S., & McCleary, K. W. (1999). A model of destination image formation. *Annals of Tourism Research., 26*, 868–897.

Bouchet, P., Bodet, G., Assollant, I., & Kada, F. (2011). Segmenting sport spectators: construction and preliminary validation of the sporting event experience search (SEES) scale. *Sport Management Review, 14*(1), 42–53.

Buonamano, R., Cei, A., & Mussino, A. (1995). Participation motivation in Italian Youth Sport. *Sport Psychologist, 9*(3), 265–281.

Chen, J. S., & Hsu, C. H. C. (2000). Measurement of Korean tourists' perceived images of overseas destinations. *Journal of Travel Research, 38*(4), 411–416.

Chen, Y., Zhang, H. & Qiu, L. (2013). A review on tourist satisfaction of tourism destinations. In *Proceedings of 2nd International Conference of Logistics, Informatics and Service Science*, pp. 593–604.

Cho, B. H., Lee, C. W., & Chon, T. J. (2004). Effect of customers' service quality satisfaction for repurchase of golf range use. *Korean Journal of Physical Education, 42*, 179–188.

Coban, S. (2012). The effects of the image of destination on tourist satisfaction and loyalty. The case of Cappadocia. *European Journal of Social Science, 29*(2), 222–232.

Court, B., & Lupton, R. A. (1997). Customer Portfolio development: Modeling destination adapters, inactives, and rejecters. *Journal of Travel Research, 36*(1), 35–43.

Hallmann, K., Kaplanidou, K., & Breurer, C. (2010). Event image perceptions among active and passive sport tourists at marathon races. *International Journal of Sports Marketing and Sponsorship, 12*, 37–52.

Hayes, A. F. (2009). Beyond Baron and Kenny: Statistical mediation analysis in the new millennium. *Communication Monographs, 76*(4), 408–420.

Hayes, A. F. (2018). Partial, conditional, and moderated mediation: Quantification, inference, and interpretation. *Communication Monographs, 85*(1), 4–40.

Huang, Y. C. (2009). Examining the antecedents of behavioral intentions in a tourism context. ProQuest Dissertations and Theses. [Online], Available at: http://search.proquest.com/docview/305111584?accountid=51189 [Accessed 6/11/2019]

Iso-Ahola, S. E. (1982). Towards a social psychology of tourism motivation- a rejoinder. *Annals of Tourism Research, 9*, 256–261.

Jago, L., Chalip, L., Brown, G., Mules, T., & Ali, S. (2003). Building events into destination branding: Insights from experts. *Event Management, 8*, 3–14.

Javier, S., & Bign, J. E. (2001). Tourism image, evaluation variables, and after purchase behavior: Inter- relationship. *Journal of Tourism Management, 22*, 607–616.

Kaplanidou, K., & Vogt, C. (2006). Do sport tourism events have a brand image? In *Proceedings of the 2006 Northeastern Recreation Research Symposium* (pp. 2–7). Newton Square, PA: U. S. Forest Service, Northern Research Station.

Kaplanidou, K., & Vogt, C. (2007). The Interrelationship between sport event and destination image and sport tourists' behaviours. *Journal of Sport and Tourism, 12*(3/4), 183–206.

Ko, Y. J., Park, H., & Claussen, C. L. (2008). Action sports participation: Consumer motivation. *International Journal of Sports Marketing and Sponsorship, 9*(2), 111–124.

Kozak, M., & Rimmington, M. (2000). Tourist Satisfaction with Mallorca, Spain, as an off- season holiday destination. *Journal of Travel Research, 38*(3), 260–269.

Krohn, B. D., & Backman, S. J. (2011). Event attributes and the structure of satisfaction: A case study of golf spectators. *Event Management, 15*(3), 267–277.

Krohn, F. B., Clarke, M., Preston, E., McDonald, M., & Preston, B. (1998). Psychological and sociological influences on attendance at small college sporting events. *College Student Journal, 32*(2), 277–288.

Lee, J., & Beeler, C. (2009). An investigation of predictors of satisfaction and future intention: Links to motivation, involvement and service quality in a local festival. *Event Management, 13*(1), 17–29.

Lee, J. S., & Kang, J. H. (2015). Effects of sport event satisfaction on team identification and revisit intent. *Sport Marketing Quarterly, 24*, 225–234.

Lee, C., Lee, Y., & Lee, B. (2005). Korea's destination image formed by the 2002 World Cup. *Annals of Tourism Research, 32*(4), 839–858.

Leong, Q. L., Shahrim, A. K. & Mohhidin, O. (2010). Relationship between Malaysian food image, tourist satisfaction, and behavioral intention. *World Applied Sciences Journal, 10*(Special Issue of Tourism and Hospitality), 164–171.

Menzies, J. L., & Nguyen, S. N. (2012). An exploration of the motivation to attend for spectators of the Lexmark Indy 300 Champ Car event, Gold Coast. *Journal of Sport and Tourism, 17*(3), 183–200.

Mouratidis, K. (2018). Sociological aspects of applying physical activities through the contribution of sport tourism and small- scale sports events. In *8th International Conference Proceedings: Effects of Applying Physical Activity on Anthropological Status of Children, Adolescents, and Adults* (pp. 463–472), Book of Proceedings, Belgrade, Serbia.

Mouratidis, K., Doumi, M., & Thanopoulos, V. (2020) Spectators' satisfaction of a small-scale sport event and intention to re-visit the sport event's destination. In V. Katsoni, T. Spyriadis (eds) *Cultural and tourism innovation in the digital era*. Springer.

Plangmarn, A., Mujtaba, B. G., & Pirani, M. (2012). Cultural value and travel motivation of European tourists. *Journal of Applied Business Research (JABR), 28*(6), 1295–1304. https://doi.org/10. 19030/jabr.v28i6.7344.

Prayag, G. (2009). Tourists' evaluation of destination image, satisfaction, and future behavioral intentions: *The case of Mauritus. Journal of Travel & Tourism Marketing, 26*(8), 836–853.

Prayag, G., & Grivel, E. (2014). Motivation, satisfaction, and behavioral intentions: segmenting youth participants at the interamnia world cup 2012. *Sport Marketing Quarterly, 23*(3), 148–160.

Rajesh, R. (2013). Impact of tourist perceptions, destination image, and tourist satisfaction on destination loyalty: A conceptual model. *PASOS. Revista de Turismo y Patrimonio Cultural, 11*(3), 67–78.

Rohn, A. J., Milne, G. R., & McDonald, M. A. (2006). Proven top- rate qualitative analysis and mixed methods research. *Sport Marketing Quarterly, 15*, 29–39.

SETE. (2019). The Contribution of Tourism to the Greek economy in 2018. (Online). Retrieved from https://www.insete.gr/Portals/0/meletes-INSETE/01/2019/2019_Tourism-Greek_Eco nomy_2017-2018_EN.pdf. Accessed on July 15, 2019.

Shank, M. D. (2001). *Sports marketing: A strategic perspective.* . Upper Saddle River, NJ: Prentice Hall.

Siri, R., Kennon, L., Josiam, B., & Spears, D., (2012) Exploring Indian tourists' motivation and perception of Bangkok, *Tourismos: An International Multidisciplinary Journal of Tourism, 7*(1), 61–79.

Sutton, W. A., McDonald, M. A., Milne, G. R., & Cimperman, J. (1997). Creating and fostering fan identification in professional sports. *Sport Marketing Quarterly, 6*(1), 15–22.

Tasci, A. D. A., Gartner, W. C., & Cavusgil, S. T. (2007). Conceptualization and operationalization of destination image. *Journal of Hospitality and Tourism Research, 31*(2), 194–223.

Tokuyama, S., & Greenwell, T. C. (2011). Examining similarities and differences in consumer motivation for playing a watching soccer. *Sport Marketing Quarterly, 20*(3), 148–156.

Wafi, A. A., Chiu, L. K., & Kayat, K. (2017a). Factors Influence visitor's experience in a small-scale sports event. *Journal of Tourism, Hospitality and Environment Management, 2*(3), 1–12.

Wafi, A. A., Chiu, L. K., & Kayat, K. (2017b). Understanding sport event visitors' motivation and satisfaction of small-scale sport event. *Journal of Tourism, Hospitality and Environment Management, 2*(3), 13–24.

Zhang, J. J., Pease, D. G., Hui, S. C., & Michaud, T. J. (1995). Variables affecting the spectator decision to attend NBA games. *Sports Marketing Quarterly, 4*(4), 29–39.

Xing, X., & Chalip, L. (2006). Effects of hosting a sport event on destination brand: A test of co-branding and match-up models. *Sport Management Review, 9*, 49–78.

Yiamjanya, S., & Wongleedee, K. (2014). International tourists' travel motivation by PUSH-PULL factors and the decision making for selecting Thailand as destination choice. *International Journal of Humanities and Social Sciences, 8*(5), 1348–1353.

Yoo, Y. S., Cho, K. M., & Chon, S. S. (2003). The effect of customer satisfaction on repurchase intention at golf practice ranges in South Korea. *Korean Journal of Sport Management, 7*, 1–13.

Yusof, A., Omar-, M. S., Shah, P. M., & Geok, S. K. (2009). Exploring small-scale sport event tourism in Malaysia. *Research Journal of International Studies, 9*, 47–58.

Strategic Negotiation Factors in Participating at Recreational Sport Activities Aiming at the Well-being and the Presentation of Perma Scale for the Greek Population

Georgia Yfantidou, Alexia Noutsou, Panagiota Balaska, Evangelos Bebetsos, Alkistis Papaioannou, and Eleni Spryridopoulou

1 Introduction

The involvement of people in modern developed societies with any form of physical and sport recreational activity is very important, as it is an individual indicator of mental, physical and psychological health but also an indicator of social well-being and progress (Conn et al., 2011). The benefits to the individual through constant engagement with physical exercise and sports activities are multidimensional (Conn et al., 2011). Thus, at the level of physical health, they are traced in the improvement of functional capacity over time (Funk et al., 2011; Murphy & Bauman, 2007), at the level of wellness through the enjoyment of hedonistic experiences (Lloyd & Little, 2010; Rupprecht & Matkin, 2012; Sato et al., 2014; Walker et al., 2011) and on a psychological level by strengthening self-esteem and sense of accomplishment (Biddle et al., 2000; Mutrie & Faulkner, 2004). At the same time, the benefits of exercise are found in the prevention of several diseases, such as obesity, cancer, diabetes and cardiovascular disease (Hamer et al., 2014; Kohl et al., 2012; Lee et al., 2012). In addition to the above, the World Health Organization states that physical activity both indoors and outdoors can be a tool for physical, mental and emotional health, to contend with the effects of the spread of COVID-19 (Bull, F. C. et al, 2020).

G. Yfantidou (✉) · A. Noutsou · E. Bebetsos · E. Spryridopoulou
Department of Physical Education and Sport Science, Democritus University of Thrace, Komotini, Greece
e-mail: gifantid@phyed.duth.gr

P. Balaska
Department of Physical Education and Sport Science, Aristotle University of Thessaloniki, Thessaloniki, Greece

A. Papaioannou
Department of Business Administration, University of Peloponnese, Kalamata, Greece

© The Author(s), under exclusive license to Springer Nature Switzerland AG 2021
V. Katsoni and C. van Zyl (eds.), *Culture and Tourism in a Smart, Globalized, and Sustainable World*, Springer Proceedings in Business and Economics, https://doi.org/10.1007/978-3-030-72469-6_5

People's participation at recreational sports also 01as a social imprint (Garrett et al., 2004). Citizens are organized into groups with the ultimate purpose of sharing experiences and making meaningful social interactions (Filo et al., 2009). Several countries have developed policies for access to sports activities for all social and age groups, with the aim of promoting the quality of life of their residents (Sallis, 2016). Given that research interest has turned to the understanding and development of internal mechanisms that favor participation in recreational sports activities for the well-being of people and foster physical activity as a vehicle for individual and social well-being (Sebire et al., 2009, 2011).

2 Literature Review

2.1 Negotiation Strategies

Negotiation strategies are defined as the internal factors that influence the formation of positive cognitive and emotional attitudes and behavioral loyalty to exercise, by avoiding and reducing the impact of restrictions on participation in sports activities of free-leisure time (Jackson, 2005; Mannell & Kleiber, 1997). These in turn help the individual to better manage negotiation strategies and promote sustainable participation at recreational sport activities (Alexandris et al., 2011, 2007; Hubbard & Mannell, 2001; White, 2008; Wilhelm et al., 2009; Balaska et al., 2019).

The concept of negotiation is based on social cognitive theory. According to this, individuals are likely to actively choose to change situations and environmental conditions that affect their behavior rather than passively accept adverse situations (Maddux, 1993). In other words, the basic concept of negotiation considers people as "active formers" instead of "passive receivers" (Mannell & Loucks-Atkinson, 2005).

2.2 Wellness

Wellness is a broad concept that has been linked to well-being, happiness, satisfaction and bliss (Seligman, 2018). Recent trends in positive psychology suggest that wellness consists of a combination of hedonistic and blissful elements (Gallagher, Lopez, & Preacher, 2009; Huppert & So, 2013; Slade, 2010). In practice, this means that in order to understand the concept, both subjective indicators (levels of positive emotions, happiness, etc.) and objective indicators (housing, level of education, life expectancy) need to be employed (Kahneman & Krueger, 2006). Powers, Dodd and Noland (2006), consider wellness as a "healthy life" that includes regular physical activity, proper nutrition, elimination of unhealthy behaviors and maintaining good emotional health.

2.3   Urban Green Spaces

According to the Council of Europe (1986), the concept of urban green spaces refers to any kind of free space, regardless of the area it occupies and changes according to time and place. However, all free public spaces of a city that host some form of vegetation and to which all citizens have access are characterized as urban green spaces. In modern cities, the role of urban green is particularly important, as it contributes to the planning of strategic actions based on the physical and mental health of citizens (Bauman et al., 2009; Bock et al., 2014). The creation of green recreational areas strengthens sports culture and sustainable engagement with physical activity (Stevinson et al., 2015), through the perspective of utilizing the natural environment, while combining the social and cultural element (Brand et al., 2014).

2.4   Aim

The purpose of this study was to examine the effect of demographic characteristics of a sample of the inhabitants of the central urban structure of Athens on negotiation strategies for participation at recreational sports and on well-being. The validity and reliability of the PERMA-profiler questionnaire were also tested in order to be used to measure the well-being of the Greek population. In addition, the dimension of "Recreational Green Attachment" as a factor of well-being was piloted, and its relationship with demographic categories was investigated, with the goal of creating green spaces that will promote sustainable engagement with recreational sports activities.

H1:   There is a statistically significant interaction of marital status and education in wellness factors.

H2:   There is a statistically significant interaction of age and gender categories on negotiation strategies factors.

H3:   There is a statistically significant interaction of education categories and marital status on negotiation strategies factors.

3   Methodology

3.1   Sample

The sample consisted of 233 inhabitants ($N = 233$) of the central urban structure of Athens, of which 67 men (28.8%) and 166 women (71.2%). The age range was located between 19 and 77 years.

3.2 Questionnaires

The questionnaire for the present survey had three parts: the demographic section, the negotiation strategies scale and the PERMA-profiler scale. At the demographic data section, a total of six questions were provided with the aim of collecting demographic data on gender, age, educational level, occupation, income and marital status. The second part was the scale of "Negotiation Strategies" for participation at sports activities. It was used the questionnaire of Balaska et al. (2019). The questionnaire consisted of a total of 33 questions and divided into 11 dimensions. The score of all the questions was counted on a seven-point Likert scale, between "1 = Never," "4 = Sometimes," "7 = Always." Finally, it was used the PERMA-profiler scale (Butler & Kern, 2016), translated and validated for the Greek population by Pezirkianidis et al. (2019). The questionnaire consisted of a total of 27 questions. For the research purpose, six questions were added to PERMA-scale and were tested regarding the commitment to green and how it contributes to well-being. The answers to the PERMA-profiler questions were scored on a seven-point Likert scale, between "Never = 1," "So and So = 4," "Always = 7."

3.3 Process

The sample questionnaires were collected in January, February and March, shortly before the COVID-19 outbreak and mandatory restrictions. The questionnaires were randomly distributed to the residents of the central urban structure of Athens in Greece and were completed once and in a single way. They were distributed electronically by sending the questionnaire to the personal e-mail of the participants. The duration of completing the questionnaire was estimated at 10 min, and the way of evaluating the answers was common for each participant. The data collected were quantitative. After the data collection, they were statistically processed with the statistical data processing software SPSS v.26.

3.4 Data Analysis

Descriptive and frequency statistics were used to provide the profile of participants. Factor analysis was performed for the wellness model with the method of exploratory factor analysis. This was followed by a reliability check of the negotiation strategies and PERMA-profiler scales, with the method of internal consistency checking the Cronbach a. To test the statistical hypotheses, it was used the multivariate analysis of variance (MANOVA).

4 Results

4.1 Demographics

The majority of the sample was 40–59 years old (49.4%), following the age group of 19–39 years old with a minor percentage (46.8%). The group over 60 had the lowest representation with 3.8%. Regarding the educational level of the sample, 39.1% were university graduates, 35.6% held a master's or doctoral degree, 13.3% were OAED, college or IEK graduates, and 12% were high school graduates. 37.8% of the sample were employed in the private sector, 30% were self-employed, and 13.7% were civil servants. The percentage of unemployed participants was 4.3%, students 5.2%, retirees at 3% and housekeepers at 2.6% (3.4% other). The annual income was divided into three scales, 47.6% of the sample had an annual income below 20,000 euro, 43.8% between 20,000 and 60,000 euro, while a percentage of 8.6% declared an income above 60,000 euro. Finally, the percentage of married amounted to 38.6%, singles made up 36.9%, domestic partnership was 16.8% and divorced 7.7%.

4.2 Validity and Reliability

Factor analysis was performed for the wellness model of PERMA with the method of exploratory factor analysis. The principal component analysis for the 27 variables revealed 5 factors that explained the 64.97% of the total variance:

- The factor "Recreational Green Attachment" consists of six questions
- The "Accomplishment—Engagement" factor consists of nine questions
- The factor "Happiness—Positive Emotions" consists of five questions
- The factor "Loneliness—Negative Emotions" consists of four questions
- The "Physical Health" factor consists of three questions (Tables 1 and 2)

Cronbach's α for all 27 questions of the PERMA-profiler questionnaire, showed high levels of reliability, $\alpha = 0.888$. The analysis revealed that the recreational green attachment is a factor with a very high reliability $a = 0.956$. In general, the reliability analysis for each dimension indicates that all dimensions showed a high degree of reliability ($\alpha = 0.681 - 0.956$), with the exception of the factor negative emotions—loneliness, which showed a relatively lower degree of reliability ($\alpha = 0.681$). However, as the study of well-being for the Greek population is at a primary stage, Cronbach's α coefficient may be acceptable at values between 0.5 and 0.6 (Tavakol & Dennick, 2011).

The structural validity of the negotiation strategies scale has been confirmed in previous research (Balaska et al., 2019). In the present study, Cronbach's α coefficient for the total of 33 questions showed very high levels of reliability $\alpha = 0.944$. The reliability analysis for each dimension indicates that all factors showed a high degree of reliability ($\alpha = 0.765 - 0.942$), with the exception of the knowledge increase

Table 1 Factor analysis for PERMA-profiler scale

Questions	Recreational green attachment	Accomplishment—engagement	Happiness—positive emotions	Loneliness—negative emotions	Physical health
R1	0.903				
R2	0.903				
R3	0.869				
R4	0.865				
R5	0.858				
R6	0.848				
A1		0.674			
A2		0.665			
A3		0.657			
A4		0.655			
A5		0.594			
A6		0.588			
A7		0.574			
A8		0.483			
A9		0.431			
P1			0.754		
P2			0.751		
P3			0.701		
P4			0.592		
P5			0.512		

(continued)

Table 1 (continued)

	Recreational green attachment	Accomplishment—engagement	Happiness—positive emotions	Loneliness—negative emotions	Physical health
N1				−0.759	
N2				−0.740	
N3				−0.591	
N4				−0.576	
H1					0.840
H2					0.837
H3					0.766
Eigenvalues	**5.130**	**4.105**	**3.842**	**2.920**	**2.843**
% of total variance	**17.691**	**14.157**	**13.249**	**10.068**	**9.804**
Total variance					**64.970**

Table 2 Reliability measures for PERMA factors

Factors	Alpha
Wellbeing	
Recreational green attachment	0.956
Accomplishment—engagement	0.885
Positive emotions—happiness	0.852
Negative emotions—loneliness	0.681
Physical health	0.867

Table 3 Reliability analysis measured for negotiation strategies factors

Negotiations strategies	Factors	alpha
	Knowledge	0.633
	Negative impact understanding non-exercise	0.891
	Self-motivation	0.830
	Enable	0.848
	Socialization	0.765
	Enhancement	0.779
	Commitment	0.844
	Create pulse	0.835
	Relations to encourage development	0.810
	Time	0.942
	Financial	0.921

dimension, which showed a relatively lower degree of reliability ($\alpha = 0.633$) (Table 3).

4.3 Demographic Effects

H1: Multivariate analysis (Manova) was used to examine the interaction of independent variables of marital status and education on the factors of dependent variable wellness. The results of the analysis showed a statistically significant interaction on the factor of recreational green attachment $F_{(9.230)} = 2.267$, $p < 0.05$. Specifically, the participants who belong to the category married and were graduates of college/OAED (Mean = 5.927) showed higher levels of attachment compared to the categories married with a master's degree (Mean = 4.467), single graduates of AEI-TEI (Mean = 5.222) and singles-high school graduates (Mean = 3.630). Continuing the analysis of the main effects, the following results were identified and are presented at Tables 4 and 5.

Table 4 Analysis of variance between education categories and wellness factors

		High school (1)	OAEΔ-college (2)	University AEI-TEI (3)	Master/PhD (4)	F	Post-hoc/categories differencies
Physical health	Mean	5.438	5.581	4.985	4.905	$F = 3.198$ $p < 0.05$	2–3, $p < 0.05$ 2–4, $p < 0.05$

H2: Multivariate analysis (Manova) was used to examine the interaction of age and gender as independent variables and the dependent factors of negotiation strategies. The results of the analysis did not show a statistically significant interaction. But, it was revealed a statistically significant main effect of the independent variable age on the factor commitment $F_{(2.233)} = 3.806, p < 0.05$. The Sidak multiple comparison test revealed statistically significant differences between the age groups 60 + (Mean = 6.083) and 40–59 (Mean = 4.709).

H3: Multivariate analysis (Manova) was used to examine the interaction of the independent variables' education and marital status on the dependent factors of nego-tiation strategies. There was no statistically significant interaction. Continuing the analysis to investigate the existence of main effects, the independent variable marital status have a significant effect on the factor of knowledge $F_{(3.233)} = 2.824, p < 0.05$. The Sidak test identified statistically significant differences between the categories of domestic partnership (Mean = 4.622) and singles (Mean = 3.828).

5 Conclusions

The present research was added to the research work that examines the multidimen-sional nature of wellness, in the light of positive psychology (Seligman 2012, 2018; Butler & Kern, 2016). The recent PERMA-profiler questionnaire (Butler & Kern, 2016) was tested with high reliability and validity values. However, it was decided to remove the elements "In general, how often do you feel positive?" and "How often do you lose track of time by doing something that pleases you?" because they showed low loads and negatively affected the validity of the questionnaire and this is in accordance with the results of the pilot study of Butler and Kern (2016) who designed the scale. At the same time, "Recreational Green Attachment" as a factor of well-being was studied on a pilot basis, with very satisfactory reliability indicators.

The housekeepers and married people presented the highest averages in terms of accomplishment—engagement factor. Married and those who live in domestic partnership showed the highest levels of recreational green attachment. Students and postgraduate or doctoral students reported the highest rates of physical health, with the unemployed reporting the lowest. Participants aged 60 and over said that they were trying to persuade themselves to exercise, which shows that interventions are needed to facilitate their access to exercise services. Finally, people in domestic partnership choose to be more informed about the health benefits of multilevel exercise.

Table 5 Analysis of variance between marital status categories and wellness factors

		Married	Divorced	Singles	Domestic partnership	F	Post-hoc/categories differencies
Recreational green attachment	Mean	5.323	5.451	4.587	5.548	$F = 3.656$ $p < 0.05$	1–3, $p < 0.05$ 3–4, $p < 0.05$
Accomplishment—engagement		5.438	5.358	4.950	5.460	$F = 3.549$ $p < 0.05$	1–3, $p < 0.05$
Positive emotions—happiness		5.563	5.092	4.8	5.66	$F = 6.697$ $p < 0.05$	1–3, $p < 0.05$ 3–4, $p < 0.05$
Negative emotions—loneliness		5.325	4.873	5.315	5.504	$F = 3.616$ $p < 0.05$	1–3, $p < 0.05$

The benefits for the participants residents of the central urban structure of Athens because of the green recreation areas were highlighted in this research, an element that is in line with the international literature (Hartig et al., 2010; Conway, 2000; Coley et al., 1997). In addition, the findings showed that participants' well-being was inextricably linked to the factors of physical health, accomplishment, recreational green attachment and happiness. Extremely important are the internal factors that motivate the individual to engage in sports and recreational activities aiming at the well-being and physical fitness. The above is in line with the research work of other researchers (Butler & Kern, 2016; Forgeard et al., 2011; Kern et al., 2015; Seligman & Csikszentmihalyi, 2000). Equally important are private and public initiatives to provide recreational sports activities for all, targeting at the benefits of the contact with the natural environment. However, the success of any proposal and measure is determined by the social impact it causes to its individuals. Therefore, it is proposed to be drawn up strategies so that citizens are informed properly about the benefits of recreational exercise in green spaces, with the ultimate goal of creating an advanced culture promoting sustainable living in a modern metropolis, such as Athens.

References

Alexandris, K., Funk, D. C., & Pritchard, M. (2011). The impact of constraints on motivation, activity attachment, and skier intentions to continue. *Journal of Leisure Research, 43*(11), 56–79.

Alexandris, K, Kouthouris, C., & Girgolas, G. (2007). Investigating the relationships among motivation, negotiation and intention to continuing participation: A study in recreational alpine skiing. *Journal of Leisure Research, 39*(4), 648–668.

Balaska, P, Yfantidou, G., Kenanidis, T., Spyridopoulou, E., & Alexandris, K. (2019). Exploring how recreational sport participants with different motivation levels use leisure negotiation strategies. *European Academy of Management*, 1039, 1–35. ISSN 2466–7498, ISBN 978–2–9602195–1–7.

Bauman, A., Bull, F., Chey, T., Craig, C., Ainsworth, B. E., Sallis, J., Bowles, H., Hagstromer, M., Sjostrom, M., Pratt, M. & The IPS Group (2009). The international prevalence study on physical activity: results from 20 countries. *International Journal of Behavioral Nutrition and Physical Activity, 6*(1), 21.

Biddle, S, Kenneth R., Fox, K., & Boutcher, S.H. (2000). *Physical Activity and Psychological Well-being*. Psychology Press.

Bock, D. G., Andrew, R. L., & Rieseberg, L. H. (2014). On the adaptive value of cytoplasmic genomes in plants. *Molecular Ecology, 23*, 4899–4911.

Brand, M., Laier, C., & Young, K. S. (2014). Internet addiction: Coping styles, expectancies, and treatment implications. *Front Psychology, 5*, 12–36.

Bull, F. C., Al-Ansari, S. S., Biddle, S., Borodulin, K., Buman, M. P., Cardon, G., & Willumsen, J. F. (2020). World Health Organization 2020 guidelines on physical activity and sedentary behaviour. *British journal of sports medicine, 54*(24), 1451–1462. https://bjsm.bmj.com/content/54/24/1451. abstract

Butler, J., & Kern, M. (2016). The Perma-Profiler: A brief multidimensional measure of flourishing. *International Journal of Wellbeing, 6*(3), 34–39.

Coley, R., Kuo, F., & Sullivan, W. (1997). Where does community grow? The social context created by nature in urban public housing. *Journal of Environment and Behavior, 29*, 468–494.

Conn, V., Hafdahl, A., & Mehr, D. (2011). Interventions to Increase Physical Activity Among Healthy Adults: Meta-Analysis of Outcomes. *American Journal of Public Health, 101,* 751–758. https://doi.org/10.2105/AJPH.2010.194381.

Conway, P. (2000). The unpopular patient revisited: characteristics or traits of patients which may result in their being considered as "difficult" by nurses. *Qualitative Evidence-based Practice Conference,* Coventry University.

Filo, K., Funk, D. C., & Hornby, G. (2009). The role of web site content on motive and attitude change for sport events. *Journal of Sport Management, 23*(1), 21–40.

Forgeard, M. J. C., Jayawickreme, E., Kern, M., & Seligman, M. E. P. (2011). Doing the right thing: Measuring wellbeing for public policy. *International Journal of Wellbeing, 1*(1), 79–106.

Funk, D., Jordan, J., Ridinger, L., & Kaplanidou, K. (2011). Capacity of Mass Participant Sport Events for the Development of Activity Commitment and Future Exercise Intention. *Leisure Sciences, 33*(3), 250–268.

Gallagher, M. W., Lopez, S. J., & Preacher, K. J. (2009). The hierarchical structure of well-being. *Journal of Personality, 77*(4), 1025–1050.

Garrett, N. A., Brasure, M., Schmitz, K. H., Schultz, M. M., & Huber, M. R. (2004). Physical inactivity: direct cost to a health plan. *American journal of preventive medicine, 27*(4), 304–309.

Hamer, M., Lavoie, K. L., & Bacon, S. L. (2014). Taking up physical activity in later life and healthy ageing: The English longitudinal study of ageing. *British Journal of Sports Medicine, 48*(3), 239–243.

Hubbard, J., & Mannell, R. (2001). Testing competing models of the leisure constraint negotiation process in a corporate employee recreation setting. *Leisure Sciences, 23*(3), 145–163.

Huppert, F. A., & So, T. T. C. (2013). Flourishing across Europe: Application of a new conceptual framework for defining well-being. *Social Indicators Research, 110*(3), 837–861.

Jackson, E. L. (2005). Impacts of life transitions on leisure and constraints to leisure. *Constraints to Leisure,* 115–136.

Kahneman, D., & Krueger, A. B. (2006). Developments in the Measurement of Subjective Well-Being. *Journal of Economic Perspectives, 20*(1), 3–24.

Kern, M. L., Waters, L. E., Adler, A., & White, M. A. (2015). A multidimensional approach to measuring well-being in students: Application of the PERMA framework. *Journal of Positive Psychology, 10*(3), 262–327.

Kohl et al., 2012. Kohl, H., Craig, C., Lambert, E.V., Inoue, S., Alkadari, J.R., Leetongin, G., & Kahlmeier, S. (Lancet Physical Activity Series Working Group) (2012). The Pandemic of Physical Inactivity: Global Action for Public Health. *The Lancet, 380*(9838), 294–305. https://doi.org/https://doi.org/10.1016/S0140-6736(12)60898-8

Lee, I. M., Shiroma, E. J., Lobelo, F., Puska, P., Blair, S. N., & Katzmarzyk, P. T. (2012). Effect of physical inactivity on major non-communicable diseases worldwide: An analysis of burden of disease and life expectancy. *The Lancet, 380*(9838), 219–229.

Lloyd, K., & Little, D. E. (2010). Self-determination theory as a framework for understanding women's psychological well-being outcomes from leisure-time physical activity. *Leisure Sciences, 32*(4), 369–385.

Hartig, T., Mang Huta, V., & Ryan, R. M. (2010). Pursuing pleasure or virtue: The differential and overlapping well-being benefits of hedonic and eudaimonic motives Pursuing Pleasure or Virtue : The Differential and Overlapping Well-Being Benefits of Hedonic. (October), M., Evans, G.W. (1991), Restorative Effects of Natural Environment Experiences. *The Astrophysical Journal, 377,* L5–L8.

Maddux, J. E. (1993). Social cognitive models of health and exercise behavior: An introduction and review of conceptual issues. *Journal of Applied Sport Psychology, 5*(2), 116–140.

Mannell, R. C., & Kleiber, D. A. (1997). *Social psychology of leisure.* State College, PA: Venture.

Mannell, R. C., & Loucks-Atkinson, A. (2005). Why don't people do what's "good" for them? Cross-fertilization among the psychologies of nonparticipation in leisure, health, and exercise behaviors. *Constraints to Leisure,* 221–232.

Μπαλάσκα, Π., Υφαντίδου, Γ., Παπαϊωάννου, Ά. & Σπυριδοπούλου, Ε. (2018). Διερεύνηση της σχέσης των στάσεων και των στρατηγικών διαπραγμάτευσης σε δραστηριότητες αθλητικής αναψυχής. *International Review of Services Management (IRSM), 1*(1), 89–95.

Murphy, N. M., & Bauman, A. (2007). Mass Sporting and Physical Activity Events—Are They 'Bread and Circuses' or Public Health Interventions to Increase Population Levels of Physical Activity? *Journal of Physical Activity & Health, 4*(2), 193–202.

Mutrie, N., & Faulkner, G. (2004). Physical activity: Positive psychology in motion. *Positive Psychology in Practice*, 146–164.

Pezirkianidis, C., Stalikas, A., Lakioti, A., & Yotsidi, V. (2019). Validating a multidimensional measure of wellbeing in Greece: Translation, factor structure, and measurement invariance of the PERMA Profiler. *Current Psychology*. https://doi.org/10.1007/s12144-019-00236-7.

Powers, S., Dodd, S., & Noland, V. (2006). *Total fitness and wellness* (brief). San Francisco, CA: Benjamin Cummings.

Rupprecht, P. M., & Matkin, G. S. (2012). Finishing the Race: Exploring the Meaning of Marathons for Women Who Run Multiple Races. *Journal of Leisure Research, 44*(3), 308–331.

Sallis, J. F., Cerin, E., Conway, T. L., Adams, M. A., Frank, L. D., Pratt, M., & Owen, N. (2016). Physical activity in relation to urban environments in 14 cities worldwide: A cross-sectional study. *the Lanset, 387,* 2163–2262.

Sato, M, Jordan, J. S., & Funk, D. C. (2014). The Role of Physically Active Leisure for Enhancing Quality of Life. *Leisure Sciences, 36*(3), 254–259.

Sebire, S., Standage, M., & Vansteenkiste, V. (2009). Examining intrinsic versus extrinsic exercise goals: Cognitive, affective, and behavioral outcomes. *Journal of Sport and Exercise Psychology, 31,* 189–210.

Sebire, S., Standage, M., & Vansteenkiste, V. (2011). Predicting Objectively Assessed Physical Activity from the Content and Regulation of Exercise Goals: Evidence for a Mediational Model. *Journal of Sport & Exercise Psychology, 33,* 175–197.

Seligman, M. E. P., & Csikszentmihalyi, M. (2000). Positive psychology: An introduction. *American Psychologist, 55*(1), 5–14.

Seligman, M. E. (2012). *Flourish: A visionary new understanding of happiness and well-being.* Simon and Schuster.

Seligman, M. (2018). PERMA and the building blocks of well-being. *the Journal of Positive Psychology, 13*(4), 333–335.

Slade, M. (2010). Mental Illness and Well-Being: The Central Importance of Positive Psychology and Recovery Approaches. *Health Service Results, 10,* 26–28.

Stevinson, C., Wiltshire, G., & Hickson, M. (2015). Facilitating participation in health-enhancing physical activity: A qualitative study of parkrun. *International Journal of Behavioral Medicine, 22*(2), 170–177.

Tavakol, M., & Dennick, R. (2011). Making Sense of Cronbach's Alpha. *International Journal of Medical Education, 2,* 53–55.

Walker, G. J., Halpenny, E., Spiers, A., & Deng, J. (2011). A prospective panel study of Chinese–Canadian immigrants' leisure participation and leisure satisfaction. *Leisure Sciences, 33*(5), 349–365.

White, D. (2008). A structural model of leisure constraints negotiation in outdoor recreation. *Leisure Sciences, 30,* 342–359.

Wilhelm, S. A., Schneider, I. E., & Russell, K. (2009). Leisure time physical activity of park visitors: Retesting constraint models in adoption and maintenance stages. *Leisure Sciences, 31*(3), 287–304.

Exploring Scuba Diving Tourism Sector in Malta and Its Sustainable Impact on the Island

Simon Caruana and Tiffany Sultana

1 Introduction

Scuba diving is a well-consolidated niche market in the Maltese Islands. As with other nature-related activities, its interaction with natural resources may affect the ecological life. There is great potential held by the scuba diving industry to carry out and encourage conservation, attract tourism, generate revenue, improve people's quality of life and promote community pride. (De Groot & Bush, 2010; Mota & Frausto, 2014; Wongthong & Harvey, 2014). Sustainability is gaining more popularity as mass tourism is leaving its effects on the environment. Small islands such as Malta are perceived as vulnerable in this aspect especially as tourism is a great dominant factor of the overall economy (Lockhart & Drakakis-Smith, 1997). Scuba diving derived from an even bigger niche, sports tourism, has been researched and considered as acquiring great economic advantages (Standeven, 1998). They also help small islands promote themselves for special interest holidays (Bull & Weed, 1999). However, good management and planning needs to be done not only in the promotion of diving but also in the conservation aspects. This niche sector is predominantly based on marine activity thus, if this deteriorates, so does the sector.

Research Objective

The scope is to put in context the number of tourists who are motivated to visit Malta due to diving and those who engage in this sport whilst on holiday, hence identifying solutions as to merge the gap, if any, between what is currently offered and how this can be excelled. Moreover, it will determine whether the existing legislation and enforcement help preserve the national resources and whether sustainability is given the due consideration.

S. Caruana (✉) · T. Sultana
University of Malta, Msida, Malta
e-mail: simon.caruana@um.edu.mt

V. Katsoni and C. van Zyl (eds.), *Culture and Tourism in a Smart, Globalized, and Sustainable World*, Springer Proceedings in Business and Economics,
https://doi.org/10.1007/978-3-030-72469-6_6

2   Scuba Diving in Maltese islands

Recreational, technical, commercial and free diving all fall under the diving category, and however, the first two types are acknowledged in this research when referring to the niche scuba diving market. Both would include:

1.   Training for recreational and/or technical diving
2.   Tours organised for certified divers—this may be either with the services of a local instructor or as an independent group (subject to local legislation)

A Malta Tourism Authority (MTA) analysis document (MTA, 2018) notes, 5.2% of total inbound tourism were motivated to visit Malta cause of diving, whereas 6.8% of total tourists participated in diving activities during their trip. Around sixty diving centres established around the Maltese islands provide service and support to the visiting divers. The instructors/guides working in the sector all occupy an instructor qualification acquired mostly from PADI, BSAC and CMAS certification agencies (PDSA, 2017; VisitMalta, 2019).

2.1   Popular Dive Sites

In the 'Master Plan to Support a Sustainable Diving Industry in Malta', commissioned by the MTA (ADI, 2011), it is reported that there are 29 shore and 15 boat dive sites, respectively. The majority situated around Sliema, St. Julians, Marsaxlokk and Marsascala area whilst others are found in more remote coastal areas such as Cirkewwa, Ghar Lapsi and Wied iz-Zurrieq. Such attractions can differ from wrecks example, the X127 Water Lighter (landing craft), Manoel Island and the P29 Patrol Boat at Cirkewwa to more remarkable less common features of marine topography example, the Mini Blue Hole at Marsascala or the Cirkewwa Arch. Numerous dive sites also offer the divers a change to wander across reef wildlife example, the Coral Gardens in St. Julians or the Cirkewwa Reef (ADI, 2011).

2.2   Data Related to Tourists Engaging in Diving on the Maltese Islands

MTA portrayed the tourists participating in diving segmented to Malta, Gozo and Comino (MTA, 2018) and comparison can be done with a prior study (MTA, 2015) noting the improvements this niche sector has obtained (Tables 1 and 2).

During the same publication, as references below, in peak season statistics excel by around 10% when compared to shoulder months (Fig. 1).

Table 1 Tourists engaging in scuba diving activity 2017

Type of trip	Estimated no. of people
Malta only	105,720
Gozo only	29,898
Comino only	1,849
Malta and Gozo	17,877
Total	155,344

MTA (2018)

Table 2 Tourists engaging in scuba diving activity 2014

2014	Number of divers
Malta	81,788
Gozo	28,346
Comino	1,625
Total	101,600

MTA (2015)

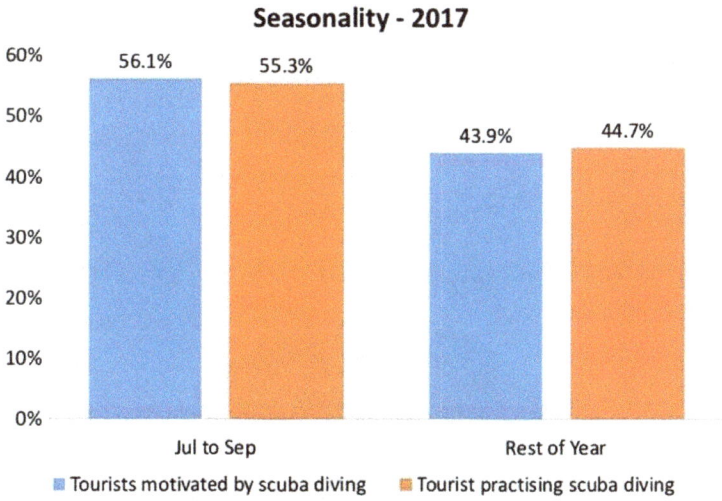

Fig. 1 Seasonality. MTA, 2018

2.3 Who is a Typical Scuba Diving Tourist?

The most common range between 45 and 54 years of age, having tertiary level of education earning an income of approximately 2,780 Euro per month (MTA, 2018). Their spending is more than the 'average' tourist as they stay longer, ten or eleven nights. They also more likely to hire cars and self-catering accommodation thus being more independent. The UK, Italy, Germany and France being amongst the top four countries where Malta gets its share. Although 71.5% of tourists engaging in this sports were first timers, 28.5% were repeated tourists. Additionally, a remarkable 95.1% said that they would recommend it to friends and relatives. (MTA, 2018) (Figs. 2 and 3).

The 'Master Plan to Support a Sustainable Diving Industry in Malta' (ADI, 2011) notes that two global destinations, namely the Red Sea/Egypt and the Maldives attract

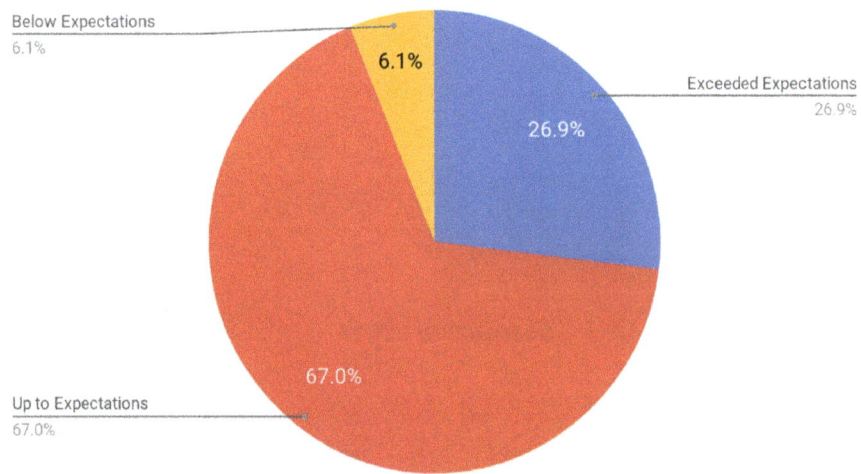

Fig. 2 Overall experience. MTA (2018)

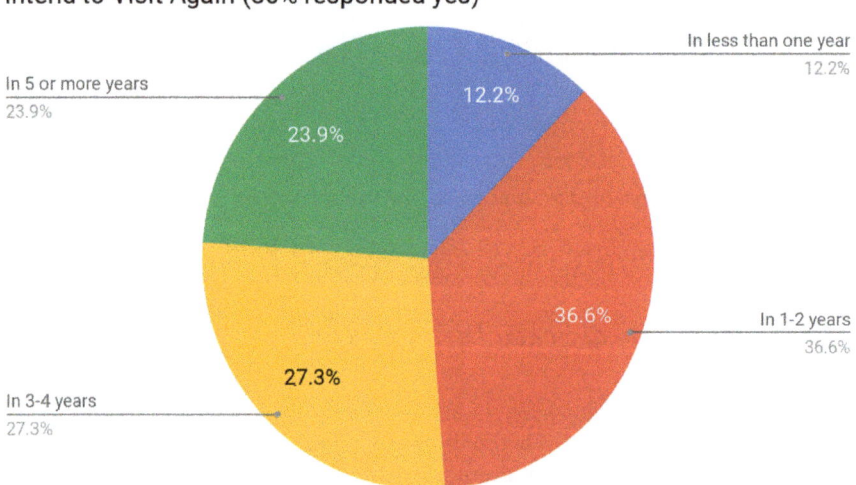

Fig. 3 Intention to visit again. MTA (2018)

many visitors from Europe itself. Within Europe, the main competitor markets are Greece, the Balearic Islands, Cyprus, Portugal, Sardinia and Corsica.

2.4 Boat Dive Sites Issues

In the 2011 master plan (ADI, 2011), it is indicated that many of the dive sites are situated in the north, west and north and south eastern coasts whilst only few situated in the south because of deeper water. This research suggests that divers and other representatives perceive overfishing and spearfishing as great problems especially at wreck dive sites. Moreover, mooring buoys and the importance of off-shore diving sites are not given enough importance especially when looking into the safety and conservation of wildlife. Finally ADI consultants (2011) also mentioned that fishing nets and anchors put divers in harm's way and could cause damage to boat dive sites in their close proximities.

2.5 Shore Dive Sites Issues

In the same master plan (ADI, 2011), divers interviewed noted that whilst it is true that shore access points are scattered over the island, parking at sheltered harbours could be intensified as when strong winds occur, it makes it difficult for divers to visit the more exposed locations. Manoel Island shore dive site was mentioned as a perfect example of sustainability issue, where although it is the second most popular shore site in central region as it gives access to the X127 Water Lighter, a marina above the shipwreck is being proposed (PA, 2003) by the Manoel island project. This will not only limit visitation but will also be less safe and poorer the quality of water.

2.6 Industry and Regulations

The diving regulations as found in the subsidiary legislation (LN 359, 2012) provide Malta with a set of laws which one needs to abide to. These focus on procedures to ensure safety of divers and details related to the requirements of diving centres in this respect. Looking into the regulations one may see various points emphasising the importance of European standard equipment, health and safety at sites, training and risk assessments. However, the conservation and sustainability of the sites during dives have remained unmentioned. In the master plan commissioned by the MTA (ADI, 2011), respondents had already highlighted these problems and stated that although environmental protection regulations are somehow visible, enforcement is neglected especially when it comes to the protection of sites and the aquatic life.

Additionally, proper management of dive sites is important, as the following issues were noted by some respondents (ADI, 2011):

1. lack of disability access,
2. poor road access,
3. lack of mooring buoys,
4. lack of public awareness,
5. lack of toilets etc.
6. lack of balance between fishing and diving.

2.7 Marine Protected Areas

The 2011 master plan (ADI, 2011) depicted that both shore and boat sites designated for diving are considered as marine protected areas. Across the world including in Malta, these are used to control fish stock exhaustion and help counteracting the mortification the marine environment is facing. Locally, these fall under the EU Council Directive 92/43/EEC on the 'Conservation of Natural Habitats and of Wild Fauna and Flora' (the Habitats Directive) of 1992 (Directive, 1992) provides a basis for designation of MPAs. According to the Maltese Environmental and Resources Authority (ERA) (2016), 35.5% of Maltese waters was published as being protected. Fourteen MPAs areas were identified of which nine of them officially designated as such in 2016.

2.8 Natura 2000 Sites

The EU Council Directive 92/43/EEC on the 'Conservation of Natural Habitats and of Wild Fauna and Flora' (Directive, 1992) provides a foundation for both land and marine locations which are perceived of having international significance. These have recognised some Special Areas for Conservation (SAC)'s as potential Natura 2000 sites. Management plans for these must be executed as part of the directive. Various shore dive access points are placed in close proximity to the sites mentioned previously thus, new developments will need permits and assessment to ensure the site remains unharmed. As of 2010, the Malta Environmenal and Planning Authority (MEPA) had issued protection to the coastline between Rdum Majjiesa and Ras ir-Raheb known as MT0000101 and later on other areas were issued protection too, such as the marine area in proximity of Ghar Lapsi and Filfla and that in the northwest of Malta. In 2008, Malta Maritime Authority had issued seven underwater area of conservation namely; The Um el Faroud in Wied Iż-Żurrieq, MV Xlendi, Cominoland, Karwela off Xatt l-Aħmar, Tug St. Michael, Tug 10 in Marsaskala, The Imperial Eagle off Qawra Point, Roži, P29 off Ċirkewwa, Blenheim Bomber off Xrobb l-Għaġin and Bristol Beaufighter off Exiles Point. This has been replaced by an updated list in which further areas were added in 2019. (TM, 2019).

3 Research Approach

An inductive approach was adopted as it enables the raw primary data to be transferred into a summary format and sets a connection between the research objectives Thomas, 2006). The findings of the research are to be extracted from the raw material gathered from the relevant participants. An inductive approach creates a framework which can be used to determine the structures of processes and experiences found in the collection phase. It is also thought that with this approach it will be possible to move from a more 'generic' idea which is the scuba diving sector as a whole, to a more confined meaning as to looking at its sustainability. It allows also observation to be made amongst the relevant stakeholders to determine common thoughts and ideas and how these differ to get a better understanding. (Thomas, 2006).

3.1 Collection of Data

There are different ways of conducting interviews being in a structured, semi-structured, unstructured, via telephone or the most common being face to face (Opdenakker, 2006). Semi-structured interviews have been chosen for the purpose of the study. Such allows reflexivity in both researcher and interviewees thus engage in more meaningful conversation. Opdenakker (2006) found that as they require concurrent discussion, voice, intonation and body language can add significance to the data being given when answering the questions.

Interviewees ranged between 20–45 min and took place either at their centres, office or a cafeteria for a more relaxed setting. Interviewees were recorded, and however, their identity is kept disclosed for confidentiality purposes.

3.2 Consulted Stakeholders

Different entities concerning the legislations, management of sites and regulations were consulted. These being the scuba diving centres as they are the prime users and sector drivers of this industry alongside, representatives from the Malta Tourism Authority (MTA), the Environmental Resources Authority (ERA) and Transport Malta (TM) who amongst other responsibilities oversees the marine transport sector.

4 Findings

4.1 Analysis of the Data Collected

A matrix analysis approach was adopted to analyse the data. Its table format permits the collection of information over a grid (matrix) facilitating data comparision and subsequent analysis (Miles, Huberman & Saldaña, 2014). From the resulting matrix, the following issues were isolated and treated in detail in the following sections.

4.2 Sustainability

The definition and meaning of sustainability was primarily discussed amongst the diving centres and authorities who all agreed that:

1.   Should be positive and/or not negative regarding the environment,
2.   Should be improving and not exploiting natural resources
3.   Generating more possible continuation of that activity in the future.

Correspondingly, one diving centre added;

> keep in consideration the environment for future and when we assess what we do and the repercussions they have both on society and environment. (sdc3 p.13).

Hence, noting also the impact on the local community. Looking into research made of how sustainability came into formation alongside comments from interviewees, the protection and conservation of natural resources for future population has been continuous (Kuhlman & Farrington, 2010). This also reflects Wongthong & Harvey's, (2014) statements that the protection of environment should be secured alongside the privacy and respect of local residents.

4.3 Sustainability in the Tourism Industry

The majority of respondents agreed that awareness is increasing even amongst business owners and local community. However, all concurred that more work is required priority needs to be emphasised further. As a matter of fact, an authority implied that;

> My personal feeling is that commercial interest are given the importance they are due and although nowadays the environment is being taken into consideration I still feel that the ratio is not yet right (A4 p.35).

A diving centre regarding the number of tourists visiting remarked that;

> I believe that a lot of the tourism which comes to Malta tends to be one hit wonder. As to looking for repeat tourists and gaining ones which keep coming year after year, I don't think much is done to achieve that on the island. (sdc2 p.8).

However, statistics representing both the diving industry and tourism in general contradict slightly such views as 28.5% of tourists participating in this sport were repeated visitors. Additionally, 80% of respondents said that they intend to visit Malta again (MTA, 2018). Correspondingly, statistics shows that over the previous year, the general inbound repeated tourists increased by 11.8%.

Continuing on the note made before the last, another authority implied that some use the term 'sustainability' to get funds for their projects and thus, is very confusing of what it truly means. However, as an authority they seek to be environmentally cautious not only with regard to scuba but also in relation to coast and rural.

The remarks made are similar to ones found in other sources, such as the 'Globe'90' conference in Vancouver, where it was depicted that tourism provides employment also in its subsidiary sectors hence, gives an opportunity to local businesses, therefore increases awareness amongst the latter. Literature commensurates with the interviewees views when mentioning the rise in sustainability awareness. On another note, (Andrades-Caldito,et al, 2014) it was implied that resistance is present as sustainable measures are recognised as costly. Although such reference was not directly made, a responded stated that;

> So, in general, people in the tourism industry are aware of sustainability issues, whether they act on them is a different story. (sdc1 p.1).

This also denotes to the comment made prior tackling that commercial interest exceeds that of the environment.

4.4   Laws and Enforcement Efforts

The lack of laws in relation to the conservation of dive sites was emphasised. Moreover, the insufficient presence of enforcement was predominant in all views made by authorities and diving centres. In particular, spearfishing, overfishing, littering and anchoring. Enforcement and compliance with fisheries regulations have been acknowledged as great issues that need to be resolved for the purpose of protecting and managing fisheries.

However, some respondents did note laws and regulations present such as the notice to mariners and 'SL.549.44'. One centre referred to a recent legislative amendment regulating diving on historical wrecks around the island. These comments corroborate a problem outlined earlier by literature that this industry is highly dependent on the quality of aquatic habitat. Hence, assuming more responsibility and attention to recycling and education is fundamental (Haddock-Fraser and Hampton, 2012).

All interviewees agreed that enforcement is lacking, there were some contrasting views in the answers given by the centres and the authorities. One of the latter implied that;

Our enforcement through the maritime enforcement unit does on a regular basis during their daily petrols, visit these sites to ensure that vessels in the area are complied with these and if there is no compliance, there are legal procedures. (A4 p.35).

Some centres agreed with that enforcement agencies are somewhat quick in their response.

However, others noted that;

Well, to be honest with you the AFM say that you should report to them, but then sometimes they say you should report to TM and then you go to TM and say you should report to MTA and vice-versa etc.etc. So they should set up a specific body that will take care of these cases etc. (sdc4 p.19).

4.5   Monitoring of Dive Sites

As remarked by Haughton (2003), there is a deficiency in resources for monitoring, surveillance and control in the particular research carried out in Caribbean islands states. All parties interviewed agreed in that they referred to the need of CCTV for better protection, use of technology, more manpower and a designated authority in charge of incidents and reports.

4.6   Implications of Scuba Diving

When reviewing the effects to the environment by scuba diving, all respondents expressed that all human activity has an effect on the natural habitat. Touching, possible collision by air tanks, 'souvenir' collection and bubbles which causes oxidation in caves. These all correspond to what Barker and Roberts (2004) and Worachananant et al., (2008) reported, all of which refer to the immediate damage done to underwater ecosystem caused by the physical interaction.

However, all interviewees complied that when outweighing the implications, the diving industry leaves more positive than negative effect on the environment. Furthermore, much of the impact were said to incidental.

Giving the number of dive centres, each one certainly do their own bit with regards to clean up of dive sites and taking litter out of the sea. They tempt to be a responsible bunch of people that care for the environment… (sdc2 p.9).

Furtherore,

a lot of people seeing what there is underwater. Most of the problems we have is because a lot of people have no clue what there is under the surface and they don't care about it sometimes not even looking at it. So, as soon as something goes under the surface it is lost. (sdc1 p.2).

Literature seems to point towards the diving industry as creating sewage pollution (Lamb et al., 2014) and dumping food, contrary to the diving centres viewpoint as they remarked of resolving such issues rather than adding on (Table 3).

Table 3 Sites in need of better management regarding fishing

Boat dive sites	Shore dive sites
HMS Hellespoint	East Reef-Wied iz-Zurrieq
St. Angelo	West Reef & Caves-Wied iz-Zurrieq
Scot Craig	Middle Reef-Ghar Lapsi
	Finger Reef & Crib-Ghar Lapsi

'Master Plan to Support a Sustainable Diving Industry in Malta' (2011).

4.7   Other Areas Impacting the Diving Industry

Spearfishing, fishing and anchoring were three areas which interviewees perceived as problematic. Although dive centres admitted that things have improved since the 'Master Plan Publication,'(ADI, 2011), the advances made were noted with regard to a recent notice to mariners (2019) and marker buoys placed at most wrecks. Whereas, more awareness needs to be placed into the marine protected areas as diving centres still reported incidents on wrecks, anchoring as it can be not only dangerous to the reefs but also to divers, and more protection to the fish. Almost all respondents agree that the main bone of contention remains that of lack of enforcement. As a matter of fact, a diving centre implied that;

> here if we really protect our reefs,….. in certain areas in Malta where its very difficult to spearfish the abundance of life both in shallow and deep is mesmerising (sdc 4 p.18).

Similarly;

> But till that happens they will continue because fish brings a lot of money and on these dive sites there is a lot of fish, probably the only place where fish is left because the rest have caught everything. It's that bad the situation, there's no fish left. (A3 p.34).

Moreover, although marine protected areas are appointed, interviewees reported an anomaly in the law as some type of fishing is still permitted. The same situation prevailed in the master plan (ADI, 2011) which noted areas where fishing management is needed and such are listed in table below. This was also verified by an authority's comment (Table 3);

> The most reports we get relates to fishing, because the conservation areas only permits certain types of fishing. (A4 p.37).

Liaising to research indicating that around the world in some coastal ecosystem, the loss can range from 1 to 10% yearly (Waycott et al., 2009).

Furthermore, literature shows that overfishing is a problem which is present in many countries. Stergiou (2002) reported that fishing predicts an extermination point which impedes dramatic effects on the aquatic ecosystem. Additionally, it is noted that (ADI, 2011) when compared to competitors, Malta is in a disadvantaged position regarding the lack of fish. Interviewees shared the same views as the latter and in return to combat such problems, an authority reported that artificial reefs have been sunk and these attract fish whilst providing them with algae.

When discussing anchoring issues, authorities have implemented marker buoys that serve also as a reference point to boaters indicating that a wreck is below. These also serve as tie points instead of having boat men anchor everyday which in return could harm the environment. Research notes that seagrasses around the world suffer from damage caused due to boat anchors. Especially in marine protected areas as these are highly frequented by boaters (Milazzo, et al., 2004). This is corroborated by one of the competent authorities:

> popular dive site, especially boat dive sites example Comino the caves, the idea and the discussion has been going on to put actual mooring block which rather than throwing an anchor everyday, everytime, a diver will go in and pass a rope through the block.(A3 p.31).

4.8   Means to Reduce Effects on the Industry

Barker & Roberts (2004) remark that when supervised dives were given, impact on reefs was found to be less. When asked how to minimise negative effects, diving centres consulted stated that training and briefing is highly beneficial, whilst also putting a carrying capacity limit in some caves if needed, self-regulation, underwater supervision and more general education.

4.9   Economic Contribution

Statistics represent that the average expenditure of a diving tourist prior to the travel and during amounts to €734.79 and €422.70, respectively (MTA, 2018). Whereas, for global tourism by market, money spent before averaged out to €609 and during their stay accounted to €339 (MTA, 2018). This is pointed out by the diving centres.

> the authorities in my opinion, are not recognising the role that diving has in the tourism sector. The investment that the authorities make in other areas is massive compared to what they invest in the diving sector. And this sector brings in a considerable amount of tourists, that spend a lot.. (sdc1 p.6).

> I mean the diving industry is larger than the English language industry. So if we are marketing ourselves as a diving destination, which we are, than we need to have the right infrastructure in place… (A3 p.32).

These reflect similar thoughts found in literature which examine the potential scuba diving possesses with regard to enhancing promotion of conservation of the sea whilst attracting more tourists, hence increasing the expenditure and also improving the community's life (De Groot and Bush, 2010; Mota and Frausto, 2014; Wongthong and Harvey, 2014).

In their linking, the economic contribution to sustainability, interviewees shared the view of Moeller et al. (2011) that indicted that 40% of total market attributing to higher expenditure value, amounted to lower environmental impact.

4.10   Management and Coordination of Dive Sites

The management and coordination of sites was another area which angered many diving centres. The absence of parking facilities was worrying alongside the need for better infrastructure on site such as toilets, showers and outdoor furniture (e.g. railings) facilitate entry and exit of the waters. Although they all agreed that improvements have been made since the master plan, (ADI, 2011), such in Cirkewwa, other sites still lacking. Furthermore, centres mentioned that equipment (e.g. ladders) is removed taken out in winter to avoid being destructed by weather, thus centres brought the argument of;

> No, they take them out specifically before winter hits so the ladders aren't washed away. So, my argument would be why don't you put them in properly to start with so they don't wash away and make them a permanent structure (sdc2 p.11).

With regard to infrastructure across all the dive sites, improvements seem to be occurring at a slow pace. Centres shared the same views when mentioning Cirkewwa as it is the main attributed location, and however, the rest of the sites reported insufficiency. Mostly, respondents mentioned Wied iz-Zurrieq as it is highly frequented and requires more attention. An interviewee remarked such sites as problematic.

> Yes, at the moment we have dive sites with completely no infrastructure whatsoever and toilet facilities. It's that bad. Example, Exiles at Sliema, Marsascala the Tug boats, Maori in Valletta, Sirens Bugibba, Qawra point. (A3 p.33).

With regard to parking, again only Cirkewwa proved as being up to standard, the rest all depicted a lack. Centres believe that as parking spaces are being reserved for other services such as taxis and electric cars, so should be the case for divers near sites. However, when mentioning these respondents were non-consistent in their answers, example a diving centre stated;

> Yes, sort of. Example in Cirkewwa we have a section only for divers, in Zurrieq as well and in Sliema we are trying. It's improving slowly. (sdc1 p.5).

Whereas another responded contradicted that;

> I actually got three points on my licence for parking on the rock and a fifty euro fine. (Parking in Exiles, Sliema) So, Cirkewwa being the main place where there is parking but everywhere else you park where you can. (sdc2 p.10).

Whilst most diving centres abide to the response made above, one interviewee noted that parking is an issue all around the island and thus has to be faced like the rest of the community. The designated authority responsible for management acknowledged such views by noting the below and adding on that parking arises issues with local councils and community area.

> We had worked on Qawra Ta' Fra ben, which are parking spaces not for divers to park but for unloading and loading so example especially divers one can unload equipment etc.and then go park somewhere else and like this at least near the dive site they can unload their tanks etc. (A1 p.26).

The need of better amenities was linked to the importance of giving a high-quality service to the tourists who participate in diving. The respondents viewpoints that correspond to research made were more conservation and refurbishment of sites will better the flow of people and hence, help with the issue of overpopulation (ADI, 2011).

4.10.1 Underwater Areas of Conservation

All respondents agreed that more areas should be acknowledged as underwater areas of conservation. However, they noted a recent 'notice to mariners no 11 of 2019' which included five additional wrecks with the aim of adding four more by the end of the year, thus upgrading on the prior 'notice to mariners no 5 of 2008'. Nevertheless, the centres implied that as mentioned prior, enforcement issues prevail. Yet, as a step forward the concerned authority had noted the below with regard to the new deep water wrecks of historical importance;

> So they can fall under the conservation areas and be dived on by anyone without specific permits. And then has put together all the other wrecks as historical wrecks which require procedure to comply with so you have to be a member of Heritage Malta, you have to have equipment installed on board (AAS) so that we can monitor and better enforce. (A4 p.38).

4.10.2 Seasonality

The interviewees acknowledged that scuba diving has a distinctive peak and low season. However, interviewees noted increase in activity in shoulder months as now the season picks up between April and May and lasts till around October. Again, there was agreement in that the low season is from the months of December till March. This is reflected in the research done by MTA (2018), depicted that from July to September, inbound tourists motivated by scuba diving and those practicing the sport represented a 10% increase over other months.

A diving centre added that seasonality is more felt in remote areas by implying that;

> Obviously peak season is the majority and big part of the business, the diving industry is highly seasonal for a diving centre based in Marsascala because we lack the walking tourists that other diving centre have (sdc3 p.15).

This is mirrored in the MTA dive travellers' report (2018), where the favoured city for accommodation purpose amongst diving tourists is highest in St. Paul's/Salina area which holds 25.2% of total number, followed by Mellieha. Sliema and St. Julians with 21.9%, 10.8% and 9.2%, respectively. Marsascala was noted to occupy 0.6% of market share.

Interviewees mentioned the increase in activity during the shoulder months, which corresponds to the data on the MTA report (2017) which indicated that good flight connections represented 42.3% regarding it as a factor which influenced the destination choice, followed by value for money. However, it is expected that the infrastructure is maintained all year round.

> But since we advertise ourselves as a yearlong destination, I keep my dive centre open all year round, I still employ all my staff and the least that I expect is that the infrastructure are well maintained and kept in place so I can do what I am selling (A3 p.34).

4.10.3 Scuba Diving as a Niche Sector

The diving master plan (ADI, 2011) concluded that Malta is highly attractive as a diving destination due to its warm waters, good visibility and underwater scenery. This is reflected in the words of a representative of one of the authority interviewed;

> What's good about Malta is that if you came for diving and it was windy cause normally that's the issue not the water being cold… You can participate in other activities…, you can visit museums etc. Also, if a family member doesn't want to do diving there are other activities which they can do. (A1 p.27).

However, as the positive attributes which make Malta attractive were depicted in the same studies discussed prior, a respondent mentioned such negatives which could affect Malta's product;

> the weather also in winter and we've had quite a terrible winter this year and also the wind constantly picks up etc. not to mention that with global warming summer has also been a turbulent one but in winter it's much more predominant. (sdc4 p.21).

4.10.4 Collaboration Between Stakeholders

Contradicting viewpoints emerged amongst respondents as centres mentioned that synergy needs to be improved amongst the relative stakeholders for better management of sites. During the studies carried out for the master plan (ADI, 2011), consultants depicted same issues. Whereas one authority noted;

> We meet on a regular basis, at least from my point of view….. so we all pretty much know each other and have each other numbers on the phone. So the collaboration is good and it will only get better over time. (A4 p.40).

5 Conclusions

5.1 Sustainability

The main arising question of this study was the relation between the diving sector and sustainability relating to laws and regulations. Analysis depicted synergy regarding the importance and idea of sustainability; however, more laws regarding protection of fish, spearfishing and anchoring were demanded. If dealt with, the sector will hold a more prospective future.

5.2 Legislations and Enforcement Powers

The issue is the lack of enforcement, thus minimising the scope and effectiveness of those already present. Whilst improvements have been done since the master plan implementation, all participants expressed the necessity for greater imposition of authorities on wrecks and dive sites. They recommend, amongst other things:

1. A sustainable diving project to assess the issues of fishing, overfishing, spearfishing and anchoring,
2. Need for better allocated marine protected areas
3. Review of conservation areas as, as remarked by an authority in the previous chapter, it still permits certain type of fishing.
4. More artificial reefs to help with fishing and also adds to the diving tourism product.

5.3 Scuba Diving as a Tourism Product

Respondents noted that the close proximity between wrecks, weather and ability to participate in other activities besides diving betters the tourism product. Yet, to promote diving as an all year round activity, dive sites are in need of better management, coordination and need to be serviced accordingly. This also entails protection of fish as this affects the attractiveness of Maltese waters which in turn with carrying capacity and enhances Malta's competitive position.

5.4 Coordination and Management of Sites

Since the master plan publication and subsequent implementation (ADI, 2011), improvements have been reported in some regards such as;

• Implementation of marker buoys,
• Notice to mariners 'no 11 of 2019'.

The latter marked new historical wrecks with the purpose of better coordination and management framework. However, a proper management programme is required for enforcement and better running of sites. This would include updating infrastructure facilities of dive sites including parking, toilets and shower facilities, but also set up marker buoys and ladders which are strong enough withstand winter weather conditions.

5.5   Synergy Between Stakeholders

From both the research conducted, it is clear that collaboration needs to be heightened. Furthermore, a representative body needs to take control of enforcement entailing reports filed and monitoring such as surveillance and patrols. As outlined in the findings, not only it is difficult to report, but no authority is taking responsibility for such. This situation had already been outlined in the master plan findings (ADI, 2011), and unfortunately, the same situation persists today.

5.6   Seasonality

Whilst research clearly suggests that seasonality is being mitigated, diving is still a seasonal sport as it relies on other elements. This puts emphasis on;

- Updating infrastructure needs and management of sites,
- Increase promotion of diving as a niche sector (as a yearly activity)

5.7   Final Remarks

This research has helped shed light on the importance of this niche product by noting its improvements and moreover, depicting the gap which requires growth and attention. 'Sustainability' has been a word used in abundance and its awareness is being not only acknowledged in literature but also in research conducted primarily. However, measures to attain such take time, resources, investments and powers. Additionally, the lack of coordination and enforcement is stagnating or not enabling diving tourism to reach its full potential. Hence, the above highlights the demand and requirements for development especially when considering the improvements made since the implementation of some of the 2011 master plan recommendations (ADI, 2011).

References

ADI Associates Environmental Consultants. (2011). *Master plan for a sustainable diving industry for Malta*. Malta Tourism Authority.

Andrades-Caldito, L., Sánchez-Rivero, M., & Pulido-Fernández, J. I. (2014). Tourism destination competitiveness from a demand point of view: An empirical analysis for Andalusia. *Tourism Analysis, 19*(4), 425–440.

Barker, N., & Roberts, C. (2004). *Scuba diver behaviour and the management of diving impacts on coral reefs* (4th ed.). Environment Department, University of York.

Bull, C., & Weed, M. (1999). Niche markets and small island tourism: The development of sports tourism in Malta. *Managing Leisure, 4*(3), 142–155. https://doi.org/10.1080/136067199375814.

De Groot, J., & Bush, S. R. (2010). The potential for dive tourism led entrepreneurial marine protected areas in Curacao. *Marine Policy, 34*(5), 1051–1059.

Directive, H. (1992). Council Directive 92/43/EEC of 21 May 1992 on the conservation of natural habitats and of wild fauna and flora. *Official Journal of the European Union, 206,* 7–50.

ERA. (2016). *Protected Areas- National.* Webpage of Environmental & Resources Authority (ERA). Retrieved from https://era.org.mt/en/Pages/Protected-Areas-National.aspx. Accessed on December 4, 2019.

Haddock-Fraser, J., & Hampton, M. P. (2012). Multistakeholder values on the sustainability of dive tourism: Case studies of Sipadan and Perhentian Islands Malaysia. *Tourism Analysis, 17*(1), 27–41.

Haughton, M. O. (2003). Compliance and enforcement of fisheries regulations in the Caribbean. Retrieved from https://aquaticcommons.org/13555/1/gcfi_54-14.pdf. Accessed on December 13, 2018.

Kuhlman, T., & Farrington, J. (2010). What is sustainability?, *Sustainability, 2*(11), 3436–3448.

Lamb, J. B., True, J. D., Piromvaragorn, S., & Willis, B. L. (2014). Scuba diving damage and intensity of tourist activities increases coral disease prevalence. *Biological Conservation, 178,* 88–96.

Lockhart, D., & Drakakis-Smith, D.W. eds. (1997). *Island tourism: Trends and prospects.* Thomson Learning.

Milazzo, M., Badalamenti, F., Ceccherelli, G., & Chemello, R. (2004). Boat anchoring on Posidonia oceanica beds in a marine protected area (Italy, western Mediterranean): Effect of anchor types in different anchoring stages. *Journal of Experimental Marine Biology and Ecology, 299*(1), 51–62.

Miles, M. B., Huberman, A. M., & Saldaña, J. (2014). *Qualitative data analysis: A methods sourcebook* (3rd ed.). Ic: SAGE Publications.

Moeller, T., Dolnicar, S., & Leisch, F. (2011). The sustainability–profitability trade-off in tourism: Can it be overcome? *Journal of Sustainable Tourism, 19*(2), 155–169.

Mota, L., & Frausto, O. (2014). The use of scuba diving tourism for marine protected area management. *International Journal of Social, Behavioral, Educational, Economic, Business and Industrial Engineering 8,* 3159–3164.

Opdenakker, R. (2008). Advantages and disadvantages of four interview techniques in qualitative research.*Forum Qualitative Sozialforschung/Forum: Qualitative Social Research, 7*(4), Art. 11.

LN 359. (2012). *Legal Notice 359 of 2012: Recreational Diving Services Regulations (PDF). Malta Travel and Tourism Services Act: Subsidiary legislation 409.13. Malta. 19 October 2012.* Retrieved from https://www.justiceservices.gov.mt/DownloadDocument.aspx?app=lom&itemid=10626. Accessed on November 23, 2019.

MTA. (2018). *The profile of diving travellers in 2017.* Research Unit, Malta Tourism Authority, Malta. Retrieved from https://www.mta.com.mt/en/file.aspx?f=31852.

MTA. (2015). *The profile of diving travellers in 2014.* Malta Tourism Authority, Research Unit.

PA. (2003). *PA case application details. Construction of breakwater to enclose marina.* Planning Authority official website, Malta. (Online). Retrieved from https://www.pa.org.mt/en/pacasedetails?CaseType=PA/03254/03. Accessed August 30, 2019.

PDSA. (2017). *Rules & Regulations.* Homepage of Professional Dive Schools Association (PDSA), (Online). Retrieved from https://pdsa.org.mt/rules-regulations/. Accessed December 6, 2018.

Standeven, J. (1998). Sport tourism: Joint marketing—A starting point for beneficial synergies. *Journal of Vacation Marketing, 4*(1), 39–51.

Stergiou, K. I. (2002). Overfishing, tropicalization of fish stocks, uncertainty and ecosystem management: Resharpening Ockham's Razor. *Fisheries Research, 55*(1–3), 1–9.

Thomas, D. R. (2006). A general inductive approach for analyzing qualitative evaluation data. *American Journal of Evaluation, 27*(2), 237–246. https://doi.org/10.1177/1098214005283748.

TM. (2019). Notice to Mariners No 8 of 2019. *Conservation Areas Around Wrecks.* Transport Malta (TM) website. (Online). Retrieved from https://www.transport.gov.mt/include/filestreaming.asp?fileid=3563. Accessed on January 20, 2020.

VisitMalta. (2019). *List of Registered Dive Centres.* VistMalta Website. (Online). Retrieved from https://www.visitmalta.com/en/dive-centres?pg=2. Accessed on December 3, 2019.

Waycott, M., Duarte, C. M., Carruthers, T. J., Orth, R. J., Dennison, W. C., Olyarnik, S., et al. (2009). Accelerating loss of seagrasses across the globe threatens coastal ecosystems. *Proceedings of the National Academy of Sciences, 106*(30), 12377–12381.

Worachananant, S., Carter, R. W., Hockings, M., & Reopanichkul, P. (2008). Managing the impacts of SCUBA divers on Thailand's coral reefs. *Journal of Sustainable Tourism, 16*(6), 645–663.

Wongthong, P., & Harvey, N. (2014). Integrated coastal management and sustainable tourism: A case study of the reef-based SCUBA dive industry from Thailand. *Ocean & Coastal Management, 95,* 138–146.

eSports Tourism: Sports Tourism in a Modern Tourism Environment

Ioannis A. Nikas and Ioulia Poulaki

1 Introduction

In the last 20 years, along with technological advances, especially in the field of software development and online services, another form of sports has developed; the case of electronic games or as it is already called the eSports. eSports is, currently, a worldwide trend. More and more disciplines of science and society are showing interest in the academic study of the phenomenon and want to be part of the eSports industry, introducing them into the culture of our daily lives. Contrary to their obvious hi-tech nature, their presence, however, is not limited to a pure digital form: Many and great eSports events are organized through the world, including real athletes and teams from around the world and sponsors supporting them, taking place in large traditional sports facilities where spectators watch these events either by filling stadiums, or by watching online these experiences, increasing the total number of enthusiasts from a few thousand to a few millions (Alden Analytics LLC—Cary, NC, 2020).

Despite the recent development of eSports, their beginning is found in the early 1970s, while the first official event took place in 1980 (Li, 2017; Funk et al., 2018). The emerging need to understand the eSports phenomenon has led the academic community to define and thoroughly study them, contributing from many different research fields (e.g. Peša et al., 2017; Ross, 2001; Pereira et al., 2019)). Their internationalization, as a result of the convergence of Western and Asian culture (Seo, 2013; Reitman et al., 2020), brought huge revenues that some cases can be compared globally with traditional sports events such as Confederations Cup and NBA Championship (Genius Works, 2019).

I. A. Nikas (✉) · I. Poulaki
Department of Tourism Management, University of Patras, Patras, Greece
e-mail: nikas@upatras.gr

On the other hand, every sporting activity in the form of an event may be also been seen as an event of tourist interest; eSports could not be an exception, so, it can be considered as a special form of tourism and, thus, an opportunity for mass movement of people (Dilek, 2019). Without a doubt, eSports can become a great tourism product proposition and a good paradigm of how a futuristic trend can create innovative consumer products.

In this direction, the main purpose of this work is the investigation of the potentials of eSports in countries where there is so far no initiative on organizing such events, despite the fact that eSports are well-known to online players and eGames enthusiasts. Furthermore, appropriate and adequate infrastructures exist and ready to be used as venues to similar events. Such a case is Greece and most of its regional cities, like the city of Patras.

This paper presents the aspects of eSports and detects their study as a tourism product through the existing literature. It also describes the potential of eSports, as well as how they can become an efficient new form of tourism and an important alternative in the tourism sector. Finally, the available infrastructure that could be used for the development of eSports events are presented and the opportunities that arise for further tourism development in Greece and especially in the city of Patras are discussed.

2 Literature Review

2.1 eSports and eSport Events

It is almost two decades that efforts are recorded aiming to define eSports and determine their context. There are indeed authors that have contributed significantly toward eSports scientific independence. Several definitions are given in this section.

One of the first efforts in eSports definition was the Wagner's (2006) that defines eSports as "an area of sport activities in which people develop and train mental or physical abilities in the use of information and communication technologies". The same author extended the prior definition postulating that eSports is formally defined as "an area of sport activities in which people develop and train mental or physical abilities in the use of information and communication technologies" (Wagner, 2007; Seo, 2013).

Later on, Hamari and Sjöblom (2017) define eSports as "a form of sports where the primary aspects of the sport are facilitated by electronic systems; the input of players and teams as well as the output of the eSports system are mediated by human–computer interfaces". Lokhman et al. (2018) use other terms for eSports, like computer sports or cybersports and define them as a video games competition that takes place in a virtual space. Additionally, for Dilek (2019) eSports, or "electronic sports" is the term used to refer to professional gaming.

In a brief summary Jenny et al. (2017) state that eSports, in the form of organized video game competitions, is also known as cybersport, virtual sport, and competitive gaming.

Unlike different computer-game practices, wherever customers could play to relish storytelling or to alleviate themselves from disappointing aspects of their everyday lives, eSports is primarily contended to enhance client talents within the use of digital technologies and taking part in competitive computer games. Consequently, such games designed for eSports should feature some objective measures of comparison that may be accustomed choose players' performances among the sport. These measures could and sometimes differ among game platforms (Seo, 2013).

Additionally, to the definitions, in the literature, efforts are underlined to present eSports as a commercial activity. For example, Parshakov and Zavertiaeva (2015), postulate that eSports have four characteristics as follows:

(a) the investment costs of eSports are lower than of traditional team sports as sports facilities for spectator gaming are much cheaper, the investment required in eSport players is less, and players can participate in multiple games on a professional level.
(b) eSports is relatively new and thus, no professional schools have been established to train players.
(c) two types of tournaments are held for most games, that is to say offline (LAN) and online, with top tournaments and events to be held offline.
(d) the rewards and money won are mostly performance-based.

Going one step further in eSports as a commercial activity in a wider service environment, Seo (2016) considers eSports as an activity of recreation, postulating that eSports tourism development in modern society as well as the market for these services are actively increasing.

The official organization of eSports events is observed in 1980, when Atari's Space Invaders Championship appears as the first major video game competition. Participants were recorded to be more than 10,000 (Li, 2017; Funk et al., 2018). From all around the world, professional video game players of both genders and young in age competing each other in such events and the winners received prizes usually from corporate sponsorships (McTee, 2014). At this point, it is worth illustrating the supply side of these events that includes mainly game developers and IP owners as well as game publishers, event organizers media platforms, commercial brands, and of course supporting services such as hardware vendors and merchandisers (OGCIO, 2017).

2.2 The Wider eSports Environment and Its Potential

Historically, eSports are characterized by developments that have significantly contributed to its evolution. Two of them concern the increase of consumer literacy and the popularity of computer games as well as the dynamic technological evolution of the Internet and digital technologies (Hartmann & Klimmt, 2006). According to Tamborini and Skalski (2006) contemporary virtual games offer substantial immersive capacities allowing many players at the same time competing each other (Chan & Vorderer, 2006). Therefore, efforts are made toward entrepreneurial and commercial activity in the context of eSports, which is defined as activity for profit. More specifically, it seems that computer games are leading the commercial sector in the context of sports services (Lokhman, et al., 2018).

Dilek (2019) characterizes eSports as a multidimensional term and a hallmark of the "experience economy" concept, based on Toffler (1970) who noted that experiential products can be of two types:

(a) Simulated environments in relevance with computers, robotics, etc.
(b) Live environments in relevance with experiential geographical hubs similar to sports, travel/tourism, and gaming events.

Newzoo published in 2016 (Holleman, 2016) the first completed report on the global eSports market in 2015 where it was estimated that eSports generates US$612 million and has 134 million viewers globally, while in 2017 the revenue generated by eSports reached the US$325 million worldwide, based on Newzoo respective report of 2018 (Newzoo, 2018) with the organization to forecast that in 2019 eSports economy will grow to US$463 million in a year over year growth of 43% with an expected audience of 131 million and 125 million occasional viewers in addition. In any case, eSports and eSports events are recording a tremendous growth in the last decade.

2.3 eSports Tourism as a Modern Tourism Form

Sports is an important activity within tourism and tourism is a fundamental characteristic of sports. It is considered that tourism and sport are related and overlap each other. Sport tourism experiences a significant increase lately in terms of academic research as well as tourism product (Gibson, 1998). Delpy (1998) defines sports tourism as the travel away from home to play sport, watch sport, or visit a sports attraction, including both competitive and non-competitive activities, while sports tourism may be divided into five main categories: (a) attractions, (b) resorts, (c) cruises, (d) tours, and (e) events.

All of the above categories are also strongly related with other tourism products. Focusing on the last category one may say that in the wider context of sports events, eSports events may be included. There are many reasons why eSports events are attractive. eSport organizations in cooperation with corporations within the computer game industry, arrange events in the form of competitions at both national and international levels (Jonasson & Thiborg, 2010). All the participants in such events, players as well as fans have the opportunity to watch the competition among the best players in the world, to meet their favorite ones and share their passion with others. Therefore, from the side of fans there is a willingness to travel long distances in order to attend eSports events. Consequently, the potential for a new tourism form rises to be called eSports tourism.

Agius (2015) stressed that the concept of eSports as a special form is without a doubt realistic for the travel industry and it may probably turn out to be lucrative with the enormous excitement that previously appeared in this field. In fact, numerous scholars have postulated that eSports are part of the area of diversion and that within the travel industry only eSports have been developed in today's society (Seo, 2016; Lokhman, et al., 2018).

In addition, Gibson (2003) postulates that sports tourism has recreational dimensions motivating individuals to move from their homes to places where they can take part or watch physical exercises, or even to visit attractions related with physical activities, while she suggests that there are three kinds of sports for travel: dynamic games, occasion sports and wistfulness sports. According to Peša et al. (2017) eSports do not concern the physical exercise of individuals in the way that traditional games do, traveling for them may be evaluated as sports tourism since it concerns big crowds globally, with Statista (Gough, 2020) to report that the Intel Extreme Masters Katowice held in Poland in 2017 achieved to be the most watched eSports event recording 46 million watchers. Moreover, the second most popular competition in terms of watchers happened to be the League of Legends World Championship that was held in the USA in October 2016, with the respective number of watchers to reach about 43 million.

Sports tourism refers to the journey made in order to participate or monitor sports activities. It is commonly accepted that there are three types of sports tourism: Sports event tourism, active sports tourism and nostalgic sports tourism (Ross, 2001).

Sports tourism is also characterized as a complete tourist experience in which sports are at the center of the process. Furthermore, sports tourism is defined as the experience of travel as a result of participation in sports activity or event (Karlis, 2002), where sporting event is the main cause of travel (Katerinopoulou, 2002). Given the above, eSports events belong under the umbrella of sports tourism. In fact, there is an obvious potential toward the touristic dimension of eSports. Additionally, the training character that a significant number of eSports events have, the idea of "edutainment" is developed driven from the combination of training and amusement (Dilek, 2019).

3 Empirical Part

The potentials of eSports have begun to emerge recently, through various aspects of our daily lives, such as academic activity, intense economic activity, geographical spread, and the actions of eSports legislation as a sport.

Academic Activity. Reitman et al. (2020) studied the literature of eSports from 2002 to the Spring of 2018. Surprisingly, 55.3% of the eSports-focused articles come from the disciplines of business, sports science, sociology, etc., contrary to the

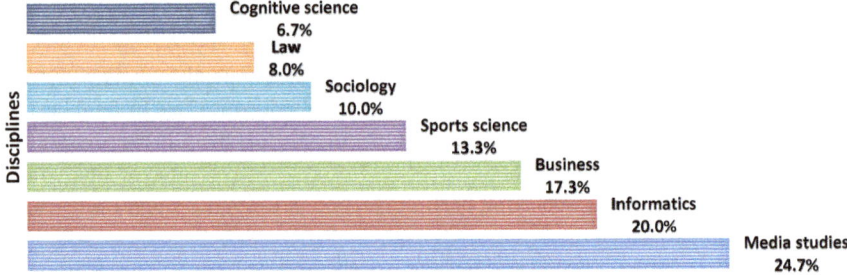

Fig. 1 Analyzing eSports corpus per area of study

common sense where one may consider that eSports should be part of media studies or computer science (Fig. 1).

The growing interest in eSports can be also seen, for example, in a Google Scholar search, where the results for a raw searching of the word "eSports" for the years 2002–2019 seem to increase significantly as we approach 2019 (Fig. 2a) (Google Scholar, 2020). This is, also, confirmed by the global trend in the search for the word "eSports" using the Google Trend tool from 2004 till September 2020 (Fig. 2b) searching worldwide (Google Trend, 2020).

Economic activity. The economic characteristics of eSports such as the low cost of the needed infrastructures (always in relation to traditional sports), the fact that

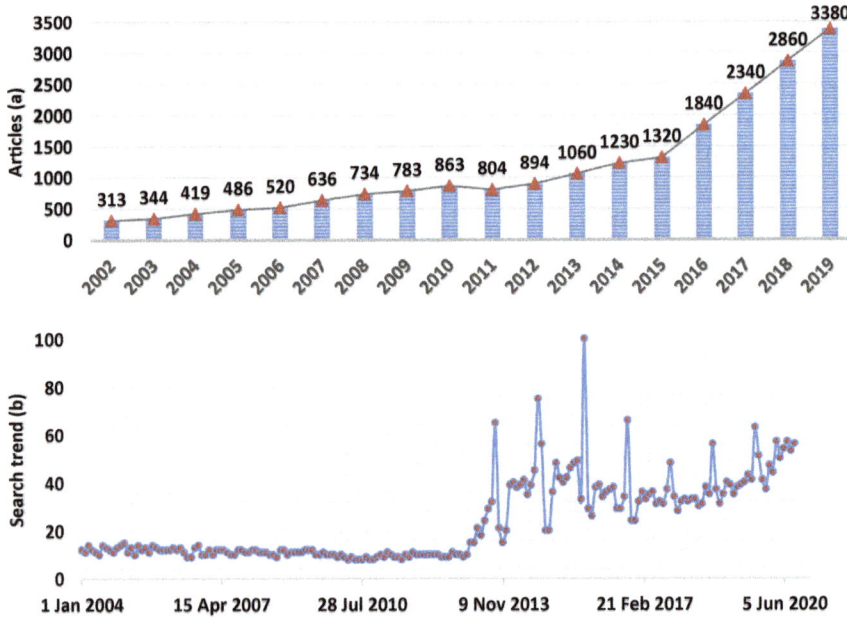

Fig. 2 Scholar search **a** and trend search **b** for the word "eSports"

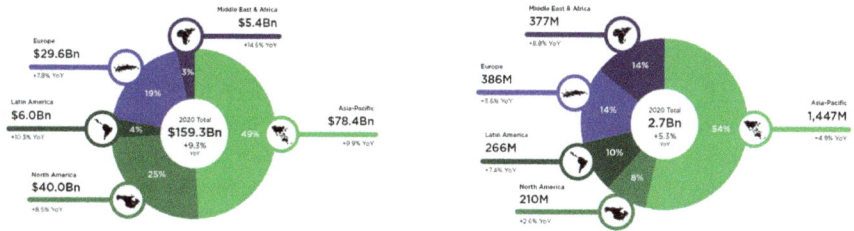

Fig. 3 Global market revenue per region and players per region

watching eSports events is cheaper for spectators, etc., gave the opportunity to create a business trend, which according to Newzoo is expected to create revenue of $ 159.3 billion in 2020 (annual growth of 9.3%) in the total toy market, with the lion's share coming from Asia and Oceania and secondarily from North America and Europe (Fig. 3) (Newzoo, 2020).

Internationalization. Revenues from eSports in geographical regions like Africa and Middle East are estimated to show an overall annual increase of 14.5%. Also, it is of particular interest the large increase in the number of eSports players from the Middle East and Africa (annual growth of 8.8%), as well as the fact that they constitute the 14% of the total number of players worldwide.

Legislation of eSports. Recently, in addition to the many common features that eSports and traditional sports have (Seo, 2013), there have been moves and efforts to establish eSports as an accepted form of sport. A milestone year in this direction was 2013, when scholarships were given from American universities regarding eSports, as well as the same year eSports athletes were recognized as professional athletes (Alden Analytics LLC—Cary, NC, 2020). Then, in 2018 one of the topics for discussion in 7th Olympic Summit was the inclusion of eSports/eGames as medal game event (International Olympic Committee, 2018), after the effort Japanese who were trying to legalize eSports and include them in the 2024 Olympic Games (Nakamura et al., 2018). Recently, the Olympic Council of Asia approved eSports as a medal game event at the Asian Games in China 2022. Finally, the International eSports Federation accepted officially eSports as a sport in more than 60 countries (Martin-Niedecken & Schättin, 2020; Alden Analytics LLC—Cary, NC, 2020; Pereira et al., 2019).

The increased popularity of eSports seems to have been significantly affected by the pandemic of Covid-19, increasing the revenues from mobile gaming, among others. Newzoo, also, prediction for 2023 estimates that the total revenue of the eSports market will reach 1.6 billion US dollars, while the size of the spectators is estimated to reach 646 million spectators worldwide (Newzoo, 2020).

3.1   The Opportunity of eSports Events and eSports Tourism

Considering all the above along with the case of Greece as a well-known tourism destination, some questions are arisen: Do the cities in Greece have the potential to take advantage of the international trend of eSports? May such a potential thrive in a country like Greece with high tourist preference but under the 4S model? Can regional cities, such as Patras in the region of Western Greece, attract and organize such events?

Greece has a significant tradition in organizing large-scale events and conference tourism infrastructures. In the first case, many of these infrastructures either remain unused for a long time period or are used for cultural events and concerts. Employing them as eSports venues will enable the better management and maintenance of these facilities, as the cost is far less than conventional sport events and the revenues seem promisingly high. Many sports venues around the world have been used to run such events (Jenny et al., 2017). Furthermore, the regional cities of Greece, such as Patras, have high-level academic infrastructure like high-speed network, conference centers, and technological know-how that could support the coverage and/or organization of eSports events. The facilities of University of Patras (along with other institutes worldwide) have already been used as venue to cover several cultural Web events, e.g. "The Met: Live in HD" live performances from the Metropolitan Opera (MET in Greece, 2020). Another good paradigm is the case of Diavlos (2020) where Web events of various themes are offered to online users. The experience from such cultural Web events can be used, as well, in covering major eSports events.

The organization of eSports events will be beneficial to the tourism sector as well: eSports may be combined with 3s tourism for the case of an event to take place in Greece or any other leisure 3s destination; the age groups of people visiting Greece will be extended; there will be a flourishing of city tourism and consequently a bigger tourism season. eSports event hosting will showcase opportunities for the development of intelligent ecosystems to create smart cities and intelligent tourism services.

Finally, Patras as a geographical center among important archeological monuments and sites in the region of Western and Central Greece, such as ancient Olympia and Delphi, with a high way connecting the city with the Greek capital Athens, and in combination with the available hotel and conference facilities has the ability to produce modern tourist products and experiences.

4   Conclusions and the Way Forward

This paper investigates the phenomenon of eSports and its potential as a tourist product, especially in countries and cities with traditional tourism product development. eSports, as a modern form of sport, can deliver to the tourism industry new products contributing toward sustainability, in the context of a new type of sports tourism form. In addition to sustainability, eSports events are a characteristic paradigm of a circular economy with significant economic prosperity, low cost, and minimal further intervention in the environment. Furthermore, given the event character of eSports and digital transformation of events emerged during Covid-19 pandemic outbreak, an interesting combination seems to be generated that concerns digital sport events leading to a new form of tourism which may be developed through eSports and may consist a recovery form of tourism during the periods of crises. Undoubtedly, further research is suggested toward this direction in order to increase the contribution of eSports market within tourism sector.

References

Agius, M. (2015). *E-sports as a niche tourist attraction : an international exploratory study.* (B.A. Thesis). L-Università ta' Malta. Retrieved from https://www.um.edu.mt/library/oar//handle/123 456789/8064

Alden Analytics LLC—Cary, NC. (2020). *The Rise of Esports.* Retrieved from Preceden: https://www.preceden.com/timelines/336488-the-rise-of-esports.

Chan, E., & Vorderer, P. (2006). Massively multiplayer online games. In *Playing video games: Motives, responses, and consequences* (pp. 77–88). Routledge. https://doi.org/10.4324/978020 3873700

Delpy, L. (1998). An overview of sport tourism: Building towards a dimensional framework. *Journal of Vacation Marketing,* 23–38. https://doi.org/10.1177/135676679800400103.

Diavlos. (2020). Retrieved from Diavlos: https://diavlos.grnet.gr/en/

Dilek, S. E. (2019). E-Sport events within tourism paradigm: A conceptual discussion. *International Journal of Contemporary Tourism Research, 3*(1), 12–22. https://doi.org/10.30625/ijctr.525426.

Funk, D. C., Pizzo, A. D., & Baker, B. J. (2018). eSport management: Embracing eSport education and research opportunities. *Sport Management Review, 21*(1), 7–13. https://doi.org/10.1016/j.smr.2017.07.008.

Gaming Bountie. (2018, January 03). *The history and evolution of eSports.* Retrieved from Medium: https://medium.com/@BountieGaming/the-history-and-evolution-of-esports-8ab6c1cf3257

Genius Works. (2019). *FIFA to Fortnite … Tencent to Twitch … How eSports is transforming the world of sport and entertainment.* Retrieved from Genius Works: https://www.thegeniusworks.com/2019/10/fifa-to-fortnite-tencent-to-twitch-how-esports-is-transforming-the-world-of-sport-and-entertainment

Gibson, H. J. (1998). Sport tourism: A critical analysis of research. *Sport Management Review, 1*(1), 45–76. https://doi.org/10.1016/S1441-3523(98)70099-3.

Gibson, H. J. (2003). Sport tourism: An introduction to the special issue. *Journal of Sport Management, 17*(3), 205–213. https://doi.org/10.1123/jsm.17.3.205

Google Scholar. (2020). *Search for the word "eSports".* Retrieved from Google Scholar: https://scholar.google.gr/scholar?q=esports&hl=el&as_sdt=0%2C5&as_ylo=2002&as_yhi=2020

Google Trend. (2020). *Trend for the word "eSports".* Retrieved from Google Trend: https://trends.google.com/trends/explore?date=all&q=esports

Gough, C. (2020). *eSports market—Statistics & Facts.* Retrieved from statista: https://www.statista.com/topics/3121/esports-market/

Hamari, J., & Sjöblom, M. (2017). What is eSports and why do people watch it? *Internet Research, 27*(2), 211–232. https://doi.org/10.1108/IntR-04-2016-0085.

Hartmann, T., & Klimmt, C. (2006). Gender and computer games: Exploring females' dislikes. *Journal of Computer-Mediated Communication, 11*(4), 910–931. https://doi.org/10.1111/j.1083-6101.2006.00301.x.

Holleman, G. (2016). *Global Esports Market Report: Revenues to Jump to $463M in 2016 as US Leads the Way*. Retrieved from Newzoo: https://newzoo.com/insights/articles/global-esports-mar ket-report-revenues-to-jump-to-463-million-in-2016-as-us-leads-the-way/

Influencer Marketing Hub. (2020). *The Incredible Growth of eSports [+ eSports Statistics]*. Retrieved from Influencer Marketing Hub: https://influencermarketinghub.com/growth-of-esp orts-stats/

International Olympic Committee. (2018). *Communique of the 7th Olympic Summit*. Retrieved from International Olympic Committee: https://www.olympic.org/news/communique-of-the-7th-oly mpic-summit

Jenny, S. E., Manning, R. D., Keiper, M. C., & Olrichd, T. W. (2017). Virtual(ly) athletes: Where eSports fit within the definition of "sport." *Quest, 69*(1), 1–18. https://doi.org/10.1080/00336297. 2016.1144517.

Jonasson, K., & Thiborg, J. (2010). Electronic sport and its impact on future sport. *Sport in Society, 13*(2), 287–299. https://doi.org/10.1080/17430430903522996.

Karlis, G. (2002). Higher education and research: Current state of condition. In *4th Annual Conference of the Athens Institute for Education and Research, May 25*. Athens, Greece.

Katerinopoulou, A. (2002). Sport tourism in Greece. *Economics and Athletics, 2*(2), 26–33.

Li, R. (2017). *Good luck have fun: The rise of eSports*. Skyhorse Publishing.

Lokhman, N., Karashchuk, O., & Kornilova, O. (2018). Analysis of eSports as a commercial activity. *Problems and Perspectives in Management, 16*(1), 207–213. https://doi.org/10.21511/ppm.16(1). 2018.20.

Martin-Niedecken, A. L., & Schättin, A. (2020). Let the body'n'brain games begin: Toward innovative training approaches in esports athletes. *Frontiers in Psychology, 11*(Article 138). https:// doi.org/10.3389/fpsyg.2020.00138

McTee, M. (2014). E-Sports: more than just a fad. *Oklahoma Journal of Law and Technology, 10*(1), Article 3. Retrieved from https://digitalcommons.law.ou.edu/okjolt/vol10/iss1/3

MET in Greece. (2020). Retrieved from MET in Greece: https://www.metingreece.com/en/hall. html

Nakamura, Y., Nobuhiro, E., & Taniguchi, T. (2018). *Shinzo Abe's Party Wants Japan Ready for Video Games in Olympics*. Retrieved from Bloomberg: https://www.bloomberg.com/news/art icles/2018-01-18/shinzo-abe-s-party-wants-japan-ready-for-video-games-in-olympics

Newzoo. (2020). *Newzoo Global Games Market Report 2020 | Light Version*. Retrieved from Newzoo : https://newzoo.com/insights/trend-reports/newzoo-global-games-market-report-2020-light-version/

Newzoo. (2018). *Newzoo Global Games Market Report 2018 | Light Version*. Retrieved from Newzoo: https://newzoo.com/insights/trend-reports/newzoo-global-games-market-report-2018-light-version/

OGCIO. (2017). *The Office of the Government Chief Information Officer (OGCIO) Report on Promotion of E-sports Development in Hong Kong*. Retrieved from https://www.ogcio.gov.hk/ en/news/publications/doc/EN_e-sports_report.pdf

Parshakov, P., & Zavertiaeva, M. (2015). Success in eSports: Does Country Matter? Available at SSRN: https://ssrn.com/abstract=2662343. https://doi.org/10.2139/ssrn.2662343

Pereira, A. M., Brito, J., Figueiredo, P., & Verhagen, E. (2019). Virtual sports deserve real sports medical attention. *BMJ Open Sport and Exercise Medicine, 5*, e000606. https://doi.org/10.1136/ bmjsem-2019-000606.

Peša, A. R., Čičin-Šain, D., & Blažević, T. (2017). New business model in the growing e-sports industry. *Poslovna izvrsnost (Business Excellence)*, 121–131. https://doi.org/10.22598/ pi-be/2017.11.2.121.

Reitman, J. G., Anderson-Coto, M. J., Wu, M., Lee, J. S., & Steinkuehler, C. (2020). Esports research: A literature review. *Games and Culture, 15*(1), 32–50. https://doi.org/10.1177/155541 2019840892.

Ross, S. D. (2001). *Developing sports tourism—An eGuide for destination marketers and sports events planners.* University of Ilinois at Urbana Champaign, National Laboratory for Tourism and eCommerce.

Seo, Y. (2013). Electronic sports: A new marketing landscape of the experience economy. *Journal of Marketing Management, 29*(13–14: Virtual Worlds), 1542–1560. https://doi.org/10.1080/026 7257X.2013.822906

Seo, Y. (2016). Professionalized consumption and identity transformations in the field of eSports. *Journal of Business Research, 69*(1), 264–272. https://doi.org/10.1016/j.jbusres.2015.07.039.

Tamborini, R., & Skalski, P. (2006). The role of presence in the experience of electronic games. In P. Vorderer, & J. Bryant (Eds.), *Playing Video Games: Motives, Responses, and Consequences* (pp. 225–240). Routledge. https://doi.org/10.4324/9780203873700

Toffler, A. (1970). *Future Shock.* Random House.

Wagner, M. G. (2006). On the Scientific Relevance of eSports. In H. R. Arabnia (Ed.), *Proceedings of the 2006 International Conference on Internet Computing & Conference on Computer Games Development, ICOMP 2006* (pp. 437--442). CSREA Press.

Wagner, M. G. (2007). Competing in metagame gamespace—eSports as the first professionalized computer metagame. In F. von Borries, S. P. Walz, & M. Böttger (Eds.), *Space Time Play* (pp. 182–185). Birkhäuser Verlag AG.

Nostalgia Sport Tourism: An Examination of an Underestimated Post-event Tourism Proposal

Ourania Vrondou

1 Introduction

Sport tourism has been enjoying significant attention amongst the scholars for more than thirty years, since Glyptis (1982) effort to mark the relation between sport and tourism forecasting a dynamic development and significant theoretical work. Two main components aiming to map the sport tourism environment those of 'energetic' and 'passive' types based on tourists' demand and behavior during holidays. 'Energetic' sport tourism has evidently produced vast amounts of research and studies due to its multi-facet character and different expressed activities occurring during holidays. Secondly, 'passive' sport tourism expresses the willingness of tourists to spectate events in combination with holidays, thus presenting an appealing proposal for different localities with important sporting activity.

Since the begging of the sport tourism concept, a small but noteworthy part of the literature identified an expression of sport-related tourism activity hidden between cultural tourism, sport fans colorful passion, and landmark importance. Redmond's (1973) early conceptual search produced the very first notion of nostalgia sport tourism. Gradually, building-specific characteristics this expression represented the vivid willingness of people to visit sites, monuments, and venues of particular sport importance and history. Early related literature uses the term 'nostalgia' sport tourism borrowing the term from the Greek 'νοσταλγία' that defines the melancholic recollection of the past with the eager to visit places of previous experience (Mega Lexicon of the Greek language, 1964). Nostalgia sport tourism despite its sentimental feel failed to attract equivalent volume of research often overlooked by literature signaling the need for further examination and magnitude estimation.

O. Vrondou (✉)
Department of Sports Management, University of Peloponnese, Sparta, Greece
e-mail: ovrondou@yahoo.gr

© The Author(s), under exclusive license to Springer Nature Switzerland AG 2021
V. Katsoni and C. van Zyl (eds.), *Culture and Tourism in a Smart, Globalized, and Sustainable World*, Springer Proceedings in Business and Economics,
https://doi.org/10.1007/978-3-030-72469-6_8

2 Theoretical Considerations

2.1 Theory Questioning the Nostalgia Sport Tourism Nature

Gibson (1998) was one of the very first authors to suggest nostalgia sport tourism as one of the three distinct pylons of sport tourism including travelers that visit sports facilities, museums, and sports camps for a recollection of their youth sports moments, heroes, and victories. She suggested that nostalgia sport tourism involves 'visiting famous sports-related attractions...visits to sports halls of fame....sports museums...and famous venues' (Gibson, 1998). More extensively, Fairley and Gammon (in Gibson, 2006) emphasize the extended value of the term encompassing not only the sport tradition, venues as landmarks and sport museums but also the 'relive of a social experience' associated with sport. Recollection of personal sport performance, romantic memories of amateur, and innocent sport years as well as participation through volunteering in sport events demonstrate a much wider platform that could generate nostalgic traveling that goes beyond visiting well-known sport facilities and venues.

With theory maturing overtime nostalgia sport tourism has evidently gained the third place amongst the three main facets of the sport tourism typology. However, doubts seem to be gathering among the theorists regarding the value and the capacity of this form to generate autonomous tourism activity with specific and willing clientele. Does limited research prove its inability to be a tourism generator? Weed and Bull (2004) strongly insisted on the less distinct character of nostalgia sport tourism. They argued that people would very unlikely travel primarily to engage to nostalgia tourism activities, suggesting that there is only a small minority that will visit a sports museum or venue as the primary reason to travel to a place thus functioning more as a supplementary activity during holidays.

Debates at the early stages of the theoretical work regarding the capacity of nostalgia sport tourism to create specific sport tourism segments revisit the core definitions of the 'tourist,' the 'excursionist,' 'sport tourist pilgrims,' 'tourism with sports content,' and 'tourists interested in sport' (Weed & Bull, 2004) leaving little room for optimism. Recorded doubts and criticisms underline the complex character of this sport tourism form. Heavily relaying on a plethora of different involved organizations and a hesitating market seeking for proof of segment interest in order to follow, development seems doubtful. Even more so, when the core product is intangible, based on the reproduction of the past with the only intervention possible being the effective exhibition and promotion.

Attempts to map the phenomenon seem to be realizing its extended nature and depart from the narrow perspective of the loving sport memories to wider conceptual paths that include the charting of tourists' holistic sport relation and experience that could be translated to specifically designed policies and market offers. From Fairley and Gammon's (in Gibson, 2006) dual conceptualization of either 'nostalgia for sport place or artifact' or 'nostalgia for social experience,' this sport tourism form has created an extended theoretical spectrum. Cho et al. (2014) attempt to clarify nostalgia produced a four-dimensional classification including experience, socialization, personal identity, and group identity. Greater emphasis is given on the 'importance of different types of experience' making the examination of nostalgia even more complex. Recent literature tends to agree on the core ingredient of a nostalgia sport tourism expression being based on 'individuals experience playing an important role in shaping their sport nostalgia' (Cho et al., 2019), thus shaping the demand for related travel mode.

Despite the considerable attention given on the examination of the different conceptual dimensions, previous or contemporary work has produced limited insight since the majority of studies gather around the notion of 'individual sport engage-ment,' thus making related theory of limited applicable value. Nostalgia sport tourism remains a wide thematic proposal with the sport glory and history predictably found in the original setting, that of sport venues and their peripheral expressions such as sport museums and exhibitions.

Expectedly, the magnitude and importance of sport have driven some parts of the market, the sports world, and the tourism business almost instinctively to design and offer sport history-related activities to attract visitors and tourists. However, one of the important reasons obstructing participation in nostalgia sport tourism was early identified in the literature when sport tourists 'are more likely to be constrained by lower levels of supply and perhaps by lack of awareness about existing opportuni-ties' (Hinch, Jackson, Hudson and Walker, in Gibson, 2006). Currently, the lack of information on actual venue visiting rates, post-events' tourism activity, and most importantly, estimated willingness to travel for sport nostalgia programs continue to leave policymakers and sport organizations as well as the tourism managers unable to discover the real visiting scale that could be generated from a significant sport incident, competition, or icon.

2.2   Sport Events and Venues as Tourism Generators

Large-scale events are fascinating because of their competition importance, venues' magnitude and architectural value, as well as the historic sport moments occurring inside the field of play. Having significant sentimental value for sport fans and spec-tators is often positioned amongst the most effective developmental vehicles for localities and nations. Similarly, large sports venues are directly benefiting from international sport excellence, records, athletes, battles, and sport heroes. Historical moments, national pride experiences, and cultural interconnected activities complete their significant importance.

The battle between the optimists and the realists is ongoing with the literature being divided into two main sides arguing over the effectiveness of large-scale events to generate tourism during or post-event. Actually, the arguments are extended to the overall ability of the events to boost economic activity after the completion of competition, let alone to stimulate tourism in the long run. Positive sides strongly support the unique power sport to transfer an appealing image for the host locality building quality perceptions to possible future visitors despite the enormous planning, organizational and construction costs involved in the preparation and materialization of a large international event.

Direct or indirect promotion of a place is the most evidenced outcome of a large event with limited criticism recorded on this notion. Literature suggests that the accommodation of numerous sport events and related activities increases local economic and social activity. The process of building or renovating sport facilities permanent or temporary for the needs of the event, apart from supporting local future sporting needs set the basis of an upgraded quality environment that could work as the stage of a new local product. There is a huge number of candidate cities bidding not only for mega events, thus are in need for superstructures, but also a variety of local, national, and international sports events that can attract the attention of the sport audience. In the case of mega events, there is enough evidence now to prove that they have become part of a more holistic strategy of nations and cities toward image alteration or regeneration. Sport venues become the catalytic factor of evident change built to last for decades with cities like Barcelona proving that they can materialize image alteration as well as long-term local regeneration in many social activity areas such as sport participation and sport-related tourism development. Sport venues become landmarks that boost a positive local image through architectural uplifting but also though the accommodation of popular sport events that match the willingness of a city to imply a new quality future for local residents (Kiuri & Teller, 2015).

Adversely, numerous studies have been fiercely criticizing the creation of an 'ice cream economy' deriving from events implying the short-term financial benefits melting soon after the completion of the event despite any other social and cultural long-term impact (Bouchon, 2017; Pilette & Kadri, 2005). After all, the event duration is very small in relation to future local development leaving behind unexploited costly 'white elephants' (sport venues) endangering the viability of the national economy overall (Giannakopoulou, 2019). Thus, paradoxically, planning for future use receives much less attention if any (Yawei, 2012).

The battle is reinforced by the lack of evaluation studies measuring post-events' impact on localities especially in terms of long-term economic returns. Political myopia excludes many types of post-event impact such as national pride uplifting, sport participation increase, cultural enhancement, and overall local regeneration focusing largely on profit being regenerated because of and due to the event. It is not only the lack of inefficient evaluation techniques and tools; it is also the quite diverse legacy aspects related to different places, diverse event scale and nature that makes measuring the impact a challenging task. Extending this notion, different consequent governments and organizers instinctively or not may not wish to measure the post-event impact in the fear of facing a negative reality. Additionally, research is obstructed when accessing information on previous events since organizing committees fade away and interest is placed on forthcoming events (Dickson et al., 2011). Thus, the difficulty to measure event's promotional impact to the city/country especially in the long run makes the task of planning post-event tourism problematic

(Cornelissen et al., 2011). Furthermore, a great lack of interest amongst local planners is noticed toward supporting large long-term research since the possibility to undertake another mega event soon seems significantly small.

International Olympic Committee realizing the increase of mismanaged post-event years produced a 12 year 'Olympic Games Impact–OGI Study' using score-cards with 126 indicators aiming to guide organizers to effective post-games management (IOC, 2010). This is an evident assumption of the need to educate localities on potentials and also reassure viability of investments post-games. Similarly, despite its legitimate role in developing their sport further, leading sport bodies can heavily influence and ensure that part of the legacy will include post-event tourism development where organizers will have to demonstrate relevant willingness and planning skills (Horne, 2017). Sport bodies but also governmental independent organizations should continue long-term post-event evaluation to ensure implementation and avoid sterile future event proposals (Dickson et al., 2011).

Literature has not overcome the obvious relation between large-scale events and tourism activity (Chappelet & Junod, 2006). As a marketing strategy, large events become an effective promotional tool through infinite hours of live TV coverage, staged to broadcast not only sport competition but also the local setting since most organizers realize the unique opportunity of showing a quality local picture to the world. Contemporary organizers face the challenge of investing in two main focal points of development under the same local roof: sport and tourism. Despite the vague picture regarding tourism activity generated around historic venues after the completion of significant sport events, there is a common appreciation on the ability of large sport events to attract attention and help generate long-term tourism strategies.

3   Methodology

Literature admits that tourism activity generated by sport events hasn't been recorded sufficiently to lead to concrete knowledge, obviously due to its difficulty in measuring tourist numbers and behaviors in the long run. The aim of the study is to dive deeper into the basis of the issue and understand the reasons that would complicate the development of post-event tourism at a policy making level.

The present study targeted experienced officials of Olympic sports international federations, Managers of Organizing Committees for Olympic Games (Athens, 2004; Beijing, 2008; Rio, 2016), Managers of Ministries supporting events and local authorities of Olympic suburbs. More than fifteen officially pre-planned interviews were employed as the tool to discover truth since knowledge derives directly from managers already involved in key bodies necessary for the completion of the games as well as the mapping of a post-event tourism development. Semi-structured interviews aimed to structure the discussion but also discover any unpredicted additional information on all the managerial and policy issues involved. Content analysis was necessary to extract information and convert it into applicable knowledge that would support the realization of post-event nostalgia tourism developments.

4 Results

4.1 Factors Obstructing Nostalgia Sport Tourism Development

Official interviews contacted with professionals having extensive experience in the field of planning and managing international events presented an unexpected content similarity that secures results' reliability. Despite the diverse nature of the sports as well as the events' organizational idiosyncrasy, retrieved information presented uniformity in terms of the prioritization of the successful sports competition and event completion as the predominant focus rather than any other local development and post-events legacy.

Sport demonstrated its powerful position amongst all sides involved in the present research being admittedly the focus of the venue creation and the athletes' representation from around the world. The importance of the sport competition especially at mega multi-sport events expectedly creates appealing sport superstructures but also enormous costs, huge organizational burden, and complexity in the whole of the local authorities and bodies in order to adjust operations to match games needs and requirements. Respondents emphasized this fact when having to work with local organizations for the preparation of Olympic competition. Critical decisions on permanent massive investments are taken under high organizational and international federations' pressure without considering future image building and thus tourism future increase.

The majority of the public bodies are recruited to facilitate large and complex parts of the event producing additional weight on their management and resources. This weigh as suggested by the respondents leaves little room to local authorities and tourism-related schemes to design future development since all resources have to be offered for the causes of the upcoming event. Similarly, the future is as distant as the event dates since the outcome is unknown and the benefits intangible. Public bodies but also private businesses are looking for event-related chances to benefit from, either through subsidies, increased resources, or enhanced image. The latter despite the obvious necessity of all parties to collaborate toward supporting the event, creates a certain grade of departmental egoism in an effort to promote each authority's importance in the successful outcome.

Respondents of the organizing committees having strong sports background strongly argued that the attention is inevitably placed on the successful completion of the competition and the total of the event procedures. More importantly, the heavy supervision by the international sport federations at all stages of the preparation leaves little room for post-event legacy plans, even toward developing sport in the future. With attention being placed on the organizational needs, respondents from different organizations agreed that other social, cultural, and tourism potentials are disregarded. Very rarely, according to their experience, policy declarations for tourism developments post-event are finally implemented, remaining political speeches, and organizers' wishful thinking (Cashman, 2006). Moreover, they strongly reminded that when the event is over, most of the involved partners are not active or change especially regarding the public bodies. Organizing committees dismantle, government changes, and professionals seek the next carrier path leading any post-event tourism proposals stand on insecure grounds.

Public bodies' representatives having experience in supporting mega events give emphasis to different political visions, constant change in governmental policy makers and their diverse personal preferences that from their point of stance suggest the most crucial factors of leading to problematic nostalgia sport tourism developments. Sadly, as suggested also in the literature (Coakley & Souza, 2013) all respondents agree, that intended political schemes regarding a sport tourism-related post-events development are not evident during the preparation and staging of the event despite its obvious potential.

In addition, industry's short-sighted view, different economic interests and diverse suggestions fail to produce a common and robust proposal that could produce viable tourism development. Limited optimism can be generated due to accidental or not production of legacy programs needing collective efforts to make them viable but at least keep suggesting that the potential remains to be exploited.

4.2   Partners' Consensus for a Holistic Approach to Tourism Development

It became evident amongst all respondents that the main obstacle for planning post-event tourism is the wideness of organizations' representation needed to map all issues involved in the process. Sport and venue managers involved in the interviews had already experienced it during the Olympic games, and many international events held locally suggesting that it is always the greatest task when undertaking a large event. The majority of related public bodies and independent organizations have to reach an unbreakable chain of communicating, facilitating, and managing all event-related issues. Partners' consensus proves to be a difficult equation when development is based on widely different strategic goals, organizational cultures, and time perspectives (Hartman & Zandberg, 2015).

Respondents admitted that if possible, sport legacy programs should be prioritized amongst organizers and sport managers as the most related and expected outcome of a mega event and as the logical exploitation of the existing and permanent venues. Sport managers eager to turn a mega event success into a local sport legacy outcome was expectedly obvious during the interviews. Similarly, international federation officials suggested that one of the crucial responsibilities of the international sport bodies is to reassure the viability and further local development of the sport post-event. International federations will be judged by their effectiveness to produce strong local legacy programs after the completion of the games turning event legacy into a political and organizational exhibition of power and influence. Sport and venue managers recalled the pressure opposed on them during the first stages of venue design by the international federations toward increased venue capacities, better and quality support areas, additional sports equipment and most crucially the permanent accommodation of their sport in the venue. Sport events, sports camps, and training programs were the developmental future that international federations strongly suggested, proving their pressure on sport legacy programs leaving little room for tourism-related proposals. The dependence of the local organizers on the international federations' legitimate event authority led to one-way sport legacy discussions.

The ability to plan for viable local tourism-related programs becomes further complex when mega events and more specifically venue planning and competition management does not include local government at any stage of the event preparation. Examining the organizational chart prior and during the event, it becomes obvious that there is no representation of the local authority at every sport/venue section. Their role is restricted to facilitating traffic, visitors' transportation, and sewage collection excluding local say from all stages of venue design and adaptation, operational planning, and viability proposals. This paradox of supporting an event that occurs in the neighborhood without been asked to contribute to the venue's future and related local development becomes one of the most crucial results of the present study directly linked with the absence of any nostalgia sport tourism future. Additionally, local communities not involved in central policy decisions are facing phenomena of elitist urban establishments and superstructures that could disorient local image away from its indigenous character being the basis for future tourism development.

Local government authorities agreed strongly on their exclusion from any step of the event planning and venue design underlining the gap between them and the government and the organizing committee. Huge superstructures are opposed on their community without any negotiations or local input for future use and development. National goals go before local say jeopardizing the future economic and social life. Vice Mayor of an Olympic suburb recalled being excluded at the initial phase of the games from any information on planning and venue designs. This paradox relation often causes local reactions, resistance, and finally a negative image for the locality. The exclusion of local say in the planning and management of the games has been recorded in the literature (Vrondou et al., 2018), and representatives provide enough proof to support it. Structurally, there is no official role for the local government in the planning of the locally accommodated competition and the venue structure. It was strongly emphasized that local sport habits and preferences were not included in government's plans thus endangering local participation post-event. The design, planning, and implementation of relevant programs in order to be viable have to be connected with existing social structures and local governing schemes (Coakley & Souza, 2013). It was strongly argued that the local government is expected to face the function, costs, and viability of a huge superstructure post-event which most probably will not match the preferences and lifestyle of the local residents.

Optimism was generated by some respondents of local authorities that appreciated the fact that despite their exclusion, the international promotion gained through the event can operate as tourism initiator. Local authorities admitted their lack of sport knowledge prior to the event in terms of the sport importance, event's promotional power, and possible legacy development but became wise enough to realize that sport can now become part of their tourism strategy and overall image enhancement. Still, they agreed that investing in sport-related tourism programs requires knowledge and advisory support in order to be materialized. Government's tourism national planning has to support local efforts to produce viable and internationally accepted proposals.

5   Conclusions

Despite the difficulty in recording tourism activity due to a sport event, tourism seems to increase the consequent years. However, there are still doubts if tourism is generated because of the event since the promotional impact of the event remains unnumbered. Evidence suggests that a considerable tourism segment remains sport oriented with a passion to visit venues and sport historical sites to revive emotions. The literature instinctively promotes sport nostalgia and its power to generate tourism activity but fails to estimate the realistic developmental dimensions and to measure the intensity of willingness to visit a sport place as the main motive to travel. Nostalgia sport tourism segment depends largely on the participation of all related partners to produce viable programs. It became evident that sport as the ultimate focus of the event and venue planning, forces tourism to become a secondary future proposal. Local legacy remains for many post-event environments a mismanaged exercise basically due to the anxiety of organizers and governors to successfully complete sports competition. Both sport bodies and governments have the responsibility to guide future post-event multi-facet legacy programs. Long-term viable nostalgia sport tourism programs are in need for wider organizational consensus to be materialized balancing different interests and perspectives on future development. International sport bodies, IOC, and related organizations should ensure that part of the legacy will include sport tourism-related development where organizers will have to demonstrate relevant commitment, planning, and viability skills in order to achieve a holistic social sustainability post-event.

References

Bouchon, F. (2017). 'Ice cream economy' and mega-events legacy, perspectives for urban tourism management. *Journal of Tourism and Hospitality, 6*(5).

Cashman, R. (2006). *The Bitter-Sweet awakening: The legacy of the Sydney 2000 Olympic Games: Walla Walla Press in conjunction with the Australian Centre for Olympic studies*. Sydney: University of Technology.

Chappelet, J. L., & Junod, T. (2006). A Tale of 3 Olympic Cities: What can Turin learn from the Olympic legacy of other Alpine cities? In *Major Sport Events as Opportunity for Development, Proceedings of the Valencia Summit* (pp. 83–89).

Cho, H., Joo, D., & Chi, C. G. (2019). Examining Nostalgia in sport tourism: The case of US college football fans. *Tourism Management Perspectives, 29,* 97–104.

Cho, H., Ramshaw, G., & Norman, W. (2014). A conceptual model for nostalgia in the context of sport tourism: Re-classifying the sporting past. *Journal of Sport & Tourism, 19*(2), 145–167. https://doi.org/10.1080/14775085.2015.1033444.

Coakley, J. J., & Souza, D. L. (2013) *Sport mega-events: Can legacies and development be equitable and sustainable? Motriz: Revista de Educacao Física, 19*(3), 580–589.

Cornelissen, S., Bob, U., & Swart, K. (2011). Towards redefining the concept of legacy in relation to sport mega-events: Insights from the 2010 FIFA World Cup. *Development Southern Africa, 28*(3), 307–318.

Dickson, T., Benson, A., & Blackman, D. (2011). Developing a framework for evaluating Olympic and Paralympic legacies. *Journal of Sport and Tourism, 16*(4), 285–302.

Fairley, S., & Gammon, S. (2006). *Something lived, something learned: Nostalgia's expanding role in sport tourism, in Gibson HG Sport Tourism—Concepts and theories*. London: Routledge.

Giannakopoulou, M. (2019). Olympic gigantism and the multifaceted concept of sports venues. In *Proceedings of the 6th International Conference on Cultural and Tourism Innovation: Integration and Digital Transition, IACuDiT*, Athens, June 2019.

Gibson, H. G. (2006). *Sport tourism—Concepts and theories*. Routledge.

Gibson, H. G. (1998). Sport tourism: A critical analysis of research sport. *Management Review, 1*(1), 45–76.

Glyptis, S. A. (1982). *Sport and tourism in Western Europe*. British Travel Educational Trust.

Hartman, S., & Zandberg, T. (2015). The future of mega sport events: Examining the 'Dutch Approach' to legacy planning. *Journal of Tourism Futures, 1*(2), 108–116.

Hinch, T., Jackson, E., Hudson, S., & Walker, G. (2006). *Understanding sport tourism socio-cultural perspectives leisure constraint theory and sport tourism, in Gibson sport tourism—Concepts and theories*. Routledge.

Horne, J. (2017). Understanding the denial of abuses of human rights connected to sports mega-events. *Leisure Studies, 37*(1), 11–21.

IOC. (2010). *Olympic Games Impact (OGI) Study, Lauzanne*.

Kiuri, M., & Teller, J. (2015). Olympic stadiums and cultural heritage: On the nature and status of heritage values in large sport facilities. *International Journal of the History of Sport, 32*(5), 684–707.

Mega Lexicon of the Greek language, 1964, Athens.

Pilette, D., & Kadri, B. (2005). *Tourisme métropolitain: Cas de Montréal*. Sainte-Foy: Presses de l'Université du Québec.

Redmond, G. (1973). A plethora of shrines: Sport in the museum and the hall of fame. *Quest, 19*, 41–48.

Vrondou, O., Kriemadis, T, Papaioannou, A., & Douvis, J. (2018). Forming policy networks between the organizing committee for olympic games and the host city. *Advances in Sport Management Journal, 1*(1), 23–41.

Weed, M., & Bull, C. (2004). *Sports tourism: Participants policy and providers*. Elsevier.

Yawei, C. (2012). Urban strategies and post-event legacy: The cases of summer Olympic cities. In *AESOP 26th Annual Congress*.

The Role of Sports Tourism Infrastructures and Sports Events in Destinations Competitiveness

Dália Liberato, Pedro Liberato, and Catarina Moreira

1 Introduction

Sport has seen strong growth in recent years (Magalhães, 2010; Ramos, 2013). It is at the end of the twentieth century, with the change in society's behaviors, with a greater concern regarding body care, physical and emotional well-being, and with the increase in the time available for the practice of sports activities, that sport manages to attract the attention of society and consolidate its place (Ramos, 2013). The sports industry plays an important role in the development of society and in its social and cultural representation is too relevant to be ignored (Bauer et al., 2008; Magalhães, 2010; Pereira, 2015). It is characterized as one of the most comprehensive and most relevant social phenomena of contemporary society, due to its universal appeal (Semedo, 2015), as it arouses the interest of millions of followers around the world, and strong emotions (Paulico, 2008). Currently, sport is lived with great intensity (Pereira, 2013) and as an open market, because it is a complex area that deals with an infinite network of relationships and thus involves a great diversity and heterogeneity of interests and actors. For this reason, the impact that sport produces have an influence on the daily life of the entire population, thus constituting a cultural model used and accepted internationally (Constantino, 2013). In addition, sport is also considered a human activity with unique particularities that involves strong emotional components, as it has the power to generate exceptional emotional responses in its audiences. Sport turns out to be a source of joy and emotion, transcending work, politics, and daily restrictions on personal and collective emotional expression (Coakley & Souza, 2013). In addition to sport, really lived with intensity by those who engage in the

D. Liberato (✉) · P. Liberato · C. Moreira
School of Hospitality and Tourism, Polytechnic Institute of Porto, Vila do Conde, Portugal
e-mail: dalialib@esht.ipp.pt

D. Liberato · P. Liberato
CiTUR Researcher (Centre for Tourism Research, Development, and Innovation), Faro, Portugal

© The Author(s), under exclusive license to Springer Nature Switzerland AG 2021   127
V. Katsoni and C. van Zyl (eds.), *Culture and Tourism in a Smart, Globalized, and Sustainable World*, Springer Proceedings in Business and Economics,
https://doi.org/10.1007/978-3-030-72469-6_9

immensity of competition through the show, the sports tourist can also enjoy a unique experience (Carvalho & Lourenço, 2009).

This study intends to be an important contribution to the scientific literature in the field of sports tourism, namely in the relationship between the Porto Football Club and Oporto tourism destination, for the definition of future strategies, regarding collaboration and strengthening partnerships in a competitive perspective for both. This study reports two thematic areas that have been very relevant to Oporto destination, as a focus of attraction and commitment: sports and tourism. However, there are scarce scientific studies that complement and analyze the role of football, in particular, the Porto Football Club and its influence on the tourism destination. In this context, and in an effort to address this issue of research scarcity, this research aims to study this segment of the tourism sector, the Sports Tourism, and to ascertain its importance and influence in Oporto destination, using the analysis of two infrastructures, relevant to the city, which are still poorly explored, the *Dragão* Stadium and the FC Porto Museum.

The general objective of this research study is to understand the role of football, in particular, of Porto Football Club, in Oporto as a tourism destination, analyzing the existence of positive relationships, such as the reasons for the trip and the assistance to a football game, with visits to the *Dragão* Stadium and the FC Porto Museum. Considering the general objective, and based on the studies of Alegrias (2017), Carvalho and Lourenço (2009), Constantino (2013), García (2006), Getz (2008), Latiesa and Paniza (2006), Magalhães et al. (2017), Manzano (2014), Rodrigues and Dávila (2007) and Stornino et al. (2016), two research hypotheses were proposed: H1: The main reasons for the trip influence the visit to the *Dragão* Stadium and the FC Porto Museum; and H2: The attendance to a football game at *Dragão* Stadium, influence the visit to the *Dragão* Stadium and the FC Porto Museum.

Gammon and Robinson (2003) and Latiesa and Paniza (2006) report that the tourists' motivations are determinant in tourism research. Although the sports tourist is motivated by the competition or by the show itself or sports event, by the performance of recreational or competitive sports activities, or by physical activity (active or passive), sport is the primary motivation of the trip. From another point of view, Ferreira and da Silva (2017) refer that the sports tourist is the individual who participates actively or passively in competitive sporting events, adding individuals who participate in or perform a sports activity in a recreational way or complement other types of tourism, whether it's beach tourism, wine, and gastronomy, or only cultural. This tourist travels outside his natural environment for any other purpose (business, leisure, personal reasons) and participates in an activity or sports context during his trip/stay (Pereira, 2016). Regarding the motivations of the sports tourist, Ferreira and da Silva (2017) and Gammon and Robinson (2003) state that the main motivation is the participation in a sporting event and the secondary motivation defines the casual attendance to the sporting event during the travel period, in which the tourist is not aware of the event, but ends up attending or participating in the activity during the trip. Thus, some areas emerge that potentially contribute to the development of sports tourism, such as adventure tourism, cultural tourism, ecotourism, business tourism, among others (Pereira, 2016). Carvalho and Lourenço (2009), Robinson and Gammon (2004) and Semedo (2015) also mention that the sports show tourist, is the individual who, during his trip attends any show or sporting event, regardless of the predominant reason for his trip. In the decision-making process of the place of destination, Latiesa, and Paniza (2006) consider that it is essential for sports tourists to take into account the motivational intensity in choosing the destination, as well as its positioning at the national and international level, not forgetting the sports offer of the destination itself. In the study by Magalhães et al. (2017), it is possible to mention cities such as Madrid, Barcelona, and Manchester, to the extent that their respective sports clubs have a significant impact on local economies, considering the entire potential of the sports club alone, assumed in the projection of its city as an interesting and attractive destination (Magalhães et al., 2017).

Regarding the destination positioning, this concept is related to the tourist mental image towards the tourism destination, some of them well highlighted as representative of sport, clearly identified with a certain sports practice and for this reason, their positioning is quite attractive for sports tourists. In short, a destination which is well positioned and contains a high motivational intensity, concerning any sport, encourages the sports tourist in the best way, because the places correctly positioned, in the face of sports offer, through infrastructure, events, or shows, tend to attract the sports tourist. In other cases, although the initial motivation of tourists is not to perform sports activities or assistance to events, many of these end up experimenting. For example, some of the tourists have as main motivation to watch sports shows, although some of them, with different motivations, take advantage of the fact that this same show is taking place in the same destination, so they watch and enjoy the opportunity. From the point of view of Carvalho and Lourenço (2009) and Robinson and Gammon (2004), there is a division between tourists who travel primarily through sport and those who consider sport as a secondary design. In this context, sports tourism can be considered as the main reason for travel or as a complement to the trip. According to Carvalho and Lourenço (2009), the relations between tourism and sport go beyond sports tourism and sports show tourism. For example, some of these relationships may be associated with other types of tourism such as cultural tourism, presenting close links with museums, or sports exhibitions. In this case, Carvalho and Lourenço (2009), and Pigeassou (2004) refer to other types of sports tourism, as the case of sports tourism culture, which refers to a more intellectual and cognitive character of sports culture, is also related to a sense of intellectual curiosity, sporting history or veneration, in which tourists make visits to attractions related to sport, sports stadiums or museums or other historical sports sites. This is how some tourist-sports attractions arise, such as stadiums, museums, theme parks, and tourist-sports events, and which encompass all major sporting events that mobilize a large number of spectators and participants (Pereira, 2016). However, there are no studies related to Porto sports club, which specifically relate the reasons for the trip with the visit to the *Dragão* Stadium and the FC Porto Museum. Carvalho and Lourenço (2009), Robinson and Gammon (2004) and Semedo (2015) report that tourists as spectators enjoy the assistance of sports shows or sporting events, regardless of the predominant reason for their trip, that is, the relations between tourism and sport go beyond sports show tourism. For example, some of these relationships may be associated with other types of tourism such as cultural tourism and close links with museums or sports exhibitions. Magalhães (2010) also mentions that this type of sporting event is outlined in advance with visibly determined objectives, but that it can associate a character not only sporting, but also cultural and social, promoting the visit to attractions related to sport.

2 Literature Review

The sport has grown, essentially, by the notoriety of the shows and events provided. This would be possible without a spectacle, but it would not be the same, because its popularization concerns its size as a spectacle. On a global scale, sport is recognized as an amazing spectacle, typical of moving clusters of people and retaining a vast number of spectators, in which they usually watch live games, or accompany them from a distance (Carvalho & Lourenço, 2009). Sport is also a universal language and contributes in many aspects in society, as the feeling of brotherhood among peoples is improved, unites people, regardless of their origin, beliefs, religion, economic status, social class or background, promotes greater social inclusion, promotes the country's image in international markets, and stimulates sports practice (Chun et al., 2005; Magalhães, 2010; Paulico, 2008; Silva, 2013). Currently, it is evident the impact and power of sports organizations on society, with emphasis on the football sport. Thus, the sport with the greatest tradition, more popular and the most public mobilizes in the globality of European countries is undoubtedly football (Cox, 1993; Ramos, 2013). Gammon and Robinson (2003) and Latiesa and Paniza (2006) report that although the sports tourist is motivated by competition or by the show itself or sporting event, by performing recreational or competitive sports activities, or by physical activity (active or passive), sport is the primary motivation of the trip, as well as that individual or group of individuals who travels to a place other than that of their habitual residence to participate actively or passively in a competitive or recreational sport. The visitor/tourist seeks experiences and novelties related to their desires, able to develop mechanisms of creativity and cultural appreciation. Therefore, tourism is no longer entirely linked to travel but also concerned with providing experiences, thus constituting the change in the offer of this type of service. In this context, the tourist experience includes the destination choice, the experience during the period on the road, the discovery and experimentation of the places visited and the final sharing of their memories during the trip, with the possibility of repeating the destination (Costa, 2015). Currently, individuals who aspire to attend and participate in sporting events also want to enjoy a pleasant experience, full of emotions and with a completely unpredictable outcome (Noite, 2013).

When sport is addressed as a factor of tourism development in the region, it is crucial to mention the central role of sports event tourism. It is well known that sport is used by several countries and cities as an instrument of tourism development, in order to increase the influx of people in sporting events. This development is thus complemented with the various promotion strategies through the tourism of sporting events. According to Carvalho and Lourenço (2009), the relations between tourism and sport go beyond sports tourism and sports show tourism. For example, some of these relationships may be associated with other types of tourism such as cultural tourism and close links with museums or sports exhibitions. In this case, Carvalho and Lourenço (2009) and Pigeassou (2004) refers to other types of sports tourism, such as "sports tourism culture" that refers to a more intellectual and cognitive character of sports culture and that may be related to a sense of intellectual curiosity, sporting history, or veneration, in which tourists make visits to attractions related to sport, stadiums or sports museums or other historical sports sites. Another example of tourism mentioned is "involvement of sports tourism" which refers to exclusive situations in the world of sport, especially the multiple travel eventualities related to sports administration or training. Through sport, in this specific case, football, FC Porto is one of the major agents responsible for the massive movement for the city of new and different audiences, national and foreign, throughout the year. Another infrastructure, the FC Porto Museum is divided into 27 thematic areas (plus a virtual and itinerant, area 28) and has a total area of 7,000 square meters of which 4,000 are exhibitions, including trophies and thematic exhibitions. The Museum uses an exhibition innovation language and strong technological dynamics with access to technological resources, capable of expressing the passion for football through its different identity representations linked to the history of the club and the memory of the fans. It also often uses new technologies for communication and dissemination of the history, events, activities, and visits of public figures (Magalhães et al., 2017). In recent years, the commitment that the Museum has been demonstrating through the programming and performance of events, actions, and activities, with the aim of bringing together, involving, reinforcing and capturing the various audiences, communities, and institutions, has been very evident. The Museum also has a strong presence in national and international fairs, such as the International Tourism Fair (FITUR), Soccerex (in England, Manchester), the Macau International Fair, and the Travel Fair. Regarding the awards received, the FC Porto museum was awarded the Innovation and Creativity Award 2015, by the Portuguese Association of Museology (APOM). The Museum also distinguished itself as the center of cultural attraction of the city of Porto, positioning itself in the tourism market, through the first place of TripAdvisor's tourist network, among the choices of museums, from the entire northern region of Portugal. In 2017, the FC Porto Museum was the first Museum of the World to be accepted as an affiliated member of the World Tourism Organization, and Porto Football Club was the first one to be represented in the UNWTO. This distinction is considered very important, as it denotes a great honor, importance, and extraordinary international recognition. The FC Porto Museum will thus share technical knowledge, about the tourism sector, with other affiliated members, organize conferences and events in collaboration with UNWTO and access privileged information, among other benefits.

3 Methodology

The general objective of this research study is to understand the role of football, in particular, of Porto Football Club, in Oporto as a tourism destination, in particular, to analyse the existence of positive relationships, such as the reasons for the trip and the assistance to a football game, with visits to the Dragão Stadium and the FC Porto Museum. Considering the general objective, two research hypotheses were proposed: H1: The main reasons for the trip influence the visit to the Dragão Stadium and the FC Porto Museum; and H2: The attendance to a football game at Dragão Stadium, influence the visit to the Dragão Stadium and the FC Porto Museum. Visioning the research objectives, a quantitative methodology was applied through the application of a questionnaire to 400 tourists/visitors in the city of Oporto.

4 Results

Hypothesis H1 The main reasons for the trip influence the visit to the Dragão Stadium and the FC Porto Museum. To study this hypothesis, will be analyzed the relationship between Questions "Main reason(s) of the trip" with Questions "Have you ever visited Dragão Stadium?", "Have you ever visited the FCP Museum?" and "If so, what are the reasons for your visit?" (Table 1).

For the reasons "Visit to family and friends", "Sports Tourism" and "City & Short Breaks", there are statistically significant differences between those who have visited the *Dragão* Stadium and those who have not visited. The percentage that indicates the travel reasons "Visit to family and friends" and "Sports Tourism" is higher for those who have visited the *Dragão* Stadium; the percentage that indicates the reason for travel "City & Short Breaks" is higher for those who have not visited the *Dragão* Stadium and the differences observed are statistically significant (Table 2).

For the reasons "Visit to family and friends", "Sports Tourism" and "Touring Cultural", there are statistically significant differences between those who have visited the FCP Museum and those who have not visited. The percentage that points out the travel reasons "Visit to family and friends" and "Sports Tourism" is higher for those who have visited the FCP Museum, the percentage that indicates the reason for travel "Touring Cultural" is higher for those who have not visited the FCP Museum, and the differences are statistically significant. There are statistically significant positive relationships between: "Sports Tourism" and "Football lover"; "Cultural *Touring*" and "Cultural interest"; "Health & Wellness" and "Cultural interest" and "Architectural interest", and "Other" and "Cultural interest", which means that those who indicate the reason for the trip also values the visit to the FC Porto Stadium (Tables 3 and 4).

There are statistically significant positive relationships between: "Visit to family and friends" and "Curiosity" and "Trophies and titles won"; "Sports Tourism" and

Table 1 Descriptive statistics and tests t: visit the *Dragão* stadium

	Question	N	% Yes	Std. deviation (%)	T	P
Holidays/leisure	No	229	55.0	49.9	−0.806	0.421
	Yes	171	59.1	49.3		
Visit to family and friends	No	229	9.2	28.9	−3.549	0.000**
	Yes	171	21.6	41.3		
Sports tourism	No	229	38.4	48.7	−5.036	0.000**
	Yes	171	63.2	48.4		
Cultural touring	No	229	17.9	38.4	1.886	0.060
	Yes	171	11.1	31.5		
Business tourism	No	229	7.9	27.0	0.779	0.436
	Yes	171	5.8	23.5		
City and short breaks	No	229	10.5	30.7	2.906	0.004**
	Yes	171	2.9	16.9		
Health and wellness	No	229	2.6	16.0	0.178	0.859
	Yes	171	2.3	15.2		
Wine and gastronomy tourism	No	229	4.8	21.4	1.283	0.200
	Yes	171	2.3	15.2		
Another	No	229	9.6	29.5	1.855	0.064
	Yes	171	4.7	21.2		

$^*p < 0.05$; $^{**}p < 0.01$
Source Own elaboration based on SPSS outputs

"Historical interest" and "Trophies and titles won"; "Cultural Touring" and "Recognition of the Museum by the UNWTO"; "Health & Wellness" and "Cultural interest" and "Historical interest" and "Trophies and titles won", which means that those who indicate the reason for the trip also value the visit to the FC Porto Museum.

Hypothesis H2 Attendance at a football match at *Dragão* stadium influences the visit to Estádio do Dragão and the FC Porto Museum. To study this hypothesis, will be analyzed the relationship between Questions "Have you ever watched a football game at *Dragão* Stadium?" and "If so, what are the reasons for the assistance?" with Questions "Have you ever visited *Dragão* Stadium?", "Have you ever visited the FCP Museum?" and "If so, what are the reasons for your visit?" (Table 5).

For Question "Have you ever watched a football game at Dragão Stadium?" and the motive "Football lover", there are statistically significant differences between those who have visited the Dragão Stadium and those who have not visited. The percentage that "Have you ever watched a football game at The Dragão Stadium?" is much higher for those who have visited the Dragão Stadium; the percentage that marks the reason "Football lover" to watch a football game is higher for those who have visited the Dragão Stadium, being the differences observed statistically significant (Table 6).

Table 2 Descriptive statistics and tests *t*: visit to the FC Porto Museum

	Question	N	% Yes	Std. deviation (%)	T	P
Holidays/leisure	No	308	54.9	49.8	−1.388	0.166
	Yes	92	63.0	48.5		
Visit to family and friends	No	308	11.0	31.4	−3.648	0.000**
	Yes	92	26.1	44.2		
Sports tourism	No	308	40.3	49.1	−6.736	0.000**
	Yes	92	78.3	41.5		
Cultural touring	No	308	17.5	38.1	2.611	0.009**
	Yes	92	6.5	24.8		
Business tourism	No	308	7.5	26.3	0.669	0.504
	Yes	92	5.4	22.8		
City and short breaks	No	308	8.4	27.8	1.683	0.093
	Yes	92	3.3	17.9		
Health and wellness	No	308	1.9	13.8	−1.293	0.197
	Yes	92	4.3	20.5		
Wine and gastronomy tourism	No	308	4.2	20.1	0.905	0.366
	Yes	92	2.2	14.7		
Another	No	308	8.4	27.8	1.308	0.192
	Yes	92	4.3	20.5		

*$p < 0.05$; **$p < 0.01$
Source Own elaboration based on SPSS outputs

For Question "Have you ever watched a football game at *Dragão* Stadium?" and the motivation "Recognition of the club", there are statistically significant differences between those who have visited the FCP Museum and those who have not visited.

The percentage that "Have you ever watched a football game at *Dragão* Stadium?" is much higher for those who have visited the FCP Museum, the percentage that marks the reason "Recognition of the club" to watch a football game is superior for those who have visited the FCP Museum, and the differences are statistically significant.

There are statistically significant positive relationships between:

- "Have you ever watched a football game at Dragão Stadium?" and "Football lover", which means that those who have already watched a football match at *Dragão* Stadium also indicates the reason for the visit to the FC Porto Stadium.

There are statistically significant positive relationships between (Table 7):

- "Club recognition" and "Recognition of the club"; "Football Lover" and "Football lover"; "Curiosity" and "Cultural interest" and "Architectural interest" and "Curiosity"; "Interaction" and "Club recognition" and "Cultural interest" and

Table 3 Spearman correlation: if so, what are the reasons for your visit? Porto FC Stadium

$N = 171$		Recognized the club	Football lover	Cultural interest	Architectural interest	Curiosity
Holidays/leisure	Correlation Coef.	0.051	−0.022	−0.016	−0.030	−0.002
	Proof value	0.504	0.772	0.839	0.696	0.975
Visit to family and friends	Correlation Coef.	0.127	−0.034	−0.024	0.118	0.130
	Proof value	0.097	0.663	0.758	0.124	0.091
Sports tourism	Correlation Coef.	−0.028	0.275(**)	−0.012	0.014	−0.136
	Proof value	0.715	0.000	0.880	0.857	0.076
Cultural touring	Correlation Coef.	−0.075	−0.079	0.169(*)	0.103	0.010
	Proof value	0.333	0.303	0.027	0.180	0.893
Business tourism	Correlation Coef.	0.013	−0.017	0.021	0.064	0.126
	Proof value	0.865	0.821	0.787	0.403	0.102
City and short breaks	Correlation Coef.	−0.026	−0.051	0.054	0.045	−0.086
	Proof value	0.740	0.505	0.484	0.560	0.261
Health and wellness	Correlation Coef.	0.008	0.007	0.171(*)	0.305(**)	0.117
	Proof value	0.916	0.931	0.025	0.000	0.128
Wine and gastronomy tourism	Correlation Coef.	−0.069	0.094	−0.092	−0.056	−0.077
	Proof value	0.368	0.222	0.229	0.464	0.316
Another	Correlation Coef.	−0.099	−0.115	0.182(*)	0.006	0.098
	Proof value	0.197	0.133	0.017	0.943	0.203

$^*p < 0.05$; $^{**}p < 0.01$
Source Own elaboration based on SPSS outputs

"Architectural interest" and "Curiosity"; "Feeling/Emotion Felt" and "Club recognition" and "Football lover" and "Architectural interest" and "Curiosity"; "Entertainment/By Spectacle" and "Cultural interest"; "Architectural interest" and "Curiosity", which means that those who indicate the reason for watching a football match at *Dragão* Stadium also indicates the reason for the visit to the FC Porto Stadium.

There are statistically significant positive relationships between (Table 8):

- "Club Recognition" and "Football lover" and "Historical interest"; "Football Lover" and "Football lover"; "Curiosity" and "Prizes awarded"; "Interaction" and

Table 4 Spearman correlation: If so, what are the reasons for your visit?: FC Porto Museum

N = 94		Recognized the club	Football lover	Cultural interest	Historical interest	Curiosity	Recognition of the Museum by the UNWTO	Trophies and titles won	Prizes awarded
Holidays/leisure	Correlation Coef.	−0.003	0.090	−0.130	0.083	0.060	−0.086	0.020	0.160
	Proof value	0.977	0.389	0.214	0.430	0.567	0.414	0.850	0.126
Visit to family and friends	Correlation Coef.	0.129	−0.076	0.036	−0.002	0.256(*)	0.082	0.267(**)	0.134
	Proof value	0.217	0.468	0.733	0.988	0.013	0.435	0.010	0.199
Sports Tourism	Correlation Coef.	0.191	0.191	0.165	0.207(*)	−0.038	−0.018	0.250(*)	0.059
	Proof value	0.067	0.067	0.114	0.046	0.718	0.866	0.016	0.577
Cultural Touring	Correlation Coef.	0.012	−0.171	−0.029	−0.204(*)	0.018	0.232(*)	−0.018	−0.146
	Proof value	0.911	0.101	0.783	0.050	0.861	0.025	0.867	0.162
Business tourism	Correlation Coef.	−0.122	0.077	0.202	0.110	0.155	0.097	0.116	0.204
	Proof value	0.243	0.462	0.052	0.294	0.137	0.356	0.268	0.050
City & Short Breaks	Correlation Coef.	−0.246(*)	−0.119	−0.020	−0.016	0.146	−0.056	−0.012	0.042
	Proof value	0.017	0.256	0.848	0.877	0.164	0.594	0.907	0.692
Health & Wellness	Correlation Coef.	−0.064	0.157	0.267(**)	0.273(**)	0.092	−0.065	0.279(**)	0.131
	Proof value	0.540	0.132	0.010	0.008	0.381	0.536	0.007	0.209

(continued)

Table 4 (continued)

N = 94		Recognized the club	Football lover	Cultural interest	Historical interest	Curiosity	Recognition of the Museum by the UNWTO	Trophies and titles won	Prizes awarded
Wine and Gastronomy Tourism	Correlation Coef.	0.110	−0.200	0.187	−0.115	−0.097	−0.045	−0.113	−0.083
	Proof value	0.294	0.055	0.073	0.272	0.353	0.665	0.283	0.432
Another	Correlation Coef.	−0.175	−0.064	0.158	−0.055	−0.024	−0.065	−0.051	0.007
	Proof value	0.093	0.540	0.131	0.599	0.822	0.536	0.628	0.949

$* p < 0.05$; $** p < 0.01$
Source Own elaboration based on SPSS outputs

Table 5 Descriptive statistics and tests t: *Dragão* Stadium

	Question	N	% Yes	Std. deviation (%)	T	P
Have you ever watched a game football stadium at Dragão Stadium?	No	229	7.0	25.5	−21.270	0.000**
	Yes	171	78.4	41.3		
Club recognition	No	16	37.5	50.0	−1.055	0.293
	Yes	134	51.5	50.2		
Football lover	No	16	50.0	51.6	−2.172	0.031*
	Yes	134	75.4	43.2		
Curiosity	No	16	25.0	44.7	1.572	0.118
	Yes	134	11.2	31.6		
Interaction	No	16	25.0	44.7	1.341	0.182
	Yes	134	12.7	33.4		
Feeling/emotion felt	No	16	12.5	34.2	−1.672	0.097
	Yes	134	32.8	47.1		
Entertainment/For the Spectacle	No	16	37.5	50.0	0.823	0.412
	Yes	134	27.6	44.9		

$^*p < 0.05$; $^{**}p < 0.01$
Source Own elaboration based on SPSS outputs

Table 6 Descriptive statistics and tests t: FC Porto Museum

	Question 15	N	% Yes	Std. deviation (%)	T	P
Have you ever watched a game football stadium at Dragão Stadium?	No	308	24.4	43.0	−11.425	0.000**
	Yes	92	81.5	39.0		
Club recognition	No	75	36.0	48.3	−3.548	0.001**
	Yes	75	64.0	48.3		
Football lover	No	75	68.0	47.0	−1.281	0.202
	Yes	75	77.3	42.1		
Curiosity	No	75	12.0	32.7	−0.244	0.808
	Yes	75	13.3	34.2		
Interaction	No	75	13.3	34.2	−0.234	0.815
	Yes	75	14.7	35.6		
Feeling/Emotion felt	No	75	24.0	43.0	−1.778	0.078
	Yes	75	37.3	48.7		
Entertainment/For the Spectacle	No	75	28.0	45.2	−0.179	0.858
	Yes	75	29.3	45.8		

$^*p < 0.05$; $^{**}p < 0.01$
Source Own elaboration based on SPSS outputs

Table 7 Spearman correlation: If so, what are the reasons for your visit?: FC Porto Stadium (*Dragão* Stadium)

		Recognized the club	Football lover	Cultural interest	Architectural interest	Curiosity
Have you ever watched to game on the Dragão?	Correlation Coef.	0.100	0.226(**)	−0.041	0.014	−0.236(**)
	Proof value	0.192	0.003	0.597	0.851	0.002
Recognition of the club	Correlation Coef.	0.552(**)	−0.111	0.051	0.127	0.113
	Proof value	0.000	0.201	0.557	0.143	0.193
Football lover	Correlation Coef.	−0.087	0.667(**)	−0.025	0.157	0.045
	Proof value	0.320	0.000	0.775	0.070	0.606
Curiosity	Correlation Coef.	0.071	0.072	0.391(**)	0.307(**)	0.516(**)
	Proof value	0.415	0.411	0.000	0.000	0.000
Interaction	Correlation Coef.	0.202(*)	0.091	0.190(*)	0.275(**)	0.344(**)
	Proof value	0.019	0.293	0.028	0.001	0.000
Feeling/Emotion	Correlation Coef.	0.222(**)	0.175(*)	0.104	0.282(**)	0.331(**)
	Proof value	0.010	0.044	0.234	0.001	0.000
Entertainment/For the Spectacle	Correlation Coef.	0.150	−0.040	0.254(**)	0.390(**)	0.303(**)
	Proof value	0.083	0.644	0.003	0.000	0.000

$^*p < 0.05$; $^{**}p < 0.01$
Source Own elaboration based on SPSS outputs

"Recognition of the Museum by OMT" and "Prizes awarded"; "Feeling/Emotion Felt" and "Football lover" and "Cultural interest" and "Trophies and titles won" and "Prizes awarded"; "Entertainment/By Spectacle" and "Curiosity" and "Trophies and titles won" and "Prizes awarded", which means that those who indicate the reason for watching a football match at *Dragão* Stadium also indicates the reason for the visit to the FC Porto Museum.

5 Conclusions

The present research had as general objective to analyse the role of football, in particular, the Porto Football Club, in Oporto as a tourism destination. Regarding the decision-making process of the place of destination, Latiesa, and Paniza (2006) state that it is essential for sports tourists to consider the motivational intensity in choosing the destination, as well as its positioning at national and international level, not forgetting the sports offer of the destination itself. It is verified that there are

Table 8 Spearman correlation: If so, what are the reasons for your visit? FC Porto Museum

		Recognized the club	Football lover	Cultural interest	Historical interest	Curiosity	Recognition of the Museum by the UNWTO	Trophies and titles won	Prizes awarded
Have you ever watched to a game on the Dragão?	Correlation Coef.	0.092	0.149	0.110	−0.013	0.084	0.053	0.146	0.081
	Proof value	0.382	0.155	0.294	0.904	0.422	0.612	0.163	0.443
Recognition of the club	Correlation Coef.	0.648(**)	0.260(*)	0.065	0.234(*)	−0.081	0.145	0.159	0.054
	Proof value	0.000	0.024	0.577	0.043	0.490	0.214	0.174	0.648
Football lover	Correlation Coef.	−0.045	0.380(**)	−0.063	0.023	−0.107	0.064	0.117	0.096
	Proof value	0.701	0.001	0.592	0.846	0.363	0.584	0.317	0.414
Curiosity	Correlation Coef.	−0.055	0.101	0.149	0.103	0.151	0.144	0.160	0.313(**)
	Proof value	0.636	0.389	0.203	0.381	0.195	0.218	0.170	0.006
Interaction	Correlation Coef.	0.133	0.123	0.188	0.148	0.200	0.256(*)	0.200	0.278(*)
	Proof value	0.254	0.294	0.107	0.207	0.085	0.027	0.085	0.016
Feeling/ Emotion felt	Correlation Coef.	0.195	0.234(*)	0.304(**)	0.088	0.121	0.131	0.608(**)	0.248(*)
	Proof value	0.094	0.043	0.008	0.452	0.303	0.261	0.000	0.032
Entertainment	Correlation Coef.	0.083	0.065	0.173	0.048	0.437(**)	0.196	0.430(**)	0.298(**)

(continued)

Table 8 (continued)

		Recognized the club	Football lover	Cultural interest	Historical interest	Curiosity	Recognition of the Museum by the UNWTO	Trophies and titles won	Prizes awarded
For the Spectacle	Proof value	0.480	0.578	0.138	0.685	0.000	0.092	0.000	0.009

*p < 0.05; **p < 0.01
Source Own elaboration based on SPSS outputs

no studies relating to the sports club FC Porto and in particular, with regard to this investigation hypothesis (H1), as well as no researches were found applied to the destination Porto and its football club FC Porto. Only studies were found that relate other destinations to their football clubs such as Barcelona, Manchester, and Madrid (Magalhães et al., 2017). There is also a notorious shortage of studies that relate the motivation of travel to a tourist destination and the visit to the equipment of the respective sports club, reasons why this original hypothesis is considered. It is also considered that it will be crucial to extend and deepen the study of the influence of the motivation of the trip on the visit to the respective sports equipment measured in H2 hypothesis. From the results obtained, it is important to value for further analysis, the relationship between the motivation "Sports Tourism" and the historical interest and trophies and titles won by the club, as well as the relationship between the motivation *"Cultural Touring"* and the visit to the Museum of FC Porto for its recognition by the World Tourism Organization, going against the studies of Carvalho and Lourenço (2009) and Pigeassou (2004).

In the studies of Carvalho and Lourenço (2009), Magalhães (2012), Pigeassou (2004), Robinson and Gammon (2004) and Semedo (2015), sporting events can associate a character, not only concerning sports, but also cultural and social, in that it promotes the visit to attractions related to sport, and can raise the sports base to cultural perspectives, regarding other existing equipment. There are no studies concerning Porto Football Club, which relate, in particular, the assistance to the football game at *Dragão* Stadium, both with its visitors and to the Museum of FC Porto. It was considered, therefore, determinant to extend and deepen the study that relates the attendance to a football game at *Dragão* Stadium, with the visit to the Stadium and the Museum of FC Porto. From the results obtained it was possible to highlight the motivations "Football Lover", "Recognition of the club" and "Feeling/Emotion felt" for the visit to the *Dragão* Stadium. Regarding the Visit to the Museum for those who attended a football game, the relationship between "Curiosity" and "Prizes awarded; "Interaction" and "Recognition of the Museum by UNWTO"; "Cultural interest" and Trophies and titles won" and "Prizes awarded".

The world's notoriety of Porto Football Club demonstrates its importance in the development and projection that the city of Oporto has had in recent years, due not only to the titles, but also to the development of the club's infrastructure heritage, the construction of the *Dragão* Stadium and the Museum (distinguished by UNWTO), and the affirmation for quality, of the FC Porto brand on a global scale, in which players, club coaches, and managers continue to be recognized by the most relevant institutions in the sporting world. Regional and local DMOs must promote partnerships and collaborative projects to identify opportunities and redefine supply, thus considering other areas of the city that are not currently visited but can take on great potential. The benefits of this initiative would be the growth of the city as a tourism destination and the reduction of the pressure of tourist demand around the historic centre of Porto. It is precisely in this challenge that sports tourism can be valued and can also contribute to this tourism growth, visioning the international destination competitiveness.

Acknowledgements The authors acknowledge the financial support of CiTUR, R&D unit funded by the FCT—Portuguese Foundation for the Development of Science and Technology, Ministry of Science, Technology and Higher Education, under the scope of the project UID/BP/04470/2020.

References

Alegrias, L. (2017). O futebol na construção das representações identitárias nos museus. *Cadernos de Sociomuseologia. Edições Universitárias Lusófonas, 54*(10), 135–162. https://revistas.ulusof ona.pt/index.php/cadernosociomuseologia/article/view/5952

Bauer, H., Stokburger-Sauer, N., & Exler, S. (2008). Brand image and fan loyalty in professional team sport: A refined model and empirical assessment. *Journal of Sport Management, 22*(2), 205–226. https://doi.org/10.1123/jsm.22.2.205

Carvalho, P., & Lourenço, R. (2009). Turismo de prática desportiva: um segmento do mercado do turismo desportivo. *Revista Portuguesa de Ciências do Desporto, 9*(2), 122–132. https://www.scielo.mec.pt/scielo.php?script=sci_arttext&pid=S1645-05232009000200014&lng=pt

Chun, S., Gentry, J., & McGinnis, L. (2005). Ritual aspects of sports consumption: How do sports fans become ritualized? *AP-Asia Pacific Advances in Consumer Research, 6*, 331–336. https://www.acrwebsite.org/volumes/11930/volumes/ap06/AP-06

Coakley, J., & Souza, D. (2013). Sport mega-events: Can legacies and development be equitable and sustainable? *Motriz: Revista de Educação Física, 19*(3), 580–589. https://doi.org/10.1590/S1980-65742013000300008

Constantino, J. (2013). Dentro e fora do estádio: o espetáculo, a globalização e o seu significado social: o espaço desportivo globalizado. *Revista USP. N° 99*, 31–44. https://doi.org/10.11606/issn.2316-9036.v0i99p31-44

Costa, C. (2015). *O turismo como arena da globalização*. Editora Observare. Janus. https://hdl.han dle.net/10071/10857

Cox, R. (1993). Annual bibliography of publications and thesis on the history of sport in Britain. *The Sports Historian, 13*, 93–114. https://doi.org/10.1080/17460269309446384.

Ferreira, E., & da Silva, L. (2017). Turismo futebolístico: Perfil e motivações do torcedor viajante que frequenta o "novo" Mineirão. *Revista Brasileira De Ciências Do Esporte, 39*(3), 298–275. https://doi.org/10.1016/j.rbce.2017.02.014.

Gammon, S., & Robinson, T. (2003). Sport and tourism: A conceptual framework. *Journal of Sport Tourism, 8*(1), 21–26. https://doi.org/10.1080/14775080306236.

Getz, D. (2008). Event tourism: Definition, evolution, and research. *Tourism Management, 29*(3), 403–428. https://doi.org/10.1016/j.tourman.2007.07.017.

García, C. (2006). La Diosa Blanca y el Real Madrid. Celebraciones deportivas y espacio urbano. *Revista de Dialectología y Tradiciones Populares, IX*(2), 191–208. https://doi.org/10.3989/rdtp.2006.v61.i2.21

Latiesa, M., & Paniza, J. (2006). Turistas desportivos: Una perspectiva de análisis. *Revista Internacional De Sociología., 64*(44), 133–149. https://doi.org/10.3989/ris.2006.i44.31.

Magalhães, M., Horta, P., Valente, L., & Costa, J. (2017). Sports museums as part of the touristic and cultural itineraries: The case of FC Porto and the Dragão stadium. *Worldwide Hospitality and Tourism Themes, 9*(6), 669–674. https://doi.org/10.1108/WHATT-09-2017-0048.

Magalhães, P. (2010). *Percepções e Práticas de Responsabilidade Social Empresarial no Futebol Profissional Português: O Caso dos Três Grandes*. Master Thesis. ISCTE Business School. Instituto Universitário de Lisboa. https://hdl.handle.net/10071/2843

Manzano, A. (2014). La comercialización del producto "turismo deportivo". *Revista Dimensión Empresarial, 12*(2), 46–58. https://www.scielo.org.co/pdf/diem/v12n2/v12n2a04.pdf

Noite, A. (2013). *Marketing Desportivo*. Master Thesis. Faculdade de Desporto da Universidade do Porto. https://repositorio-aberto.up.pt/handle/10216/70113

Paulico, F. (2008). Marketing Desportivo no Pódio. *Gestin: Revista Científica da Escola Superior de Gestão do Instituto Politécnico de Castelo Branco*. Ano VII, nº 7, 113–121. https://hdl.han dle.net/10400.11/251

Pereira, F. (2015). *A Importância da Marca nas Organizações Desportivas: O Caso Benfica*. Master Thesis. Universidade Europeia - Laureate International Universities. https://hdl.handle.net/10400. 26/9643

Pereira, M. (2016). *Dragon Force International Clinics – do Planeamento à Concretização de um Programa Internacional de Futebol*. Master Thesis. Faculdade de Desporto da Universidade do Porto. https://sigarra.up.pt/reitoria/pt/pub_geral.pub_view?pi_pub_base_id=135622

Pereira, P. (2013). *Marketing Desportivo Digital: A importância do marketing digital para os clubes desportivos – Estudo de Caso F.C. Porto*. Master Thesis. Escola de Economia e Gestão. Universidade do Minho. https://hdl.handle.net/1822/28437

Pigeassou, C. (2004). Contribution to the definition of sport tourism. *Journal of Sport Tourism, 9*(3), 287–289. https://doi.org/10.1080/1477508042000320205.

Ramos, H. (2013). *A identidade da marca versus imagem: Caso Futebol Clube do Porto*. Master Thesis. Faculdade de Economia da Universidade do Porto. https://repositorio-aberto.up.pt/han dle/10216/70777

Robinson, T., & Gammon, S. (2004). A question of primary and secondary motives: Revisiting and applying the sport tourism framework. *Journal of Sport Tourism, 9*(3), 221–233. https://doi.org/ 10.1080/1477508042000320223.

Rodrigues, P., & Dávila, J. (2007). *Turismo desportivo: benefícios da generalização da participação*. In 8º congresso da Associação Portuguesa de Gestores de Desporto. Aveiro. https://hdl.handle. net/10198/7535

Semedo, M. (2015). *Avaliação do potencial do turismo marítimo-desportivo em Cabo Verde: Uma análise a partir da população residente*. Master Thesis. Escola Superior de Educação do Instituto Politécnico de Coimbra. https://hdl.handle.net/10400.26/13429

Silva, P. (2013). *Avaliação da qualidade do serviço prestado num estádio de futebol: o caso do F.C. do Porto*. Master Thesis. IPAM - The Marketing School. Escola Superior do Porto. https://hdl. handle.net/10400.26/15360

Stornino, C., Chagas, M., Moutinho, M., & Leite, P. (2016). A Nova Recomendação da UNESCO sobre Museus Colecções sua Diversidade e Função Social. *Informal Museology Studies*, nº 13, 1–51. https://ces.uc.pt/myces/UserFiles/livros/1097_arecomenda%E7%E3ounesco2015.pdf

Sustainable Tourism

Case Study Protocol for the Analysis of Sustainable Business Models

Joaquin Sanchez-Planelles and Marival Segarra-Oña

1 Introduction

During the last years, a growing number of companies around the world have implemented different environmental practices due to the consumers behaviour's changes, stakeholders' pressures and ESG requirements. Companies adopt those practices on different areas and levels of their companies: into their strategy, business model and processes (Eccles et al., 2014; Segarra-Oña et al., 2016). Furthermore, the progressive rate of adoptions of these practices amongst existing companies and the generation of new sustainable business models has lead academics to focus their efforts on finding out which practices can lead to a performance improvement for companies (Ioannou & Serafeim, 2019).

The increasing concern about sustainable development has motivated the creation of a large body of research called corporate sustainability that only during 2019 lead to the publication of 3.338 research papers (from Web of Science records). Although those publications have the corporate sustainability topic in common, they are focused on several different units of analysis. For this reason, and heading the same way as research in management, we have considered necessary to review this type of literature in order to identify which are the best practices or combination of practices that work as causal mechanisms to generate improvement for the companies that implement them. Therefore, we are working on a project with the aim to create a Theory of Sustainability according to the theory building process proposed by Carlile and Christensen (2005) (see Table 1).

According to these authors, the development of a theory about sustainability needs to carry out a careful review of the phenomena through the literature review, but also examining outcomes extracted from corporations via a case study procedure.

J. Sanchez-Planelles (✉) · M. Segarra-Oña
Universitat Politècnica de València, Valencia, Spain
e-mail: joasanpl@ade.upv.es

© The Author(s), under exclusive license to Springer Nature Switzerland AG 2021
V. Katsoni and C. van Zyl (eds.), *Culture and Tourism in a Smart, Globalized, and Sustainable World*, Springer Proceedings in Business and Economics, https://doi.org/10.1007/978-3-030-72469-6_10

Table 1 Phases of the process of building a descriptive theory

Descriptive theory		
Phase 1	Phase 2	Phase 3
Observation and description of the phenomena (creation of 'constructs')	Attributes categorisation of the phenomena	Statements of correlation between categories

Adapted from Carlile and Christensen (2005)

Table 2 Steps of the research process that frames the elaboration of this case study protocol

Research process					
Step 1	Phase 2	Phase 3	Phase 4	Phase 5	Phase 6
Observation based on literature review	Construction of concepts about knowledge bodies about corporate sustainability	Draft of the sustainable theory based on the concepts identified that will be in a descriptive stage	Creation of a case study protocol for retrieving data from companies about the integration of sustainability in the organisation	Identification of associations between concepts through the application of the case study protocol in sustainable companies and literature review	Transition from a transition stage to normative stage of the sustainable theory

Several authors (Yin, 1993; Johnson et al., 1999; Hillebrand et al., 2001; Myers, 2013) state that case study research is an appropriate methodology to identify the existence or absence of phenomenon under specific conditions, especially for matters related with social sciences (Table 2).

Hence, the aim of this chapter is to provide academics with a case study protocol embedded into the case study research about the Theory of Sustainability which will allow them to gather data about the implementation of sustainability among corporations and its outcomes.

According to Johnson et al. (1999), case research studies are suitable for testing theory in social sciences whether replace statistical correlations with logical argumentation. So the logical argumentation plays an essential role for supporting causal relationships for the application of the theoretical generalisation in order to expect outcomes on those cases or companies with similar structure.

1.1 Building a Sustainability Theory

The sustainability theory has the aim to prove the following hypothesis:

The companies that develop specific sustainable practices, which target certain areas of the business, can get better results in terms of turnover, profitability and/or performance.

The observation process through a careful review of the literature has led to the identification of four bodies of knowledge about the type of research about on corporate sustainability and its implications in corporations that have been named in the following concepts (Sanchez-Planelles, et al. 2021) (Table 3).

This concepts' clarification is a useful framework for developing a case study protocol as they allow researchers to envisage the companies from a broad point of view (holistic sustainability) to an operational level detail (sustainable operations). In addition, taking into account those concepts during the case study process makes it easier to keep a flow of questions in order to get all the relevant data from the interviewee. Therefore, with the aim of moving forward with the Theory of Sustainability, and once the observation process is finished (it has been deployed through a literature review), the next stage of the theory building process will consist in the identification of correlations between concepts and outcomes (Carlile & Christensen, 2005), which will eventually lead to the conclusion of the stage of building a descriptive theory.

The process of correlations' identification will be developed based on a combination of the literature review of papers about sustainable-oriented innovation and

Table 3 Definition of the concepts of the sustainability theory

Concept	Description
Holistic Sustainability Porter and Kramer (2011), Nidumolu et al. (2009), Ioannou and Serafeim (2019)	Policies with a long-term vision and a broad perspective that encompasses sustainable actions for reshaping the interaction of the company with its stakeholders
Sustainable Business Models Lüdeke-Freund (2010), Schaltegger et al. (2016), Bocken et al. (2014)	Business model that creates competitive advantage through superior customer value and contributes to a sustainable development of the company and society
Sustainable Methodologies Joyce and Paquin (2016), França et al. (2016), Bocken et al. (2013), Rodríguez-Vilá and Bharadwaj (2017)	Methodologies and tools designed for managers to improve the company's performance and sustainability
Sustainable Operations Segarra-Oña (2012), Cheng et al. (2014)	Activities and business processes that reduce the environmental impact that only involve specific areas of the organisation (i.e. product development, waste management, eco-innovation, etc.)

Source Own elaboration based on the academic literature

business case studies. The end of that stage will allow us to advance to the prescriptive phase, which will be specifically focused on analysing the causal mechanism inductor of the result.

The hypothesis states that there are groups of environmental practices that can be gathered according to their characteristics and the proper combination of them at different levels of the organisations can lead to better results. On one hand, the unit of analysis of this research extends to those private organisations that have implemented sustainable practices in a successful way, so they are regarded as environmentally friendly companies by prestigious independent organisations. On the other hand, this protocol has been designed for developing a multiple case study.

2 Methodology

This case study protocol has been designed according to the guidelines stated by Yin (2003), and it has been deployed a literature review of other case study protocols throughout the selection of the next keywords and data strings: protocol case research, case study research and sustainability in Web of Science or Scopus.

In addition, a deep comprehension of the integration of sustainability among the business model needs researchers to know what the company offers to the market and how the activities and strategy are executed for offering value to its customers (Esty & Winston, 2006). Therefore, another literature review has been deployed for retrieving those frameworks used in the field of management that can help to know how sustainable practices create value through the value chain. The keywords selected for this research were business framework, canvas, value chain and analysis tool.

Then, the techniques incorporated in this protocol had to match the following criteria:

- Should be widely used in the management field for facilitating its used by researchers with different backgrounds and experience.
- These techniques should offer enough broadness to reflect the sustainable practices carried out by the companies that were selected for the study.

There is a discussion between defenders of that just one case is enough to achieve a theoretical generalisation through a logical argumentation (Yin, 2013) between those that state that replications will increase the reliability of the research and its results (Hillebrand et al., 2001). For this theory building process, we have assumed to replicate this process through this case study protocol in several firms of the following types of businesses in order to cases be considered structurally similar valid:

- B2B business.
- B2C business.
- Product sellers.
- Services sellers.

2.1 Case Study Protocol

Yin (1993) defends the use of a case study protocol for establishing the criteria of the data collection. The protocol designed for this purpose combines two main sources of data: interviews with a variety of informants and the use of documentation.

Interviews play an essential role for the case study protocol because they are the main source of qualitative data in the field (Walsham, 2006). Interviews will be semi-structured, open-ended questionnaires and documentations will need to be scrutinised in order to avoid a desired image projected by the company. The combinations of these techniques will allow the triangulation that confers reliability to the study.

However, before kicking off each interview, researchers will need to retrieve information about the company and also about its market niche or sector. Following Porter's (1985) guidelines, in order to analyse a company researchers need to analyse the sector where the company competes and the competitive positioning of the company.

The following sections show the process of design of this case study protocol, which includes the interview's content and also the frameworks that will help researchers to gather information and structure the amount of data obtained during the fieldwork. Once the fieldwork is finished, data retrieved from the study cases will be analysed individually and a cross-case analysis will be conducted to highlight differences and similarities between cases through the triangulation methodology (Steckler et al., 1992; Patton, 1990).

2.2 Case Study Protocol Design

According to Yin (2003), a case study protocol should include an overview of the project that frames the protocol, field procedures, case study questions and a guide for the case study report. Once the company has been assigned, the researcher needs to look for information and data about the company. Drawing the business model canvas (Osterwalder & Pigneur, 2010) and Porter's value chain (Porter, 1985) will be useful tools for helping the researcher to learn what kind of business the company is and its market, which will also help to address better questions during the interview. Then, Porter's theory of Shared Value (2011) will show how the company creates value through its sustainable practices. In Table 4, there is a summary with the needed information.

2.3 Interview's Content

Case study questions have been designed as open-ended questions through a semi-structured interview.

Table 4 Information of the company that needs to be retrieved by the researcher before the interview

Business model		Value chain	
How the companies do business and what its value proposition is		Analysis of the degree of integration of sustainability in each activity of the company and the interrelation between activities	
Revenue model	Cost structure	Primary activities	Support activities
Target customers of the company, how it establishes relationships with them and its revenue streams	Main resources and activities needed to offer the value proposition to customers and their costs	Sustainable practices implemented in the logistics, production and how they are exposed through marketing and sales department	Degree of awareness about environment by the executives and rest of the staff. Research programmes that promote sustainability and purchasing based on sustainable premises
	Shared Value		
	Level of alignment between the way the company offers products / services and its relationship between the stakeholders		
	Internal	External	
	Products and services offered to customers and practices focused on an efficient use of resources	Channels for delivering products and services to customers and the relationship of the company with regional businesses, clusters, regulatory environment and institutions	

Adapted from Osterwalder and Pigneur (2010), Porter (1985) and Porter and Kramer (2011)

The content of the interview designed via this case study protocol has been designed after the analysis of sustainable methodologies showed in the next sections and concepts from the Theory of Sustainability presented above.

Nowadays, sustainability is a transversal idea that influences all the levels of organisations (Porter & Kramer, 2014; Ioannou & Serafeim, 2019), therefore researchers deploying a case study about sustainability need to understand all the interactions of the company between stakeholders, its strategic and organisational structure and the relationship between its employees.

For this reason, and according to the concepts about the Theory of Sustainability, the content of the interview of this protocol asks questions in order to get information in a holistic way about the presence of sustainability among the activities developed by the company.

The structure of the interview has two kind of ways to retrieve information:

- A set of questions that include the source of that information and some tips for helping researchers to obtain the data.

- Sustainable methodologies which are frameworks that can extract data from the company filling the required fields by the researcher.

The set of questions (see the Annex) have come out as outcomes from the concepts presented above. There are questions designed for understanding how the company deals with sustainable policies from a holistic point of view (stakeholders relationships, strategic plan or decision-making process). Another group of questions aims to figure out the type and the main points of the sustainable business model (if the company can be considered as a sustainable business model as a whole or if only certain departments of it are sustainable). Additionally, other bunch of questions uncovers the level of environmental awareness of the business processes. Lastly, there are questions that seek to know if managers and employees have used any methodology for implementing sustainable practices.

Furthermore, the sustainable methodologies that need to be used for completing this protocol are based on frameworks that should be filled by the researcher with the late interviewee's validation.

2.3.1 Holistic Sustainability

The first part of the interview seeks to understand the strategy and the purpose of the interviewee's company. It is necessary to know what motivated board members to adopt sustainable practices and if the company can be considered as a purpose-driven company. Then, once the strategy and purpose have been defined, the researcher needs to know how executives include environmental criteria on their daily decision-making process in order to know if the mission, vision and objectives are coherent with the strategy execution. Should there be a gap between the company's statement and its execution, this needs to be pointed out in the protocol.

Other reasons that determine the adoption of sustainable practices by companies have to do with the market they compete in. So, this protocol also takes in account aspects from the Stakeholders' Theory (Freeman, 1984) and the level of convergence of sustainable practices of market competitors (Ioannou & Serafeim, 2019) with the aim of identifying if the basis of the competition of the market are being modified because of the development of sustainable practices.

2.3.2 Sustainable Business Model

Companies can turn their business into a sustainable business model or even create sustainable business models within the company as a new business line (Schaltegger et al., 2016). The questions of this are focused to examine if the company under study can be considered as a truly sustainable business model or if some of its areas are developing sustainable practices, even they if they operate as silos.

Sustainable business models can be classified according to the following list (Technical Secretary of the Eco-innovation Laboratory, 2016):

- Circular economy.
- Sustainable production.
- Servitisation.
- Sustainable consumption.

The strategy deployed by the company will determine the definition of the business model and the resource allocation process (Bower, 1986). Hence, the strategy and business model of the company will determine which products can be launched to the market and what kind of market niches satisfy (Christensen & Johnson, 2009). During this step of the protocol, the researcher has to understand how the company is able to offer a superior value to customers by improving the society (Schaltegger et al., 2016). So, if the channels developed for delivering products and services to the customers have sustainable attributes, these need to be examined. In addition, the process in which sustainable practices and values are communicated to employees, customers and stakeholders needs to be recorded.

Finally, in order to know how the company deals with its supply chain and providers, questions about the green certificates or other measures related to the company's partners and providers will be required.

2.3.3   Sustainable Operations

The concept of sustainable operations refers to those activities with a specific scope that are not transversal to several departments. Usually, these are activities that reduce the environmental impact caused by logistics, production and supply chain management.

In the field of management, there are methodologies that are used for the identification of the activities that take part in the business model. One of the most used frameworks for analysing companies for the last years has been the business model canvas (Osterwalder & Pigneur, 2010). Nonetheless, Joyce and Paquin (2016) adapted the business model canvas adding two more layers with the aim to explore sustainable-oriented innovation activities among the business model. The first layer explores the environmental life cycle, and the second layer shows the social stakeholder relationships of the organisation (Chairul, 2019).

Another framework widely used on the field of management is the value chain designed by Porter (1985). The theory of shared value has become a well-known strategy (2013) associated with corporate sustainability and sustainable development. Porter's value chain can be a useful methodology in order to identify how companies can create value through sustainability. Therefore, identifying in the value chain the activities with a sustainable component and their relationship between others can show us the sources of sustainable value generation. Therefore, the analysis of these frameworks has led to the generation of questions about the activities, processes and operations carried out by sustainable companies.

Moreover, sustainable operations also refer to the activities that reduce the environmental impact of the supply chain management and the development of eco-innovative products.

2.3.4 Sustainable Methodologies

Another body of knowledge identified during the study has been the generation, test and improvement of methodologies to help managers to adopt sustainable practices or turn traditional business models into sustainable business models (Sanchez-Planelles & Segarra-Oña, 2019a, 2019b). This concept includes methodologies widely accepted like the life cycle assessment (Guinee, 2002) and methodologies or frameworks like the previously exposed 'Triple-Layered Canvas' exposed above (Joyce and Paquin, 2016). In addition, it is necessary to know if the company has implemented any green certificate (e.g. ISO 14001, EMAS or ecologic labels).

2.3.5 Corporate Sustainability Stages

Environmental practices carried out by companies have evolved during last years and can be grouped according to their similarities (Segarra-Oña et al., 2012). The sustainable strategies' evolution is represented in Table 5. During last decades, companies have dealt with environmental protection in different ways. Some decades ago, public administration established environmental requirements that companies had to accomplish, and currently, there are companies launching sustainable business models or even creating new business platforms. So, the evolution of the corporate sustainability can be classified in four stages according to the environmental policies deployed by companies (Nidumolu et al., 2009; Sanchez-Planelles & Segarra-Oña, 2019a, 2019b):

This classification might be used as a roadmap to suggest the following steps that should be adopted by managers of the company studied. Existing companies can also launch new business units based on the sustainable business models principles or even transform its business into a sustainable business model.

Table 5 Evolution of corporate sustainability

Evolution of corporate sustainability				
Stage 1	Stage 2	Stage 3	Stage 4	Stage 5
Taking advantage of compliance opportunities	Developing a sustainable value chain	Implementing eco-innovative practices	Generating sustainable business models	Creation of new business platforms

Adapted from Nidumolu (2009), Sánchez-Planelles and Segarra-Oña (2019)

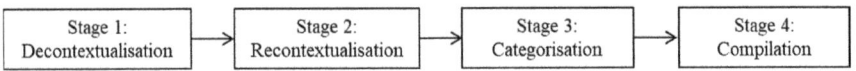

Fig. 1 An overview of the five stages of the process of a qualitative content analysis (Bengtsson, 2016)

3 Data Analysis and Results

The data obtained from interviews needs to be examined using the content analysis technique for drawing realistic conclusions.

Krippendorff (2004) defined content analysis as 'a research technique for making replicable and valid inferences from texts (or other meaningful matter) to the contexts of their use' (p. 18). Downe-Wambolt (1992) underlines that content analysis is more than a counting process, as the goal is to link the results to their context or to the environment in which they were produced: 'Content analysis is a research method that provides a systematic and objective means to make valid inferences from verbal, visual, or written data in order to describe and quantify specific phenomena' (p. 314).

Then, methodology moves forward through the following steps (Fig. 1):

- Decontextualisation: The data from the interviews needs to be broken down in 'meaning units'. A meaning unit is the smallest unit that contains some of the insights required to perform the research. Each identified meaning unit is labelled with a code, which should be understood in relation to the context.
- Recontextualisation: After the meaning units have been identified, it will be checked if all the aspects of the content have been covered in relation to the aim.
- Categorisation: Extended units will be condensed. Then, themes and categories will be identified. Categories must be rooted in the data from which they arise.
- Compilation: During this phase, the essence of the studied phenomenon will be found. The results will be presented as a summary of themes, categories/sub-themes and sub-categories/sub-headings.

At the end of those phases, the new findings will need to be checked in order to know how they correspond to the literature and identify if there is any issue.

4 Empirical Application of the Case Study Protocol in a Grocery Retailing Company

The first test of this Case Study Protocol for the development of the sustainability theory has been deployed in a Spanish grocery retailing company based in Valencia (Spain) called Consum.[1]

---

[1]https://www.consum.es/.

Consum is a firm organised as a cooperative. It was founded in 1975, and it has been growing steadily reaching 755 supermarkets (447 owned by Consum and 283 franchised). Its turnover was 2.9356 millions of euros and employed 16.031 people in 2019. During the research process, all the sustainability memories published by the company were analysed (from 2007 to 2018), it was retrieved information from the website and also from press notes that have been released in national and regional mass media. The interview was performed with a member of the Corporate Social Responsibility member on 27th, March and took 51 min.

The following table shows the information about the sustainable practices developed by Consum according to the data structure proposed by this case study protocol (Table 6).

4.1 Recommendations

According to the corporate sustainability classification, the ESG department of Consum needs to set objectives that seek to fully develop sustainable practices that lead the company to move from the stage of sustainable value chain to the stage of sustainable business model based on the sustainable production archetype. So, it will need to develop the following sustainable practices:

- Increase of the number of ecologic products offered in the supermarkets.
- Application of the life cycle assessment of the suppliers' products. After a careful analysis, implementation of practices that allow an environmental impact reduction (e.g. lightening the packaging, energy efficiency processes during the production, etc.).
- Establishment of purchasing policies that gives weight to the sustainable attributes of potential suppliers.
- Continuous implementation of energy efficiency practices in the supermarkets, logistics centres and offices.
- Trainings to employees to increase their environmental awareness.
- Marketing campaigns for positioning the company as a first mover implementing sustainable practices among the retail sector.

5 Conclusions, Limitations of the Study and Future Research

This protocol seeks to retrieve sufficient information to allow researchers to understand how companies can create value through sustainability. The protocol is composed by three phases:

Table 6 Information about sustainable practices developed by Consum according to the case study protocol structure

Business model		Value chain		Shared Value	
Chain of supermarkets that offers a wide range of local products, especially promoting fresh products, leading this market niche		The key activities of Consum are the transportation of the products from suppliers to supermarkets via logistics centres, the chain of supermarket and its advertisement activities		Consum has deployed sustainable practices from an internal point of view (e.g. reducing environmental impact generated by its operations) and from an external point of view (e.g. selling ecologic products)	
Revenue model	Cost structure	Primary activities	Support activities	Internal	External
• Incomes generated by customers' purchases in supermarkets • Customers interact with Consum through supermarkets' employees and social networks • Main channels based on goods transportations to supermarkets and home shopping	• Costs come from goods acquisition, supermarket maintenance and labour costs • Logistics plays a key role in order to optimise delivery and transportation of products in order to save costs • Assets like supermarkets, logistics centres and vehicles	• Consum reduces the environmental impact applying sustainable practices in the logistics activities (goods transportation) and implementing energy efficiency activities in supermarkets • Consum has a programme for recovering waste generated in the logistics flows	• There is a ESG department that establishes objectives for being matched by the managers based on the materiality index • ESG department works for offering ecologic products, increasing the employees awareness about environment and establishing relationship with stakeholders based on sustainable concerns	Consum has focused its efforts to become sustainable in the reduction of emissions during goods transportation and its carbon footprint improving the energy efficiency of supermarkets	The company is increasing the number of ecologic products offered to customers and has strong relationships with NGO's for donating food and beverages for non-resources families

(continued)

Table 6 (continued)

Stage of the corporate sustainability evolution

STAGE 1—Compliance	STAGE 2—Sustainable Value Chain	STAGE 3—Eco-innovation	STAGE 4—Sustainable Business Model	STAGE 5—New Business Platforms
• Consum accomplishes the environmental requirements	• 83.6% of carbon footprint reduction since 2015 (through the application of energy efficiency practices) • 99% of national suppliers, 66% of them are placed near the facilities • 98% of renewable electricity consumption • LED lighting installations • Logistics programs for optimising routes and reducing the journeys (TEO and Nodriza)	• 239 ecologic products, 57 of them of the Consum brand • Fleet of 319 eco-efficient vehicles • Reduction of the environmental impact of the packaging of the Consum brand products' through the reduction of grammages and implementation of sustainable materials	• The internal business unit 'Residuo Cero' (Zero Waste) recovers 99% of packaging waste, containers, pallets and other waste that are generated from the Consum's logistics centres • Consum has donated 6.900 Tn of products (packaged products, from the deli, meat, fruit, vegetable, sweets and dairy products) have been donated via the program 'Profit'	• At the moment, there are no evidences of the advance of Consum forward a new business platform

Sustainability theory concepts

(continued)

Table 6 (continued)

CONCEPT 1—Holistic Sustainability	CONCEPT 2—Sustainable Business Model	CONCEPT 3—Sustainable Methodologies	CONCEPT 4—Sustainable Operations
• There is a ESG Department that establishes objectives depending on the results of the materiality matrix • Objectives about sustainability are communicated to the rest of the company's departments • The main reporting index used for monitoring the evolution of these objectives is the Global Reporting Initiative (GRI) • Board members drive employees and associates to become more sustainable and adopt these initiatives, however, the providers that offer products in the Consum's supermarkets are not audited or elected according to their sustainable practices	• Although Consum cannot be considered as a sustainable business model, the 'Zero waste' project for the waste reduction through its recover can be classified as a circular economy business model • Consum's headquarters, supermarkets and logistics centres have installed energy efficiency practices like LED lighting and refrigerant gases with 0 global warming potential	• For the implementation of sustainable practices, managers have not implemented any methodology, neither methodologies published by academic researchers nor widely known methodologies like life cycle assessment	• Consum has been replacing its vehicles for eco-efficiency vans and trucks • Two projects that aim to reduce the amount of kilometres of the trucks during the transport of goods (from suppliers to logistics centres and from logistics centres to supermarkets) have been developed (Nodriza and TEO)

- Data extraction of the company from public sources that allow the researcher to understand the business model and value chain and also get some information about the sustainable practices.
- Interview with the company's employees with a set of questions classified depending on the unit of analysis of each one and the type of outcomes expected to obtain.
- Use of the content analysis methodology and data triangulation between the cases studied.

In addition, the use of this protocol seeks to standardise the research process for analysing how companies implement environmental practices and compare them with other companies (from the same sector or from other sectors). This protocol has a wide scope in order to understand the sustainable positioning of the company in its market and how the market values the sustainable attribute. So, the main limitation is the difficulty to get details of specific business units. This might lead to miss some information, especially from the operative level with the aim of learning about the integration of sustainability practices from a strategic point of view.

Additionally, a proper use of this protocol will need from researchers to have basic knowledge about managerial and strategic concepts like business model canvas (Osterwalder & Pigneur, 2010), value chain (Porter, 1985) or the Theory of Stakeholders (Freeman, 1984).

Acknowledgements We acknowledge Mr. Elías Amor Montiel from the ESG Department of Consum for his help and collaboration in the interview and providing us with data.

Appendix

Questions about Holistic Sustainability

Checking the company's sustainable policies against these questions will allow to identify how board members deal with the subject matter and know what kind of strategies is executing the company.

Question 1: What motivated the company to become sustainable? What is the company's purpose?

Source of data:

- CEO (Chief Executive Officer)
- CSO (Chief Strategy Officer)
- CSO (Chief Sustainability Officer)

Sample strategies:

- Mission, vision and values statements from website.
- Media press.

Question 2: How is the market dealing with sustainability? Is this company a leader, follower or laggard on the implementation of sustainable practices?

Source of data:

- Product manager
- Product manager assistant
- Sales manager
- Catalogue
- Products portfolio

Sample strategies:

- List the direct competitors of the company.
- Identify which attributes about sustainability their products have.
- Identify which attributes about sustainability those companies have.
- Gather data about when sustainable practices were announced by direct competitors in order to determine which one is the leader, follower and laggard.

Question 3: Analysis of the influence of sustainability in the decision-making process. Are environmental criteria taken into account during the decision-making process? List the environmental criteria used for the decision-making process. Is the company's statement about sustainability aligned with the decision-making process?

Source of data:

- CEO (Chief Executive Officer)
- CSO (Chief Strategy Officer)
- CSO (Chief Sustainability Officer)

Sample strategies:

- Identify the decision-making process established by the board members of the company.
- Estimate how much environmental concerns are taken into consideration during the decision-making process.
- Create a framework or diagram of the decision-making process.

Question 4: What is the process for detecting market needs focused on sustainable attributes?

Source of data:

- CMO (Chief Marketing Officer)
- Sales Manager
- CSO (Chief Sustainability Officer)

Sample strategies:

- Identify if there is a department focused on detecting market trends.
- Gather data about providers that offer services related to markets analysis, consumer studies, etc.

Question 5: Analysis of the relationship with stakeholders and the influence of sustainability in the relationship between company and stakeholders.

Source of data:

- CEO (Chief Executive Officer)
- CSO (Chief Strategy Officer)
- CSO (Chief Sustainability Officer)

Sample strategies:

- Determine what kind of relationships has the company with:

  - Capital market
  - Suppliers
  - Networks and associations
  - Policymakers
  - Research
  - Mass media
  - Business partners
  - Local stakeholders
  - Civil society and NGO's
  - Employers
  - Customers

Question 6: Identify if the company's board members establish environmental goals for the short, medium and long term.

Source of data:

- CEO (Chief Executive Officer)
- CSO (Chief Strategy Officer)
- CSO (Chief Sustainability Officer)

Sample strategies:

- UN Sustainable Development Goals
- Materiality matrix

Questions about Sustainable Business Models:

Checking the company's business model against these questions will allow to identify how the company creates superior value to customers improving the society and reducing the environmental impact.

Question 7: Identify which activities generate value through sustainability and determine flows of value through activities.

Source of data:

- CEO (Chief Executive Officer)
- CSO (Chief Strategy Officer)
- CSO (Chief Sustainability Officer)

Sample strategies:

- Draw the Porter's value chain and complete each activity
- Draw the value flows between activities from the value chain.

Question 8: Classify the sustainable business model developed by the company: circular economy, sustainable production, servitisation and sustainable consumption.

Source of data:

- CEO (Chief Executive Officer)
- CSO (Chief Strategy Officer)
- CSO (Chief Sustainability Officer)

Sample strategies:

- Draw the flows of inputs and outputs that take part in the business.
- Create an organisational chart of the company that shows the different business lines and possible sustainable business models within the company (e.g. circular economy procedures to revalorise waste).
- Whom does the business model supervise?
- What customer segment does the business model target? Is it targeting external or internal customers?

Question 9: Are eco-friendly products and/or services addressed to a specific market niche or are they launched to broad customer segments?

Source of data:

- Sales manager
- Product manager

Sample strategies:

- Reports about the market sector.
- News or press notes published in mass media.

Question 10: How does the company informs or communicates the sustainable practices to customers, users and other groups of interest?

Source of data:

- Chief of Staff
- CSO (Chief Sustainability Officer)

Sample strategies:

- Identify the channels used to deliver information: videos, seminars, conferences, online courses, short sessions, etc.

Question 11: Does the company consider the degree of sustainability of its providers or partners? If does, what are those criteria?

Source of data:

- Purchasing manager
- CSO (Chief Sustainability Officer)

Sample strategies:

- Identify the most key partners and providers of the company.
- Examine what criteria o requirements need to be matched in order to work with the company. For instance: ISO 14001, EMAS, green certificates, eco certificates, etc.

Questions about Sustainable Operations.

Checking the company's operations against these questions will allow to identify how business processes from the operational level might reduce the environmental impact.

Question 12: List the products and/or services which incorporate eco-innovative attributes.

Source of data:

- Product manager
- Product manager assistant
- Catalogue
- Products portfolio

Sample strategies:

- Draw a chart with the products and services managed by business line.
- Retrieve information about eco-innovative technologies and practices developed for the last three years.

- Check the eco-innovative practices that have been integrated in the company's products or services.

Question 13: What decision-making process or criteria is considered by the company to invest resources in the development and release of eco-innovative products / services?

Source of data:

- R&D manager
- Product manager
- Product manager assistant
- Catalogue
- Products portfolio

Sample strategies:

- Retrieve information about eco-innovative technologies and practices developed for the last three years and identify characteristics which are similar between each other.
- Identification of the customers' needs that try to solve the eco-innovative products / services.

Question 14: Identify the business areas that create value through sustainable activities.

Source of data:

- CEO (Chief Executive Officer)
- CSO (Chief Strategy Officer)
- CSO (Chief Sustainability Officer)

Sample strategies:

- Complete the business model canvas (Osterwalder & Pigneur, 2010).
- Complete the triple-layered business model canvas (Joyce & Paquin, 2016).

Question 15: Are the channels to deliver products and services to your customers sustainable?

Source of data:

- CLO (Chief Logistics Officer)
- CSO (Chief Sustainability Officer)

Sample strategies:

- Identify the channels used to deliver products and services to customers: vehicle fleet, logistics, shops, offices, stores, etc.

- Identify which eco-innovations or sustainable practices have been implemented in each channel. For instance, energy efficiency processes in cooling systems, eco-innovative trucks, etc.

Question 16: Have the company implemented any measure to reduce the environmental impact of its assets? For instance, energy efficiency measures in offices and production plants, emissions-reduction devices, etc.

Source of data:

- Production Manager
- Chief of Maintenance
- CSO (Chief Sustainability Officer)

Sample strategies:

- Identify the strategic assets of the company. For instance, production centre, factory, stores, shops, offices, vehicle fleets, fields, etc.
- Examine what environmental improvements have been implemented recently in the facilities and assets. For instance, acquisitions of eco-innovative production systems, installation of solar panels, implementation of sustainable architecture principles in the company's buildings, etc.

Questions about Sustainable Methodologies.

These questions will show if methodologies designed for implementing sustainable practices among companies are widely used by managers and which of them are the most commonly applied.

Question 17: Identify if the company has any green certificate (e.g. ISO 14,001, EMAS, BREAM, Ecologic label, etc.) or if it is working to achieve one.

Source of data:

- CSO (Chief Sustainability Officer)
- CEO (Chief Executive Officer)

Sample strategies:

- Information retrieved from website.

Question 18: Examine if the company works with any international standard to report its sustainable practices.

Source of data:

- CEO (Chief Executive Officer)
- CSO (Chief Strategy Officer)
- CSO (Chief Sustainability Officer)

Sample strategies:

Some of the most common international standards for measuring the implementation of sustainable practices are:

- GRI (Global Reporting Initiative)
- Rainforest Alliance
- ISO 26000

Question 19: Did the company use any framework or methodology to implement sustainable practices?

Source of data:

- CSO (Chief Sustainability Officer)

Sample strategies:

Some of the most common sustainable methodologies and frameworks are:

- Triple-Layered Canvas
- Framework for Strategic Sustainable Development
- RESTART
- Shareholder-value framework
- Value Mapping Tool

Questions about the Evolution of the Corporate Sustainability.

This question will classify the company in the stage of the corporate sustainability evolution and will enlighten the potential practices that might be deployed in order to move forward to a sustainable business model or a new business platform.

Question 20: According to the different stages of the corporate sustainability evolution, in which stage does the company fit?

Source of data:

- Sustainability memories of the company.
- CSO (Chief Sustainability Officer)

Sample strategies:

- Classify the company according to the following corporate sustainability stages:

  - Compliance
  - Sustainable Value Chain
  - Eco-innovative practices
  - Sustainable Business Model
  - New business platforms

References

Bengtsson, M. (2016). How to plan and perform a qualitative study using content analysis. *NursingPlus Open, 2*(2016), 8–14.

Bocken, N., Short, S., Rana, P., & Evans, S. (2013). A value mapping tool for sustainable business modelling. *Corporate Governance, 13*(5), 482–497.

Bower, J. L. (1986). *Managing the resource allocation process: A study of corporate planning and investment.* Harvard Business School Press.

Carlile, P., & Christensen, C. (2005). The cycles of theory building in management research.

Chairul, F. (2019). Business development of coffee farmers group using triple layered business model canvas. *GATR Journals, Global Academy of Training and Research (GATR) Enterprise.* https://EconPapers.repec.org/RePEc:gtr:gatrjs:jber182.

Cheng, C. C. J., Yang, C.-L., & Sheu, C. (2014). The link between eco-innovation and business performance: A Taiwanese industry context.

Christensen, C., & Johnson, M. (2009). What are business models, and how are they built? Harvard Business School Module Note 610–019, August 2009. (Revised May 2016).

Downe-Wambolt, B. (1992). Content analysis: Method, applications and issues. *Health Care for Women International, 13,* 313–321.

Eccles, R. G., Ioannou, I., & Serafeim, G. (2014). The impact of corporate sustainability on organizational processes and performance. *Management Science, 60*(11), 2835–2857.

Esty, D. C., & Winston, A. S. (2006). *Green to gold: How smart companies use environmental strategy to innovate, create value, and build competitive advantage.* Yale University Press.

França, C. L., Bromana, G., Karl-Henrik, R., Basile, B., & Trygg, L. (2016). An approach to business model innovation and design for strategic sustainable development. *Journal of Cleaner Production, 140*(Part 1), 155–166.

Freeman, R. E. (1984). *Strategic management: A stakeholder approach.* Pitman.

Guinee, J. B. (2002). Handbook on life cycle assessment operational guide to the ISO standards. *International Journal of Life Cycle Assessment, 7,* 311.

Hillebrand, B., Kok, R., & Biemans, W. G. (2001). Theory-testing using case studies: A comment on Johnston, Leach, and Liu. *Industrial Marketing Management, 30,* 651–657.

Ioannou, I., & Serafeim, G. (2019). *Corporate sustainability: A strategy?* (January 1, 2019). Harvard Business School Accounting & Management Unit Working Paper No. 19–065.

Johnston, W. J., Leach, M. P., & Liu, A. H. (1999). Theory testing using case studies in business-to-business research. *Industrial Marketing Management, 28,* 201–213.

Joyce, A., & Paquin, R. (2016). The triple layered business model canvas: A tool to design more sustainable business models. *Journal of Cleaner Production, 135*(2016), 1474–1486.

Krippendorff, K. (2004). *Content analysis: An introduction to its methodology.* Sage.

Myers, M. D. (2013). *Qualitative research in business & management* (2nd ed.). Sage.

Nidumolu, R., Prahalad, C. K., & Rangaswami, M. R. (2009). Why sustainability is now the key driver of innovation. In *Harvard Business Review September Issue.*

Osterwalder, A., & Pigneur, Y. (2010). *Business model generation: A handbook for visionaries, game changers, and challengers.* Wiley.

Patton, M. Q. (1990). *Qualitative evaluation and research methods.* Sage.

Porter, M. E. (1985). *Competitive advantage. Creating and sustaining superior performance* (p. 557). Free Press.

Porter, M. E., & Kramer, M. R. (2011, January–February). Creating shared value. *Harvard Business Review, 89*(1–2), 62–77.

Rodríguez-Vilá, O., & Bharadwaj, S. (2017, September–October). Competing on social purpose. *Harvard Business Review.*

Sanchez-Planelles, J., & Segarra-Oña, M. (2019a). Conference paper: Modelos de negocio sostenibles y su implementación en el mercado. II Congreso Iberoamericano AJICEDE. Valencia, 28–29th November.

Sanchez-Planelles, J., & Segarra-Oña, M. (2019b). Reshaping Business Models with an Environmental Perspective. Corporate Social Responsibility in the Manufacturing and Services Sectors, EcoProduction. Springer Nature 2019.

Sanchez-Planelles, J., & Segarra-Oña, M. Conference paper: Modelos de negocio sostenibles y su implementación en el mercado. II Congreso Iberoamericano AJICEDE. Valencia, 28–29th November.

Sanchez-Planelles, J., Segarra-Oña, M., & Peiro-Signes, A. (2021). Building a Theoretical Framework for Corporate Sustainability. *Sustainability, 13*(1), 273.

Schaltegger, S., Hansen, E., & Lüdeke-Freund, F. (2016). Business models for sustainability: Origins, present research, and future avenues. *Organization & Environment., 29*, 3–10.

Secretaría Técnica del Laboratorio de Ecoinnovación. (2016). Informe de Tendencias #1: Modelos de Negocio Ecoinnovadores. Available from https://www.laboratorioecoinnovacion.com/informes_de_tendencias.

Segarra Oña, M., Merello Giménez, P., Segura Maroto, M., Peiró Signes, A., & Maroto Álvarez, M. (2012). Proactividad medioambiental en la empresa: Clasificación empírica y determinación de aspectos clave. *Tec Empresarial, 6*(1), 35–48.

Segarra-Oña, M., Peiró-Signes, A., Mondéjar-Jiménez, J., & Sáez-Martínez, F. J. (2016). Friendly environmental policies implementation within the company: An ESG ratings analysis and its applicability to companies environmental performance enhancement. *Global NEST Journal, 18*(4), 885–893.

Steckler, A., McLeroy, K. R., Bird, S. T., & McCormick, L. (1992). Toward integrating qualitative and quantitative methods: An introduction. *Health Education Quarterly, 19*(1), 1–8.

Walsham, G. (2006). Doing interpretive research. *European Journal of Information Systems, 15*, 320–330.

Wohlin, C. (2014). Guidelines for snowballing in systematic literature studies and a replication in software engineering. In EASE '14.

Yin, R. (1993). *K Applications of case study research*. Sage.

Yin, R. K. (2013). Validity and generalization in future case study evaluations. *Evaluation, 19*(3), 321–332. https://doi.org/10.1177/1356389013497081.

Yin, R. K. (2003). *Case study research: Design and methods*. Sage.

Examining the Relationship Between Tourism Seasonality and Saturation for the Greek Prefectures: A Combined Operational and TALC-Theoretic Approach

Thomas Krabokoukis, Dimitrios Tsiotas, and Serafeim Polyzos

1 Introduction

Tourism seasonality is a phenomenon referring to the uneven distribution of tourism demand along the year timeframe, for a given destination (Butler, 2001; Batista e Silva et al., 2019). This phenomenon has substantial differences between countries and is rather complex, having both temporal and spatial dimensions (Romao & Saito, 2017; Batista e Silva et al., 2019; Tsiotas et al., 2020). Despite that seasonality affects almost every destination, it appears more intense in countries with mass tourism, such as in Mediterranean, where the summer aspects of the tourism product prevail (Corluka et al., 2016). Seasonality sets all environmental aspects (natural, economic, cultural, structured, and anthropogenic) under great pressure during peak months, in which the tourism carrying capacity of destinations is violated (Martin et al., 2019). This causes tourism companies to face high fixed costs to meet the functional needs required in the peak season, thereby increasing the average company cost and reducing its overall profitability (Corluka et al., 2016; Polyzos, 2019). Also, this phenomenon considerably affects the labor configuration because it offers short-term employment opportunities and therefore not high levels of specialization, education, and training (Lee et al., 2008; Polyzos, 2019).

In literature, tourism seasonality is studied by various methodologies depending on the geographical and socioeconomic framework of the destinations and on the availability of temporal data (Ferrante et al., 2018; Martin et al., 2019). Temporal

T. Krabokoukis (✉) · D. Tsiotas · S. Polyzos
Department of Planning and Regional Development, University of Thessaly, Volos, Greece
e-mail: tkrabokoukis@uth.gr

D. Tsiotas
Department of Regional and Economic Development, Agricultural University of Athens, Amfissa, Greece

© The Author(s), under exclusive license to Springer Nature Switzerland AG 2021
V. Katsoni and C. van Zyl (eds.), *Culture and Tourism in a Smart, Globalized, and Sustainable World*, Springer Proceedings in Business and Economics, https://doi.org/10.1007/978-3-030-72469-6_11

approaches usually focus on measurement methods of seasonality, while socioeconomic and spatial approaches deal with issues such as over-tourism (Choe et al., 2019), competitiveness (Liu et al., 2018; Gomez-Vega & Picazo-Tadeo, 2019), tourism in neighboring areas (Romao et al., 2017), society, economy, and environment (Ferrante et al., 2018; Batista e Silva et al., 2019; Polyzos, 2019; Tsiotas et al., 2020). On the one hand, the measurement of seasonality is usually implemented by examining a particular variable (e.g., overnight stays), at a specific time-unit (e.g., months), regardless of their temporal pattern (Porhallsdottir & Olafsson, 2017; Ferrante et al., 2018). Some of the classic seasonal measures are the seasonality range (Karamustafa & Ulama, 2010), seasonality ratio (Polyzos, 2019), and coefficient of seasonal variation (Tsiotas et al., 2020) provide information on the range of seasonal factors, over periods, within the year, or on their volatility at a given time-point (De Cantis et al., 2011; Lo Magno et al., 2017). Some other more composite indicators that are used are the Gini coefficient and the Theil index (Kulendran & Wong, 2005; Duro, 2016; Fernandez-Morales et al., 2016; Porhallsdottir & Olafsson, 2017), the Relative Seasonality Index—RSI (Lo Magno et al., 2017; Ferrante et al., 2018) and synthetic index DP_2 (Martin et al., 2019), which are considered as more effective and elegant (e.g., they can be decomposed into components) but sometimes are also more demanding to compute.

On the other hand, tourism seasonality is affected by a combination of natural (e.g., sea, forests, climate, weather) (Butler, 2001), institutional (e.g., holiday period, travel patterns, planned cultural events, national days) (Butler, 2001; Lee et al., 2008; Pegg et al., 2012; Fang & Yin, 2015; Duro & Turrion-Prats, 2019), and other sociocultural factors (Almeida & Kastenholz, 2019), such as fashion and traditions (Hylleberg, 1992; Butler, 1994; Ruggieri, 2015; Fernandez-Morales et al., 2016), religious, cultural, social and sports factors (Lee et al., 2008; Rossello & Sanso, 2017), the type of tourist product (Cuccia & Rizzo, 2011), the market structure (Fernandez-Morales et al., 2016), the accessibility (Lundtorp et al., 1999), and the configuration of local economies (Duro & Turrion-Prats, 2019). These factors fairly create complex interactions between them (Lee et al., 2008; Charles et al., 2013).

In the extent to which tourism is related to regional development, seasonality can induce economic and social imbalance to regional economies, depending on the ability of destinations to manage the unevenly distributed tourism demand along the year timeframe (Polyzos, 2019). Within this framework, tourism seasonality can be seen as an aspect of the regional problem, which has become the subject of research for many recent studies (Connell et al., 2015; Corluka et al., 2016; Cisneros-Martinez et al., 2017; Batista e Silva et al., 2019; Martin et al., 2019). Moreover, seasonality can further relate to tourism carrying capacity and saturation because according to the intensity and concentration of tourism demand can induce consequent uneven pressures in the natural, economic, cultural, and structured environment (Coccosis & Mexa, 2004; Jurado et al., 2012). However, the relation between tourism seasonality and tourism saturation has not yet been studied in a comprehensive context (Polyzos, 2019; Tsiotas et al., 2020) because it currently mainly focuses on the qualitative assessment of this relation.

Aiming to serve the demand of integration about the quantitative conceptualization of seasonality, this paper aims to quantitatively detect relations between tourism seasonality and saturation for the case of Greece. This is done by using classification and linear regression techniques applied to a major set of tourism variables having regional configuration. In particular, the analysis compares variables configured by seasonality measures (indices) and by tourism saturation coefficients computed according to the Tourism Area Life Cycle (TALC) theory. TALC theory interprets tourism saturation of destinations in terms of the intensity of their tourism demand (Polyzos et al., 2013), and provides information useful to tourism policy (Polyzos et al., 2013) because it models evolution of tourist destinations within the context of product life cycle used in marketing (Butler, 2006; Candela & Figini, 2012). Generally, TALC theory describes that each tourism area passes through certain stages as the number of visitors increases, which are the exploration, involvement, development, consolidation, stagnation, and the decline (or rejuvenation) stage (Polyzos et al., 2013). This approach allows determining the level of saturation of tourist destinations and thus to detect cases capable of further touristic development (Polyzos et al., 2013).

The study focuses on the case of Greece, which is a coastal country with a complex geomorphology mixing mountainous, land, coastal, and insular areas, a major tourism specialization, and a composite tourism product (Kalantzi et al., 2016), which is diversely distributed along with the various tourism destinations of the country (Tsiotas, 2017; Polyzos, 2019). In Greece, more than 55 km^2 mountainous areas, more than 16,000 km of coastline and more than 1350 islands, islets, and rocky islands, of which over 230 are inhabited (Tsiotas, 2017). For the year 2019, according to the Greek Tourism Confederation (SETE), the overall contribution of the tourism sector on GDP reached 20.8%, with total foreign arrivals (without the arrivals from cruises) at 31.3 million visitors and the 56% of these arrivals take place in July–August–September (Tsiotas et al., 2020). The further purpose of the paper is to contribute toward integrating the conceptual framework of tourism seasonality by proposing a quantitative framework for incorporating the spatial and the temporal aspect in the study of tourism seasonality. The common consideration of tourism seasonality and saturation is expected to provide an aggregate approximation of the spatiotemporal conceptualization of the phenomenon.

The remainder of this paper is organized as follows: Sect. 1 is a brief literature review discussing measurement, temporal, and spatial assessment of tourism seasonality. Section 2 describes the methodological framework of the study, the available data, and the variables participating in the analysis. Section 3 presents the results and discusses them within the context of regional science and tourism development. Finally, in Sect. 4 conclusions are given.

2   Methodology and Data

The methodological framework of the study builds on correlation analysis applied to a pair of regional variables, the first expressing tourism seasonality and the second expressing tourism saturation. The first variable of tourism seasonality (RSI) is configured by computing the Relative Seasonality Index (S) proposed by Lo Magno et al. (2017). This index is defined within the context of the transportation problem by minimizing the cost of eliminating seasonality by transferring units from high- to low-season periods (Lo Magno et al., 2017; Ferrante et al., 2018). The mathematical expression of the RSI is described by the formula:

$$S(\mu, C) = \sum_{i \in A} \sum_{j \in B} c_{ij} x_{ij} / \mu \cdot \max_{i \in M} \left\{ \sum_{j \in M} c_{ij} \right\} \tag{1}$$

where x_i is the ith observation of a time-series x (expressing a tourism-variable), μ is the average value of the available observations, C is the total cost for eliminating seasonality, A is the set of high-season time periods, B is the set of low-season time periods, and M is the set of all possible observed time-patterns.

The second variable of tourism saturation (RST) is configured by computing the coefficient of Tourism Area Life Cycle (TALC). In general, TALC theory conceptualizes a tourism destination as a market specialized to tourism and describes its evolution as a sequence of stages (exploration, involvement, development, consolidation, stagnation, and decline or rejuvenation) changing as the number of visitors increase (Polyzos et al., 2013). By using a logistic curve to model the TALC process, the magnitude of coefficient $r(t)$ expresses the speed at which a tourism destination reaches saturation. The TALC coefficient of saturation (RST) can be expressed through by the formula (Polyzos et al., 2013):

$$RST(t) = (\ln(N(t)) - \ln(A)) / t \tag{2}$$

where $N(t)$ is the value of a time-variant variable expressing tourism demand (usually it refers to the number of visitors) that is defined at time t and A is a fitting coefficient depending on the starting conditions of the curve.

Both variables $RSI = \{S_i \mid i = 1, ..., 51\}$ and $RST = \{r(t)_i \mid i = 1, ..., 51\}$ are computed on data referring to the monthly number of overnight stays (including both foreign and domestic visitors) per Greek prefecture, for the period 1998–2018. The available data were granted upon request by the Hellenic Statistical Authority (ELSTAT, 2019) to be used under an exclusive license, for the purpose of this study. Each variable has length 51, namely including scores corresponding to the 51 Greek prefectures.

After configuring the available variables RSI and RST, a multi-level correlation analysis is further applied. At the first level, the scatterplots $RST * RSI$ (Norusis, 2011) of these variables are constructed, where the available 51 cases (prefectures) are grouped into quadrants (Low Saturation-Low Seasonality, Low–High, High-Low, and High- High) defined by the average lines per axis. The spatial distribution of the grouping is mapped. At the second level, bivariate parametric and non-parametric coefficients of correlation are computed on the pair of (RST, RSI) variables. Parametric computations are made by using the *Pearson's bivariate coefficient of correlation* (Norusis, 2011; Devore & Berk, 2012; Walpole et al., 2012), which is defined as:

$$r_{XY} = \frac{\text{cov}(X, Y)}{\sqrt{\text{var}(X)} \cdot \sqrt{\text{var}(Y)}} \tag{3}$$

where $\text{cov}(X, Y)$ is the *covariance* of variables X, Y and $\sqrt{\text{var}(\cdot)}$ is the sample standard deviations. Pearson's coefficient of correlation ranges within the interval $[-1, 1]$ and detects linear relations when $|r_{XY}| \rightarrow 1$ (Devore & Berk, 2012). On the other hand, non-parametric computations are made by using *Kendall's tau and Spearman rho rank correlation coefficients* (Norusis, 2011; Walpole et al., 2012). *Kendall's tau* correlation coefficient is computed on the number of concordant and discordant pairs according to the expression (Norusis, 2011):

$$t = ((\text{number of concordant pairs}) - (\text{number of discordant pairs}))/\binom{n}{2} \tag{4}$$

where n is the number of items included in the variable. The Spearman's correlation coefficient r_s is computed on the standard formula of the Pearson's correlation coefficient where the rank-variable $rnk(x)$ and $rnk(y)$ with the rank of magnitude of each element are used instead of the values of variables x, y. Both Kendall's tau and Spearman's rho coefficients of correlation range within the interval $[-1, 1]$, describing a perfect linear relation (positive or negative) in cases they equal to one.

At the final level of analysis, a sequence of Pearson's correlation coefficients $P = \{r_1, r_2, ..., r_n\}$ is computed by excluding from the set of Greek regions outlier cases, which are defined by variant confidence intervals for the linear regression slope. In particular, a univariate linear regression model is applied to the pair-set $RSI * RST$, of the form (Norusis, 2011; Walpole et al., 2012):

$$y = bx + c \overset{y=RSI}{\underset{x=RST}{\Rightarrow}} RSI = b \cdot RST + c \tag{5}$$

where b is the regression coefficient (or slope of the regression line) estimated under the Ordinary Least Square algorithm (Walpole et al., 2012). Interval estimations of the regression slope can be made by using the formula (Norusis, 2011; Walpole et al., 2012):

$$b_a = b \pm t_{1-a/2, n-2} \cdot sb \tag{6}$$

where $t_{1-a/2, n-2}$ is the value of the Student's t-distribution computed for $a\%$ confidence interval (CI) and $n-2$ degrees of freedom (Walpole et al., 2012). Cases exceeding the zones defined by the $a\%$ confidence intervals for the linear regression slope, where $a = 99, 95, 90, 85, ..., 50$ are considered as outliers and thus they are omitted from the sample. This approach aims to detect the minimum number of outlier cases so that the correlation $r(RSI, RST)$ to be considered as high ($r > 0.75$).

3   Results and Discussion

The results of the first part of the analysis, which builds on correlation scatterplots $RSI * RST$, are shown in Fig. 1. In this figure, the relevant map shows the spatial distribution of the Greek regions according to the quadrant grouping defined by the RSI and RST average reference lines. As it can be observed, an arc of LL (Low Saturation-Low Seasonality) cases is configured in the mainland Greece, which is composed by the prefectures of Rodopi (1), Drama (2), Evros (3), Xanthi (5), Imathia (7), Kilkis (8), Pella (9), Serres (11), Kozani (13), Grevena (14), Kastoria (15), Florina (16), Arta (18), Larissa (21), Karditsa (22), Trikala (24), Viotia (26), Evritania (28), Fokida (29), and Aitoloakarnania (35). A final LL case can be found in the center of the region of Peloponnesus and it regards the prefecture of Arkadia (37), which is the one with the least coastline length in this region. According to this map, the LL behavior seems to be attributed to geographical centrality and to mainland formation.

On the other hand, the pattern of the spatial distribution of the HH (High Saturation-High Seasonality) cases appears more as a matter of insularity and coastal morphology of the Greek regions. In particular, the HH prefectures of Kavala (4), Pieria (10), and Halkidiki (12) are arranged along the coastal forehead of the metropolitan (with population of over a million people) prefecture of Thessaloniki (6), similarly to the HH prefectures of Evia (27), Argolida (38), and Korinthia (39), which are arranged along the coastal forehead of the metropolitan (with population over 3 million people) prefecture of Attiki (42). In addition, the HH prefectures of Kerkyra (30), Zakynthos (31), Kefallonia (32), Ilia (36), and Messinia (41) are

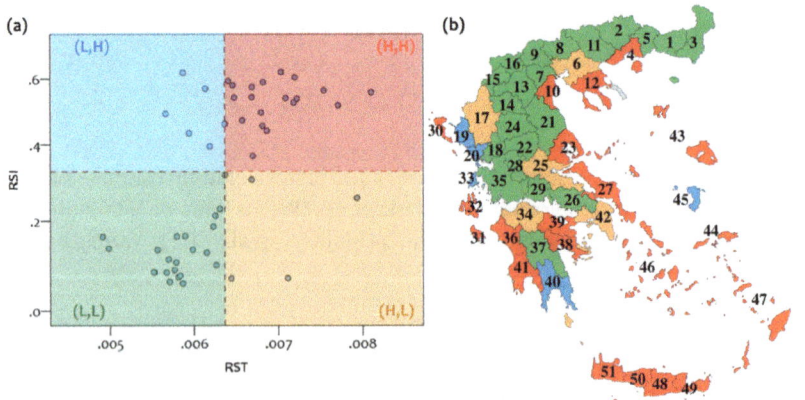

Fig. 1 The spatial Greek prefectures according to their seasonality (RSI) and level of saturation (TALC)

arranged along the forehead of the Ionian Sea, and the prefectures of Magnessia (23) and Evia (27) are arranged along the coastal forehead of the central Aegean sea. Also, a HH behavior appears for the prefectures of Lesvos (43), Samos (44), Cyclades (46), Dodecanese (47), which are located at the central and south Aegean, and the prefectures of Heraklion (48), Lasithi (49), Rethymno (50), and Hania (51), which are located at the south Aegean. This spatial imbalance of the HH cases complies with other literature findings (Tsiotas et al., 2020) describing the sea-driven (3S) configuration of the tourism product in Greece.

Next, the prefectures of Thessaloniki (6), Ioannina (17), Fthiotida (25), Achaia (34), and Attiki (42) are HL cases described by high saturation but low seasonality. The spatial pattern configured by these prefectures is rather scattered but they have a functional similarity due to their central roles in their regions (e.g., the prefecture of Thessaloniki is the capital region of north Greece, Ioannina is the center of the region of Epirus, Fthiotida is the center of the region of Sterea Ellada, Achaia is the center of Peloponnesus, and Attiki is the metropolitan region of the country). Finally, the prefectures of Thesprotia (19), Preveza (20), Lefkada (33), Lakonia (40), and Chios (45), are LH cases described by low saturation and high seasonality. The spatial distribution of these prefectures also configures a scattered pattern in the map and their common attribute is related to their peripheral location (all these prefectures are the most distant in their regions).

At the next step, a (parametric and non-parametric) bivariate correlation analysis is applied to the variables RST and RSI, the results of which are shown in Table 1. As it can be observed, the correlation between tourism saturation and seasonality is highly significant (implying that is less than 1% possibility to be a matter of chance) but the value of coefficient is not very high (ranging between 0.418 and 0.601), implying not a strong linear relation between the variables.

Within this context, the third part of the analysis attempts to examine which individual cases can be considered as outliers in the linear relation between tourism saturation and seasonality. The analysis builds on confidence intervals constructed for the linear regression slope between these variables and the results are shown in Fig. 2.

Table 1 Results of the parametric and non-parametric correlation analysis

	Formula	y-variable: TALC Coefficient (R_{st})	
		Statistics	R_{st}
x-y variable: RSI	Pearson's (r)	Corr. Coefficient	0.594[**]
		Sig. (2-tailed)	0.000
		N	51
	Kendall's tau (b)	Corr. Coefficient	0.418[**]
		Sig. (2-tailed)	0.000
		N	51
	Spearman's (ρ)	Corr. Coefficient	0.601[**]
		Sig. (2-tailed)	0.000
		N	51

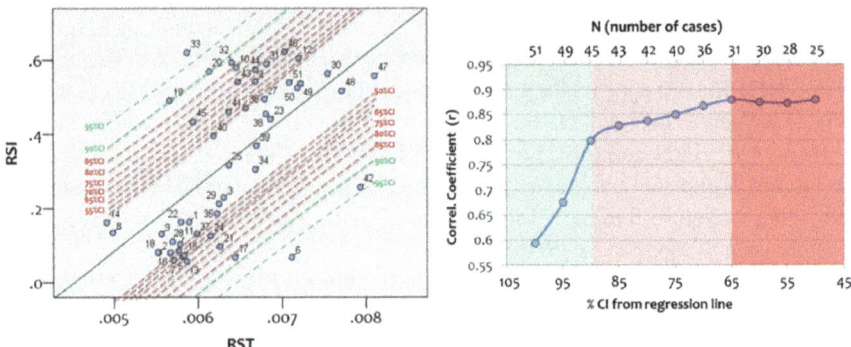

Fig. 2 Gradual confidence intervals (ranging from confidence level $a = 50\%$ up to $a = 99\%$) for the linear regression slope of the bivariate relation between tourism saturation (*RST*) and seasonality (*RSI*). On the left, the confidence interval bounds are shown per confidence level (%CI) in the *RST* * *RSI* scatterplot. On the right, the diagram shows the gain in the Pearson's correlation coefficient per %CI. From the point of 90% CI and so, the correlation coefficient is subjected to diminishing growth

As it can be observed in Fig. 2, by excluding six outlier prefectures, the Pearson's correlation coefficient ($r_{90} = 0.798$) grows almost 30% relative to the initial value ($r_{100} = 0.594$). This case (corresponding to 90% CI) is the one with the highest marginal growth (~18%). After this point (90% CI), the correlation coefficient is subjected to diminishing growth. Especially, from the point of 65% CI and so the marginal growth is negative, implying that no more cases should be further omitted. In detail, the labeled cases that are omitted in each case of confidence level are shown in Table 2.

According to Table 2, the cases that should be omitted to gain the highest marginal growth in the correlation coefficient are the prefectures of Thessaloniki (6), Thesprotia (19), Preveza (20), Kefallonia (32), Lefkada (33), and Attiki (42). In this case, the correlation coefficient grows from $r_{100} = 0.594$ to $r_{90} = 0.798$. The prefectures can be classified into two categories; the first includes the metropolitan prefectures of Athens and Thessaloniki, which are megacities in terms of population size, for the scale of Greece (Tsiotas & Polyzos, 2013). As previously shown, these prefectures belong to the HL class described by high saturation but low seasonality. The second group includes the coastal prefectures of Thesprotia (19) and Preveza (20) and the island prefectures of Kefallonia (32) and Lefkada (30), which are all located at the Ionian Sea (west) side. According to the analysis shown in Fig. 1, all these prefectures belong to the LH quadrant except Kefallonia (32), which belongs to the HH. Low levels of saturation are probably related to restricted infrastructure capacity and accessibility of these prefectures that consequently constraints their tourism capacity (Polyzos, 2019). Overall, by excluding the pair of metropolitan prefectures and these four prefectures belonging to the Ionian Sea, the correlation between tourism seasonality and saturation turns to high levels, implying that seasonality is a factor driving tourism saturation in the majority of the Greek prefectures, which is obviously related to the sea-driven (3S) configuration of the tourism product in Greece.

Table 2 Cases (prefectures) omitted from the initial set per confidence level and the respective correlation coefficients[a] computed for the variables *RST* and *RSI*

CI_95 (0.676)		CI_90 (0.798)		CI_85 (0.828)		CI_80 (0.837)		CI_75 (0.85)	
6	THESSALONIKI	6	THESSALONIKI	6	THESSALONIKI	6	THESSALONIKI	6	THESSALONIKI
33	LEFKADA	19	THESPOTIA	10	PIERIA	10	PIERIA	10	PIERIA
CI_50 (0.88)		20	PREVEZA	17	IOANNINA	17	IOANNINA	17	IOANNINA
4	KAVALA	32	KEFALLONIA	19	THESPOTIA	19	THESPOTIA	19	THESPOTIA
5	XANTHI	33	LEFKADA	20	PREVEZA	20	PREVEZA	20	PREVEZA
6	THESSALONIKI	42	ATTIKI	32	KEFALLONIA	21	LARISSA	21	LARISSA
7	HMATHIA	**CI_55 (0.873)**		33	LEFKADA	32	KEFALLONIA	32	KEFALLONIA
10	PIERIA	4	KAVALA	42	ATTIKI	33	LEFKADA	33	LEFKADA
12	HALKIDIKI	5	XANTHI	**CI_60 (0.875)**		42	ATTIKI	42	ATTIKI
15	KASTORIA	6	THESSALONIKI	4	KAVALAS	**CI_65 (0.88)**		43	LESVOS
16	FLORINA	7	IMATHIA	5	XANTHIS	4	KAVALA	44	SAMOS
17	IOANNINA	10	PIERIA	6	THESSALONIKHS	5	XANTHI	**CI_70 (0.867)**	
18	ARTA	12	HALKIDIKI	7	HMATHIAS	6	THESSALONIKI	6	THESSALONIKI
19	THESPOTIA	15	KASTORIA	10	PIERIAS	7	IMATHIA	10	PIERIA
20	PREVEZA	17	IOANNINA	12	CHALKEDIKHS	10	PIERIA	12	HALKIDIKI
21	LARISSA	18	ARTA	15	KASTORIAS	12	HALKIDIKI	17	IOANNINA
24	TRIKALA	19	THESPOTIA	17	IOANNINON	15	KASTORIA	19	THESPOTIA
28	EVRYTANIA	20	PREVEZA	18	ARTAS	17	IOANNINA	20	PREVEZA
31	ZAKYNTHOS	21	LARISSA	19	THESPOTIAS	19	THESPOTIA	21	LARISSA
32	KEFALLONIA	24	TRIKALA	20	PREVEZAS	20	PREVEZA	31	ZAKYNTHOS

(continued)

Table 2 (continued)

CI_95 (0.676)		CI_90 (0.798)		CI_85 (0.828)		CI_80 (0.837)		CI_75 (0.85)	
6	THESSALONIKI	6	THESSALONIKI	6	THESSALONIKI	6	THESSALONIKI	6	THESSALONIKI
33	LEFKADA	28	EVRYTANIA	21	LARISSIS	21	LARISSA	32	KEFALLONIA
35	AITOLOAKARNANIA	31	ZAKYNTHOS	24	TRIKALON	24	TRIKALA	33	LEFKADA
36	ILIA	32	KEFALLONIA	31	ZAKYNTHOS	31	ZAKYNTHOS	42	ATTIKI
37	ARKADIA	33	LEFKADA	32	KEFALLONIAS	32	KEFALLONIA	43	LESVOS
42	ATTIKI	37	ARKADIA	33	LEFKADOS	33	LEFKADA	44	SAMOS
43	LESVOS	42	ATTIKI	42	ATTIKHS	42	ATTIKI	45	CHIOS
44	SAMOS	43	LESVOS	43	LESVOU	43	LESVOS	46	CYCLADES
45	CHIOS	44	SAMOS	44	SAMOU	44	SAMOS		
46	CYCLADES	45	CHIOS	45	CHIOU	45	CHIOS		
		46	CYCLADES	46	KEEKLADON	46	CYCLADES		

[a]Pearson correlation coefficients computed according to relation (3) for the variables of tourism saturation and seasonality, on the sequential sets shown in the scatterplot of Fig. 2

4   Conclusions

This paper examined the relationship between tourism seasonality and tourism saturation. The proposed analysis was built on correlation analysis to classify (into quadrants defined by the average lines per axis) the pair of regional variables of the Greek prefectures, for the period 1998–2018. The resulting four groups (Low Saturation-Low Seasonality, Low-High, High-Low, and High-High), were examined in terms of their geographical characteristics. In particular, the LL (Low Saturation-Low Seasonality) group is configured in mainland Greece. The HH (High Saturation-High Seasonality) cases appear more as a matter of insularity and coastal morphology of the Greek regions. The other two groups include the fewest regions. The HL (High Saturation-Low Seasonality) cases have a functional similarity due to their central roles in their regions, while the common attribute of LH (Low Saturation-High Seasonality) regions is related to their peripheral location (all these prefectures are the most distant in their regions). According to the (parametric and non-parametric) bivariate correlation analysis that was applied, the correlation between tourism saturation and seasonality was found highly significant, but the value of coefficient was not very high. By excluding the outliers from the linear relation between tourism saturation and seasonality, the correlation between tourism seasonality and saturation turned to high levels and had the maximum marginal gain. The outliers were the pair of metropolitan prefectures (Attiki, Thessaloniki) and four prefectures belonging to the Ionian Sea (Thesprotia, Preveza, Kefallonia, Lefkada). The overall analysis showed that seasonality is a factor driving tourism saturation in the majority of the Greek prefectures, which is related to the sea-driven (3S) configuration of the tourism product in Greece.

Acknowledgements This research is co-financed by Greece and the European Union (European Social Fund—ESF) through the Operational Programme "Human Resources Development, Education and Lifelong Learning 2014–2020" in the context of the project "Analysis and methodological approach of tourism seasonality for the Greek regions" (MIS 5048961).

Operational Programme
Human Resources Development,
Education and Lifelong Learning

Co-financed by Greece and the European Union

Appendix

See Appendix Table 3.

Table 3 Numerical results of the TALC coefficients and relative seasonality index (RSI) per Greek region

Code	Prefecture	Code	Prefecture	Code	Prefecture	Code	Prefecture
1	Rodopi	14	Grevena	27	Evia	40	Lakonia
2	Drama	15	Kastoria	28	Evritania	41	Messinia
3	Evros	16	Florina	29	Fokida	42	Attiki
4	Kavala	17	Ioannina	30	Corfu	43	Lesvos
5	Xanthi	18	Arta	31	Zakynthos	44	Samos
6	Thessaloniki	19	Thesprotia	32	Kefallonia	45	Chios
7	Imathia	20	Preveza	33	Lefkada	46	Cyclades
8	Kilkis	21	Larissa	34	Achaia	47	Dodecanese
9	Pella	22	Karditsa	35	Aitoloakarnania	48	Heraklion
10	Pieria	23	Magnessia	36	Ilia	49	Lasithi
11	Serres	24	Trikala	37	Arkadia	50	Rethymno
12	Halkidiki	25	Fthiotida	38	Argolida	51	Hania
13	Kozani	26	Viotia	39	Korinthia		

References

Almeida, A., & Kastenholz, E. (2019). Towards a theoretical model of seasonal tourist consumption behavior. *Tourism Planning and Development, 16*(5), 533–555.

Batista e Silva, F., Kavalov, B., & Lavalle, C. (2019). *Socio-economic regional microscope series— Territorial patterns of tourism intensity and seasonality in the EU*. Publications Office of the European Union, Luxembourg.

Butler, R. W. (1994). Seasonality in tourism: Issues and implications. In: A. Seaton (Ed.), *Tourism: The state of the art*. Chichester: Wiley.

Butler, R. W. (2001). Seasonality in tourism: Issues and implication. In T. Baum, & S. Lundtorp, (Eds.) *Seasonality in tourism* (pp. 5–21). Oxford: Elsevier Ltd.

Butler, R. W. (2006). *The tourism area life cycle* (vol. 1), *Applications and modifications*. Channel View Publications.

Candela, G., & Figini, P. (2012). *The economics of tourism destinations (Springer texts in business and economics)*. Berlin: Springer.

Charles-Edwards, E., & Bell, M. (2013). Seasonal flux in Australia's population geography: Linking space and time. *Population, Space and Place, 21*(2), 103–123.

Choe, Y., Kim, H., & Joun, H. (2019). Differences in tourist behaviors across the seasons: The case of Northern Indiana. *Sustainability, 11*(16), 1–16.

Cisneros-Martinez, J., McCabe, S., & Fernandez-Morales, A. (2017). The contribution of social tourism to sustainable tourism: A case study of seasonally adjusted programs in Spain. *Journal of Sustainable Tourism, 26*(1), 85–107.

Coccossis, H., & Mexa, A. (2004). *The challenge of tourism carrying capacity assessment: Theory and Practice*. Aldershot: Ashgate Publishing Ltd.

Connell, J., Page, S., & Meyer, D. (2015). Visitor attractions and events: Responding to seasonality. *Tourism Management, 46*(1), 283–298.

Corluka, G., Mikinac, K., & Milenkovska, A. (2016). Classification of tourist season in coastal tourism. *UTMS Journal of Economics, 7*(1), 71–83.

Cuccia, T., & Rizzo, I. (2011). Tourism seasonality in cultural destinations: Empirical evidence from Sicily. *Tourism Management, 32*(3), 589–595.

De Cantis, S., Ferrante, M., & Vaccina, F. (2011). Seasonal pattern and amplitude—Logical framework to analyze seasonality in tourism: An application to bed occupancy in Sicilian hotels. *Tourism Economics, 17*(3), 655–678.

Devore, J., & Berk, K. (2012). *Modern mathematical statistics with applications.* London: Springer.

Duro, A. (2016). Seasonality of hotel demand in the main Spanish provinces: Measurements and decomposition exercises. *Tourism Management, 52*(1), 52–63.

Duro, J., & Turrion-Prats, J. (2019). Tourism seasonality worldwide. Tourism Management. *Perspectives, 31*(1), 38–53.

Fang, Y., & Yin, J. (2015). National assessment of climate resources for tourism seasonality in China using the tourism climate index. *Atmosphere, 6*(1), 183–194.

Fernandez-Morales, A., Cisneros-Martinez, J. D., & McCabe, S. (2016). Seasonal concentration of tourism demand: Decomposition analysis and marketing implication. *Tourism Management, 56*(1), 172–190.

Ferrante, M., Lo Magno, G., & De Cantis, S. (2018). Measuring tourism seasonality across European countries. *Tourism Management, 68*(1), 220–235.

Gomez-Vega, M., & Picazo-Tadeo, A. (2019). Ranking world tourist destinations with a composite indicator of competitiveness: To weigh or not to weigh? *Tourism Management, 72*(1), 281–291.

Hellenic Statistical Authority—ELSTAT. (2019). Number of monthly overnight stays in the Greek prefectures for the period 1998–2018. Data granted upon request by ELSTAT (www.statistics.gr) to be used under an exclusive license, for the purpose of this study.

Hylleberg, S. (1992). *General introduction in Modelling seasonality.* Oxford: Oxford University Press.

Jurado, E., Tejada, M., Almeida Garcia, F., Cabello Conzalez, J., Cortes Macias, R., Delgado Pena, J., Fernandez Gutierrez, F., Gutierez Fernandez, G., Luque Gallego, M., Malvarez Garcia, G., Marcerano Gutierrez, O., Navas Cocha, F., Ruiz de la Rua, F., Ruiz Sinoga, J., & Solis Beccerra, F. (2012). Carrying capacity assessment for tourist destinations. Methodology for the creation of synthetic indicators applied in a coastal area. *Tourism Management, 33*(1), 1337–1346.

Kalantzi, O., Tsiotas, D., & Polyzos, S. (2016). The contribution of tourism in national economies: Evidence of Greece. *European Journal of Business and Social Sciences (EJBSS), 5*(5), 41–64.

Karamustafa, K., & Ulama, S. (2010). Measuring the seasonality in tourism with the comparison of different methods. *EuroMed Journal of Business, 5*(2), 191–214.

Kulendran, N., & Wong, K. (2005). Modeling seasonality in tourism forecasting. *Journal of Travel Research, 44*(2), 163–170.

Lee, C., Bergin-Seers, S., Galloway, G., O'Mahony, B., & McMurray, A. (2008). *Seasonality in the tourism industry—Impacts and strategies.* Australia: Sustainable Tourism Cooperative Research Center.

Liu, Y., Li., Y., & Parkpian, P. (2018). Inbound tourism in Thailand: Market form and scale differentiation in ASEAN source countries. *Tourism Management, 64*(1), 22–36.

Lo Magno, L., Ferrante, M., & De Cantis, S. (2017). A new index for measuring seasonality: A transportation cost approach. *Mathematical Social Sciences, 88*(1), 55–65.

Lundtorp, S., Rassing, C., & Wanhill, S. (1999). The off-season is "no season: The case of the Danish Island of Bornholm. *Tourism Economics, 5*(1), 49–68.

Martin, J., Fernandez, J. A., & Martin, J. (2019). Comprehensive evaluation of the tourism seasonality using a synthetic DP2 indicator. *Tourism Geographies, 21*(2), 284–305.

Norusis, M. (2011). *IBM SPSS statistics 19.0 guide to data analysis.* Upper Saddle River, NJ: Prentice Hall.

Pegg, S., Patterson, I., & Gariddo, P. V. (2012). The impact of seasonality on tourism and hospitality operations in the alpine region of New South Wales, Australia. *International Journal of Hospitality Management, 31*(3), 659–666.

Polyzos, S. (2019). *Regional development.* Athens: Kritiki.

Polyzos, S., Tsiotas, D., & Kantlis, A. (2013). Determining the tourism development capabilities of the Greek regions, by using TALC theory. *Tourisms: An International Multidisciplinary Journal of Tourism, 8*(2), 159–178.

Porhallsdottir, G., & Olafsson, R. (2017). A method to analyze seasonality in the distribution of tourists in Iceland. *Journal of Outdoor Recreation and Tourism, 19*(1), 17–24.

Romao, J., Guerreiro, J., & Rodrigues, P. (2017). Territory and sustainable tourism development: A space-time analysis on European regions. *The Region, 4*(3).

Romao, J., & Saito, H. (2017). A spatial analysis on the determinants of tourism performance in Japanese Prefectures. *Asia-Pacific Journal of Regional Science, 1*(1), 243–264.

Rossello, J., & Sanso, A. (2017). Yearly, monthly and weekly seasonality of tourism demand: A decomposition analysis. *Tourism Management, 60*(1), 379–389.

Ruggieri, G. (2015). Island tourism seasonality. In E. Pechlaner, & E. Smeral (Eds.), *Tourism and leisure: Current issues and perspectives of development*. Wiesbaden: Springer Gabler.

Tsiotas, D. (2017). The imprint of tourism on the topology of maritime networks: Evidence from Greece Anatolia. *An International Journal of Tourism and Hospitality Research, 28*(1), 52–68.

Tsiotas, D., Krabokoukis, T., & Polyzos, S. (2020). Detecting interregional patterns in tourism seasonality of Greece: A principal components analysis approach. *Regional Science Inquiry, 2*, 91–112.

Walpole, R. E., Myers, R. H., Myers, S. L., & Ye, K. (2012). *Probability & statistics for engineers & scientists* (9th ed.). Prentice Hall: New York, NY.

Sustainable Tourism Development in the Ionian Islands. The Case of Corfu Island

Konstantinos Mouratidis

1 Introduction

Tourism has been formally recognized as a major socio-economic sector and the third largest export category worldwide. In 2019, international tourist arrivals grew 3.8% compared to the previous year and destinations worldwide received around 1.5 billion arrivals (54 million more than 2018). Taking into consideration that total exports earnings from international tourism activity reached US$ 1.7 trillion, or almost 5 billion a day on average, in 2018 (UNWTO), it should be mentioned that forecasts for 1.6 billion international arrivals and US$ 2 trillion receipts from international tourism for 2020 may not be verified due to the pandemic crisis and adverse effects of COVID-19 on tourism industry worldwide. Under these aspects, this manuscript analyzed a priori the major trends of tourism in Greece, Ionian Islands region, and Corfu Island over the last decade, and attempts to explore the emergence of sustainable tourism development for Corfu, based on special interest tourism and alternative tourism products that should be considered as a key axis of local tourism development, providing a pattern of tourism development aimed to the balance among the local community, economy and the environment of the area. The methodology used is the critical review of the literature and the bibliographic references described more of an illustration of the topic. Results and discussion conclude this summary reporting the parameters for sustainable tourism development in Corfu, as the success and the implementation of sustainable development programs can lead to the protection of the environment and cultural characteristics of each region and determine the long-run benefits of the sustainability agenda.

K. Mouratidis (✉)
Tourism Planning, Management & Policy, University of the Aegean, Chios, Greece
e-mail: mouratidis.konstantinos@outlook.com.gr

© The Author(s), under exclusive license to Springer Nature Switzerland AG 2021
V. Katsoni and C. van Zyl (eds.), *Culture and Tourism in a Smart, Globalized, and Sustainable World*, Springer Proceedings in Business and Economics,
https://doi.org/10.1007/978-3-030-72469-6_12

1.1   Characteristics and Figures of the Tourism Sector in Greece

Greece remains one of the most mature destinations in Southern Mediterranean Europe and has reported increased trends both in arrivals, receipts, and overnights for the period 2015–2018 (Table 1). According to SETE, Greece reached 30.1 million incoming tourists, in 2018 (+10.8% over 2017) and the country's tourism receipts were slightly €15.7 billion (+11.7% compared to the previous year). Contrastingly, the average expenditure per trip decreased from 580€ in 2015 to 520€ in 2018 and average expenditure per night reached 89.75€ in 2018, 6.5€ lower regard to relative numbers of 2015, while the length of stay fluctuated approximately in same levels. However, Greek tourism product has to overcome seasonality and unhooked of principle model of mass tourism. Regrettably, in 2018, it is estimated that 80.2% of arrivals and 84.4% of revenues are recorded from April to September, demonstrating the seasonality of demand and bringing several problems to destinations and accommodations. In terms of tourist accommodation establishments, Greece accounts for almost 38.000 official accommodation facilities (hotels and rented rooms), in 2018, with a total capacity of approximately 1,275,000 beds on which they recorded 227

Table 1   Data related to Greek tourism demand 2015–2018

	2015	2016	2017	2018
Arrivals of incoming tourists (in millions)	23,599,455	24,799,349	27,194,185	30,122,781
Tourism receipts (in million Euros)	13,679,194,302	12,749,275,919	14,202,462,080	15,633,185,629
Overnight stays (Eurozone countries)	82,220,559	84,650,448	92,050,905	103,278,212
Overnight stays (European and non-eurozone countries)	50,621,734	56,017,893	57,075,366	60,940,582
Overnight stays (Non-European countries)	52,184,852	49,733,478	60,728,886	62,793,178
Overnight stays (Total)	185,027,145	190,401,819	209,855,157	227,011,971
Occupancy (Yearly)	49.1%	50.1%	51.9%	52.7%
Average length of stay (Days)	7.8	7.7	7.7	7.5
Average expenditure per night (in Euros)	96.25	90	85.75	89.75
Average expenditure per trip (in Euros)	580	514	522	520

Source Bank of Greece & INSETE Intelligence, Hellenic Statistical Authority—Author's Editing

million overnight stays. Hence, low occupancy rates—especially in off-seasons—remain a major issue for national tourism policy. In 2018, the highest occupancy rate was recorded during August (77%) and the lowest during November (19.2%).

According to 2019 Travel & Tourism Competitiveness Index (TTCI), which is calculated by World Economic Forum (WEF), Greece ranked at 25th position among 140 countries, 25th based on natural and cultural resources, 13th based on the prioritization of Travel & Tourism and 18th regard to its tourism service infrastructure. Greece's GDP in current prices, in 2018, amounted to €184,714 million (+2.5% compared to 2017 recorded €180,218 million), while the direct contribution of tourism to the formation of GDP amounted to 11.7% or almost €21.6 billion. Taking into account the multiplier benefits of tourism activity, the total (directly and indirectly) contribution of tourism amounts between €47.4 and €57.1 billion, (i.e., ranged between 25.7% and 30.9% of GDP), displaying the high dependence of Greek economy on tourism sector. It is indicative that the economy of three island regions depends to a significant extent on tourism, as the sector's contribution to regional GDP amounts to 47.2% in Crete, 71.2% in the Ionian Islands, and 97.1% in South Aegean. These regions have one of the highest per capita GDPs in the country, arguing that tourism leads to an improvement in the living standards of residents. According to the World Travel & Tourism Council (WTTC), one-quarter of all employment, in 2018, in Greece is based on Tourism (988.6 k jobs), while Greek Tourism outpaced EU's regional Travel and Tourism growth rate of 2.4%. Both EU and Greece's wider economies grew at a rate of 2%, but the Greek travel sector leads ahead of regional averages.

The structure of international tourist arrivals and receipts recorded that the vast majority of incoming tourists came from North and Central European countries (i.e., Germany, U.K., France) as those markets considered also "big spenders" of the national tourism product. Cumulatively, for the years 2015–2018, the share of the German source market remains steady in the first place, both in arrivals (13.3%) and receipts (17.6%) of the total market, followed by the UK (10.6% and 14.2%, respectively). Namely, the French market placed 5th (5.5%) in terms of arrivals and 3rd (7.2%) regarding tourism revenues, followed by the U.S.A (4th) and Italy (5th) (6.3% and 5.8% respectively) concerning tourism receipts for the same period.

1.2 Characteristics and Figures of the Tourism Sector in the Ionian Islands and Corfu

Ionian Islands region is one of the thirteen administrative regions of Greece and consists of a complex of islands divided into five regional units (i.e., Corfu, Ithaca, Lefkas, Zante, and Cephalonia). Regions population in 2011 was 207,855 (−1.5% compared to 2001) and its GDP for 2018 was €3270 million, while the GDP per capita for 2016 was €15,182, which was slightly lower than the national median of €16,378 for the same year. The Ionian Islands considered attractive worldwide tourist

Table 2 Data related to Ionian Islands region tourism demand 2016–2019

	2016		2017		2018		2019	
	Total	%[a]	Total	%[a]	Total	%[a]	Total	%[a]
Arrivals of incoming tourists (in millions)	2457.1	8.7	2966.3	9.6	3162.3	9.1	3047.8	8.3
Tourism receipts (in million Euros)	1503.6	11.8	1774.9	12.5	1691.1	10.8	1911.2	10.8
Overnight stays (Total)	21,493.2	11.3	24,943.2	11.9	24,761.8	10.9	23,744.4	10.2
Average length of stay (Days)	8.7		8.4		7.8		7.8	
Average expenditure per night (in Euros)	70		71.2		68.3		80.5	
Average expenditure per trip (in Euros)	611.9		598.4		534.8		627.1	

Source Border Research of Bank of Greece, INSETE Intelligence–Author's Editing
[a]% of Greece's Total

destinations and the airports of Corfu, Zante, and Cephalonia recorded 2,644,287 international arrivals for 2019, as well as Corfu airport received 1,457,420 (49.5% of the region) being 6th airport by several international arrivals nationwide (Table 5). Tourism demand has grown in the Ionian Islands region for 2016–2019, in terms of international arrivals and tourism revenues, as well as region have reported a slight decrease trend in overnights stays for the same period. According to Bank of Greece Border Research, in the Ionian Islands region, the average expenditure per trip increased from 611.9€ in 2016 to 627€ in 2019 and average expenditure per night reached 80.5€ in 2019, 10.5€ higher regard to relative numbers of 2016, while the length of stay decreased from 8.7 in 2016 to 7.8 in 2019 (Table 2).

However, the Ionian Islands tourism product is linked with the supply of a single product, generally sea- and sports-related holidays, and the presence of a primarily local or domestic tourism market. In terms of tourist accommodation establishments, the Ionian Islands region account for 5242 official accommodation facilities (hotels and rented rooms), in 2018, with a total capacity of 160,687 beds in which they recorded 10,231,098 overnight stays (Table 3). In 2018, the occupancy rate of bed places ranged to 63% for Hotels Establishments of the Ionian Islands region and 65% for Corfu's Hotels accommodations. According to SETE, in 2018, the Ionian Islands region being 6th most visited of thirteen regions of Greece (7.2% of international arrivals in Greece) and 6th among country's regions in terms of overnight stays (8.9% of Greece's total overnights). Region's structure of international tourist arrivals and receipts displayed that the vast majority of incoming tourists originated from North Europe (i.e., U.K, Germany) and neighbor (i.e., Italy) countries. Cumulatively, for 2016–2019, the share of the U.K market remain steadily in the first place, both in arrivals (32%) and receipts (36.4%) of the total market, followed by Italy in terms of arrivals (10.7%) and German in terms of revenues (11%).

Table 3 Accommodation capacity of Corfu, Ionian Islands Region, and Greece

		Hotel capacity					Rented rooms capacity	
		2015	2016	2017	2018	2019	2018	2019
Corfu	Units	400	394	394	402	404	1625	1625
	Rooms	23,584	23,772	23,772	24,180	24,691	8233	8233
	Beds	44,709	45,139	45,638	46,948	48,139	21,636	21,636
Ionian Islands	Units	931	930	933	965	980	4277	4254
	Rooms	47,631	47,888	48,121	50,066	51,445	25,954	25,816
	Beds	91,555	92,350	93,440	98,223	101,405	62,464	62,124
Greece	Units	9757	9730	9783	9873	9971	28,074	27,705[a]
	Rooms	406,200	407,146	414,127	425,973	433,689	188,853	186,585[a]
	Beds	784,315	788,553	806,045	835,773	856,347	438,936	433,626[a]

Source Hellenic Chamber of Hotels & INSETE Intelligence–Author's Editing
[a]*n* except rented rooms capacity figures of Western Macedonia Region

Located in Ionian Archipelago, Corfu Island, the 4th biggest (in population size) and 7th (inland size) among Greek islands and the 13th biggest in the Mediterranean Sea has already raised its tourism development in a period with undefined orientation for the country's tourism policy. Historically, Corfu's tourism development is based on two phases; in the first (until 1975), Island attracted tourists of high and medium incomes and in the second (since 1975) considered a pole of attraction for mass tourism of medium and low incomes (Tsartas, 1995). In Corfu, the spread of tourism activity started initially around the historical center of the municipality and northern side of the island to expand after the 1980s and in the southern (i.e., Kavos) and eastern coasts (Nisaki-Messonghi) of the island. Corfu's natural environment (i.e., coasts, green areas) and preserved structured environment (i.e., Old city of Corfu, which, in 2007, was added to UNESCO World Heritage List, following a recommendation by International Council on Monuments and Sites (ICOMOS)) defined the characteristics of its tourism product, which, unfortunately, until today, cannot be unhooked from the mass "Sun lust" tourism model and is characterized by high seasonality, as well as hospitality services offered mainly by Small and Medium-sized Tourism Enterprises (SMTE) (Table 4).

Studying the period 2015–2018, arrivals and overnights of foreigners and Greek tourists in Corfu's hotel accommodations highlighted an increasing trend (Table 5), while the reported numbers of international and domestic air arrivals (Table 6) present the seasonal tourism activity of the island and defines that Corfu is still considered as a summer seaside tourist destination. The international air arrivals for tourist season (June–September) of 2018 and 2019 represented 78% and 77.4%, respectively of international air arrivals per each year.

The income per capita is of the highest in the country because the local economy is dependant on tourism, while Corfu's tourism development is dominated by Tour Operators and considered in the stage of stagnation (Pappas, 2005). This dominant

Table 4 Structure of hotel establishments in Corfu and Ionian Islands 2011–2019

Hotels	Ionian Islands Region			Corfu		
	Units		Percentage change (%)	Units		Percentage change (%)
	2011	2019		2011	2019	
5*	23	62	169.5	14	25	78.5
4*	96	157	63.5	50	54	8
3*	203	251	23.6	83	100	20.5
2*	514	437	−15	183	175	−4.3
1*	78	73	−6.4	56	50	−10.7
Total	914	980	7.2	386	404	4.7

Source Hellenic Chamber of Hotels–Author's Editing

Table 5 Arrivals, overnights, and occupancy in Corfu's' hotels establishments

Year	2015	2016	2017	2018
Arrivals of foreigners	695,294	751,222	859,791	865,724
Arrivals of Greeks	75,868	92,324	91,954	101,565
Arrivals (Total)	771,162	843,546	951,745	967,289
Overnights of foreigners	4,193,280	4,660,605	4,949,987	5,036,660
Overnights of Greeks	245,076	304,074	284,455	310,519
Overnights (Total)	4,438,356	4,964,679	5,234,442	5,347,179
Occupancy (%)	62.8	66	66.3	65

Source Hellenic Statistical Authority & INSETE Intelligence–Author's Editing

Table 6 International and domestic air arrivals in Corfu

		2018	2019
Domestic air arrivals	(January–December)	165,039	166,553
	(June–September)	75,298	73,858
International air arrivals	(January–December)	1,502,305	1,457,420
	(June–September)	1,173,302	1,127,703
Total air arrivals	(January–December)	1,667,344	1,623,973
	(June–September)	1,248,600	1,201,561

Source Civil Aviation Authority–Author's Editing

pattern of tourism development proved to be problematic already from the early '80s, as it led to gigantism of the arbitrary construction of rented rooms and environmental pollution from the uncontrolled increase of tourist infrastructures (i.e., hotels, restaurants, etc.) (Tsartas, 1995). Pappas (2005) emphasized that the uncontrolled tourist infrastructure combined with the lack of urban and regional planning created a deterioration of green areas in the city of Corfu and that coastal zone is partially polluted. Taking into consideration that Corfu's tourism development depends on the availability of natural and/ or anthropogenic resources on the island, several environmental impacts have been emerged, especially during the seasonal- summer period where the consumption from locals and tourists is maximized, making it imperative to implement sustainable tourism practices for the development of Corfu tourism product.

2 Literature Review

2.1 Emergence of Sustainable Tourism Development

By the early 1990s, *sustainable tourism* was gaining ground among academics and practitioners to describe the range from the most rudimentary forms of alternative tourism to the most intensive manifestation of mass tourism. Hence, the idea of sustainable tourism has emerged as a primary objective of the global tourism sector since the mid-1990s as international governments and organizations focused their interest on destinations' natural and socio-cultural environment (Weaver, 2006). As a term, sustainable tourism development describes a specific type of tourism development that is balanced in the local, social, economic, cultural, and environmental structure of each tourist destination, while configuring conditions (e.g., services, infrastructure, etc.) for its continuous feedback, based into principal parameters displayed in Table 7.

Sustainable development promotes the pattern of tourism development that respects the destination's "tourism carrying capacity" (TCC) without any deterioration of its environmental identity (Kalafatis et al., 2003). TCC emerged as a remarkable scientific concept in the 1970s and 1980s, while practitioners and scholars proposed several definitions in broader literature (Coccossis & Parpairis, 1992; Chamberlain, 1997; Middleton & Hawkins, 1998; Coccossis & Mexa, 2004).

Table 7 Principal parameters of sustainable tourism development

• Special planning for tourism development aiming the balance among the society, the economy, and the environment
• Strengthening all measures (local development initiatives, functional interconnections between different sectors of the economy, research, education, marketing) that contribute to the feedback processes of the development
• The special institutional framework that will promote sustainable tourism development processes and local participation
• Promoting measures and policies that contribute to the protection and promotion of the local, natural, and structured environment
• Using Special and Alternative forms of tourism as a key axis of local tourism development

Source Coccossis et al. (2011), p. 93

World Tourism Organization (UNWTO, 1981) defined carrying capacity as "the maximum number of people that may visit a tourist destination at the same time, without destroying the physical, economic and socio-cultural environment and an unacceptable decrease in the quality of visitors' satisfaction", while this definition also adopted by MAP's Priority Actions Programme (PAP). TCC should be considered more as a strategic tool or index for sustainability, which will demonstrate the limit of tourism development in destinations to reduce its negative imprints on natural resources, socio-cultural patterns and land uses in local communities (Miller, 2001; Coccossis & Mexa, 2004). Thus, in the context of integrated and sustainable development, tourism becoming more and more a crucial factor for the development of Ionian Islands, as long as it is prior explored from a different perspective and removed from mimetic reproductions of an unbalanced and not eternally effective mass tourism model (Kalafatis et al., 2003), which is dominating Greek Tourism product already from the post-war period and has raised over time the popularity of country's island destinations (i.e., Rhodes, Mykonos, Corfu) for massive organized vacation tourism. Although investigating the assessment of tourism carrying capacity requires special studies that should take into account qualitative factors (i.e., a need for environmental and cultural heritage protection, local customs and traditions, etc.), a first approach to the issue for Ionian Islands region and Corfu can be achieved displaying some relevant quantitative indicators (Table 8). Corfu, in 2011, ranked in 2nd place both in Tourism Function Index (TFI) and Tourism Intensity Ratio (TIR), while Lefkas recorded the lowest TFI and Cephalonia and Ithaca the lowest TIR. According to Pearce (1987) classification regarding TFI rates, Ionian Islands Region, in total, as well as Zante and Corfu are classified in areas where tourism is considered a key sector for their development, while Lefkas, Cephalonia, and Ithaca are classified in areas where tourism considered important, but not the key sector for their development.

Table 8 Tourism carrying capacity index (2011)

Index	Ionian Islands Region	Corfu	Zante	Cephalonia and Ithaca	Lefkas
Tourist Function Index (TFI) (beds/population × 100)	43	42	71	27	23
Tourism Intensity Ratio (TIR) (tourist arrivals/population × 100)	561	581	889	302	335

Source Hellenic Chambers of Hotels and I.T.E.P (2016)–Author's Editing

2.2 Factors Underlying the Importance of Special Interest Tourism (S.I.T) and Alternative Tourism Products

By the early 1970s and 1980s, significant changes, which concerned initially tourists' motivations linked to nature, environment, local traditions, culture, sports, etc. (Tsartas, 1996; Middleton & Hawkins, 1998) have been recorded in tourism demand, and structurally ultimately affect the characteristics of tourism development at a local level and the formulation of alternative development patterns in tourist destinations (Coccossis & Tsartas, 2019). The monoculture of Greek tourism in the dominant "sun and sea" vacation model, the limited average length of stay of foreign tourists in hotel accommodation, and the intense seasonality of demand for the Greek tourist product, which present high rates of concentration mainly during the summer season in mature destinations which dominated by the pattern of unsustainable mass tourism are some of the crucial factors that underlined the need to promote sustainable tourism development programs. In this context, especially after the '80s, the development of Special Interest Tourism (SIT) and alternative tourism products are used as part of a two-goal program, on the one hand, offering a greater number of specialized services that expect the demand of special groups of tourists, and on the other hand, conceptualizing a development model that is balanced in the existing structure, by improving the pre-existing situation (i.e., destinations with massive organized tourism), or by forming a new productive structure in the tourism sector where infrastructures and services of special and alternative forms of tourism will dominate (Tsartas, 2004; Coccossis & Tsartas, 2019). Hence, Special Interest Tourism and Alternative tourism considered more relevant forms of engagement with the idea of sustainability and were conceived to encompass products and activities that were thought to be more appropriate than conventional mass tourism (Weaver, 2006). For discussion purposes, characteristics and criteria that underline Special Interest Tourism and Alternative tourism as ideal types for sustainable tourism development can be grouped into the four categories of attractions, accommodation, market type, and tourists' motivations and autonomous types of travels.

2.2.1 Attractions

By the early 1990s, destination attractions are in growing demand by non-local visitors, who recognized natural and man-made attractions as key elements of destination's environment which, individually and combined, serve as the primary motivation for holiday tourism visits (Middleton & Hawkins, 1998). According to Tsartas and Sarantakou (2014), SIT market assists practitioners to understand that everything and different things can be a tourism resource if there is a demand, such as cycling and running tours, wineries tours, and local gastronomy, music festivals, thematic parks for children, etc. The growth of SIT and alternative tourism market has been extremely linked with the preference of modern-day visitors for "authentic" cultural, historical and natural attractions that are perceived to capture a destinations' unique sense of place and allow for interactions between visitors and locals (Weaver, 2006).

2.2.2 Accommodation

Accommodation is, by definition, an integral part of the tourism product and may not necessarily be part of the environmental attractions or quality of the places in which it is located (Middleton & Hawkins, 1998). Alternative tourism accommodations tend to be small-scale facilities (i.e., houses of locals, guesthouses, monasteries and retreats, tents, youth hostels, dormitories, etc. (Weaver, 2006). In contrast to the non-local corporate ownership of large hotels and resorts, SIT and alternative tourism hospitality services are offered mainly by Small and Medium-sized Tourism Enterprises (SMTE), as those controlled mainly by locals, or by the local community more generally.

2.2.3 Market Type and Tourists' Motivations

According to Tsartas and Sarantakou (2014), studying tourists' motivations assists to understand the Special Interest Tourism market segmentation and thus the demand for exclusive tourism products. Alternative tourists are free and independent travelers (*FITs*) who avoid high volume package tour arrangements and prefer to travel as individuals or in small groups, often in the off-season period, tend to explore more authentic places than well-known "tourist destinations" (Weaver, 2006). Also, visitors started to avoid the mass marketing system and started seeking more autonomous types of travel and preferred choosing individual offbeat places, travel timetable, accommodation type, travel means, prices, and costs, attractions visited, etc. (Tsartas, 1996; Weaver, 2006; Coccossis et al., 2011; Tsartas & Sarantakou, 2014). Thus, a significant turn both in offer and demand was based on multiple-motive new types of customers, which expect to satisfy different needs, tourism services, and infrastructures in the visited destination. In Europe and North America, customers' interest in the heritage, social customs, locals' way of life, and culture of the visited destination are revealed as principal motivators for international travels and visits to cities, towns, and rural areas away from traditional beach resorts, while customers' interest in particular ecotourism and "green" products express notoriously their attention in and demand environmental quality (Middleton & Hawkins, 1998). It is worth mentioning that modern-day tourists, although traveling based on a strong special motivation, often operate in the destination area multidisciplinary, choosing activities and services both of different special and alternative tourism, but also services of conventional mass tourism.

2.2.4 Autonomous Types of Travels

In contrast to conventional mass tourism products, a key feature of special and alternative tourism products is the provision of a significant degree of freedom and ultimately the direct and indirect participation of tourists in the design of the tourist product (Coccossis et al., 2011). The rapid growth of the Internet has contributed also heavily to this trend, providing information, different prices, booking options, determining, in parallel, that the organization of travel is no longer an exclusive hypothesis for Tour Operators and Travel agencies (Tsartas & Sarantakou, 2014). Internet and Web sites open up potentially limitless access on a global basis, as destination information systems bring smaller business into information networks (Poon, 1993; Middleton & Hawkins, 1998), as the importance of independent travel and online bookings is increasing and desire for personalization is becoming more pronounced (i.e., business trip, visiting relatives, etc.) (Gudurash, 2014). The differentiation provided by special and alternative tourism products matches customer segmentation and promotes destination qualities, as modern-day tourists tend to be more sophisticated and travel-experienced visitors who customized their travel arrangements and choose different types of travels during the year.

3 Methodology

The methodology used in the current manuscript is a critical review of the literature. The author has used a variety of sources and information based on popular online bibliographic databases (i.e., Science Direct, Research Gate, etc.) and scientific search engines such as Google Scholar. The types of bibliographic sources included in the manuscript are articles published on scientific books, academic journals, conference proceedings, thesis, online sites and journals, and several studies from national and international organizations. In particular, a significant source of information has been the Greek Tourism Organization and SETE, which both provide online databases with statistical figures regarding tourism in Greece and its regions. The selection criteria of these bibliographic references were based on the relevance to the manuscript's topic and this survey described more of an illustration of the topic.

4 Discussion

Taking into consideration that the development of tourist activity in island complexes and coastal areas is extremely linked to the traditional massive organized vacation tourism product, the turn to more sustainable development of tourism and more autonomous patterns of travel has been gaining more and more ground over the last 40 years. Sustainable tourism development should be based on the quality upgrade and enrichment of the Greek tourism product through the upgrade of infrastructure and services related to SIT Market, to build a quality market that will gradually formulate and contribute to the upgrade of the profile of Greek tourism. The Greek Ministry of Tourism to diversify the Greek tourism product underlines the significance of special interest tourism and alternative tourism products, which allow the spread of tourist activity throughout the year. Additionally, Greek National Tourism Organization (GNTO) focused on the quality improvement of tourism infrastructure and hospitality services and its strategic plans attempt to promote the hospitable image of Greek tourism worldwide.

In Corfu Island, in particular, the development of tourism up to date has created mature tourist areas, and/ or many areas have been driven into a crisis because having exceeded their tourism carrying capacity. Corfu's tourism development is linear and one-dimensional along the coasts. The hotels of Corfu are located mainly on its eastern coasts, from the area of Messonghi to Kouloura, while on the other coasts of the island there are various zones of tourist development. The over-concentration of tourism activity in Corfu, since the early '70s, following the construction of hotel infrastructure to serve the rapidly growing demand, which is synonymous with mass demand for services and infrastructure associated with the dominant 3s model of tourism development, has been created the first environmental problems and transform specific zones to tourist areas with large hotels (i.e., Palaiokastritsa, Glyfada, Ermones, Livadi, Benitses, Roda), altering the character of these areas, whilst the strong presence of Small and Medium-sized Tourism Enterprises providing low to middle-level hospitality services, is also obvious in other areas (i.e., Kavos, Kassiopi, Acharavi, etc.). The special planning and implementation of a sustainable tourism development model, with the participation of all stakeholders (i.e., local authorities, private and public sector executives, local community, etc.), is the most appropriate way to formulate a long-term effective strategy, where the promotion of "locality", as a key parameter of sustainable tourism, will affect positively the boost of special and alternative tourism products. This trend, consequently, will contribute to the qualitative upgrade of the image of Corfu's tourism, to the improvement of hospitality services provided to local and foreign visitors, and will allow the diffusion of the benefits of tourism development in the local community. Hence, sustainable tourism development, using special and alternative forms of tourism considered the most appropriate model for local tourism development. This type conceptualizes the key figures of an effective tourism strategy, which based on the specifics, needs, possibilities, and weaknesses of Corfu's tourism product can define the development course of the island and its reintegration into the global tourist market by providing a complex of high-quality tourism products and services. Principal parameters for special planning of sustainable tourism development in Corfu are presented in Table 9.

Table 9 Principal parameters of sustainable tourism development in Corfu

Parameter	Implementation axes
Infrastructures	• Investments in basic infrastructure (i.e., airports, ports, marinas, roads, water supply, renewable energy sources, telecommunications, etc.)
	• Investments in special infrastructure (i.e., leisure and sports facilities, hiking trails, cycling, horseback riding, etc.) to improve the quality of life of locals and visitors and attract high tourist income
	• Investments in tourist facilities with special uses (i.e., Hydrotherapy centers, swimming pools, etc.)
	• Upgrading the hospitality infrastructure of the monastic complexes for overnight and feeding pilgrims and religious tourists
	• Upgrading the hospitality infrastructure of the existing farmhouses, ranch, wineries for overnight and feeding "green" tourists, eco-tourists, wine tourists, etc.
Environment	• Protection and promotion of the natural environment (coasts, mountainous areas, etc.) and classification of certain areas as "nature reserve" and areas of "special natural beauty"
	• Improving accessibility in these areas (i.e., hiking trails, special routes for cycling, horseback riding, etc.)
	• Protection of terrestrial ecosystems from arbitrary activities (i.e., hunting, free access of wheeled vehicles, the illegal opening of the road network, fires, etc.)
	• Protection of marine ecosystems from marine pollution due to the use of the sea as a recipient of domestic and industrial waste
	• Solid waste management and problem-solving related to ungoverned and sprawling residential development
	• Protection from aesthetic pollution, either due to a structured space or due to ways of promoting services (advertisements, buildings, etc.)
Accommodations	• Monitoring and controlling the quality of hotels' accommodation, operating both on a seasonal and/ or yearly basis
	• Institutionalizing a type of catalog/ list that will segment the types of accommodation (i.e., Hotels, apartments, rooms for rent, camping, etc.), as those will be subject to an on-site inspection by competent authorities for their suitability and operation levels
	• Investing in a small number of high-quality accommodations in selected areas (i.e., in mountainous and/or less developed tourist areas of the island)
	• Investing in existing types of accommodation to provide more high-quality products and services
Services	• Continuous and quality training programs for employees of the tourism sector, and public and private sector executives (i.e., participation in national/ international congresses, conferences, and seminars for the development of sustainable tourism at national/ international level, synergies with Ionian University for exploitation the graduates of tourism studies, etc.)

(continued)

Table 9 (continued)

Parameter	Implementation axes
	• Development of new high-quality tourism products and services based on the use of skills and innovative practices (i.e., maritime and yachting tourism, cultural tourism, etc.), promoted mainly by modern methods of marketing
	• Upgrading the communication channels and visitors' information sources (on-line sources, printed materials, media shows, travelogues, broadcasts, etc.)
	• Organizing a small number of high qualities and international recognition of cultural, artistic, and sports events throughout the year

Source Author's Editing

5 Conclusions

The rising volume of Greek tourism is made up of a variety of natural, cultural, and historical attractions spreading throughout the country's regions. Corfu Island is considered a well-known seaside leisure destination that offers sea, sun, Mediterranean climate, as well as a rich multi-cultural heritage and historic and cultural monuments. The tourism sector is still a pillar of support for the local economy, while demand for special and alternative tourism products is still growing, but at a slow pace. For sustainable tourism development, however, several tourism destinations, such as Corfu, an island that can provide alternative and more sustainable tourism solutions, besides the traditional unsustainable mass tourism infrastructure, offer services that delivered modern-day tourists' expectations and needs for new tourism products. Taking into consideration the long-run benefits of the sustainability agenda, a bottom-up approach that involved the different characteristics of those products can be systemically developed at the local level by elected local governments and public sector managers who recognized or supported sustainable goals. Nevertheless, the general picture about the emergence of sustainable tourism development in the Corfu Island and Ionian Islands Region could be derived from the present analysis, which discussed the principal parameters for special planning of sustainable tourism development on the island. Corfu can be unhooked by the massive pattern of tourism development and/ or tend to be more competitive into the global tourist market by providing a complex of high quality special and alternative tourism products and services. As tourism, compared to the other economic sectors, is considered a dynamic sector of the local economy, local, regional and international stakeholders should provide, implement and encourage sustainable development programs aiming at protecting the environmental and historical- cultural characteristics of each region.

References

Chamberlain, K. (1997). *Carrying capacity, UNEP industry, and environment 8 (January–June 1997)*. Paris: UNEP.

Coccossis, H., & Parpairis, A. (1992). Tourism and the environment: Some observation on the concept of carrying capacity. In H. Briassoulis, & J. van der Straaten (Eds.), *Tourism and the environment: Regional, economic and policy issues*. Dordrecht Kluwer Academic Publishers.

Coccossis, H., & Mexa, A. (2004). *The challenge of tourism carrying capacity assessment: Theory and practice*. Basingstoke, Hampshire: Ashgate.

Coccossis, H., & Tsartas, P. (2019). *Βιώσιμη Τουριστική Ανάπτυξη και Περιβάλλον, 2η Έκδοση*. Athens: Kritiki (In Greek).

Coccossis, H., Tsartas, P., & Gkrimpa, E. (2011). *Ειδικές και Εναλλακτικές Μορφές Τουρισμού: Ζήτηση και Προσφορά Νέων Προϊόντων Τουρισμού*, Athens: Kritiki (in Greek).

Guduraš, D. (2014). *Economic crisis and tourism: Case of the Greek tourism sector. Ekonomska misao i praksa, 2*, 613–632. Retrieved from: https://hrcak.srce.hr/130855. Accessed April 18, 2019.

Hellenic Chamber of Hotels. (2020). *Δυναμικότητες*. Retrieved from: https://www.grhotels.gr/category/epicheirimatiki-enimerosi/statistika/dynamikotites/. Accessed April 18, 2019.

Hellenic Chambers of Hotels and I.T.E.P. (2016). *Tourism characteristics of Ionian Islands Region*. Retrieved from: https://hotelmag.gr/wp-content/uploads/2016/05/Ionia-Nisia_Tourism_Characteristics_180516.pdf. Accessed April 18, 2019.

Hellenic Statistical Authority. (2018). *Αφίξεις και Διανυκτερεύσεις στα Καταλύματα Ξενοδοχειακού Τύπου και Κάμπινγκ: Έτος 2017*. Retrieved from: https://www.statistics.gr/documents/20181/658333dd-a13f-48f0-9023-433d0721d8cb. Accessed April 18, 2019.

Hellenic Statistical Authority. (2019). *Αφίξεις και Διανυκτερεύσεις στα Καταλύματα Ξενοδοχειακού Τύπου και Κάμπινγκ: Έτος 2018*. Retrieved from: https://www.statistics.gr/documents/20181/1fc93e78-6c15-7551-3995-67fe03c67e1c. Accessed April 18, 2019.

Kalafatis, Th., Pakos, Th., & Skountzos, Th. (2003). *Επτάνησα 20ος-21ος Αιώνας*. Οικονομία, Δημογραφία, Περιβάλλον και Πολιτισμός, Athens: Panteion University (In Greek).

Middleton, V. T. C., & Hawkins, R. (1998). *Sustainable Tourism: A Marketing Perspective*. Butterworth-Heinemann.

Miller, G. (2001). The development of indicators for sustainable tourism: Results of a Delphi survey of tourism researchers. *Tourism Management, 22*, 351–362.

Pappas, N. (2005). Mediterranean tourism: Comparative study of urban Island host destinations. *International Journal of Tourism Research, 7*(4–5) (International Conference on Tourism Development and Planning, 11–12 Jun 2005, Patras, Greece).

Pearce, D. (1987). *Tourism today: A geographical analysis. Longman scientific and technical*. Wiley.

Poon, A. (1993). *Tourism, technology, and competitive strategies*. CAB International.

SETE. (2019). *The contribution of tourism to the Greek economy in 2018*. Retrieved from: https://www.insete.gr/Portals/0/meletes-INSETE/01/2019/2019_Tourism-Greek_Economy_2017-2018_EN.pdf. Accessed July 15, 2019.

Tsartas, P. (1995). *Οι Κοινωνικές Επιπτώσεις του Τουρισμού στους Νομούς Κέρκυρας και Λασιθίου*. EKKE (In Greek).

Tsartas, P. (1996). *Τουρίστες, Ταξίδια, Τόποι: Κοινωνιολογικές Προσεγγίσεις στον Τουρισμό*. Exandas (In Greek).

Tsartas, P. (2004). *Επτά Βασικές προϋποθέσεις για μια βιώσιμη τουριστική ανάπτυξη, Η ΚΑΘΗΜΕΡΙΝΗ*. Retrieved from: https://www.kathimerini.gr/197313/article/oikonomia/ellhnikh-oikonomia/epta-vasikes-proupo8eseis-gia-mia-viwsimh-toyristikh-anapty3h. Accessed June 20, 2020.

Tsartas, P., & Sarantakou, E. (2014). From tourism to "tourisms": Destination management the key factor, Voland 2014. In *1st International Conference "Volcanic Landscapes"*, Santorini.

UNEP/MAP/PAP. (1997). Guidelines for carrying capacity assessment for tourism in mediterranean coastal areas, Priority Action Programme, Regional Activity Centre, Split.

UNWTO. (1981). *Saturation of tourist destinations.* Report of the secretary-general.

UNWTO. (2001). *Tourism 2020 vision vol. 7 global forecast and profiles of market segments.* Retrieved from: https://www.e-unwto.org/doi/pdf/10.18111/9789284404667. Accessed April 18, 2019.

UNWTO. (2020). *World tourism barometer*, vol. 18, Issue 1 January 2020. Retrieved from: https://www.e-unwto.org/doi/pdf/10.18111/wtobarometereng.2020.18.1.1. Accessed April 18, 2019.

Weaver, D. (2006). *Sustainable tourism.* Routledge.

World Economic Forum (WEF). (2019). *The travel & tourism competitiveness report 2019.* Retrieved from: https://www.weforum.org/reports/the-travel-tourism-competitiveness-report-2019. Accessed April 18, 2019.

World Travel & Tourism Council (WTTC). (2019). *Economic impact 2019, Greece.* Retrieved from: https://wttc.org/en-gb/Research/Economic-Impact. Accessed April 18, 2019.

Sustainable Tourism; Vector of the Social and Solidarity Economy: Case of Region Souss Massa, South of Morocco

Asma Edaoudi, M'bark Houssas, and Abdelhaq Lahfidi

1 Introduction

Sustainable tourism is defined according to the World Tourism Organization (WTO) as *"tourism that takes full account of its current and future economic, social and environmental impacts, while meeting the needs of visitors, professionals, and environment and host communities"*. Sustainable tourism is applicable to all tourist sites on the condition of respecting the balance between environmental, economic, and socio-cultural aspects. This ensures the viability of the environments in question.

The notion of sustainable tourism developed since the 1990s. In the year 1992, at the Rio Summit, the terms of sustainable development and tourism were associated, but true tourism practice was only mentioned for the first time in 1995, during the conference Lanzarote Sustainable Tourism World Conference, which defines the principles of sustainable tourism.

Subsequently, in 1999, the general assembly of the WTO adopted the world code of ethics for tourism which declined various recommendations to guide the actors of the territory. It is about promoting sustainability through a balance between environment, economy, social, and culture (Luciani, 2016). It was only in 2002, at the World Summit on Sustainable Development in Johannesburg that sustainable tourism was recognized as a management strategy for the protection of natural resources. The culmination of this recognition takes place 10 years later, in 2012, at the Rio + 20 conference, during which new reflections on the sustainability of economy, generated by sustainable tourism, were evoked as an instrument of developing the territory. In addition, in 2004, the principles of sustainable tourism were updated by the Sustainable Development Committee. Since Cop 21 in Paris, in 2015, sustainable development and therefore sustainable tourism is no longer just concepts but have become meanings of implementing actions (Ruiz, 2014). The year 2017 is even more

A. Edaoudi (✉) · M. Houssas · A. Lahfidi
Ibn Zohr University, ENCG Agadir, Agadir, Morocco

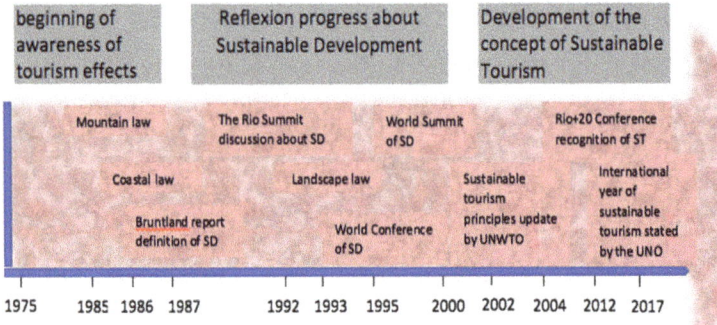

Fig. 1 Evolution frieze of the theory concerning the sustainability to tourism

promising for sustainable tourism because the United Nations declares it the international year of sustainable tourism for development. The aim of this proclamation is to develop sustainable tourism internationally and therefore promote comprehension between all people, acceptance of different civilizations, and increase the appreciation of the values inherent in various cultures. This year allows *"To increase the contribution of the tourism sector to the pillars of sustainability"* and therefore draw attention to the dimensions and repercussions of this sector, which are often undervalued. This promotion of tourism sustainability in 2017 follows the declaration at Rio + 20 conferences *that states: well-designed and well-organized tourism can contribute to dimensions of sustainable development and the creation of jobs and commercial outlets.* "Fig. 1 shows all these thoughts evolving in the direction of developing a sustainable tourism practice".

2   Literature Review

2.1   *Sustainability and Sustainable Tourism Development*

Sustainability is one of the newest degree subjects that attempts to bridge social science with civic engineering and environmental science with the technology of the future. As a set of goals sustainability describes the desired conditions of the environment and the ability of humans to receive benefits directly and indirectly from it, in the present as well as in the future. As practices and behaviors, sustainability describes human actions that support and enhance the human well-being derived through interaction with the environment, and which support the ability of human society to interact with the environment in ways that discourage the reduced benefits.

Sustainable tourism development has attracted significant attention in many scientific studies particularly in tourism studies and has been one of the very fast-growing areas of tourism studies research since the late 1980s. According to Buckley (2012) the specific term "sustainable tourism" was first used almost two decades ago. During the first decade, basic frameworks from backgrounds in tourism, economics, and environmental management were studied. The second decade yielded a number of reconceptualization and a series of critiques including Sharpley (2000), Gossling (2002), Liu (2003), Saarinen (2006), Lane (2009b), and Liu et al. (2013). According to Bramwell and Lane, the two greatest founders of these concepts in the tourism industry, sustainable tourism emerged in part as a negative and a reactive concept in response to the many tourism issues, such as environmental damage and serious impacts on society and traditional cultures (Bramwell & Lane, 1993). Gradually, tourism development has been seen as a solution capable of creating positive changes through the ideas of sustainable tourism. Sustainable tourism has played an important role in identifying ways to secure positive benefits, as well as the established approaches of regulation and development control (Bramwell & Lane, 2012).

There are a large number of definitions of sustainability and sustainable development. The best-known definition of sustainable development is *"development that meets the needs of the present without compromising the ability of future generations to meet their own needs"* (WCED, 1987). This definition implies the connections between economic development, environmental protection, and social equity, each element reinforcing the other. The World Tourism Organization (WTO, 2001) defined sustainable development as follows:

> Sustainable tourism development meets the needs of present tourists and host regions while protecting and enhancing opportunities for the future. It is envisaged as leading to management of all resources in such a way that economic, social and aesthetic needs can be fulfilled while maintaining cultural integrity, essential ecological processes, biological diversity and life support systems.

Liu et al. (2013) highlights the precise definition of "sustainability", which implies the significant role of states in preparing a steady progress in life conditions for generations to come; "sustainable development" is more process-oriented and associated with managed changes that cause improvement in conditions for those involved in such development; and "sustainable tourism" is defined as all types of tourism that are compatible with or contribute to sustainable development. "Sustainable tourism" requires both the sustainable growth of tourism's contribution to the economy and society and the sustainable use of resources and the environment, which will be gained by a deep understanding and proper management of tourism demand (Liu et al., 2013). Liu and Seaton (1994) defined tourism development as a dynamic process of matching tourism resources to the demands and preferences of actual or potential tourists.

The concept of sustainability is a wide approach everybody is talking about in a period when environmental problems caused by various human activities are requiring serious solutions. As it is well known, the concept found its roots in the United Nations' 1987 Brundtland Commission Report *"Our Common Future"* and even earlier in the 1980s World Conservation Strategy. Starting from a 'pure' ecologically-based concept in the 1970s and in the World Conservation Strategy, it transformed very quickly into a more comprehensive socio-economic approach.

> In essence, sustainable development is a process of change in which the exploitation of resources, the direction of investments, the orientation of technological development and institutional change are all in harmony and enhance both current and future potential to meet human needs and aspirations. (ibid., p. 46)

Sustainable Development, is the development which "meets the needs of the present without compromising the ability of future generations to meet their own needs" (Eber, 1992: 1). Thus, some researchers suggest that there may be a symbiotic relationship between Tourism and the environment (Mathieson & Wall, 1989: 102). It is also a development that is likely to achieve lasting satisfaction of human needs and improvement of the quality of human life (Allen, 1980). It helps to meet the needs of the present generation without compromising the ability of future generations to meet their own needs (Brundtland Commission, 1987), It continues indefinitely because it is based on the exploitation of renewable resources and causes insufficient environmental damage for this to pose an eventual limit (Allaby, 1988).

The basic idea of sustainable development is simple in the context of natural resources (excluding exhaustibles) and environments: the use made of these inputs to the development process should be sustainable. through time. If we now apply the idea to resources, sustainability ought to mean that a given stock of resources— trees, soil quality, water, and so on—should not decline (Markandya & Pearce, 1988). The sustainable development concept constitutes a further elaboration of the close links between economic activity and the conservation of environmental resources. It implies a partnership between the environment and the economy, within which a key element is the legacy of environmental resources that is not "unduly" diminished (OECD, 1990).

Sustainable Tourism, "is the ideal model of Tourism which is able to operate the system, by creating a profitable and 'healthy' environment from the Tourism industry, during the time" (Harrison, 1996: 35–41). The purpose of sustainable tourism is to make a balance between protecting the environment, establishing social justice, maintaining cultural integrity and promoting economic benefits and meeting the needs of the entire population in terms of improved living standers both in the short and long term (Liu et al., 2013). It has played an important role in identifying ways to secure positive benefits, as well as the established approaches of regulation and development control (Bramwell & Lane, 2012). Operates in harmony with the local environment, community, and culture, so it can become create permanently benefits and minimize the negative effects of development. Achieving sustainability depends on a balance of private initiative, economic instruments and regulation, translating global principles into focused local action, and new public–private sector delivery mechanisms. This may bear a new and necessary tourism culture that focuses on the environment as a valid raw subject for sustainable tourism development.

The United Nations' Environment Programme and World Tourism Organization say sustainable tourism should:

- Make use of environmental resources in a way that maintains essential ecological processes and helps to conserve the natural heritage and biodiversity.

- Respect the authenticity of host communities, conserve the cultural heritage and traditional values, and contribute to inter-cultural understanding and tolerance.
- Ensure viable, long-term economic operations, providing benefits across the community—including employment and income-earning opportunities.

2.2 Social and Solidarity Economy

The social and solidarity economy (SSE) is a concept that refers to enterprises and organizations, in particular cooperatives, mutual benefit societies, associations, foundations, and social enterprises, which specifically produce goods, services, and knowledge while pursuing economic and social aims and fostering solidarity.

In recent years, there has been an increasing number of organizations in the productive and social, and personal services industries that are based on principles of cooperation, self-management, and free association. Indeed, the expansion of these kinds of organizations has led to programmes and actions in both the public and private sectors and consideration of actively promoting them as a means of generating income and improving the quality of life (Morais & Bacic, 2009).

The social and solidarity economy (SSE) refers to specific forms of organizations and enterprises, the most common of which are cooperatives, mutual societies, associations, community organizations, social enterprises, and some foundations. The SSE offers many advantages to address social, economic, political, and environmental challenges worldwide, including social cohesion, empowerment, and the recognition of a plural economy (Fonteneau et al., 2010). It is undoubtedly a sector worth studying in greater depth, bearing in mind its contribution in dealing with the present socio-economic reality and in creating green jobs, income, social inclusion, and environmental awareness.

In this context, many programmes aim to combat inequality, reduce poverty and, as a consequence, improve the standard of living for a large part of the population.

3 Database

The economic contribution of tourism activity in Morocco is certainly important. It is increasingly considered by senior officials as a strategic sector to support economic development. But what about the environmental and social costs as well as for sustainable development? The main objective of this article is to explore the aspects that make tourism a driving force for sustainable development in Morocco and especially in the region of Souss Massa. Sustainable tourism activity can be transposed to all tourism activities, it acts on the behavior of consumers and producers in the sector to ensure respect for the environment and the resources they use. The aim of this practice is to educate the population to produce and consume sustainably throughout the supply and demand chain. Thus, strategies and policies are necessary for all the actors of the territory.

Due to the existence of problems such as lack of information concerning sustainable tourism in the region Souss Massa, and in order not to place any restriction on our sample, we have tried to develop our own database by combining several sources of information. Our goal is to have as much information as possible on as many companies are aware of the importance of sustainability and adopt it as a strategy in their activities. We took a sample of 25 companies, hotels, council offices, and public bodies in the region of Souss Massa. We questioned them whether it was by distributing surveys or establishing interviews, we also used the computerized database of some public bodies and the annual activity reports of a few companies.

4 Methodology

We adopt here a quantitative and qualitative approach where we have based our paper on several research works, distributing paper questionnaires, online surveys, and establishing interviews.

The target population is made up of establishments in Souss Massa whose main activity is associated with one or other of the following five sectoral groups and associated with tourism (Fig. 2). In the Fig. 3 the percentage of establishments survived for companies/corporations is 40%.

Juridical form	Establishments survived	%
Compagnies/Corporations	10	cc
Individual companies	7	28%
A non-profit organizations	5	20%
corporation partnership)	2	8%
Cooperative	1	4%
Total	25	100%

The survey plan included mixed data collection (web and telephone) and aimed to complete 10 interviews.

The sample of 25 units was processed in two phases:

Phase 1:

All companies with an email address were invited to answer the questionnaire on Mars 6 and 7, 2020.

All companies without an email address were included in the telephone collection from Mars 7, 2020.

Juridical form	Establishments survied	%
Compagnies/ Corporations	10	cc
Individual companies	7	28%
A non-profit organizations	5	20%
corporation partnership)	2	8%
Cooperative	1	4%
Total	**25**	**100%**

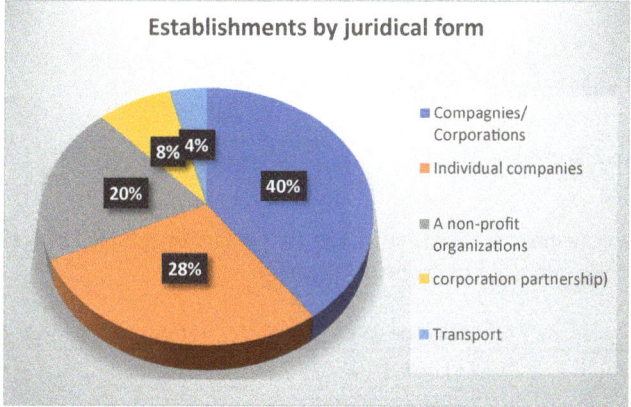

Fig. 2 Establishments by juridical form

Phase 2:

Companies that did not respond to the web component have been added to the telephone component as of Mars 15, 2020. It was initially known that the desired number of interviews in the transport sector would be unachievable. It was agreed to distribute these "missing" interviews to the other sectors.

The original questionnaire was provided by the client, then revised, translated.

The telephone version of the questionnaire was tested with 4 respondents on February 25 and 26, 2020. Changes made to the questionnaire following the test were approved by the client. Interviews were conducted in French or Arabic, at the choice of the respondent. They lasted an average of 20 min on the phone and 15 min on the web.

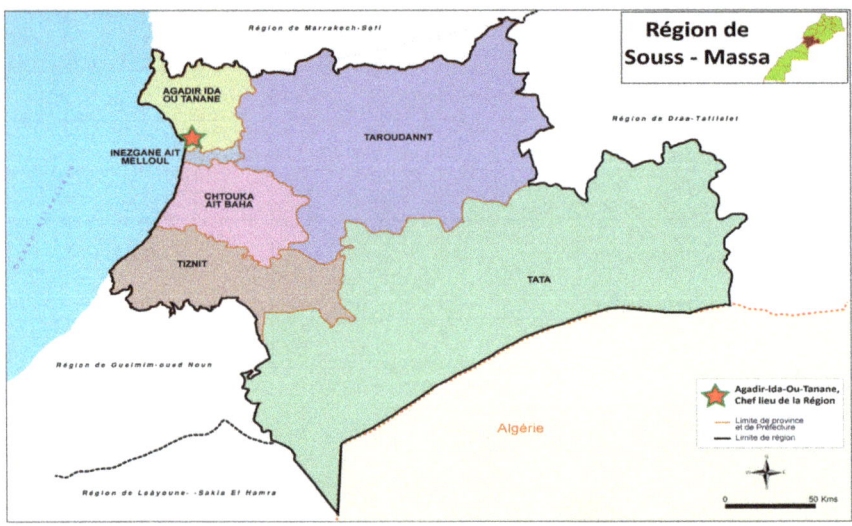

Fig. 3 Region map of Souss-Massa

The data collection was carried out in two ways: telephone collection and web collection. Like indicated in the board below:

	Collection date	Response rate (%)
Telephone	7 Mars 2020	20.3
Web	15 Mars 2020	12

The data was first validated in order to eliminate any errors, outliers or other anomalies. The data was weighted for all eligible and ineligible respondents, by simple extrapolation to the joint distribution (geographic area and sector).

5 Results

5.1 Sustainable Development Within the Region of Souss Massa

The region of Souss-Massa is the first seaside resort of Morocco, and it is renowned for its western atmosphere. It is also the first tourist hub in the country and has the first fishing port of Morocco. It is bordered to the west by the Atlantic Ocean, to the north by the region of Marrakech-Safi, East by the Drâa-Tafilalet region, to the south-east by Algeria, and to the south-west by the region of Guelmim-Oued Noun. The region is bordered on the north by the mountain range of the High Atlas, following the valley of Wadi Souss, it is crossed in the center by the Massa river and the Anti-Atlas, and bordered on the south by the Drâa River.

It counted in the 2014 census approximately 2,676,847 inhabitants spread over an area of 30,321 km². The region is made up of mountain ranges, plains, and plateaus and surrounded by both the desert and the Atlantic Ocean. It has a park that is frequented by 46 species of mammals, 40 species of reptiles and amphibians, and 9 species of fish, more than 275 species of birds observed, and many species of Lepidoptera. There are 300 species of plants, 13 of which are endemic to the southwest of Morocco. The biodiversity of the Souss-Massa region is particularly interesting for the combination of Palearctic and Afro-tropical species but also for a very interesting endemic compontent.

The pre-Saharan Souss-Massa Region south of the High Atlas Mountains covers two types of regional destinations, coastal tourism along the Atlantic Ocean (especially around Agadir) and rural destinations of mountains and valleys south of the High Atlas. Almost every regional destination in Morocco has seen smaller private tourism initiatives which tend to focus on sustainability (at least in a broader sense). However, the RDTR (The Rural Tourism Development Network) in the Souss-Massa Region appears to be the most substantial private initiative in Morocco taking a sustainable tourism perspective.

A comprehensive study on sustainable tourism activities in the Souss-Massa Region undertaken between 2012 and 2014 revealed that, among the traditional hotel sector (mostly run by national and international hotel chains) in the region, only Club Robinson (owned by the tour operator TUI) had a significant orientation towards the goals of sustainable development (Boudribili, 2014, p. 69 et seq. & p. 90 et seq.). Apart from this beach resort, all the other hotels and holiday resorts in the study fulfilled only the (few) legal requirements concerning the impact of their business on the environment, avoiding any further commitment to the idea of sustainability.

A second example of good practices in the Souss-Massa Region was an owner-operated lodge, the Ecolodge Atlas Kasbah. This location focuses on rural tourism and practises a comprehensive sustainability-oriented approach with a broad environmental management plan concerning water, energy, and waste management. Apart from that, social aspects concerning the integration of the local population and awareness-raising among visitors are included as well, following a broader Corporate Social Responsibility approach. The sustainability orientation is combined with a high level of quality of the product, including different activities ranging from conventional excursions to workshops on traditional crafts (e.g., pottery, soap making). While the sustainability orientation of Club Robinson limits itself to rather isolated internal activities, the owner of the Atlas Kasbah can be seen as one of the rare examples where individual activities make a broader impact.

The owner of the Atlas Kasbah has been one of the key persons since 2011, when a few tourism professionals and academics founded the RDTR with the encouragement of the regional council. Its target has been to bring together small tourism stakeholders in the rural milieu of the region in order to structure and organize the rural tourism sector, organize and manage all common interests and promote the practices of rural tourism (promotion, communication, marketing assistance), promote the exchange of good practices, ensure the quality of rural tourism products, participate in the development of sustainable and responsible rural tourism, and build partnerships with regional, national and international tourism stakeholders" (RDTR & Afkar, 2013, p. 6, translation A. Kagermeier).

The primary purpose of the network is to strengthen the performance of small tourism activities by building a network focusing on capacity building and providing the service orientation necessary to successfully participate in the tourism market—which is marked by a high degree of competition. Apart from the development and promotion of the rural tourism product itself, the sustainability orientation has been an integral part of the RDTR's mission from the very beginning.

At the same time, a number of marketing activities were started—especially using internet and social media tools (RDTR & Afkar, 2013), which are easily accessible, and thus have low barriers to entry and do not result in high monetary costs. In spite of these various activities, the number of members stagnated, so the optimistic hope that membership would rise from 70 to an estimated number of 200 by 2013 (RDTR, 2012, p. 4) was never fulfilled.

It is difficult to evaluate the reasons for the stagnation of the members. Perhaps the focus of the key players exercising leadership in the network has been a little bit too much oriented towards the representative function of conferences. It also must be noted that the spatial focal point of the network lies in the immediate surroundings of Agadir, so perhaps the intensity of interaction decreases significantly with the distance from the spatial (and functional) core node of the network, even if formally each of the provinces in the Souss-Massa Region were to send an equal number of delegates.

In order to strengthen the competitiveness of tourism in the region as well as the sustainability orientation in 2012, the RDTR began working in cooperation with the University Ibn-Zohr in Agadir to establish a Charter of Quality and the Environment in Tourism (QET). The general meeting agreed on this charter in 2013 (Boudribili, 2014, p. 121). The charter opened the way for a classification scheme focusing on two core aspects: product quality and sustainability, including classical environmentally-oriented aspects, as well as attention to human resources, the local population, raising visitor awareness, and other aspects related to an integrated understanding of sustainability. Four different grades are possible, ranging from a "basic" level to an "excellent" level. Two-thirds of the classification points deal with quality aspects and one-third with sustainability aspects (Boudribili, 2014, p. 124). The classification scheme has so far been open to RDTR members only (thus functioning as a point of added value for members and with the intention of stimulating interest in joining the network). Of the 48 RDTR members owning an accommodation establishment or a restaurant, 44 have been evaluated for classification. Only two establishments were excluded from the classification outright (Boudribili, 2014, p. 127), which might be interpreted as a selective interest in the network by those owners already showing a minimum level of interest in quality and sustainability.

However, it can also be seen as ambivalent if the classification institution and the classified establishments are in a specific relationship. At the same time, only 14% of the theoretically possible points have to be fulfilled to be classified at the basic level (Boudribili, 2014, p. 124). This seems to be a relatively low barrier to entry, attracting even establishments with rather moderate quality and sustainability ambitions hoping to further orient their business on issues of sustainability in the future as a result of an initial basic classification. The process-oriented intention becomes visible by the fact that only four establishments (among them, of course, the Ecolodge Atlas Kasbah as the best-ranked accommodation) were classified as "excellent", thus signaling to others that future improvements might lead to a higher classification (Boudribili, 2014, p. 127).

Even if the classification approach of the RDTR started out as quite compelling, a follow-up visit in 2016 revealed the fact that continuous quality improvement and a focus on a sustainability approach could not be achieved by the NGO. After the first classification assessment, no additional repeated surveys have been conducted. At the same time, the idea of continuous quality and sustainability-oriented training has also not been realized. A lack of human capacity as well as other interests on the part of the leading members might be a preliminary explanation for the lack of continuity in the approach of the RDTR.

6 Conclusions

This paper has presented a review of the literature on sustainable tourism definitions and its application in the Souss Massa region. This review attempts to create a window of opportunity to help researchers' and practitioners' efforts and also to meet their requirements for easy access to sustainable tourism publications. Adding that both English and French journals are considered in this research work.

This research has some limitations. The limitation is that a part of the data used in this review is collected from scholarly journals, which exclude conference proceeding papers, master's dissertations, doctoral theses, textbooks, and unpublished working papers in the Sustainable Tourism literature. This may imply that this review is not thorough; however, it provides a comprehensive review since the majority of papers published by scholarly journals are included. Therefore, this paper offers to academic researchers and practitioners a framework for future research.

References

Aall, C. (2014). Sustainable tourism in practice: Promoting or perverting the quest for a sustainable development? pp. 1–22.

Andreas Kagermeier (Trier), Lahoucine Amzil (Rabat), & Brahim Elfasskaoui (Meknes). (2018). *Governance aspects of sustainable tourism in the Global South—Evidence from Morocco.*

Allaby, M. (1988). Macmillian dictionary of the environment. London: Macmillian.

Bramwell, B., & Lane, B. (1993). Sustainable tourism: An evolving global approach. *Journal of Sustainable Tourism, 1*(1), 1–5.

Bramwell, B., & Lane, B. (1999). Sustainable tourism: Contributing to the debates. *Journal of Sustainable Tourism, 7*(1), 1–5.

Bramwell, B., & Lane, B. (2008). Priorities in sustainable tourism research. *Journal of Sustainable Tourism, 16*(1), 1–4.

Bramwell, B., & Lane, B. (2009). Economic cycles, times of change and sustainable tourism. *Journal of Sustainable Tourism, 17*(1), 1–4.

Bramwell, B., & Lane, B. (2010). Sustainable tourism and the evolving roles of government planning. *Journal of Sustainable Tourism, 18*(1), 1–5.

Bramwell, B., & Lane, B. (2011). Crises, sustainable tourism and achieving critical understanding. *Journal of Sustainable Tourism, 19*(1), 1–3.

Bramwell, B., & Lane, B. (2012). Towards innovation in sustainable tourism research? *Journal of Sustainable Tourism, 20*(1), 1–7.

Bramwell, B., & Lane, B. (2013). Getting from here to there: Systems change, behavioural change and sustainable tourism. *Journal of Sustainable Tourism, 21*(1), 1–4.

Buckley, R. (2012). Sustainable tourism: Research and reality. *Annals of Tourism Research, 39*, 528–546.

Camargo, B., Lane, K., & Jamal, T. (2004). Environmental justice and sustainable tourism: The missing cultural link, 1–12.

Eber, S. (1992). Beyond the Green Horizon. *Principles of Sustainable Tourism: A discussion paper*. London: World Tourism Concern and Wide Fund for Nature.

El Boudribili, Y. (2014). Mise en œuvre du management environnemental pour le développement touristique durable du territoire de la région Souss-Massa-Draa (Maroc): une approche globale et des solutions locales. Trier (Dissertation, retrieved from: ubt.opus.hbz-nrw.de/volltexte/2014/900/).

El Boudribili, Y., Kabbachi, B., & Kagermeier, A. (2012). Environmental management and sustainability in the hospitality business: The case of the Ecolodge Atlas Kasbah Agadir, Southwest Morocco. In A. Kagermeier, & J. Saarinen (Hrsg.) *Transforming and managing destinations: Tourism and leisure in a time of global change and risks* (S. 353–364). Mannheim.

Fonteneau, et al. (2010). *Social and solidarity economy: Building a common understanding*. CIRIEC/EESC and ITC/ILO, Turin.

Gossling, S. (2002). Global environmental consequences of tourism. *Global Environmental Change, 12*, 283–302.

Gossling, S. (2003). Market integration and ecosystem degradation: Is sustainable tourism development in rural communities a contradiction in terms? *Environment, Development and Sustainability, 5*, 383–400.

Harrison, D. (1996). Sustainability and tourism: Reflections in a muddy pool. In *Sustainable Tourism in Islands and Small States: Issues and Policies*, Vol. 1, pp. 69–89. London: Cassell.

Lane, B. (2009b). Thirty years of sustainable tourism. In S. Go¨ssling, C. M. Hall, & D. B.Weaver (Eds.), *Sustainable tourism futures* (pp. 19–32).

Liu, Z. (2003). Sustainable tourism development: A critique. *Journal of Sustainable Tourism, 11*, 459–475.

Liu, Z. H., et al. (1994). Tourism development—A systems analysis. In A. V. Seaton (Ed.), *Tourism: The state of the art* (pp. 20–30). Chichester: Wiley.

Liu, G., Liu, Z., Hu, H., Wu, G., & Dai, L. (2008). The impact of tourism on agriculture in Lugu Lake region. *International Journal of Sustainable Development & World Ecology, 15*, 3–9.

Liu, J., Ouyang, Z., & Miao, H. (2010). Environmental attitudes of stakeholders and their perceptions regarding protected area-community conflicts: A case in China. *Journal of Environment Management, 91*, 2254–2262.

Liu, C. H., Tzeng, G. H., & Lee, M. H. (2012). Improving tourism policy implementation—The use of hybrid MCDM models. *Tourism Management, 33*, 413–426.

Liu, C. H., Tzeng, G. H., Lee, M. H., & Lee, P. Y. (2013). Improving metro–airport connection service for tourism development: Using hybrid MCDM models. *Tourism Management Perspectives, 6*, 95–107.

Lütteken, A., & Hagedorn, K. (1999). *Concepts and issues of sustainability in countries in transition—An institutional concept of sustainability as a basis for the network*. Humboldt University of Berlin-Department of Agricultural Economics and Social Sciences.

Royaume Du Maroc- Ministère de l'intérieur- Direction Générale des Collectivités Locales. (2015). La Région de Souss Massa (Monographie Générale), P 3 on 62.

Markandya, & Pearce, D. (1988). Natural environments and the social rate of discount. Project Appraisal, Vol. 3 (No. 1).

Miller, G., & Twining-Ward, L. (2005). Monitoring for sustainable tourism transition, the challenge of developing and using indicators book.

Ministry of Tourism, Handicraft and Social Economy. www.tourisme.gov.ma.

Morais, L., & Bacic, M. (2009). *Solidarity economy and public policies in Brazil: Challenges, difficulties and opportunities in a world undergoing transformation*. Seville: International Congress CIRIEC.

Niranjan Ray Netaji Mahavidyalaya. (2017). Business infrastructure for sustainability in developing economies, India.

Organization for Economic Cooperation and Development (OECD). (1990). ISSUESPAPERS: On Integrating Environment and Economics. Paris.

RDTR (= Réseau de Développement Touristique Rural). (2012). Plan d'action & Réalisations. Agadir.

RDTR (= Réseau de Développement Touristique Rural). (2013). La charte Qualité et Environnement en Tourisme (QET). Agadir.

RDTR (= Réseau de Développement Touristique Rural) & Afkar Consultance. (2013). Promotion du tourisme rural au moyen des réseaux sociaux: cas du Réseau de Développement Touristique Rural. Agadir.

Saarinen, J. (2006). Traditions of sustainability in tourism studies. *Annals of Tourism Research, 33*, 1121–1140.

Sharpley, R. (2000). Tourism and sustainable development: Exploring the theoretical divide. *Journal of Sustainable Tourism, 8*(1), 1–19.

Stelios, B., & Melisidou, S. (2010, November). "Globalization," tourism research institute. *Journal of Tourism Research, 1*(1), 33–53.

Worldwide Tourism Organization [WTO]. (2003). www.world-tourism.org.

WCED. (1987). *Our common future*. Oxford: Oxford University Press.

WTO. (2001). The concept of sustainable tourism. https://www.world-tourism.org/sustainable/concepts.htm.

Zolfani, S. H., Sedaghat, M., Maknoon, R., & Zavadskas, E. K. (2015). Sustainable Tourism: A comprehensive literature review of frameworks and applications (pp. 1–30). Publication details, including instructions for authors and subscription information: https://www.tandfonline.com/loi/rero20.

Overtourism and Tourism Carrying Capacity: A Regional Perspective for Greece

Sophia Panousi and George Petrakos

1 Introduction

Despite the current health crisis with the outbreak of Covid19 pandemic, world tourism has recorded unprecedented growth during the last decade. Tourism worldwide has proven to be one of the world's highest priority industries and employers. Its resilience until recently to serious external negative factors helped many countries to overcome the economic crisis minimizing the casualties for their economies.

For decades, the primary rationale in tourism development was increasing the numbers of tourist arrivals to a destination. Improved living conditions and technology innovations have been key drivers of the tourism growth. Increased disposable income, wealthier retirees in a very good health made tourism services accessible to a larger share of the world population. Technological development, the rise of cruise and low-cost carriers boosted international mobility, whereas product diversification and innovation transformed a trip into a life-changing experience.

Greece over the last decade has witnessed a striking expansion of its tourist flows. During the last 10-year period 2010–2019, the number of inbound travelers increased by 127%, while the number of non-residents' overnight stays declined by 69%. This difference can mainly be attributed to the decline of the average duration of stay by

---

This paper was prepared before the outbreak of the CoViD-19 pandemic. Travel and tourism, in contrast to the resilience shown against the economic crisis, is currently one of the most affected sectors by this unprecedented global health, social and economic crisis. It remains to be seen how the world tourism will address this challenge and overcome the sharp decline expected in 2020.

S. Panousi · G. Petrakos (✉)
Research Institute for Tourism, Athens, Greece
e-mail: petrakos@panteion.gr

G. Petrakos
Department of Public Administration, Panteion University, Athens, Greece

© The Author(s), under exclusive license to Springer Nature Switzerland AG 2021
V. Katsoni and C. van Zyl (eds.), *Culture and Tourism in a Smart, Globalized, and Sustainable World*, Springer Proceedings in Business and Economics,
https://doi.org/10.1007/978-3-030-72469-6_14

26%. Finally, at the same period, tourism receipts increased by 89%, while the non-residents' expenditure per journey reduced by 17% (Bank of Greece, 2020). Inbound tourism in Greece, especially at the second half of the current decade, was affected by the world tourism trend, according to which tourists prefer to take shorter and more frequent trips.

At regional level in 2019, the three most popular Greek island regions, i.e., Crete, Ionian Islands, and South Aegean Islands, account for the 42% of inbound tourism arrivals, corresponding to 52% of the overnight stays and to 60% of total tourism receipts. Furthermore, these three regions manifest particularly high seasonality. In Crete, 52% of non-residents' overnight stays occur at the third quarter of the year, while the corresponding percentages for the South Aegean Islands and the Ionian Islands are 62% and 68%, respectively.

During the recent economic crisis that seriously afflicted Greece, tourism sector has proven to be one of the pillar sectors that helped the Greek economy to mitigate the negative impacts of the crisis (Petrakos et al., 2020). The beneficial effects of tourism were even more evident at regional level. Increased volumes of tourists will further consolidate the role of tourism as a driver of development and diversification for local and regional economies.

However, the continuous uncontrolled growth of tourism is accompanied by serious problems affecting regions specialized in leisure industry. Specifically, problems concerning demographic congestion, high consumption of natural and energy resources, ecological impacts affecting the landscape, the cultural and historical sites and even the moral values and customs of the indigenous populations are some of the emerging factors that degrade and deteriorate the very same assets that attract visitors.

Today more than ever, the imperative objective of tourist activity stands out: sustainability. Consequently, coping with the phenomenon of overtourism and simultaneously achieving its sustainable development have created an urging need for better and more effective planning, development, and management of the tourist activities. While the direction is set, the viability of tourism development in the long run depends on the approach that the public and private stakeholders will adopt to tap opportunities and respond to challenges.

Since the objective of contemporary tourism is sustainability, there is an emerging need for constructing and evaluating measurable indicators using open data. The evaluation of those indicators at a regional level will help us to refrain from subjective conceptions and to develop a better understanding of both overtourism and tourism sustainability.

The rest of the paper is organized as follows. Section 2 provides the relative literature review. Section 3 discusses the methodology used for the construction of the composite indicators and the feasibility of their application to the regional level. Section 3 presents the results of the analysis using real data for all the 13 Greek regions between 2016 and 2019. Section 4 summarizes the main findings and conclusions regarding the practical implementation of the proposed indicators on a regional level as a measurement of tourism intensity of a destination and as a useful tool for planning tourism policies toward sustainability. Finally, in Sect. 5, the references of the paper are listed.

2   Literature Review

The explosive growth that tourism experienced in recent years intensified an old problem and brought to the center of the academic community the issue of "Overtourism." Overtourism is a new term to describe an old problem, mainly the situation where too many tourists visit a specific destination at a certain period inflicting negative impacts of all types on the local community. In other words, we can say that overtourism is the tourism pressure exerted on a destination.

Tourists overflowing destinations causing all kinds of transformations on the site have been recorded for well over a century. Complains about mass tourism and the impact on destination have been recorded in Venice in the mid-nineteenth century (Butler, 2006). In the 1960s, authors discussed the ways in which tourism negatively affected destinations causing discomfort and nuisance to locals. Doxey's irridex model (1975) is one of the first academic papers in which negative reactions of local populations have been mentioned against the over flooding numbers of visitors. Butler's tourist life cycle (1980) and Pizam's Attitude-Index (1978) for describing the social costs to destination communities are some of the fundamental works referring to the continuing decline in the environmental quality and therefore to the attractiveness of many tourist areas. In many cases, these problems have been exacerbated by the seasonal nature of tourism (Milano et al., 2019).

In the 1980s, discussion regarding the carrying capacity of a destination moved this debate forward. The concept of carrying capacity evolved from a neo-Malthusian perspective of resource limitations. The idea was to find the "magical number" of tourists that can visit a destination without causing serious negative consequences of any kind (Van der Borg et al., 1996). For many years, carrying capacity became a popular concept for estimating the negative impacts of tourism. However, the usefulness of the proposed indices for measuring the carrying capacity of a destination has been seriously questioned. As Koens et al. (2018) observed in Rosenow and Pulsipher (1979), they had coined the term "visitor overkill" to describe three main causes negatively effecting a destination because of tourism activities. Visitor behavior, timing, concentration are some of the factors as important as tourism numbers. Saveriades (2000) advocates that each destination can sustain a specific level of acceptance of tourist development and use, beyond which further development can result in sociocultural deterioration or a decline in the quality of the experience gained by visitors. While the impact on the physical environment can somehow be measured, it is far more difficult to measure the impacts on social environment, as it is a subjective concept, depending on ever-changing individuals, with disparate tolerance levels and different interests.

Turning away from the search of an intrinsic and elusive numerical carrying capacity that is ultimately based on unrealistic assumptions and hidden value judgements, authors suggested alternative approaches by developing planning frameworks. Such planning frameworks include the Limits of Acceptable Change (LAC) (McCool, 1989, 1994; Stankey et al., 1985), Visitor Impact Management (Graefe et al., 1990), Visitor Activity Management Planning (VAMP) (Nilsen & Grant, 1998), and the Tourism Optimization Management Model (TOMM) (Manidis Roberts Consultants, 1997). These frameworks are decision-making frameworks and not a scientific theory and this is the major difference from the carrying capacity approach. The benefit of the debates around impact-based approaches is that the emphasis has shifted from numbers to one that is based more on perceived benefits and disadvantages.

From the late 1990s onwards, the emphasis of work on dealing with tourism impacts shifted toward adopting the notion of a less intervening policy, advocating the concept of "allowing the market to act as a form of governance" (Hall, 2011). One of the main flaws of such theories, for which they have been criticized, is that they put too much responsibilities onto stakeholders who lack the resources or the knowledge to act in a sustainable way (Koens & Thomas, 2016; Leslie, 2012).

In recent years, the stunning developments in technology, the growth of low-cost carriers, the influence of social media, as well as the emergence of new inexpensive accommodation offers through Airbnb and other online platforms have exacerbated the problem of overtourism, especially in popular and renown destinations. Veiga et al. (2018) advocate that increased tourism demand has contributed to social and environmental unsustainability in tourism. They also claim that overtourism is strictly connected to the problem of mass tourism saturation of many destinations worldwide, diminishing the quality of life of residents and creating negative experiences for tourists. Butler (2018) advocates that the uncontrolled growth in the number of tourists in many destinations has resulted to the rise of anti-tourism movements from permanent residents. Furthermore, he distinguishes between the concept of overtourism and that of over-crowding, clarifying that overtourism is the situation in which tourists exert extra burden to local services and facilities and at the same time become a serious inconvenience for permanent residents. Croce (2018) proposes that the economic development should take into consideration and prioritize the conservation, preservation, and protection of cultural and natural environment that are the lifeblood of tourism and ultimately averting tourism from becoming its main cause of disruption.[1]

Nevertheless, the development of tourism inevitably induces changes on the social and environmental character of a destination. Therefore, we believe that we must have reliable and impartial measurements that will help us validate possible processes of sustainability or unsustainability of a territory. Having as a starting point two indices proposed by Manera and Valle (2018) for measuring tourism intensity worldwide, we adapted them so to become applicable to the case of Greek regions, as the authors

---

[1]For further reading on implications and future perspectives of overtourism see Cappochi et al. (2019).

noted "…the indices and the methodology we have constructed would be equally applicable to a more regional-scale analysis."

3   Methodology and Indices Presentation

3.1   Description and Formulation of the Indices

Despite the importance of tourism to national, regional, or local economies, there is no consensus among researchers on how to define and measure tourism intensity, as clearly was pointed out in Manera and Valle (2018). They have noticed that there are not accurate indicators to measure tourism intensity and, above all, there are no composite indicators for this purpose. Therefore, they proposed two specific instruments for measuring tourism intensity: The Tourism Intensity Index (TII) and the Tourism Density Index (TDI). In a preceding study, Manera and Navinés (2018) provided a tourism intensity index that was then applied to 18 insular economies throughout the world.

Four variables are used to formulate the TII: (i) the number of tourist arrivals at a destination, (ii) the destination's population, (iii) tourism revenues and the GDP of each country under consideration.

After TII has been calculated, the TDI can be obtained by multiplying the demographic component of the TII by the population ratio per km^2 of the country.

The four variables that formulate those indices were taking values from relative data published by the World Tourism Organization (UNWTO) and by the World Bank (WB).

The proposed indices have three basic attributes. Firstly, they introduce in a single synthetic index, four determining variables that are closer to the economic reality of the destination in consideration. Secondly, they facilitate their application on international, national, and regional scale enabling intra-destinations comparisons. Thirdly, they use homogenized data, leading to more robust results which open up new pathways for research, not only in the field of tourism, but also for studying economic structures (for an interesting intra-regional analysis in Spain, see Pérez-Dacal et al. (2014).

The initial research hypothesis was that tourism intensity mainly affects insular economies and by constructing tourism intensity indices research would help assessing the level of a destination's sustainability. The results of their analysis showed that these measurable indicators allowed for country classification free from subjective conceptions about the phenomenon of sustainability.

As it was mentioned previously, the two proposed indices are the Tourism Intensity Index (TII) and the Tourism Density (TD). For calculating the TII, they used demographic and economic variables obtained from World Tourism Organization (WTO) for all countries and not from the Official Statistical Authorities of each country. This ensured the homogeneity of the variables for drawing up the index.

They defined the TII as follows:

$$\text{Tourism Intensity Index}^i = \sqrt{\frac{\frac{T_i}{P_i}}{\frac{T_w}{P_w}} \times \frac{\frac{TR_i}{GDP_i}}{\frac{TR_w}{GDP_w}}} \times 100$$

where T is the number of tourists, P is the population, TR is tourism revenues, GDP is the gross domestic product with subscript (i) for the specific country and (w) for the world. Although they have used the WTO database as their source, they have experienced missing values issues for certain countries and years. In order to cope with this methodological problem, they focused on the trends of the series.

While carrying out the calculation of the TII, they added a new geographical aspect: The Tourism Density (TD) of each country, defined as the number of tourists per km^2.

$$\begin{aligned} TD &= \frac{\text{Inbound Tourism} + \text{Domestic Tourism}}{\text{Population}} \times \frac{\text{Population}}{\text{km}^2} \\ &= \frac{\text{Inbound Tourism} + \text{Domestic Tourism}}{\text{km}^2} \end{aligned}$$

By multiplying the demographic component of the TII with the population ratio per km^2, the authors produced the Tourism Density Index. Apart from the WTO database, they derived data regarding the population and the area in km^2 for each country from the World Bank database.

Calculating the two indices for all the countries between 1995 and 2015, they acquired the ranking of the countries from greater to lower tourism intensity, according to the mean value of TII over the period under consideration. The countries were classified into four groups according to the magnitude of their TII:

1	Very high TII:	mean > 1500
2	High TII:	500 < mean < 1500
3	Medium TII:	100 < mean < 500
4	Low TII:	mean < 100

Furthermore, they distinguished between the two components of the TII: the demographic component $\left(\frac{T_i}{P_i}\right)$ and the economic component $\left(\frac{TR_i}{GDP_i}\right)$. The higher the demographic component, the bigger the tourism pressure exerted on the country. The higher the economic component, the greater the dependency from tourism of the country.

Next the authors proceeded to the calculation of the Tourism Density for all the countries for the same period. Again, they acquired a new classification of all countries, ranking them according to the magnitude of the mean value of TD for the period. The formation of the four groups is now based on the number of tourists per km^2.

1	Very high TD:	mean > 10,000 tourists/km^2
2	High TD:	1000 < mean < 10,000 tourists/km^2
3	Medium TD:	300 < mean < 1000 tourists/km^2
4	Low TD:	mean < 300 tourists/km^2

The findings from their analysis corroborated their initial hypothesis, that island countries are more susceptible to tourism pressures and much more dependent on tourism. These classifications can be proven very useful for planning tourism policies in order to eliminate negative externalities and to boost positive ones.

3.2 Adaptation of the Indices to the Case of Greek Regions

Although the above described indices were originally designed to be used at national level, they could equally be applicable for regional analysis. In this section, we will present the adaptations we propose, so these indices could be applied to measure the tourism intensity in the 13 regions of Greece. Instead of comparing the four major tourism variables of a country to the corresponding world variables, we compare the regional variables to the total Greece ones. Also, we have substituted the number of tourists by the number of overnight stays, since we believe that the later variable is more robust and better represents the tourism burden of a region.

The demographic (population) and economic (GDP) variables are obtained from the Hellenic Statistical Authority (EL.STAT.) while tourism revenues and overnight stays of non-residents are obtained from the Bank of Greece (BoG). Available tourism regional data from BoG exist for the period from 2016 to 2019.

Having selected the variables and the data sources, the TII is defined as follows:

$$\text{Tourism Intensity Index}^r = \sqrt{\frac{\frac{N_r}{P_r}}{\frac{N_{gr}}{P_{gr}}} \times \frac{\frac{TR_r}{GDP_r}}{\frac{TR_{gr}}{GDP_{gr}}}} \times 100$$

where N is the number of overnight stays of non-residents, P is the population, TR is tourism revenues, GDP is the gross domestic product with subscript (r) for the region and (gr) for the country (Greece).

Regarding the computation of the Tourism Density, we confined our analysis using only the number of inbound overnight stays, since there are no equivalent data available in regional level for domestic tourism.[2]

---

[2] Besides, domestic overnight stays in hotels and short-stay rentals in 2018, constituted only 7% of inbound overnight stays.

If the demographic component of the TII is multiplied by the population ratio per km^2, the Tourism Density (TD) of the region is obtained.

$$TD = \frac{\text{Inbound Overnight Stays}}{\text{Population}} \times \frac{\text{Population}}{\text{km}^2}$$

$$= \frac{\text{Inbound Overnight Stays}}{\text{km}^2}$$

Implementing the TII on regional level for Greece, we defined three groups of regions according to the mean value of the index for the period 2016–2019.

1	High TII:	mean > 100
2	Medium TII:	50 < mean < 100
3	Low TII:	mean < 50

Respectively, the three groups of regions according to the mean value of Tourism Density for the period 2016–2019 are formed as follows:

1	High TD:	mean > 1000
2	Medium TD:	500 < mean < 1000
3	Low TD:	mean < 500

Next, we have applied the above-mentioned methodology to the Greek regional data available for the 2016–2019, obtaining the results presented in the following section.

4   Results

Having defined the contents of the two indices, the calculation of the TII for all the regions of Greece between 2016 and 2019 is presented below. The regions were ranked from greater to lower tourism intensity, depending on the mean value of the indicator over the 4-year period and were classified according to whether they had a high TII (mean greater than 100), a medium TII (mean greater than 50 and lower than 100), or a low TII (mean lower than 50) (Table 1).

According to the mean value of TII, the 13 regions of Greece are classified into three almost equal in number of regions groups. The values and the trends of TII for the top group over the studied time period are shown in Fig. 1.

We see that the regions in the group with the highest TII are the most prominent and renowned island regions of Greece and the mainland region of Central Macedonia, where the popular destinations of Chalkidiki are situated.

In Table 2, the evolution of the TII is given for each year of the 4-year period 2016–2019 for the 13 regions of Greece.

Table 1 Ranking of 13 administrative regions of Greece according to TII mean value

Region	TII mean value 2016–2019	TII mean value range
Southern Aegean	741.51	100+
Ionian Islands	611.39	100+
Crete	376.63	100+
Central Macedonia	102.41	100+
Northern Aegean	73.69	50–100
Epirus	62.89	50–100
Eastern Macedonia and Thrace	56.81	50–100
Peloponnese	54.52	50–100
Thessaly	37.20	>50
Attica	34.59	>50
Western Greece	27.11	>50
Central Greece	22.44	>50
Western Macedonia	21.53	>50

Data source BoG, EL.STAT

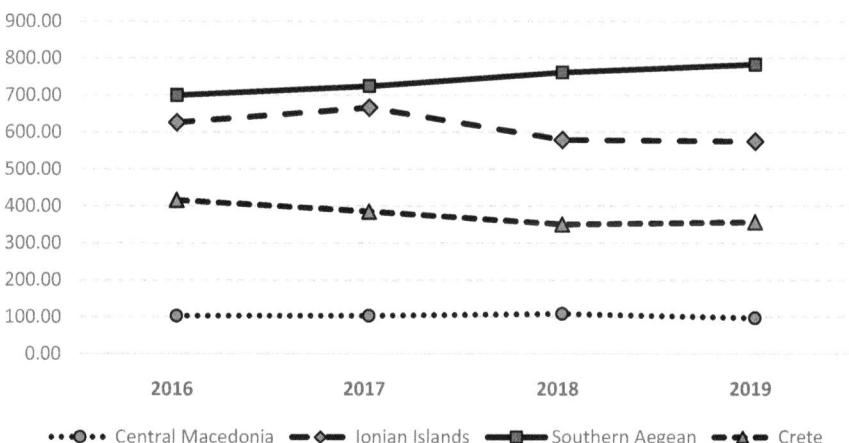

Fig. 1 High tourism intensity index evolution

From the four regions with high TII (Table 1), only the index of the Southern Aegean Region has recorded continuous growth during the 4-year period in consideration.

In Table 3, the TII components for the first and the last year of the period in consideration are broken down into their demographic and economic aspects.

The higher the demographic index, the higher the pressure exerted on the region. Respectively, the highest the economic component of the index, the greater the dependency on tourism of the region.

Table 2 Evolution of the tourism intensity index, 2016–2019

Region	2016	2017	2018	2019
Eastern Macedonia and Thrace	54.06	48.76	53.34	71.06
Central Macedonia	102.43	102.81	107.88	96.51
Western Macedonia	26.97	15.32	21.26	22.59
Thessaly	42.35	37.43	34.16	34.86
Epirus	67.96	62.06	60.68	60.88
Ionian Islands	626.02	665.90	579.02	574.64
Western Greece	24.09	23.35	28.81	32.18
Central Greece	19.26	17.75	29.58	23.17
Peloponnese	56.38	47.38	61.89	52.40
Attica	32.52	35.16	34.75	35.94
Northern Aegean	72.12	81.59	76.91	64.14
Southern Aegean	699.42	723.58	760.66	782.39
Crete	415.35	384.68	350.35	356.15

Data source BoG, EL.STAT

Table 3 Basic components of the tourism intensity index, 2016 and 2019

Region	2016		2019	
	$\frac{N_r}{P_r}$	$\frac{TR_r}{GDP_r}$ (%)	$\frac{N_r}{P_r}$	$\frac{TR_r}{GDP_r}$ (%)
Eastern Macedonia and Thrace	8.96	4.2	16.96	6.1
Central Macedonia	19.29	7.0	21.78	8.7
Western Macedonia	5.39	1.7	5.69	1.8
Thessaly	7.02	3.3	6.80	3.7
Epirus	10.75	5.5	11.99	6.3
Ionian Islands	104.26	48.1	116.47	58.0
Western Greece	4.10	1.8	6.91	3.1
Central Greece	3.32	1.4	5.36	2.0
Peloponnese	9.91	4.1	11.26	5.0
Attica	6.55	2.1	9.09	2.9
Northern Aegean	12.50	5.3	13.13	6.4
Southern Aegean	119.47	52.4	154.55	81.0
Crete	62.32	35.4	68.13	38.1

Data source BoG, EL.STAT

As shown in Table 3, from the demographic point of view, the greatest pressure is calculated in the regions of Southern Aegean and Ionian Islands for both reference years. At the economic level, again the region of Southern Aegean reveals the greatest dependency on tourism followed by the region of Ionian Islands. Moreover, the dependency on tourism of the Southern Aegean region rises sharply between 2016 and 2019, whereas the corresponding rise of the index for the regions of Ionian Island and Crete is substantially milder.

Next, we calculate the Tourism Density (TD) for the 13 Administrative regions of Greece for the same 4-year period and rank the regions based on the mean value of the indicator. The regions were classified according to the magnitude of the index into groups with high TD (mean greater than 1000 overnight stays per km^2), medium TD (mean less than 1000 overnight stays per km^2 and over 500 overnight stays), and small TD (mean less than 500 overnight stays per km^2).

In contrast to the TII, the three groups include different number of regions. Five regions fall into the group with the high TD, only two regions fall into the second group with the medium TD, and six regions fall into the last group with the low TD.

As shown in Table 4, the regions of Ionian Islands and the Southern Aegean occupy the two first places with the highest TD mean value, only they have interchanged places compared to those in Table 1. However, the third place is occupied by the region of Attica, revealing the tourism pressure that this region receives.

In Fig. 2, the evolution of Tourism Density for the period 2016–2019 can be observed for the five regions of the group with the high TD. The TD for the Ionian Islands region records a downwards trend from 2018 to 2019, attributable to a 3%

Table 4 Ranking of 13 administrative regions of Greece according to TD mean value

Region	TD mean value 2016–2019	TD mean value range
Ionian Islands	10385.43	1000+
Southern Aegean	9182.69	1000+
Attica	8029.84	1000+
Crete	5036.58	1000+
Central Macedonia	2196.70	1000+
Northern Aegean	804.36	500–1000
Eastern Macedonia and Thrace	501.10	500–1000
Epirus	425.68	>500
Peloponnese	419.09	>500
Thessaly	371.30	>500
Western Greece	314.06	>500
Central Greece	177.94	>500
Western Macedonia	143.95	>500

Data source BoG, EL.STAT

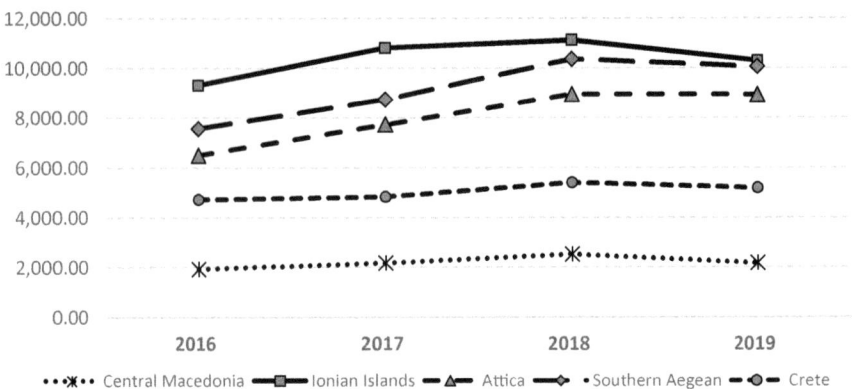

Fig. 2 High tourism density index evolution

decline of overnight stays in 2019 in relation to 2018. The same but less eminent trend is also reflected in the evolution of the TD of the other four regions.

In Table 5, the evolution of the TD is given for each year of the 4-year period 2016–2019 for the 13 regions of Greece.

The island regions of Greece record the highest TD, since over 50% of total overnight stays are realized into these regions. The regions with the highest percentage changes of TD are the emerging regions that are newly discovered by tourists.

Table 5 Evolution of the tourism density index, 2016–2019

Region	2016	2017	2018	2019	% change 2019/16
Eastern Macedonia and Thrace	382.38	382.86	520.80	718.36	88
Central Macedonia	1931.31	2168.01	2518.08	2169.38	12
Western Macedonia	156.09	90.87	168.06	160.80	3
Thessaly	364.80	358.09	414.04	348.27	−5
Epirus	393.56	395.87	478.39	434.91	11
Ionian Islands	9316.51	10812.14	11120.75	10292.33	10
Western Greece	241.56	248.33	367.18	399.15	65
Central Greece	118.66	129.44	272.15	191.52	61
Peloponnese	371.87	336.58	550.47	417.45	12
Attica	6504.36	7730.28	8948.80	8935.92	37
Northern Aegean	640.82	838.69	981.38	756.54	18
Southern Aegean	7566.42	8741.96	10363.98	10058.40	33
Crete	4723.79	4830.94	5402.50	5189.08	10

Data source BoG, EL.STAT

5 Conclusions

The primary objective of sustainable tourism development concerns enhancing the welfare of those affected by it, ensuring the preservation of the local community's cultural and natural heritage, as well as the perceptions of tourists.

It is very important being able to estimate the tourism burden exerted on a destination, as it can affect the well-being of both residents and tourists. The compound Tourism Intensity Index consisting of four determining variables: the GDP, tourism revenues, the population, and the number of overnight stays are a very efficient and useful tool for tourism analysis. Despite its descriptive character, the TII enables classifying a destination according to its magnitude, a fact that can be proved very useful when it comes to determining the type of policies necessary to eliminate negative externalities and promote the positive ones. The TII is complemented by the Tourism Density. Both methodologies lead to different rankings that can be used accordingly by decision makers for tourism policy planning.

The insular regions of Greece recorded the highest TII and TD, a conclusion in line with the results of Manera and Valle for the island economies worldwide. The tourism dependency of these regions has recorded remarkable increase in 2019 in respect to 2016, rendering them more vulnerable to tourism turmoil.[3]

Furthermore, the percentage changes of TD in 2019 over 2016 reveal the emerging regions of Greece. The higher the change, the higher the recognition of the new destination.

The changes in the tourism sector in a globalized world are overwhelming. The tourist numbers rise rapidly, seriously affecting countries, nations, regions specialized in leisure industry. Problems of demographic congestion, high consumption of natural and energy resources, environmental and cultural impacts arise by the day and greater immense, forcing policymakers to take urgent actions to avert and, if possible, invert the negative impacts.

The proposed indices are impartial and unbiased and can be a very useful tool for designing tourism policies aiming at correcting externalities because of advancing of mass tourism and therefore ensure the environmental sustainability of a destination. Implementing a management strategy that specifically identifies these conditions and establishes explicit standards of quality will be more efficacious than relying on numerical carrying capacities. It behoves all stakeholders involved in tourism industry to admit the existence of the problem of overtourism and take proper actions before permanently damaging tourism, before, as they say, "killing the goose that lays the golden eggs."

If you cannot measure it, you cannot manage it, and the indices presented in this paper help us measure tourism intensity and density. However, we must be aware that measurement is always partial and cannot constitute the be-all and end-all of knowledge of the phenomena investigated. The pursuit of constructing other indices is constant and continuous in the research community.

---

[3] The current pandemic had much more severe repercussions on the tourism of the island regions of Greece.

References

Bank of Greece. (2020). The border survey. https://www.bankofgreece.gr/en/statistics/external-sec tor/balance-of-payments/travel-services

Benner, M. (2019). From overtourism to sustainability: A research agenda for qualitative tourism development in the Adriatic, Heidelberg University. MPRA Paper No. 92213. Retrieved from https://mpra.ub.uni-muenchen.de/92213/. Accessed the 19 July 2020.

Butler, R. W. (Ed.). (2006). *The tourism area life cycle* (Vol. 1). Channel view publications.

Butler, R. W. (1980). The concept of tourism area cycle of evolution: Implications for management of resources. *Canadian Geographer, 24,* 5–12.

Butler, R. W. (2018). Challenges and opportunities. *Worldwide Hospitality and Tourism Themes, 10,* 635–641.

Capocchi, A., Vallone, C., Pierotti, M., & Amaduzzi, A. (2019). Overtourism: A literature review to assess implications and future perspectives. *Sustainability,* 11, 3303. https://doi.org/10.3390/su11123303

Croce, V. (2018). With growth comes accountability: Could a leisure activity turn into a driver for sustainable growth. *Journal of Tourism Futures, 4.* https://doi.org/10.1108/JTF-04-2018-0020.

Doxey, G. V. (1975). A causation theory of visitor-resident irritants, methodology and research inferences: The impact of tourism. In *Sixth Annual Conference Proceedings of the Travel Research Association,* San Diego, CA, pp. 195–168.

Graefe, A. R., Kuss, F. R., et al. (1990). *Visitor Impact management: A planning framework.* National Parks and Conservation Association.

Hall, C. M. (2011). A typology of governance and its implications for tourism policy analysis. *Journal of Sustainable Tourism, 19,* 437–457.

Koens, K., Postma, A., & Papp, B. (2018). Is Overtourism overused? Understanding the impact of tourism in a city context. *Sustainability, 10,* 4384.

Koens, K., & Thomas, R. (2016). "You know that's a rip-off": Policies and practices surrounding micro-enterprises and poverty alleviation in South African township tourism. *Journal of Sustainable Tourism, 24,* 1641–1654.

Koens, K., Postma, A., & Bernadett, P. (2018). Is overtourism overused? Understanding the impact of tourism in a city context. *Sustainability, 10,* 4384. https://doi.org/10.3390/su10124384.

Leslie, D. (Ed.). (2012). *Responsible Tourism: Concepts, theory and practice.* CABI. ISBN 978-1-84593-987.

Manera, C., & Navinés, F. (2018). La Indústria Invisible, 1950–2016. El Desenvolupament del Turisme a L'economia Balear; Lleonard Muntaner Editor: Palma, Spain.

Manera, C., & Valle, E. (2018). Tourist Intensity in the World, 1995–2015: Two measurement proposals. *Sustainability, 10,* 4546. https://doi.org/10.3390/su10124546.

Manidis Roberts Consultants. (1997). *Developing a tourism optimization management model (TOMM), a model to monitor and manage tourism on Kangaroo Island, South Australia.* Surry Hills, NSW: Manidis Roberts Consultants.

McCool, S. F. (1989). *Limits of acceptable change: Some principles. Toward serving visitors and managing our resources.* Waterloo, ON: Tourism Research and Education Centre.

McCool, S. F. (1994). Planning for sustainable nature-dependent tourism development: The Limits of Acceptable Change system. *Tourism Recreation Research, 19*(2), 51–55.

McCool, S. F., & Lime, D. W. (2001). Tourism carrying capacity: Tempting fantasy or useful reality? *Journal of Sustainable Tourism, 9,* 372–388.

McCool, S., & Lime, D. (2001). Tourism carrying capacity: Tempting fantasy or useful reality? *Journal of Sustainable Tourism, 9.* https://doi.org/10.1080/09669580108667409

Milano, C. (2017). *Overtourism and Tourismphobia: Global trends and local contexts.* Ostelea School of Tourism and Hospitality: Barcelona, Spain

Milano, C., Novelli, M., & Cheer, J. M. (2019). Overtourism and tourismphobia: A journey through four decades of tourism development, planning and local concerns. *Tourism Planning and Development, 16*(4), 353–357.

Nilsen, P., & Grant, T. (1998). A comparative analysis of protected area planning and management frameworks. In S. F. McCool & D. N. Cole (Eds.), *Proceedings—Limits of Acceptable Change and Related Planning Processes: Progress and Future Directions* (pp. 49–57). USDA Forest Service Rocky Mountain Research Station.

Pérez-Dacal, D., Pena-Boquete, Y., & Fernández, M. (2014). Measuring tourism specialization: A composite indicator for the Spanish Regions. *Alma Tourism, 5,* 35–73.

Petrakos, G., Beneki, Ch., & Panousi, S. (2020). *The Greek hotel sector during economic crisis.* RIT Publication.

Pizam, A. (1978). Tourism's impacts: The social costs to the destination community as perceived by its residents. *Journal of Travel Research, 16,* 8–12.

Rosenow, J. E., & Pulsipher, G. L. (1979). *Tourism the good, the bad, and the ugly.* Media Productions & Marketing.

Saveriades, A. (2000). Establishing the social tourism carrying capacity for the tourist resorts of the east coast of the Republic of Cyprus. *Tourism Management, 21,* 147–156.

Stankey, G. H., Cole, D. N., et al. (1985). *The limits of acceptable change (LAC) system for wilderness planning.* USDA Forest Service Intermountain Research Station.

Van der Borg, J., Costa, P., & Gotti, G. (1996). Tourism in European heritage cities. *Annals of Tourism Research, 23,* 306–321.

Veiga, C., Santos, M. C., Águas, P., & Santos, J. A. C. (2018). Sustainability as a key driver to address challenges. *Worldwide Hospitality and Tourism Themes, 10,* 662–673.

Tourism Transportation Services Provided on the Principle of Sharing Economy

Radka Marčeková, L'ubica Šebová, Kristína Pompurová, and Ivana Šimočková

1 Introduction

One of the new forms of the transportation services is the offer, which works on the principle of sharing economy and allows passengers to use vacancies in various means of transportations (most often cars) that would not otherwise be occupied, travel cheaper, faster and more environmentally friendly.

We understand the sharing economy as a system in which the equity or services are shared by private persons, resp. as a business model that is applied on the basis of sharing existing resources free of charge or for a consideration and usually via the Internet.

2 Literature Review

According to the Brinke (1999), the transportation is the ensemble of transportation services that result in purposeful movement of people, goods and energy in space and time. Transportation services are one of the most important components of tourism if we assume that every visitor needs to get to the destination at time (Horner & Swarbrooke, 2003).

As experts agree that without the possibility of accessibility, tourism cannot take place (Prideaux, 2000, In Van Truong & Shimzut, 2017), we must understand accessibility in a broader context, not only as transport accessibility of the destination (and

R. Marčeková (✉) · L. Šebová · K. Pompurová · I. Šimočková
Department of Tourism, Faculty of Economics, Matej Bel University, Banská Bystrica, Slovakia
e-mail: radka.marcekova@umb.sk

its quality, speed, reliability and affordability), but also as a possibility of transportation directly in the destination, which influences the choice of tourism destination (Boopen, 2005 in Van Troung & Shimizu, 2017).

Quality transportation infrastructure is not only remedy for the future development of the destination, but it is a key factor in motivating visitors to the destination and thus contributes to increasing competitiveness and the development of tourism in the destination (Borovská, 2019).

Transportation has a significant impact on tourism demand, affecting tourism revenues, relative prices of goods and services and of course, the availability of destinations (Khadaroo & Seetanah, 2007). The provision of transportation services is also subject to external factors that we cannot influence and that significantly affect its quality (weather, insufficient airport capacity, etc.).

Transportation must also develop sustainably. Reckless development of transportation and transportation services can be a threat to the destination, especially in terms of limited capacity, increasing the level of competitiveness of destinations, environmental problems (Vaněček, 2008) and also limited human resources (Navickas and Malakauskaite, 2009, In Khan et al., 2017).

The digital age on the threshold of the fourth industrial revolution also brings with it new forms of economy and thus an extension of the existing forms of flexible work, which are also reflected in transportation. The phenomenon is the sharing economy, which has expanded especially in recent years (Barancová, 2017).

Numerous authors (Allen & Berg, 2014; Botsman, 2015; Botsman & Rogers, 2011; Dervojeda et al., 2013 and others) define the sharing economy as an economic system of decentralized networks and markets which allows the value of unused assets to be used by interconnecting needs and resources in a way that omits traditional intermediaries.

Acquier, Daudigeos and Pinkse (2017) explain that this is a rapidly expanding system of services in the open market, often provided by private individuals on the basis of existing demand and on the principle of temporary use of assets. The difference compared to the traditional economy is that while in traditional markets consumers buy the products they subsequently own, in the sharing economy, suppliers temporarily share their assets with consumers, either for free or for consideration, which can be financial or non-financial (Dervojeda et al., 2013; Ikkala & Lampinen, 2015; Hamari et al., 2016; Juul, 2017; Ranjbari et al., 2018). It is possible to share almost anything for the pre-agreed time, for example, tangible assets, services, skills, competencies.

The most frequently applied model of the sharing economy is so-called the peer-to-peer model (P2P, Fig. 1), in which individuals offer and request products, and another subject, the online platform, acts as an intermediary between them (Rifkin, 2015; Tussyadiah and Personen, 2015).

A further application of the model is real sharing and commercial sharing (B2C) (Codagnone & Martens, 2016; Analysis of the sharing economy in an environment of the SMEs, 2018). The platforms of genuine sharing of goods and services include platforms on the principle of individuals in a particular community of people who put it into practice in the form of non-financial performance. Commercial (B2C)

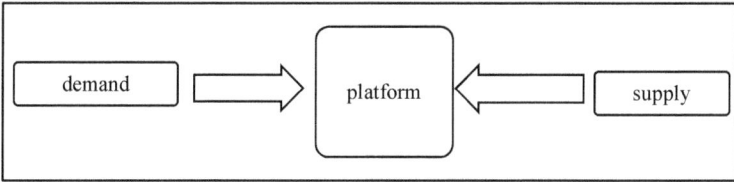

Source: Demary, 2014.

Fig. 1 Model of shared economy (P2P). *Source* Demary, 2014

sharing includes activities in the field of sharing economy provided by businesses for financial reward.

Although the sharing economy is understood in the presence as the new model and is often discussed (after entering the English term sharing economy into Google's search engine on July 27th, 2020, 950,000,000 search results were found), its roots are in history. However, the trend is considered to be the emergence of online platforms that mediate the sharing of property and services not only in the local but also in the global market among individuals who do not know each other, do not live in the same city, in the same state or on the same continent (Kostakis & Bauwens, 2014). Online platforms act on the market as intermediaries who earn a commission for mediation. The sharing economy has significantly changed the value chain (Demary, 2014 in Roblek et al., 2016). It has a positive but also a negative impact on the tourism. On the one hand, it provides easy access to a wide range of services, which are often of better quality and affordability than their traditional equivalents; on the other hand, it provides scope for unfair competition, reduces employment security (based on non-payment of health and social insurance), avoids taxes and, according to some authors, poses a threat to safety, health and hygiene standards (Goudin, 2016) and does not contribute to tax into the state budget.

The most frequently used modes of transport based on the principle of sharing economy can include carsharing, carpooling, ridesharing, lift-sharing, bikesharing, motorcyclesharing and, at present, more alternative options of the sharing of parking spaces, the use of boats, catamarans and other vessels (boat-sharing) and the joint use of private aircraft (flightsharing) or helicopters (helisharing). The most important role in tourism is played by the joint use of cars and rides in them and subsequently the joint use of bicycles.

The subject of joint use can be not only tangible assets, the "rental" of a vehicle without a driver, but also a service that is directly linked to the participation of the owner/lessor of the vehicle (usually the driver). In the article, we will continue to focus only on the common use of the seats in cars as part of transportation services, as we consider them important for tourism.

Transportation services intermediaries in a sharing economy are platforms that are the basis in the digital economy for creating new values, and at the same time, they represent a virtual space (market) where "sellers" and "buyers" meet (Accenture,

2016). According to Chovanculiak (2018), platforms "sell" transaction cost reductions. They effectively connect demand with supply and create the possibility of direct payments. According to the European Commission (2016), the platforms help to connect providers willing to lease their assets with users who are interested in the temporary use of these assets. Dervojeda et al. (2013) state that service providers in the sharing economy temporarily provide their assets/resources to users free of charge or for a fee. In the article, we identify with the mentioned authors and understand the importance of the sharing economy in this context.

3 Methodology

The research object of the paper is the supply of the transportation services in the tourism. The subject of paper is the sharing economy and its application in tourism.

The aim of the paper is to analyze the public's awareness of transport services provided on the principle of sharing economy, their usage and motivation of potential providers of such services.

To achieve the goal of the paper, we set two research questions:

Q1: What motivates and demotivates potential providers of transportation services to use and provide platforms operating on the principle of the sharing economy?
Q3: What needs to be changed in Slovakia in order to expand the possibilities of offering transportation services in tourism on the principle of sharing economy?

The acquisition of primary data is based on the questionnaire survey method. We have distributed the questionnaire in printed but also in electronic form. Electronic distribution has taken place through direct e-mails to foreign contacts with a request for further distribution on social network site Facebook in groups "Foreigners in Bratislava", distribution through own contacts (January–March 2019) and electronically through Matej Bel University in Banská Bystrica students, who carried out the survey in their surroundings and family environment (January–June 2020). Questionnaires distribution ran mostly online because of the pandemic situation due to the disease Covid-19 and strict quarantine measures in Slovakia. The questionnaire was prepared in two language versions (English and Slovak version), divided into several sections.

In the first part of the questionnaire, we have found out information about the number of motor vehicles in the respondent's household, frequency and reason for their use. We were interested in how the respondents perceive the amount of costs related to the operation of their own motor vehicle and whether they plan to buy a new one in the future.

The second part of the questionnaire related to the awareness about sharing economy among respondents, third part was aimed at identifying the motivations, which would affect respondents in the phase of deciding to become providers of transportation services. In the case of respondents who have already used such transport services, we tried to find out what are the biggest benefits of providing these services for them. The last part of the questionnaire examined socio-demographic data of respondents (age, sex, social status, country/region). The results of the questionnaire survey were processed using a spreadsheet program MS Excel.

In the article, we use theoretical methods of scientific work such as abstraction, content-causal analysis, synthesis, induction and deduction, method of comparison and analogy.

The basic sample for the primary survey is individuals older than 18 years from Slovakia and abroad. The research sample on the demand side is individuals over the age of 18, willing to participate in the survey, selected by an accessible selection (727 respondents).

When processing the paper, we use theoretical research methods, sociological methods of data collection and exploratory statistics, as well as correlation analysis in their processing.

We use exploratory statistics to express absolute and relative abundances when evaluating a questionnaire survey and correlation analysis (calculation of the degree of tightness of the dependence of monitored variables) to determine the dependence between the monitored variables. We are finding out a relationship between the age of respondents and awareness of the sharing economy, between the age of respondents and the use of transportation services in the sharing economy and between the age of respondents and the potential interest in providing transportation services on the principle of sharing economy in the future. The reason for detecting dependencies between variables is the assumption that in Slovakia, where research was mostly conducted, the sharing economy is only in the initial phase of development and its awareness, as well as its use and interest in using, respectively, providing of such services is affected by the age of the population.

We consider for important to determine the impact (direct or indirect impact) and its intensity (strong, average, weak impact), or to identify variables that are not affected. To verify the agreement of the detected state and the assumption (we always examine the assumption that the characters A and B are independent) we use the method of Spearman's nonparametric correlation coefficient, which also verifies the degree of intensity of the dependence of the examined characters. We verify the dependence of the variables using the statistical software IBM SPSS Statistics, version 25. In the case of the detected influence, we verify whether it is a direct or indirect influence and we determine its strength (strong, average, weak). Spearman's coefficient of rank correlation confirms (or rejects) the dependence of the observed features and proves the strength of their dependence. It can range from negative values -1 to positive $+1$. A value of -1 represents the highest negative and a value of $+1$ represents the highest positive dependence. If the dependence coefficient ranges from 0 to 0.3, we speak of a weak dependence, in the interval from 0.3 to 0.6, we speak about medium dependence, and in the interval from 0.6 to 1.0, we say about strong dependence.

4 Results

Transport services based on the principle of economy of joint use are represented by many platforms that enable interconnection between providers and users of these services. Internationally the platforms Lyft and Didi are mostly common. In the paper, we devote ourselves to the analysis of the platforms BlaBlaCar, Uber and Bolt, which belong to the most widespread and most used in Europe and are available also in Slovakia (Table 1).

4.1 Awareness of the Transportation Services on the Principle of Sharing economy and Their Use

Awareness of transportation services provided on the principle of sharing economy and the motivation of potential providers of transportation services were examined by a questionnaire survey. All graphs and tables were created based on the results of the survey and show the situation in Slovakia and abroad. We reached together 320 people during the period from January 18th to March 12th, 2019, by the questionnaire survey, with 268 respondents who completed the questionnaire, of which 166 respondents completed the questionnaire in Slovak version and 102 respondents completed the questionnaire in English version. The return rate of the questionnaires in the mentioned period reached 83.75%. The survey then continued in the period from March 1st to June 30th, 2020, exclusively online. A total of 727 respondents took part in the survey, of which 625 were citizens of the Slovak Republic and 102 were respondents from abroad. Of the total number of respondents, 63.5% were women and 36.5% were men. There is a disparity in the participation of Slovak women who were more willing to participate in the survey than men, the proportion of men and women from abroad was balanced (Table 2).

We can confirm that the survey was mostly interesting for the young and middle generation, for who the use of new ways of transportation is simple and attractive (Table 3).

Table 1 Indicators comparison of the examined platforms

Platform	Number of users (mil)	Number of conductors (mil)	Number of countries	Average driving time (km)
BlaBlaCar	70	0.1	22	300
Bolt	25	0.5	30	0.2
Uber	75	3	65	8

Note 1—The BlaBlaCar platform does not differentiate users and service providers in the statistics, 2—The Bolt platform does not record statistics on the average driving distance
Source Own elaboration based on www.blablacar.com, www.uber.com, www.bolt.com, 2019

Table 2 Gender of respondents

Sex	Slovakia		Abroad	
	Absolute number	Relative number (in %)	Absolute number	Relative number (in %)
Women	412	56.7	50	7.0
Men	213	29.3	52	7.0
Total	625	86.0	102	14.0

Source Own elaboration, 2020

Table 3 Age structure of respondents

Age group	Slovakia		Abroad	
	Absolute number	Relative number (in %)	Absolute number	Relative number (in %)
18–25 years	223	35.7	31	30.4
26–35 years	99	15.8	31	30.4
36–45 years	92	14.7	16	15.7
46–55 years	113	18.1	17	16.7
56–65 years	50	8.0	5	4.9
66 years and older	48	7.7	2	1.9
Total	625	100	102	100

Source Own elaboration, 2020

In terms of social status, employed respondents predominated. The group created 48.13% of all respondents (Table 4). Almost a third of the respondents were students, self-employed and entrepreneurs made up 11.19% of respondents and other groups were pensioners, the unemployed and mothers on maternity leave (8.21%).

The survey covered respondents from all regions of Slovakia, the English version was filled by respondents from Germany (19), Austria (12), Slovakians living abroad (11), respondents from France (6), Poland (6), Czech Republic (5), Spain (5), Mexico (4), Vietnam (4), England (2), El Salvador (2), Ghana (2), Portugal (2), Italy (2), Azerbaijan (1), Bulgaria (1), Denmark (1), Netherlands (1), Kazakhstan (1), Colombia (1), Latvia (1), Hungary (1), Moldova (1), Russia (1), Scotland (1), Tunisia (1), Turkey (1), Ukraine (1), USA (1) and Venezuela (1). The number of respondents is not sufficient to evaluate the differences in the context of respondents' nationality, but it is interesting to compare the opinions of a group of foreign respondents as a whole with the opinions of domestic respondents.

In the first question of the questionnaire, we found out how many cars the respondents own in the household, after deducting company cars. The survey involved 536 respondents who own at least one motor vehicle (Table 5).

Other questions were intended only for the owners of at least one car, therefore only 536 respondents answered them (including 469 domestic and 67 foreign respondents). From the results of the survey, we have found out that 48.6% of domestic

Table 4 Social status of respondents

Status	Slovakia		Abroad	
	Absolute number	Relative number (in %)	Absolute number	Relative number (in %)
Employed	289	46.2	54	52.9
Unemployed	12	1.9	5	4.9
On maternity/parental leave	18	2.9	0	0
Student	177	28.3	24	23.5
Retiree	71	11.4	4	3.9
Other (self-employed, entrepreneur)	58	9.3	15	14.7
Total	625	100	102	100

Source Own processing, 2020

Table 5 Number of cars owned by respondents

Number of cars	Slovakia		Abroad	
	Absolute number	Relative number (in %)	Absolute number	Relative number (in %)
One	313	50.1	33	32.3
Two	116	18.5	27	26.5
Three or more	40	6.4	7	6.9
No	156	25.0	35	34.3
Total number/share of respondents	625	100	102	100

Source Own elaboration, 2020

respondents and 50.8% of foreign respondents use the car on daily basis (Table 6). Frequent use of the cars could be an incentive to create a project similar to BlaBla-Lines (www.blablalines, 2019), which has been operating successfully in France for more than three years.

In the next question, we have found out what the main reason for using a car is (Table 7). 11.5% of domestic and 9% of foreign respondents use the car especially for long distances outside the place of permanent/temporary residence, which can be understood as the potential for the sharing economy in transport (BlaBlaCar).

Similarly, we can perceive more than potentially the high percentage of people who need to go to work every day by car. Parking in the cities is an increasing problem and driving via transportation platforms could help solve this problem with fewer cars on the roads and their better exploitation.

Table 6 Frequency of use of the car by its owner

Frequency	Slovakia		Abroad	
	Absolute number	Relative number (in %)	Absolute number	Relative number (in %)
Daily	228	48.6	34	50.8
Several times a week	163	34.8	22	32.7
Occasionally (at least once a month)	44	9.4	6	9.0
Exceptionally (at least once a year)	34	7.2	5	7.5
Total number/share of respondents	469	100	67	100

Source Own elaboration, 2020

Table 7 The main reason for using the car

The reason	Slovakia		Abroad	
	Absolute number	Relative number (in %)	Absolute number	Relative number (in %)
Journey to/from work	174	37.1	37	55.2
Shopping	73	15.6	13	19.4
Free time	136	29.0	10	14.9
Long distances	54	11.5	6	9.0
Other	32	6.8	1	1.5
Total	469	100	67	100

Source Own elaboration, 2020

The following question concerned the average annual costs for operating and maintaining a car. We found out how these costs perceived by car owners (Table 8). We were slightly surprised by the perception of costs in 61% of domestic respondents and almost 63% of foreign respondents, who perceive the costs for operating the vehicle as average high. We can state that for some of the respondents, the car is still a luxury. Transport services in a sharing economy allow service providers to share operating costs, in particular fuel costs, and for respondents who considered the costs to be high, this could in future reduce cost perception to at least a reasonable cost level.

We also examined whether drivers usually travel alone or with other passengers (Table 9). We found that almost half of the domestic respondents usually travel with at least one passenger. We can state that this is a positive result, as at least one other

Table 8 Perception of the average annual costs for operating a car

Costs	Slovakia		Abroad	
	Absolute number	Relative number (in %)	Absolute number	Relative number (in %)
High	174	37.1	20	29.8
Reasonable	286	61.0	42	62.7
Low	9	1.9	5	7.5
Total number/share of respondents	469	100	67	100

Source Own elaboration, 2020

Table 9 The average number of people in cars

Number of persons	Slovakia		Abroad	
	Absolute number	Relative number (in %)	Absolute number	Relative number (in %)
Alone	170	36.2	25	37.3
With one passenger	223	47.5	28	41.8
With two or more passengers	76	16.3	14	20.9
Total number/share of respondents	469	100	67	100

Source Own elaboration, 2020

place besides the driver is usually used in the cars of the respondents. We assume that our survey involved 384 respondents (53%) aged 18 to 35 years, who can be characterized as drivers who usually do not have a family of their own yet and usually travel only in the company of a partner.

We consider the answers on the following question in the survey to be crucial, because the ownership of a motor vehicle is a basic precondition for the provision of transportation services in the sharing economy. The question was answered by the respondents who did not own any motor vehicle at the time of the survey (156 domestic and 35 foreign respondents). We asked them if they planned to buy a car in the future (Table 10). We perceive positively the answers of respondents from Slovakia, where 25% of respondents plan to buy a vehicle, 35.3% consider buying depending on the financial situation. Nearly 40% of domestic respondents were not considering the car purchase. This finding can be attributed to the number of students who participated in the survey. It is generally known that students in Slovakia usually do not have enough funds to buy a vehicle while studying at university. Another important factor is that students in Slovakia have the opportunity to travel by train

Table 10 Planning a car purchase in the future

Plan	Slovakia		Abroad	
	Absolute number	Relative number (in %)	Absolute number	Relative number (in %)
Yes, I plan to	39	25	12	34.3
Maybe, depending on the financial situation	55	35.3	14	40.0
No, I do not plan to	62	39.7	9	25.7
Total number/share of respondents	156	100	35	100

Source Own elaboration, 2020

for free, so their need to own a car is more a question of comfort or building their own image.

In the questionnaire, we asked respondents if they knew the concept of sharing economy (Table 11). Given the age structure of the respondents, we conclude that the awareness of the sharing economy is not high enough. We assumed that the age groups in younger and middle productive age know this concept, respectively, they even use some of the services that are part of the sharing economy. Foreign respondents have a much higher awareness than domestic respondents.

We were interested whether the respondents heard about the concepts of carsharing, carpooling, which indicate sharing of the car or places in them. In a comprehensive way, we can say that about half of the respondents knew the terms and understood their meaning.

Table 11 Awareness of the sharing economy

Plan	Slovakia		Abroad	
	Absolute number	Relative number (in %)	Absolute number	Relative number (in %)
I know the concept	229	36.6	68	66.6
I heard the term, but I do not understand it	219	35	17	16.7
I do not know the term	177	28.4	17	16.7
Total number/share of respondents	625	100	102	100

Source Own elaboration, 2020

Table 12 Usage of the transportation platforms

Usage	Slovakia		Abroad	
	Absolute number	Relative number (in %)	Absolute number	Relative number (in %)
Yes, I use it as a user (passenger)	196	31.4	60	58.8
Yes, I use it as a driver	12	1.9	3	2.9
Yes, I use both (the user and the driver)	21	3.4	5	4.9
No, I do not	237	37.9	23	22.5
No, I do not use it, but I plan to use it	159	25.4	11	10.9
Total number/share of respondents	625	100	102	100

Source Own elaboration, 2020

The following question in the second part of the questionnaire examined in more detail whether the respondents had ever used the ride via the Uber, Taxify, BlaBlaCar platforms (Table 12). We focused only on these three platforms providing carsharing, which are the most known in Slovakia.

Another question examined the perception of the greatest benefits for users who have ever used platform for ridesharing. Out of the total number of 625 domestic and 102 foreign respondents, 297 respondents have already tried transportation services on the principle of the sharing economy (Table 13). We conveniently evaluate the question in absolute terms and rank the advantages in the table in descending order from those that were evaluated with the highest number of points in the category "most important" and "very important".

The respondents saw the greatest importance of the sharing economy in saving money (270 respondents appointed this factor as the most important or important) and in saving time (239 respondents perceive it as the most important or important). At present, a person is forced to use his time as efficiently as possible, as evidenced by the results of the survey. Financial benefits, time reduction and flexible travel are crucial factors in deciding whether to use public transportation services in the city, long-distance bus or railway lines or carsharing. Users see potential advantages of ridesharing in the possibility of traveling at such time and place, where public transportation is not available.

It is interesting to observe that respondents usually do not primarily use transportation services in the sharing economy due to the development of social ties and new contacts, just as they are not primarily motivated to learn about the local authentic culture when choosing the sort of transport.

Table 13 The advantages of the transportation services on the principle of sharing economy from the perspective of platform users

Benefits	The most important	Important	Moderately important	Less important	Not important
Saving money	173	97	23	3	1
Save time when traveling	111	128	42	9	7
Possibility to travel outside the reach of public transportation	123	109	57	6	2
An ecological way of transport	104	94	62	28	9
Possibility of social life (zero tolerance of alcohol for drivers in Slovakia)	89	94	91	19	4
Greater credibility than the classic taxi service	82	98	92	20	5
Opportunity to develop social ties	61	112	69	36	1
Possibility to get new contacts	72	89	76	34	26
Authenticity of local culture	43	115	78	42	19

Source Own elaboration, 2020

4.2 Motivation of Potential Providers of Transportation Service on the Principle of Sharing Economy

In the next part of the survey, we were interested in the potential of providing transportation services in the sharing economy. We asked respondents if they were interested in providing services in this way in the future. The question was aimed on all respondents who have not so far used the transportation services on the principle of sharing economy, or they plan to use them in the future (396 domestic and 34 foreign respondents). We found out that more than half of the respondents might be interested in providing such services under certain conditions (Table 14).

We asked all 396 domestic and 34 foreign respondents what would motivate them to become providers of transportation services on the principle of sharing economy.

Table 14 Interest in providing transportation services in the sharing economy in the future

Interest	Slovakia		Abroad	
	Absolute number	Relative number (in %)	Absolute number	Relative number (in %)
Yes, I am interested	109	27.5	15	44.1
No, I'm not interested	162	40.9	10	29.4
Perhaps	125	31.6	9	26.5

Source Own elaboration, 2020

We rank the answers according to the degree of the highest agreement with the stated statements in descending order and present them in absolute terms (Table 15).

We found out that the most motivating was the possibility of gaining financial profit (364 respondents strongly agreed or agreed). Other important motivating factors are more efficient car usage (308 respondents strongly agreed or agreed) and the

Table 15 Incentive factors for potential transportation service providers on the principle of sharing economy

Motivation	I strongly agree	I agree	I do not know	I do not agree	I strongly disagree
Possibility of gaining financial profit	228	136	53	9	4
More efficient use of the car	137	171	118	2	2
Allocation of car operating costs	143	162	121	3	1
Opportunity to gain new experience	135	161	64	27	43
The possibility of developing communication skills	123	152	104	24	27
Helping others who do not own a car	118	141	138	25	8
Possibility to use free time effectively	112	139	127	29	23
Opportunity to meet new people	114	129	93	56	38

Source Own elaboration, 2020

distribution of costs for the operation of the vehicle, with which up to 305 respondents expressed their consent. Financial motivating factors are therefore a priority.

The survey confirmed that for some respondents, all the factors offered are a potential motivation for providing transportation services on the principle of sharing economy. However, if we focused on the statements with which the greatest number of respondents expressed disapproval, these are by the large number of respondents the statements that represent discomfort in leaving their own comfort zone: the opportunity to meet new people, the opportunity to gain new experiences, develop communication skills, but also the opportunity to use free time effectively. The connection between these statements of the respondents is probably to be found in considering the diversity of human characters (Stelmack & Stalikas, 1991).

In the next question, we focused on examining the demotivating factors for respondents' decision regarding the provision of transportation services in the sharing economy. As in the previous question, this concerned respondents who stated that they would be potentially interested in providing such services (430 respondents). The answers are part of Table 16, expressed in absolute and relative numbers in descending order of importance. As the respondent also had the opportunity to mark 3 answers at the same time, the resulting number of answers exceeds the total number of respondents.

As the survey was conducted during the pandemic situation related to Covid-19 and part of it took place during the period of strict quarantine in Slovakia, we were interested in how the respondents perceive the potential impact of a pandemic on the development of the sharing economy. The question of how the current epidemiological situation in the world affects supply and demand for services provided on the principle of sharing economy was answered by 459 respondents (online responds). We found out that up to 266 respondents (58%) are convinced that interest in providing

Table 16 Disadvantages of providing transportation services on the principle of sharing economy from the point of view of potential providers

Disadvantages	Absolute number	Relative number (in %)
Loss of privacy	300	69.8
Impairment of car	265	61.6
Concerns about own safety	200	46.5
Insufficient legislative protection	132	30.7
The need to adapt to the customers	113	26.3
Lack of time and flexibility	90	20.9
Stress from contact with strangers	74	17.2
Stress from communication with customers	37	8.6
Other (customer liability concerns, lack of information)	2	0.5

Note Respondents had the opportunity to mark more than one answer at a time, so the relative number exceeds 100
Source Own elaboration, 2020

these services will decline due to the epidemiological situation. Another 93 respondents (20.3%) believe that the interest in using these services will decrease because they perceive the situation sensitively and lose interest in services, the use of which they do not perceive as sufficiently safe. On the other hand, exactly 100 respondents (21.7%) think that the situation will gradually calm down and transportation services, which are part of the sharing economy, will maintain a growing trend, as they allow individualization of customer requirements. Given the lower number of people who come to mutual personal contact by providing these services compared to using the services of public transportation, the transportation services provided on the principle of the sharing economy are perceived as a safer alternative.

4.3 The Impact of the Age Category of Individuals on the Awareness, Use and Interest in the Provision of Transportation Services in the Sharing Economy

We have found out the relationship between age of respondents and other variables. Spearman's rank correlation coefficient confirmed in the researched period a weak indirect relationship between the age of respondents, and the awareness of the sharing economy (Table 17).

The result can be interpreted as with a higher age of respondents increases the likelihood of their lower awareness of transportation services provided on the principle of sharing economy. Many individuals in the older productive or in the post-productive age probably do not have sufficient information about the possibilities of services that they can use, they do not know the concept of sharing economy, or they heard already the term, but did not know its closer meaning.

Spearman's rank correlation coefficient also confirmed in the researched period moderately strong indirect correlation between the age of respondents and the use of transportation services on the principle of sharing economy (Table 18).

Table 17 Matrix of the correlation coefficient of age and awareness of individuals about transportation services on the principle of sharing economy

Researched variables	Age
Awareness of transportation services on the principle of sharing economy	−0.228

Source Own elaboration according to the results of IBM SPSS Statistics, version 25, 2020

Table 18 Matrix of the correlation coefficient of age and the use of transportation platforms in the sharing economy

Researched variables	Age
Use of transportation platforms	−0.315

Source Own elaboration according to the results of IBM SPSS Statistics, version 25, 2020

Table 19 Matrix of the correlation coefficient of age and interest in providing transportation services in the sharing economy

Researched variables	Age
Interest in providing transportation services in the sharing economy	−0.428

Source Own elaboration according to the results of IBM SPSS Statistics, version 25, 2020

This means that the older the respondents are, the less likely they use the available transportation platforms, either as passengers or as service providers.

At the same time, Spearman's rank correlation coefficient confirmed a moderately indirect relationship between the age of respondents and their interest in providing transport services on the principle of sharing economy (Table 19).

The result can be interpreted that the decrease in the age of respondents in the survey also decreased interest in providing transportation services based on the principle of sharing economy. The factors we have identified as the most important (especially financial motivation) were not so important for respondents in older productive age as for respondents in younger productive age.

5 Conclusion

The results of the questionnaire survey confirmed some of the expectations we obtained after processing the theoretical basis concerning the supply of the transportation services in tourism in the sharing economy. From the survey results, the strongest incentives for potential transportation service providers are financial motives—the possibility of financial profit, possibility of more efficient use of cars and the possibility of sharing the car operational costs. At the same time, however, the survey revealed several respondents who considered the new social contacts to be demotivating and perceived them as disadvantages of providing such services. This finding can be considered as partly contradictory, because it does not correspond with the basic characteristics of sharing economy (European Parliament, 2007, Dervojeda et al., 2013; European Commission, 2016; Barancová, 2017; Stevenson, 2017) which speaks about temporary use of assets by individuals, which is mediated through information technologies and is based on the need to survive and maintain social contacts. The results of questionnaire survey have in fact found out that communication and meeting new people were not sufficiently motivating for the potential service providers.

From the results of the survey, we observe that the awareness of the sharing economy in Slovakia is not yet high enough. Respondents hesitate also about the possibility to try the services. In particular, domestic respondents still prefer ownership rather than sharing and are hesitant mainly due to a lack of trust in service providers and security considerations.

We see the potential for tourism mainly in the offer of transportation services when traveling within the city (Uber, Bolt). Using this method of traveling is easy for tourists thanks to information technologies that allow online payment by credit card. Long-distance traveling (BlaBlaCar) remains a typical example of the sharing economy.

Cities not only in Slovakia need to be aware of the importance of the sharing economy in the present time and what its contribution can be in the future, in order to be able to make the most of it. A crucial element is a step needed to be done by the state that would allow or at least simplify the business in the field of sharing economy. Likewise, the authorities that make up the country's tourism policy must respond quickly to the current situation and create an appropriate legislative framework that can use shared services on behalf of tourism and increase the confidence of providers but also users of transportation services on the principle of sharing economy.

Today, cities attract more and more people not only because of the job opportunities, but also because of better opportunities for social life. In terms of higher concentration of the population and the number of goods that these inhabitants own, platforms represent an effective solution for the problems of the cities how to fight the deteriorating traffic situation or lack of the parking spaces.

Acknowledgements This research was funded by the Scientific Grant Agency of the Ministry of Education, Science, Research and Sport of Slovak Republic VEGA, grant number 1/0368/20 "Sharing economy as an opportunity for sustainable and competitive development of tourist destinations in Slovakia".

References

Accenture. (2016). *The platform economy: Innovation from the outside-in*. https://www.accenture.com/us-en/insight-digital-platform-economy.

Acquier, A., Daudigeos, T., & Pinkse, J. (2017). Promises and paradoxes of the sharing economy: An organizing framework. *Technological Forecasting and Social Change, 125*, 1–10. https://www.sciencedirect.com/science/article/pii/S0040162517309101.

Allen, D., & Berg, Ch. (2014). *The sharing economy. How over-regulation could destroy an economic revolution*. https://collaborativeeconomy.com/wp/wp-content/uploads/2015/04/Allen-D.-and-Berg-C.2014.The-Sharing-Economy.-Institute-of-Public-Affairs.-.pdf.

Barancová, H. (2017). Collaborative economy-digital age and their reflections for further development of the labor law. In J. Pichrt, et al., *Sharing economy-a shared legal problem?* (p. 336). Wolters Kluwer. ISBN 978-80-7552-874-2.

Boopen, S. (2005). Transport capital as a determinant of tourism development. A time series approach. https://mpra.ub.uni-muenchen.de/25402

Borovská, T. (2019). The offer of tourism transportation services provided on the principles of sharing economy in Slovakia and abroad. Matej Bel University in Banská Bystrica, Faculty of Economics.

Botsman, R. (2015). *Defining the Sharing Economy: What is Collaborative consumption and What isn't*. Fast company article. https://www.fastcompany.com/3046119/defining-the-sharing-economy-what-is-collaborative-consumption-and-what-isnt.

Botsman, R., & Rogers, R. (2011). *What's Mine Is Yours: The Rise of Collaborative Consumption* (p. 304). Harper Business Print. ISBN: 9780007395910.

Brinke, J. (1999). *Introduction to the geography of the transportation* (p. 112). Karolinum. ISBN: 80-7184-923-5.

Chovanculiak, R. (2018). *A sharing economy would help the development of the eastern Slovakia, there is a potential.* https://www.finreport.sk/lidri/robert-chovanculiak-zdielana-ekonomika-by-pomohla-rozvoju-vychodneho-slovenska-je-tam-potencial/.

Codagnone, C., & Martens, B. (2016). *Scoping the Sharing Economy: Origins, Definitions, Impact and Regulatory Issues.* JRC Technical Reports. Institute for Prospective Technological Studies Digital Economy. Working Paper 2016/01. ISSN 1831-9408.

Demary, V. (2014). *Competition in the sharing economy.* IW policy paper, No. 19/2015. Provided in Cooperation with German Economic Institute (IW), Cologne. https://www.econstor.eu/bitstream/10419/112778/1/830325093.pdf.

Dervojeda, K. et al. (2013). *The sharing Economy: Accessibility Based Business Models for Peer-to-Peer Markets.* European Commision: Business Inovation Observatory Contract No 190/PPENT/CIP/12/C/N03C01 Case study 12.

European Commision. (2016). *Europe and agenda for the colaborative economy–supporting analysis.* 2016. Commision Staff Working Document Ref. Ares (2016)2562059.

European Parliament. (2007). *A renewed EU Tourism Policy: European Parliament resolution from 29th of November 2007 on a renewed EU Tourism Policy: Towards a stronger partnership for European Tourism.* Retrieved from https://www.europarl.europa.eu/sides/getDoc.do?pubRef=-//EP//TEXT+TA+P6-TA-2007-0575+0+DOC+XML+V0//SK#def_1_1.

Goudin, P. (2016). *The cost of non-Europe in the sharing economy. economic, social and legal challenges and opportunities.* European Parliament. https://www.europarl.europa.eu/RegData/etudes/STUD/2016/558777/EPRS_STU(2016)558777_EN.pdf.

Hamari, J., Sjöklint, M., & Ukkonen, A. (2016). The Sharing Economy: Why People Participate in Collaborative Consumption. *Journal of the Association for Information Science and Technology, 67,* 2047–2059. https://doi.org/10.1002/asi.23552. https://www.researchgate.net/publication/255698095_The_Sharing_Economy_Why_People_Participate_in_Collaborative_Consumption.

Horner, S., & Swarbrooke, J. (2003). Tourism, accommodation and hospitality, the usage of the free time: applied marketing of services (p. 488). Grada. ISBN: 8024702029.

Ikkala, T., & Lampinen, A. (2015). Monetizing Network Hospitality: Hospitality and Sociability in the Context of Airbnb. In *CSCW'15 Proceedings of the ACM 2015 conference on Computer supported cooperative work* (pp. 1033–1044). ACM. ISBN: 978-1-4503-2922-4.

Juul, M. (2017). *Tourism and the sharing economy.* European Parliament. https://www.europarl.europa.eu/thinktank/en/document.html?reference=EPRS_BRI(2017)595897.

Khadaroo, J., & Seetanah, B. (2007). Transport infrastructure and tourism development. *Annals of Tourism Research, 34*(4), 1021–1032. ISSN: 01607383. https://linkinghub.elsevier.com/retrieve/pii/S0160738307000837.

Khan, R. et al. (2017). Travel and tourism competitiveness index: The impact of air transportation, railways transportation, travel and transport services on international inbound and outbound tourism. *Journal of Air Transport Management* [online], 125–134 [cit. 2019–02–11]. ISSN 09696997. https://linkinghub.elsevier.com/retrieve/pii/S0969699716301351.

Kostakis, V., & Bauwens, M. (2014). *Network Society and Future Scenarios for a Collaborative Economy* (p. 97). Palgrave Macmillan. ISBN 978-1-137-40689-7.

Navickas, V. & Malakauskaite, A. (2009).The possibilities for the identification and evaluation of tourism sector competitiveness factors. *Engineering Economics, 61*(1), 37–44.

Prideaux, B. (2000). The role of the transport system in destination development. *Tourism Management, 21*(1), 53–63. https://doi.org/10.1016/S0261-5177(99)00079-5

Ranjbari, M., Morales-Alonso, G., & Carrasco-Galleg, R. (2018). Conceptualizing the sharing economy through presenting a comprehensive framework. *Sustainability, 10,* 7. ISSN 2071-1050.

Rifkin, J. (2015). *Zero marginal cost society* (p. 446). Reprint version. Griffin. ISBN 978-1137280114.

Roblek, V., Meško Štok, Z., & Meško, M. (2016). Complexity of a sharing economy for tourism and hospitality. In *Conference: Tourism and hospitality industry trends and challenges 2016* (pp. 374–387). https://doi.org/10.13140/RG.2.1.3000.2165. https://www.researchgate.net/public ation/301612962_Complexity_of_a_sharing_economy_for_tourism_and_hospitality.

Slovak Business Agency, (2018). *Analysis of the sharing economy in the environment of the small and medium enterprises.* https://www.sbagency.sk/sites/default/files/6_analyza_zdielanej_ekonomiky_v_prostredi_msp.pdf.

Stelmack, M., & Stalikas, A. (1991). Galen and the humour theory of temperament. *Personality and Individual Differences, 12*(3), 255–263. ISSN: 01918869. https://linkinghub.elsevier.com/retrieve/pii/019188699190111N.

Stevenson, A. (2017). *Oxford dictionary of English* (3rd ed.). Oxford University Press. ISBN 978-0199571123.

Tussyadiah, P., & Personen J. (2015). Impacts of Peer-to-Peer Accommodation Use on Travel Patterns. *Journal of Travel Research, 55*(8), 1022–1040. Article first published online: October 12, 2015; Issue published: November 1, 2016. ISNN 1552-6763. https://doi.org/10.1177/004728 7515608505.

Vaněček, D. (2008). *Logistics.* Jihočeská Univerzita, Ekonomická fakulta, 2008. ISBN 978-80-7394-085-0.

Van Truong, N., & Shimizu, T. (2017). The effect of transportation on tourism promotion: Literature review on application of the computable general equilibrium (CGE) model. *Transportation Research Procedia* [online]. *2017*, 3096–3115 [cit. 2019–02–11]. https://linkinghub.elsevier.com/retrieve/pii/S2352146517306439.

Integration of Sustainable Practices in Firms: The Specifics of the Tourism, Leisure and Hospitality Sectors

Inés Díez Martínez and Ángel Peiró Signes

1 Introduction

Historically, tourism and related activities have had a great impact on societal and natural environments. From exploiting the potential of different areas and creating new infrastructure to the development of new services and leisure alternatives, this sector has been changing landscapes and producing high energy, waste and water footprint all around the world.

Currently, many sustainable practices are put in place to reduce this footprint. Some of these practices involve infrastructure changes such as reshaping facilities with the use of eco-design of the integration of water-efficient technologies. Other initiatives are related to the improvement of current processes, reducing a companies' footprint through the integration of locally sourced ingredients or the improvement of the recyclability of products. Finally, brand-new business models that have sustainability in their core are also arising, such as the development of new activities attractive to responsible tourists.

This paper will investigate and examine existing approaches and trends related to sustainable models in firms, focusing on the identification of sustainable practices applied in firms related to the tourism, leisure and hospitality sectors.

2 Background

To perform research and improve the understanding of the development and integration of sustainable practices, it is worthy to first take a step back and look into the meaning of sustainable development.

I. D. Martínez (✉) · Á. P. Signes
Universitat Politècnica de València, Valencia, Spain

Sustainable development is often described using the definition from the 1987 Brundtland Commission Report, where it is described as the "development that meets the needs of the present without compromising the ability of future generations to meet their own needs" (Roostaie et al, 2019). Within the context of this definition, it is also commonly accepted that sustainability has three main dimensions, which are environment, social and economic (Strezov et al., 2016; Torugsa et al., 2013; Hansmann et al., 2012). Consequently, a sustainable business model or practice would also be expected to target the same objectives and dimensions to be aligned with sustainability principles, considering environmental, social and economic matters. As shown in Fig. 1 and described by Irsan and Utama (2019), these dimensions interrelate, and it is when the combination of all is achieved that we could consider a system to be sustainable. When these dimensions are all integrated, they would make a business model sustainable, which can be defined as bearable, viable and equitable (Slocum, 2015).

Starting with the environmental dimension, by itself, it refers to all the aspects related to ecologic matters, such as the use of resources, emissions and biodiversity. These shall be combined with the social and economic dimensions to achieve sustainability. As discussed by Barile et al. (2018), when looking into the social aspect, a practice that integrates environmental and social aspects could be considered "bearable". By all means, when the environmental dimension is interlinked with the social aspect, the outcome is the creation of policies to ensure environmental justice and conservation. Following Fig. 1 and moving clockwise to the social dimension, this dimension covers aspects related to human rights, labor conditions and social inclusion. Thus, when interrelated with the economic dimension, the result is "equitable", referring to fair trade, business ethics and workers benefits (Irsan & Utama, 2019). Finally, the third pillar, economic, is focused on revenue and profit generation, growth

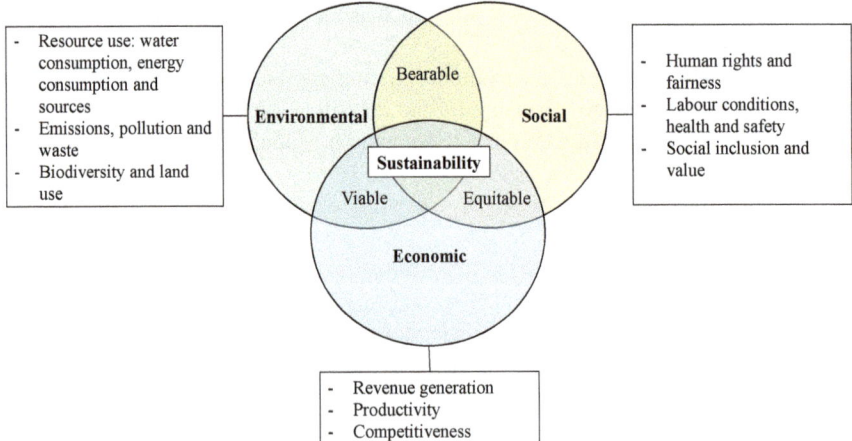

Fig. 1 Three dimensions of sustainability (adapted from Irsan and Utama (2019), Barile et al. (2018) and Slocum (2015))

and productivity and competitiveness including cost-saving initiatives. Closing the circle, a sustainable practice should also be "viable", which is obtained when as per the combination of the economic and environmental pillars (Irsan & Utama, 2019). "Viable" would refer to elements such as energy efficiency, incentives/subsidies for the use of greener technology and renewable fuels, etc. An outline of these three pillars and their interactions is shown in Fig. 1.

To summarize, a sustainable business model or practice would be expected to target as much as possible these dimensions alongside the combined products obtained when these three pillars interact. As discussed by Nicholas and Thapa (2010) these pillars are not meant to be mutually exclusive but the opposite, they would be expected to have a synergistic effect and be mutually reinforcing.

3 Methodology

Once the main characteristics of what would be the framework of a sustainable business model are understood, it is possible to proceed and investigate what would be their integration in the framework of the tourism, leisure and hospitality industry. To perform this investigation, this paper performs a systematic analysis of the existing literature. This study conducted this type of review since it is a rigorous standalone review in which the methodological approach could be followed by others interested in the topic. Additionally, a systematic review has an exhaustive scope in which a high volume of documents can be processed, to ensure that all pertinent material is included (Okoli, 2015; Reim et al., 2015).

Consequently, a systematic review would be an appropriate method to examine sources that consolidate a high volume of information such as research databases. Accordingly, a database was selected to perform the review. In this case, the chosen database is Scopus (www.scopus.com). As analyzed by Aksnes and Sivertsen (2019), and Aghaei Chadegani et al. (2013), Scopus can be considered as a suitable option, since this database would be able to provide comprehensive coverage of the world's scientific and scholarly literature.

The objective of the search is focused on understanding the integration of sustainable business models from the tourism, leisure and hospitality sector. In Scopus, the category of "Tourism, Leisure and Hospitality" is under the research area "Business, Management and Accounting", therefore we would limit the search to this category. As per the search string, it would include the term sustainability along with the three pillars discussed in the previous section as follows: *SUBJAREA(BUSI) ALL ("SUSTAINABILITY") AND title-abs-key (*sustainability* OR *sustainable* OR *environment* or *social* or *economic*).* The output of this search was 80,614 documents. The high volume of results evidenced the need to fine-tune the search and ensure its focus on the tourism, leisure and hospitality sectors. The methodology followed to refine the search consists of the three steps proposed by Reim et al. (2015), which are indicated in Fig. 2.

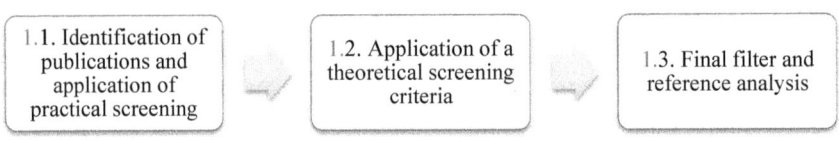

Fig. 2 Methodology steps followed to refine results

3.1 Identification of Publications and Application of Practical Screening

To filter the high volume of results to find those relevant for the screening, the first step would be to identify results that are linked to the industry of the research, in this case, "Tourism, leisure and hospitality". By doing so, we have limited the results to those journals that are most relevant under this category.

To find the most relevant journals, the Scimago (www.scimagojr.com) ranking was used, and the top 5 journals, with the highest SJR ranking in 2019, in the "Tourism, Leisure and Hospitality Management" category were selected. This step is focused on the identification of publications to ensure the quality of the review, in this case by aiming at those journals that would be the most relevant to the scope of the search. The top 5 journals for the category are shown in Table 1.

Additionally, since the focus is on understanding the latest state of the art, within this framework, articles older than 2017 will not be reviewed to keep the focus on the most recent studies. The final research string considering the above was: *SUBJAREA(BUSI) ALL ("SUSTAINABILITY") AND title-abs-key(*sustainability* OR *sustainable* OR *environment* or *social* or *economic*) AND (LIMIT-TO (DOCTYPE,"ar")) AND (LIMIT-TO (PUBYEAR,2020) OR LIMIT-TO (PUBYEAR, 2019) OR LIMIT-TO (PUBYEAR, 2018) OR LIMIT-TO (PUBYEAR,2017)) AND (LIMIT-TO (EXACTSRCTITLE, "Tourism Management") OR LIMIT-TO (EXACT-SRCTITLE, "Annals Of Tourism Research") OR LIMIT-TO (EXACTSRCTITLE, "International Journal Of Hospitality Management") OR LIMIT-TO (EXACTSRC-TITLE, "International Journal Of Contemporary Hospitality Management") OR LIMIT-TO (EXACTSRCTITLE, "Journal Of Travel Research")).*

Table 1 Journals selected for the systematic review

Rank	Source id	Title	SJR	SJR Quartile	H index
1	16547	Tourism Management	3068	Q1	179
2	14813	Journal of Travel Research	3014	Q1	122
3	30718	Annals of Travel Research	2228	Q1	158
4	28686	International Journal of Hospitality Management	2217	Q1	106
5	144742	International Journal of Contemporary Hospitality Management	2203	Q1	76

3.2 Application of a Theoretical Screening Criteria

Since the objective of the review is focused on sustainable business models, articles would need to target practices including one or several sustainability dimensions, or ideally, all of them. For articles to be included in the analysis, the abstract needed to provide insights about the articles, by meeting the criteria below:

1. Articles need to describe eco-innovation and sustainable development practices and business models within the hospitality, tourism and leisure firms, including best practices in this field.
2. Articles need to address how a given practice addresses the sustainability dimensions, the sustainable challenge to overcome and how the practice in place is providing a solution for the sustainable challenge.
3. Articles need to include information about the specifics of the type of companies for which these business models are applied (e.g., region, company size (medium/small family-owned hotels and hostels and/or chain-brand hotels, etc.)) in order to provide insights of how a given practice is implemented.

3.3 Final Filter and Reference Analysis

In this final stage, those articles that met the inclusion criteria defined above were downloaded and read to perform an analysis of their content.

A summary of the overall process is described in Fig. 3.

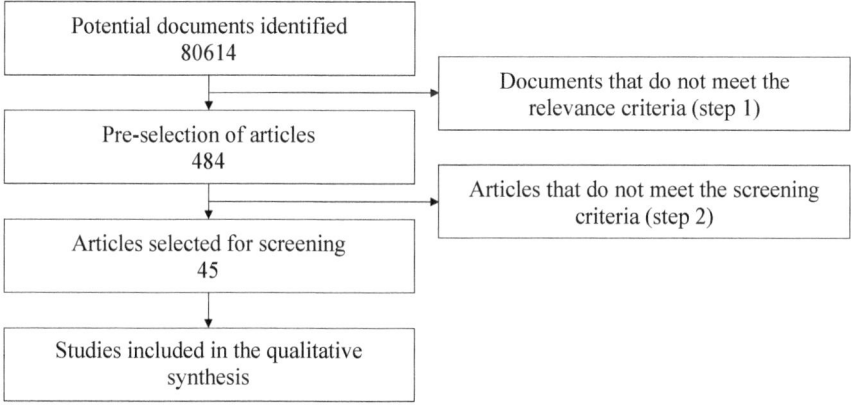

Fig. 3 Flow diagram of the review steps

4   Results and Discussion

Once the methodology is specified, the list of selected articles to be included in the qualitative synthesis is shown in Table 2.

Additionally, Table 3 provides an overview of the distribution of results per journal. As evidenced in Table 3, all the journals defined during step 1 of the methodology have been found relevant, since articles were selected across the five different journals with valuable information concerning the search. Results also show that the journal with the highest ranking, "Tourism Management" is also the journal where more results that were relevant to sustainability practices have been found.

Overall, among the different journals and papers presented above, several business models and practices to improve sustainability in the industries are discussed. The most relevant sustainable practices found through the articles selected are displayed in Table 4. Sustainable practices have been classified according to the main pillar or combination of pillars that they address.

The main business practices seem to have a focus on either environmental or social pillar, with economic aspect acting as a positive reinforcement to strengthen the integration of the practice within the firm. On the environmental side, there seem to be two main objectives: either the creation of a more environmentally friendly service as proposed by Ballantyne et al. (2018), Huang and Liu (2017), Lee and Jan, (2018) and Lucrezi et al. (2017); or, as discussed by Abaeian et al. (2019), Anastasiadou and Vettese (2019), Cha et al. (2018) Gössling et al. (2019), or the improvement in the efficiency of the resource use (e.g., for energy and water) along with waste reduction.

With the focus on creating an environmentally friendly service, practices commonly target the learning experience of the visitor. As discussed by Ballantyne et al. (2018), this is done by using a values-based approach to visitor interpretation, which refers to the integration of environmental learning and values, promoting self-reflection of visitors, which in many cases is poor and shows a need for environmental education. Its added value is that it promotes environmental learning and enhances the post-visit environmental behavior of visitors. This practice could be established for different company types and sectors, from the case of wildlife tourism, zoos and aquariums (Ballantyne et al., 2018) to scuba diving tourism (Lucrezi et al., 2017). Additionally, environmentally friendly services could also be targeted through the community, as discussed by Batle et al. (2018) by collaborating with the community addressing environmental issues together, an idea also supported by Batle et al. (2018), Ertuna et al. (2019), Raub and Martin-Rios (2019) and Hang et al. (2020) and connecting with locals farms and promoting environmentally friendly agricultural methods.

Regarding the improvement of efficiency in the resource use, articles propose using alternative energy systems (Batle et al., 2018) and reducing energy and water consumption (Abaeian et al., 2019; Cha et al., 2018). A clear example is a case described by Cha et al. (2018), related to private clubs, where maintaining golf courses is an activity that often exceeds the required water and chemicals. Energy

Table 2 List of selected articles

Author (year)	Title
Abaeian et al. (2019)	Motivations of undertaking CSR initiatives by independent hotels: a holistic approach
Agapito (2020)	The senses in tourism design: a bibliometric review
Akbar and Tracogna (2018)	The sharing economy and the future of the hotel industry: transaction cost theory and platform economics
Alderighi and Gaggero (2019)	Flight availability and international tourism flows
Aleshinloye et al. (2020)	The influence of place attachment on social distance: examining mediating effects of emotional solidarity and the moderating role of interaction
Anastasiadou and Vettese (2019)	"From souvenirs to 3D printed souvenirs". Exploring the capabilities of additive manufacturing technologies in (re)-framing tourist souvenirs
Arbogast et al. (2020)	Using social design to visualize outcomes of sustainable tourism planning: a multiphase, transdisciplinary approach
Asmelash and Kumar (2019)	Assessing progress of tourism sustainability: Developing and validating sustainability indicators
Avila-Foucat and Rodríguez-Robayo (2018)	Determinants of livelihood diversification: The case wildlife tourism in four coastal communities in Oaxaca, Mexico
Ballantyne et al. (2018)	Visitors' values and environmental learning outcomes at wildlife attractions: Implications for interpretive practice
Batle et al. (2018)	Environmental management best practices: Towards social innovation
Bertella (2020)	Re-thinking sustainability and food in tourism
Bianchi (2018)	The political economy of tourism development: A critical review
Boley and Woosnam (2021)	Going global or going local? Why travelers choose franchise and independent accommodations
Cabral and Chiappetta Jabbour (2020)	Understanding the human side of green hospitality management
Camilleri and Neuhofer (2017)	Value co-creation and co-destruction in the Airbnb sharing economy

(continued)

Table 2 (continued)

Author (year)	Title
Cantele and Cassia (2020)	Sustainability implementation in restaurants: A comprehensive model of drivers, barriers, and competitiveness-mediated effects on firm performance
Cha et al. (2018)	Adoption of sustainable business practices in the private club industry from GMs and COOs' perspectives
Chan et al. (2020)	What hinders hotels' adoption of environmental technologies: A quantitative study
Chen (2019)	Hotel chain affiliation as an environmental performance strategy for luxury hotels
Chi et al. (2020)	Wellness hotel: Conceptualization, scale development and validation
Craig and Feng (2018)	A temporal and spatial analysis of climate change, weather events and tourism businesses
Cui et al. (2020)	Moral Effects of Physical Cleansing and Pro-environmental Hotel Choices
Dogru et al. (2019)	Climate change: Vulnerability and resilience of tourism and the entire economy
Dolnicar (2020)	Designing for more environmentally friendly tourism
Ertuna et al. (2019)	Diffusion of sustainability and CSR discourse in hospitality industry: Dynamics of local context
Garay et al. (2019)	Sustainability-oriented innovation in tourism: An analysis based on the decomposed theory of planned behavior
Gerdt et al. (2019)	The relationship between sustainability and customer satisfaction in hospitality: An explorative investigation using eWOM as a data source
Gössling et al. (2019)	Towel reuse in hotels: Importance of normative appeal designs
Horng et al. (2017)	Developing a sustainable service innovation framework for the hospitality industry
Huang and Liu (2017)	Moderating and mediating roles of environmental concern and ecotourism experience for revisit intention
Jones and Wynn (2019)	The circular economy, natural capital and resilience in tourism and hospitality

(continued)

Table 2 (continued)

Author (year)	Title
Khatter et al. (2019)	Analysis of hotels' environmentally sustainable policies and practices: Sustainability and corporate social responsibility in hospitality and tourism
Lee and Jan (2018)	Ecotourism behavior of nature-based tourists: An integrative framework
Lithgow et al. (2019)	Exploring the co-occurrence between coastal squeeze and coastal tourism in a changing climate and its consequences
Lucrezi et al. (2017)	Scuba diving tourism systems and sustainability: Perceptions by the scuba diving industry in two Marine Protected Areas
Martinez-Martinez et al. (2019)	Knowledge agents as drivers of environmental sustainability and business performance in the hospitality sector
Ng et al. (2017)	Seeking tourism sustainability–A case study of Tioman Island, Malaysia
Pulido-Fernández and Cárdenas-García (2020)	Analyzing the bidirectional relationship between tourism growth and economic development
Raub and Martin-Rios (2019)	"Think sustainable, act local"–a stakeholder-filter-model for translating SDGs into sustainability initiatives with local impact
Schliephack and Dickinson (2017)	Tourists' representations of coastal managed realignment as a climate change adaptation strategy
Su et al. (2019)	Livelihood sustainability in a rural tourism destination—Hetu Town, Anhui Province, China
Vogt et al. (2020)	Designing for quality of life and sustainability
Zhang and Zhang (2018)	Perception of small tourism enterprises in Lao PDR regarding social sustainability under the influence of social network
Zhang et al. (2020)	Sociocultural Sustainability and the Formation of Social Capital from Community-based Tourism

efficiency is also proving to be relevant to the branding value of operators as discussed by Chen (2019). In this aspect, is it also key to increase the knowledge inside the organization, gearing toward making smart environmental choices, with initiatives such as green training (Chan et al., 2020; Cabral & Chiappetta Jabbour, 2020) and create knowledge agents within the organization (Martinez-Martinez et al., 2019).

On the social side, initiatives are predominantly focused on either the local communities and/or the firms' staff. With the focus on the local communities, research

Table 3 Articles selected grouped by journal

Journal	Number of articles selected
Tourism Management	15
International Journal of Hospitality Management	9
International Journal of Contemporary Hospitality Management	8
Journal of Travel Research	7
Annals of Travel Research	6

proposes engaging with the well-being of local communities as described by Vogt et al. (2020) and increasing the employment of residents (Zhang & Zhang, 2018). This last initiative is also proposed with a special focus on those unemployed or less fortunate within the community (Abaeian et al., 2019). Also, Su et al. (2019) propose to merge tourism with other income sources to increase the whole livelihood of the community. With the focus on the staff, initiatives such as the ones described above are directly linked with the employment of locals and helping those in need within the community. Finally, firms also could integrate initiatives to increase the well-being of the employees (Abaeian et al., 2019).

5 Conclusion

Sustainability considerations are key in the development of new practices and innovation for those firms belonging to the tourism, leisure and hospitality sectors. Sustainability practices are put in place in firms of all sizes and services, as well as regions all around the world.

Sustainability is integrated into practices that consider and create synergies between its three dimensions; environmental, social and economic. Practices related to environmentally friendly services and products, resource efficiency and sustainable management. Furthermore, many companies are opting for educating clients and visitors on eco-friendly values, which has proven to be an efficient method and advisable practice to change perspectives and influence eco-friendly conducts, encouraging a more sustainable society.

Additionally, practices often consider the well-being of society as well as the local communities. Often, practices promote the interaction between firms and locals, encouraging employment and communications between the two parties. Furthermore, there are practices where the social side is focused on the well-being of staff. They do so by improving the working conditions of those who are employed by the firms. Besides, companies also promote internal change through green training of employees and similar initiatives.

Table 4 Sustainable practices found during the research

Sustainable practice	Source	Environmental	Social	Economic
Adapt outdoor tourism to long-term weather conditions and weather events	Craig and Feng (2018)	X		X
Adopt a managed realignment to address coastal change (i.e., erosion)	Schliephack and Dickinson (2017)	X		X
Adopt families with no jobs and invite people less fortunate at the community	(Abaeian et al., 2019)		X	
Building social networks and the provision of training and employment of locals	Zhang, and Zhang (2018)		X	
Collect rainwater for reuse	Abaeian et al. (2019)	X		X
Connect with local farms and ecological and slow food associations	Batle et al. (2018)	X	X	X
Create an experimental agricultural program and use alternative agricultural methods	Batle et al. (2018)	X	X	
Create the role of knowledge agent within the firm	Martinez-Martinez et al. (2019)	X		
Designing places that yield high quality of life (QOL) for residents	Vogt et al. (2020)		X	
Engage with the local community (e.g., to address ecological ans social issues through community and local venues)	Batle et al. (2018) Raub and Martin-Rios (2019) Zhang et al. (2020)	X	X	
Have a home-grown herb garden	Abaeian et al. (2019)	X		
Integrate environmental and/or social learning and values, promoting self-reflection of visitors	Ballantyne et al. (2018) Huang and Liu (2017) Lee and Jan (2018) Lucrezi et al. (2017)	X	X	
Merge tourism with other income sources, enhancing overall livelihood sustainability	Su et al. (2019)		X	X
Motivate consumers to engage in pro-environmental travel choices through physical cleansing	Cui et al. (2020)	X		

(continued)

Table 4 (continued)

Sustainable practice	Source	Environmental	Social	Economic
Organize activities to improve staff's well-being	Abaeian et al. (2019)		X	
Perform green training	Cabral and Chiappetta Jabbour (2020) Chan et al. (2020)	X		
Provide newspaper only upon request	Abaeian et al. (2019)	X		X
Recover traditional harvesting and hunting methods	Batle et al. (2018)	X		X
Recycle cooking oil	Abaeian et al. (2019)	X		X
Reduce chemicals alongside allergies/reactions in staff members and clients	Cha et al. (2018)	X	X	
Reduce food waste (reduce stock, donate excess food, etc.)	Cha et al. (2018)	X	X	X
Reduce water usage with flow regulators and low flow alternatives	Abaeian et al. (2019) Cha et al. (2018)	X		X
Reduce water usage with the reuse of towels and bed linen	Abaeian et al. (2019) Gössling et al. (2019)	X		X
Use alternative energy-saving systems (e.g., light on a timer, motion sensors, etc.)	Abaeian et al. (2019) Batle et al. (2018) Cha et al. (2018) Chen, (2019)	X		X
Use recycled materials and products (e.g., recycled furniture, toilet paper, etc.)	Abaeian et al. (2019)	X		

Overall, research shows a wide variety of best practices that could be put in place to improve the sustainability performance of those firms that are linked to the hospitality, leisure and tourist sectors. In most cases, either the environmental and/or social pillars are addressed. However, research shows that there is limited discussion on the economic pillar of sustainability, which could be typically addressed through cost-saving initiatives within the scope of the practices found.

References

Abaeian, V., Khong, K. W., Yeoh, K. K., & McCabe, S. (2019). Motivations of undertaking CSR initiatives by independent hotels: A holistic approach. *International Journal of Contemporary Hospitality Management, 31*(6), 2468–2487. https://doi.org/10.1108/IJCHM-03-2018-0193.

Agapito, D. (2020). The senses in tourism design: A bibliometric review. *Annals of Tourism Research, 83*. https://doi.org/10.1016/j.annals.2020.102934.

Aghaei Chadegani, A., Salehi, H., Yunus, M., Farhadi, H., Fooladi, M., Farhadi, M., & Ebrahim, N. A. (2013). A Comparison between two main academic literature collections: Web of science and Scopus databases. *Asian Social Science, 9*(5), 18–26.

Akbar, Y. H., & Tracogna, A. (2018). The sharing economy and the future of the hotel industry: Transaction cost theory and platform economics. *International Journal of Hospitality Management, 71*, 91–101. https://doi.org/10.1016/j.ijhm.2017.12.004.

Aksnes, D.W., & Sivertsen, G. (2019). A criteria-based assessment of the coverage of Scopus and web of science. *Journal of Data and Information Science, 4*(1). https://doi.org/10.2478/jdis-2019-0001.

Alderighi, M., & Gaggero, A. A. (2019). Flight availability and international tourism flows. *Annals of Tourism Research, 79*. https://doi.org/10.1016/j.annals.2018.11.009.

Aleshinloye, K. D., Fu, X., Ribeiro, M. A., Woosnam, K. M., & Tasci, A. D. A. (2020). The influence of place attachment on social distance: Examining mediating effects of emotional solidarity and the moderating role of interaction. *Journal of Travel Research, 59*(5), 828–849. https://doi.org/10.1177/0047287519863883.

Anastasiadou, C., & Vettese, S. (2019). "From souvenirs to 3D printed souvenirs". Exploring the capabilities of additive manufacturing technologies in (re)-framing tourist souvenirs. *Tourism Management, 71*, 428–442. https://doi.org/10.1016/j.tourman.2018.10.032.

Arbogast, D., Butler, P., Faulkes, E., Eades, D., Deng, J., Maumbe, K., & Smaldone, D. (2020). Using social design to visualize outcomes of sustainable tourism planning: A multiphase, transdisciplinary approach. *International Journal of Contemporary Hospitality Management, 32*(4), 1413–1448. https://doi.org/10.1108/IJCHM-02-2019-0140.

Asmelash, A. G., & Kumar, S. (2019). Assessing progress of tourism sustainability: Developing and validating sustainability indicators. *Tourism Management, 71*, 67–83. https://doi.org/10.1016/j.tourman.2018.09.020.

Avila-Foucat, V. S., & Rodríguez-Robayo, K. J. (2018). Determinants of livelihood diversification: The case wildlife tourism in four coastal communities in Oaxaca, Mexico. *Tourism Management, 69*, 223–231. https://doi.org/10.1016/j.tourman.2018.06.021.

Ballantyne, R., Hughes, K., Lee, J., Packer, J., & Sneddon, J. (2018). Visitors' values and environmental learning outcomes at wildlife attractions: Implications for interpretive practice. *Tourism Management, 64*, 190–201. https://doi.org/10.1016/j.tourman.2017.07.015.

Batle, J., Orfila-Sintes, F., & Moon, C. J. (2018). Environmental management best practices: Towards social innovation. *International Journal of Hospitality Management, 69*, 14–20. https://doi.org/10.1016/j.ijhm.2017.10.013.

Barile, S., Quattrociocchi, B., Calabrese, M., & Iandolo, F. (2018). Sustainability and the viable systems approach: opportunities and issues for the governance of the territory. *Sustainability, 10*, 790.

Bertella, G. (2020). Re-thinking sustainability and food in tourism. *Annals of Tourism Research.* https://doi.org/10.1016/j.annals.2020.103005.

Bianchi, R. (2018). The political economy of tourism development: A critical review. *Annals of Tourism Research, 70*, 88–102. https://doi.org/10.1016/j.annals.2017.08.005.

Boley, B. B., & Woosnam, K. M. (2021). Going global or going local? Why travelers choose franchise and independent accommodations. *Journal of Travel Research, 60*(2), 354–369. https://doi.org/10.1177/0047287520904786.

Cabral, C., & Chiappetta Jabbour, C. J. (2020). Understanding the human side of green hospitality management. *International Journal of Hospitality Management, 88*. https://doi.org/10.1016/j.ijhm.2019.102389.

Camilleri, J., & Neuhofer, B. (2017). Value co-creation and co-destruction in the airbnb sharing economy. *International Journal of Contemporary Hospitality Management, 29*(9), 2322–2340. https://doi.org/10.1108/IJCHM-09-2016-0492.

Cantele, S., & Cassia, F. (2020). Sustainability implementation in restaurants: A comprehensive model of drivers, barriers, and competitiveness-mediated effects on firm performance. *International Journal of Hospitality Management, 87.* doi:https://doi.org/10.1016/j.ijhm.2020. 102510.

Cha, J., Kim, S. J., & Cichy, R. F. (2018). Adoption of sustainable business practices in the private club industry from GMs and COOs' perspectives. *International Journal of Hospitality Management, 68,* 1–11. https://doi.org/10.1016/j.ijhm.2017.08.014.

Chan, E. S. W., Okumus, F., & Chan, W. (2020). What hinders hotels' adoption of environmental technologies: A quantitative study. *International Journal of Hospitality Management, 84.* https://doi.org/10.1016/j.ijhm.2019.102324.

Chen, L. (2019). Hotel chain affiliation as an environmental performance strategy for luxury hotels. *International Journal of Hospitality Management, 77,* 1–6. https://doi.org/10.1016/j.ijhm.2018. 08.021.

Chi, C. G., Chi, O. H., & Ouyang, Z. (2020). Wellness hotel: Conceptualization, scale development, and validation. *International Journal of Hospitality Management.* https://doi.org/10.1016/j.ijhm. 2019.102404.

Craig, C. A., & Feng, S. (2018). A temporal and spatial analysis of climate change, weather events, and tourism businesses. *Tourism Management, 67,* 351–361. https://doi.org/10.1016/j.tourman. 2018.02.013.

Cui, Y., Errmann, A., Kim, J., Seo, Y., Xu, Y., & Zhao, F. (2020). Moral effects of physical cleansing and pro-environmental hotel choices. *Journal of Travel Research, 59*(6), 1105–1118. https://doi. org/10.1177/0047287519872821.

Dogru, T., Marchio, E. A., Bulut, U., & Suess, C. (2019). Climate change: Vulnerability and resilience of tourism and the entire economy. *Tourism Management, 72,* 292–305. https://doi. org/10.1016/j.tourman.2018.12.010.

Dolnicar, S. (2020). Designing for more environmentally friendly tourism. *Annals of Tourism Research, 84.* https://doi.org/10.1016/j.annals.2020.102933.

Ertuna, B., Karatas-Ozkan, M., & Yamak, S. (2019). Diffusion of sustainability and CSR discourse in hospitality industry: Dynamics of local context. *International Journal of Contemporary Hospitality Management, 31*(6), 2564–2581. https://doi.org/10.1108/IJCHM-06-2018-0464.

Garay, L., Font, X., & Corrons, A. (2019). Sustainability-oriented innovation in tourism: An analysis based on the decomposed theory of planned behavior. *Journal of Travel Research, 58*(4), 622–636. https://doi.org/10.1177/0047287518771215.

Gerdt, S., Wagner, E., & Schewe, G. (2019). The relationship between sustainability and customer satisfaction in hospitality: An explorative investigation using eWOM as a data source. *Tourism Management, 74,* 155–172. https://doi.org/10.1016/j.tourman.2019.02.010.

Gössling, S., Araña, J. E., & Aguiar-Quintana J. T. (2019). Towel reuse in hotels: Importance of normative appeal designs. *Tourism Management, 70,* 273–283. https://doi.org/10.1016/j.tourman. 2018.08.027.

Hansmann, R., Mieg, H. A., & Frischknecht, P. (2012). Principal sustainability components: empirical analysis of synergies between the three pillars of sustainability. *International Journal of Sustainable Development & World Ecology, 19*(5), 451–459. https://doi.org/10.1080/13504509. 2012.696220.

Horng, J., Liu, C., Chou, S., Tsai, C., & Chung, Y. (2017). From innovation to sustainability: Sustainability innovations of eco-friendly hotels in Taiwan. *International Journal of Hospitality Management, 63,* 44–52. https://doi.org/10.1016/j.ijhm.2017.02.005.

Huang, Y. C., & Liu, C. H. (2017). Moderating and mediating roles of environmental concern and ecotourism experience for revisit intention. *International Journal of Contemporary Hospitality Management, 29.* https://doi.org/10.1108/IJCHM-12-2015-0677.

Irsan, I., & Utama, M. (2019). The political law on coal mining in the fulfilment of people's welfare in Indonesia. *Sriwijaya Law Review, 3*(1), 11. https://doi.org/10.28946/slrev.Vol3.Iss1.202.pp1 1-25.

Jones, P., & Wynn, M. G. (2019). The circular economy, natural capital and resilience in tourism and hospitality. *International Journal of Contemporary Hospitality Management, 31*, 6, 2544–2563. https://doi.org/10.1108/IJCHM-05-2018-0370.

Khatter, A., McGrath, M., Pyke, J., White, L., & Lockstone-Binney, L. (2019). Analysis of hotels' environmentally sustainable policies and practices: Sustainability and corporate social responsibility in hospitality and tourism. *International Journal of Contemporary Hospitality Management, 31*(6), 2394–2410. https://doi.org/10.1108/IJCHM-08-2018-0670.

Lee, T., & Jan, F. H. (2018). Ecotourism behavior of nature-based tourists: An integrative framework. *Journal of Travel Research, 57*(6), 792–810. https://doi.org/10.1177/0047287517717350.

Lithgow, D., Martínez, M. L., Gallego-Fernández, J. B., Silva, R., & Ramírez-Vargas, D. L. (2019). Exploring the co-occurrence between coastal squeeze and coastal tourism in a changing climate and its consequences. *Tourism Management, 74,* 43–54. https://doi.org/10.1016/j.tourman.2019.02.005.

Lucrezi, S., Milanese, M., Markantonatou, V., Cerrano, C., Sarà, A., Palma, M., & Saayman, M. (2017). Scuba diving tourism systems and sustainability: Perceptions by the scuba diving industry in two marine protected areas. *Tourism Management, 59,* 385–403. https://doi.org/10.1016/j.tourman.2016.09.004.

Martinez-Martinez, A., Cegarra-Navarro, J. G., Garcia-Perez, A., & Wensley, A. (2019). Knowledge agents as drivers of environmental sustainability and business performance in the hospitality sector. *Tourism Management, 70,* 381–389. https://doi.org/10.1016/j.tourman.2018.08.030.

Nicholas, L., & Thapa, B. (2010). Visitor perspectives on sustainable tourism development in the Pitons management area world heritage site, St. Lucia. *Environment, Development and Sustainability, 12,* 839–857. https://doi.org/10.1007/s10668-009-9227-y.

Ng, S. I., Chia, K. W., Ho, J. A., & Ramachandran, S. (2017). Seeking tourism sustainability–A case study of Tioman island, Malaysia. *Tourism Management, 58,* 101–107. https://doi.org/10.1016/j.tourman.2016.10.007.

Okoli, C. (2015). A guide to conducting a standalone systematic literature review. *Communications of the Association for Information Systems, 37.* https://doi.org/10.17705/1CAIS.03743.

Pulido-Fernández, J. I., & Cárdenas-García, P. J. (2020). Analyzing the bidirectional relationship between tourism growth and economic development. *Journal of Travel Research.* https://doi.org/10.1177/0047287520922316.

Raub, S. P., & Martin-Rios, C. (2019). "Think sustainable, act local"–a stakeholder-filter-model for translating SDGs into sustainability initiatives with local impact. *International Journal of Contemporary Hospitality Management, 31*(6), 2428–2447. https://doi.org/10.1108/IJCHM-06-2018-0453.

Reim, W., Parida, V., & Örtqvist, D. (2015). Product-Service Systems (PSS) business models and tactics–a systematic literature review. *Journal of Cleaner Production, 97,* 61–75. https://doi.org/10.1016/j.jclepro.2014.07.003.

Roostaie, S., Nawari, N., & Kibert, C. J. (2019). Sustainability and resilience: A review of definitions, relationships, and their integration into a combined building assessment framework. *Building and Environment, 154,* 132–144. https://doi.org/10.1016/j.buildenv.2019.02.042.

Schliephack, J., & Dickinson, J. E. (2017). Tourists' representations of coastal managed realignment as a climate change adaptation strategy. *Tourism Management, 59,* 182–192. https://doi.org/10.1016/j.tourman.2016.08.004.

Slocum, S. L. (2015). The viable, equitable and bearable in Tanzania. *Tourism Management Perspectives, 16,* 92–99. https://doi.org/10.1016/j.tmp.2015.07.012.

Strezov, V., Evans, A., & Evans, T. J. (2016). Assessment of the economic, social and environmental dimensions of the indicators for sustainable development. *Sustainable Development, 55*(3), 242–253. https://doi.org/10.1002/sd.1649.

Su, M. M., Wall, G., Wang, Y., & Jin, M. (2020). Livelihood sustainability in a rural tourism destination-Hetu town, Anhui province, China. *Tourism Management, 71,* 272–281. https://doi.org/10.1016/j.tourman.2018.10.019.

Torugsa, N. A., O'Donohue, W., & Hecker, R. (2013). Proactive CSR: An empirical analysis of the role of its economic, social and environmental dimensions on the association between capabilities and performance. *Journal of Business Ethics, 115,* 383–402. https://doi.org/10.1007/s10551-012-1405-4.

Vogt, C. A., Andereck, K. L., & Pham, K. (2020). Designing for quality of life and sustainability. *Annals of Tourism Research, 83.* https://doi.org/10.1016/j.annals.2020.102963.

Zhang, L., & Zhang, J. (2018). Perception of small tourism enterprises in Lao PDR regarding social sustainability under the influence of social network. *Tourism Management, 69,* 109–120. https://doi.org/10.1016/j.tourman.2018.05.012.

Zhang, Y., Xiong, Y., Lee, T. J., Ye, M., & Nunkoo, R. (2020). Sociocultural sustainability and the formation of social capital from community-based tourism. *Journal of Travel Research.* https://doi.org/10.1177/0047287520933673.

Trending World Changes

Environmentally Friendly Tourists In Morocco

Charef Kenza and D. M. Bark Houssas

1 Introduction

The travel and tourism industry is placed among the largest industries in the world. However, the degrading effects of tourism have become a big concern and need to be addressed quickly. From this perspective, the concept of sustainable tourism has emerged to reduce the negative effects of tourism activities, which has become almost universally accepted as a desirable and politically appropriate approach to tourism development. The ecological sustainability of tourism has been researched on a larger scale. The central aim of sustainable tourism research has been to identify how an economically capable tourism industry can be developed and maintained at a destination while minimizing environmental impacts and in so doing, preserve the destination's natural and cultural resources for both residents and future generations of tourists.

Many procedures to expand the ecological sustainability of destinations have been proposed by researchers, and they address the supply side of tourism. Supply-side measures take the tourists as given and try to modify their behavior once at their destinations.

Examples are regulations imposed on businesses and initiatives to educate tourists and stimulate pro-environmental behavior. This research work seeks to study the progress of research on ecofriendly tourists by conducting structured examination of peer reviewed journal articles in recent years and to identify the key disciplines, journals, articles, and authors. It investigates the current situation of what tourist behavior can be considered as sustainable and little consensus about who environment-friendly tourists are. This study reviews theoretical and empirical studies by tourism researchers and explores work done on environmentally friendly behavior. Throughout this paper, we will use the term EFTs to describe tourists

C. Kenza (✉) · D. M. Bark Houssas
National School of Business ENCG, EDMP, LARGE, Agadir, Morocco

© The Author(s), under exclusive license to Springer Nature Switzerland AG 2021
V. Katsoni and C. van Zyl (eds.), *Culture and Tourism in a Smart, Globalized,
and Sustainable World*, Springer Proceedings in Business and Economics,
https://doi.org/10.1007/978-3-030-72469-6_17

with a low environmental footprint at the destination. The term ecotourist refers to a subgroup of EFTs that engages in "responsible travel that conserves natural environs" (The International Ecotourism Society, 2002).

2 Literature Review

2.1 The Brief Review of Sustainable Tourism Development

Sustainable tourism development has attracted significant attention in many scientific studies particularly in tourism studies and has been one of the very fast-growing areas of tourism studies research since the late 1980s. There are many definitions of sustainability and sustainable development. The best-known definition of sustainable development is "development that meets the needs of the present without compromising the ability of future generations to meet their own needs" (WCED, 1987). This definition implies the connections between economic development, environmental protection, and social equity, each element reinforcing the other. The World Tourism Organization (WTO, 2001) defined sustainable development as follows:

> Sustainable tourism development meets the needs of present tourists and host regions while protecting and enhancing opportunities for the future. It is envisaged as leading to management of all resources in such a way that economic, social and aesthetic needs can be fulfilled while maintaining cultural integrity, essential ecological processes, biological diversity and life support systems.

Kotler et al. (2010) define sustainable tourism as a concept of tourism management that anticipates the problems that can occur when the operating capacity of tourist destinations is exceeded. This capacity is estimated in relationship to the environmental impacts. Williams and Shaw (1998) explain that sustainable or sustainable development meets the needs of the present without compromising the ability of future generations to meet their needs. Middleton (1998) defines sustainable tourism as the accomplishment of a combination of number and type of tourists, the effect of accumulation of activities in a given destination and the actions to maintain the businesses can continue in the foreseeable future without damaging the quality of the environment in which these activities are based. Collin (2002) argues that sustainability can only be achieved with an appropriate number of tourists and a destination capacity.

Frey and George (2009) classified sustainable tourism as follows:

- Responsible tourism: According to Spenceley et al. (2002), it consists in delivering better holiday experiences to tourists and good business opportunities to enjoy a better quality of life by increasing socio-economic benefits and improving the management of natural resources.
- Ethical tourism: Weeden (2001) determines that ethical tourism goes beyond the three principles of sustainability; tourists and tourism service providers must assume their moral responsibility.

- Cultural tourism: Tourism that respects the cultural and human heritage of places and people.
- Alternative tourism: According to Krippendorf (1987), alternative tourists seek to distance themselves as much as possible from mass tourists; they must have a unique and authentic tourist experience with interaction with the local population and the environment.

2.2 Sustainable Development

Sustainable development is a new concept of development that focuses on meeting the needs of the present generation, without compromising the needs of future generations. The assumption on which sustainable development is based is that the quality of the environment is better if preserved by economic management (Bauernhansl et al., 2012; Farid, 2011). Sustainable development involves the interaction of three or more dimensions: an economic, social, and environmental dimension. Concrete actions relating to one or more dimensions must be implemented in order to guarantee sustainable growth. One of the first initiatives to study the future of humanity was the Meadows report, which considered five factors: exponential population growth, exponential growth in the need for food, exponential growth in industry and pollution, industry and the exponential decrease in non-regenerable natural resources, (Duguleana, 2002; Lampa et al., 2013). From the equations formulated, future extrapolations were made (based on data up to 1970) on the future of humanity. The report was very pessimistic and stated that our civilization will not survive beyond 2050. Then, there were improvements on various input parameters: food, mineral reserves, reduction of pollution, and reduction population growth (Sinclair, 2011; Zaccai, 2012). Then, research began to move toward complementary analysis, located between economic growth and the environment. The concept of "steady state economics," introduced by Professors Daly and Cobb, constitutes an element of further consolidation of the concept of sustainable development.

In their work "For the Common good" (1989), these authors advance the idea according to which the company must be directed toward the systems which are based on renewable energies and natural resources, without forcing the economic development and, therefore, without destroying the resources of the environment. The Rio de Janeiro Declaration (June 1992), signed at the Conference on Environment and Development, includes 27 principles for sustainable development and an action plan known as Agenda 21 (Teodosiu and Bârjoveanu, 2011). This plan includes principles which concern the acceleration of sustainable development in developing countries, the fight against poverty, the control of demographic dynamics, the health of populations, the integration of the environment in decision making, the atmosphere, planning, and management of resources (Peng et al., 2011; Sedlacek, 2013). Berca (2001) considers that "this controversial concept of sustainable development, in itself, whatever the definition and whatever its name, corresponds to an objective reality of the evolution of the living world on our planet."

Sustainable development represents the change in a lifestyle, of behaviors which avoid various destructive actions for the environment, but which have an obvious importance on the social and economic environment (Sinclair, 2011, Munitlak Ivanovic et al., 2009).

The desire to discover the effective method in order to achieve sustainable development according to the sectors of activity is an illusion because sustainability is only possible if it is present in all areas of human activity; therefore, the action of ensuring the sustainability of one activity may be incompatible with the sustainability of another activity (Heikkurinen and Bonnedahl, 2013).

The concept of sustainable development has evolved and now covers three perspectives: economic, social, and ecological.

- The economic approach is based on the concept of the maximum flow of income which can be generated by the maintenance of the stock of values (or capital) which produced these benefits.
- The ecological approach to sustainable development has in mind the biological and physical stability of systems. The viability of subsystems is of increasing importance because these subsystems are essential to the stability of the entire system. A key aspect is biological diversity (Bon, 2009; Van Der Yeught, 2009).
- The social and cultural approach seeks to maintain the stability of social and cultural systems, including through the reduction of destructive conflicts.

We can observe that sustainable development is based on reasonable management and equity in the process of distributing the benefits that result from the use of resources. Tourism has to choose, like any other entity, strategies which allow the realization of its gains thanks to available resources, taking into account the opportunities as well as the threats which concern the modifications of the environment (Lijing et al., 2011; Hunter 1997; Inskeep, 1994). The tourism industry was one of the first to adopt the Agenda 21 in 1996, the 20's witnessed the evolution of the code of ethics. Tourism development must be based on the criteria of sustainability, which implies that tourism must be ecological in the long term, economically sustainable, socially equitable, and ethical for local authorities (Cahndralal, 2010; Buckley, 2012).

2.3   Stakeholders in Sustainable Tourism

With the increasing economic importance of tourism, given the employment it generates and the fact that tourism business activity is conducted in places that belong to local society, these businesses owe society, the natural environment, and other elements in the surroundings, a certain responsibility, which is where stakeholder theory comes into play (Aguera, 2013). The concept of "stakeholders" gained wide acceptance with Freeman's (1984) book "Strategic Management: A Stakeholder Approach," a fact that is widely recognized by researchers (Donaldson and Preston, 1995; Mitchell et al., 1997; Jawahar and Mc Laughlin, 2001). Stakeholders in tourism, generally refer to those groups or individuals who are associated with tourism development initiatives and therefore can affect or are affected by the decisions and activities concerning those initiatives (Waligo et al., 2013).

The most popular definition of a stakeholder is the one presented by Freeman 1984 "A stakeholder is an individual or a group of individuals who can affect or be affected by the achievement of organizational goals." This vision of the company as a field of cohesion between different actors is found both in stakeholder theory and in the resource-based theory.

Wernerfelt (1984) was the first one to put out a factor of company profitability that is other than the products. He presented in his article in 1984 a first formulation of the resource-based theory. Indeed, each company has its own resources which gives it a competitive advantage. Barney (1991) develops the theory and wants to make it more operational. It sets up criteria that make a resource a permanent competitive advantage: first of all, a resource that creates value, a resource that cannot be imitated, and finally cannot be substituted. Subsequently, Maradok (2001) proposes an operational way for the selection and exploitation of resources.

3 Methodology

The presented research aims at understanding past and current research, creating some direction for future studies, and therefore advancing the application of sustainable development in the tourism industry. For this reason, it was decided to investigate as many articles as possible in order to discover several areas of the sustainable tourism domain, which was necessary to ensure the reliability and representativeness of the results. It also should be noted here that the number of citations and the popularity of publishers are the most significant criteria for publication selection to clarify the authenticity of them. A descriptive bibliography study was conducted. Journal articles with a high impact on the scientific community were downloaded from different online databases: ProQuest, Cairn, and Sciences direct. All articles published after 2000 were included in the review if they contained either a definition, an operationalization, or an empirical profile of EFTs. Based on these criteria, 29 articles were selected.

This literature review was undertaken in four steps: First collecting the evidence and information contained in each article and then coding reviewed articles with respect to the information contained there, data entry, and last analysis. Content analysis is adopted to identify categories and produce descriptive information on the content of previous research (Silverman, 1997). The scholarly papers are sorted by year of publication, publication journal, subject area, authors' nationality, and region of focus. It is believed that these aspects can provide information on theorical and empirical studies done by tourism researchers on environmentally friendly behavior. Moreover, it is expected that such an analysis uncovers the potential gaps in the literature and identifies future research opportunities.

Table 1 Definition of EFT's

Definitional feature	Percentage of articles (%)
Natural location	57
Cultural interactions	25
Conservation of nature	12
Protection of nature	6

4 Definitions, Operationalization and Descriptions of EFTs

4.1 Definitions of EFTs

Eight definitions of EFTs were identified. The following table combines those eight factors ordered by the percentage of articles that used each of the listed components as part of their definition. Definitions regarding nature were frequently included with almost two thirds of all articles using this as a characteristic. The second most common feature was "cultural interactions," and "conservation of nature" is only mentioned in 12% of the articles and protection of nature' by 6% of the studies included nature protection in their definition, leaving 25% which do not include any aspect of nature protection as part of their definition. All articles indicated, either explicitly or implicitly, that ecological sustainability or protection was an important element of the study. The same goes for the definitional component "Learning about nature" which is the second most frequently used.

The prominence of nature-based definitions is a result of the dominance of ecotourism studies in terms of the demand-side or customer-oriented view of ecologically sustainable tourism.

Research into EFTs in other than nature-based contexts is needed to learn more about the personal trait of environmentally friendly behavior which may help all sustainable tourism destinations, not only those offering their natural assets as the main tourist attraction (Table 1).

4.2 Operationalization of EFTs

The operationalization of EFTs is the most important item for the purpose of evaluating the state of knowledge because it provides an insight of what was measured empirically. The significant impact of different Operationalizations of a concept has been demonstrated by Tao, Eagles and Smith (2004).

59% of the studies included an empirical research; they all aimed at profiling EFTs, either by describing the EFT sample under study or by comparing it to a reference group. The most common means by which EFTs were reviewed were to distribute a questionnaire through ecotourism operators (Khan, 2003; Wight,

1996). This approach considers that all tourists that choose ecotourism operators are environmentally friendly in their behavior.

Another group of researchers used respondents' expressed interest in nature-based or ecotourism activities as the selection criterion. Activities were general, for example, interest in traveling to a destination for the purpose of outdoor recreation (Pennington-Gray & Ker Stetter, 2002) or wanting to undertake a trip to increase understanding and appreciation of nature (Blamey & Braithwaite, 1997). This approach is limited by the fact that they implicitly assume that an interest in outdoor recreation and wanting to understand nature are indicative of pro-environmental behavior.

Eagles (1992) used members of organizations with pro-environmental aims as distributors for the surveys. This operational approach clearly orientates respondents toward pro-environmental attitudes to begin with. Tao et al. (2004) compared profiles of visitors to a national park in Taiwan, who perceived themselves as ecotourists, with those who complied with three criteria (learning about nature, wilderness setting, and spending a substantial proportion of the trip in the park), finding significantly different results. They raised a concern about cumulative knowledge of EFTs, in fact no cumulative knowledge about EFTs can develop if each study uses either a different or unknown rule for empirical measurement.

4.3   Characteristics of EFTs

The bibliography study produced 14 characteristics used to outline EFTs. These characteristics fall into four categories: socio-demographic factors, behavior-related characteristics, travel motivations, and other characteristics. Table 2 below gives a summary of EFT's characteristics. As can be seen, the only characteristic that has been studied repeatedly is the tourist's level of education. Almost half of the studies incorporated age and a third of all studies included interest in learning and income.

While the results regarding to income, education, and interest in learning have been consistent across the studies, pointing to higher educated tourists with an interest in learning and higher income levels, the results with respect to age are contradictory: five studies concluded that EFTs are middle aged, and two studies come to the conclusion that they are older tourists. "Environmental concern" was examined as a

Table 2 Operationalization of EFTs

Characteristics of EFTs	Percentage (%)
Higher/tertiary education	50
High income	31
Age	45
Environmental concern	19
High environmental awareness	13

potentially useful characteristic of EFTs in only three studies. Similarly, "environmental awareness" was investigated in only two studies, each of which concluded that this was an important characteristic. Health concerns, physical activity, adventure seeking, occupation, and willingness to forget comfort were only investigated by one study each. It is evident from the analysis of EFT characteristics, derived from previous research, that only a few characteristics have been studied extensively.

4.4 Sustainable Tourist Behavior in Morocco

This article aimed to present a typology of ecofriendly behavior in Morocco to understand the share of Moroccan consumers interested in this new way of traveling. Sustainable tourism in Morocco operates as follows: regular services open to all, for example, the use of labor and local products or respect to traditions. The offer of solidarity, fair trade, or ecotourism holidays primarily concerns markets for foreign tourists and is offered by agencies specializing in this niche, it emerges that sustainable tourism remains a marginal concept in our society; it lacks the credibility of professional actors, a price policy for sustainable travelers and tourists' awareness of environmental benefits.

Also, the concept of regional identity appears to be a particularly important and useful indicator. Carrus et al. (2005) found regional identity to play a major role in environmental behavior. The tourism implications of these findings essentially put forward the hypothesis that environmentally friendly behavior will decrease with lower levels of regional identification by tourists.

Sustainable tourism appears to be a key factor for the success territorial development. From mass tourism to sustainable tourism, Morocco is now being built as a global tourist destination that integrates its regions by enhancing all its geographic and historical wealth. Sustainable development offers the possibility to advance a territory with new practices. These consider both economic imperatives, but also respect local populations while seeking to minimize the effect on the environment.

5   Conclusion

The point of this study was to evaluate the state of knowledge about EFTs in order to provide destination management with an additional tool to reduce the ecological impact of the tourism industry without necessarily sacrificing tourism revenues. The state of knowledge was assessed by reviewing definitions, operationalizations, and empirical profiles of EFTs as well as investigating the value of contributions outside the field of tourism research.

The methods used to operationalize EFTs revealed different approaches to measure EFTs. They have several limitations which suggest that further research into this market segment is necessary. First, the study of EFTs has typically focused on ecotourism, assuming that individuals who take an interest in nature impact the environment to a lesser extent than other tourists.

References

Eagles, P.F. (1992). The travel motivations of Canadian ecotourists. *Journal of Travel Research.*

Eagles, P. F., & Cascagnette, J. W. (1995). Canadian ecotourists: Who are they? Tourismecotourists. *Tourism Analysis, 9,* 1–13.

Frey, N., & George, R. (2010). Responsible tourism management: The missing link between business owners' attitudes and behaviour in the Cape Town tourism industry. *Tourism Management, 31*(5), 621–628. Horses for old courses. Questioning the limitations of sustainable tourism to supply-driven.

Peeters, P., & Schouten, F. (2006). Reducing the ecological footprint of inbound and transport to Amsterdam. *Journal of Sustainable Tourism, 14*(2), 157–171. *Recreation Research, 20*(1), 22–28

Tao, C.-H., Eagles, P. F., & Smith, S. L. (2004). Implications of alternative definitions of tourism *13*(6), 346–565.

Tourism and Contact Tracing Apps in the COVID-19 Era

Agisilaos Konidaris, Ourania Stellatou, Spyros E. Polykalas, and Vicky Katsoni

1 Introduction

In the era of the COVID-19 pandemic, it is well known that personal hygiene and wearing a mask are the two best strategies to stop the spread. To facilitate these basic strategies, health authorities around the world have also adopted mass testing and contact tracing. Contact tracing (World Health Organization, 2020) comes into play right after a case has been confirmed. This process is usually carried out in a non-digital way. People, known as contact tracers, interview a confirmed case and then communicate with all persons identified by them as close contacts during the past few days. Classic old-school contact tracing has been around for a long time and is fairly easy to implement in small or rural areas. In these areas, everybody knows everybody else, and it is easy for someone infected by the virus to identify everyone that he or she has come in contact with during the past few days. Things tend to be more complex in large cities, because it is not easy for someone to identify people that they have come in contact with. There are hundreds of potential contacts who they do not know and cannot identify. Contact tracing in tourism destinations faces a similar, if not more complex, situation. The process can become a nightmare even in small tourism destinations. These destinations, usually islands, have more tourists than residents during the summer. Tourists continuously move around and remain at a tourist destination for about 6–7 days on average. How can classic contact tracing be carried out in an area where no one knows anyone, and people constantly arrive and leave? This situation is the exact opposite of the bubble concept (McLaws, 2020) that provides some kind of health certainty when complemented with testing. It is

A. Konidaris (✉) · O. Stellatou · S. E. Polykalas
Digital Media and Communication Department, Ionian University, Kefalonia, Greece
e-mail: konidaris@ionio.gr

V. Katsoni
Tourism Management Department, University of West Attica, Athens, Greece

© The Author(s), under exclusive license to Springer Nature Switzerland AG 2021
V. Katsoni and C. van Zyl (eds.), *Culture and Tourism in a Smart, Globalized, and Sustainable World*, Springer Proceedings in Business and Economics,
https://doi.org/10.1007/978-3-030-72469-6_18

obvious that tourism destinations are the ideal setting for digital contact tracing. The development and the deployment of Contact Tracing Apps (CTAs henceforward) has been a rocky road. The lack of regulation and global standards, the different development and data handling approaches adopted, the fragmented and country-centered app development approaches and some development failures have not enabled CTAs to really take off.

The only coherent and global initiative in the field of CTAs is the development of the Exposure Notifications System API by tech giants Google and Apple (Google, 2020). The API that was introduced in the end of May 2020 provides a base on which decentralized CTAs can be developed. Decentralized are Apps that do not require the storage of personal user data on servers, thus substantially limiting data protection concerns. Even though the two tech giants developed an API that works on both Android and iOS devices, several countries and US states are still very reluctant to adopt it. Currently, several CTAs have been developed by different countries. Only a fragment of these apps uses the API provided by Google/Apple. Most of them cannot interconnect and exchange data. Users are still very reluctant to download them. This grim situation is far from what tourism would require.

In this paper, we argue that the development of a pan-European or even global decentralized and open-source contact tracing app would be a very effective tool for stopping the spread of the coronavirus while at the same time keeping tourism alive.

2   Related Work

Coronavirus disease 2019 (COVID-19), caused by severe acute respiratory syndrome–coronavirus 2 (SARS-CoV-2), has the clear potential for a long-lasting global pandemic, high fatality rates, and incapacitated health systems. Until vaccines are widely available, the only available infection prevention approaches are case isolation, contact tracing and quarantine, physical distancing, decontamination, and hygiene measures. (Ferretti et al., 2020).

According to Ozturk et al. (2020), Yu et al. (2020) and Ferretti et al. (2020) (as cited in Rizzo, 2020) the SARS-CoV-2 pandemic will be remembered as the first to be tackled "technologically": in a few months, the major social networks have implemented information centers on the disease and neural networks, supercomputers and mathematical models have been developed for various purposes, including detection of COVID-19 cases, rapid screening of drugs, and contact tracing.

The importance of contact tracing has been particularly emphasized by the World's Health Organization Director-General in March (as cited in Salathé et al., 2020) by saying "You cannot fight a fire blindfolded. And we cannot stop this pandemic if we do not know who is infected.

Use of technology as a means of mass surveillance has been successfully demonstrated in some countries that responded well to the pandemic (Sen-Crowe et al., 2020). Health professionals have long considered conventional mapping, and more recently geographic information systems (GIS), as critical tools in tracking and combating contagion (Boulos & Geraghty, 2020).

A recent study by Nuzzo et al. (2020) concluded that wide adoption of digital contact tracing can mitigate infection spread similar to universal shelter-in-place, but with considerably fewer individuals isolated. It was reported in literature that as long as the tracing delay is distinctively shorter than the latency period resp. the infectious period, it seems to be better to put effort in the detection of more contacts or index cases. (Muller et al., 2016). Access to testing should therefore be optimized, and mobile app technology might reduce delays in the contact tracing process and optimize contact tracing coverage (Kretzschmar et al., 2020).

Four categories of applications (apps) are being proposed for digital contact tracing. Technical specifications aside, they vary from one another by the degree of invasiveness in terms of privacy (Sinha et al., 2020). Data first or privacy first? Different approaches developed by different governments and corporations cover a wide range of the possible spectrum. (Fahey et al., 2020). It is notable that the most strong data first approaches have been adopted by South Korea and Taiwan as well as China, Iran and Quatar, while both France and the UK have also resisted the privacy first approach (O'Neill et al., 2020 as cited in Fahey, 2020). According to Rowe (2020) in France, as in many countries, the idea of building a mobile app that would inform a smartphone user if she/he crossed the way of contagious individuals has led the French government to add this protection method to the traditional contact tracing method by human investigation. In Asia, such apps have been seen as a successful part of the tracing and testing suppression strategy, but in Europe and elsewhere, privacy considerations are seen as vital. A particular issue for the UK is that, unlike most other European countries, it initially chose to go with a centralized design. The UK's choice of architecture has since been revisited so that contact alerts are only set off by actual positive testing—but this design choice may also yet be rolled back or reconsidered once more. Interestingly, Norway, which also chose to go with a centralized architecture and added collection of location data, has now been told by its privacy regulator that this design is not legally acceptable, as of June 2020 (Darbyshire, 2020). Germany deployed technological solutions, including an anonymized and decentralized contact tracing app (Reintjes, 2020).

As the contact tracing applications are developing there are several ethical and practical issues, such as privacy, questions about efficacy, lower user adoption rates, and concern by some public health experts that mobile apps might distract resources from the core work of conventional contact tracing (Ivers et al., 2020). According to Parker et al. (2020), a profoundly important ethical question presented by this technology relates to the problem of how and whether societies can find ways to benefit from the potential of algorithmic approaches to improve public and individual health, while also ensuring that the legacy of the deployment of these technologies does not impact negatively on future generations.

Researchers at the University of Oxford released a report (University of Oxford, 2020) that found that 60% of the population in the UK would need to install a contact tracing app in order for it to be effective. Another possible problem is that app users may troll the system by falsely claiming symptoms and their trace will be difficult as decentralized apps only store anonymous data temporarily, without collecting location. Also, range of Bluetooth can vary greatly depending on how people hold their phones, and whether they are indoors or outdoors. People in different rooms could be unnecessarily flagged as having had contact.

A further potential issue is the quality of the data. Researchers have reported that many apps being considered would record contacts only every 5 min, which might mean infectious contacts are missed. Other key issues include the level of trust between citizens and governments, how privacy is preserved, whether apps are kept voluntary and how to also protect people who might not be able to install an app—a group that is likely to include many vulnerable older people (Vaughan, 2020). This is also highlighted by Rizzo (2020) as these apps would have a greater epidemiologic importance for older subjects, who are however less inclined to technology and more suspicious on a leak of personal data.

3   The Current State of Contact Tracing Apps

CTAs for smartphones have emerged as a significant technological solution with potential to aid the fight against COVID-19. From the beginning, CTA development has not taken travel into account. Every CTA has been implemented for use inside a country by its citizens. This can be viewed as somewhat logical when considering the time frame when these apps were designed. It was a time of lockdown when air travel had been restricted and tourism was not even considered. Even so, CTA development did not consider the foreseeable future and the world ended up with the situation depicted in the following table. It shows the fragmented and independent approaches taken by countries around the world (Council of Europe, 2020).

Table 1 clearly reveals the proprietary approach to CTAs taken by countries worldwide during the summer of 2020. There were 86 apps already implemented and only 11 used the Exposure Notification System API. In addition, only 30 apps were open-source. This particular snapshot of the state of CTAs shows that at the moment it is extremely difficult to use CTAs for travel purposes.

Even though it is obvious that contact tracing only works when tackled uniformly, almost every country tried to develop its own proprietary CTA.

Recently there have been initiatives to somewhat resolve the lack of interoperability in the European Union (European Commission, 2020). Even so, these initiatives are not yet operable. They are only a set of guidelines that have a long way to go until apps become interoperable in the EU.

Table 1   Current state of CTAs around the world

Countries developing apps	Total apps related to COVID-19	Apps using Google/Apple API	Apps provided under an open-source license
44	86	11	30

Table 2 Survey demographics

Gender	%
Men	60
Women	39.1
Other	0.9
Total	100
Age	%
18–25	16.5
26–35	21.7
36–45	47
46–55	7
Over 55	7.8
Total	100
Education	%
Secondary	8.7
Graduate	59.1
Postgraduate/Doctorate	32.2
Total	100

4 Research Methodology

We conducted an electronic survey among 542 Greek residents between 3/7/2020 and 21/7/2020, based on voluntary sampling, in order to determine the level of knowledge, the attitude and willingness to install CTAs should they be available. We then compare our results to similar survey results in other countries. In our study, we are particularly interested in the relation between CTA usage and tourism. In Table 2, we present the demographic breakdown of the respondents, and in Table 5, we present our results in a way that shows which demographic group has the most positive view about contact tracing apps.

The demographic data of our study are shown in Table 2.

5 Preliminary Results

In this section, we present our initial findings of the survey based on the respondent's answers. From these results, we extract further user grouping information that we then use for the creation of Table 2.

5.1 Relation of Respondents to Tourism

47.9% of respondents live on Greek islands and thus are in some way related to tourism. Also 43.6% of respondents consider their occupation either directly or indirectly related to tourism.

5.2 Knowledge of CTAs

Only 41% said they know what CTAs are. 27.3% answered that they do not know what CTAs are and the rest (31.6%) are not sure.

5.3 Personal Data Privacy

One of the most important and controversial issues of CTAs is that of personal data handling and protection. We asked if respondents generally considered personal data protection before installing apps on their smartphones. 53.8% answered that they always or usually pay attention to how an app handles personal data. On the other hand, 24.8% of respondents said they never or that they usually don't pay any attention to how apps handle data. 21.4% were unsure.

5.4 Willingness to Learn More About CTAs

79.5% of respondents said they would welcome more information on how CTAs and the Google/Apple API work. It is obvious that respondents need more information on how CTAs work and how the Google/Apple Exposure Notification System handles data.

5.5 CTA Usefulness in Stopping the Spread and Health Assurance

48.7% believe that CTAs can play an important role in stopping the spread of COVID-19. 41% are not sure and 10.3% believe that CTAs cannot play a role in stopping the spread.

On the other hand, a lot more people (64.1%) think that CTAs can play an important role in the reassurance of residents on health safety in areas of the country that are considered tourist resorts.

5.6   CTA Installation When Traveling Abroad

51.3% of respondents would install a CTA if it was an optional recommendation while traveling to another country. 35.1% are not sure if they would install the optional app and 13.7% would not install the App.

5.7   Necessity of a CTA in Greece

Even without adequate information on how CTAs work, the majority of respondents (66.7%) said that Greece should develop a CTA for the COVID-19 virus. Only 7.7% reject the idea and 25.6% are unsure.

5.8   Intention to Install a CTA in Greece

While 66.7% of respondents think a CTA should be developed in Greece, only 42.7% are positive they would install the app should it exist. 11.1% would not install the App. 37.6% would probably install the CTA (coming to an intent to install of 80.3%). The rest are unsure. There was no significant variance between those living on islands to those living in large cities. Positive answers were equally distributed.

6   Comparison to Other Countries

In this section, we compare our results in Greece to similar surveys in other countries. The comparison is made on whether a country should implement a CTA and whether citizens would install a CTA should it be available in the country.

6.1   Favorability Toward Developing a CTA in the Country

As shown in Table 3, 66.7% of respondents in Greece are in favor of the development of a CTA. This result is very close to the survey (Survation-Open Knowledge Foundation, 2020) that found the favorability to be 65% in the UK. The UK survey was

Table 3 Favorability toward developing a CTA in the country (UK results from Survation—Open Knowledge Foundation, 2020)

Greece	66.7%
UK	65%

conducted in May 2020 about 2 months prior to ours. It is obvious that citizens are not opposed to the development of a CTA in their country. Two out of three citizens are favorable.

6.2   Intent to Install a CTA if Available

As shown in Table 4, the question of developing a CTA is a lot different than actually installing the CTA. In the UK the intent to install is 12% lower than favorability in developing. In Greece, we found only 42.7% very likely to install a CTA should it be available but also 37.6% likely to install. If these percentages were to be added we would be looking at 80.3% of respondents in Greece leaning favorably toward CTA installation. This of course is an extraordinary result that needs to be looked at by the Greek health authorities. It is also extraordinary when compared to other countries where the installation intent is between 47% (in Germany) and 57% (in Singapore).

In any case, respondents in Greece show that the existence of a CTA would be welcomed and would be installed by the majority of respondents.

7   Analysis of Favorability Toward CTAs

In Table 5, we present the positive opinions expressed in our survey by the demographic groups identified. The goal of the table is to depict which respondent demographic has the most positive views about CTAs and that is why we have called it the "Positive Response Table". We also calculated averages and standard deviation of user responses. High averages imply high adoption of CTAs. Low standard deviation values imply high consensus on a matter. Table rows provide percentages of positive answers to questions about CTAs. Table columns provide percentages of positive answers to questions within specific demographic groups. High averages in rows show positive responses to questions on CTA acceptance and adoption. Low standard deviations in rows show consensus in answers among different demographic groups. High averages in columns show high general acceptance of CTAs among specific demographic groups. Low standard deviation in table columns shows

Table 4 Intent to install a CTA if available (McCarthy (2020), Roper (2020), Opinium (2020), Martelli-Banégas (2020), Ho (2020))		
	UK	53%
	FRANCE	49%
	GREECE	80.3% (42.7 Very Likely—37.6% Likely)
	SINGAPORE	57%
	GERMANY	47%
	USA	50%

Table 5 Positive response table

	GOOD knowledge of contact tracing apps	LIVE in tourist destinations	OCCUPIED in the tourism sector	NOT occupied in the tourism sector	Highly educated	Basic education	Aged 18–35	Aged over 36	MEN	WOMEN	Concerned about data privacy	Average	Standard deviation
Would it be useful for the country (GREECE) to create a contact tracing app?	81.3	52	68.6	66.6	66	72	68.4	68	67.4	65.7	76.2	66.7	7.26
Would you install a contact tracing app should it be available in GREECE?	62.2	42.9	45.1	44.4	43.4	36.4	35.6	47.2	43.5	42.9	41.3	42.7	6.94
Would you install a contact tracing app should it be a recommendation when traveling abroad?	81.3	48.2	68.6	66.7	50	63.6	46.7	54.2	52.2	50	76.2	51.3	12.08

(continued)

Table 5 (continued)

	GOOD knowledge of contact tracing apps	LIVE in tourist destinations	OCCUPIED in the tourism sector	NOT occupied in the tourism sector	Highly educated	Basic education	Aged 18–35	Aged over 36	MEN	WOMEN	Concerned about data privacy	Average	Standard deviation
do you think contact tracing apps can provide a sense of health safety in tourist areas?	79.2	64.3	68.6	64.8	65.1	54.5	68.9	61.1	65.2	62.9	71.4	64.1	6.25
Do you think contact tracing apps can play a significant role in reducing the spread of COVID-19?	68.8	48.2	52.9	50	50	36.4	46.7	50	50	47.1	60.3	48.7	8.16
Average	74.6	51.	60.8	58.5	54.9	52.6	53.3	56.1	55.7	53.7	65.1		
Standard deviation	8.64	8.05	11.08	10.53	10.09	16.01	14.76	8.47	10.26	10.03	14.8		

high consistency in answers about CTAs within a specific demographic group. The positive response table enabled us to arrive to the following important conclusions:

1. The user demographic with the most positive view and acceptance of contact tracing apps is the group of people with the most knowledge about the apps (74.6% positive average responses) followed by users concerned about data privacy (65.1% on average) and users occupied in the tourism sector (60.8% on average).

   Key Takeaways 1: Users that have been educated on how CTAs work as well as data privacy issues are more favorable to CTA use. Respondents working in the Tourism sector are among those with high acceptance of CTAs. On the other hand, people that live in tourist destinations do not seem to realize the importance of CTAs as much (only 51,1% positive views on average) and are the demographic group that are most consistent in all their answers.

2. Users that live in tourist destinations, those that have received only basic education and those aged 18–35 hold the least favorable views on CTAs.

   Key Takeaways 2: It is obvious that we have an oxymoron here, since users that work in the tourism sector are favorable to contact tracing apps and those that live in tourist destinations are less reluctant to adopt them. This could be due to the time of the survey. At the time (late July 2020) Greece had opened to tourism only for 20–25 days and COVID-19 numbers were very low throughout the country. Users working in the tourism sector had already realized the importance of CTAs but those just living in tourist areas had not yet realized their importance since coronavirus cases were still very few in their areas.

3. The user groups that were not consistent in their responses about CTAs (high standard deviation in their responses) were those with basic education, those aged 18–35 and those concerned about data privacy.

   Key takeaways 3: These results show that young users, not highly educated that are concerned about data privacy tend to be the demographic groups that a most unsure about the adoption of CTAs.

4. The issue concerning CTAs that finds the most target groups in favor are the usefulness of a contact tracing app for Greece (66.7% positive responses on average), and that contact tracing apps can provide a sense of health safety in tourist areas (64.1% on average).

   Key takeaways 4: Most demographic groups sense the usefulness of CTAs in Greece and their importance for health safety in the country. Less consensus among target groups can be found when asked to install a CTA on their smartphones and the significance of CTAs in reducing the spread. These are issues where demographic groups are split.

5. The basic issue that user groups do not have a consistent opinion on is whether they would install a CTA should it be a recommendation when traveling abroad.

   Key takeaways 5: It is obvious that the issue of installing a CTA is very confusing to the user groups and there is still no wide consensus on this issue.

8 Recommendations Based on the Analysis

In order to summarize our research in Greece, we provide the following recommendations to some key research questions:

1. **What should a country do to increase the number of people that would install a CTA?**
   Recommendation: Countries need to educate citizens, so they know exactly what CTAs are and how they work. It is obvious, based on our research, that citizens tend to adopt CTAs based on the level of knowledge about them. The more they know the easier it is for citizens to adopt and install CTAs. Furthermore, citizens seem to be eager to acquire more information about CTAs.

2. **Should Greece create a CTA?**
   Recommendation: Greeks show a wide acceptance of the idea that the country should create a CTA. Almost 67% of respondents in our survey agree that the country should create a CTA. Based on the above, Greece should implement a CTA.

3. **What demographic groups are more reluctant to adopt a CTA in Greece?**
   Recommendation: Young citizens aged 18–35 with basic education are the least favorable demographic group toward CTA implementation and adoption. These are the most reluctant citizens and they are the first that need to be educated in order to ensure high adoption of a CTA should it be implemented.

4. **What is the issue about CTAs on which citizens are most skeptical?**
   Recommendation: It is obviously the issue whether CTAs can play a significant role in reducing the COVID-19 spread. We observe that most demographic groups are still not completely convinced about the significance of CTAs in the battle against COVID-19. It would be a good idea to inform and educate users through case studies of CTA implementations in other countries and their results.

5. **Are people living or working in Greek tourist destinations convinced that CTAs can be beneficial?**
   Recommendation: Both demographic groups are favorable to the adoption of CTAs. The group that is more favorable is people working in the tourism sector. We propose educating citizens in tourist areas about the benefits of installing CTAs for their own well-being but also for the well-being of the destination.

9 Proposals for the Google/Apple Exposure Notification API

The Google/Apple Exposure notification API (Google, 2020) is a decentralized implementation that enables users to be directly notified on their smartphones when they have come in close contact with someone that later tested positive for COVID-19. It is not in the scope of this paper to analyze the Google/Apple API. We refer to this API only as an excellent base on which a Travel CTA can be implemented. Since the API is globally available to almost all smartphones on the planet it would be an excellent solution for a borderless CTA implementation. In this section, we propose an addition to the API in order for it to become more effective for Tourism.

9.1 Proposal 1

The API itself or the CTA built upon the API should be fed information by a central information database that contains travel destination-specific coronavirus information. Once a user is notified that they could be infected they should also be notified of where the closest testing center is at that destination. This "after notification" information is essential to a traveler at a foreign destination. It is also absolutely vital that this information is provided through the CTA, on the spot, so that the user does not need to interact with anybody prior to his/her test. This central information database can be maintained by the World Tourism Organization (UNWTO) and updated by local destination tourism and health authorities.

9.2 Proposal 2

Our second recommendation for the Google/Apple API is the addition of a feature that will enable the API to be locally trained about its accuracy. A basic problem of the API is that it largely depends on the quality of the smartphone Bluetooth signal and thus sometimes does not provide accurate notifications. This means that some notifications issued may be false. The API can also provide slightly different results on smartphones from different vendors. We propose that users will be able to inform the CTA whether the "possibly infected" notification that they received was actually true or false. This can be done by connecting the CTA with rapid test results. The new Abbott rapid test, for example, provides a QR code that when scanned by an Abbott App provides proof of testing and proof of health. If the CTA could scan such QR codes, then it would be informed that a user was not actually infected even though the API issued a notification. This proposal may lead someone to think that its implementation could compromise personal privacy or even the decentralized nature of the API. This is not true. We do not propose that any data leave the user's smartphone. We only propose an additional local (on the smartphone itself) mechanism through which users can inform their CTA whether the notifications received were valid or not. If this is made possible then every user's smartphone can be self-trained to provide accurate notifications after an initial training period of possible mistakes.

10 Benefits of Digital Contact Tracing for Tourism

Travel is inherently bound to mobility. When we travel, we usually move away from our residence, we are mobile for a long time, we meet a lot of people and visit a lot of crowded places. The definition of travel itself (Goeldner et al., 2006) contains the notion of mobility.

Tourism and travel can surely benefit from CTAs, but not in the way that they are implemented today. Tourism needs borderless CTAs as opposed to country-specific CTAs. Contact tracing needs to be globally enabled for it to work in Tourism and Travel. The current state of CTAs does not make them useful for Tourism. We propose the implementation of a pan-European or even Global CTA especially for Travel purposes. This means that every time someone needs to travel to another country, he/she will be able to use this CTA. The app will enable borderless contact tracing. This application can be easily implemented by using the globally available Google/Apple notification API. If a Travel specific borderless CTA was implemented, Travel and Tourism would largely benefit. Especially if the aforementioned recommendations were included in the CTA. The main benefits would be:

1. Contact tracing would be enabled on a global scale with all the benefits that this would bring to the reduction of COVID-19 cases.
2. Travelers would feel some health assurance at the destination since they would be notified and guided by the CTA.
3. Health information and alerts would be easily conveyed to travelers through the CTA.
4. Many people already infected but not showing symptoms would not travel to destinations.
5. Travelers returning from destinations would not be required to quarantine at all or the quarantine period would be significantly reduced.

There would also be significant psychological benefits for travelers and destination residents. In the end, a CTA adopted by most travelers and residents would contribute to the avoidance of future lockdowns and preventive measure applications.

11 Conclusions

In this paper, we have presented the close connection between CTAs and Tourism. We have presented the current fragmented and country-specific approach that the implementation of CTAs has adopted today. This approach is not in favor of the re-ignition of tourism worldwide. The fear of COVID-19 has stopped millions of people from traveling. The existence of a worldwide decentralized, open-source CTA and its wide adoption would surely benefit tourism and play the role of a reassuring measure for potential tourists. Our survey in Greece has provided a lot of insights on how countries should approach different demographic groups in order to convince them that CTAs are useful. We also concluded that most users are already in favor of CTAs in general but also in favor of CTA roll out in Greek tourist destinations. Greece should develop a CTA because the wide majority of citizens is in favor, and there is a high possibility that it would be installed by citizens. Finally, in this paper, we conclude that as in many other aspects of life, knowledge is very powerful. The most favorable demographic group to CTAs in our survey were people with good knowledge of what CTAs accomplish and how they work.

References

Boulos, K., & Geraghty, E. (2020). Geographical tracking and mapping of coronavirus disease COVID-19/severe acute respiratory syndrome coronavirus 2 (SARS-CoV-2) epidemic and associated events around the world: How 21st century GIS technologies are supporting the global fight against outbreaks and epidemics. *International Journal of Health Geographics, 19*, 8. https://doi.org/10.1186/s12942-020-00202-8.

Council of Europe. (2020). Contact Tracing Apps. https://www.coe.int/en/web/data-protection/contact-tracing-apps.

Darbyshire, T. (2020). Do we need a Coronavirus (safeguards) act 2020? Proposed legal safeguards for digital contact tracing and other apps in the COVID-19 crisis. https://doi.org/10.1016/j.patter.2020.100072.

European Commission. (2020). Coronavirus: Member states agree on an interoperability solution for mobile tracing and warning apps. https://ec.europa.eu/commission/presscorner/detail/en/ip_20_1043.

Ferretti, L., Wymant, C., Kendall, M., Zhao, L., Nurtay, A., Abeler-Dörner, L., Parker, M., Bonsall, D., & Fraser, C. (2020). Quantifying SARS-CoV-2 transmission suggests epidemic control with digital contact tracing. *Science, 368*, eabb6936. https://doi.org/10.1126/science.abb6936.

Fahey, R.A., Hino, A. (2020). COVID-19, digital privacy, and the social limits on data-focused public health responses. *International Journal of Information Management.*

Goeldner, C. R., & Ritchie, B. J. R. (2006). *Tourism: Principles, practices, philosophies* (12th ed.). Wiley.

Google. (2020). Exposure notifications: Using technology to help public health authorities fight COVID-19. https://www.google.com/covid19/exposurenotifications/.

Ho, K. (2020). Singaporeans divided on tracking token. https://sg.yougov.com/en-sg/news/2020/06/18/singaporeans-divided-tracking-token/.

Ivers, L.C., & Weitzner, D.J. (2020). Can digital contact tracing make up for lost time? https://doi.org/10.1016/S2468-2667(20)30160-2.

Kretzschmar, M. A., Rozhnova, G., Bootsma, M. C. J., Boven, M., Wijgert, J. H. H. M., & Bonten, M. J. M. (2020). Impact of delays on effectiveness of contact tracing strategies for COVID-19: A modelling study. *Lancet Public Health.* https://doi.org/10.1016/S2468-2667(20)30157-2.

Martelli-Banégas, D., Desreumaux, M., Favré, T. (2020). Corporate Harris Interactive, COVID-19 L'OBSERVATOIRE, Questions spécifiques: perceptions de l'application STOPCOVID et regards sur l'enjeu du partage des données personnelles. https://harris-interactive.fr/wp-content/uploads/sites/6/2020/05/Observatoire_DATA_PUBLICA-Sondage_HI_mai_2020_vDEF.pdf.

McLaws, M.L. (2020). What is the COVID 'bubble' concept, and could it work in Australia? The Conversation, Academic rigour, journalistic flair. Retrieved August 31, 2020 from, https://theconversation.com/what-is-the-covid-bubble-concept-and-could-it-work-in-australia-144938. Accessed August 31, 2020, at 10:00.

McCarthy, N. (2020). How many Germans plan to install the Corona App? https://www.statista.com/chart/22017/share-of-people-who-would-use-a-coronavirus-app-in-germany/

Müller, J., & Koopmann, B. (2016). The effect of delay on contact tracing. *Mathematical Biosciences, 282*(2016), 204–214.

Nuzzo, A., Tan, C.O., Raskar, R., De Simone, D.C., Kapa, S., & Gupta, R. (2020). Universal shelter-in-place vs. advanced automated contact tracing and targeted isolation: A case for 21st-century technologies for SARS-CoV-2 and future pandemics. In *Mayo Clinic Proceedings*. https://doi.org/10.1016/j.mayocp.2020.06.027.

Opinium.co.uk. (2020). Public opinion on coronavirus. https://www.opinium.co.uk/wp-content/uploads/2020/05/VI-05-05-2020-Observer-Data-Tables.xlsx.

Parker, M. J., Christophe, F., Abeler-Dörner, L., & Bonsall, D. (2020). Ethics of instantaneous contact tracing using mobile phone apps in the control of the COVID-19 pandemic. *Journal of Medical Ethics, 2020*(46), 427–431. https://doi.org/10.1136/medethics-2020-106314.

Reintjes, R. (2020). Lessons in contact tracing from Germany: Germany built on existing local infrastructure to get ahead of the covid-19 pandemic. *BMJ, 369*, m2522. https://doi.org/10.1136/bmj.m2522.

Rizzo, E. (2020). COVID-19 contact tracing apps: The 'elderly paradox.' *Public Health, 185*(2020), 127.

Rowe, F. (2020). Contact tracing apps and values dilemmas: A privacy paradox in a neoliberal world. *International Journal of Information Management* (in press).

Roper, W. (2020). Americans split on contact tracing app. https://www.statista.com/chart/21573/contact-tracing-app-adoption/.

Salathé, M., Althaus, C. L. B., Neher, R., Stringhini, S., Hodcroft, E., Fellay, J., Zwahlen, M., Senti, G., Battegay, M., Wilder-Smith, A., Eckerle, I., Egger, M., & Low, N. (2020). COVID-19 epidemic in Switzerland: On the importance of testing, contact tracing and isolation. *Swiss Medical Weekly, 150*, w202205.

Sen-Crowe, B., McKenney, M., Elkbuli, A. (2020). Utilizing technology as a method of contact tracing and surveillance to minimize the risk of contracting COVID-19 infection. *American Journal of Emergency Medicine*. https://doi.org/10.1016/j.ajem.2020.07.003.

Sinha, P., Paterson, A. E. (2020). Contact tracing: Can 'Big tech' come to the rescue, and if so, at what cost? *E Clinical Medicine*. https://doi.org/10.1016/j.eclinm.2020.100412

Survation-Open Knowledge Foundation. (2020). Topical poll. https://cdn.survation.com/wp-content/uploads/2020/05/07085617/Survation-Open-Knowledge-Foundation-1.xlsx.

University of Oxford. (2020). Digital contact tracing can slow or even stop coronavirus transmission and ease us out of lockdown. https://www.research.ox.ac.uk/Article/2020-04-16-digital-contact-tracing-can-slow-or-even-stop-coronavirus-transmission-and-ease-us-out-of-lockdown. Accessed May 20, 2020.

World Health Organization. (2020).What is contact tracing? https://www.who.int/news-room/q-a-detail/coronavirus-disease-covid-19-contact-tracing.

Tourism Customer Attitudes During the COVID-19 Crisis

Ivanka Vasenska and Blagovesta Koyundzhiyska-Davidkova

1 Introduction

At the end of 2019 and the beginning of 2020, the data of the global economic scholars and in particular those of the World Tourism Organization (WTO, 2020), demonstrated a bright development of the tourism sector. However, the world of the economy was disrupted by an unnatural disaster COVID-19 (Mostafanezhad, 2020). Consequently, the tourism sector was among the first and the most effected economic sectors and shattered by this disaster. Therefore, the authors were provoked to elaborate a research regarding Bulgarian tourism customer attitudes during the COVID-19 crisis. Moreover, the aim of the report is to establish to what extend tourism customers in Bulgaria are willing to keep their travel patterns comparing their established attitudes towards their new perception following the unfolding crisis. For this reason, we have prepared and conveyed a survey questionnaire conducted among 121 individual respondents. Following the global trend among economic researchers for digitalized analysis implementation on a global level, the authors decided to provide a new perspective towards survey statistical analysis. Furthermore, the elaborations are made by the computer language Python and a Web-based interactive computing environment for creating documents—Jupyter Notebook.2020

At the beginning, for the rest of the world the first 2019 cases of the SARS-CoV-2, officially reported by the Chinese government to the World Health Organization–WHO, (2020) on December 31, 2019, was only a local outbreak, similarly to the SARS-CoV epidemic started in Asia during 2002 (World Health Organization, 2020). Nonetheless, this outbreak, causing the COVID-19 virus, turned out to be the biggest pandemic since the Spanish flu and AIDS. Infected travellers carried the virus abroad to locations such as Toronto in Canada, Hong Kong, Singapore, Taiwan and Vietnam.

I. Vasenska (✉) · B. Koyundzhiyska-Davidkova
South-West University "Neofit Rilski", Blagoevgrad, Bulgaria
e-mail: ivankav@abv.bg

© The Author(s), under exclusive license to Springer Nature Switzerland AG 2021
V. Katsoni and C. van Zyl (eds.), *Culture and Tourism in a Smart, Globalized, and Sustainable World*, Springer Proceedings in Business and Economics,
https://doi.org/10.1007/978-3-030-72469-6_19

These locations recorded the highest numbers, but there were isolated cases found elsewhere in 29 countries altogether. Following this, there was a common pattern of reaction among official institutions and private enterprises which included the gathering and communication of information, marketing aimed at reassurance, efforts to sell to domestic markets, price cutting, a search for cost savings and greater efficiency, rationalization, capacity reduction and staff redundancies. Unfortunately, today we observe a different case scenario, and the epidemic of COVID-19 became a pandemic. What is more, a few voices of global leaders were raised seeking to lay accusations towards WHO for delayed pandemic announcement. Thus, in our opinion, this delay had reflected global governance decisions towards respective measures curbing COVID-19 outbreak. At the same time, the opinion of Kristalina Georgieva, the Director of International Monetary Fund (IMF), follows a conference call of G20 Finance Ministers and Central Bank Governors that there should not be "trade-off between saving lives and saving livelihoods". She continued with the very punctual statement: "this is a false dilemma … given that this is a pandemic crisis, victory against the virus and the protection of human health are necessary for economic recovery" (International Monetary Fund, 2020). Following the above statement, in response to the rapid spread of the Coronavirus (COVID-19), with ten thousands of deaths and intensive-care hospitalizations, a large number of regions and countries have been put under lockdown by their respective governments. Policy makers are confronted in this situation with the problem of balancing public health considerations, with the economic costs of a persistent lockdown (Gros et al., 2020).

2 Literature Review

In scientific literature, there is no unified definition of what crisis should consist of. On one side, Marshal Sahlins (1972) describes how crises may be revelatory as they may lay bare the structural contradictions of the modes of production that can no longer be ignored. On the other hand, Keown-McMullan (1997) define the crisis as a triggering event causing significant change or having the potential to cause significant change, the perceived inability to cope with this change and a threat to the existence of organizations. However, according to Hermann (1972) and Henderson (2007), a crisis includes unexpectedness, urgency and danger, yet they all have unique characteristics in generally. While Roitman (2013) highlights how crisis narratives also produce meaning and initiate critique of a given condition. Observers on the COVID-19 crisis have addressed its potentially transformational role. Thomas Friedman (2020), for instance, suggested that "There is the world B.C.—Before Corona—and the world A.C.—After Corona". At the same time, economists downplay the pandemic as a purely natural event that has originated outside of the economic system and, thus, has nothing to do with economic structures (Nowlin, 2017). Afar from the immediate health crisis, COVID-19 is basically a crisis of economized societies rooted in the growth-paradigm (Ötsch, 2020). Moreover, governments worldwide reacted to the crisis based on prognoses GDP shortfalls and steep increases in unemployment with "rescue packages" and "shock therapies" on an unprecedented scale (Gretzel et al., 2020).

One additional point of view on crises is that they are precipitated by catalysts powerful enough to undermine structures and modes of operation, with repercussions for the profitability of commercial ventures which might even be destroyed (Shrivastava & Mitroff, 1987). Lives and individual and all stakeholders' reputations may be put in peril, thereby eroding people morale. Participants are taken by surprise and have little time to make difficult decisions in an atmosphere of tension and instability. Crises also reach a crucial point when change, for the better or worse, is unavoidable, and the experience may prove beneficial for people and organizations (Prideaux, 2003).

For Selbst (1978), the nature of the crisis can be explained as "any act or omission which hinders the development of (organizational) functions of acceptable approaches for achieving the objectives of the organization or its viability or evaluation ..." In this definition, which will be discussed more fully in detail below, there are two dimensions of the crisis situation, explaining the differences between crisis and disaster and the consequences of these two situations concerning the activities of organizations and communities. One of which, according to Faulkner (2001), relates to relevant events associated with the organization and sees them as a result of its operation. The second dimension relates to the fact that the event in question must have detrimental or negative impacts on the organization as a whole or on individuals in it. Precisely that arises particularity in linguistic and convocational aspect. Moreover, researchers as Selbst (1978) were stressing mainly on the negative and destructive consequences of such event, without observing it as a turning point or opportunity. On the other hand, Faulkner (2001) has pointed out that it should be given another definition of crisis, which Fink (1986) mentioned—in Webster's dictionary, which rather refers to events with an effect of "turning point for better or worse". Namely in this context, it has been observed transforming connotations in relation to crises and disasters, and each event in the context of a particular situation leads to a potential positive and negative consequences. In terms of the conceptual basis and terminology apparatus that statement made by Faulkner (2001) draws attention. It is a result of his collaboration with Russell on the topic of turbulence (Faulkner & Russell, 2001), chaos and complexity may be perceived as context of which the crisis can arise. Whereupon, the crisis may be interpreted as consequence of failure of governance rather than cause of external factors.

According to the World Tourism Organization (2013), the crisis is an unexpected event that affects trust that travellers have towards the destination and hinders its normal development and functioning. As noted by Ritchie et al. (2004), over the years, a number of authors have been trying to formulate a definition of crisis which assists its understanding. In their research paper of "Crisis Communication and Recovery for the Tourism Industry: Lessons from the 2001 Foot and Mouth Disease Outbreak in the United Kingdom", they have presented basic definitions according to which the crisis is "a disorder that affects physical system as a whole and threatens its main manifestations; its subjective sense of self, its existential core".

In this regard, the measures recommended by the World Health Organization (WHO), (2020) and implemented by the governments of the countries affected by COVID-19 to maintain social distance are of a great importance for the protection of the physical self of the human individuals. Social distancing interventions can be effective against epidemics but are potentially detrimental for the economy. Businesses that rely heavily on face-to-face communication or close physical proximity when producing a product or providing a service are particularly vulnerable (Koren & Peto, 2020). Consequently, during crisis, in order to achieve higher efficiency and create competitive advantages, tourism businesses should use marketing as the main tool to identify market opportunities (Kyurova, 2013). On the other hand, the tourism industry is one of the most sensitive, susceptible and vulnerable industries to crisis (Santana, 2004). It is strongly affected by crisis events resulting in negative tourist perceptions (Pforr & Hosie, 2009). Moreover, it has been one of the major beneficiaries of the wealth created by these revolutions and by 2019 generated 10.3% of global GDP (The World Travel & Tourism Council (WTTC), 2020a, b), with year-on-year growth only briefly interrupted by crisis events such as SARS, the 9/11 terrorist attack on the USA and the Global Financial Crisis of 2007/08. Recovery occurred within a neoliberal global economic production system where continued economic growth was viewed as more important than the long-term impacts on the global environment. However, recovery from COVID-19 (caused by severe acute respiratory syndrome 2 (SARS-CoV-2)) is unlikely to follow the pattern of earlier post-crisis recovery (Prideaux et al., 2020). Regarding Bulgarian tourism industry data, WTTC reports 10.8% contribution to the country's GDP, which is with -2.5% for the previous period (The World Travel & Tourism Council (WTTC), 2020a, b).

The tourism industry is acknowledged to be exposed to crises or disasters (Cró & Martins, 2017). This is because tourism is impacted by many external factors, including political instability, economic conditions, the environment and weather (Okumus et al., 2005). The susceptibility of tourism industries has also been recognized by industry bodies and agencies. As a result, a number of reports, templates and toolkits to help industry prepare and respond to crises and disasters have been published (Ritchie & Jiang, 2019). Wherefore, the tourism industry's relative immaturity and dramatic expansion are noteworthy. Tourism has a long history, but modern mass tourism and the industry which supports it date from after the Second World War. International arrivals rose from 25 million in 1950 to 760 million by 2004, although geographical imbalances persist and most tourists and their spending circulate within the developed world (World Tourism Organization (WTO), 2005). The World Tourism Organization (WTO) predicts that there will be 1.6 billion international tourists in 2020 (World Tourism Organization (WTO), 1999) and domestic tourism, estimated to be 10 times greater in volume, must not be overlooked (Weaver & Lawton, 2002). Growth has been fuelled by behavioural approach that includes both sociological phycological aspects (Kliestikova et al., 2019a, b), new product development, aggressive marketing and intense global competition.

Globalization of tourist markets has increased over recent years (Levitt, 1983), contributing to the escalation of global risks. Tourism is viewed as one of the activities most susceptible to global risk factors (Ritchie, 2004). Tourism, especially international tourism, is highly sensitive to safety and security issues (Pizam & Mansfeld, 1996). It is one of the activities most liable to changes in the world stage that may produce modifications in tourist behaviour (Coshall, 2003; Dimanche & Leptic, 1999; Levantis & Gani, 2000; Pizam & Mansfeld, 1996). Safety concerns strongly influence tourists' decision-making processes (Beirman, 2002; Crompton & Ankomah, 1993; Fesenmaier, 1988; Moutinho, 2000; Woodside & King, 2001; Woodside & Lysonski, 1989). Travellers select destinations that best match their needs, offer the most benefits and have the lowest possible costs or risks. If a tourist feels insecure or threatened at a specific destination, an overall negative impression is likely to result (George, 2003). Consequently, destinations perceived as being safer may be preferred, and those perceived as risky or unsafe may be rejected (Beirman, 2002; Gu & Martin, 1992; Mansfeld, 1996; Sönmez, 1998).

Looking to the future, the WTO was not alone in its optimistic forecasts, and analysts concur that tourists demonstrate considerable resilience in the face of setbacks. Any downturns ahead for whatever reason, including crises, were discern to quickly overpass. However, there is also an appreciation that tourists can be fickle in their decision-making and behaviour (Henderson, 2007). Bauer (1967) was one of the first to suggest that customer attitudes are an act of risk, because any individual action of consumption is associated with uncertainty, implying unanticipated and possibly unpleasant consequences. In this concept, risk may be related to latent, rather than actual, individual or sets of conditions which can become crises if realized and of sufficient gravity. Hence, risk assessment is a key factor when planning for crisis, involving the anticipation of what might go wrong and identification of the reasons for divergences from expectations (Henderson, 2007). Most researchers have studied perceived risk, rather than objective or real risk, because ultimately, perceived risk determines behaviours (Bauer, 1967; Budesco & Wallstein, 1985). Several studies analyse the concept of perceived risk in tourism (Bodosca et al., 2014; Moutinho, 1987; Rohel & Fesenmaier, 1992). In the tourism context, risk is defined as the uncertainty experienced by tourists during the purchasing and consumption process of travel services and destination choice (Tsaur et al., 1997). Risk is also identified as a fundamental concern of international travellers (Yavas, 1990). However, the past several decades of study on risk perception reveal difficulties in operationalizing this concept (Rohel & Fesenmaier, 1992; Jacoby & Kaplan, 1972; Kaplan et al., 1974) suggest a multi-attribute model of risk, where several product attributes are judged separately by each customer. When the customer perceives a failure in one of the product attributes, this leads to a generalized feeling of loss. Kliestikova et al. (2019a, b) evaluate except the product attributes also benefits, imageries and attitudes.

According to Henderson (2007), however, there are fundamental differences between tourist products and other customer goods which induce the probability of crisis. Tourism industry has to transport people to the primary place of consumption, to accommodate and entertain them upon arrival, and therefore, the transport itself is one phase of production. These visits could be spoiled, and tourist safety could be compromised by variety of issues which could or could not be by culpability of those making the travel arrangements. The examples can be numerous and include transport accidents, hotel fires, street riots and multiple force majeure consequences which may result in travellers' victimization by criminals. Destinations where there are doubts about safety and security are near to a crisis of a tarnished image, official warnings against travel and a decline in visitors (Henderson, 2007).

It is necessary to remark that the Coronavirus (COVID-19) pandemic has triggered an unprecedented crisis in the tourism economy, given the immediate and immense shock to the sector. On a global scale, all tourism enterprises were forced to close for operation, and their staff were laid off. Moreover, all stakeholders in tourism industry had to respond with intensified marketing, price discounting and cost cutting, and several businesses demanded assistance from government in order to alleviate their financial pickle. Revised OECD estimates on the COVID-19 impact point to 60% decline in international tourism in 2020 (The Organisation for Economic Co-operation and Development (OECD), 2020). This could rise to 80% if recovery is delayed until December. International tourism within specific geographic regions, for example, the European Union, is expected to rebound first. Hotels, restaurants and attractions were among the first to swallow the bitter reality. Domestic tourism, which accounts for around 75% of the tourism economy in OECD countries, is expected to recover more quickly, similarly to already observed previous aftershock crisis waives. It offers the main chance for driving recovery, particularly in countries, regions and cities where the sector supports many jobs and businesses. In a similar vein, calls for global unity among tourism actors have proliferated from the UNWTO (2020) Global Tourism Crisis Committee which calls for a collective response to not only recover, but "grow back better" to all stakeholders in global tourism.

As summarized above, many articles are related with the effects of the global crisis on tourism industry in the scientific literature. Although the presented study is similar to them in some perspectives, but still, it offers a different point of view in terms of used methods and inferred analyses.

3 Methodology

The current study aims to establish Bulgarian tourists' attitudes towards travelling before and after the COVID-19 crisis. The assessment of Bulgarian travellers is based on results of a survey questionnaire conducted among 121 individual respondents for the period March–May 2020. For the data processing, statistical methods are implemented through computer language Python in a Web-based interactive computing environment for creating documents—Jupyter Notebook.

Statistical methods such as arithmetic mean "average" of data, sample standard deviation of data, single mode "most common value" of discrete or nominal data and median "middle value" of data are used for the analysis and evaluation of the results of the survey. The average value in its essence is a generalizing numerical characteristic of qualitatively homogeneous aggregates. It expresses the general, typical meaning of a given feature of the population as a whole. The average value represents the regular in the population and shows the central trend in it (Николова, 2013). When using the standard deviation, it is established to what extent each possible result differs from the expected mean value, using the root mean square value as the form of averaging (Колева & Касабова, 2016). The most common value is the meaning of the feature in which there is the greatest concentration of units or how often a given meaning of the feature participates in the population. In the middle value, the population has to be arranged in ascending or descending order of N units (Николова, 2013).

In the assembly, summarization and analysis of empirical data, it is inevitable to use the tools provided by the statistical methodology (Ламбова, 2003). Quantitative features are directly observable and allow unambiguous reflection through a numerical relational system, the operation of which consists in the registration of numerical quantities using appropriate measuring instruments, while qualitative features are in fact characterized by meanings which cannot be primarily measured. Through a number system, therefore they cannot be registered directly through measuring instruments as they represent verbal categories. For this reason, we should choose a scale to help us accurately represent our qualitative traits in quantitative.

The scale we apply in questionnaires research is the Likert scale, which is a type of psychometric scale often applied in psychological surveys. It was developed and named by organizational psychologist Rensis Likert in 1932 (Likert, 1932). One of the most widely applied tools in psychological research is self-disclosure inventories. Participants are required to state their level of agreement according a 5-point scale. Such scale is often applied to assess personality, attitudes and behaviour. In order to develop a questionnaire survey and data process, the conventional Likert scale usually has the following format: from "complete disagreement" to "complete agreement".

4 Results

COVID-19 has illustrated the fragility of life, but the same understanding has yet to be applied to addressing tourism industry. Several recent studies (Scott et al., 2010; Perriman et al., 2010) have shown that the preferences and the attitude of tourists after the crisis have changed. The tendency is to make use of the Internet tools and digitalized sources in order to avoid person-to-person contact complying with the forced self-isolating measures, thus limiting virus contamination risk.

The impact of pandemic COVID-19 shocks has reflected in the global tourism industry and has generated some considerable turbulence. The "new normal" has influenced to great extension the customer behaviour worldwide, and their confidence is changing on a daily basis. Consequently, the main differences can be seen in income disposal, the duration of the vacation, buying more private labels, entertaining at home more frequently or eating out and meeting with relatives and friends in a small percentage.

In connection with a more complete and accurate disclosure of the problem, we believe that it is necessary to outline the profile of respondents in terms of their gender, age, education, occupation and personal income.

The results of the survey show that the number of female respondents prevails—79.3%. As regards to the age range of participants, it was found that the biggest share of respondents (38.8%) is of age between 30 and 39 years old. It is worth to notice that the shares of participants at the age of 40–49 and up to 29 are also quite considerable. The majority of respondents have a higher degree (70.2%). The people interviewed followed a variety of occupations, the most of which state that they are employees (48.2%). As evident from Fig. 1, the biggest percentage of the respondents has personal income from 561 to 999 BGN (Fig. 1).

Of a high importance for the survey analysis is the residential status of the respondents due to the fact that from 13 March the Bulgarian government declared state of emergency and all Bulgarian citizen were obliged to keep strict self-isolation. It

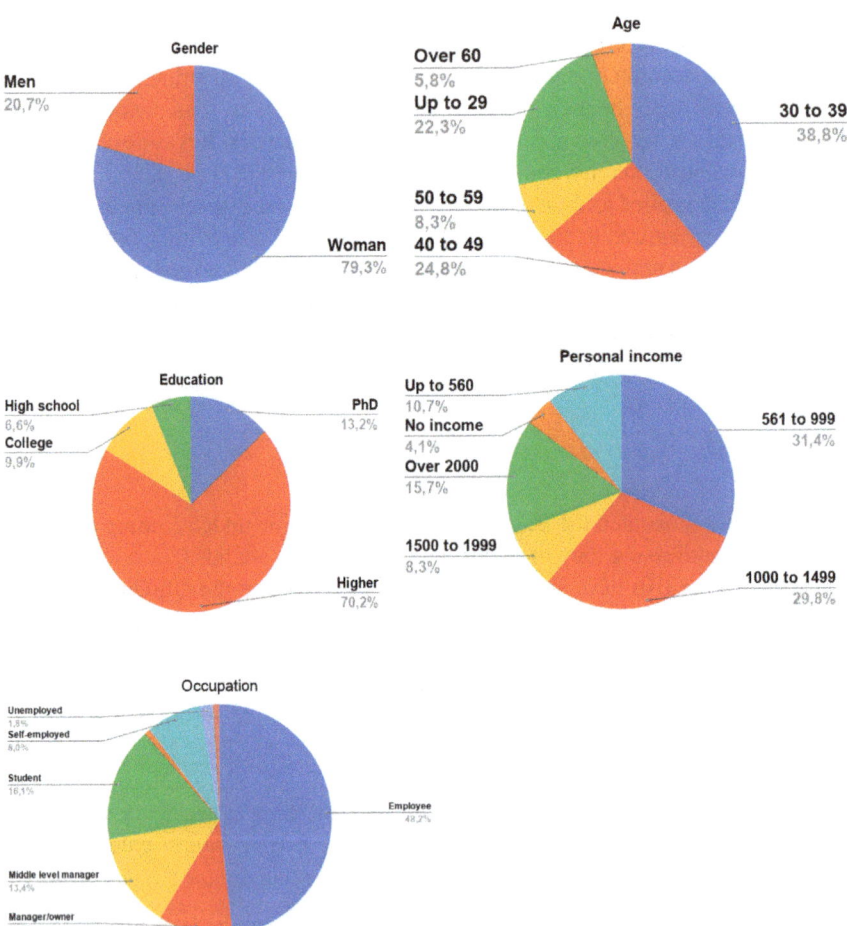

Fig. 1 Profile of respondents by gender, age, education, occupation and personal income. *Source* author's own survey; less than 1% do not appear on the figure

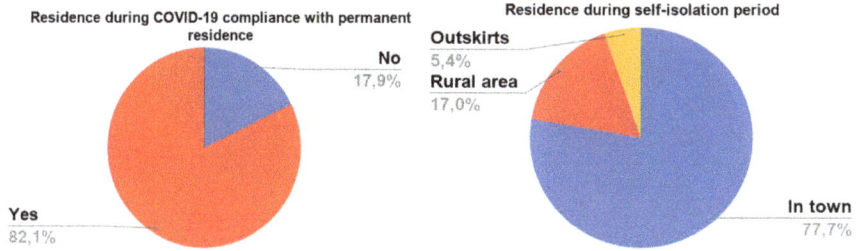

Fig. 2 Residence during COVID-19. *Source* author's own survey

can be observed from Fig. 2 that the biggest part of the interviewed residence during COVID-19 complies with their permanent residence (82.1%) which can lead us to the conclusion that the Bulgarian citizens travelled less and thus limited COVID-19 spreading risk. This can be confirmed by the highest percentage of respondents (77.7%) self-isolating in towns.

Questions with strong significance which are aiming to determine to what degree tourism customer attitudes before the COVID-19 have changed during the crisis. According to the respondents replies, applying estimations with Python via Jupyter Notebook, we calculated arithmetic mean "average" of data.

As evident from Fig. 3, the participant's attitudes towards the activities during the self-isolation according to the mean comparisons demonstrate that they miss being with relatives and friends the most. At the meantime, the mean value shows that their free movement is with high estimation.

Of interest are the questions related to the frequency of travel before and after COVID-19 (Fig. 4).

Regarding the frequency of travel, the results of the survey show that the COVID-19 factor has discouraged the respondents and they are hesitant in making travel decisions at this stage. This statement is confirmed by the registered responses of respondents who do not intend to travel at this stage, would wait to see what future measures will be taken or would travel at a given opportunity. On the other hand, the share of respondents who would take a trip for tourism once a month is not small, despite COVID-19 (17.9%).

```
In [14]: tourism_data.mean()

Out[14]: to be with relatives and friends    3.661157
         to shop                             3.661157
         to sport in nature                  2.834711
         use beauty centers services         2.553719
         use SPA services                    2.016529
         to eat out                          3.198347
         to travel for tourism               3.198347
         to travel to work                   2.991736
         attend cultural events              3.057851
         dtype: float64
```

```
In [15]: tourism_data.mean()

Out[15]: to move freely                         3.727273
         to be with relatives and friends       3.743802
         to shop                                2.735537
         to sport in nature                     2.727273
         to use beauty centers services         2.355372
         to use SPA services                    1.975207
         to eat out                             3.049587
         to travel for tourism                  3.570248
         to travel for work                     2.710744
         to attend cultural events              3.347107
         dtype: float64
```

Fig. 3 Mean of customer attitudes before and during the COVID-19. *Source* author's own calculation

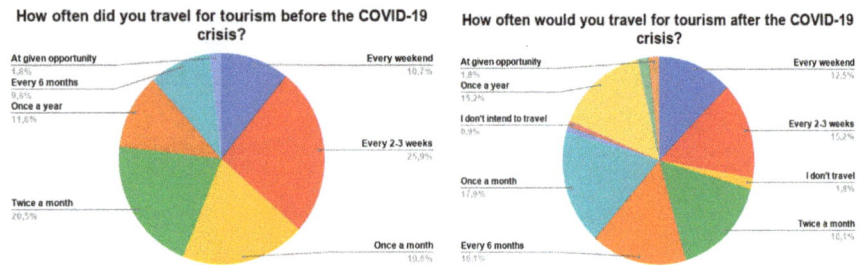

Fig. 4 Travel frequency before and after the COVID-19. *Source* author's own survey

We should mention the next couple of questions concerning the type of tourism undertaken before and after the COVID-19, visualized on Fig. 5.

In connection to the type of tourism undertaken by the interviewed, it can be observed an increase of the share that is willing to undertake sea, mountain and alternative tourism. While, some of the interviewed are hesitant to undertake any type of tourism at this stage or to travel at all.

As regards to the destination choice (Fig. 6), research results demonstrate, opposite to the expected shit, the respondents state that they are willing to travel to far destinations.

The interviewed preferences about vacation booking after COVID-19 can lead us to the assumption that they consider that it is less risky to use Internet, online agencies, applications and reserving by phone (Fig. 7).

Overall, the results of the research reveal that tourists attitudes slightly change during and after COVID-19 crisis. Their choices regarding tourism have changed by

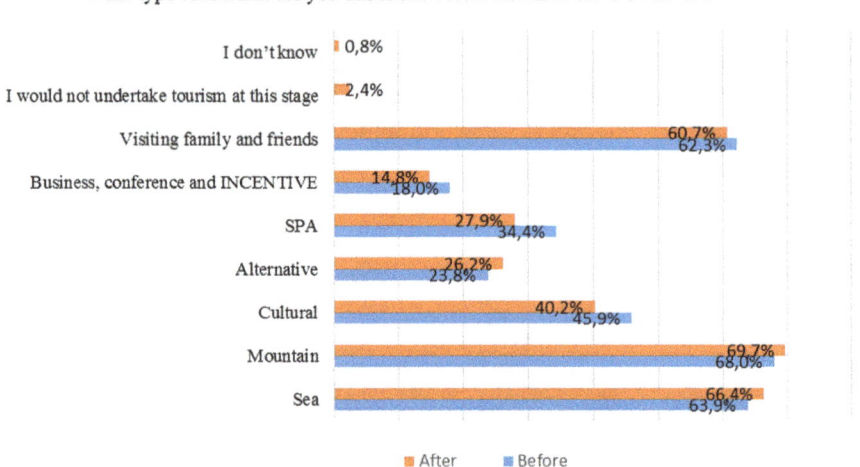

Fig. 5 Type of tourism undertaken before and after the COVID-19. *Source* author's own survey

Which destinations did you visit the most before and after the COVID-19?

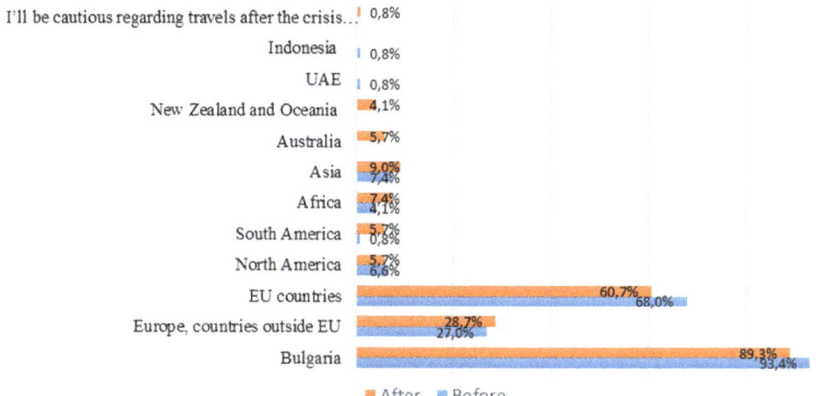

Fig. 6 Most visited destinations before and after the COVID-19. *Source* author's own survey

How did you book your vacation before and after the COVID-19?

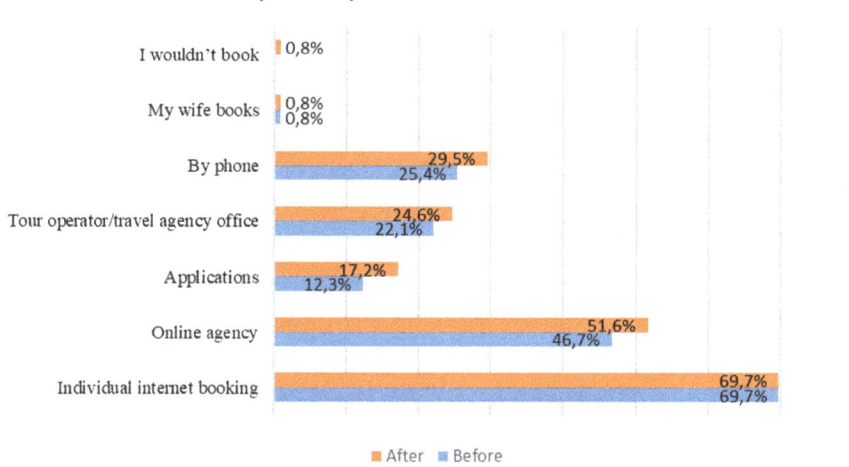

Fig. 7 Vacation booking before and after the COVID-19. *Source* author's own survey

the COVID-19 factor, and in our opinion, this is a direct consequence of the risks arising by dangerous exposure on that same factor. Moreover, such slight changes can be explained by the fact that no matter of residence, class, social status and religion people's health and safety stays first and economic measures should come second.

5 Conclusion

COVID-19 will reshape tourism as we previously knew it. Post-COVID-19 recovery of the tourism industry will be tied in the short term to the rate of global economic recovery (Prideaux et al., 2020). Yet, while there are reasons to be hopeful, who will benefit from this restructuring is still an unsettled question. There are currently more than hundred articles addressing what COVID-19 means for tourism. Many are focusing on the impacts of COVID-19 on the industry as well as how tourism can be reimagined and enacted in more sustainable and resilient ways.

All tourism industry stakeholders have to consider the personal characteristics of its customers, because they influence risk perception and their decision-making processes, such as income, travel motivations, previous contact with risks and nationality (Seabra et al., 2013). Also, earnings management is proven to be well used in Bulgaria, especially in the last 10 years (Durana et al., 2020) and will be more than necessary to prevent risks of bankruptcies in the industry. Elaborations on the issue strike also live the impression that disasters and crises could be considered as a catalyst for positive change in the destination. Moreover, among them stands out the topic of the potential created by risk or crisis for positive change through orientation towards sustainable development, product repositioning or for better respond. In this connection, Faulkner (2001) highlights crises and risks ability to act as a turning point for destinations and businesses. Furthermore, in our opinion, all tourism industry stakeholders should consider their strategies under the condition of risk and crisis in order to obtain efficient long-term results, better sustainable management for retaining customers and employees in the shifting conditions of economic, social and business environment.

References

Bauer, R. (1967). Consumer behavior as risk taking. In D. Cox (Ed.), *Risk taking and information handling in consumer behavior* (pp. 23–33). Cambridge: Harvard University Press.

Beirman, D. (2002). Marketing Of tourism destinations during a prolonged crisis. *Journal of Vacation Marketing, 8*(2), 167–176.

Bodosca, Ş, Georgica, G., & Puiu, N. (2014). Tourist consumption behaviour before and after the crisis from 2008. *Procedia Economics and Finance, 16*, 77–87. https://doi.org/10.1016/s2212-5671(14)00777-1.

Budesco, D., & Wallstein, T. (1985). Consistency in interpretation of probabilistic phrases. *Organizational Behavior and Human Decision Processes, 36*(3), 391–405.

Coshall, J. (2003). The threat of terrorism as an intervention on international travel. *Journal of Travel Research, 42*(1), 4–12.

Cró, S., & Martins, A. M. (2017). Structural breaks in international tourism demand: Are they caused by crises or disasters? *Tourism Management, 63*, 3–9.

Crompton, J., & Ankomah, P. (1993). Choice set propositions in destination decisions. *Annals Of Tourism Research*, 461–475.

Dimanche, F., & Leptic, A. (1999). New Orleans tourism and crime: A case study. *Journal of Travel Research, 38*(1), 19–23.

Durana, P., Valaskova, K., Chlebikova, D., Krastev, V., & Atanasova, I. (2020). Heads and tails of earnings management: quantitative analysis in emerging countries. *Risks, 8*(57), 1–20. https://doi.org/10.3390/risks8020057.

Faulkner, B. (2001). Towards a framework for tourism disaster management. *Tourism Management, 22,* 135–147.

Faulkner, B., & Russell, R. (2001). Tourism in the twenty-first century: Reflections on experience. In B. Faulkner, G. Moscardo, & E. Laws (Eds.), *Turbulence, chaos and complexity in tourism systems: A research direction for the new Millennium* (pp. 328–349). London: Continuum.

Fesenmaier, D. (1988). Integrating activity pattern into destination choice models. *Journal of Leisure Research, 20*(3), 175–191.

Fink, S. (1986). *Crisis management: Planning for the inevitable.* New York: Amacom.

Friedman, T. L. (2020). We need herd immunity from trump and the coronavirus. *The New York Times.* Retrieved from https://www.nytimes.com/2020/04/25/opinion/coronavirus-immunity.

George, R. (2003). Tourists' perceptions of safety and security while visiting cape. *Tourism Management, 24*(3), 575–585.

Gretzel, U., Fuchs, M., Baggio, R., Hoepken, W., Law, R., Neidhardt, J., & Xiang, Z. (2020). E-Tourism beyond Covid-19: A call for transformative. *Information Technology & Tourism, 22,* 187–203. https://doi.org/10.1007/s40558-020-00181-3.

Gros, C., Valenti, R., Valenti, K., & Gros, D. (2020). *Arxiv.Org.* Retrieved from Corell University, https://arxiv.org/abs/2004.00493v1?utm_source=feedburner&utm_medium=feed&utm_campaign=feed%3a+coronavirusarxiv+%28coronavirus+research+at+arxiv%29.

Gu, Z., & Martin, T. (1992). Terrorism, seasonality, and international air tourist. *Journal of Travel and Tourism, 1*(1), 3–15.

Henderson, J. C. (2007). *Tourism crises: Causes, consequences and management.* Butterworth–Heinemann, Elsevier Inc. ISBN: 978-0-7506-7834-6.

Hermann, C. (1972). *International crises: Insights from behavior research.* New York: The Free Press.

International Monetary Fund. (2020). *Press Release No. 20/98.* Retrieved from International Monetary Fund, https://www.imf.org/en/news/articles/2020/03/23/pr2098-imf-managing-director-statement-following-a-g20-ministerial-call-on-the-coronavirus-emergency?fbclid=iwar2hzwzn6du2pwyi9sp47sqg-kzee24esyp1remcnq06j4pg6ceikrmwhfs.

Jacoby, J., & Kaplan, L. (1972). The components of risk perception. *Proceedings of the 3rd annual conference* (pp. 382–393). Association for Consumer Research.

Kaplan, L., Szybillo, G., & Jacoby, J. (1974). Components of perceived risk in product purchase: A cross-validation. *Journal of Applied Psychology, 59*(3), 287–291.

Keown-Mcmullan, C. (1997). Crisis: When does a molehill become a mountain. *Disaster Prevention and Management, 6*(1), 4–10. https://doi.org/10.1108/09653569710162406.

Kliestikova, J., Durana, P., & Kovacova, M. (2019). Naked consumer's mind under branded dress: Case study of Slovak Republic. *Central European Business Review, 8*(1), 15–32. https://doi.org/10.18267/j.cebr.208.

Kliestikova, J., Kovacova, M., Krizanova, A., Durana, P., & Nica, E. (2019). Quo Vadis brand loyalty? Comparative study of perceived brand value sources. *Polish Journal of Management Studies, 19*(1), 190–203. https://doi.org/10.17512/pjms.2019.19.1.14.

Koren, M., & Peto, R. (2020). *Arxiv.Org.* Retrieved from Cornell University, https://arxiv.org/pdf/2003.13983.pdf.

Kyurova, V. (2013). The role of marketing in anti-crisis management in hotel business. *Economics and Management, 1,* 18–23.

Levantis, T., & Gani, A. (2000). Tourism demand and the nuisance of crime. *International Journal of Social Economics, 27,* 959–967.

Levitt, T. (1983). The globalization of marketing. *Harvard Business Review, 7,* 92–102.

Likert, R. (1932). A technique for the measurement of attitudes. In R. S. Woodworth (Ed.), *Archives of psychology* (pp. 5–55). New York: New York University.

Mansfeld, Y. (1996). Wars, tourism and the "Middle East" factor. In A. Pizam & Y. Mansfeld (Eds.), *Tourism, crime and international security issues* (pp. 265–278). New York: Wiley.

Mostafanezhad, M. (2020). Covid-19 is an unnatural disaster: Hope in revelatory moments of crisis. *Tourism Geographies*, 1–7. https://doi.org/10.1080/14616688.2020.1763446.

Moutinho, L. (1987). Consumer behavior in tourism. *European Journal of Marketing, 21*(10), 5–44.

Moutinho, L. (2000). *Consumer behavior* (pp. 41–79). New York: Cabi Publishing.

Nowlin, C. (2017). Understanding and undermining the growth paradigm. *Dialogue, 56,* 559–593.

OECD The Organisation For Economic Co-Operation And Development . (2020). *Tourism Policy Responses To The Coronavirus (Covid-19)*. Retrieved from Policy Responses To Coronavirus (Covid-19), https://www.oecd.org/coronavirus/policy-responses/tourism-policy-respon ses-to-the-coronavirus-covid-19-6466aa20/.

Okumus, F., Altinay, M., & Arasli, H. (2005). The impact of Turkey's economic crisis of February 2001 on the tourism industry in Northern Cyprus. *Tourism Management, 26*(1), 95–104.

Ötsch, W. (2020). What type of crisis is this? The coronavirus crisis is a crisis of the economized society. *Lecture at the Topical Lecture Series Of Cusanus Hochchule Für Gesellschaftsgstaltung.*

Perriman, H., Ramsaran-Fowdar, R., & Baguant, P. (2010). The impact of the global financial crisis on consumer behaviour. In *Proceedings of annual london business research conference* (pp. 1–14). World Business Institute.

Pforr, C., & Hosie, P. (2009). *Crisis management in the tourism industry.* Farnham, UK, Surrey: Ashgate Publishing.

Pizam, A., & Mansfeld, Y. (1996). *Tourism, crime and international security issues.* Chichester: Wiley.

Prideaux, B. (2003). The need to use disaster planning frameworks to respond to major. *Journal of Travel and Tourism Marketing, 3*(1), 281–298.

Prideaux, B., Thompson, M., & Pabel, A. (2020). Lessons from Covid-19 can prepare global tourism. *Tourism Geographies*, 1–12. https://doi.org/10.1080/14616688.2020.1762117.

Ritchie, B. (2004). Chaos, crises and disasters: a strategic approach to crisis. *Tourism Management, 25*(6), 669–683.

Ritchie, B. W., & Jiang, Y. (2019). A review of research on tourism risk, crisis and disaster management: Launching the annals of tourism research curated collection on tourism risk, crisis and disaster management. *Annals of Tourism Research, 79,* 1–15. https://doi.org/10.1016/j.annals.2019.102812.

Ritchie, B. W., Dorrell, H., Miller, D., & Miller, G. (2004). Crisis communication and recovery for the tourism industry: Lessons from the 2001 foot and mouth disease outbreak in the United Kingdom. *Journal of Travel & Tourism Marketing, 15*(2–3), 199–216.

Rohel, W., & Fesenmaier, D. (1992). Risk perceptions and pleasure travel: An exploratory analysis. *Journal of Travel Research, 2*(4), 17–26.

Roitman, J. (2013). *Anti-Crisis.* Duke University Press.

Sahlins, M. (1972). *Stone age economics.* Chicago: Aldine Atherton Inc.

Santana, G. (2004). Crisis management and tourism. *Journal of Travel and Tourism Marketing, 15*(4), 299–321.

Scott, N., Laws, N., & Prideaux, B. (2010). *Safety and security in tourism.* London: Routledge.

Seabra, C., Dolnicar, S., Abrantes, J., & Kastenholz, E. (2013). Heterogeneity in risk and safety perceptions of international tourists. *Tourism Management, 36,* 502–510.

Selbst, P. L. (1978). *Containment and control of organizational crises. Management handbook for public administrators.* New York: Van Nostrand.

Shrivastava, P., & Mitroff, I. (1987). Strategic management of corporate crises. *Columbia Journal of World Business,* 5–11.

Sönmez, S. (1998). Tourism, terrorism and political instability. *Annals of Tourism, 25*(2), 416–456.

Tsaur, S., Tzeng, G., & Wang, K.-C. (1997). Evaluating tourist risks from fuzzy. *Annals of Tourism Research, 24*(4), 796–812.

Weaver, D., & Lawton, L. (2002). *Tourism Management.* Sydney: Wiley.

Woodside, A., & King, R. (2001). An updated model of travel and tourism purchase consumption. *Journal of Travel and Tourism Marketing, 10*(1), 3–27.

Woodside, A., & Lysonski, A. (1989). A general model of traveler destinations choice. *Journal of Travel Research, 27*(4), 8–14.

World Health Organization. (2020). *Coronavirus disease (Covid-2019) situation reports.* Retrieved from World Health Organization: https://www.who.int/docs/default-source/coronaviruse/situat ion-reports/20200121-sitrep-1-2019-ncov.pdf?sfvrsn=20a99c10_4.

World Travel & Tourism Council (WTTC). (2020a, March 31). *Bulgara, 2020 annual research: Key highlights.* Retrieved from Country/Region Data: https://wttc.org/research/economic-impact.

World Travel & Tourism Council (WTTC). (2020b, March 31). *Economic impact reports.* Retrieved from The World Travel & Tourism Council (WTTC): https://wttc.org/research/economic-impact.

World Tourism Organization (WTO). (1999). *Tourism 2020 vision: Executive summary updated.* Madrid: World Tourism Organization.

World Tourism Organization (WTO). (2005). *Tourism highlights. edition 2004.* World Tourism Organization.

World Tourism Organization. (2013). *Unwto annual report 2012.* World Tourism Organization. Retrieved from https://cf.cdn.unwto.org/sites/all/files/pdf/annual_report_2012.pdf.

World Tourism Organization. (2020). https://www.unwto.org/statistics. Retrieved from World Tourism Barometer: https://www.unwto.org/world-tourism-barometer-n18-january-2020.

Yavas, U. (1990). Correlates of vacation travel: Some empirical evidence. *Journal of Professional Services Marketing, 5*(2), 3–18.

Колева, Р., & Касабова, С. (2016). Статистически Методи За Анализ На Бизнес Риска. *Икономика, 21*(1), 93–104.

Ламбова, М. (2003). *Измерването - Неглижираният Проблем При Емпирични Изследвания, Осъществявани С Помощта На Статистически Инструментариум.* Варна: Ик "Стено".

Николова, Н. (2013). *Статистика. Обща Теория.* София: Авангард Прима.

Solo Female Travelers as a New Trend in Tourism Destinations

Lubica Sebova, Kristina Pompurova, Radka Marcekova, and Alica Albertova

1 Introduction

The demand in tourism is characterized by several new trends in present time. Tourism demand is strongly influenced by demographic, environmental, technological changes and changes in the values of society and lifestyle (Alejziak, 2007). One of the new trends of the twenty-first century is independent travel, which is also significantly expanding among women. Women are increasingly emancipated, longing for rest, relax but also for discovery and adventure. The usual stereotype of traveling with the whole family ceases to apply in Slovakia as well, and the demand for women's trips or traveling alone is growing, sometimes even without the company of another person. In terms of participation in tourism, many women travel abroad, due to the usual opinion about better services and their lower prices.

In order for Slovakia to become attractive destination for emerging market segments, which include solo female travelers, it is necessary to know the motives and needs of women. The study of demand represents a starting point for the creation of desired products, which can stimulate an increase in the independent participation of women in tourism in Slovakia. Making the supply of products in Slovakia more attractive can thus reduce the tendency to prefer visiting destinations abroad.

The expected benefit of the paper is an increase in awareness of the mentioned segment and its requirements in Slovakia, which may have a positive impact on the participation of the segment of solo female travelers in domestic and inbound tourism in Slovakia especially after Coronavirus crisis.

L. Sebova (✉) · K. Pompurova · R. Marcekova · A. Albertova
Matej Bel University, Banska Bystrica, Slovakia
e-mail: lubica.sebova@umb.sk

© The Author(s), under exclusive license to Springer Nature Switzerland AG 2021
V. Katsoni and C. van Zyl (eds.), *Culture and Tourism in a Smart, Globalized, and Sustainable World*, Springer Proceedings in Business and Economics,
https://doi.org/10.1007/978-3-030-72469-6_20

2 Literature Review

Women's participation in outdoor leisure activities was limited before 1950. Since then, however, the proportion of working and childless women has been increasing. Women are becoming more independent and are more likely to engage in outdoor recreational activities, such as traveling (Marzuki et al., 2012). As a result of these societal changes, women spend more and more time without their parents, partners, children and other family responsibilities (Franzi & DeCarlo, 2018). Today, women are increasingly active in participating in tourism, with a variety of opportunities, resources and possibilities for spending their spare time. They are thus considered as an influential market segment within the tourism sector (Wilson & Little, 2005). They increasingly travel alone, with their girlfriends, sisters and daughters, as opposed to traveling with family (Franzi & DeCarlo, 2018).

Solo traveling women are an increasingly important segment of tourism. This fact manifests itself as a growing trend of independent travel, especially among female tourism visitors (Ministry of Transport and Construction of the Slovak Republic, 2019). However, they cannot be considered as a completely separate segment, as they can also be part of a new segment of active seniors, millennials, "singles" and "dinkies" or the middle class, it means other segments that create new trends in tourism through their participation in tourism.

According to Booking.com (2014), independent travel is on the rise, with most individual passengers being women. There are several approaches to the concept of independent or solo travel (Jafari et al., 2016). According to Cargill (2009), it is not a travel without company in the sense of moving from one place to another. He says that it is an individual participation in tourism, when a person spends a significant amount of time without direct company at the place where he has traveled.

Today, women have greater opportunities for acceptance in the society, traveling and related financial security. Nevertheless, compared to men, they continue to face several restrictions affecting participation in tourism, which result from their greater vulnerability (Junek et al., 2006). In terms of independent travel of women, therefore, we do not see the independent, individual participation of women in tourism literally. Concerns about one's own safety, language barrier, poor orientation in space and other aspects determine the usual perceptions of women's independent travel. It can be an initial phase that takes place independently—individual decision-making and purchase of the trip (product, service), followed by mutual acquaintance and joint experience of several participants of the trip (consumers of the product, service) at the destination (www.women-traveling.com).

Another type of women's independent travel is jointly planned and realized participation in tourism, where there are two women in a mutual family or friendly relationship (Fiala, 2017). Even in this case, the condition of traveling without several family members or a male partner remains.

Thanks to traveling without a partner, women enjoy their independence, a break from everyday worries and feel more confident. This is a new phenomenon that is expected to remain in tourism (Kow, 2017; Booking.com, 2014; Wilson & Little, 2005).

Gucik et al. (2010, pp. 109–110) talk about changing "the social status of women who become independent, do not plan a family and travel more and more". Junek et al. (2006) add to the mentioned societal changes that more women are currently divorcing. Gucik et al. (2010) emphasize the high proportion of women in the segment of business travelers and consider it necessary to create a suitable product for this segment.

In terms of travel for pleasure, women are more adventurous than men (Davidson & McKercher; Swarbrooke et al. in Wilson & Little, 2005) and incline to eco-tourism (Weaver in Wilson & Little, 2005).

Women who currently take part in independent tourism on their own come from different age groups. There is a significant increase in demand among visitors over the age of 50, with this segment being made up mainly of women. Some of them are interested in an easier form of adventurous experience, and many women in this and younger age categories prefer exotic destinations (African safaris) and seek cultural experiences.

Service providers are also responding to the women's demand, with a rapid increase in the number of travel agencies focusing exclusively on female clients in countries such as the USA and Canada (Junek et al., 2006).

Women can now create a clear and consistent picture of their needs for safety, comfort, power and pampering. Thanks to this knowledge, for example, hotel managers (as well as other service providers) have an excellent opportunity to offer women more than just standard hotel equipment. Today, they can provide this dynamic and growing market segment with an experience that is better than their original expectations (expanded product) (Brownell, 2011).

According to Qu et al. (In Truong et al., 2018), identifying the characteristic and unique elements of a destination supply is the first step in creating a marketing strategy that can encourage visitors to choose a destination. According to Gucik et al. (2010), however, it is important to realize that a destination cannot offer every-thing to everyone, but it is more efficient to aim on what a selected target group of customers expects, whose needs and wishes they want and can satisfy.

Borovsky et al. (2008) understand the product of tourism in a broader and narrower sense, when the product in a broader sense is considered to be the destination and its overall impact on the visitor. In the narrower sense, these are the products satisfying the partial needs of visitors, which are a condition for their basic needs fulfillment such as accommodation and meals. The destination can be described as a product, while the product is composed of many sub-products (Jakubiková, 2006). From the point of view of the product of the destination, it is necessary to emphasize the importance of market segmentation (Mariani et al., 2014). Masip (2006) speaks about value creation at the destination, which can be determined by specialization through the creation of products focused on specific market segments.

3 Methodology

The subject of the research is solo female travelers, whose travel represents a new trend in tourism. The paper contains an analysis of the requirements and preferences of women in terms of independent travel and a comparison of the current range of products for this segment in Slovakia and abroad.

We are looking for answers to following research questions:

Q1: What are the specifics of the solo female travelers' segment?
Q2: What are the requirements and preferences of women in terms of independent traveling?
Q3: What tourism products are desired by solo female travelers in Slovakia?

We use secondary and primary sources by paper processing. Secondary sources are mostly from the literature from domestic and foreign authors. By the definition and characteristics of the solo female travelers' segment, the resources are mainly from foreign online publications and articles, as reflecting the current state of research problems.

In the context of primary research, we realized questionnaire survey, which helped us to find out the demand of women in the context of their independent traveling in the Slovak Republic. The questions in the questionnaire focused on preferences, motives, requirements for the offer of destination products, experience with traveling abroad, perceived shortcomings in measures related to tourism and other aspects in the context of women's independent participation in tourism in Slovakia (Motusova, 2019).

The questionnaire contained 16 questions. The questionnaire was compiled in the Slovak language, as our goal was to obtain answers from women who are of Slovak origin or live in Slovakia or know its tourism offer thanks to traditional cultural and historical relationships (and at the same time speak Slovak). We created the questionnaire using the online tool Survio and distributed it through social networks, by e-mails and in person (to the women older than 60 years). We obtained 355 answers in total. The survey was realized in the period from January to February 2019. For the data processing obtained from the questionnaire survey, we used MS Excel software and SPSS statistics. The results are clarified in the tables.

When elaborating the paper, we used selected mathematical-statistical methods and theoretical methods of research. The theoretical methods of research were methods of abstraction, analysis and synthesis and induction and deduction. We used also the method of comparison when comparing the offer and measures of destinations abroad and in Slovakia.

4 Results

Destinations abroad and in Slovakia respond to the needs and requirements of the segment of women traveling alone in different ways. The offer of individual providers of products and services, which in aggregate represents the secondary supply of the destination, does not always satisfy the demand of individual travelers. Therefore, it is necessary to know specific preferences of solo female travelers, which we researched through questionnaire survey, which points to the current demand of solo female travelers in Slovakia.

At present, women's interest in various types of tourism and related products and services oriented to this way of participating in tourism is growing. Based on the results of the questionnaire survey, we found out that almost half (48.2%) of respondents are between 22 and 26 years old. This is a category of women belonging to the so-called Generation Y—currently and in the near future a significant market segment, to which it will be necessary to pay adequate attention. The next most numerous age group in our survey was women aged 18–21 years (15.5%), then aged 27–35 (13.8%) and 36–49 (13.8%) years. 7.6% of respondents are 50–60 years old, and the remaining 1.1% of respondents are 61 years of age or older.

In terms of social status, 46% of respondents stated that they were employed, 27.4% were students and 21.5% were studying and working at the same time. The other respondents were unemployed (2%), housewives or on maternity leave (1.7%), and 1.4% of women were retired.

We also surveyed the highest level of education achieved, with 57.5% of respondents having a university degree, 39.7% having a full secondary education and only 2.8% having lower education. Given the number of our respondents belonging to the generation of millennials, the above-mentioned socio-demographic structure reflects the fact that this generation belongs to the highly educated, which can be reflected in the expectation and search for sophisticated products.

The first question of the questionnaire asked how often women participate in tourism on their own. Although as we already stated in the theoretical background the joint participation of two women in tourism can be considered as independent travel of women, there is no strict quantitative definition of the concept of solo female travelers. Therefore, with regard to the Slovak practice, we chose as a maximum in the questions of the questionnaire up to three women to qualify for the condition of solo travelers, and it means our respondent was traveling alone or was accompanied by maximum of two women.

Based on the results of the questionnaire survey, we processed a table (Table 1), which informs about how often the respondents participate in tourism alone or in the company of no more than two women. We found out that the largest share of women (35.2%) indicated option 2–3 times a year, and option 4 and more times were chosen by 32.4% of respondents. 16.3% of respondents travel individually only one a year, and 16.1% of women do not travel solo. This is also the group which can be approached to start traveling independently by ensuring suitable conditions

Table 1 Frequency structure of women's independent travel

Frequency/year	1 time	2–3 times	4 and more times	I do not travel alone
Number of answers	58	125	115	57
Share in total number in%	16.3	35.2	32.4	16.1

Source Own elaboration, 2019

Table 2 Frequency structure of independent travel of women in Slovakia

Frequency/year	0–1 times	2–3 times	4 and more times
Number of answers	127	115	48
Share in total number in %	43.8	39.7	16.5

Source Own elaboration, 2019

and creating the desired products. When completing the questionnaire, this group of women was to continue with 12th question.

The aim of the second question was to find out how often women independently participate in tourism in Slovakia. The structure of the answers is recorded in Table 2. From 290 relevant answers, we could find out that most respondents (43.8%) participate on tourism in Slovakia independently 0–1 time a year and 39.7% participate on tourism in Slovakia 2–3 times a year.

We also asked our respondents if they preferred traveling abroad to traveling in Slovakia by their solo travels. Based on a comparison of the number of answers obtained ("yes" or "no"), we can assess that most respondents (77.1%) would prefer to travel abroad. The group of women who expressed the largest preference to travel abroad was between the age of 22 and 26 years. These negative findings represent the basis for a more detailed analysis of the perception of the shortcomings of Slovakia's supply focused on a selected market segment.

It is obvious that supply in foreign countries is very attractive for many respondents and therefore in the next question we focused on specific aspects that respondents perceive as the benefits of traveling outside Slovakia. Here are four basic answer options. 55.3 % of responses show that the rest of the world is preferred because of the variety of choices concerning the offer, while a much lower share of the overall responses indicated a higher quality of services (18.4%), other (individual) response (15.6%), better prices (7%) and good experience of friends (3.7%). Up to 38 respondents used the free space to write their own answer, which was a significant contribution to our research. Based on prepared possibilities as well as individual answers, we can deduce that the tourism offer abroad is attractive for respondents mainly due to the desire to get to know something new what they have not already seen, experienced or cannot find in Slovakia (differences in the primary offer). In addition, women are increasingly longing for adventure, which they associate with possible experiences from visiting abroad. We also connect the tendencies to travel solo abroad with the possibilities that people have today in terms of transportation (cheap

Table 3 Number of nights usually spent in accommodation facility in Slovakia during an individual trip

Number of nights	1–2	3–4	5–6	7 and more	I do not travel alone in Slovakia
Number of answers	195	49	8	3	39
Share in total number in %	66.3	16.7	2.7	1.0	13.3

Source Own elaboration, 2019

air tickets, building transport infrastructure, free movement of people, etc.). Based on the answers, the challenge for Slovakia as a destination remains to improve the quality of services provided. However, we evaluate positively that the respondents do not consider the prices of tourism products and services to be the basic difference between Slovakia and foreign countries.

If the respondents travel alone, we were interested how many nights they used to spend in an accommodation facility in Slovakia. After excluding six irrelevant answers, we collected 294 answers, from which we conclude that the respondents usually spend a weekend (66.3%) or an extended weekend (16.7%) in the accommodation facility. Only 13.3% of the answers indicate that the respondents do not travel individually in Slovakia. This is the number that could be reduced by creating relevant products associated with the introduction of support measures for solo women's travel in Slovakia. We have summarized the results in Table 3.

Since tourism is significantly affected by seasonality, we were interested in which time of the year women most often travel independently in Slovakia. With this question, we also wanted to find out the potential for increasing participation in tourism during the off-season (spring, autumn) in Slovakia through individual women's trips. Respondents could have chosen only one option. It turned out that summer is the most attractive period of time for traveling for respondents (36.9%) in terms of independent travel in Slovakia, but we consider the fact that 35.2% of answers show an inclination to year-round travel as a positive result. Respondents travel at least in the fall (5.7%), spring (4.7%) and in winter (4.4%). 13.1% of respondents do not travel in the sense of solo traveling. As a significant proportion of respondents travel throughout the year, we consider women's independent travel to be a possible determinant of the development of off-season tourism products in the country and a way to reduce fluctuations in the frequency of participation in tourism.

The subject of the seventh question was the preferred types of tourism by the selected segment in Slovakia. Each respondent could indicate up to two answer options or even enter her own answer. We interpret the results using a table (Table 4). The largest number of respondents chose cultural tourism (32.1%), followed by sports and adventure tourism (22.6%), health tourism (20.9%) and business tourism (13.4%). Thirty-nine respondents do not travel alone in Slovakia, and we also received some individual answers concerning visiting family and acquaintances and shopping tourism. Despite the dominance of the choice of cultural tourism (visit to cultural and pilgrimage monuments, museums and galleries, organized events, etc.), we state that women in Slovakia are becoming more and more brave and adventurous. Many

Table 4 Types of tourism in
which self-employed women
most often participate in
Slovakia

Type of tourism	Number of answers	Share of total responses in %
Sports/adventure	91	22.6
Commercial	54	13.4
Health/wellness	84	20.9
Cultural	129	32.1
I do not travel alone in Slovakia	39	9.7
Other: Shopping tourism	2	0.5
Visit of family and friends	3	0.8

Source Own elaboration, 2019

continue to look for rest, spa and wellness care, which we combine with Slovakia's excellent potential for providing services of this kind and at the same time with the work and other workload of today's women and the health problems of older women.

In the next question, we were asking about the motivation and reasons for solo traveling. Almost half of the respondents (49.3%) identified discovery and knowledge for their main motivation, and 23% of the answers indicate a desire to get rid of everyday stereotypes. These data agree with the identified preference for sports, adventure and activities related to cultural tourism. To the reasons belong also the rest from work (11.1%), rest from home duties (6.8%), further currently working duties (6.4%) and other reasons (3.4%). According to our findings women often cannot find a company and so travel solo.

Many aspects can influence the (potential) visitor when choosing a destination or the services provided in it, not only for the purpose of independent participation in tourism. In our survey, we gave women space to identify a maximum of the two factors that most influence their choice of destination (or its product). Despite the fact that none of the offered answers was significantly dominant among the respondents, most of the answers received the option related to the unique offer of the destination (29.1%). This was followed by a good price (23.7%), reviews published on the Internet (19.4%), recommendations from friends (14.3%), previous personal experience (9.6%), personalized answers (3%), and the least are respondents influenced by advertising (0.9%). Other responses included decision-making based on the availability of cost-effective airline tickets, knowledge of the place and the fact that it is a place not yet visited by the person.

As we mentioned earlier, most respondents prefer independent participation in tourism abroad, not in Slovakia. Our research was focused on improving products and conditions for solo female travelers in Slovakia, so we used the tenth question to find out why the women do not travel alone in Slovakia as much as they would like. Respondents were able to indicate a maximum of two options, and the results

Table 5 Reasons for not participating in independent travel in Slovakia

The answer	Number of answers	Share in total number in %
An uninteresting offer for me	38	9.8
High prices for similar services compared to abroad	98	25.2
Security risks	28	7.2
I do not have time	36	9.3
I am satisfied, I travel as much as I wish	84	21.6
The time and money that I can afford to use for the holiday, I use for the holiday with family or partner	99	25.4
Other	6	1.5

Source Own elaboration, 2019

are shown in table (Table 5). 25.4% of all marked answers represent the fact that many respondents with a certain free time fund and financial resources use them as a priority for holidays spent with family or partner. Independent travel is thus in the background. The similar result (25.2% of responses) relates to the perception of the price level in Slovakia, which was perceived for the same or similar services by the respondents as higher compared to abroad. 21.6% of responses were from women who are satisfied with how much they travel independently in the country. Less times, respondents chose options related to the uninterestingness of the Slovak offer (9.8%), lack of time (9.3%), security risks of solo travel (discomfort in public transport, on the streets, etc.—7.2%) and other answers (1.5%).

In the next question, the solo female travelers had to rank the aspects of the offer for this segment in Slovakia from the most important to the least important according to them. We have identified seven basic aspects. We found out that the complex product and all services pre-arranged (accommodation, meals, transportation, program) was most often placed in the first place by the respondents. It means that the work of travel agencies is still important today, as well as the creation of packages of services, for example, in accommodation facilities. The second most relevant aspect was most often considered to be the possibility of electronic and Internet participation in the creation of the service, and the third was the advantageous price. The fourth most important aspect for the respondents was most often identified as the aspect related to the possibility of being addressed—to hear about the product through engaging advertising or another tool of marketing communication. Attention of the service provider to the individual relationship with the customer through adapting the offer to the individual requirements so the customer can feel exceptional is on the fifth place. Respondents mostly identified the existence of a brand as the least important factor of the offer, to which they could automatically return thanks to the proven quality and the exclusive focus on female clients (existence of a women's hotel or travel agency).

In addition to the questions addressed to women, who are already solo travelers, we focused in the 12th and 13th questions of the questionnaire survey on the potential demand of women who are not yet individually traveling. We asked these respondents to rank six aspects, from the one that would motivate them the most to travel independently, to the least motivating. The survey has shown that a larger and more interesting range of products and services for single women would motivate many of respondents the most. The second biggest motivator was most often considered to be lower prices of products and services aimed at female clients, as the third biggest motivator most respondents described the reduction of safety risks associated with, for example, discomfort in public transport, on the streets, etc. This was followed by aspects related to more free time, funds set aside for holidays and, finally, the requirement for more time for the possibility of self-regeneration or arranging a women's stay, business trip or trip provided (paid for) by the employer. Based on these results, we believe that the creation of interesting products at adequate prices would create such an offer for women traveling in Slovakia, which would convince women who usually use their time and funds only for other types of participation in tourism. The challenge remains to reduce the perceived discomfort or fear that makes women prefer to travel with a male or at least a larger group of people (family, friends).

At the same time, we wanted to get information from women who do not yet travel alone, about the type of tourism that would be the most attractive to them if they decided to travel individually. The 54 answers show that almost half (44.4%) of the respondents would prefer a certain form of health tourism (wellness). The slightly lower response rate (24%) speaks about the attractiveness of sports and adventure activities. Cultural tourism would be primarily chosen by 16.6% of respondents, business by 13%, and one respondent (1.9%) gave the answer "recreation".

5   Conclusion

Through the results of a questionnaire survey, we responded to all stated research questions. We found out that our female respondents tend to participate independently in tourism, usually several times a year abroad and in Slovakia. However, most women prefer to visit destinations abroad. This is attractive for them, especially in terms of a unique primary offer, a more diverse range of products and services, while the relevant aspects influencing the decision to participate in tourism include a favorable price offer and reviews mediating the experience of other customers. When respondents travel solo in Slovakia, they most often look for activities related to cultural tourism. However, there is also considerable interest in sports, adventure and health tourism, because as we have found out, women like to discover, recognize and look for forms of escape from the stereotype. The existence of a complex product, electronic and Internet participation in the creation of the service and the price are the three most important aspects of the tourism offer for respondents.

This knowledge should be applied by analyzing the products and conditions that women currently have. In present time in Slovakia, the accommodation packages of accommodation facilities that are aimed at female clients are mostly focused on women's wellness and beauty rides. The advantage of these products is their complex nature and the frequent possibility of creating online reservations, which is in line with the demand of our respondents. We consider the weak diversity of these packages to be a shortcoming in terms of the dominant focus on providing beauty and SPA procedures. Another negativity for the overall image of Slovakia as a destination for solo female travelers may be the primary focus of providers of accommodation packages on the offer of accommodation in double rooms. For a single woman traveling alone, this can represent undesirable financial costs associated with occupying a room by only one person or choosing a single room at mostly disproportionately high prices.

Solo female travelers from our research travel mostly all year round or as a matter of priority in the summer, so the provision of occasional packages is not sufficient. However, the offer is often aimed at more solvent clients and at least two customers traveling together. The challenge thus remains the attractiveness and adaptation of product packages for lower income groups of women as well as for women traveling without any company.

In Central Europe, we find many creative modifications of SPA and wellness packages as well as packages with cultural and sports elements. There are interesting activities of some Czech tour operators, who significantly specialize their products to satisfy the demand of female customers. Although we have found some interesting products for solo female travelers in the offer of travel agencies in Slovakia, these products do not support domestic tourism but, on the contrary, the departure of women abroad. In the analysis of service providers abroad, we also encountered the application of customization, where the female customer is allowed to partially adapt the service provided to her own ideas and wishes.

There are also many organized events abroad aimed on the women. There are also already some in Slovakia, but potential still exists. The challenge is also to offer the desired comfort and safety in transportation for solo female travelers in Slovakia, and the positive example is in rail transport in the neighboring Czech Republic, Austria or Germany. Especially for younger respondents, who have limited financial resources, but still want to travel, it can be attractive, safe and at the same time economically advantageous to stay in women's hostels. There is no such type of accommodation in Slovakia, so we could learn from the examples from other countries.

Acknowledgements This research was funded by the Scientific Grant Agency of the Ministry of Education, Science, Research and Sport of Slovak Republic VEGA, grant number 1/0368/20 "Sharing economy as an opportunity for sustainable and competitive development of tourist destinations in Slovakia".

References

Alejziak, W. (2007). Megatrends and challenges in the development of national and international tourism policy. *Economic Review of Tourism, 40*(1), 3–20.

Booking.com. (2014). Holidays with me, myself and I give women a self-esteem boost. Retrieved from https://news.booking.com/holidays-with-me-myself-and-i-give-women-a-self-esteem-boost/. Accessed May 15, 2020, at 10:00.

Borovsky, J., Smolkova, E., & Ninajova, I. (2008). *Tourism: Trends and perspectives*. Bratislava: IURA Edition.

Brownell, J. (2011). Creating value for women business travelers: Focusing on emotional outcomes. Retrieved from https://scholarship.sha.cornell.edu/chrpubs/10/?utm_source=scholarship.sha.cor nell.edu%2Fchrpubs%2F10&utm_medium=PDF&utm_campaign=PDFCoverPages. Accessed May 15, 2020, at 10:30.

Cargill, G. (2009). What is solo travel? Retrieved from https://solofriendly.com/what-is-solo-tra vel/. Accessed May 15, 2020, at 11:30.

Custódio Santos, M. et al. (2016). Tourism services: Facing the challenge of new tourist profiles. *Worldwide Hospitality and Tourism Themes, 8*(6), 654–669. https://doi.org/10.1108/WHATT-09-2016-0048

Fiala, V. (2017). You can experience unforgettable wellness stays in Slovakia too. Retrieved from https://www.pisem.sk/2017/09/04/aj-na-slovensku-mozete-zazit-nezabudnutelne-wellness-pob yty/. Accessed May 16, 2020, at 12:00.

Franzi, Ch. O., & DeCarlo, D. (2018). Two disruptive trends to watch in the tourism industry. Retrieved from https://www.camoinassociates.com/two-disruptive-trends-watch-tou rism-industry. Accessed May 16, 2020, at 11:00.

Gucik, M., et al. (2010). *Tourism management*. Banska Bystrica: Slovak-Swiss Tourism.

Jafari, J. et al. (2016). *Encyclopedia of tourism* (p. 1168). Springer International Publishing. ISBN: 978-3-319-01383-1.

Jakubikova, D. (2006). Marketing management of tourism destination. Retrieved from https://www.cestovni-ruch.cz/skolstvi/destinace.php. Accessed May 17, 2020, at 10:00.

Junek, O., Binney, W., & Winn, S. (2006). All-female travel: What do women really want? *Tourism: An International Interdisciplinary Journal, 54*(1), 53–62.

Kow, N. (2017). Travel trends that will drive the tourism industry in 2018 and 2019. Retrieved from https://www.trekksoft.com/en/blog/9-travel-trends-that-will-drive-the-tourism-industry-in-2019. Accessed May 17, 2020, at 10:15.

Mariani, M. et al. (2014). Managing change in tourism destinations: Key issues and current trends. *Journal of Destination Marketing & Management, 2*(4), 269–272. ISSN 2212-571X.

Marzuki, A., Chin, T. L., & Razak A. A. (2012). What women want: Hotel characteristics preferences of women travellers. Retrieved from https://www.intechopen.com/books/strategies-for-tourism-industry-micro-and-macro-perspectives/what-women-want-hotel-characteristics-preferences-of-women-travellers. Accessed May 16, 2020, at 14:00.

Masip, J. D. (2006). Tourism product development: a way to create value, the case of La Vall de Lord. Retrieved from https://www.esade.edu/cedit2006/pdfs2006/papers/tourism_product_dev elopment_dds__esade_3r_may_2006.pdf. Accessed May 17, 2020, at 14:55.

Ministry of Transport and Construction of Slovak Republic. (2019). Actualization of marketing strategy of tourism section for years 2019–2020. Retrieved from https://www.mindop.sk/minist erstvo-1/cestovny-ruch-7/legislativa-a-koncepcne-dokumenty/koncepcne-dokumenty/aktualiza cia-marketingovej-strategie-na-roky-2019-2020-19258. Accessed May 14, 2020, at 10:00.

Motusova, T. (2019). *Solo traveling women as a new segment in tourism in Slovakia and abroad*. Banska Bystrica: Faculty of Economics, MBU.

Solo Women Traveling. (2019). Retrieved from https://www.women-traveling.com/. Accessed May 19, 2020, at 14:25.

Truong, T. L. H., Lenglet, F., & Mothe, C. (2018). Destination distinctiveness: Concept, measurement, and impact on tourist satisfaction. *Journal of Destination Marketing and Management, 2018*(8), 214–231. https://doi.org/10.1016/j.jdmm.2017.04.004.

Wilson, E., & Little, D. E. (2005). A 'relative escape'? The impact of constraints on women who travel solo. *Tourism Review International, 2005*(9), 155–175. https://doi.org/10.3727/154427205 774791672.

Strategic Hotel Management in the "Hostile" International Environment

Ioannis Rossidis, Dimitrios Belias, and Labros Vasiliadis

1 Introduction

An important aspect of management today is how the organizations will be able to tackle with the various challenges and issues which can be found on their external environment. This means that organizations must be able to plan and develop a strategy which will allow them to respond on those challenges. This is very important if we consider that the environment where organizations today operate tend to be even more fragile and uncertain. Of course, strategy supposed to be the ability to respond in an effective way upon those uncertainties, but what we see with emerging crises, such as the case of the COVID-19, some crises are not easy to be predicted and even worst the organizations have to face a dead-end. Indeed, for organizations today, drawing up the right strategy can be the difference between the survival and failure of an organization. An important element is the fact that for many years the strategy has been the subject of analysis by large organizations. But now the strategy is used in any kind of organization, whether it is a multinational company or a cultural organization (Thompson et al., 2012). In any case, the value of the strategy is that it allows an organization to combine its resources to produce competitive advantages that will allow it to perform better than the competition, and therefore to have a competitive advantage (Ansoff et al., 2018).

The concept of strategic management is used on almost all forms of organizational operations, including the tourist industry. Indeed, tourism is a fragile industry which is

I. Rossidis (✉)
University of the Peloponnese, 22100 Tripolis Campus, Greece
e-mail: i.rossidis@uop.gr

D. Belias
Department of Business Administration, University of Thessaly, 41500 Larissa, Geopolis, Greece

L. Vasiliadis
General Department, National and Kapodistrian University of Athens, Athens, Greece

© The Author(s), under exclusive license to Springer Nature Switzerland AG 2021   325
V. Katsoni and C. van Zyl (eds.), *Culture and Tourism in a Smart, Globalized,
and Sustainable World*, Springer Proceedings in Business and Economics,
https://doi.org/10.1007/978-3-030-72469-6_21

often vulnerable on changes which happen on the external environment. For example, a financial crisis or an serious political crisis (including terrorism) is affecting the operation of tourist businesses; especially of Hotels (El-Said & Elmakkawy, 2017). Such an example is the case of COVID-19 where tourism in a global scale has stopped its operations, while even if a destination is open for tourists, the hotels must be able to tackle serious issues such as the drop on demand and what to do if one of their guests is found positive on COVID-19. This is just an example of how the hotels are in a very fragile situation and they have to be able to adjust their strategies so to say competitive (González-Rodríguez et al., 2018).

The subject of this paper is to examine the perspectives of using effective strategic management in the hotel industry in order to overcome the extended difficulties of the hostile international environment. The current paper has a significant value due of the fact that it is important to understand how hotels can react in a very hostile international environment and to emerge with some strategies which will help them to get out of any crisis that they have to face. Having in mind the latest developments with the case of COVID-19, it is important to note that we need to better understand how hotels shall draw their strategy in such a hostile environment. This is a research which is a literature review. The authors have made a wide research on online databases such as EBSCO and SCOPUS by using related keywords. Special attention was given so to include the latest updates.

2 Literature Review

2.1 Definition of Corporate Strategy

Businesses focus on three levels of strategy. The first level of strategy is called corporate or business strategy and is related to the corporate vision and mission of the company and the achievement of synergies so that with the available resources of the business unit, actions be carried out that will create value for the company itself. In this way, the company can gain the opportunity with its corporate vision and mission to focus on its desired plan for the future, focusing on the products and services it wants to promote and bringing its image to the attention of those interested (Ansoff et al., 2018).

The corporate vision and the mission of the company are key elements for the corporate strategy and based on them the strategy called business unit strategy is applied in the second level and is related to the competition of the company, the achievement of competitive advantage and the exploitation of business opportunities for the development of new products and services and the allocation of resources within the business unit.

At the third level, one can find the operational strategy which is related to the way in which the unit or organization effectively implements the strategies selected at the previous two levels so that in combination with the resources and procedures followed to be able to achieve its business vision (Ansoff et al., 2018).

The business strategy or corporate strategy concerns the entire business unit and aims to create efficient synergies that will benefit the business. The corporate strategy should aim at the optimal coordination of all actions so that the business unit achieves the highest possible performance. The corporate strategy is related to the effort made by the business unit to expand the resources it uses and its fundamental capabilities in order to achieve a competitive advantage that will make it unique compared to its competitors. The corporate strategy therefore defines the broader field in which the company operates and is determined by the management itself which decides where it is appropriate to focus more so that the most efficient way of allocating resources creates the aforementioned synergies within the business unit.

The corporate strategy is a delicate handling process that involves the management of the business unit and includes the initiatives such as the supply of raw materials to hotels as well as the product and service line provided to the customer in order to lead the business unit to be able to make critical decisions to achieve organizational goals. That is why the corporate strategy should be taken into account and in line with the individual strategies followed by the business unit so that all strategies as a whole benefit its corporate strategy and can support it at the business unit level (Diffley et al., 2018).

Thus if, for example, a hotel business is having trouble providing its individual services, this could jeopardize its broader business strategy and therefore the malfunction should be taken into account and the management should coordinate the individual departments so that it is not driven in a business strategic crisis (González-Rodríguez et al., 2018).

2.2 Generic Strategies

As mentioned above, there are three levels of strategy formation, the business-corporate level, the level of the business unit strategy with the competitive strategies to achieve competitive advantage and the level of the individual functions of the business unit. In order to analyze the competitive strategies, the concept of competitive advantage should be presented where the company competes with the other companies in the context of activities provided in the same market and can supply the same consumers so that he eventually chooses on of the two companies. This makes the business able to achieve greater efficiency, something that will lead it to a competitive advantage (Köseoglu et al., 2016).

Based on Porter's generic strategies, there are two main types of competitive advantage with which the hotel can increase its profits, either by reducing its operating costs or by increasing its total revenue. Cost leadership, on the one hand, is the ability to produce and supply products at the lowest possible cost in the market, and diversification is the product's supply in such a way that the hotel provides unique features that consumers are willing to pay more for in order to acquire them.

Thus, the hotel should take into account the specific two categories of competitive advantage and make its final decisions based on the dominant choices regarding the strategy to be followed, whether it is a strategic cost leadership strategy, a differentiation strategy or a focus strategy with differentiation or with cost leadership (El-Said & ElMakkawy, 2017).

All three options when adopted by a company have the effect of bringing the business closer to the other companies in a different way, especially given the organizational structure and the skills it is called upon to develop. Thus, in order to formulate the appropriate strategy, the business unit should start its strategy by answering the questions as to who are the consumers to whom it is addressed, i.e., its scope and what are the needs of consumers, i.e., what are the unique capabilities that it must develop.

In order for the company to develop a competitive advantage, this can be done through the development of superior quality or the improvement of efficiency and innovation as well as the response to consumer needs. In order for the company to succeed in the above-mentioned areas, it must make decisions about the skills it intends to develop but also about the way in which it will organize and correlate the specific skills. The ability to develop superior quality as well as innovation and meeting the needs of consumers are realized through the adoption of the differentiation strategy. On the other hand, the development of efficiency also leads to a competitive advantage through the cost leadership strategy (Köseoglu et al., 2016).

Initially for the cost leadership strategy, it is a systematic effort on the part of companies to maintain the cost of producing and selling products and services at a lower price than the competitors in order to achieve a competitive advantage. This is a strategy that is suitable for those companies that provide services in a market seeking to reduce costs while due to the large size of the target market the availability of low-cost products can create excess capacity and automated production processes (Köseoglu et al., 2018).

However, it is important to note that a cost leadership process cannot ignore differentiation as if the company's product is considered equal to the competitors its price will have to be much lower to reveal its cost advantage.

As for the differentiation strategy, it applies to those companies that aim to gain a competitive advantage through the uniqueness of the products and services they provide to the customer. The specific advantage created by the production and distribution of products and services, is perceived by consumers as something qualitative and at the same time unique. The differentiation strategy is adopted by companies that price their services higher than the market price due to the difference in quality compared to competitors. Of course, the price is much higher than the cost of the cost leader examined in the previous paragraph.

Businesses that implement a differentiation strategy have high prices on their products and services, but consumers are willing to pay for the difference in price, believing that the product they are buying is also a symbol of prestige. Therefore, companies that implement this strategy gain a competitive advantage by offering unique products that have unparalleled quality in relation to the quality of their competitors and at the same time fall into innovative elements that meet higher consumer needs (Moutinho & Vargas-Sanchez, 2018).

In the case of hotels where the variation focuses on the quality of customer service, these services should be of a high standard even after the customer service is provided. Thus, the hotel should achieve a difference in its services and even in its accompanying features. This means, for example, that in addition to the main services provided to the customer, there should be high-level cleaning services as well as customer service support services by telephone.

At the same time, the hotel must understand the importance of diversifying the services it provides and be able to apply it to all dimensions of the services provided to the customer so that it cannot be copied by competitors. However, emphasis should be placed on the final price of the hotel product provided to the customer, but there should be no reduction in costs to the detriment of quality but strictly only to serve the customer. Thus, differentiation should lead the hotel to great functionality through improved performance and innovation with the addition of hotel features that the customer will choose when he closes the service package he wants (Lasserre, 2017).

In the focus strategy, however, the company can choose elements from the differentiation strategy but also from the cost leadership strategy, focusing on customer satisfaction on the one hand and on market satisfaction on the other, which is their main difference.

In order for the company to achieve the implementation of the focus strategy, it must use the data it has gathered for each target market and have some peculiarities that will make the consumer choose it over the competitors. Therefore, the focus strategy can be on the one hand, a low-cost strategy or a diversification strategy that will address a specific market segment that will meet its needs by choosing the highest quality.

2.3 Schools of Thought on Strategic Management

Strategic management has been in the field of culture for a few years now. The strategy is a point of reference for the whole organization, from its base to the highest levels of its leadership. In any case, the executives of an organization should be aware of the resources they have and have in their hands in combination with the changes in the external environment, so that they can design a plan that will give answers to the challenges and opportunities it creates the environment. The ultimate goal of the strategy is to create a competitive advantage. This means that an organization will be able to organize a strategy that will allow it to outperform its competitors. This performance can be achieved in three ways. The first is through the diversification strategy, where an organization produces a competitive advantage by differing from the whole competition, or through the cost leadership strategy where the organization becomes more competitive by having the lowest cost while a third strategy is that of focus, where an organization focuses—through differentiation or cost leadership

(Gassmann et al., 2016). In each case, the organization must choose which strategy to approach with competitive advantages that can ensure its viability.

As far as contemporary organizations are concerned, McKiernan (2017) refers to four models that an organization who operates in a very competitive environment can adopt. The first model is the classic school. In this case, the strategy is designed by the top management of the organization. The main goal is the financial viability of the organization through profit maximization. Any design is detailed and rational. Hence, this is a traditional model where the organizations are trying to optimize their operations and at the same time to maximize their potential to generate profit. This actually the classic school of thought, which is that an organization aims into producing profits.

The second model is the evolutionary thinking. In this case, this model refers to the fact that contemporary companies are not in a stable environment but on the contrary the environment is constantly changing. This means that an organization cannot have a stable strategy, as in the classical school, but a strategy that is constantly changing so that it can adapt to the changes in the environment (McKiernan, 2017). Indeed, the previous model had a key weakness which was that it regarded the environment as being stable. However, it must be understood that today's organizations have to operate in a very difficult and fragile environment, therefore strategy—to a large extent—is the response of organizations towards changes in a very hostile environment. This school of thought comes closer to the case of contemporary hotel management, since the hotels are operating on most of the cases in an international hostile environment (Yoshikuni & Albertin, 2018).

The third model is the procedural approach. In this case, the strategy is not the result of planning by the top executives of the organization. Instead, it is a process involving executives and employees from all levels of the organization with the goal of having a strategy that is acceptable to all audiences, including employees (Gassmann et al., 2016). This is another approach where the decisions are made not form the top executives and managers. Instead, the decision making is a process of generating ideas which starts from the bottom and goes up to the upper layers of the hierarchy. According to Collis (2016) this is a strategy which is often used from startups and high-tech organizations. This is due of the fact that those organizations have an important pool of experts and employees who have high skills and knowledge; therefore, it is important for the top management to consult them before taking any decision. On the contrary, the traditional models claimed that the upper layers must take the decisions. However, things have changed. Today's employees are flexible and they have the ability to generate new ideas, innovations and changes which lead into the sustainable competitive advantage that their companies are looking for.

Finally, the systemic approach perceives the organization as an open system. This means that it is in constant interaction with the elements of the environment such as the social, political, cultural and economic environment. This means that an organization follows in each case a different strategic approach from that of its competitors due to the peculiarities that develop in its environment (McKiernan, 2017). The open systems approach fits with the operation of a hotel in a hostile environment. In this case, the members of the organization are able to understand the changes made on the environment and to adjust the organization's settings upon those changes. Such a case can be on the accommodation industry. The hotels are part of the wider environment not only of the tourist industry, but also of the destination. Their strategy to a large extent depends on the ability of the destination to become more competitive while the strategy of the hotel depends on what happens on the destination. Hence, a hotel is not a "stand-alone" organization but rather it is part of the wider environment where the hotel interacts with its environment (Okumus et al., 2019).

In recent years, organizations have been moving towards a systemic and procedural approach. The procedural approach allows the creation of a flexible strategy which is accepted by the whole organization while the systemic one allows the organization to be integrated in the changes that exist in the environment and to take advantage of any opportunities presented but also to be able to successfully respond to resulting changes.

2.4 Strategic Management in Hotels

Strategic management is often involved with strategy even though they are two separate concepts. However, the two concepts have a similar name but a different purpose. Strategy on the one hand is a comprehensive plan to achieve organizational goals while strategic management is a continuous management process that aims to formulate the appropriate strategy to approach the right business opportunities that will lead the hotel to further profitability (Rothaermel, 2016).

It is a fact that with the right use of strategic management, some hotels can become successful compared to their competitors, something that makes them automatically lucky. This may be due to the appropriate mix of products and services they provide but in other cases this is not enough. However, all hotels that use the appropriate strategic management are interested in the stakeholders related to the hotels. These bodies are groups of people who influence and are influenced by the hotel and have an excellent relationship with it. Thus, a successful implementation of the strategic management process allows the company to analyze and learn from its external environment by setting long-term and short-term goals and formulating its strategy in such a way as to satisfy key stakeholders (Lasserre, 2017).

The strategic management process helps companies recognize what they want to achieve and how to achieve results that are valuable. The importance of this challenge is greater today than ever before. The globalized economy, an economy in which products or services flow freely between nations, is constantly pushing businesses to become more competitive. Thus, companies with increased competitiveness must offer products or services that have greater value to customers, thereby gaining competitive advantage that is reflected in profits above the industry average (Ansoff et al., 2018).

3 Methodology

The methodology used in the present paper is the critical review of the literature.

The sources of relevant literature investigation derived from popular online bibliographic databases, such as Science Direct, Emerald, EBSCO host and scientific search engines such as Google Scholar and Scirus. General search engines such as Google have also been examined.

The types of bibliographic sources included in the research are articles published on scientific journals, books, conference proceedings, company papers and studies, white papers, online sites and online journals. The selection criteria of these literature sources were based on the relevance to the topic of the paper, and this research is not exhaustive.

4 Discussion—Strategic Hotel Management in the "Hostile" International Environment

It is important to understand the environment where hotels operate today. The hotel industry is a very sensitive industry on changes which may occur on the macro-environment. An example is the 9/11 where suddenly the airliners stopped their flights and the whole industry went off for a couple of weeks. The outcome was that a significant number of tourist businesses, mostly associated with flights, went bankrupt or they had to merge or downsize. On this case, we have sudden changes on the environment that a tourist company will have to accept and take its measures (Valente & dos Anjos, 2017).

From the above, it is understood that a crucial issue on the strategic management of hotels is how to deal with a crisis and to adjust. Whether if this is a change on consumer behavior or a change in the tax system, on all cases a hotel must convince its stakeholders that it is able to cope with those changes and to continue serving its customers. Of course, implement such strategies and dealing with a crisis is not an easy task. According to Köseoglu et al. (2018) many hotels have to face several barriers such as their limited resources, the resistance from the employees on changes but also the fact that on many cases a hostile environment means that the fate of the hotel is not on the hands of its decision makers but on the hands of someone else. At this case we can refer on the example of COVID-19 where the countries had to shut down their economies. This is a case where the measures included not only a halt on international travel but also the closure of hotels, while often hotels are on the spotlight of mass media as being the place where an outbreak may take place. For example, Hoefer et al. (2020) have examined the case of COVID-19 outbreak in Spain where the first outbreak occurred to a hotel in Tenerife which had tourists from Italy. The hotel went into a quarantine for 14 days and this was a decision taken not from the management of the hotel but from the government. Therefore, this is justifying the fact that on such cases the hotel is losing its ability to control its management and the decisions taken. Instead, it has to follow the guidelines and orders given from the government. The same happened with other hotels in the area of Tenerife which had to go into a quarantine, while at the same time many bookings were cancelled or postponed.

Overall, in tourism shocks is not something rare. Instead, it is something which happens frequently and it must be taken as part of the strategy. On many cases, destinations and hotels are affected from extreme events such as terrorist attacks, bushfires or cyclones. This means that the hotel management must be ready to take decisions and adjust its strategy into such situations (Ritchie & Jiang, 2019). One of those shocks is the COVID-19 according to Dolnicar and Zare (2020) has affected the strategy of the hotels in three different ways. The first one is the overall economic shock where according to UNWTO (2020) the tourist arrivals are expected to drop by 20 to 30%, though it is too early to claim if this is argument is valid or not since it seems that the drop on arrivals can be even worst. The second impact is the reduction on the global economic growth. Businesses have closed down or they have shut down for few weeks or months. This has resulted not only in a deep recession but also in an increase in unemployment and uncertainty in the markets. This surely affects the behaviour of tourists who are discouraged from travelling—especially abroad—not only because of the restriction and the fear that they may get sick but also because they may not have the financial ability to travel. Finally, the third shock is the structural changes made on the industry and changes on how the hotels operate. An example are the hygiene protocols for COVID-19 that result in more bureaucracy and higher costs (Dolnicar & Zare, 2020).

Hence, at this point, the management of the hotel will have to decide on which strategic decisions to make. From the findings of the above paragraph, it is already mentioned the fact that in such emergencies the hotel is losing its autonomy on the decision-making process. The lockdown and the use of special measures such as social distancing, hygiene protocols and the use only of the 50% of the capacity of the hotel are decisions which are imposed from the government leading into a very hostile environment, while on other hand the intermediaries—mostly the tour operators—are out of business or they are pressing for re-negotiation of the contracts. Surely, this is a very hostile environment for a hotel. At this case, the strategic approach of the hotel can be to downsize, which is to reduce the use of resources including its human resources or to try to re-focus its strategy aiming into new segments or trying to create a turnover culture (Abo-Murad & Abdullah, 2019).

The research made from Laws and Priedeaux (2017) and Giannopoulou (2019) refers on several ways that a hotel can choose so to maintain its competitiveness in a time of crisis. Those include the following strategies:

- **Downsizing**: This is about reducing the resources used. On many cases, this may include the staff layoff in order to reduce the salary costs and overall the operation costs
- **Change on markets**: Since the targeted markets are in a crisis, the hotel can turn its focus into other segments. On some cases, a crisis can be a good reason for

re-positioning the hotel upon the market. Also, the hotel can focus also on special markets such as conferences, events, etc.

- **Price offers**: This is a situation where the hotel can use an effective pricing strategy where it will try to maintain the interest of its customers. For example, it can offer an extra day for free on loyal customers or to try to make reductions so to increase demand.

- **Expansion of tourist period**. This concerns mostly destinations such as in Greece, where there is the limitation of the tourist period. At this case, the hotel can seek ways to expand the tourist period.

- **To shut down the operations for a short period of time**. During the crisis of the COVID-19, it was noticed that several hotels have close down, while from what it is noticed from the market several large hotels will remain closed during 2020 so to avoid either the high costs of taking measures such as social distancing, employing extra personnel for the implementation of the safety protocols, etc. or because the management regards 2020 as a lost season.

The current crisis of COVID-19 has created a new reality. Since it is an ongoing crisis and there are not so many researches, it is not easy to conclude on which is the most effective strategy to deal with such situation. Bagnera et al. (2020) have remarked that most of the large hotel chains closed their hotels and they are waiting to see what will happen next, often without having a certain strategy on their mind. It is just that the management of the hotel is in a "stuck in the middle situation" where they are not sure what it may happen on the next day. The current strategy used is to shot down the hotels and wait for when the market will open again or to operate the hotels on the minimum resources, including available rooms.

5 Conclusion

The hotel industry is very sensitive on the crisis, and it has high uncertainty. On this case, there is a number of strategies that a hotel can adopt so to cope with the uncertainties and the crises made from a hostile international environment. What is more important on this paper is that there is a field where future research will have a significant impact. This is how the hotels can respond on the COVID-19 crisis. This is an ongoing phenomenon, and we still are not sure on which strategy can be the most effective. For this reason, there is a recommendation for a future research which will be on how hotels have dealt with the COVID-19 crisis. This can include a qualitative research along with a case study approach, which will produce useful results and insights.

References

Abo-Murad, M., & Abdullah, A. K. (2019). Turnover culture and crisis management: Insights from Malaysian hotel industry. *Academy of Strategic Management Journal.* Retrieved from https://www.abacademies.org/articles/turnover-culture-and-crisis-management-insights-from-malaysian-hotel-industry-8031.html. Accessed June 10, 2020.

Ansoff, H. I., Kipley, D., Lewis, A. O., Helm-Stevens, R., & Ansoff, R. (2018). *Implanting strategic management.* NJ: Springer.

Bagnera, S. M., Stewart, E., & Edition, S. (2020). Navigating hotel operations in times of COVID-19. *Boston Hospitality Review.* Retrieved from https://www.iacconline.org/docs/Navigating-Hotel-Operations-Corona.pdf. Accessed June 10, 2020.

Collis, D. (2016). Lean strategy: Start-ups need both agility and direction. *Harvard Business Review, 94*(3), 62–69.

Diffley, S., McCole, P., & Carvajal-Trujillo, E. (2018). Examining social customer relationship management among Irish hotels. *International Journal of Contemporary Hospitality Management.*

Dolnicar, S., & Zare, S. (2020). COVID19 and Airbnb–Disrupting the disruptor. *Annals of Tourism Research.* Retrieved from https://www.ncbi.nlm.nih.gov/pmc/articles/PMC7242963/. Accessed June 10, 2020.

El-Said, O., & ElMakkawy, M. (2017). Evaluating the implementation of strategic management practices in Egyptian five-star hotels: An exploratory study. *International Journal of Heritage, Tourism and Hospitality, 11*(2), 35–56.

Gassmann, O., Frankenberger, K., & Sauer, R. (2016). Leading business model research: the seven schools of thought. In *Exploring the field of business model innovation* (pp. 7–46). Palgrave Macmillan.

Giannopoulou, F. (2019). Crisis management within the hotel industry: The case of luxury hotels and resorts in popular Greek tourism destinations. In *Strategic innovative marketing and tourism* (pp. 1075–1083). Springer.

González-Rodríguez, M. R., Jiménez-Caballero, J. L., Martín-Samper, R. C., Köseoglu, M. A., & Okumus, F. (2018). Revisiting the link between business strategy and performance: Evidence from hotels. *International Journal of Hospitality Management, 72,* 21–31.

Hoefer, A., Pampaka, D., Wagner, E. R., Herrera, A. A., Alonso, E. G. R., López-Perea, N., & Gallo, D. N. (2020). Management of a COVID-19 outbreak in a hotel in Tenerife, Spain. *International Journal of Infectious Diseases., 96,* 384–386.

Köseoglu, M. A., Ross, G., & Okumus, F. (2016). Competitive intelligence practices in hotels. *International Journal of Hospitality Management, 53,* 161–172.

Köseoglu, M. A., Yazici, S., & Okumus, F. (2018). Barriers to the implementation of strategic decisions: Evidence from hotels in a developing country. *Journal of Hospitality Marketing and Management, 27*(5), 514–543.

Lasserre, P. (2017). *Global strategic management.* Macmillan International Higher Education.

Laws, E., & Prideaux, B. R. (2017). A study of crisis management strategies of hotel managers in the Washington, DC metro area. In *Tourism crises* (pp. 103–118). Routledge.

McKiernan, P. (2017). *Historical evolution of strategic management, volumes I and II.* London: Routledge.

Moutinho, L., & Vargas-Sanchez, A. (Eds.). (2018). *Strategic management in tourism.* Cabi: CABI Tourism Texts.

Okumus, F., Altinay, L., Chathoth, P., & Koseoglu, M. A. (2019). *Strategic management for hospitality and tourism.* London: Routledge.

Ritchie, B. W., & Jiang, Y. (2019). A review of research on tourism risk, crisis and disaster management: Launching the annals of tourism research curated collection on tourism risk, crisis and disaster management. *Annals of Tourism Research, 79,* 102812.

Rothaermel, F. T. (2016). *Strategic management: concepts* (Vol. 2). McGraw-Hill Education.

Thompson, A., Peteraf, M., Gamble, J. & Strickland, A. (2012). *Crafting and executing strategy– Concepts and cases, International Edition*. McGraw Hill

UNWTO. (2020). UNWTO International tourist arrivals could fall by 20–30% in 2020. https://www.unwto.org/news/international-tourism-arrivals-could-fall-in-2020. Accessed June 5, 2020.

Valente, S. B., & dos Anjos, B. G. F. (2017). Strategic management of event spaces in hotels: a multiple-case study in the city of São Paulo. *TURyDES: Revista Turismo y Desarrollo Local, 10*(23), 1–9.

Yoshikuni, A. C., & Albertin, A. L. (2018). The Effects of strategic is on firm performance: An empirical study of the three-way interaction investigation of turbulent scenario. *Journal of Public Administration and Governance, 8*(4), 20.

Strategic Human Resource Management in the International Hospitality Industry. An Extensive Literature Review

Ioannis Rossidis, Dimitrios Belias, and Labros Vasiliadis

1 Introduction

The contemporary, highly competitive business environment has rendered the strategic orientation of organizations imperative and thus shifting towards the consolidation of strategic management. Strategic management is a key factor in a company's survival as it provides the guidance needed to accomplish its mission and vision, determining the right conditions for developing competitive advantages by understanding and identifying opportunities, weaknesses, and threats (Papadakis, 2016). The multidimensional value of management strategy refers to the efficient use of resources, the creation of competitive advantages, reduction of uncertainty, creation of a general management discipline, and setting of standards regarding the final results. Strategic management is intended to support organizations on many levels, as it contributes decisively to unity in the direction of administrative action, aiming to achieve specific goals (Kreitner, 2009). In recent years, a differentiated approach has emerged, combining aspects of human resource management strategy and management.

Modern businesses are called upon to deal with a highly competitive environment that is governed by intense change and high uncertainty. These conditions render the adoption of a strategic orientation by incorporating strategic management principles, increasingly paramount. Strategic management theories have recognized the importance of internal functions, organizational structure, and, of course, how to

I. Rossidis (✉)
University of the Peloponnese, Tripolis, Greece
e-mail: i.rossidis@uop.gr

D. Belias
Department of Business Administration, University of Thessaly, Larissa, Geopolis, Greece

L. Vasiliadis
General Department for National and Kapodistrian University of Athens, Athens, Greece

© The Author(s), under exclusive license to Springer Nature Switzerland AG 2021
V. Katsoni and C. van Zyl (eds.), *Culture and Tourism in a Smart, Globalized, and Sustainable World*, Springer Proceedings in Business and Economics,
https://doi.org/10.1007/978-3-030-72469-6_22

use resources as potentially significant sources of competitive advantage. In recent years, the strategic dimension has tended to be transferred to human resource management as one of the most crucial factors in achieving strategic goals. The partnership between management strategy and human resources management can lead to high efficiency and effectiveness of the business as the prospect of employees working to attain strategic goals, is formed. Of particular interest is the value of strategic management in the highly competitive tourism industry and in particular in the hotel industry. The peculiarity of the hotel industry is that the final product produced by the hotels refers to the service experiences, which largely depend on the result of the work of the employees and the general interaction with the customers (Tracey, 2014). Therefore the quality of services provided and customer satisfaction (which results from the services received) and ultimately the efficiency and competitiveness of hotel businesses, depends to a large extent on the performance of human resources (Baum, 2015). Based on this view, it becomes clear that hotel companies have a strong interest in finding the administrative tools that will help push employees to meet their strategic goals. The importance of the human factor in fulfilling the mission and approaching the vision led to the development of human resource management strategy with emphasis on hospitality industry. This approach refers to identifying the right strategies that will stimulate the utilization of the human factor in such a way that will ultimately spur the formation of competitive advantages between hotel companies (Madera et al., 2017). Thus, human resource management strategy in the hospitality industry undertakes the exploration of strategic tools that will bring high levels of efficiency and competitiveness (Buller & McEvoy, 2012). At this point, it should be noted that although strategic human resource management is an indisputable necessity for modern organizations, in the hotel industry (although one of the most competitive sectors—due to its globalized dimension), it has not yet attracted the attention and commitment to its promotion (both in practice and in theory/academia) (Kaufman, 2012). The strategic dimension of human resource management in the hotel industry can be introduced in any of the individual functions of HR management (attraction, selection, training, development, leadership, motivation, communication, maintenance, performance management, etc.) based on improving the company's performance.

2 Literature Review

2.1 Definition of Strategic Human Resource Management

Strategic Human Resource Management (SHRM) has emerged in the 1980s. According to Wright and McMahan (1992, p. 298): SHRM can be defined as *"the design of planned facilities and human resources activities. In order to allow the company to achieve its goals"*. The analysis of this definition brings up four key aspects of its conceptual dimension (Altarawneh & Aldehayyat, 2010).

1. It focuses on the human factor as the key to developing competitive advantage.
2. The individual functions of human resource management are the required administrative tools through which the company can claim competitive advantage.
3. The strategic plan determines how the human resource management will be aligned with the company's strategic goals.
4. Human resources integrated into a defined administrative and organizational framework act in order to achieve strategic goals.

Currently, literature defines SHRM as an ongoing effort to align human resource management policies with its business strategy (Boxall et al., 2007). Through these approaches, human capital is upgraded to the decisive factor in the success of a business. Its overall purpose is to create a clear target for the use of the human factor in order to serve the strategic needs of the company (Hall et al., 2009).

Strategic Human Resources Management focuses on the relationship between a company's HRM systems and a company's performance (Huselid & Becker, 2010). A significant number of studies in recent years highlight the impact of the HRM strategy on corporate performance (Combs et al., 2006; Paauwe, 2009). Adapting the functions of human resource management in the direction defined by strategic planning can help achieve significant competitive advantages. In particular, this adjustment may refer to the following (Huselid & Becker, 2010; Rossidis et al., 2020, Aspridis et al., 2018):

- Attracting and selecting the right staff to help serve specific strategic missions
- Formulation of incentive policies and remuneration systems that link employees' personal benefit to the organization's strategic goals
- Formation of focused communication and other policies in order to clarify the form and importance of the company's strategy, to strengthen organizational commitment, to deal with possible resistance to change, to strengthen the dynamics of teams, etc.
- Laying the foundation of strategically targeted organizational culture with an emphasis on serving the goals of organizations
- Development of training and development strategies in order for employees to cultivate the skills/knowledge/training they need in order to serve the strategic plans of the companies
- Cultivate leading executives who will instill the company's vision in employees and inspire them to better serve the goals of the organization
- Maintaining and supporting employees who are considered important in promoting competitive advantages
- Management of employee performance in order to ensure the required performance that will lead to the fulfillment of strategic objectives
- Utilization of individual and organizational knowledge in order to develop competitive advantages.

2.2 Structure of Strategic Human Resource Management

As mentioned above, the human resources strategy and management partnership can contribute to particularly high levels of efficiency, effectiveness, and competitiveness, since this approach adjusts the way staff is managed to meet strategic goals. The bibliographic analysis identified seven key areas of SHRM configuration: (1) an analysis of the need for human resource management strategy and its adaptation to the administrative system, (2) a shift from commitment to human resource management to human utilization, fulfilling the objectives of the organization, (3) processing the elements that govern the human resources of the organization and the structure of the HR system, (4) expanding the scope of application of SHRM in all operations of the company, (5) activating HR to the direction of the strategic objectives, (6) measurement of SHRM results and (7) evaluation of methodological issues.

Careful research on HRM issues led to the distinction between micro (functional) and macro (strategic) areas (Huselid & Becker, 2010). Until recently, scientific research in human resource management focused on the impact of HRM practices on employees at each level. On the other hand, the modern approach of human resource management strategy examined the impact of HRM practices/methods on human resources at the macroeconomic level in relation to the performance of the organization (Wright & Boswell, 2002). Research into human resource management strategy examines HRM's functions in comparison to the organization's financial and business results (market share, quality indicators, revenue/expenditure, profitability, etc.).

According to Gannon et al. (2015), the analysis of the theory of human resource management strategy can be carried out through the elaboration of three basic approaches.

i.   The first approach investigates the case of best-practice management strategy in which organizations try to create competitive advantages by developing "high-performance" human resource management practices (Pfeffer, 1998). This approach, however, presents the disadvantage that it is particularly difficult to agree on the type and content of high-performance HRM practices. The question, therefore, arises as to which HRM practice can be considered high performance; From business to business, from industry to industry, and from country to country, the definition of high-performance HRM practices can vary considerably (Kaufman, 2014). It is a fact, of course, that through systematic investigation and analysis, practices have been conceived which under certain conditions can be defined as high efficient (since of course they are analyzed for a specific species, industry, country, etc.) (Boxall & Purcell, 2011).

ii. The second approach of human resource management strategy refers to the analysis of strategies that best suit each case given that business positions and strategies are the ones that determine human resource management policies. Under the "best fit" approach, a number of models and theories have been developed linking the formulated strategy to the applied HRM practices, taking on strategic features. In this case, however, there are disadvantages as there is excessive attention to the external environment and in the effort to create competitive advantages, tending to present great similarities with other companies in the industry, practicing the same process of analysis and strategy (Boxall & Purcell, 2011).

iii. The third approach is based on the effort to develop a competitive advantage by leveraging valuable, rare, non-substitute (human) resources (Morris et al., 2006). According to this approach, employees themselves can be the source of competitive advantage (Marchington et al., 2003). Human resources can be a valuable asset, creating added value to organizations. A prerequisite is the existence of the required strategic orientation for their utilization. In this context, this approach is embellished and richly enriched by the collaboration of modern management disciplines such as Knowledge Management, Talent Management, Organizational Learning, etc.

Through the analysis that has been presented, it appears that each of the above approaches has value (but also individual delays) which can significantly contribute to the improvement of business performance. In their practical application, companies usually avoid using such an approach. They usually resort to combination options that have elements from more than one approach (most often from all three approaches).

2.3   Implementation of Strategic Human Resource Management in Hospitality Industry

A current trend in modern business is the transition of human resource management to management with a more strategic dimension, a trend which is gradually observed in the companies in the hospitality industry. Employee development with a view to providing a high level of service (and therefore achieving the objectives of the hosting business) requires culture-oriented training in the expectations of visitors, appropriate recruitment process, organizational commitment. The strategic management of human resources in hosting companies aims to shape employees with a focus on problem-solving, the development of common values, the pursuit of organizational goals, the coexistence with the demands of the company, etc. The modern approaches of human resource management undertake the creation of psychological bonds; the satisfaction of employees' expectations; their fair treatment; the selection of the most suitable for each job; the targeted training and development of employees by providing them with all the necessary knowledge for the successful execution of their duties; the formation of appropriate communication channels in order to convey timely and correct messages and information both from top to bottom and from bottom to top, ultimately looking forward to the formation of dedicated employees who will be committed to fulfilling the strategic goals (Taylor & Finley, 2008). Orientation in strategic management helps to formulate a comprehensive program to maximize hospitality.

According to Graham and Lennon (2002), strategic HRM greatly affects corporate performance. Their research explored the impact of human and social capital on workers' behavior. Their research led to the conclusion that the strategic role of human resource management influences customer behavior through various HR functions, such as incentive and payroll policies, targeted education/training, development programs, and staff selection practices. Respectively, according to the research of Úbeda-García et al. (2013), targeting Spanish hotels, the hotels that applied human resource management strategy practices (focusing on adapting human resource management policies to their strategic targets) presented higher performance results compared to hotels choosing not to adopt similar practices. Karatepe (2013) examined the effect of HRM practices on work performance and customer service at a similar level. His research found that HRM practices (education, empowerment, incentive and pay policies, employee participation) affected work performance and customer service. According to the study by Altarawneh and Aldehayyat (2010) the strategic management of human resources in the hotel industry in Jordan is based on the following aspects: clearly defined HRM strategic plans, strategic role for HR managers, top management commitment and support, line manager support and partnerships, cost-effectiveness evaluation, organizational culture, environmental scanning. The practice of the HRM strategy deals with the Porter theory model in an effort to develop a competitive advantage. In SHRM, competitive advantage will be achieved by pushing employees to develop specific behaviors appropriate to each competing strategy. The relevant literature identified that the model, strategy, and policy of human resource management are fueled and developed by environmental forces. In this context, the corporate competitive strategy should be aligned with human resource management policy.

In the context of human resource management strategy, human resources are aligned with business strategies. According to Kearns (2004), human resources can be divided into two categories, one in which employees are considered resources and one in which employees can be seen as a source of competitive advantage. In the second case, human resources management becomes a strategic partner responsible for getting the maximum value from employees. According to this approach, the human element is a determining factor in the fulfillment of strategic goals.

3 Methodology

The methodology used in the present paper is the thorough literature review, followed by its critical analysis. The sources of literature review derived from high-quality scientific journals, books, conference proceedings, business reports, etc. The authors have used a variety of sources based on their accessibility provided by the affiliated institutions. The selection criteria of these literature sources were based on the relevance to the topic of the paper, the date of the publishing (focusing on the most recent research), and the validity of the bibliographic sources.

4 Discussion—The Added Value of Strategic Human Resources Management

The growing intensity of competition in the hotel industry outlines the need to implement a human resource management strategy. This perspective is based on the fact that the high performance of HRM practices is strongly correlated with corporate performance in the hospitality industry. The HRM practices presented above, refer to human resource management policies designed to pursue and achieve specific strategic objectives. Under these circumstances, human resource management strategy tends to be an integral part of a comprehensive business strategy. The intensification of the use of SHRM practices in the hospitality industry is identified by a number of administrative actions in this direction, such as focusing on the correlation between work objectives/requirements and employee skills, effective management of diverse human resources (with different ages, educational background, professional experience, nationality, etc.), the development of incentive and communication policies in order to push employees to achieve strategic goals, etc. (Combs et al., 2006). According to Harvey et al. (2000) strategic human resource management systems can "become firm strategic resources that expand the range of corporate strategic choices leading to a sustained competitive advantage".

The modern analysis of human resource strategic management is characterized by new ideas and practices that reinforce the emerging theoretical framework. According to Hall et al. (2009) the knowledge formed by the thorough review in this field in recent years focuses on the following:

- Highlight the strong correlation between human resource management and corporate performance.
- Focus on the conditions that must be in place to adopt SHRM and the particular impact of the external environment.
- Emphasis on the development of strategies at the level of human resources.
- Concern for the implementation of strategies at the level of human resource management; the practical use of the human factor for the implementation of strategic objectives.
- Utilization of individual and organizational knowledge to formulate better human resource management strategies aiming at developing competitive advantages.
- Focus on modern issues of human resources through which added value can be developed (e.g. administration of diverse human resources, utilization of individual talents/skills, etc.).

Despite the fact that many studies have been done on SHRM in recent years, there are still several areas that need significant research. Particularly controversial is the issue of measuring and evaluating the results of SHRM. The issue of determining the expected in relation to the results achieved raises the issue. A business can create HR policies that may seem to fit its strategic goals. However, the implementation method may not be able to lead to the desired results, according to their intended purpose. Another issue is that some companies are developing multiple strategies in a large number of business units, making it quite difficult to develop consistent human resource management strategies. Finally, a particularly topical issue (as mentioned above) is the partnership of knowledge management principles with human resource management strategy, focusing on human capital and trying to extract (through organizational learning mechanisms) elements of knowledge, experience, know-how, information, etc. with the aim of better strategic utilization of the human factor (Rossidis & Belias, 2020).

5 Conclusion

This article focuses on the bibliographic analysis of published studies on HRM strategy in the hospitality and tourism industry (Myers, 2013) presenting the way and contemporary practices by which the strategic dimension of human resource management can contribute to the fulfillment of strategic plans and therefore directly or indirectly influence business performance. The literature was examined through content analysis by identifying and examining related research topics. Through this study, the importance of human resource management strategy for the competitiveness of hosting companies was presented, linking modern human resource management practices to the effort to achieve strategic goals. Companies in the hospitality and tourism industry operating in a highly competitive environment are oriented towards the constant search for competitive advantages in order to gain competitive advantages. Focusing on the strategic management of the human factor can help provide better quality services by rendering it easier for businesses to respond to their role, thus increasing their business turnout. The strategic utilization of human resources as developed through the range of theories, creates the prospect of development of the hospitality industry, aiming at achieving higher levels of efficiency and effectiveness.

References

Altarawneh, I., & Aldehayyat, J. (2010). Strategic human resources management (SHRM) in Jordanian hotels. *International Journal of Business and Management, 6*(10), 242–255.

Aspridis, G., Tselios, D., & Rossidis, I. (2018). *Business Communications*. Athens: Kritiki publishing.

Belias, D., Velissariou, E., & Rossidis, I. (2018). The contribution of HRM on the development of effective organizational culture in hotel units—The case of Greek hotels. In *"Exploring smart tourism: The cultural and sustainability synergies" Springer proceedings in business and economics*.

Baum, T. (2015). Human resources in tourism: Still waiting for change? A 2015 reprise". *Tourism Management, 50,* 204–212.

Boxall, P., Purcell, J., & Wright, P. (2007). Human resource management: Scope, analysis and significance. In P. Boxall, J. Purcell, & P. M. Wright (Eds.), *The handbook of human resource management* (pp. 1–16). Oxford: Oxford University Press.

Buller, P., & McEvoy, G. (2012). Strategy, human resource management and performance: Sharpening line of sight. *Human Resource Management Review, 22*(1), 43–56.

Boxall, P., & Purcell, J. (2011). *Strategy and Human Resource Management* (3rd ed.). Basingstoke: Palgrave Macmillan.

Combs, J., Liu, Y., Hall, A., & Ketchen, D. (2006). How much do high-performance work practices matter? A meta-analysis of their effects on organizational performance. *Personnel Psychology, 59*(3), 501–528.

Hall, M., Hall, C., Andrade, L., & Drake, B. (2009). Strategic human resource management; The evolution of the field. *Human Resource Management Review, 19,* 64–85.

Harvey, M., Novicevic, M., & Speier, C. (2000). Strategic global human resource management: The role of inpatriate managers. *Human Resource Management Review., 10*(2), 153–175.

Huselid, M., & Becker, B. (2010). Bridging micro and macro domains: Workforce differentiation and strategic human resource management. *Journal of Management, 37*(2), 421–428.

Gannon, J., Roper, A., & Doherty, L. (2015). Strategic human resource management: Insights from the international hotel industry. *International Journal of Hospitality Management, 47,* 65–75.

Graham, M., & Lennon, J. (2002). The dilemma of operating a strategic approach to human resource management in the Scottish visitor attraction sector. *International Journal of Contemporary Hospitality Management, 14*(5), 213–220.

Karatepe, O. (2013). High-performance work practices and hotel employee performance: The mediation of work engagement. *International Journal of Hospitality Management, 32,* 132–140.

Kaufman, B. (2014). *The Development of Human Resource Management across Nations: Unity and Diversity.* Cheltenham: Edward-Elgar.

Kaufman, B. (2012). Strategic human resource management research in the United States: A failing grade after 30 years? *The Academy of Management Perspectives, 26*(2), 12–36.

Kearns, P. (2004). How strategic are you? The six "killer" questions. *Strategic HR Review, 3*(3), 20–23.

Kreitner, R., & Kinicki, A. (2009). *Organizational behavior.* New York: McGraw Hill.

Madera, J., Dawson, M., Guchait, P., & Belarmino, A. (2017). Strategic human resources management research in hospitality and tourism A review of current literature and suggestions for the future. *International Journal of Contemporary Hospitality Management, 29*(1), 48–67.

Marchington, M., Carroll, M., & Boxall, P. (2003). Labour scarcity, the resource-based view, and the survival of the small firm: A study at the margins of the UK road haulage industry. *Human Resources Management, 13*(4), 5–22.

Morris, S., Snell, S., & Wright, P. (2006). A resource-based view of international human resources: Towards a framework of integrative and creative capabilities. In G. K. Stahl & I. Bjorkman (Eds.), *Handbook of research in international human resource management*, Edward Elgar, Cheltenham, UK.

Myers, M. (2013). *Qualitative research in business and management.* London: Sage.

Papapdakis, V. (2016). *Business strategy.* Athens: Benou publications.

Pfeffer, J. (1998). The human equation: Building profits by putting people first. Harvard Business School Press, Boston, MA; Rossidis, I., Belias, D., & Aspridis, G. (2020). *Change management and leadership.* Athens: Tziola publishing.

Rossidis, I., & Belias, D. (2020). Combining strategic management with knowledge management. Trends and international perspectives. *International Review of Management and Marketing, 10*(3), 39–45.

Taylor, M., & Finley, D. (2008). Strategic human resource management in U.S. Luxury resorts-a case study. *Journal of Human Resources in Hospitality & Tourism, 8*(1), 82–95.

Tracey, J. (2014). A review of human resources management research: The past 10 years and implications for moving forward. *International Journal of Contemporary Hospitality Management, 26*(5), 679–705.

Úbeda-García, M., Marco-Lajara, B., García-Lillo, F., & Sabater-Sempere, V. (2013). Universalistic and contingent perspectives on human resource management: An empirical study of the Spanish hotel industry. *Journal of Human Resources in Hospitality & Tourism, 12*(1), 26–51.

Wright, P., & McMahan, G. (1992). Theoretical perspectives for strategic human resource management. *Journal of Management, 18*(2), 295–320.

Wright, P., & Boswell, W. (2002). Desegregating HRM: A review and synthesis of micro and macro human resource management research. *Journal of Management, 28*(3), 247–276.

From Mass Tourism and Mass Culture to Sustainable Tourism in the Post-covid19 Era: The Case of Mykonos

Konstantinos Skagias, Labros Vasiliadis, Dimitrios Belias, and Papademetriou Christos

1 Introduction

One of the key issues that tourism has to face in the twenty-first century is the impact of mass tourism and how it affects destinations. Indeed, the concept of mass tourism has helped many communities to overcome their financial problems, but in the long run it can create several problems, including a disruption on the way that local people live and a negative impact on the social and cultural values of the destination (Nukoo & Gursoy, 2017). The opportunities and threats which derive from mass tourism depend on how the destination perceives tourism and what resources it has. For example, Lundberg & Ziakas (2018) refer to the case of destinations such as Mykonos and Ibiza where mass tourism is related toh mass culture. Those destinations developed their destination image during the past years based in a number of assets including large parties and gatherings which had a tremendous impact on mass culture. Nonetheless, this image is many cases does not represent the reality of a destination. Destinations like Mykonos have become the "mecca" of mass culture, with emphasis given on rave and music culture, something which shifts away from the image of the destination from the reality of its beautiful beaches, the hospitality of locals, etc. The same happens with other popular destinations, like Goa and Ibiza (Rowe, 2006). Of course, this raises many questions as far as it concerns how those destinations shall develop

K. Skagias
Aegean Islands, Greece

L. Vasiliadis
National and Kapodistrian University of Athens, Athens, Greece

D. Belias (✉)
University of Thessaly, 41500 Larissa, Geopolis, Greece
e-mail: dbelias@pe.uth.gr

P. Christos
Neapolis University, Paphos, Cyprus

V. Katsoni and C. van Zyl (eds.), *Culture and Tourism in a Smart, Globalized, and Sustainable World*, Springer Proceedings in Business and Economics,
https://doi.org/10.1007/978-3-030-72469-6_23

347

in the future. For example, many Greek destinations have shifted from mass tourism to sustainable and alternative tourism with success (Belias et al., 2018). However, in the case of Mykonos the destination's brand name is associated with luxurious lifestyle, parties, and with mass culture. Hence, changing this image means that it may change the tourist product offer of Mykonos and its attraction on the segments that it has already attracted with its offerings.

The current situation with the Covid19 crisis has resulted in the fact that the summer of 2020 and overall the future of tourism is uncertain. Though that it is early to make conclusions, there is evidence that Greece will hardly have the 20% of tourism inflows compared to 2019. Surely, this means that the whole tourist concept must change but also that this is the time for some destinations to re-brand themselves. The packed beaches and nightclubs will not be part of the 2020s tourist offer, which means that tourist destinations may have to shift away from mass tourism and mass culture, into a more elegant and niche marketed product. For this reason, the aim of this paper is to discuss how a popular destination, such as Mykonos, can shift away— due to the Covid19 crisis—from the model of mass tourism into a sustainable model; hence to contribute to the academia by investigating a topic which will be the focus of many upcoming researches.

This is a literature review. Hence, the paper will rely on publications that were retrieved from EBSCO, SCOPUS, etc. The authors used keywords such as "mass tourism and Mykonos". Emphasis was given on using the most recent publications and the ones which were from valid journals.

2 Literature Review

2.1 The Mass Tourism Model

In recent decades, Greece has undergone a significant change in its tourism. This has led scientists to explore the particular features and capabilities that the country presents by supporting mass tourism. Thus there is interest in the way tourism is developed, which lies in the fact that tourism is an important sector of the economy that contributes greatly to economic growth and employment (Claver-Cortés et al., 2007). This means that tourism is a sector that is considered a locomotive for the Greek economy, something that makes it highly important for the data of Greece. The trends that tourism follows in Greece show differences in relation to other countries as over the years they are called to meet the needs of visitors who know Greece closely (Belias et al., 2017; Garau-Vadell et al., 2018).

The continuous development of tourism in developed and developing countries according to Hernández et al., (2016) is something that makes it an integral part of the Greek tourism product and international specialization. Naturally, Greece, like any other country, organizes its own strategy to highlight their place in order to attract more visitors (Belias et al., 2017). However, there are some problems that arise and which limit tourism development such as seasonality and some extraordinary events such as Covid-9 that has now appeared and which has significantly reduced the tourist product which was expected for the year 2020–2021.

The flourishing of Greek tourism began in 1960 with the main reasons leading the first tourists in Greece being the archeological sites of Greece combined with the possibility of rest for a few days near the Greek coast. Already at that time, intermediaries in Europe had begun to look for some destinations alternatively, compared to Spain due to the difficulties that existed in the prices and quality of the product provided, and thus the tourist product of Greece emerged. Greece is a very attractive tourist destination for the summer months due to its rich natural resources and the combination of sun and sea, however, there are some issues that need to be addressed in order to further develop tourism (Moussa, 2017; Harisson and Sharpley, 2017).

Hence tourism according to Jovicic (2016) plays a leading role in the development process, contributing to the total income and employment and the attractive elements that the tourist product has, based on which tourism is promoted, are the natural beauties, the landscapes, the prices of products and services, museums and monuments as well as the natural and cultural environment.

Also, the main elements of Greek tourism are the natural peculiarities that Greece has as well as the low prices that its entertainment has while its weakness is the sports sector and often the hotel comfort. On the other hand, Greece compared to other Mediterranean countries is the first in terms of entertainment, although the tourist product it offers in entertainment is not the best possible. However, Greece lags behind other competing Mediterranean countries in terms of infrastructure and does not have a good position in terms of railways and air transport (Belias et al., 2017).

After the war, the rise of Europe and Greece with the gradual increase in family income saw an increase in living standards and the establishment of a holiday and the development of transport and communication, something that led the peoples of Central Europe to Greece's shores, something that made the tourist product of Greece widely known. Thus, the institution of organized vacations was developed and mass tourism was created (Hernández et al., 2016). Mass tourism during the post-war period was the main model of tourism development which for many years did not allow the creation of other tourism standards. The model of mass tourism is associated with the development of popular tourist destinations to this day, despite the fact that alternative tourism is being promoted today. The concept of mass tourism is associated with activities that offer western-style amenities and simple activities and as a form of tourism is characterized by the sun and the sea (Harrison & Sharpley, 2017).

The response to this tourism model depends on how each country handles the organized tourist packages offered. These tourist packages are created and promoted by international travel agencies that direct the demand to specific tourist destinations. Most travel agencies come from Europe and the United Kingdom as well as France and Germany and are a dynamic business. Destinations that invest in mass tourism often have negative effects due to the concentration of infrastructure and tourists in a specific area where the bearing capacity has been exceeded (Brondoni, 2016).

Thus, mass tourism appears aggressive towards other wealth-producing resources and causes changes in the natural and structured environment, as a result of which the good morals of the societies living in a country are offended. In addition, a major problem caused by mass tourism is the seasonality that is in demand, whether it is summer or winter. The areas that adopt this specific tourism model are places that have rich environmental and cultural resources, but which are gradually being transformed into destinations with a specific advertising image both domestically and internationally (Claver-Cortés et al., 2007).

Tourism shows its preference in some areas and not in all. Thus, some tourist destinations literally "sink" in the summer, while other places are ignored. The result of this picture is that inequalities are created in the distribution of national income by region. Excessive, one-sided devotion to a certain type of activity carries too many risks, especially when it comes to the economy (Martínez-Garcia et al., 2017). Thus, in tourism, the tourist destinations that are based exclusively or mainly on tourism are endangered.

Another area affected by tourism is society in a broad context. Tourism according to Brondoni (2016) helps to keep the population in the urban centers, to bring the peoples closer, to discover new cultures and additional cultural activities. If tourism had not developed in Greece, many of the existing monuments and archeological sites would have been lost. Tourism is very important for many regions. It also is especially important for remote areas and areas of Greece with a long tradition, such as in Karpathos.

More to that, economic growth, which is directly and indirectly due to tourism, reducing unemployment has contributed to the reduction of social phenomena due to it, such as immigration, crime, drugs, etc. (Jovicic, 2016). With tourism, mainly the outgoing people approach each other through the constant contact of the inhabitants of the tourist places with the visitors who come from other countries or regions. They understand that everyone is a member of a larger community.

Tourism contributes positively to rising inflation (rising prices and goods). It is known that where there is increased demand in relation to supply, the prices of goods and services increase, however, having adverse effects on the living standards of residents, especially in the case of mass tourism (Garau-Vadell et al., 2018). Equally unfavorable is the rise in the value of land. The increased prices due to increased taxes and duties reduce the demand for Greek tourism, thus preventing Europeans from choosing Greece for their holidays and reducing the competitiveness that burdens Greek tourism the most.

2.2 Mass Culture

The western world, characterized by abundance, is a world in which mass culture has emerged. Mass culture, which is linked to the elimination of popular rural and traditional urban culture, has marginalized authentic art and popular culture and has been constructed in a way that fills the free time of modern man according to

Brantlinger (2016). At the same time, mass culture has affected the way man work, and has affected tourism as a whole. The genesis of mass culture has been associated with capitalism since the late nineteenth century until the early twentieth century, when the tendency to maintain established things emerged (Aramberri, 2017).

The protagonists of today's division of power and property have invented mass culture to exercise control over people's consciousness. The products of this culture are goods that are imposed through advertising, shaping the aesthetic standards that are not related to aesthetics but to economic and industrial interests and the stock market. Thus the imposition of the aesthetics of mass culture has affected culture as a whole as it has affected the world in general by overriding classical values. A key feature of mass culture is standardization. Standardization lies in the endless repetition of the same issues and techniques and significantly influences innovation in culture while highlighting specific fashions that are shamelessly exploited elements of popular culture that are brutally detached from their original contexts (Aramberri, 2017). This diversity according to Raitz (2001) which is false and shows that this culture cannot exist without an attack on the people, is a sign of modernity. It is essentially a falsification of individualization that is achieved through the excessive use of consumption and which depends on technology and consumerism in general. The standardization, however, concerns not only the products of mass culture but also the reactions of the individual himself to this culture. Thus the individual is influenced by the comments and criticisms presented by newspapers, magazines, radio, and television and which are imposed by interpreting reality in a very specific way.

Mass culture is associated with the industrial mass lifestyle and the distribution of cultural products as well as the urbanization and consumerism that have imposed uniformity and standardization on social life and human behavior. In this way, the masses adapt to a new reality, something that creates incalculable damage to civilization. In essence, mass culture is the idea of a particular culture that is produced by the masses and consumed by the masses with specific centers that follow, as mentioned above, one-way transmissions despite the need for communication between producer and recipient, resulting in a degraded culture (Butcher, 2003).

As Brantlinger (2016) claims, mass culture is associated to a quantitative determination, that is, a culture that is consumed by many and is addressed to many, which in turn has a qualitative definition, stating an inferior and degraded culture. Mass culture, which is the result of the industrialization of Western society according to the views of its conservative critics, is responsible for the erosion of moral values and the undermining of social hierarchy. On the other hand, the leftist critique considers that mass culture is separated from popular culture and has a purely commercial character with low quality being a product of cultural imperialism. However, there is also the pluralistic approach in which mass culture is the popular culture that is addressed to everyone as a whole and meets the needs of people today.

Thus mass culture is linked to industry and industrial culture which in turn is linked to technology. Millions of people are involved in this mass culture, which means that some production processes are necessary due to the needs that must be met by the goods that are sold. Technology in mass culture is gaining strength over society, a force of the economically stronger (Brondoni, 2016). Technological rationality is the rationality of domination today with a compelling character of an alienated society. So cars, movies, and theater today have the level of mass culture that is constantly being promoted. In fact, mass culture is characterized by the international interests in which the liberated business system has moved. Everything today is being rebuilt with new constructions similar to those presented in trade fairs, which praises technical progress and requires the individual to consume in order to acquire this new product (Harrison & Sharpley, 2017).

In this culture, the person is constantly consuming. The goal of imperialism is for the individual to constantly work to consume. The individual receives mortgages as well as consumer loans which he will spend on his vacation or on consumer goods acquiring specific products against the price imposed by the monopoly of capitalism, feeling that it is a unit against the absolute power of capital (Butcher, 2003).

The person in capitalism according to Chaney, (2002) seeks work and pleasure in a well-organized complex of fancy products that give the person the belief that he has a real identity while in fact his identity is completely false. Under the monopoly of capitalism, the lines of mass culture are all the same. The man who is in mass culture is the same as everyone else in a monopoly and is left to his own devices without seeking the truth further away. Thus, in this power, cinema and art are not recommended as art forms but as a business which must justify the garbage, it deliberately produces.

2.3   From Mass Culture and Mass Tourism to Sustainable Tourism

As noted in the introduction of this paper, mass tourism is often associated with mass culture. For example, Ibiza built its brand image based on the large raves and music gatherings of the 90s, while it was the key destinations of hippies during the late 60s and 70s and till today its brand image is heavily related with pop and rave music, clubbing and parties (Miternique, 2018). The same happens with Berlin where its annual rave parade and the large clubs have attracted a huge volume of tourists (Garcia, 2016). Hence, there are well-known destinations which have relied on mass culture—on most cases on huge raves and clubbing culture—so to promote tourism. However, this model of mass tourism has been subject of criticism.

Sarantakou (2020) has examined the case of some popular Greek destinations, including Mykonos. The sudden increase of tourist inflows has created several opportunities for the local societies along with welfare but it has raised questions about the sustainability of those destinations. For example, Mykonos is portrayed as a luxurious—lifestyle destination well-known for its nightlife. This brand image has alienated the traditional tourists of the island but also many potential tourists, such as families and middle-income tourists, who consider Mykonos as a very expensive destination and also not a destination for relaxing holidays but rather as a destination for partying and lifestyle. Of course, this is away from the truth and the reality of

Mykonos. Nonetheless, it is important to note that tourists are consuming experiences and images, hence the perceptions about the destination image will have a crucial role when a tourist will take his/her decisions (Boavida-Portugal et al., 2017).

During the past decades, the case of sustainable tourism has been developed, as a way to ensure that tourism development can co-exist along with the natural environment and the culture of a destination. One of the negative consequences of mass tourism was the negative impact on the local societies and on natural environment. The outcome is a market segment of tourists who respect the physical environment and they are enjoying their co-existence with the local society (Belias et al., 2017). The outcome is the model of sustainable tourism is a very successful way of tourist development which grants with value the destinations and it respects the local culture. Hence, a destination can create wealth without relying either on the hordes of mass tourists or on mass culture. Such as an example is provided from Pozoukidou et al. (2017) who refer to the case of Syros which has used its intangible cultural elements so to promote its tourism, including the opera festival which happens every two years and the rebetiko music heritage. Those are just some of the elements that shape the tourist offering of the island. The outcome is that it has become a popular destination for high-end tourists while at the same time it is accessible to middle-income tourists. Syros is an example of how high culture can become a mean of tourist development in a sustainable way.

After having made the necessary investigation, it is essential to examine how the Covid-19 crisis can affect the way that destinations are marketed in relation with sustainable tourism. This will happen in the next chapter by examining the case of Mykonos.

3 Methodology

The methodology used in the present paper is the critical review of the literature.

The sources of relevant literature investigation derived from popular online bibliographic databases, such as Science Direct, Emerald, EBSCO host, and scientific search engines such as Google Scholar and Scirus. General search engines such as Google have also been examined.

The types of bibliographic sources included in the research are articles published on scientific journals, books, conference proceedings, company papers and studies, white papers, online sites, and online journals. The selection criteria of these literature sources were based on the relevance to the topic of the paper and this research is not exhaustive.

4 Discussion—The Case of Mykonos in the Post-covid19 Era

Tourism development in Greece has decisively influenced some of the elements of the social structure that are considered representative of stability such as social relations and the participation of residents in activities that highlighted the cultural peculiarities of an area. Unrestrained and unplanned tourism development has brought about a sharp urbanization of these societies. In fact, in surveys conducted in Serifos and Ios, the inhabitants of these islands reported that (a) economic interests, as a result of mass tourism, caused envy and competition among the inhabitants and (b) the lack of old intimacy and solidarity had negative consequences. consequences for social cohesion (Tsartas, 2010).

The result is that in recent years, these societies have increasingly sought an alternative model of economic development that respects their cultural heritage. In many cases, the societies themselves turned against mass tourism and shifted their focus to the milder forms of tourism and in many cases to cultural tourism. Turning to cultural tourism has many benefits, especially in areas with significant cultural capital. First of all, the residents themselves, as connoisseurs of their tradition and culture, are able to understand and adapt the tourist packages of their areas in order to focus on their cultural richness. Also, cultural tourism has as a point of reference tourists who have not only high incomes but also a high level of education which allows them not only to be able to immediately understand the cultural wealth of a destination but also to respect it (Lesli, 2010). One of the main problems of mass tourism is that these tourists often not only do not respect or understand the value of the culture that a destination has but often also engage in acts such as the destruction of monuments or generally behaviors that show ignorance and arrogance. In relation to the culture of a destination.

The impact on the cohesion of the local community is also significant. The fact that cultural tourism activities are carried out in a way that respects the political heritage of a place, results in a reduction in tensions within society and a significant acceptance of tourism by the local community. Also, cultural tourism significantly helps to increase the knowledge that the inhabitants of a society have about the cultural heritage of their place (Sigala, 2019). Miller's (2001) research showed that in all areas where cultural tourism developed, there was an increase in the activities and interest of young people in the cultural goods of their place. So it is considered that cultural tourism has only to offer positively to a destination that will decide to turn to this form of development.

In the case of Mykonos, the examined island has developed its destination branding based on a luxurious profile which includes elements of mass culture such as clubbing culture and a lifestyle-nightlife which attracts the interest of many young persons but also of the elite. However, there is a question on whether this is sustainable and what can be the impact of COVID-19 crisis.

Regarding the case of COVID-19, it is too early to make conclusions. What we know from our experience is that tourism in Greece and elsewhere will face a very difficult and dramatic year, maybe the worst ever. Researches such as Gössling et al., (2020) and Karim et al. (2020) along with the report made from UNTWO (2010) indicate not only a dramatic loss on tourist inflows all over the world but also that for some destinations this may mean that the whole tourist year can be lost. Mykonos has already suffered from the dramatic reduction of tourism demand, while a key issue is how mass tourist can exist with the measures against COVID-19. An example is that as soon as Mykonos opened for inbound tourists from Greece, one of the first news which circulated on national but also on international media was a number of parties which took place resulted in the closure of a very popular and high-profile night club due of violations of COVID-19 measures, while this has damaged the profile of the island. According to the Greek chamber of tourist professionals (SETE, 2020), which is the largest in Greece, claims that it is very important to focus on how to protect the brand image and the reputation of Greece in the post-covid era. It is easily understood, based on SETE (2020)s interview and on Renaud (2020), that there two key issues that destinations such as Mykonos would have to consider for the future. The first one, that is, the reputation will have a crucial role, where the tourists will prefer to make their holidays on destinations where they feel secure and safe from Covid-19. This does not mean only to have limited or not cases of Covid-19—actually it is expected that all of the well-known tourist destinations will face cases of Covid-19—but also that the destination has taken the necessary measures so to protect the tourist. The second issue is that for 2020 and probably 2021 the destinations must not rely on mass tourism. Actually, there may not be mass tourism at all till 2022.

From the above, it is understood that a destination such as Mykonos for the upcoming months, even years, can not rely on mass tourism and on massive parties and clubbing culture. In order to survive and to stay with competition, it is necessary to change the way that it understands how Mykonos can develop. If we consider that the future of tourism is about less tourists who seek a sustainable and safe destination, then Mykonos and its tourist businesses would have to adjust. This requires to focus on a sustainable tourist model which will not rely on endless parties—something with the Covid-19 is too risky for the tourists and for the reputation of the destination—but rather on a quite and tranquil beaches and focus on high culture which it may be music concerts or cultural exhibitions.

Of course, the above model may mean less tourists but not necessarily less income. Instead, it can attract both families and high-end tourists but at the same time to avoid the hordes of party-goers and of those who are coming only for the club culture. Surely, this will result in major changes on the current businesses but still, Mykonos will have to change its mindset and to become a safe haven and an affordable destination for the tourist who will struggle with the economic impact of the covid 19 crisis. Hence, Mykonos and its tourist businesses must change their mindset so to make a tourist marketing offer which will include the following:

- Develop with emphasis on quality
- Focus on tourist segments that will respect the natural beauties and the culture of the destination

- Comply with Covid-19 regulations such as on social distancing
- Shift the destination brand from a mass tourist destination to a sustainable one.
- Adjust the prices on the current financial situation so as to make it affordable for everyone
- Shift from clubbing culture into a high culture with a focus on art exhibitions.

The above are some of the initiatives that Mykonos can take so to change its tourist product features which will help her to adjust into the post-covid19 era.

5 Conclusions and Recommendations

The paper examined the case of Mykonos and it needs to do so to retain its competitive advantage in the post-covid19 tourist industry. The key conclusion of this paper is that the covid-19 crisis is a good chance to reconsider the tourist product of the island. This will include a shift from mass tourism and mass culture into a sustainable tourist destination which will focus on high culture and safety for the tourists. For this reason, the local authorities along with the tourist professionals must take a number of decisions so to change the profile of Mykonoς and to make it sustainable for the future.

It is also important to remark the lack of such research, which is reasonable since the covid 19 crisis is an ongoing phenomenon. For this reason, the author suggests a future research with a case study in a selected destination, such as Mykonos, which will monitor the progress made and which policies—practices were implemented.

References

Aramberri, J. (2017). Mass tourism does not need defending. In D. Harrison & R. Sharpley (Eds.), *Mass Tourism in a Small World* (pp. 15–27). Wallingford: CABI.

Belias, D., Velissariou, E., Kyriakou, D., Varsanis, K., Vasiliadis, L., Mantas, C., & Koustelios, A. (2018). Tourism consumer behavior and alternative tourism: the case of agrotourism in Greece. In *Innovative Approaches to Tourism and Leisure* (pp. 465–478). Springer, Cham.

Belias, D., Velissariou, E., Kyriakou, D., Vasiliadis, L., Roditis, A., Varsanis, K., & Koustelios, A. (2017, June). The differences on consumer behavior between mass tourism and sustainable tourism in Greece. In *5th International Conference on Contemporary Marketing Issues ICCMI*, June 21–23, 2017 (p. 176). Thessaloniki, Greece

Boavida-Portugal, I., Ferreira, C. C., & Rocha, J. (2017). Where to vacation? An agent-based approach to modelling tourist decision-making process. *Current Issues in Tourism, 20*(15), 1557–1574.

Brantlinger, P. (2016). Bread and circuses: Theories of mass culture as social decay. Cornell University Press.

Bray, R., & Raitz, V. (2001). *Flight To the Sun: The Story of the Package Holiday Revolution.* London: Continuum.

Brondoni, S. M. (2016). Global tourism management. Mass, experience and sensations tourism. *Symphonya. Emerging Issues in Management, 1*, 7–24.

Butcher, J. (2003). *The moralisation of tourism: Sun, sand ... and saving the world?* London: Routledge.

Chaney, D. (2002). *Cultural change and everyday life.* London: Palgrave.

Claver-Cortés, E., Molina-Azorı, J. F., & Pereira-Moliner, J. (2007). Competitiveness in mass tourism. *Annals of Tourism Research, 34*(3), 727–745.

Garau-Vadell, J. B., Gutierrez-Taño, D., & Diaz-Armas, R. (2018). Economic crisis and residents' perception of the impacts of tourism in mass tourism destinations. *Journal of Destination Marketing & Management, 7,* 68–75.

Garcia, L. M. (2016). Techno-tourism and post-industrial neo-romanticism in Berlin's electronic dance music scenes. *Tourist Studies, 16*(3), 276–295.

Gilmore, J. (2017). Quality and quantity in tourism. *Journal of Hotel Business Management, 6*(1), 1–3.

Gössling, S., Scott, D., & Hall, C. M. (2020). Pandemics, tourism and global change: a rapid assessment of COVID-19. *Journal of Sustainable Tourism.* Retrieved on June 30, 2020, from https://www.tandfonline.com/doi/pdf/10.1080/09669582.2020.1758708

Harrison, D., & Sharpley, R. (Eds.). (2017). Mass tourism in a small world. CABI.

Harrison, R., Newholm, T., & Shaw, D. (Eds.). (2005). *The Ethical Consumer.* London: Sage.

Hernández, J. M., Suárez-Vega, R., & Santana-Jiménez, Y. (2016). The inter-relationship between rural and mass tourism: The case of Catalonia, Spain. *Tourism Management, 54,* 43–57.

Higgins-Desbiolles, F. (2020). Socialising tourism for social and ecological justice after COVID-19. *Tourism Geographies.* Retrieved on June 30, 2020, from https://www.tandfonline.com/doi/full/10.1080/14616688.2020.1757748

Jovicic, D. (2016). Cultural tourism in the context of relations between mass and alternative tourism. *Current Issues in Tourism, 19*(6), 605–612.

Karim, W., Haque, A., Anis, Z., & Ulfy, M. A. (2020). The movement control order (mco) for covid-19 crisis and its impact on tourism and hospitality sector in malaysia. *International Tourism and Hopitality Yournal, 3*(2), 1–7.

Kourgiantakis, M., Apostolakis, A., & Dimou, I. (2020). COVID-19 and holiday intentions: The case of Crete, Greece. Anatolia, 1–4.

Leslie, D. (2010). The European Union, sustainable tourism policy and rural Europe. In *Sustainable Tourism in Rural Europe* (pp. 59–76). Routledge.

Lundberg, C., & Ziakas, V. (2018). *The Routledge handbook of popular culture and tourism.* Routledge.

Martínez-Garcia, E., Raya, J. M., & Majó, J. (2017). Differences in residents' attitudes towards tourism among mass tourism destinations. *International Journal of Tourism Research, 19*(5), 535–545.

Miller, G. (2001). The development of indicators for sustainable tourism: Results of a Delphi survey of tourism researchers. *Tourism Management, 22*(4), 351–362.

Miternique, H. C. (2018). The Ibiza's nightlife as a bend from marginalization to tourism centrality. In *Nature, Tourism and Ethnicity as Drivers of (De) Marginalization* (pp. 109–118). Springer, Cham.

Moussa, M. (2017). Constructing tourism in Greece in 50s and 60s: The Xenia Hotels Project. *Journal of Tourism and Research* 1, 17.

Nunkoo, R., & Gursoy, D. (2017). Political trust and residents' support for alternative and mass tourism: An improved structural model. *Tourism Geographies, 19*(3), 318–339.

Pozoukidou, G., Linaki, E., & Planner, C. (2017). Tangible and intangible cultural assets as means for sustainable urban planning and place making. The case of Ano Syros, Greece. In *Proceedings of International Conference on Changing Cities III: Spatial, Design, Landscape & Socio-economic Dimensions (No. IKEECONF-2017-361).* Graphima Publications.

Renaud, L. (2020). Reconsidering global mobility–distancing from mass cruise tourism in the aftermath of COVID-19. *Tourism Geographies.* Retrieved on June 30, 2020, from https://www.researchgate.net/profile/Luc_Renaud2/publication/341322194_Reconsidering_global_mobility_-_distancing_from_mass_cruise_tourism_in_the_aftermath_of_COVID-19/links/5edce8d24585

1529453fd3e7/Reconsidering-global-mobility-distancing-from-mass-cruise-tourism-in-the-aft ermath-of-COVID-19.pdf

Rowe, D. (2006). Leisure, mass communications and media. In *A handbook of leisure studies* (pp. 317–331). Palgrave Macmillan, London.

Sarantakou, E. (2020). Mechanisms for the formation of tourism organization models in Greece through a comparative analysis of ten Greek destinations' development. In *Destination Management and Marketing: Breakthroughs in Research and Practice* (pp. 560–573). IGI Global.

SETE. (2020). Priority is to save our reputation. Retrieved June 29, 2020, from https://www.ant 1news.gr/Economy/article/572454/sete-proteraiotita-gia-fetos-i-diafylaxi-tis-fimis-tis-xoras

Sigala, M. (2019). The synergy of wine and culture: The case of Ariousios wine, Greece. In *Management and marketing of wine tourism business* (pp. 295–312). Palgrave Macmillan, Cham.

Tsartas, P. (2010). *Sustainable tourism* (Greek). Athens: Kritiki.

UNWTO International tourist arrivals could fall by 20–30% in 2020. Retrieved on June 5, 2020, from https://www.unwto.org/news/international-tourism-arrivals-could-fall-in-2020

Smart Analysis of Volatility Visualization as a Tool of Financial and Tourism Risk Management

Ani Stoykova and Mariya Paskaleva

1 Introduction

Risk affects the investment decisions of economic agents, especially when its size and manifestations deviate from their expectations. This idea is promoted in Hayek (1945), Keynes (1936), Akerlof and Shiller (2010) studies under the title "animal spirits". Nowadays, animal spirits describe the psychological and emotional factors that drive investors to take action when faced with high levels of volatility in the capital markets.

Many researchers claim that capital markets are a tool for measuring capital dynamics in the country (Ang & Longstaff, 2011). On the other hand, sovereign credit instruments are affected by systematic credit risk. According to the Portfolio Theory, the asset value can be calculated as a payment of the asset value, with a discount using the appropriate stochastic discount factor (SCF). All necessary adjustments to asset pricing must be included in the SDF. According to the Capital Asset Pricing Model (CAPM) (Bodie et al., 2010), SDF is connected with an increase in the marginal utility of consumption and is expressed as a linear function of a group of risk factors.

The tourism industry can be considered as an important sector for accelerating economic growth, increasing efficiency, and economic recovery. The tourism sector is influenced by risk factors and the effect of these factors is included in the returns of tourism listed companies. According to Keller and Bieger (2010) supply-side trends influence largely the tourism phenomenon. Additionally, supply-side trends accelerate the structural change of tourism-related industries and tourism destinations. The development of the financial markets leads to the formation of investment potentials for specific forms of tourism infrastructure. The emergence of leasing facilities has pushed investment in cruise ships and airplanes.

A. Stoykova (✉) · M. Paskaleva
South-West University "Neofit Rilski", Blagoevgrad, Bulgaria
e-mail: ani_qankova_st@abv.bg

© The Author(s), under exclusive license to Springer Nature Switzerland AG 2021
V. Katsoni and C. van Zyl (eds.), *Culture and Tourism in a Smart, Globalized, and Sustainable World*, Springer Proceedings in Business and Economics,
https://doi.org/10.1007/978-3-030-72469-6_24

According to financial risk management theories, the Efficient market hypothesis (EMH) is of primary importance because it can also be considered as an element in building an Early Warning System for an upcoming financial crisis. After the global financial crisis of 2008, the following question was raised: "Do unusual levels of financial market volatility imply an increased likelihood of a subsequent financial crisis?" (Danielsson et al., 2016). Volatility is important for option traders because it affects options prices. Generally, higher volatility makes options more valuable, and vice versa.

There are two types of volatility: realized and implied. Realized volatility reflects the historical price and fluctuations of the asset. Implied volatility is always forward-looking. It is the expected volatility from now until the option's expiration. The volatility index (VIX) is one of the most popular measurement tools for stock market volatility. VIX measures the 30-day volatility implied by the S&P 500 stock index option prices. Market risk can be low when volatility is low. However, low volatility could be a catalyst for market participants to take on more risk, making the financial system more fragile. This is defined as a phenomenon known as the instability/volatility paradox (volatility paradox).

In this article, we aim to examine the stock market dynamics of Bulgaria (SOFIX), France (CAC 40), Germany (DAX), The United Kingdom (FTSE 100), Belgium (BEL-20), Bulgaria (SOFIX), Romania (BET), Greece (ATHEX20), Portugal (PSI-20), Ireland (ISEQ-20), Spain (IBEX35), and USA (S&P 500) on one hand and the dynamics of GEPU, VIX, and S&P 500 on the other hand. We observe the performance of S&P 500, Stoxx 600, and Stoxx Europe 600 Travel and Leisure price index. The analyzed period is 2003–1016. The results show that studied stock indices tend to move synchronously during the examined period. We register a high level of volatility for the period 2007–2011. The first major bottom of VIX is in December 2007. We can conclude that there is a strong correlation between VIX, GEPU, and S&P500 measuring "investor risk appetite". GEPU fluctuates around consistently high levels since mid-2011 until the beginning of 2013. The dynamics of GEPU and VIX are not synchronized for the period 2013–2016. In the crisis period, travel and leisure stock performance is lower because of increasing global uncertainty.

2 Literature Review

Ang and Longstaff (2011) find that U.S. systemic sovereign credit risk is highly correlated with European systemic credit risk. They prove that both are strongly related to financial market variables such as stock returns. What is more, U.S. systemic credit risk is significantly negatively related to changes in the VIX index. Thus, as markets become more volatile, the credit risk of the U.S. Treasury improves.

Whaley (2000) analyzes the relation between stock market returns and changes in VIX, the relationship is asymmetric (e.g. Fleming et al., 1995; Giot, 2005). Also, Whaley (2000) proves that VIX is more a barometer of investors' fear (investor sentiment) of the downside risk and it is a barometer of investors' excitement (or greed) in a market rally.

Shaikh and Padhi (2015) examine implied volatility as the investor fear gauge or/and forward-looking expectation of future stock market volatility within emerging markets setting-India VIX. The results show that VIX is the gauge of investor fear, wherein the expected stock market volatility rises when the given market is declined. It is also proven that expected volatility is being an unbiased estimate of the actual return volatility (30-calendar days); hence, during the market turmoil VIX likely to be biased.

Here, we have to note that the volatility of capital markets has a direction. Volatility falls when markets rise and volatility rises as markets fall. The relationship has the following direction—from the capital market to volatility, and not vice versa. Low volatility usually reflects an extended bull market. In other words, volatility is a lagging indicator. Additionally, bear markets follow bull markets. Consequently, the longer the bull market, the higher the risk of a bear market and the simultaneous increase in volatility. These asymmetric relationships between volatility and stock return are empirically proven (Bates, 2000; Bollerslev & Zhou, 2006; Dennis et al., 2006; Ederington & Guan, 2010; Frijns et al., 2010; Giot, 2005; Schwert, 1989, 1990; Kownatzki, 2006; Moghaddam et al., 2019).

A systematic basis for the asymmetric relationship between implied volatility and stock market returns was first provided by Schwert (1989, 1990) and Fleming et al. (1995), who find a significant negative and asymmetric relationship between volatility and stock returns. Crisis periods often precede periods of unusually low volatility. Baker et al. (2016) and Gulen and Ion (2016) prove that high stock volatility is related to high political uncertainty, reduced investment, production, and employment.

Simeonov (2020) concludes that the global financial crisis of 2008 has a significant and lasting negative impact only on the price component of the stock exchange profiles, while the stock exchange activity of the studied exchanges remains completely unaffected. Albulescu (2020) search for the effect of official announcements regarding new cases of infection and death ratio of COVID-19 on the financial markets volatility index (VIX). Their results show that: the spread of coronavirus increases the financial volatility. The persistence of COVID-19 might generate a new episode of international financial stress.

Baker et al. (2016) develop a simple, transparent, scalable method for constructing newspaper-based Equity Market Volatility (EMV) trackers. Implementing the method using eleven major U.S. newspapers, our EMV tracker moves closely with the VIX and with realized volatility on the S&P 500. Their results reveal Monetary Policy and Tax Policy to be the most important policy-related sources of stock market volatility, followed by our aggregated Regulation category. The contribution of specific policy categories to stock market volatility fluctuates markedly over time. Hedström et al. (2020) find that the geopolitical risk has no impact on either the return or volatility spillovers for 10 emerging capital markets. However, the general stock market risk (VIX) is connected to individual market volatilities.

Mora and Sethapramote (2019) examine the spillover effects of global financial uncertainty (VIX index), the Global Economic Policy Uncertainty (GEPU) index, and other global risk factors, i.e. oil prices and gold prices, on the Stock Exchange of Thailand (SET). The empirical results show that both GEPU and VIX indices have significant impacts on the returns of the stock market in Thailand before 2010. After 2010, the effects of GEPU on the stock return are not statistically significant. Yu et al. (2018) investigate how Global Economic Policy Uncertainty (GEPU) drives the long-run components of volatilities and correlations in crude oil and U.S. industry-level stock markets. They find that GEPU is positively related to the long-run volatility of Financials and Consumer Discretionary industries.

Saffet et al. (2019) prove that the change in VIX index had permanent causal relationship towards the changes in the tourism indices of China, Denmark, Greece, Italia, Spain, Swedish, Turkey, and the UK, but the causality towards the changes in the index of Finland was found to be temporary. In addition, they find that there was a long-term relationship between the variables and that the increase in VIX caused a decrease in the return of tourism indices.

3   Methodology and Data

In this study, we explore eleven EU Member States—France, Germany, The United Kingdom, Belgium, Bulgaria, Romania, Greece, Portugal, Ireland, Italy, and Spain, and the USA. The variables that we use, represent the capital market indexes for these countries: France (CAC 40), Germany (DAX), The United Kingdom (FTSE 100), Belgium (BEL-20), Bulgaria (SOFIX), Romania (BET), Greece (ATHEX20), Portugal (PSI-20), Ireland (ISEQ-20), Italy (FTSEMIB), Spain (IBEX35), and the performance of S&P 500, Stoxx Europe 600 and Stoxx Europe 600 Travel and Leisure (SXTP). The proxy for the tourism industry is the Stoxx Europe 600 Travel and Leisure Price index. We choose the EU countries listed above based on the following criteria: countries with a developed capital market, the values of which CDS during the crisis of 2008 has not suffered significant changes (UK, Germany, France, and Belgium); countries with relatively developing capital markets (*emerging markets*), which CDS spreads grow immediately after the crisis, but their values gradually decrease during the debt crisis (Bulgaria, Romania); countries with emerging capital markets which CDS spread reaches peak values—"problem countries" (*distressed countries*) (Greece, Portugal, Ireland, Italy, and Spain). A country's index data is obtained from the internet sources of their capital markets. The date is with monthly frequency. The explored period is March 2003–June 2016. We divide the explored period of two sub-periods: Period 1–crisis period (March 2003–September 2011) and Period 2 (September 2011–June 2016). The division above was made based on the peak values of VIX and GEPU in 2011, which were significantly higher than in 2008.

The data used in this study is the following: monthly values of the studied stock indices, VIX, GEPU for the period 03.03.2003–01.07.2016 and we calculate the return of these variables:

$$r_t = \ln\left(\frac{P_t}{P_{t-1}}\right) \tag{1}$$

where:

r_t the return of the explored variable at time t;
PI_t the value of the variable at time t;
PI_{t-1} the value of the variable at time $t1$.

Using the natural logarithm of price ratios corresponds to the financial and mathematical aspect since they cannot take negative values and must be log-normally distributed. By calculating return in this way, we examine the gradual change of stock index values, VIX values, and GEPU values over two adjacent periods t and $t - 1$, rather than their specific values. We should make a note here that the European capital markets close several hours before the U.S. market closes, and therefore they can reflect the dynamics of the VIX to the last and include it in the value of their indices, namely, the closing values are those used in this study. The US market continues to operate and accumulate information content, and after the European markets have closed.

We use Global Economic Policy Uncertainty Index (GEPU). The GEPU Index is a GDP-weighted average of national EPU indices for 21 countries: Australia, Brazil, Canada, Chile, China, Colombia, France, Germany, Greece, India, Ireland, Italy, Japan, Mexico, the Netherlands, Russia, South Korea, Spain, Sweden, the United Kingdom, and the United States. Each national EPU index reflects the relative frequency of own-country newspaper articles that contain a trio of terms about the economy (E), policy (P), and uncertainty (U). In other words, each monthly national EPU index value is proportional to the share of own-country newspaper articles that discuss economic policy uncertainty in that month.

To examine the dynamics of GEPU and stock volatility, we use the volatility index (VIX). VIX is a popular measure of the stock market's expectation of volatility based on S&P 500 index options. It is calculated and disseminated on a real-time basis by the CBOE (Chicago Board Options Exchange) and is often referred to as the fear index or fear gauge. Low VIX values do not necessarily indicate that there is impending financial stress. A high level of VIX suggests more fear. Volatility often measured as the standard deviation of historical returns and it is used as a proxy for risk (Markowitz, 1952).

In 1993 Chicago Board Options Exchange (CBOE) introduces the volatility index, also known as VIX. After the recent 2008 financial crisis, financial media regularly report on VIX dynamics along with stock market indices dynamics. Whaley (1993) suggests that the VIX provided a "reliable estimate of expected short-term market volatility. Additionally, Whaley (2009) argues that the main attraction was its forward-looking nature, "measuring volatility that investors expect to see". We can assume that VIX is the investors' sentiment index and it is the barometer of future stock market risk. In general, VIX is constructed using observed option prices. The market participant buys call/put options to hedge/trade the volatility, and the same observed option price is used to derive VIX in real-time (Shaikh & Padhi, 2015). Additionally, Whaley (2000) points out that a high level of VIX is observed due to the high degree of market turmoil.

The STOXX Europe 600, also called STOXX 600 (SXXP), is a stock index of European stocks designed by STOXX Ltd. This index has a fixed number of 600 components representing large, mid, and small capitalization companies among 17 European countries, covering approximately 90% of the free-float market capitalization of the European stock market (not limited to the Eurozone). The countries that make up the index are the United Kingdom (comprising around 27% of the index), France, Germany, and Switzerland (accounting for around 15% of the index each), as well as Austria, Belgium, Denmark, Finland, Ireland, Italy, Luxembourg, the Netherlands, Norway, Poland, Portugal, Spain, and Sweden. The STOXX Supersector indices, including Stoxx Europe 600 Travel and Leisure price index, track supersectors of the relevant benchmark index. There are 19 supersectors according to the Industry Classification Benchmark (ICB). Companies are categorized according to their primary source of revenue. The supersector of travel and leisure is also available.

In order to test the impact of financial crisis on the tourism industry, we divide the explored period of three sub-periods: pre-crisis period 03.03.2003–29.12.2006; crisis period (02.01.2007–28.12.2012) and post-crisis period 03.01.2013–30.06.2016). The financial crisis had a negative impact on the stock markets, investors' risk behavior, and tourism industry. This is proved by the drop in their average annualized returns. Post-crisis period is characterized by a slow recovery, but the values of the explored indices are lower than the values in the pre-crisis period. We should focus our attention on the fact that VIX and GEPU—the global measures of economic and investment uncertainty increase their values in crisis period. The investors' decisions for their investments and stock portfolio is in accordance with risk and returns. The increasing global economic uncertainty corresponds to a decrease in the tourism development (Table 1).

Table 1 Descriptive statistics on tourism, market returns, VIX, and GEPU (*Source* Authors' calculations)

Index	Pre-crisis period		Crisis period		Post-crisis period	
	Average return (%)	Standard deviation (%)	Average return (%)	Standard deviation (%)	Average return (%)	Standard deviation (%)
S&P 500	5.21	18.01	3.85	12.85	4.15	13.85
Stoxx Europe 600	4.09	15.24	3.21	12.05	4.08	11.56
Stoxx Europe 600 Travel and Leisure	4.98	15.03	2.95	11.59	4.25	14.89
VIX	4.18	14.58	7.21	25.12	5.28	15.23
GEPU	4.07	14.28	6.18	20.17	5.86	16.46

4   Analysis and Results

Figure 1 shows that studied stock indices tend to move synchronously and the only exception of this market co-movement is Belgian BEL 20. Our findings reconfirm the analysis made by Peneva (2020). All examined stock indices, except for the Belgian index, reach peak values before 2007 with a subsequent decrease in their values after that. This co-movement of stock indices must be related to identical external shocks to which the studied economies are subjected (Gerunov, 2014). However, the VIX index still significantly affects the return and volatility of the examined stock market.

Figure 2 presents the dynamics of S&P 500 and VIX for the period 2003–2016. We establish a high level of volatility for the period 2007–2011. We register the first major bottom of VIX is just before the start of the global financial crisis of 2008 (December 2007). VIX registered very low values during Period 2, which is characterized by low volatility. This low volatility is considered a prerequisite for the upcoming financial crisis by several economists and analysts. Considering the above analysis, the studied sub-periods are defined as a period of high volatility-Period 1 and a period of long-term low volatility-Period 2.

Predicting crises is inherently challenging because these are rare events and data is often not enough. Huang et al. (2019) believe that data (news) related to investor behavior can be useful to researchers and policymakers for multiple applications. First, the fact that investor sentiment indices tend to spike and/or trends up ahead of financial crises (e.g., Brazil 1999) or periods of severe economic stress (e.g., Turkey 2018) suggests that "news-based" indicators could potentially improve performance of traditional forecasting models. Second, this investor sentiment dataset could be used to examine potential similarities and/or differences in cross country business and

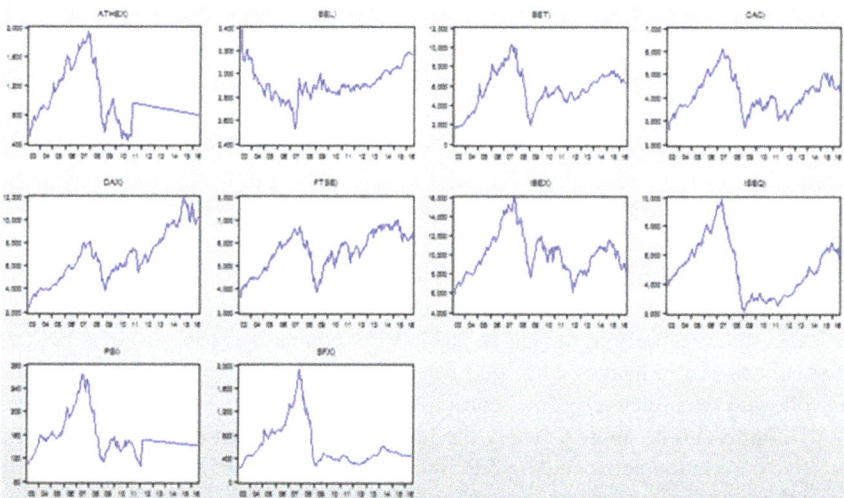

Fig. 1 Dynamics of stock indices for the period 2003–2016 (*Source* Authors' elaboration)

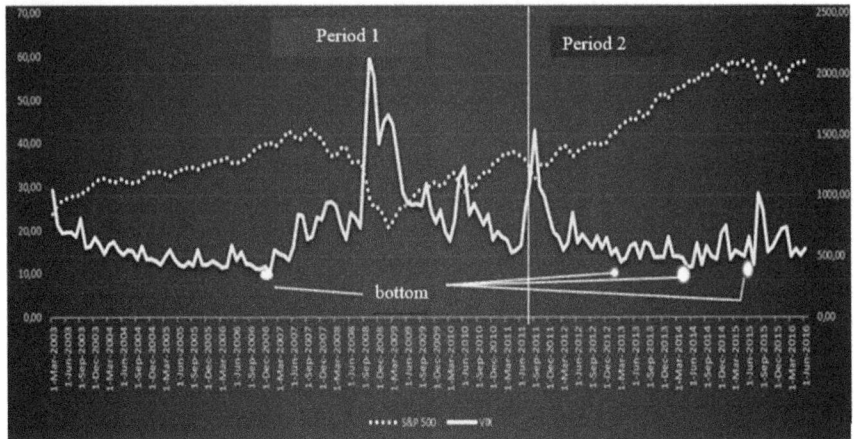

Fig. 2 Dynamics of S&P 500 и VIX for the period 2003–2016 (*Source* Authors' elaboration)

customer sentiment, and subsequent economic outcomes, when there are common global shocks.

The outlook for the economy and investment markets is uncertain, given the huge political changes we are seeing. Also, most assets are expensive and are most likely priced to offer sub-normal potential returns. History shows us that periods of crisis often precede periods of unusually low volatility. This is reconfirmed by the dynamics of VIX and GEPU, shown in Fig. 3 for the examined period 2003–2016.

GEPU index rapidly increases as a response to the US invasion of Iraq in 2003; the global financial crisis of 2008; the European migrant crisis and fear about China's economy at the end of 2015; and The Brexit referendum in June 2016. GEPU fluctuates around consistently high levels since mid-2011 until the beginning of 2013. This period is characterized by recurring debt and banking crises in the Eurozone, intense battles over fiscal and health policy in the United States, and the transition to General leadership in China. The average value of GEPU index is 60% higher during the period July 2011–August 2016 than in the previous fourteen years period. What is more, the average value of GEPU index in July 2011 is 22% higher than in 2008–2009, when policymakers faced the worst economic crisis since the Great Depression of the 1930s. These results suggest that policy-related issues have become a major source of economic uncertainty in recent years.

Figure 3 shows two high values for VIX, coinciding with the mortgage crisis in 2008, and instability in the US in 2011. VIX dynamics present cyclical periods of small and large changes with irregular intervals. Sudden upward spikes of VIX are followed by a relatively slow return to the average value. The maximum values of VIX appear to be approximately the same as relative market lows, indicating a negative correlation between the S&P 500. Low values of VIX does not necessarily mean that severe financial stress is unlikely to occur.

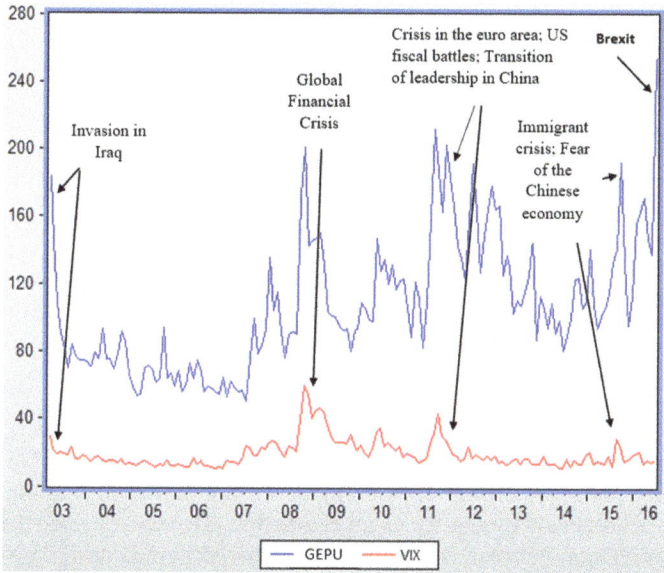

Fig. 3 Dynamics of GEPU and VIX for the period 2003–2016 (*Source* Authors' elaboration)

We can make two important remarks here. First, there is historically a strong relationship between the two variables—GEPU and VIX. Second, the dynamics of GEPU and VIX are not synchronized for the period 2013–2016. Figure 3 can be used to target the vulnerability of capital markets, given that based on such measures, capital markets do not reflect deteriorating conditions. However, like understanding the relationship between equity markets and volatility, the question of causality is important. GEPU index is based on the relative frequency of newspaper articles containing the terms "economy", "politics" and "uncertainty", i.e. it reflects the mood of the General audience, but does not show us definitely what caused this feeling. Comparing Figs. 1 and 3, in the period 2006–2007, we can point out that the studied stock indexes, except for BEL 20, register peak values and low volatility, registered by the dynamics of VIX and GEPU.

Figure 4 reveals the relationship between VIX, GEPU, and S&P500. We can conclude that there is a strong correlation between VIX and S&P500, which in most cases move "against each other" (Fig. 4). What is more, we also register a strong correlation between VIX, GEPU, and S&P500 measuring "investor risk appetite".

5 Conclusion

During periods of high volatility, financial markets are relatively more efficient. Volatility falls when markets rise and rise as markets fall. This paper focuses on

Fig. 4 3-D visualization of relationship between 3-D VIX, GEPU and S&P500 (*Source* Authors' calculations based on IBM Watson Studio)

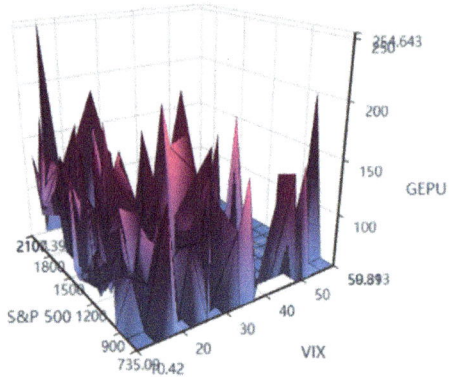

examining the dynamics of European capital markets, S&P 500, GEPU, and VIX. The analyzed period is 2003–1016. The results show that studied stock indices tend to move synchronously during the examined period. We can conclude that there is a strong correlation between VIX, GEPU, and S&P500 measuring "investor risk appetite". GEPU fluctuates around consistently high levels since mid-2011 until the beginning of 2013. The dynamics of GEPU and VIX are not synchronized for the period 2013–2016.

During periods of high volatility, financial markets are relatively more efficient. Volatility falls when markets rise and rise as markets fall. It was also shown that the expected volatility is an objective assessment of the actual volatility of earnings and, therefore, during the market turmoil, VIX is likely to react hastily, which in turn correlates with investor nervousness and brings potential profit to the options seller. Long periods with low volatility can further reduce correlations, contributing to further risk. This pro-cyclical behavior increases investors 'risk of losing a systematic shock when spikes in volatility and correlations in asset returns return to historical levels. Low volatility correlated with poor market performance can directly affect market risk. During this period, investors underestimate the likelihood of a possible upcoming spike in volatility and financial distress. The tourism sector is accepted as a fundamental industry for economic growth. We prove that during unstable period—crisis period—travel and leisure stock performance lower because of increasing global uncertainty. We established that the European tourism industry recovers quickly during post-crisis period.

References

Akerlof, G. A., & Shiller, R. J. (2010). *Animal spirits: How human psychology drives the economy, and why it matters for global capitalism.* Princeton University Press.

Albulescu, C. (2020). Coronavirus and financial volatility: 40 days of fasting and fear (pp. 1–7). Retrieved from https://arxiv.org/ftp/arxiv/papers/2003/2003.04005.pdf

Ang, A., & Longstaff, Francis A. (2011). *Systemic Sovereign Credit Risk: Lessons from the U.S. and Europe.* Retrieved from https://ssrn.com/abstract=1815502 or /https://doi.org/10.2139/ssrn. 1815502

Baker, S. R., Bloom, N., & Davis, S. J. (2016). Measuring economic policy uncertainty. *The Quarterly Journal of Economics, 131,* 1593–1636.

Bates, D. S. (2000). Post-'87 crash fears in the S&P 500 futures option market. *Journal of Econometrics, 94*(1–2), 181–238.

Bodie, Z., Kane, A., & Marcus, A. (2010). *Essentials of investments* (5th ed.). McGraw-Hill Irwin.

Bollerslev, T., & Zhou, H. (2006). Volatility puzzles: a simple framework for gauging return-volatility regressions. *Journal of Econometrics, 131*(1e2), 123–150.

Danielsson, J., Marcela, V., & Ilknur, Z. (2016). Learning from history: Volatility and financial crises review of financial studies, Forthcoming; FEDS Working Paper No. 2016–093. Retrieved from https://ssrn.com/abstract=2872651 or https://doi.org/10.17016/FEDS.2016.093

Dennis, P., Mayhew, S., & Stivers, C. (2006). Stock returns, implied volatility innovations, and the asymmetric volatility phenomenon. *Journal of Financial and Quantitative Analysis, 41*(2), 381–406.

Ederington, L. H., & Guan, W. (2010). How asymmetric is U.S. stock market volatility? *Journal of Financial Markets, 13*(2), 225–248.

Fleming, J., Ostdiek, B., & Whaley, R. E. (1995). Predicting stock market volatility: A new measure. *Journal of Futures Markets, 15*(3), 265–302.

Frijns, B., Tallau, C., & Tourani-Rad, A. (2010). The information content of implied volatility: Evidence from Australia. *Journal of Futures Markets, 30*(2), 134–155.

Gerunov, A. (2014). Linkages between public sentiments and stock market dynamics in the context of the efficient market hypothesis. *Economic and Social Alternatives, 3,* 58–71 (in Bulgarian).

Giot, P. (2005). Relationships between implied volatility indexes and stock index returns. *The Journal of Portfolio Management, 31*(3), 92–100.

Gulen, H., & Ion, M. (2016). *Policy Uncertainty and Corporate Investment, 29,* 523–564.

Hayek, F. A. (1945). The use of knowledge in society. *American Economic Review, 54,* 519–530.

Hedström, A., Zelander, N., Junttila, J., & Uddin, G. (2020). Emerging market contagion under geopolitical uncertainty. *Emerging Markets Finance and Trade, 56*(6), 1377–1401. https://doi.org/10.1080/1540496X.2018.1562895.

Huang, C., Simpson, S., Ulybina, D., & Roitman, A. (2019). News-based sentiment indicators, IMF Working Paper No. 19/273. Retrieved from https://ssrn.com/abstract=3523146

Imlak, S., & Puja, P. (2015). The implied volatility index: Is 'investor fear gauge' or 'forward-looking'? *Borsa Istanbul Review, Research and Business Development Department, Borsa Istanbul, 15*(1), 44–52.

Keller, P., & Bieger, T. (2010). *Managing change in tourism: Creating opportunities—Overcoming obstacles* (p. 6). Schmidt: Publisher.

Keynes, J. M. (1936). T*he General Theory of Employment, Interest and Money* (pp. 161–162). London: Macmillan.

Kownatzki, C. (2006). How good is the VIX as a predictor of market risk? *Journal of Accounting and Finance, 16*(6), 39–60.

Markowitz, H. (1952). Portfolio selection. *The Journal of Finance, 7*(1), 77–91.

Moghaddam, M. D., Liu, Z., & Serota, R. A. (2019). Distribution of historic market data ¨C implied and realized volatility. *Applied Economics and Finance, Redfame Publishing, 6*(5), 104–130.

Mora, A., & Sethapramote, Y. (2019). Spillover effects of global financial uncertainty and economic policy uncertainty on stock exchange of Thailand. *Chiang Mai University Journal of Economics, 23*(21), 1–26.

Peneva, A. (2020). Analysis of the relationships between capital and foreign exchange markets, Godishen almanah "Nauchni izsledvaniya na doktoranti", XII, Book 15 (In Bulgarian).

Saffet, A., İlker, K., & Hakan, Y. (2019). Does VIX scare stocks of tourism companies? *Letters in Spatial and Resource Sciences, Springer, 12*(3), 215–232.

Schwert, G. W. (1989). Why does stock market volatility change over time? *Journal of Finance,* *44*(5), 1115–1153.

Schwert, G. W. (1990). Stock volatility and the crash of '87. *Review of Financial Studies, 3*(1), 77–102.

Shaikh, I., & Padhi, P. (2015). The implied volatility index: Is 'investor fear gauge' or 'forward-looking'?, Borsa Istanbul Review. *Research and Business Development Department, Borsa Istanbul, 15*(1), 44–52.

Scott R. B., Bloom, N., Davis, S., & Kost, K. (2019). Policy news and stock market volatility, NBER Working Papers 25720, National Bureau of Economic Research, Inc. (pp. 1–53).

Simeonov, S. (2020). Analiz na aktivnostta na osnovnite Iztochnoaziatski fondovi borsi (v perioda 2007–2019), e-Journal VFU, 13, 1–25.

Whaley, R. E. (1993). Derivatives on market volatility: Hedging tools long overdue. *Journal of Derivatives, 1,* 71–84.

Whaley, R. E. (2000). The investor fear gauge. *The Journal of Portfolio Management, 26*(3), 12–17.

Whaley, R. E. (2009). Understanding the VIX. *The Journal of Portfolio Management.*

Yu H., Fang L., & Sun B. (2018). The role of global economic policy uncertainty in long-run volatilities and correlations of U.S. industry-level stock returns and crude oil. *PLoS ONE, 13*(2): e0192305. https://doi.org/10.1371/journal.pone.0192305

The Productivity Puzzle in Cultural Tourism at Regional Level

Eleonora Santos, Inês Lisboa, Jacinta Moreira, and Neuza Ribeiro

1 Introduction

Tourism is an important driver of economic growth via tax revenues, infrastructure development and job creation (Pulido-Fernández et al., 2013). As a result, tourism is regarded as a priority area within the scope of the research and innovation strategy for smart specialization (RIS3) in Portugal (Ramos & Rosa, 2018; Marques et al., 2019). In addition, European regions highly specialized in tourism have been reported as having more difficulties to return to a growth path after negative external shocks from international crisis (Milio, 2014). This issue is echoed in the spatial distribution of the clusters relating tourism firms' performance to productivity (Romão & Nijkamp, 2019).

Productivity is linked to the notion of efficiency that translates into firms getting the most from their scarce resources. Thus, increased efficiency can lead to higher economic growth (Hall & Williams, 2019).

However, regional disparities in productivity greatly challenge the overall regional equity. As a result, a primary concern for policymakers is how to balance productivity growth which can be achieved, to some extent, by promoting tourism. In this context, because human capital is related to the implementation of competitive strategies based on differentiation (Úbeda-García et al., 2014; Yang & Cai, 2015), labour productivity can be seen as a relevant indicator of tourism competitiveness (Cvelbar et al., 2016).

In a territory such as Portugal, which combines the diversity of natural resources, with a unique architectural heritage and a mild climate, cultural tourism activities can be important catalysts for the social development and recovery of regions weakened from the social and economic point of view. Cultural tourism is increasingly

E. Santos (✉) · I. Lisboa · J. Moreira · N. Ribeiro
CARME – Centre of Applied Research in Management and Economics, Polytechnic Institute of Leiria, Leiria, Portugal

© The Author(s), under exclusive license to Springer Nature Switzerland AG 2021
V. Katsoni and C. van Zyl (eds.), *Culture and Tourism in a Smart, Globalized, and Sustainable World*, Springer Proceedings in Business and Economics,
https://doi.org/10.1007/978-3-030-72469-6_25

understood as an instrument for preserving cultural heritage. Though, the dynamics of tourism is characterized by a heterogeneous spatial distribution that concentrates around clusters of regions (Kang et al., 2014; Majewska, 2015). Yet, Guedes and Jiménez (2015) suggest that organized tourist programs based on cultural heritage break reduce to some extent the asymmetry of the spatiality of the Portuguese tourism model that historically was concentrated around the Algarve.

Previous research for Portugal (Santos, 2020) suggests that firms operating in cultural tourism activities have higher average profits, make a more efficient usage of investors' funds and display a better liquidity position. However, the analysis across business cycles indicates that these firms are more vulnerable to periods of crisis and expansion than the remaining tourism firms. Also, the analysis for larger firms shows that cultural tourism firms display better financial structures.

In the last two decades of the XX century, the great importance of tourism on local and national economies has been acknowledged (Church et al., 2000). Nevertheless, the increasingly policy awareness has not been accompanied by research on the organizational and structure aspects of the supply side of tourism. Three major reasons can explain the lack of research in this field: first, because the tourism industry has specificities, the theoretical framework has been relatively isolated from mainstream economic geography; second, the fragmentation of the production and the fact that tourism activities fall into several sectors (such as transport, retailing and catering) creates major difficulties regarding the collection and systematization of secondary statistics, which have not been compensated by primary data collection at firm level.

This paper fills the gap by using secondary data at firm level to calculate the spatial distribution of labour productivity in Portugal and to compare it with that of the distribution of changes in tourism demand (guests) across NUTS II regions.

In what follows, Sect. 2 presents the empirical literature. Section 3 describes and analyses the trends in spatial distribution of tourism productivity, and Sect. 4 provides the concluding remarks.

2 Literature Review, Capitalize the First Letter of Every Word in the Title

This paper relates to two strands of literature: Cultural tourism and the supply side (production) of Tourism. Concerning the topic of cultural tourism, several authors created various definitions of cultural tourism, among which Richards's definition (2002, 2007) is one of the most popular, describing cultural tourism as the movement of people towards cultural attractions, somewhere other than their usual place of residence, in order to find information and knowledge to fulfil their own cultural demands.

Cultural tourism firms are a sustainable pro-poor tourism initiative by engaging local communities in various tourism activities and allowing them to earn an income from tourism (Anderson, 2015). Thus, cultural tourism is a key tourism typology incorporated in worldwide national and regional policies. Its popularity among policymakers is due mainly to the potential to attract high-quality, high-spending tourists and at the same time provide economic support for culture (Richards, 2018). For example, Gomes et al. (2006) in their analysis of cultural and artistic entities in Portugal denote that, by and large, regional tourism entities invest on cultural tourism due to their local development potential. The importance of cultural tourism is also associated with its potential in reducing rigid seasonal cycles, since it has the potential to increase the positive effects of tourism development at the regional level (Bonet, 2003; Cuccia & Rizzo, 2011; Figini & Vici, 2012). In this context, areas far from the main tourist routes or that do not have the traditional resources to seduce tourists (beach and surf, monuments and other symbolic attractions) have the potential to attract tourists and develop economically via the conservation and restoration of their cultural heritage (European Commission, 1995 in Bonet, 2003).

In previous studies, it was generally assumed that cultural attractions appeal more to higher socio-economic groups with higher cultural capital. Accordingly, most cultural tourists are regarded and described as "up-scaled" (mature aged with high education and high income earnings) and female (Craik, 2002). This is corroborated by the research of ATLAS (2007), showing that the average consumption of cultural tourists is 10% higher than the average consumption of leisure tourists. However, this basic profile of cultural tourists does not seem to reflect the progressively diversified group of cultural tourist attractions (e.g. art galleries, opera, amusement parks, history museums, music concerts, etc.) along with the changing cultural tastes in post-modern society (Bryson, 1996; DiMaggio & Mukhtar, 2004; Kim et al., 2007). Furthermore, not all visitors are cultural tourists, and the motivation among cultural tourists is very different. The average cultural tourist looks for a mix of cultural activities, entertainment and relaxation, and not only for the traditional "high culture" products (Jovicic, 2016).

In recent years, cultural tourism activities have become increasingly important for visited regions given its economic implications, and therefore, different forms of tourism have grown in popularity and have captured practitioners' interest (Bell et al., 2007). This recognition creates opportunities to extend the existing knowledge about the impacts of tourism activities by approaching these activities from a supply perspective. Understanding and researching in tourism is critical to destinations' and businesses' marketing strategy tourism management and tourism sustainable development. Accordingly, this paper focuses on the supply side (production) of tourism.

In the last two decades of the XX century, the great importance of tourism on local and national economies has been acknowledged (Church et al., 2000). Nevertheless, the increasingly policy awareness has not been accompanied by research on the organizational and structure aspects of the supply side of tourism.

Two major reasons can explain the lack of research in this field: first, because tourism industry has specificities, the theoretical framework has been relatively isolated from mainstream economic geography; second, the fragmentation of the production and the fact that tourism activities fall into several sectors (such as transport, retailing and catering) creates major difficulties regarding the collection and systematization of secondary statistics, which have not been compensated by primary data collection at firm level. For this reason, recently most of supply side research on tourism use surveys and questionnaires. For example, Lundberg and Fredman (2012) provide an exploratory analysis of business success factors and constraints among nature-based tourism entrepreneurs in Sweden. Data collected in 2009 include "life-history" interviews, a telephone survey of 176 entrepreneurs and follow-up critical incident interviews. The results show that internal factors are more common for business success, while external factors dominate among the constraints. Among the 26 success items studied, management, access to natural resources and lifestyle are considered the most important. Low profitability, lack of capital, regulations, infrastructure and taxes are given the highest weights among the constraints.

Rauken et al. (2010) study the effects of weather changes on small- and medium-sized enterprises (SMEs) in the tourism and hospitality industry as perceived by firms in two coastal areas in Norway. The data are derived from two-stage semi-structured interviews with industry representatives. Results indicate that weather and weather changes do not stand out as being a major concern among the respondents, although it is acknowledged that some types of weather, notably precipitation and low visibility, can negatively affect businesses. At the same time, the operators are familiar with combining outdoor 'recreation and unpredictable weather', meaning that the weather is just not "bad enough" to be concerned with, especially given tourist expectations of weather in these locations.

Fredman et al. (2012) investigate to what extent nature-based tourism companies in Sweden depend upon natural environments and facilities, open access and exclusive rights to natural resources. They use a mixed method approach, including qualitative data from interviews and quantitative data from a subsequent telephone survey to 11 nature-based tourism companies in Sweden. Findings suggest that open access is much more important than exclusive rights, while naturalness and facilities represent important attributes.

Kastenholz et al. (2012) scrutinize the rural tourism experience offered by a small village in Central Portugal. Based on interviews, the authors analyse the experience of tourists and residents, the interactions between different stakeholders, impacts and marketing implications. Different stakeholders were interviewed, including four local supply agents, during March and April 2011. Supply agents identified constraints to local development, namely the decrease of population, the bureaucracy involved in investing in tourism and the lack of investment, particularly regarding stores offering local products. These agents confirm that the village's tourist demand is mostly domestic, composed largely of same-day visitors, although international tourists also stand out. They identify nature, local heritage and the paragliding festivals as the tourists' main motivations.

Wang and Lyons (2012) use a qualitative approach to examine the values that underpin commercial adventure tourism supply. A sample of 21 commercial adventure tour operators in national parks in New South Wales (NSW), in Australia, was selected from three distinct regions representing the diversity of operational and geographical characteristics of adventure tourism supply. In-depth interviews were conducted that explored not only the business approaches that managers took but also their personal histories and experiences in protected areas. The results highlight that while business-related values dominate decision-making among operators, they are tempered by personal values associated with conservation and recreation.

Jones et al. (2015) attempt to identify the critical success factors that have shaped the Napa Valley rise and growth and its unique competitiveness in both the domestic and global marketplaces for wine tourism destinations. The selection of interviewed participants was based on their organizational experiences and expertise as well as recommendations from industry leaders. A 12-question, open-ended survey instrument for qualitative interviews was designed. Semi-structured interviews were conducted from August 2012 to January 2013 with ten suppliers of the Napa wine tourism experience. Findings reveal an aligned marketing effort and strategic partnerships among suppliers.

Shan and Marn (2013) examine and identify the major critical success factors that are crucial in the successful development of the tourism industry of Penang Island. Interview questions were built upon the theory as proposed by Baker (1998). The interview sessions have been carried out through face-to-face interviews with qualified managerial-level local authority and stakeholders from the five major components of tourism suppliers proposed by Gunn and Vars. The respondents were free to provide insights, opinions and comments that are relevant to the area of the research topic that is deemed informative for the researcher. The findings suggest that the authority and relevant stakeholders in the industry should provide greater emphasis and effort in developing competence and strengths in the seven identified factors in order to improve overall destination competitiveness locally as well as globally.

McNicol and Rettie (2018) present results of commercial tour operator's perspectives of environmental supply in two Canadian Rocky Mountain national parks. They use a four-stage mixed methodology including one-on-one field interviews and surveys targeting commercial tour guiding businesses operating within Banff and Jasper national parks, in 2011–2015. Each interview included formal questions, ranging from a description of the types of activities included during tours to actual or perceived limitations placed on operations due to managers' considerations about environmental supply. Results reflect 17 randomly sampled formal interviews and 41 completed online surveys with representatives of tour companies operating within Banff or Jasper national parks. They conclude that tour operators must supply quality visitor experiences while operating within a different policy environment than required by other tourism operators providing guided tours outside of park boundaries.

3   Trends in Spatial Distribution of Tourism Productivity in Portugal, 2015–2018

According to the data from the Tourism Satellite Account (TSA), released by the National Statistics Institute (INE), tourism consumption in the territory and economy reached 29.8 million euros, which represented an increase in nominal terms of 7.7% relative to the previous year. The estimated consumption of tourism in 2018 represented 14.6% of the Gross Domestic Product (GDP) and 8% of the Gross Value Added (GVA).

The Gross Value Added by Tourism (VABGT) reached 14.1 million euros and increased by 8.0% in nominal terms, while the GVA of the national economy grew only 4.0%.

In 2017, tourism employed 413,567 individuals, representing 9.0% of the national employment, increasing 8.7%, against the increase at national level (3.4%). Moreover, the TSA data for 2017 show that tourism contributes to 21.7% of the total national exports of services. The highest share of tourists' expenses comes from restaurants (26.9%), accommodation (25.7%) and in transport (21.3%).

Figure 1 shows the growth rate of tourism demand (%) across Portuguese NUTS II regions. In 2015, the regions with major increases in touristic demand were Azores, Centre and Alentejo. In 2016, the Islands registered the highest rate followed by the central region; while in 2017, the Centre region showed the highest increase in terms of number of guests, followed by Azores and Alentejo. Finally, in 2018, the regions of Azores and North, followed by Alentejo, show higher rates of growth of touristic demand, measured by the number of guests. These facts show the differences regarding tourism demand among regions, calling the need to better understand the impact of productivity at a regional context.

Using panel data from SABI database and the National Tourism Registry, the sample to calculate labour productivity comprises 2866 firms operating in cultural tourism over a period of 4 years (2015–2018), in a total of 11,464 observations (see

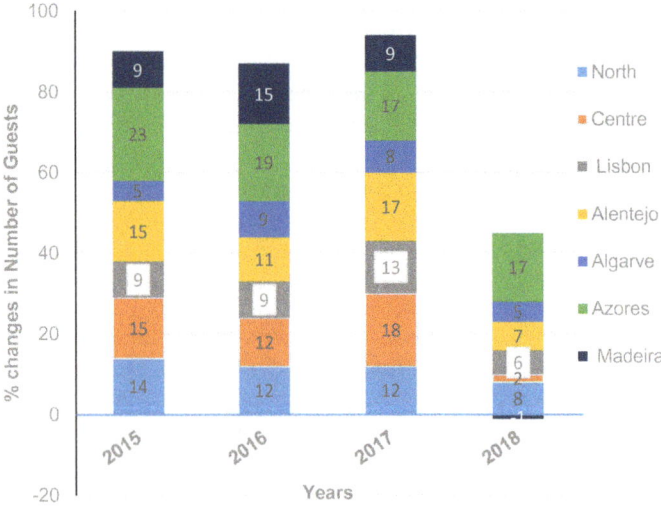

Fig. 1 Growth rate of tourism demand (%) by NUTS II regions, 2015–2018. *Source* Authors calculations from PORDATA

Table 1 Regional distribution of firms in the sample

Region	# firms	% in total
North	616	21
Centre	191	7
Lisbon	1489	52
Alentejo	105	4
Algarve	260	9
Azores	84	3
Madeira	121	4

Source Authors' calculations

Table A1 in the appendix for basic statistics). The number of employees and turnover are provided by SABI financial reports. Labour productivity is calculated as the ratio between turnover and the number of employees.

On the basis of data published in SABI, the regional distribution of firms and its share on total number of firms in the sample is calculated and listed in Table 1. Thus, there is a dominance of Lisbon region regarding the number of cultural tourism firms (52%) followed by the Northern region (21%). This is not surprising bearing in mind that Lisbon is the capital and Porto is the second most important city in Portugal. The Islands (Azores and Madeira) capture only 7% of cultural tourism firms.

Table 2 exhibits the annual average turnover by region. The highest average turnover is shown in firms in Algarve, followed by the Northern region and Madeira. The central region displays the lowest values of average turnover, but also only 7% of firms over the total sample.

According to Fig. 2, during the period from 2016 till 2018, there was a declining trend regarding the average turnover in all regions. The exceptions are the regions of Azores, Alentejo and Madeira in 2016–2017 and the Northern region in 2016. A possible explanation lies in the higher growth rates of tourism demand in these regions, measured by changes in the number of guests.

Regarding the annual average number of employees (see Table 3), the Algarve region shows the higher value of five employees, followed by the Northern region with four employees. The regions of Lisbon, Alentejo and Centre have an average of two employees by cultural tourism firm. In the Islands, cultural tourism firm has, on average, three employees.

Table 2 Annual average turnover by NUTS II regions

Year	North	Centre	Lisbon	Alentejo	Algarve	Azores	Madeira
2015	239	109	140	98	292	97	117
2016	255	98	128	108	296	112	123
2017	240	88	111	113	290	130	128
2018	220	83	102	103	250	122	118

Source Authors' calculations

Fig. 2 Changes in average turnover by NUTS II regions. *Source* Authors calculations

Table 3 Annual average employment by region

Year	North	Centre	Lisbon	Alentejo	Algarve	Azores	Madeira
2015	4	3	2	2	5	2	3
2016	4	2	2	2	5	3	3
2017	4	2	2	2	5	3	3
2018	4	2	2	2	5	3	3

Source Authors' calculations

According to Table 4, the average labour productivity increased by 1.6% in 2016 and by 2.1% in 2017. However, it declined in 2018 by nearly 4%.

Table 5 shows the regional distribution of average labour productivity. The average labour productivity is similar across regions; notwithstanding, the maximum values are very uneven. The Lisbon region displays the highest values for the maximum labour productivity, followed by Azores. In the case of Lisbon, the maximum labour productivity is expected due to the higher number of firms located in the region; however, in Azores, the maximum labour productivity may be related to other factors such as the growth rate of tourism demand (measured by the number of guests) in 2015–2018. The Northern region ranks in third place but with a maximum labour productivity that is less than 1/3 of those recorded by Lisbon and Alentejo. Cultural tourism firms that exhibit higher values of labour productivity operate in amusement and recreational activities. Regarding average labour productivity, the regions of

Table 4 Average labour productivity and annual growth rate, 2015–2018

Year	Average labour productivity	Annual growth rate
2016	41.6	1.6
2017	42.5	2.1
2018	40.8	−3.9

Source Authors' calculation on Stata 13.0

Table 5 Average and maximum labour productivity, by NUTS II regions, 2015–2018

Region	Average labour productivity	Maximum labour productivity
North	38	522
Centre	38	494
Lisbon	45	1818
Alentejo	40	284
Algarve	34	474
Azores	44	1704
Madeira	40	371

Source Authors' calculation on Stata 13.0

Fig. 3 Changes in tourism demand and average labour productivity by NUTS II regions. *Source* Authors calculations

Lisbon and Azores show the highest values, and Algarve displays the lowest value in 2015–2018.

As expected, regional inequality of maximum labour productivity in cultural tourism firms is shown to be much greater than for that of average labour productivity across NUTS II regions.

Crossing data on changes of tourism demand (measured by the growth rates of the number of guests) and changes in average labour productivity by region (Fig. 3), it appears that both variables change in same direction in 2016 in the Centre, Alentejo and the Islands. However, in 2017 only in the region of Azores, both variables change in same direction; while in 2018 that happens only for the Northern region of the country. Thus, these results as well those of Pearson correlation show that there is no correlation between changes in tourism demand and labour productivity. As a result, policies designed to attract tourists to certain regions would probably fall short on contributing to increase firms' productivity.

Regarding the implications for tourism management, in some Portuguese regions, if cultural tourism managers want to increase the labour productivity, they should reduce the number of employees. In particular in the Northern and Algarve regions, firms report the higher levels of average turnover but not the higher levels of labour productivity because they maintain high levels of employment as well. The case of Algarve is a major source of concern since it records the lowest average labour productivity, even though cultural tourism firms exhibit the highest levels of average turnover. Thus, if cultural tourism firms in the Algarve regions pursuit levels of labour productivity similar to those recorded in Lisbon and Azores, they must reduce the personnel in nearly half to keep the number of workers in 2–3 individuals.

4   Conclusion and Policy Implications

Productivity measurement in tourism has been raising increasing interest in academia and recognized as critically important, since it is an important indicator for understanding the strengths and weaknesses of the tourism sector. This paper adopts a regional view in order to follow an exploratory analysis of the productivity of tourism firms in Portugal. The specific focus is upon understanding the regional distribution of labour productivity among cultural tourism firms.

This research is innovative for four reasons: first, it contributes to the literature of tourism from the supply side; secondly, by using firm level data from SABI database, it helps to fill the gap regarding the collection and use of tourism secondary data at firm level; thirdly, it calculates the labour productivity for a special type of tourism, the cultural tourism. Calculating productivity of tourism industry causes difficulties related to data collection and systematization; fourthly, it focuses on all regions of Portugal contrary to the tradition of focusing only one ad hoc region, usually one with greater tourist influx, without regional comparisons and, hence, with no practical implications for regional policy. Indeed, research on tourism would gain from regional asymmetries analysis especially regarding productivity. This is so because productivity is a notion closely related to competitiveness and living standards of local communities. In this context, cultural tourism is a type of tourism that can potentially mitigate the regional asymmetries towards income. In fact, resources in cultural tourism are more democratic, in the sense that, unlike other types of tourism, such as nautical, it is not limited to a region's specific natural resource.

Results indicate that, while in many advanced economies, tourism is increasingly becoming a knowledge-based activity, in Portugal, it is clearly a place-based activity. The pattern of distribution of productivity across NUTS II regions shows that current firms' performance is not associated to the dynamics of neighbouring regions but to the resources and conditions available in each region.

Thus, results can increase the knowledge of tourism managers, not only because results show that more employees do not mean more productivity, but also conclusions are different from those of advanced economies in tourism. Indeed, the results provide insights when it comes to design the priorities of tourist development for different regions. Tourism management includes a reduction of employment in cultural tourism firms located in the Northern and Southern (Algarve) regions of Portugal since in these regions the higher levels of average turnover are being obtained with resource to a higher number of (apparently) unproductive personnel. Hence, this paper suggests a trade-off between employment and firms' efficiency, which is uncovered by the results on the spatial distribution of labour productivity. However, at macro level, government and local authorities should evaluate the gains and costs of such measures in terms of regional competitiveness and regional employment.

As stated by Ferreira (2010), cultural policies in Portugal have changed from strategies based in infrastructures, subsidization of artistic production and the promotion of equal opportunities to measures supporting cultural and creative businesses, linking subsidies and incentives and encouraging the use of culture, as an element of regional identity and a factor of regional differentiation.

Tourism industry has different challenges regarding productivity, and it is often reported as having low productivity when compared to other industries. Accordingly, cultural strategies have consisted in some cases of precarious forms of labour. Finally, by identifying the areas where tourism management needs to be improved, this paper provides suggestions on measures to increase regional productivity and thus regional competitiveness. National and regional tourism planning needs to be carried out because tourism makes a significant contribution to the sustainable economic development of regions. By helping the local community to increase employment and income, preserve historical and cultural monuments, strengthen local identity and the established way of life, cultural tourism plays a key role in economic and sociocultural sustainability.

Acknowledgements This paper is financed by National Funds of the FCT—Portuguese Foundation for Science and Technology within the project "UIDB/04928/2020"

References

Anderson, W. (2015). Cultural tourism and poverty alleviation in rural Kilimanjaro, Tanzania. *Journal of Tourism and Cultural Change, 13*(3), 208–224.

ATLAS. (2007). ATLAS cultural tourism survey—Summary report. Retrieved April 14, 2011, from www.tram-research.com/atlas

Bell, S., Tyrvainen, L., Sievanen, T., Probstl, U., & Simpson, M. (2007). Outdoor recreation and nature tourism: A European perspective. *Living Reviews in Landscape Research, 1*(2), 2. https://doi.org/10.12942/lrlr-2007-2

Bonet, L. (2003). Cultural tourism. In R. Towse (Ed.), *A Handbook of Cultural Economics* (pp. 187–193). Cheltenham, UK: Edward Elgar.

Bryson, B. (1996). "Anything but heavy metal": Symbolic exclusion and musical dislikes. *American Sociological Review, 61*(5), 884–899.

Church, A., Ball, R., Bull, C., & Tyler, D. (2000). Public policy engagement with British tourism: The national, local and the European Union. *Tourism Geographies, 2*(3), 312–336.

Craik, J. (2002). The culture of tourism. In Rojek, C. & Urry, J. (Eds.), *Touring cultures: Transformations of Travel and Theory* (pp. 123–146). Routledge.

Cuccia, T., & Rizzo, I. (2011). Tourism seasonality in cultural destinations: Empirical evidence from Sicily. *Tourism Management, 32*(3), 589–595.

DiMaggio, P., & Mukhtar, T. (2004). Arts participation as cultural capital in the United States, 1982–2002: Signs of decline? *Poetics, 32*(2), 169–194.

European Commission. (1995). The role of the European Union in the promotion of tourism, Brussels, EC. In Bonet, L. & Towse, R. (Ed.), *Cultural Tourism. A Handbook of Cultural Economics.* Edward Elgar: Cheltenham, UK, 2003 (pp. 187–193).

Ferreira, A. (2010). Comunidades criativas e desenvolvimento, jornad—A criatividade empresarial como caminho para a criatividade. CRIA Divisão de Empreendedorismo e Transferência de Tecnologia da Universidade do Algarve, 20 de Abril de 2010.

Figini, P., & Vici, L. (2012). Off-season tourists and the cultural offer of a mass-tourism destination: The case of Rimini. *Tourism Management, 33*(4), 825–839.

Fredman, P., Wall-Reinius, S., & Grundén, A. (2012). The nature of nature in nature-based tourism. *Scandinavian Journal of Hospitality and Tourism, 12*(4), 289–309.

Gomes, R. T., Lourenço, V., & Martinho, T. D. (2006). Entidades Culturais e Artísticas em Portugal. Centro Cultural de Cascais.

Guedes, A. S., & Jiménez, M. I. M. (2015). Spatial patterns of cultural tourism in Portugal. *Tourism Management Perspectives, 16,* 107–115.

Hall, C. M., & Williams, A. M. (2019). *Tourism and Innovation.* New York, NY: Routledge.

Jones, M. F., Singh, N., & Hsiung, Y. (2015). Determining the critical success factors of the wine tourism region of Napa from a supply perspective. *International Journal of Tourism Research, 17*(3), 261–271.

Jovicic, D. (2016). Cultural tourism in the context of relations between mass and alternative tourism. *Current Issues in Tourism, 19*(6), 605–612.

Kang, S., Kim, J., & Nicholls, S. (2014). National tourism policy and spatial patterns of domestic tourism in South Korea. *Journal of Travel Research, 53*(6), 791–804.

Kastenholz, E., Carneiro, M. J., Marques, C. P., & Lima, J. (2012). Understanding and managing the rural tourism experience—The case of a historical village in Portugal. *Tourism Management Perspectives, 4,* 207–214.

Kim, H., Cheng, C. K., & O'Leary, J. T. (2007). Understanding participation patterns and trends in tourism cultural attractions. *Tourism Management, 28*(5), 1366–1371.

Knežević Cvelbar, L., Dwyer, L., Koman, M., & Mihalič, T. (2016). Drivers of destination competitiveness in tourism: A global investigation. *Journal of Travel Research, 55*(8), 1041–1050.

Lundberg, C., & Fredman, P. (2012). Success factors and constraints among nature-based tourism entrepreneurs. *Current Issues in Tourism, 15*(7), 649–671.

Majewska, J. (2015). Inter-regional agglomeration effects in tourism in Poland. *Tourism Geographies, 17*(3), 408–436.

Marques, A. V., Marques, C., Braga, V., & Marques, P. M. (2019). University-industry technology transfer within the context of RIS3 North of Portugal. *Knowledge Management Research & Practice, 17*(4), 473–485.

McNicol, B., & Rettie, K. (2018). Tourism operators' perspectives of environmental supply of guided tours in national parks. *Journal of Outdoor Recreation and Tourism, 21,* 19–29.

Milio, S. (2014). Impact of the economic crisis on social, economic and territorial cohesion of the European Union (Vol. 1). Brussels: Directorate-General for Internal Policies, Policy Department B (Structural and Cohesion Policies).

Pulido-Fernández, J. I., Cárdenas-García, P. J., & Villanueva-Álvaro, J. J. (2013). On the role of environmental sustainability in the transformation of tourism growth into economic development. *Environmental Engineering & Management Journal (EEMJ), 12*(10), 2009–2018.

Ramos, A. S., & Rosa, F. (2018). Empreendendo descoberta inteligente: Uma abordagem aos modelos de implementação da especialização regional em Portugal. *Public Policy Portuguese Journal, 3*(1), 57–74.

Rauken, T., Kelman, I., Steen Jacobsen, J. K., & Hovelsrud, G. K. (2010). Who can stop the rain? Perceptions of summer weather effects among small tourism businesses. *Anatolia, 21*(2), 289–304.

Richards, G. (2002). Tourism attraction systems: Exploring cultural behavior. *Annals of Tourism Research, 29*(4), 1048–1064.

Richards, G. (2007). *Cultural tourism: Global and Local Perspectives.* New York: Psychology Press.

Richards, G. (2018). Cultural tourism: A review of recent research and trends. *Journal of Hospitality and Tourism Management, 36,* 12–21.

Romão, J., & Nijkamp, P. (2019). Impacts of innovation, productivity and specialization on tourism competitiveness—A spatial econometric analysis on European regions. *Current Issues in Tourism, 22*(10), 1150–1169.

Santos, E. (2020). Do cultural tourism firms perform better than their rivals?. In Rocha, Á., Abreu, A., de Carvalho, J., Liberato, D., González, E., & Liberato, P. (Eds.), Advances in Tourism, Technology and Smart Systems. Smart Innovation, Systems and Technologies (Vol. 171). Springer: Singapore.

Shan, B. Y., & Marn, J. T. K. (2013). Perceived critical success factors (CSFS) for the tourism industry of Penang Island: A supply perspective. *Interdisciplinary Journal of Contemporary Research in Business, 4*(9), 495–510.

Úbeda-García, M., Cortés, E. C., Marco-Lajara, B., & Zaragoza-Sáez, P. (2014). Strategy, training and performance fit. *International Journal of Hospitality Management, 42,* 100–116.

Wang, P. Y., & Lyons, K. D. (2012). Values congruity in tourism and protected areas policy: Evidence from adventure tourism supply in New South Wales Australia. *Annals of Leisure Research, 15*(2), 188–192.

Yang, Z., & Cai, J. (2015). Do regional factors matter? Determinants of hotel industry performance in China. *Tourism Management, 52,* 242–253.

Factors Affecting Destinations

Capital Structure Determinants of Greek Hotels: The Impact of the Greek Debt Crisis

Panagiotis E. Dimitropoulos and Konstantinos Koronios

1 Introduction

The debates on capital structure have lasted for six decades since the irrelevance propositions provided by Modigliani and Miller's (1958) claiming that the cost of capital is independent of a firm's capital structure, which is conducted without considering the effects of tax, bankruptcy cost, and agency cost. Considering the divergence and conflict of interests among shareholders, managers, and creditors, Myers' (1984) introducing agency cost into the trade-off theory and predicting an optimal capital structure where the tax benefits of increased leverage are offset by the corresponding bankruptcy costs and agency costs. Nevertheless, pecking order theory argues that firms prefer to use internal resources such as retained earnings to finance their investment rather than issue debt, and equity financing might be the last choice (Myers, 1977; Myers & Majluf, 1984). Similar to the conflicts between theoretical studies, the empirical findings are also contradictory, supporting either trade-off theory or pecking order theory or even both of them (Psillaki & Daskalakis, 2009; Kyriazopoulos, 2017; Chatzinas & Papadopoulos, 2018).

The onset of Greek debt crisis can be traced to the US sub-prime crisis in 2007 (Arghyrou & Tsoukalas, 2011). Wickens (2017) attributes this crisis to discretionary policy and the failure of tax revenues to match expenditures, while Tsoukis et al. (2017) find that the rapid rise in costs, loss of competitiveness, rise in consumption, decline in savings, the government budget deficits, demographic changes, and the unsustainable public funding of the pension system all are the causes of the crisis.

P. E. Dimitropoulos (✉)
Department of Sport Organization & Management, University of Peloponnese, Sparta, Greece
e-mail: dimitrop@uop.gr

K. Koronios
Department of Accounting and Finance, University of Peloponnese, Kalamata, Greece

© The Author(s), under exclusive license to Springer Nature Switzerland AG 2021   387
V. Katsoni and C. van Zyl (eds.), *Culture and Tourism in a Smart, Globalized, and Sustainable World*, Springer Proceedings in Business and Economics,
https://doi.org/10.1007/978-3-030-72469-6_26

Greece is severely suffered in this crisis, having its rating in 2015 dropped to B- (S&P), Caa3 (Moody's), CCC (Fitch). Moreover, its GDP shrank about 25% over the period of 2010–2015, accompanied with nearly 25% unemployment (roughly 60% youth unemployment) and high ratios of debt to GDP, which were 177% for public and 142% for foreign debt. Also, the capital market was frozen and bank lending mitigated, facts that affected the capital structure of Greek firms (Chatzinas & Papadopoulos, 2018). Furthermore, the Gini coefficient of Greece climbed to approximately 0.34–0.345 in 2013, with the increase of relative poverty rate from nearly 0.12–0.15 (Economides et al., 2017; Tsoukis et al., 2017).

The Greek hotel industry remains an important pillar for economic development since it is considered as the "hard" industry of the Greek economy. According to Dimitropoulos (2018) and Dimitropoulos et al. (2019), hotels' long-term liabilities increased during the crisis period and several hotel firms faced significant problems in raising financing from various sources. This fact has a significant effect on the hotel's capital structure decisions and their ability to finance daily operations and future investment projects. Pacheco and Tavares (2017) argue that the intensive fixed assets structure of hotel firms may influence their capital structure because of the enhanced fixed costs they face and their volatile revenue streams they generate. This may lead hotels to utilize more internal sources of capital instead of lending but on the contrary, the fact that their capital is invested on fixed assets that could be collateralized for assessing bank lending, they could resolve to external financing. Maças Nunes and Serrasqueiro (2017) corroborate the above arguments by documenting that external financing is even more difficult for small- and medium-sized hotels due to their high volatility of financial results. This fact could have major importance for hotels especially during times of financial frictions.

This paper aims to investigate the firm-specific determinants of capital structure of Greek non-listed hotel firms before and after the period of debt crisis. In addition, this paper intends to study whether the firm-specific determinates of capital structure and their effects on leverage will differ across those two sub-periods. The first motivation of this paper is the special macroeconomic environment of Greece during the period of crisis, since several studies find that the macroeconomic factors such as economic crisis could affect firm's capital structure (Daskalakis et al., 2017; McNamara et al., 2017; Ahsan et al., 2016). Greece might suffer the most severe consequences of this crisis compared to other Eurozone countries and the period of 2010–2016 seems to be the most severe time for the Greek economy. The lack of literature is also the motivation of this paper, for the reason that most of the previous studies focus on firms in UK or the USA and seldom consider Greece.

Furthermore, studies published on the Greek business setting either investigate the behaviors of small- and medium-sized enterprises (SMEs) or select a different period of time, most of which do not focus on firm-specific factors (Noulas and Genimakis, 2011; Balios et al., 2016). Moreover, the current study extends the previous work by Chatzinas and Papadopoulos (2018) and Kyriazopoulos (2017) by including a largest sample (almost the majority) of non-listed, non-financial corporations in Greece and not focusing only on listed firms as the previously mentioned studies. Moreover, we include a large sample period even before the US financial crisis of 2007–2008, and thus, we are able to extract more efficient inferences regarding the impact of firm-specific determinants on the capital structure. Finally, this is the first study in the Greek hotel industry that tries to assess the impact of the sovereign debt crisis on the capital structure decisions of hotels.

This paper has five sections. A detailed literature review of the determinants of capital structure including both theoretical researches and empirical studies will be provided in the following section followed by the research design and data selection explained in Sect. 3. The results and discussions will be demonstrated and analyzed in Sect. 4. Finally, an overall conclusion will be summarized in the last section.

2   Literature Review and Testable Hypothesis

There are several empirical researches that focused on the determinants of capital structure. Kayo and Kimura (2011) study the hierarchical determinants of capital structure after analyzing the effects of time-level, firm-level, industry-level, and country-level determinants of leverage, finding that 78% of firm leverage can be explained by time-level and firm-level factors. Also, Psillaki and Daskalakis (2009) claim that it is not country-specific but firm-specific factors that explain the differences in leverage choices. Thus, this paper will focus on the firm-level determinants of capital structure. Following previous studies by Chen (2004), Balios et al. (2016) and Moradi and Paulet (2015), size, growth opportunities, profitability, earning volatility, asset tangibility, non-debt tax shields and tax are selected as the possible leverage determinants in the Greek hotel business setting.

2.1   Size

A positive relationship between firm size and leverage is expected since large firms are considered to be more diversified and less likely to go bankrupt resulting in lower bankruptcy cost (Ozkan, 2001). At the same time, the transaction costs associated with debt issuing in large firms are expected to be lower compared to other firms. Thus, according to the trade-off theory, large firms might have a high leverage level (Jensen & Meckling, 1976; Myers, 1984), which complies with several empirical researches (Psillaki & Daskalakis, 2009; Abor & Biekpe, 2009). However, large firms may also prefer to issue shares because of their low information asymmetry, considering the pecking order theory, which may result in a low leverage level (Rajan & Zingales, 1995). This opinion is supported by Titman and Wessels' (1988) empirical evidence. Nevertheless, small sized firms are more vulnerable during times of financial turmoil. According to Hoang et al. (2018), smaller firms during crisis are not able to finance their operations from external sources and they resort more often to internal financing. So based on the above discussion, it is expected that firms' size will have a positive impact on total and long-term debt during the crisis period.

H1: During the crisis period, firm size is positively associated with total and long-term debt.

2.2 Growth Opportunities

The theoretical predictions about the relationship between growth opportunities and leverage are conflicting. According to the trade-off theory, firms with growth potentials are riskier since they have a higher bankruptcy cost and agency cost of debt, and thus, they might find it difficult to finance operations through debt and thus tend to have lower leverage (Myers 1977, 1984). Nevertheless, the pecking order theory suggests a positive relationship between growth opportunities and leverage (Myers & Majluf, 1984). Yang et al. (2010) and Rajagopal (2011) argue that both negative and positive relationship exist between growth potential and leverage. Of course, during times of financial distress and recession growth is deteriorated and those firms that have growth potential could use that as a potential signal to gain access to financial markets. Hoang et al. (2018) provide evidence of a positive association between debt and growth opportunities during crisis. However, the opposite association could also be true since firms with high growth opportunities may be perceived as riskier and thus they can choose to re-pay a part of their debt obligations during crisis and finance their operations from internally generated funds (Daskalakis et al., 2017). So the second research hypothesis is stated in the null form.

H2: During the crisis period, growth opportunities will be associated with total and long-term debt.

2.3 Profitability

The relationships between profitability and leverage under different theories are conflicting. There might be a positive relationship between the variables according to the trade-off theory. Firms with high profitability are expected to have more capacity to benefit from the tax shield of debt because of their high profit and the corresponding low bankruptcy cost (Jensen & Meckling, 1976; Myers, 1984). However, the pecking order theory predicts a negative relationship between profitability and leverage (Myers & Majluf, 1984). Firms with high profitability seems to have adequate retained earnings to meet their investment needs, and thus, they will probably prefer financing through internal sources rather than seek external financing due to latter's high information costs and transaction costs (Myers, 1984; Myers & Majluf, 1984). Especially, during crisis periods profitability is restrained by market recession and decreased demand, making it even more difficult for firms to obtain external financing (Hoang et al., 2018; Daskalakis et al., 2017) and finance their operations through long-term debt. In consequence, firm's profitability is expected to be negatively associated with firm leverage:

H3: During the crisis period, profitability will be negatively associated with total and long-term debt.

2.4 Earning Volatility

Earning volatility refers to the risk and uncertainty of the future income streams of the firm (i.e., the higher the earning volatility, the riskier the firm is). The trade-off theory suggests a negative relationship between earning volatility and leverage since firms with high earning volatility are expected to be more likely to go bankrupt and thus have a high cost of financial distress, which forcing the firms to reduce their debt level (Bradley et al., 1984; Myers, 1984). Specifically, during crisis periods riskier firms may be more inclined to transfer their risk to external finance providers so they tend to borrow more (Daskalakis et al., 2017). Nevertheless, the majority of the previous works find a negative relationship between earning volatility and leverage (Banerjee et al., 2004; Kim et al., 2006; Psillaki & Daskalakis, 2009; Rajagopal, 2011; Choi & Richardson, 2016). Thus, fourth research hypothesis is stated as follows:

H4: During the crisis period, earnings volatility will be associated with total and long-term debt.

2.5 Asset Tangibility

Asset tangibility seems to be positively related to debt according to the trade-off theory since tangible assets can be used as collateral for creditors to borrow or issue debt (Miller, 1977; Myers, 1984). During crisis periods, the information asymmetry between borrowers and lenders increases and so tangible assets are more crucial for receiving external financing. Hoang et al., (2018) provide evidence of a positive impact of tangible assets on total and long-term leverage during the crisis period in France. The positive relationship is supported by several studies, such as Hovakimian et al., (2001), de Jong et al. (2008), Frank and Goyal (2003), Yang et al. (2010). Hence, based on the above discussion we state the fifth research hypothesis as follows:

H5: During the crisis period, asset tangibility will be positively associated with total and long-term debt.

2.6 Non-debt Tax Shields

Non-debt tax shields mainly refer to certain items with tax deductibility, such as depreciation, amortization, investment tax credits, and research and development expenses. Non-debt tax shields can be seen as the substitutes for tax deductibility of debt since these items have been deducted before calculating the taxable income, which means that they can also reduce firms' taxable income like debt (DeAngelo & Masulis, 1980). Consequently, according to the trade-off theory, firms having large non-debt tax shields might have lower leverage due to the remaining relatively less deductible tax shields (Modigliani & Miller, 1963; Myers, 1984). A recent study by Daskalakis et al. (2017) in Greece pointed to a negative impact of NDTS on long-term debt during the Greek crisis period because on that period firms' tax burden increased (due to enhanced tax rates) and profits decrease and thus NDTS were the only alternative in order to reduce their taxable income and their tax burden. Consequently, a negative relationship between a firm's non-debt tax shields level and its leverage level is predicted according to the following hypothesis:

H6: During the crisis period, net debt tax shields will be negatively associated with total and long-term debt.

2.7 Taxation

According to the trade-off theory, there might be a positive relationship between taxes and leverage since firms paying large corporate taxes are motivated to issue more debt to take advantage of the benefit of tax deductibility of debt (Modigliani & Miller, 1963; Myers, 1984). However, both the tax rate and profitability should be considered on the basis of the pecking order theory. If firms face a high corporate tax rate (aka, during the Greek crisis period), the retained earnings might be less and firms are forced to issue debt to meet their investment needs, which indicates a positive relationship between tax and debt level (Myers & Majluf, 1984). While the high corporate taxes might also represent a high profitability, which suggests a negative relationship between the variables (Myers & Majluf, 1984; Cespedes et al., 2010), therefore, the final research hypothesis is stated as follows:

H7: During the crisis period, income taxes will be associated with total and long-term debt.

3   Data Selection and Research Design

The sample of this study includes all private, non-listed hotel corporations registered in the Greek chamber of trade during the period 2003–2016. All data were extracted from ICAP database. We collected data from firms' annual financial statements which closed their fiscal year on December and have full financial data for at least four consecutive years. This procedure resulted in a usable unbalanced sample of 120 hotel firms summing up to 362 firm-year observations.

Following previous empirical studies (Chen, 2004; de Jong et al., 2008; Psillaki & Daskalakis, 2009; Kayo & Kimura, 2011; Balios et al., 2016; Moradi & Paulet, 2015), two dependent variables are employed in the study. Total leverage (TLEV) and long-term leverage (LLEV) are utilized as the proxies of firm's capital structure. Total leverage (TLEV) is defined as the ratio of the book value of long-term debt plus interest-bearing short-term debt to the book value of total assets. This study also adopts long-term leverage (LLEV) to research the relationship between long-term leverage level and the determinants of capital structure. The natural logarithm of total assets is assigned as a proxy for the size of the firm (SIZE) since the transformation of natural logarithm might scale down the amount of total assets. Growth opportunities (GROWTH) is defined as the annual percentage change in the book value of total assets which is also adopted by Moradi and Paulet (2015). Profitability (PROF) is measured as the ratio of earnings before interest and tax (EBIT) to the book value of total assets. The standard deviation of the first difference in the ratio of EBIT to the book value of total assets is employed as the proxy for earning volatility (EVOL).

The ratio of the book value of fixed assets to the book value of total assets is adopted as the proxy for asset tangibility (TANG). Non-debt tax shields (NDTS) is defined as the ratio of the book value of depreciation and amortization to the book value of total assets. The firm's tax level (TAX) is considered as the ratio of the book value of total income taxes to the book value of total assets (Huang & Song, 2006).

Complying with most of the previous studies on capital structure, panel data will be estimated through the ordinary least squares (OLS) and multivariate linear regression model to investigate the firm-specific determinates' impact on firm's capital structure decision. The basic regression model can be specified as follows:

$$
\begin{aligned}
\text{LEV}_{it} = {} & \beta_0 + \beta_1 \text{SIZE}_{it} + \beta_2 \text{GROW}_{it} + \beta_3 \text{PROF}_{it} + \beta_4 \text{EVOL}_{it} \\
& + \beta_5 \text{TANG}_{it} + \beta_6 \text{NDTS}_{it} + \beta_7 \text{TAX}_{it} + e_{it}
\end{aligned}
\tag{1}
$$

where i denotes the firm dimension and t denotes the time dimension. The dependent variable LEV (including TLEV_{it} and LLEV_{it}) represents the leverage level of the ith firm in tth year, βk is the regression coefficient for the kth independent variable, SIZE, GROWTH, PROF, EVOL, TANG, NDTS, and TAX, represent the size, grow opportunities, profitability, earnings volatility, asset tangibility, non-debt tax shields, and tax level of the firm i during period t, respectively, $\varepsilon_{i,t}$ denotes the error term for the ith firm in t year. Model (1) will be estimated by random effect since the Breusch and

Pagan Lagrangian multiplier test for random effects produced a highly significant chi-square value suggesting that random effects is the most efficient estimation method.

4 Empirical Results

The descriptive statistics of the sample variables are demonstrated in Table 1. As can be seen in the table, the average total leverage level (TLEV) of the Greek hotel firms is about 36% and seems to have decreased after the crisis period (2010–2016) up to 34.6%. This means that more than one-third of the firms' assets are financed through debt. However, total leverage (TLEV) increased after the crisis to 28.3% form 25.4% before the crisis. This might mainly due to the debt crisis in Greece which increases the needs of financing, and those needs are expected to be covered by external funds. In terms of the independent variables, the average GROWTH and PROF are −23.1% and −38.8%, respectively, for the whole period, and after the crisis, both variables were significantly deteriorated. Finally, NDTS has increased after the crisis indicating that the sample firms reported higher levels of depreciations. The rest of the variable do not present significant fluctuations between the two sub-periods. The Pearson correlations results of the sample variables are shown in Table 2. As seen in the table, TLEV seems to have a negative and significant relationship to GROWTH, PROF, NDTS, and TAX, while LLEV was positively associated with NDTS. The rest of the correlation coefficients are highly significant and with economic meaning. In order to examine the multicollinearity among the independent variables for the total sample, the variance inflation factors (VIFs) have been performed and the results of VIF for every variables is less than 3, which indicates that there might not be sever multicollinearity among the explanatory variables. Meanwhile, the result of Durbin-Watson test is around 2, meaning autocorrelation in this regression model might be not significant. The model is also tested for heteroscedasticity through plotting method and the outcomes together with the residual statistics show a satisfactory result.

The regression results from the estimation of model (1) using the TLEV and LLEV as the dependent variables are demonstrated in Tables 3 and 4. The only significant independent variables were GROWTH, TANG, NDTS, and TAX. Specifically, GROWTH produced a negative and significant coefficient for the crisis period on both TLEV and LLEV. These findings verify H2 and suggest that high growth opportunities firms may be perceived as riskier, and thus, they can choose to re-pay a part of their debt obligations during crisis and finance their operations from internally generated funds (Daskalakis et al., 2017). Moreover, TANG produced a positive and significant impact on TLEV and LLEV for both sub-periods yielding support to H5. Firms with high asset tangibility might borrow more, resulting in TANG's positive influence on LLEV and TLEV. This result supports the trade-off theory and agency cost theory and complies with the previous studies of Frank and Goyal (2003), Yang et al. (2010), Kyriazopoulos (2017). Furthermore, a negative and significant statistical relationship between NDTS, LLEV, and TLEV is evidenced for the whole period

Table 1 Descriptive statistics of full sample variables, before and after the Greek debt crisis

Variables	Full sample				Pre-crisis				Post-crisis			
	Mean	St. dev	Min	Max	Mean	St. dev	Min	Max	Mean	St. dev	Min	Max
TLEV	0.360	0.211	−0.001	1.208	0.382	0.205	−0.002	1.208	0.346	0.197	0.002	1.194
LLEV	0.271	0.887	−0.001	26.536	0.254	0.210	−0.001	1.019	0.283	1.140	0.000	26.536
SIZE	14.021	1.766	4.639	23.419	14.145	1.608	8.528	23.367	13.951	1.846	4.639	23.419
GROWTH	−0.231	13.933	−1008.43	7.413	0.066	0.672	−11.255	6.395	−0.386	17.176	−1008.43	7.413
PROF	−0.388	27.597	−2257.45	33.319	0.044	0.797	−13.525	33.319	−0.633	34.566	−2257.45	18.697
EVOL	6.712	6.614	0.354	19.244	3.646	4.325	0.354	14.154	8.463	7.044	1.044	19.244
TANG	0.326	0.325	0.000	1.139	0.343	0.326	0.000	1.139	0.312	0.324	0.000	1.191
NDTS	0.788	32.074	0.000	22.927	0.280	0.476	0.000	8.081	1.170	42.445	0.000	22.927
TAX	0.032	0.044	−0.149	0.928	0.034	0.042	−0.149	0.326	0.031	0.046	−0.088	0.928

Table 2 Pearson correlation coefficients of sample variables

Variables	TLEV	LLEV	Size	Growth	Prof	Evol	Tang	NDTS	Tax
TLEV	1								
LLEV	0.617*** (0.001)	1							
SIZE	0.004 (0.902)	−0.151*** (0.001)	1						
GROWTH	−0.080** (0.050)	−0.477*** (0.001)	0.061*** (0.001)	1					
PROF	−0.353*** (0.001)	−0.416*** (0.001)	0.053*** (0.001)	0.667*** (0.001)	1				
EVOL	−0.025 (0.500)	0.018 (0.575)	−0.031** (0.011)	−0.017 (0.223)	−0.014 (0.242)	1			
TANG	0.059 (0.112)	0.001 (0.963)	0.299*** (0.001)	0.016 (0.294)	0.012 (0.352)	−0.009 (0.482)	1		
NDTS	−0.066* (0.078)	0.432*** (0.001)	−0.061*** (0.001)	−0.598*** (0.001)	−0.698*** (0.001)	0.013 (0.350)	−0.009 (0.500)	1	
TAX	−0.328*** (0.001)	−0.162*** (0.001)	−0.338*** (0.001)	−0.021 (0.248)	0.066*** (0.001)	−0.03*** (0.023)	−0.195*** (0.001)	0.016 (0.389)	1

***, **, *, indicate statistical significance at the 1%, 5%, and 10% significance level, respectively. P-values in the parentheses

Table 3 Panel random effects regression results on the determinants of total leverage before and after the crisis (robust standard errors)

Variables	Full sample			Pre-crisis			Post-crisis		
	Coef	z-test	p-value	Coef	z-test	p-value	Coef	z-test	p-value
Constant	0.426**	2.09	0.037	0.255	0.91	0.360	0.343	1.61	0.107
SIZE	−0.003	−0.26	0.794	0.015	0.87	0.384	−0.002	−0.13	0.894
GROWTH	0.001	0.14	0.890	−0.032	−0.93	0.354	−0.042**	−2.35	0.019
PROF	0.063	0.25	0.799	−0.449	−1.00	0.317	0.165	0.35	0.727
EVOL	−0.001	−1.51	0.132	−0.001	−0.49	0.623	−0.001	−0.06	0.948
TANG	0.107*	1.68	0.093	−0.068	−0.99	0.321	0.165***	2.85	0.004
NDTS	−0.293***	−3.03	0.002	−0.408***	−3.36	0.001	−0.193***	−3.24	0.001
TAX	−1.791*	−1.86	0.063	−0.545	−0.36	0.722	−1.581	−1.25	0.211
Wald−x^2	33.20***			38.40***			34.11***		
R^2	0.087			0.217			0.109		
Observations	362			141			221		

***, **, * indicate statistical significance at the 1% and 5% significance level respectively. Dependent variable TLEV

Table 4 Panel random effects regression results on the determinants of long-term leverage before and after the crisis (robust standard errors)

Variables	Ful samplel			Pre-crisis			Post-crisis		
	Coef	z-test	p-value	Coef	z-test	p-value	Coef	z-test	p-value
Constant	0.233	1.26	0.207	0.089	0.42	0.676	0.333*	1.88	0.060
SIZE	−0.003	−0.29	0.774	0.006	0.49	0.622	−0.013	−1.14	0.253
GROWTH	−0.010	−0.50	0.619	0.074*	1.69	0.091	−0.058***	−3.43	0.001
PROF	−0.157	−0.95	0.342	−0.625	−1.21	0.226	−0.130	−0.78	0.437
EVOL	−0.001	−0.84	0.399	0.001	0.43	0.670	0.001	0.73	0.466
TANG	0.214***	3.38	0.001	0.185***	2.75	0.006	0.248***	5.64	0.001
NDTS	−0.242***	−5.49	0.001	−0.225**	−2.05	0.040	−0.221***	−6.25	0.001
TAX	−0.312	−0.40	0.686	0.501	0.31	0.758	0.230	0.29	0.770
Wald-x^2	47.05***			23.18***			78.91***		
R^2	0.127			0.121			0.129		
Observations	362			141			221		

***, **, * indicate statistical significance at the 1%, 5% and 10% significance level respectively. Dependent variable LLEV

and also before and after the crisis. This result leads us to accept H6 and complies with the expectation of the trade-off theory, claiming that non-debt tax shields can negatively affect the firms leverage level since they can be seen as the substitutes for tax deductibility of debt, which might result in relatively less deductible tax shields remaining for debt (Myers, 1984). Finally, TAX is negatively related to TLEV but marginally significant at the 10% level. This finding is consistent with the prediction of trade-off theory, focusing on the full utilization of debt's tax deductibility and the expectation of pecking order theory, claiming the possible external financing need caused by the reduce of retained earnings with the increase of tax level (Myers & Majluf, 1984). This result yields support to H7. The rest of the independent variables were insignificant within conventional levels so H1, H3, H4, and H5 were not supported.

5 Conclusion

This paper studies the possible hotel firm-specific determinants of capital structure during the pre- and post-debt crisis periods. A multivariate panel regression model is employed using a panel dataset of 8529 Greek firms during the period of 2003–2016. The results indicate that asset tangibility is directly and positively related to total leverage and long-term leverage during the pre- and post-crisis periods, while growth opportunities, non-debt tax shields and tax payments are negatively and significantly impacting total and long-term leverage during the crisis period. This study has important implications for regulators and investors since it extends our understanding regarding the behavior of Greek hotel corporations and how they adjust their financing needs before and during a period of financial turmoil. Regulators could use the findings of the current study in order to adjust state policies (tax policy, financing policy, etc.) for assisting firms to access financing in difficult times, while investors should consider the differential impact of capital structure determinants during troubled times when receiving their financing and investment decisions.

Nevertheless, it should be noted that this paper has some limitations. Firstly, since the data used in this study are all at book value extracted from financial statements, if firms utilize off-balance sheet financing instruments, the capital structure might be influenced and such effects might be difficult to recognize. Meanwhile, this paper is bounded by the inherent drawbacks existing in any regression analysis such as parameter instability. For future research, empirical studies can focus on the determinants of firm-specific capital structure of Greek small and very small hotel firms and their effects on leverage during such period of debt crisis. Furthermore, having studied the relationship between capital structure and its firm-specific determinants, the future researchers are expected to study how to achieve optimal capital structure. Finally, it will be also interesting to consider the ownership structure of hotels (family ownership or dispersed ownership) and how it affects capital structure decisions under a volatile financial environment.

References

Abor, J., & Biekpe, N. (2009). How do we explain the capital structure of SMEs in sub-Saharan Africa? *Journal of Economic Studies, 36*(1), 83–97. https://doi.org/10.1108/014435809 10923812.

Ahsan, T., Wang, M., & Qureshi, M. A. (2016). How do they adjust their capital structure along their life cycle? An empirical study about capital structure over the life cycle of Pakistani firms. *Journal of Asian Business Studies, 10*(3), 276–302. https://doi.org/10.1108/JABS-06-2015-0080.

Arghyrou, M., & Tsoukalas, J. (2011). The Greek debt crisis: Likely causes, mechanics and outcomes. *The World Economy, 34*(2), 173–191. https://doi.org/10.1111/j.1467-9701.2011.013 28.x.

Balios, D., Daskalakis, N., Eriotis, N., & Vasiliou, D. (2016). SMEs capital structure determinants during severe economic crisis: The case of Greece. *Cogent Economics & Finance, 4*(1), 1145535. https://doi.org/10.1080/23322039.2016.1145535.

Banerjee, S., Heshmati, A., & Wihlborg, C. (2004). The dynamics of capital structure. *Research in Banking and Finance, 4*(1), 275–297.

Bradley, M., Jarrell, G., & Kim, E. (1984). On the existence of an optimal capital structure: Theory and evidence. *The Journal of Finance, 39*(3), 857–878. https://doi.org/10.1111/j.1540-6261.1984. tb03680.x.

Cespedes, J., Gonzalez, M., & Molina, C. (2010). Ownership and capital structure in Latin America. *Journal of Business Research, 63*(3), 248–254. https://doi.org/10.1016/j.jbusres.2009.03.010.

Chatzinas, G., & Papadopoulos, S. (2018). Trade-off vs. pecking order theory: Evidence from Greek firms in a period of debt crisis. *International Journal of Banking, Accounting and Finance, 9*(2), 107–191. https://doi.org/10.1504/IJBAAF.2018.092133

Chen, J. (2004). Determinants of capital structure of Chinese-listed companies. *Journal of Business Research, 57*(12), 1341–1351. https://doi.org/10.1016/S0148-2963(03)00070-5.

Choi, J., & Richardson, M. (2016). The volatility of a firm's assets and the leverage effect. *Journal of Financial Economics, 121*(2), 254–277. https://doi.org/10.1016/j.jfineco.2016.05.009.

Daskalakis, N., Balios, D., & Dalla, V. (2017). The behaviour of SMEs' capital structure determinants in different macroeconomic states. *Journal of Corporate Finance, 46,* 248–260. https://doi.org/10.1016/j.jcorpfin.2017.07.005.

de Jong, A., Kabir, R., & Nguyen, T. (2008). Capital structure around the world: The roles of firm- and country-specific determinants. *Journal of Banking & Finance, 32*(9), 1954–1969. https://doi.org/10.1016/j.jbankfin.2007.12.034.

DeAngelo, H., & Masulis, R. (1980). Optimal capital structure under corporate and personal taxation. *Journal of Financial Economics, 8*(1), 3–29. https://doi.org/10.1016/0304-405X(80)900 19-7.

Dimitropoulos, P. (2018). Profitability determinants of the Greek hospitality industry: The crisis effect. In V. Katsoni & K. Velander (Eds.), *Innovative Approaches to Tourism and Leisure* (pp. 405–416). Switzerland: Springer. https://doi.org/10.1007/978-3-319-67603-6_31

Dimitropoulos P., Vrondou, O., & Koronios, K. (2019). Earnings predictability of the Greek hospitality industry during the crisis. In V. Katsoni & M. Segarra-Oña (Eds.), *Smart Tourism as a Driver for Culture and Sustainability* (pp. 647–658). Switzerland: Springer. https://doi.org/10. 1007/978-3-030-03910-3_43

Economides, G., Papageorgiou, D., & Philippopoulos, A. (2017). The Greek great depression: a general equilibrium study of its drivers. In *Political Economy Perspectives on the Greek Crisis Introduction* (pp. 205–221). Cham: Palgrave Macmillan.

Frank, M., & Goyal, V. (2003). Testing the pecking order theory of capital structure. *Journal of Financial Economics, 67*(2), 217–248. https://doi.org/10.1016/S0304-405X(02)00252-0.

Hoang, T. H. V., Gurău, C., Lahiani, A. L., & Seran, T.-L. (2018). Do crises impact capital structure? A study of French micro-enterprises. *Small Business Economics, 50,* 181–199. https://doi.org/ 10.1007/s11187-017-9899-x.

Hovakimian, A., Opler, T., & Titman, S. (2001). The debt-equity choice. *Journal of Financial and Quantitative Analysis, 36*(1), 1–24. https://doi.org/10.2307/2676195.

Huang, G., & Song, F. (2006). The determinants of capital structure: Evidence from China. *China Economic Review, 17*(1), 14–36. https://doi.org/10.1016/j.chieco.2005.02.007.

Jensen, M., & Meckling, W. (1976). Theory of the firm: Managerial behavior, agency costs and ownership structure. *Journal of Financial Economics, 3*(4), 305–360. https://doi.org/10.1007/978-94-009-9257-3_8.

Kayo, E., & Kimura, H. (2011). Hierarchical determinants of capital structure. *Journal of Banking & Finance, 35*(2), 358–371. https://doi.org/10.1016/j.jbankfin.2010.08.015.

Kim, H., Heshmati, A., & Aoun, D. (2006). Dynamics of capital structure: The case of Korean listed manufacturing companies. *Asian Economic Journal, 20*(3), 275–302. https://doi.org/10.1111/j.1467-8381.2006.00236.x.

Kyriazopoulos, G. (2017). Corporate governance and capital structure in the periods of financial distress: Evidence from Greece. *Investment Management and Financial Innovations, 14*(1), 254–262.

Maçãs Nunes, P., & Serrasqueiro, Z. (2017). Short-term debt and long-term debt determinants in small and medium-sized hospitality firms. *Tourism Economics, 23*(3), 543–560. https://doi.org/10.5367/te.2015.0529.

McNamara, A., Murro, P., & O'Donohoe, S. (2017). Countries lending infrastructure and capital structure determination: The case of European SMEs. *Journal of Corporate Finance, 43*, 122–138. https://doi.org/10.1016/j.jcorpfin.2016.12.008.

Miller, M. (1977). Debt and taxes. *The Journal of Finance, 32*(2), 261–275. https://doi.org/10.1111/j.1540-6261.1977.tb03267.x.

Modigliani, F., & Miller, M. (1958). The cost of capital, corporation finance and the theory of investment. *American Economic Review, 48*(3), 261–297. Retrieved from https://www.jstor.org/stable/1809766

Modigliani, F., & Miller, M. (1963). Corporate income taxes and the cost of capital: a Correction. *American Economic Review, 53*(3), 433–443. Retrieved from https://www.jstor.org/stable/1809167

Moradi, A., & Paulet, E. (2015). A causal loop analysis of the austerity policy adopted to address the Euro crisis—Effects and side effects. *International Journal of Applied Decision Sciences, 8*(1), 1–20. https://doi.org/10.1504/IJADS.2015.066552

Myers, S. (1977). Determinants of corporate borrowing. *Journal of Financial Economics, 5*(2), 147–175. https://doi.org/10.1016/0304-405X(77)90015-0.

Myers, S. (1984). The capital structure puzzle. *The Journal of Finance, 39*(3), 574–592. https://doi.org/10.1111/j.1540-6261.1984.tb03646.x.

Myers, S., & Majluf, N. (1984). Corporate financing and investment decisions when firms have information that investors do not have. *Journal of Financial Economics, 13*(2), 187–221. https://doi.org/10.3386/w1396.

Noulas, A., & Genimakis, G. (2011). The determinants of capital structure choice: Evidence from Greek listed companies. *Applied Financial Economics, 21*(6), 379–387. https://doi.org/10.1080/09603107.2010.532108.

Ozkan, A. (2001). Determinants of capital structure and adjustment to long run target: Evidence from UK company panel data. *Journal of Business Finance & Accounting, 28*(1–2), 175–198. https://doi.org/10.1111/1468-5957.00370.

Pacheco, L., & Tavares, F. (2017). Capital structure determinants of hospitality sector SMEs. *Tourism Economics, 23*(1), 113–132. https://doi.org/10.5367/te2015.0501.

Psillaki, M., & Daskalakis, N. (2009). Are the determinants of capital structure country or firm specific? *Small Business Economics, 33*(3), 319–333. https://doi.org/10.1007/s11187-008-9103-4.

Rajagopal, S. (2011). The portability of capital structure theory: Do traditional models fit in an emerging economy? *Journal of Finance and Accountancy, 5*, 1–17.

Rajan, R., & Zingales, L. (1995). What do we know about capital structure? Some evidence from international data. *The Journal of Finance, 50*(5), 1421–1460. https://doi.org/10.1111/j.1540-6261.1995.tb05184.x.

Titman, S., & Wessels, R. (1988). The determinants of capital structure choice. *The Journal of Finance, 43*(1), 1–19. https://doi.org/10.1111/j.1540-6261.1988.tb02585.x.

Tsoukis, C., Bournakis, I., Christopoulos, D., & Palivos, T. (2017). Introduction. *Political Economy Perspectives on the Greek Crisis* (pp. 3–40). Cham: Palgrave Macmillan.

Wickens, M. (2017). A macroeconomic perspective on the Greek debt crisis. *Political Economy Perspectives on the Greek Crisis* (pp. 157–175). Cham: Palgrave Macmillan.

Yang, C., Lee, C., Gu, Y., & Lee, Y. (2010). Co-determination of capital structure and stock Returns-A LISREL approach. *The Quarterly Review of Economics and Finance, 50*(2), 222–233. https://doi.org/10.1016/j.qref.2009.12.001.

Dr. Panagiotis E. Dimitropoulos is a member of teaching staff of the Department of Sport Organization and Management, University of Peloponnese. His research interests are on accounting and financial management. He has published in referred academic journals like British Accounting Review, Corporate Governance: An International Review, Journal of Economic Behavior and Organization, European Sport Management Quarterly and other.

Human Resource Empowerment and Employees' Job Satisfaction in a Public Tourism Organization: The Case of Greek Ministry of Tourism

Alkistis Papaioannou, George Baroutas, Ioulia Poulaki, Georgia Yfantidou, and Alexia Noutsou

1 Introduction

Human resource empowerment has been the focus of various researches and studies and has been defined in various ways. Among these, it has been reported as the transfer of decision-making and responsibility from managers to employees as well as the support of employees in order to become more effective in their work (Byham & Cox, 1989).

Overall, we can consider that empowerment of human resources is the process of encouraging and rewarding the initiative of employees. A basic condition is that these employees are willing and able to take on and handle additional responsibility and power, as well as they are able to make good decisions and execute them effectively (Chelladurai, 1999). It essentially consists of processes, methods, tools and techniques that have been developed in the context of human resource development, motivation, job planning and delegation (Kriemadis & Papaioannou, 2006).

Human resource empowerment has been defined as the process that allows employees to make decisions and take responsibility for their actions (Kriemadis & Papaioannou, 2006). Human resource empowerment is considered as a vital strategy for service success and improvement (Sarason, 1992) and the critical role of empowerment in promoting staff effectiveness has been highlighted in the literature (Blase

A. Papaioannou (✉)
University of Peloponnese, Kalamata, Greece
e-mail: alkistisp@uop.gr

G. Baroutas
Hellenic Open University, Patra, Greece

I. Poulaki
University of Patras, Patras, Greece

G. Yfantidou · A. Noutsou
Democritus University of Thrace, Komotini, Greece

© The Author(s), under exclusive license to Springer Nature Switzerland AG 2021
V. Katsoni and C. van Zyl (eds.), *Culture and Tourism in a Smart, Globalized, and Sustainable World*, Springer Proceedings in Business and Economics, https://doi.org/10.1007/978-3-030-72469-6_27

& Blase, 1996; Crow & Pounder, 2000; Wilson & Coolican, 1996; Papaioannou et al., 2009; Papaioannou et al., 2012). The most important benefits of its implementation in the workplace, as presented by the literature (Papaioannou et al., 2009), are the following: increased productivity, increased organizational performance, improved service quality and customer satisfaction, increased employee loyalty, increased work/organizational commitment, increased employee satisfaction, increased employee motivation and efficiency and increased trust between employees and management.

Moreover, in the service sector should be the most in need of the Human resource empowerment as both the customers and the employees are involved the same time in the production of the service (Alabar & Abubakar, 2013).

Regarding the job satisfaction of the human resources, the literature has (Hoppock, 1935) defined it as the combination of psychological, physical and environmental conditions that lead the individual to express satisfaction with his job. According to (Spector, 1985), it is identified with the degree of fulfillment of the individual's needs in the workplace as well as with the fulfillment of his goals, while (Wright & Davis, 2003) have stated that satisfaction concerns the evaluation of the individual as to what he is seeking from his job and what he is ultimately receiving. (Wright, 2006) concluded that job satisfaction became the most widely used behavioral measure for organizational research.

Job satisfaction has been the subject of research for many experts (Koustelios, 2001) from different sciences in the field of organizational and industrial psychology, as well as it is considered one of the most popular fields and the most frequently researched variable (Wright, 2006). Despite the different approaches, everyone agrees that this is a multidimensional conceptual construction which is influenced in various ways by many individual elements (Koustelios, 2001).

One of the most commonly used definitions of job satisfaction defines it in terms of how people feel about their job and its various aspects (Spector, 1985, 1997) and has to do with the extent to which people like it or dislike it. A widely accepted definition, according to Baron (1986), could be the following which defines it as "the positive and negative attitudes of the individual, about his job" (Koustelios, 2001).

In conclusion, job satisfaction is a very important variable, as it has consequences for both the employee and the organization on a larger scale, and is certainly an important part of the overall satisfaction that a person receives in his life.

Empirical research on human resource empowerment and job satisfaction has been conducted in the field of service sector, where employees empowerment is usually associated with their participation in decision making and their increased professional satisfaction (Rinehart & Short, 1994; Rice & Schneider, 1994; Wu & Short, 1996; Cypert, 2009; Papaioannou, 2009; Bogler & Nir, 2010, Papaioannou et al., 2017a, b).

Nevertheless, research regarding human resource empowerment and job satisfaction of employees in a Greek public tourism organization is still limited.

The primary objectives of this research were: (a) to determine the extent to which human resource empowerment is applied to the Greek Ministry of Tourism, (b) to examine the extent to which the employees' job satisfaction was existed in this public tourism organization and (c) to investigate the relationship between human resource empowerment and employees' job satisfaction.

2 Literature Review

Regarding human resource empowerment in the tourism sector Al-Ababneh et al. (2017) examined the effect of empowerment on job satisfaction in five-star hotels in Jordan. The results pointed out that human resource empowerment had significant effects on job satisfaction.

Furthermore, in an additionally study was investigated the influences of managers' and supervisors' commitment and satisfaction through high-commitment HR practices on organizational performance in the hospitality industry in Spain. The findings proposed that managers' commitment and satisfaction did not lead to improved organizational performance, while supervisors' commitment and satisfaction led to better economic outcomes. In conclusion, it seems that the use of high-commitment HR practices is essential in order to improve their employees' commitment and job satisfaction in the hospitality industry which possibly results in better customer service, allowing hotels to ensure improved economic outcomes (Domínguez-Falcón et al., 2016).

In addition, one more study placed under investigation the effect of employee empowerment on employee satisfaction and service quality, and the influence of employee satisfaction on service quality, in 20 different financial institutions in Bangladesh. The results revealed that employee satisfaction and service quality significantly be supported by employees empowerment, and satisfied employees delivered improved service quality (Ukil & Ullah, 2016).

Accordingly, in an additional study were investigated the impact of employee empowerment, training and teamwork on job satisfaction in higher education sector in northern Malaysia using an online survey. The results indicated that employee empowerment and training as well as teamwork had an important impact on job satisfaction (Hanayshaa & Tahir, 2016).

Furthermore, one more study placed under investigation human resource empowerment and job satisfaction in sport divisions of the Higher Military Educational Institutions (HMEI) and higher schools of Armed Forces (SF) in Greece. The findings pointed out that human resource empowerment was implemented in a small extent by the sport divisions of HMEI and schools of SF, while the overall job satisfaction of employees was above average. Also there were found significantly positive correlations among the key factors of human resource empowerment and physical education employees' job satisfaction (Kaniadakis et al., 2017).

In another study was investigated the causal relationships among employee empowerment, ethical climate, organizational support, top management commitment and employee job satisfaction in companies of South Africa. The findings of this study showed that all the above mentioned variables had a favorable effect on job satisfaction (Chinomona et al., 2017).

In addition, in another study was examined the degree to which the empowerment of human resources is applied in the secondary education, and the correlation between the human resource empowerment and employees' job satisfaction. The results of the research showed that the empowerment of human resources is applied to a large extent in the secondary education, while positive and important correlations were found between the empowerment of human resources and the employees' job satisfaction (Papaioannou et al., 2017a, b).

Moreover, in an additional study was explored the relationship between employee empowerment and job satisfaction in the urban Malaysian context. The findings supported that employee empowerment positively affects the job satisfaction, while there were found important dissimilarities in employee empowerment and job satisfaction between higher and lower echelons of the organizational hierarchy (Idris et al., 2018).

In conclusion, the finding that empowered employees enjoy high job satisfaction is common in a large number of studies. From the very first attempts to understand employee behavior to the most recent human resource studies, the results supported the same conclusion. A review of the Greek and World literature have shown that the empowerment of human resources leads to an increase in job satisfaction in the service sector such as: the educational sector, the service sector, the banking sector, and the hospitality industry.

3   Methodology

Sample

The research was addressed to all full-time staff of the Central Service of the Greek Ministry of Tourism (250 people were employed during the time of the research according to the Human Resources Director of the Greek Ministry of Tourism) and involved 250 experienced managers and employees who were the most knowledgeable regarding personnel management issues and who can respond as accurately as possible. Responses were received from 142 managers and employees of the aforementioned organization with a 56.8% response rate. They all had full time experience. According to the findings of this research 35% of the participants were males and 65% were females with a mean age of 52 (sd = 7). The educational level of the participants was grouped into three categories: 52.4% had post graduate level education, 33.3% had university level education, and 14.3% had elementary/high school education. 40% of the participants were managers and 60% of the participants were heads of divisions. By reference to the work experience of the research participants, 69.6% had over 15 years, 24.8% had ten to fifteen years, 3.2% had seven to ten years and 2.4% had three to seven years work experience.

Questionnaires

For the purpose of this study two questionnaires, were used: (a) the questionnaire of human resource empowerment (developed by Vogt & Murrell, 1990 and modified and simplified by Kriemadis, 2011; Papaioannou, 2011) using a five point Likert scale and (b) the questionnaire of employees' job satisfaction (developed by Warr et al., 1979), using a seven point Likert scale. The scales regarding human resource empowerment represented three different managerial styles which were: (a) The "empowering style" of management: the manager has a managerial style that reflects the manager who creates and shares power, (b) The "middle-ground style" of management: combines the two styles of controlling and empowering in equal proportions and (c) The "controlling style" of management: the manager is concerned with control or seldom shares, creates or empowers subordinates. The questionnaire included six units based on the following key factors of human resource empowerment:

"Management—information/communication system skills".
"Decision-making and action-taking skills".
"Project-planning, organizing, and system-integration skills".
"System-evaluation and internal-control skills".
"Leadership, motivation, and reward-systems skills".
"Selection, placement, and development of people skills".

The next section of the questionnaire, which refers to employees' job satisfaction, included the following 10 closed-ended questions: (a) "Physical working conditions", (b) "Freedom to choose your own method of working", (c) "Your colleagues and fellow workers", (d) "Recognition you get for good work", (e) "Amount of responsibility you are given", (f) "Your remuneration", (g) "Opportunity to use your abilities", (h) "Your hours of work", (i) "Amount of variety in your job", (j) "Taking everything into consideration, how do you feel about your job" and (k) "Overall employees' job satisfaction".

The final part of the questionnaire included 4 questions in relation to demographic characteristics of the respondents such as: the gender of the research participants, the educational level of the participants, the position of responsibility and the work experience of the respondents.

The content validity of the questionnaire was determined by a panel of experts consisting of academics and practitioners in human resource management and tourism management research. The reliability of the scale was found to be $\alpha = 0.88$.

Procedure

In cooperation with the Human Resources Director of the Greek Ministry of Tourism, questionnaires were distributed to the Central Service of the Ministry of Tourism, both in hard copy and in electronic form, using the capabilities provided by the google forms service. The employees of the Ministry of Tourism participated in the research voluntarily, while the completion of the questionnaire was anonymous. The time required to complete the questionnaire was approximately 10 min.

Data Analysis

Standard descriptive statistics including mean, percentages and standard deviation were used to answer the first and second research question. Data were also examined for normality, using normal probability plots and the Kolmogorov–Smirnov test. Since were indicated serious deviations of normality ($p < 0.05$), non-parametric statistical tests of Spearman's rho, Mann Whitney and Kruskal Wallis were performed. Research questions three was answered by using Spearman Correlation, using SPSS software (version 22).

4   Results

Data of the managers and heads of divisions' responses were analyzed to gather information specific to the extent to which human resource empowerment is applied to the Greek Ministry of Tourism. From the analysis it is shown that the managers and heads of divisions used 61.7% "empowering style", 12.2% "controlling style" and 26.1% "middle-ground style" (combines the two styles of controlling and empowering in equal proportions) in two key factors of human resource empowerment: (a) "Decision-making and action-taking skills" and (b) "Leadership, motivation, and reward-systems skills". Furthermore, they used 61.9% "empowering style", 22.5% "middle-ground style" and 15.6% "controlling style" in the key factor: "Selection, placement, and development of people skills".

In relation to the key factor "system-evaluation and internal-control skills" the managers and heads of divisions used 60.7% "empowering style", 24.5% "middle-ground style" and 14.8% "controlling style". Moreover, concerning the key factor: "Total for all managerial functions" they used 60.1% "empowering style", 33.5% "middle-ground style" and 6.4% "controlling style". Concerning the key factor "Project-planning, organizing, and system-integration skills", the managers and heads of divisions used 60% "empowering style", 25.5% "middle-ground style", and 14.5% "controlling style". Finally, in the key factor "Management—information/communication system skills", the managers and heads of divisions used 59.2% "empowering style", 26.9% "middle-ground style" and 13.9% "controlling style" (Fig. 1).

Table 1 shows that the overall managers' and heads of divisions' job satisfaction in the Greek Ministry of Tourism was above average (M = 4.43, TA = 1.18). More analytically, 76.9% of managers and heads of divisions were at least little satisfied with their colleagues and fellow workers, 62.7% were satisfied with their hours of work, 61.9% were satisfied with the amount of responsibility they are given, 58.0% were satisfied with the freedom to choose their own method of working, 57.1% were satisfied with their physical working conditions, 53.9% were satisfied with how they feel about their job, 53.1% were satisfied with the recognition they get for good work, 49.2% were satisfied with the amount of variety in their job, 46.9% were

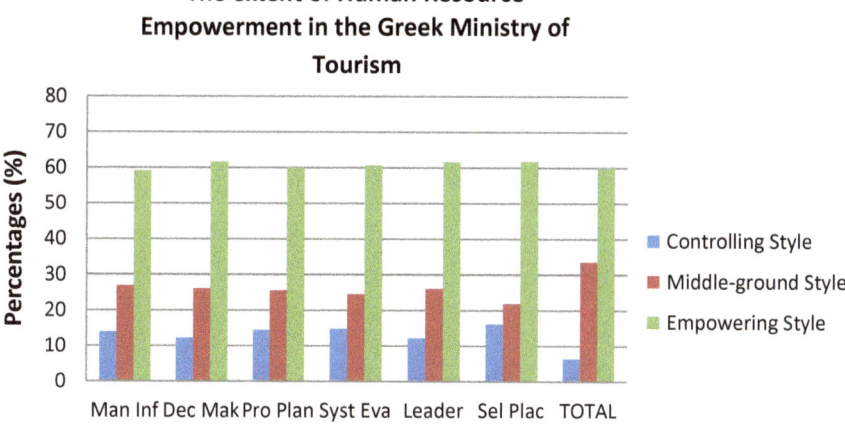

Fig. 1 The extent of human resource empowerment in the Greek Ministry of Tourism

satisfied with the opportunity to use their abilities and 26.9% were satisfied with their remuneration.

The results of the study indicated that there were significant and positive relationships between all the factors of human resource empowerment (Total for all managerial functions) and managers' and heads of divisions' job satisfaction in the Greek Ministry of Tourism, as shown in the Table 2. More specifically, the results showed that the human resource empowerment had a significant and positive relationship with: (a) "Physical working conditions" ($r = 0.372, p = 0.002$), (b) "Freedom to choose your own method of working", ($r = 0.557, p = 0.000$), (c) "Your colleagues and fellow workers" ($r = 0.368, p = 0.002$), (d) "Recognition you get for good work" ($r = 0.500, p = 0.000$), (e) "Amount of responsibility you are given" ($r = 0.496, p = 0.000$), (g) "Opportunity to use your abilities" ($r = 0.400, p = 0.000$), (h) "Your hours of work" ($r = 0.289, p = 0.023$), (i) "Amount of variety in your job" ($r = 0.442, p = 0.000$), (j) "Taking everything into consideration how do you feel about your job" ($r = 0.598, p = 0.000$) and (k) "Overall employees' job satisfaction" ($r = 0.545, p = 0.000$). On the contrary, there was no significant relationship between all the factors of human resource empowerment ("Total for all managerial functions") and the parameter of managers' and heads of divisions' job satisfaction: (f) "Your remuneration" ($r = 0.148, p = 0.107$) (Table 2).

Table 1 The extent of managers' and heads of divisions' Job satisfaction in the Greek Ministry of tourism

	Totally dissatisfied		Very dissatisfied		Little satisfied		Neither satisfied nor satisfied		Little satisfied		Very satisfied		Totally satisfied		M	TA
	N	%	N	%	N	%	N	%	N	%	N	%	N	%		
Physical working conditions	11	8.7	9	7.1	22	17.5	12	9.5	29	23.0	34	27.0	9	7.1	4.40	1.75
Freedom to choose your own method of working	8	6.3	14	11.1	9	7.1	22	17.5	39	31.0	30	23.8	4	3.2	4.40	1.58
Your colleagues and fellow workers	1	0.8	9	7.1	3	2.4	16	12.7	25	19.8	48	38.1	24	19.0	5.34	1.43
Recognition you get for good work	12	9.5	10	7.9	16	12.7	21	16.7	29	23.0	28	22.2	10	7.9	4.34	1.74
Amount of responsibility you are given	2	1.6	13	10.3	11	8.7	22	17.5	29	23.0	42	33.3	7	5.6	4.72	1.48
Your remuneration	22	17.5	27	21.4	22	17.5	21	16.7	24	19.0	8	6.3	2	1.6	3.24	1.62
Opportunity to use your abilities	12	9.5	15	11.9	21	16.7	19	15.1	32	25.4	20	15.9	7	5.6	4.05	1.71

(continued)

Table 1 (continued)

	Totally dissatisfied		Very dissatisfied		Little satisfied		Neither satisfied nor satisfied		Little satisfied		Very satisfied		Totally satisfied			
	N	%	N	%	N	%	N	%	N	%	N	%	N	%	M	TA
Your hours of work	2	1.6	9	7.1	9	7.1	27	21.4	22	17.5	41	32.5	16	12.7	4.94	1.50
Amount of variety in your job	10	7.9	8	6.3	19	15.1	27	21.4	25	19.8	30	23.8	1	5.6	4.33	1.64
Taking everything Into consideration, how do you feel about your job?	6	4.8	9	7.1	13	10.3	30	23.8	29	23.0	31	24.6	8	6.3	4.52	1.53
Overall employees' job Satisfaction	9	7	13	10	8	6.5	23	18.3	39	31.0	30	24	4	3.2	4.43	1.18

Table 2 Significance of the relationship between the dimensions of employees' job satisfaction and the factors of human resource empowerment

	Physical working conditions	Freedom to choose your own method of working	Colleagues and fellow workers	Recognition you get for good work	Amount of responsibility you are given	remuneration	Opportunity to use your abilities	Hours of work	Amount of variety in your job	How do you feel about your job?	Overall employees' job satisfaction
Management—information/communication system skills	0.178 ($p = 0.267$)	0.452** ($p = 0.000$)	0.358** ($p = 0.002$)	0.452** ($p = 0.000$)	0.297* ($p = 0.016$)	0.059 ($p = 0.493$)	0.286* ($p = 0.041$)	.289** ($p = 0.008$)	0.287** ($p = 0.008$)	0.467** ($p = 0.000$)	0.364** ($p = 0.001$)
Decision-making and action-taking skills	0.245 ($p = 0.054$)	0.482** ($p = 0.000$)	0.283 ($p = 0.058$)	0.378** ($p = 0.002$)	0.489** ($p = 0.000$)	0.083 ($p = 0.587$)	0.385** ($p = 0.002$)	0.018 ($p = 0.983$)	0.384** ($p = 0.004$)	0.392** ($p = 0.000$)	0.382** ($p = 0.001$)
Project-planning, organizing, and system-integration skills	0.388** ($p = 0.001$)	0.582** ($p = 0.000$)	0.349** ($p = 0.004$)	0.458** ($p = 0.000$)	0.595** ($p = 0.000$)	395** ($p = 0.005$)	0.478** ($p = 0.000$)	0.286** ($p = 0.009$)	0.438** ($p = 0.000$)	0.588** ($p = 0.000$)	0.620** ($p = 0.000$)
System-evaluation and internal-control skills	0.487** ($p = 0.000$)	0.699** ($p = 0.000$)	0.354** ($p = 0.002$)	0.535** ($p = 0.000$)	0.587** ($p = 0.000$)	0.122 ($p = 0.317$)	0.410** ($p = 0.000$)	0.365** ($p = 0.005$)	0.578** ($p = 0.000$)	0.668** ($p = 0.000$)	0.558** ($p = 0.000$)
Leadership, motivation, and reward-systems skills	0.332** ($p = 0.005$)	0.549** ($p = 0.000$)	0.279* ($p = 0.026$)	0.488** ($p = 0.000$)	0.478** ($p = 0.000$)	0.155 ($p = 0.156$)	0.445** ($p = 0.000$)	0.422** ($p = 0.000$)	0.387** ($p = 0.001$)	0.665** ($p = 0.000$)	0.550** ($p = 0.000$)
Selection, placement, and development of people skills	0.236 ($p = 0.057$)	0.429** ($p = 0.000$)	0.398** ($p = 0.000$)	0.445** ($p = 0.000$)	324** ($p = 0.005$)	0.084 ($p = 0.619$)	0.369** ($p = 0.002$)	0.295** ($p = 0.008$)	0.359** ($p = 0.000$)	0.478** ($p = 0.000$)	0.435** ($p = 0.000$)
Total for all managerial functions	0.372** ($p = 0.002$)	0.557** ($p = 0.000$)	0.368** ($p = 0.002$)	0.500** ($p = 0.000$)	0.496** ($p = 0.000$)	0.148 ($p = 0.107$)	0.400** ($p = 0.000$)	.289** ($p = 0.023$)	0.442** ($p = 0.000$)	0.598** ($p = 0.000$)	0.545** ($p = 0.000$)

5 Conclusion

Based on the findings of this study the human resource empowerment process is applied to a mediocre extent by the Greek Ministry of Tourism (60%). This consists a unique and original finding since there is no related supporting literature.

These findings suggested that the managers of the Greek Ministry of Tourism were partially aware of new management techniques such as team working, devolved management, performance appraisals, etc. At the same time, the top management of the Ministry induced the human resources to some extent to set an example and to collaborate with each other. Also, the availability of non-confidential information to people was limited. In addition the managers in the organization had partly knowledge of their performance. Finally the findings showed that the managers of the organization did not give enough support to the devolution of power to take decision and responsibilities in their working field (Papaioannou & Kriemadis, 2017b).

Based on the results of the study (Table 2), it was found that managers' and heads of divisions' job satisfaction was above average ($M = 4.43$, $TA = 1.18$). The findings of this research are in agreement with the literature in the higher educational sector, indicating that employees in the public sector are broadly satisfied with their job (Kaniadakis et al., 2017; Eyupoglu & Saner, 2009).

Additionally, it was found that there is a significant positive relationship between managers' and heads of divisions' job satisfaction and human resource empowerment. These results are in accordance with previous research conducted in the tourism sector, in the service sector, in the health care sector and in the higher educational sector (Gill et al., 2010; Kaniadakis et al., 2017; Lee et al., 2006; Morrison et al., 1997; Ugboro & Obeng, 2000) and displays that the extent of human resource empowerment which managers exerted in the Greek ministry of tourism is linked to their employees' job satisfaction.

The findings of the particular study have implications for the development and use of the human resource empowerment process in the Greek Ministry of tourism. Given that the human resource empowerment is a factor of strategic importance which increases the employees' job satisfaction in the public tourism sector, the managers of Greek Ministry of tourism can be encouraged to implement it in a greater extent. This could happen by developing an empowering culture based on principals such as (Vogt & Murrell, 1990; Papaioannou et al., 2018):

- Delegation of authority and responsibilities to employees in their working field
- Implementation of evaluative systems of performance
- Create a motivational climate which assists and supports employees
- Continuous education and training of employees
- Participative planning process.

The literature in the area of human resource empowerment in the tourism public sector is limited. It seems that there is an ever increasing concern in this field, and further studies could turn out to be advantageous. Based on the findings of this study, the following recommendations are offered for future research:

- Follow-up studies should be carried out in the same sample over three to five years, to investigate possible changes in the utilization of the human resource empowerment process.
- A comparative research could also be conducted between a public tourism organization and a large private tourism enterprise. These comparisons would be useful in order to find out similarities and differences in human resource empowerment process and employees' job satisfaction.
- There could also be a study to measure the impact of human resource empowerment regarding the citizens' satisfaction.

The study yields interesting insights. However, possible limitations of this research need to be acknowledged. A total of 142 (56.8%) managers and employees responded to the survey. According to research methods bibliography a response rate of 50% or more is generally considered an acceptable response rate in the surveys (Babbie, 2015). Bearing that in mind, it appears that the results of the study could be generalized to the target population (Greek Ministry of tourism). The present study was demarcated at the Greek Ministry of Tourism. Data for this study were only collected from the Greek Ministry of Tourism and there was no attempt to generalize this information to the whole public sector. The study was also demarcated by the questionnaire, which was aimed to collect data on human resource empowerment and employees' job satisfaction at the Greek Ministry of Tourism. The following were acknowledged as the limitations of the study:

- There was no proof of the objectivity, of the participants when responding to the questionnaire
- There was no proof of the level of participants' comprehension about human resource empowerment terms.

References

Al-Ababneh, M., Al-Sabi, S., Al-Shakhsheer, F., & Masadeh, M. (2017). The influence of employee empowerment on employee job satisfaction in five-star hotels in Jordan. *International Business Research, 10*(3), 133–147.

Alabar, T. T., & Abubakar, H. S. (2013). Impact of employee empowerment on servic and quality— An empirical analysis of the Nigerian banking industry. *British Journal of Marketing Studies, 1*(4), 32–40.

Babbie, E. R. (2015). *The practice of social research.* Nelson Education, Wadsworth.

Blase, J., & Blase, J. (1996). Facilitative school leadership and teacher empowerment: Teachers' perspectives. *Social Psychology of Education, 1,* 117–145.

Bogler, R., & Nir, A. (2010). The importance of teachers' perceived organizational support to job satisfaction. What's empowerment got to do with it? *Journal of Educational Administration, 50*(3), 287–306.

Byham, W. C., & Cox, J. (1989). *Zapp! The lightening of empowerment.* Development Dimensions International Press.

Chelladurai, P. (1999). *Management of human resources in sport and recreation.* Human Kinetics Publishers.

Chinomona, E., Popoola, B. A., & Imuezerua, E. (2017). The influence of employee empowerment, ethical climate, organisational support and top management commitment on employee job satisfaction. A case of companies in the Gauteng province of South Africa. *Journal of Applied Business Research (JABR), 33*(1), 27–42.

Crow, M. G., & Pounder, D. G. (2000). Interdisciplinary teacher teams: Context, design, and process. *Educational Administration Quarterly, 36,* 216–254.

Cypert, C.B. (2009). Job satisfaction and empowerment of Georgia high school career and technical education teachers. Dissertation, Thesis (Ed. D.).University of Georgia.

Domínguez-Falcón, C., Martín-Santana, J. D., & De Saá-Pérez, P. (2016). Human resources management and performance in the hotel industry. *International Journal of Contemporary Hospitality Management.*

Eyupoglu, S. Z., & Saner, T. (2009). The relationship between job satisfaction and academic rank: A study of academicians in Northern Cyprus. *Procedia Social and Behavioral Sciences, 1,* 686–691.

Gill, A., Flaschner, A. B., Shah, C., & Bhutani, I. (2010). The relations of transformational leadership and empowerment with employee job satisfaction: A study among Indian restaurant employees. *Business and Economics Journal, 18,* 1–10.

Hanaysha, J., & Tahir, P. R. (2016). Examining the effects of employee empowerment, teamwork, and employee training on job satisfaction. *Procedia-Social and Behavioral Sciences, 219,* 272–282.

Hoppock, R. (1935). *Job satisfaction.* Harper.

Idris, A., See, D., & Coughlan, P. (2018). Employee empowerment and job satisfaction in urban Malaysia. *Journal of Organizational Change Management.*

Kaniadakis, A., Papaioannou, A. & Kriemadis, T. (2017) Human resource empowerment and job satisfaction in sport divisions of the Higher Military Educational Institutions and higher schools of Armed Forces. *Sport Tourism and Leisure Magazine, 11*(b), 12–46.

Kriemadis, T., & Papaioannou, A. (2006). Empowerment methods and techniques for sport Managers. *Choregia. Sport Management International Journal, 2*(1–2), 117–133.

Kriemadis, A. (2011). Marketing plan and human resource empowerment as contributing factors of the competitiveness of professional sports (football & basketball clubs). Proceedings the 19th International Congress on Physical Education and Sport, ICPES, Komotini.

Koustelios, D. A. (2001). Personal Characteristics and job Satisfaction of Greek teachers. *The International Journal of Educational Management, 15*(7), 354–358.

Lee, Y., Nam, J., Park, D., & Lee, K. (2006). What factors influence customer-oriented prosocial behavior of customer-contact employees? *Journal of Services Marketing, 20*(4), 251–264. https://doi.org/10.1108/08876040610674599.

Morrison, R., Jones, L., & Fuller, B. (1997). The Relation Between Leadership Style and empowerment on Job Satisfaction of Nurses. *the Journal of Nursing Administration, 27*(5), 27–34.

Papaioannou, A. (2009). *Human resource empowerment applied to Greek sport Organizations.* Unpublished doctoral dissertation, University of Peloponnese, Sparta.

Papaioannou, A., Kriemadis, T., Alexopoulos, P., Vrondou, O., & Kartakoullis, N. (2009). The relationship between human resource empowerment and organizational performance in football clubs. *International Journal of Sport Management, Recreation and Tourism, 4,* 20–39.

Papaioannou, A. (2011). Human resource empowerment in football & basketball clubs. In *Proceedings of the 19th international congress on physical education and Sport, ICPES,* Komotini.

Papaioannou, A., Kriemadis, T., Alexopoulos, P., & Vrondou, O. (2012). An analysis of human resource empowerment and organizational performance in Greek sport federations. *World Review of Entrepreneurship, Management and Sustainable Development, 8,* 439–455.

Papaioannou A., Kriemadis A., Kourtesopoulou A., Sioutou A. & Avgerinou V. (2017a). Sport business excellence: A systemic approach. In *2017 HSSS 13th national and international conference of Hellenic society for systemic studies, systemic organizational excellence* (pp. 83–85). HSSS.

Papaioannou A., Kriemadis Th., Koronios K., & Sioutou A. (2017b). Empowerment of human resources in the educational sector: The case of the secondary education school units of the region of Peloponnese. In *Proceedings of the 2nd Panhellenic educational conference*. Gythio, Laconia, Marios (2017).

Papaioannou, A., Sioutou, A., & Kriemadis, T. (2018). An investigation of human resource empowerment in municipal sport organizations. *International Review of Services Management, 1*, 23–39.

Rice, E. M., & Schneider, G. T. (1994). A decade of teacher empowerment: An empirical analysis of teacher involvement in decision making, 1980–1991. *Journal of Educational Administration, 32*, 43–58.

Rinehart, J. S., & Short, P. M. (1994). Job satisfaction and empowerment among teacher leaders, reading recovery teachers, and regular classroom teachers. *Education, 114*, 570–580.

Sarason, S. B. (1992). *The predictable failure of educational reform: Can we change course before it's too late?* Jossey-Bass.

Spector, P. E. (1985). Measurement of human service staff satisfaction: Development of the job satisfaction survey. *American Journal of Community Psychology, 13*(6), 693–713.

Spector, P. E. (1997). *Job satisfaction: Application, assessment, causes and consequences*. Sage.

Ugboro, I. O., & Obeng, K. (2000). Top management leadership, employee empowerment, job satisfaction in TQM organizations: An empirical study. *Journal of Quality Management, 5*, 247–272.

Ukil, M. I., & Ullah, M. S. (2016). Effect of occupational stress on personal and professional life of bank employees in Bangladesh: Do coping strategies matter. *Journal of Psychological and Educational Research, 24*(2), 75.

Vogt, F. J., & Murrell, L. K. (1990). *Empowerment in organizations: How to spark exceptional performance*. University Associates Inc.

Warr, P., Cook, J., & Wall, T. (1979). Scales for the measurement of some work attitudes and aspects of psychological well-being. *Journal of Occupational Psychology, 52*, 129–148.

Wilson, S., & Coolican, M. J. (1996). How high and low self-empowered teachers work with colleagues and school principals. *Journal of Educational Thought, 30*, 99–118.

Wright, B., & Davis, B. (2003). Job satisfaction in the public sector: The role of the work environment. *The American Review of Public Administration, 33*(1), 70–90.

Wright, T. A. (2006). The emergence of job satisfaction in organizational behavior. A historical overview of the dawn of job attitude research. *Journal of Management History, 12*(3), 262–277.

Wu, V., & Short, P. M. (1996). The relationship of empowerment to teacher job commitment and job satisfaction. *Journal of Instructional Psychology, 25*, 85–89.

Baron, R. S. (1986). Distraction-conflict theory: Progress and problems. *Advances in experimental social psychology, 19*, 1-40.

The City of Thessaloniki as a Culture Tourism Destination for Israeli Tourist

Efstathios Velissariou and Iliostalakti Mitonidou

1 Introduction

The scope of the paper is to demonstrate the importance of Thessaloniki for the Jews and to explore the motives of the visit, the characteristics and the behavior of the Israeli tourists in Thessaloniki. Through this research, it is intended to draw important conclusions that will help to improve the organization, the development and the promotion of the city of Thessaloniki, as a cultural destination for the Jews. The city of Thessaloniki was a refuge for the persecuted Jews of Europe for more than 20 centuries and especially during the period 1492–1943, consolidating in the city many historical centers of the Diaspora, resulting in the creation of a large Jewish community.

The paper first presents the city of Thessaloniki, but also the importance of the city for the Jews, as well as the main cultural elements related to the history and the presence of the Jews in Thessaloniki.

The research part presents the results of a survey of Israeli tourists that took place from November 2019 to March 2020. The survey initially concerned the demographic characteristics and motivations of the respondents. Then the way of organizing the journey, the type of accommodation, and the length of stay of the tourists were investigated. The activities during the stay of the tourists as well as the points of interest visited were also recorded. Finally, the evaluation of Thessaloniki as a tourist and cultural destination was requested, as well as the formulation of proposals by the visitors for the improvement of the tourist product of the city.

E. Velissariou (✉) · I. Mitonidou
University of Thessaly, Larissa, Greece
e-mail: belissar@uth.gr

V. Katsoni and C. van Zyl (eds.), *Culture and Tourism in a Smart, Globalized, and Sustainable World*, Springer Proceedings in Business and Economics, https://doi.org/10.1007/978-3-030-72469-6_28

2   City Tourism Und Cultural Tourism

According to the European Commission, (2000), city tourism is divided on the basis of the purpose of travel, in leisure tourism, cultural tourism, conference tourism, and sports tourism. In the context of leisure tourism in a city, experiences related to tourism, leisure, and entertainment are produced and consumed (Page and Hall, 2003). During a visit or stay in a city, the majority of trips include some kind of cultural activities. Therefore, very often cultural tourism and city tourism considered as similar. Often these terms are considered identical in the literature (Chevrier & Clair-Saillant, 2006). Visitors to a city for a conference, during their stay combine other forms of tourism, such as cultural tourism, gastronomic tourism, etc. (Law, 1993). To be noted that a city break trip is often a complementary holiday trip where tourists spend some days to discover the beauties of a city (Dunne et al., 2010). Of course, it should be also noted that the activities that interest city tourists are mainly of cultural interest, with visits to museums, visits to historical sites, and clearly to attractions (D'Hauteserre, 2000). In addition to these activities, they show particular interest in leisure activities, such as shopping, night out, watching a sporting event, a concert or theater, attending conferences and exhibitions, or even visiting a relative or friend (Kokkosis et al., 2011).

Cultural tourism is considered the oldest form of tourism. A visit to historical monuments, museums, cultural landscapes, attending special events, and happenings has always been a subset of the overall tourist experience (McKercher & Du Cros, 2002).

Researchers and tourism operators, in the late 1970s, began to realize that many tourists traveled in order to understand the culture of a place destination and to learn about its cultural heritage. With this finding, cultural tourism began to be recognized as a special and alternative category of tourism (Tighe, 1986). Cultural tourism or cultural heritage tourism places particular emphasis on the preservation, restoration, and utilization of cultural resources in order to create tourist flows. Thus, it is classified as a special form of tourism as it focuses on quality and sustainable development (Cater, 1987; Gould, 1994).

Curiosity but also the desire to learn and get to know others, to explore but also to increase our own experiences, is what we would call the "cultural" version of tourism (Herreman, 2003). Bonink (1992) makes two main distinctions. The first distinction is the "monuments and sites." It tries to describe the type of cultural attractions that attract cultural tourism tourists. The second distinction is a conceptual approach that is more specific and tries to outline the motivations and activities of cultural tourism tourists that lead them to ultimately visit monuments and sites of cultural interest.

Cultural products and the identity of the cultural tourist have a very common pattern, according to which the cultural tourist (Silberberg, 1995), particularly: (1) Earns and spends more money during his travels, while staying longer space in one place. (2) It is more likely to make purchases. (3) Its educational level is higher in relation to the wider tourist public. (4) It mostly belongs to a larger age group (target group), and the largest percentage are women.

The high educational level of cultural tourists is confirmed by Association for Tourism and Leisure Education (ATLAS), which conducted a series of surveys from 1992 to 2001 in Europe. According to Hughes (2002), the cultural tourist does not belong to a homogeneous mass market, but to a heterogeneous market that have different characteristics and needs (Stylianou & Lambert, 2011). Hughes (2002) initially divides the cultural tourist into two types: "core", refers to the tourist who travels to a destination to get to know and experience the culture of the place and its people, and to a "peripheral", refers to the tourist traveling for other reasons. He then divides the main cultural tourist into primary and multi-primary, while Silberberg differentiates the cultural tourist into highly motivated and motivated in part, respectively. Similarly, it divides the regional cultural tourist into incidental and accidental, while Silberberg divides it into adjunct and accidental, respectively (Stylianou & Lambert, 2011). McKercher and Du Cros (2002) have identified five categories of cultural tourists: purposeful cultural tourist, sightseeing cultural tourist, serendipitous cultural tourist, casual cultural tourist, and incidental cultural tourist. Most distinctions are made based on the degree to which the tourist is involved with cultural tourism.

Religiously motivated visitors should be added to the cultural heritage visitors. According to Rinschede (1992), religious tourism is the form of tourism, the participants of which move either partially or exclusively for religious reasons. In fact, he considers that religious tourism is a subcategory of cultural tourism. In the case of religious heritage visitors, tourists usually travel in groups with a specific religious inclination, however, they combine the trip of religious interest with other tourist activities such as visits to archeological sites, markets, etc. This is how the concept of religious tourist is formed, which is contrasted by the pilgrim (Vukonic, 1996).

3 The City of Thessaloniki. Main Characteristics and Tourism

Thessaloniki is located in Northern Greece, in the region of Central Macedonia and is the second largest city in Greece in terms of area and population. It is the seat of the region of Central Macedonia, but also of the Decentralized Administration of Macedonia—Thrace. The urban complex of Thessaloniki has an area of 111,703 Km^2 and consists of seven Municipalities, while in the suburbs there are four other municipalities. According to the 2011 census, 1.02 million people live in the urban complex, with the municipality of Thessaloniki (in the center) having 325.2 thousand inhabitants (Hellenic Statistic Authority, 2011).

Thessaloniki is a transport hub and gateway from the Balkans with easy accessibility, as it has the second largest airport in the country, one of the largest ports, connection to the railway network of Greece and the Egnatia highway. Thessaloniki International Airport in 2018 made 1.11 million domestic arrivals and 2.08 million arrivals from abroad. A total of 6.89 million passengers were handled (arrivals and departures) in 2019 (Skg-airport, 2020). Thessaloniki is home of three Universities and has the largest University campus in Greece, but also the largest exhibition center in Greece, the "Thessaloniki International Exhibition & Congress Center." The above infrastructures complement the existence of a number of cultural and archeological monuments, included in the "List of World Cultural Heritage" of UNESCO. The existence of great historicity in the city of Thessaloniki is the result of a combination of ancient Greek, Byzantine, Roman, Turkish, and Israeli cultural elements.

Regarding the hotel capacity, Thessaloniki had in 2019 a total of 145 hotels with a capacity of 15,339 beds, of which 50.7% belong to the highest categories of 4 and 5 stars, demonstrating the high level of services in the city. Also, according to the Thessaloniki Hoteliers Association, it is estimated that there are about 2000 properties available in the field of short-term rent, but which are not available all year round (Thessaloniki Hotel Association, 2020).

In the year 2019, 2,442,050 overnight stays were recorded in Thessaloniki, of which 1,119,582 or 45.8% concerned overnight stays of Greek travelers, demonstrating the important place that Thessaloniki holds in the economic and social life of Greece. In recent years (2018 and 2019), overnight stays of tourists from Israel occupy the first place, followed by overnight stays from travelers coming from Cyprus, Germany, and the USA. As shown in Fig. 1, after 2016, Israeli overnight stays increased rapidly. The passengers from Israel at the airport of Thessaloniki "Macedonia" were also increased. More specifically, in the year 2017, about 55,412 passengers were handled, in the year 2018 a total of 124,232 passengers, and in 2019 187,029 passengers (Fraport Greece, 2020).

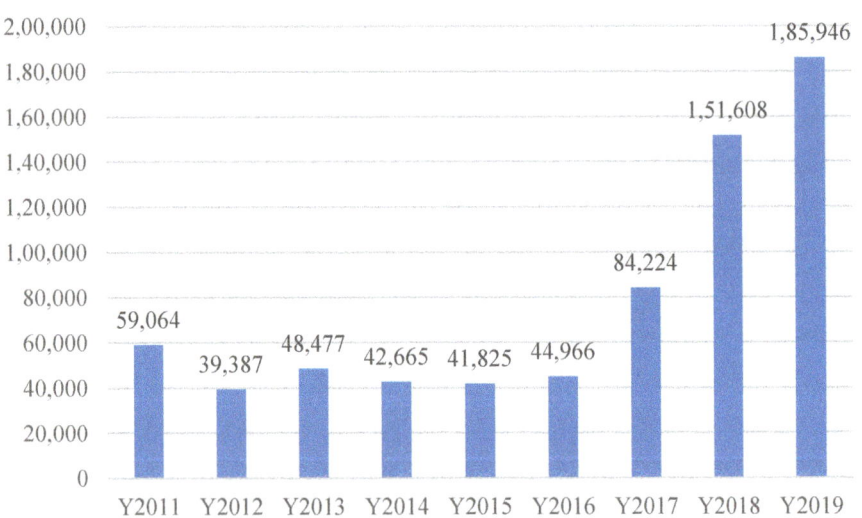

Fig. 1 Overnight stays by Israelis in the city of Thessaloniki (2011–2019). *Source* Thessaloniki Hotels Association (2020)

4   The History of Thessaloniki Over the Last 23 Centuries

Thessaloniki was founded in 316 BC by Cassander, general of Philip II, who named the city after his wife, the sister of Alexander the Great. Due to its great geographical location, but also the policy of the successors of Alexander the Great to establish cities in key positions, in order to facilitate the communication of the rest of the world with the Macedonian state, the region developed rapidly into an urban center, attracting many inhabitants of the surrounding area. The organized port of Thessaloniki was necessary, since trade and communications with distant places were growing rapidly. The development of the city was rapid during the second century BC, so walls and a citadel were built (City of Thessaloniki, 2017a).

In 149 BC, the Macedonian State was occupied by the Romans (Roman era) and divided into four provinces, one of which became the capital Thessaloniki. The city retains its administrative autonomy and is developing quite rapidly, while it is not long before it becomes the most important city in Macedonia. The construction of Via Egnatia (146–120 BC), connecting Dyrrachium to Evros, helped raise Thessaloniki to a major commercial, cultural, and military center. The fourth century was one of great changes for the city, characterized by the pre-eminence of Christianity, beginning around the visit of Paul the Apostle to the city. By the end of the fifth century, the Roman city had been transformed into a center of Christianity (City of Thessaloniki, 2017b).

In 413 AD, the holy temple of Agios Dimitrios, the patron saint of Thessaloniki, was built, and for the following centuries, the enemies made attacks in order to conquer the city. From the seventh to the eleventh century, the city of Thessaloniki experienced many developments. The conquest of Thessaloniki by the Normans in 1185 and its domination by the Latins, for about twenty years, after 1204, were events that temporarily stopped the evolution of its civilization, but without interrupting it permanently. The historical character of Thessaloniki is undoubtedly linked to its Byzantine life. The walls and extant inscriptions record the tumultuous history of the city. The walled city and its monuments could reasonably be described as an open museum.

During the Ottoman period (1430–1912 AD), most of the Christian temples, as well as the catholic monasteries of the city, were converted into mosques. At the same time, mosques, the covered market, and the baths were built, and new complexes were built throughout the city; while at the same time, the existing water supply system of the city was expanded with underground and above-ground cisterns. Thessaloniki, around 1500 AD, acquires a multicultural and multi-religious character, which it maintains until its liberation, due to the arrival of Jewish refugees from Spain (Vakalopoulos, 1983).

During the twentieth century, Thessaloniki undergoes many changes. More specifically, in 1912 Thessaloniki was liberated and incorporated to the Greek state. In 1917, an entire cultural heritage is engulfed in the flames of a great fire; while over 70,000 people are left homeless, all their belongings are lost. After the signing of the Treaty of Lausanne (1923) and the decision to exchange populations, thousands of refugees from Central Asia are gathering. While the city is being emptied of Muslims, the refugees from Central Asia come to settle in temples, in abandoned camps, but also in the settlements burned by the fire, and Thessaloniki becomes the Capital of Refugees (Karathanasis, 2012).

During World War II, Thessaloniki suffered because of the German occupation, as large numbers of Thessaloniki Jews were taken to Nazi concentration camps. However, during the post-war period, Thessaloniki again became a pole of attraction.

In the 90s, the eastern bloc collapsed, and Thessaloniki received a large wave of immigrants; while in 1997, Thessaloniki was declared "European Capital of Culture," and in 2012, it celebrated a whole century since its liberation, while in 2014, "European Youth Capital."

5   The Jews of Thessaloniki

Thessaloniki, called also "Jerusalem of the Balkans," was once home to the largest Jewish community in the world (Devin, 2016), a fact that justifies the numerous books and publications referring to the Jews of Thessaloniki. According Hagouel (2013), the history of the Jewish presence in Salonika may be partitioned in four chronological periods (A) From ancient times to 1492 CE, (B) From 1492 to the occupation of Salonika by the German Armed Forces in April 9, 1941, (C) The period of the German Occupation (April 9, 1941 to October 30, 1944), and (D) From the date of liberation to the present.

It is speculated that around 140 BC, the first Jews arrived in Thessaloniki, who came from Egypt and more specifically from Alexandria. "Etz Achaim," in other words "Tree of Life," was the oldest synagogue in Thessaloniki, according to Jewish tradition (Jewish community of Thessaloniki, 1997a). Jews from all over Europe had settled in Thessaloniki from time to time, establishing their synagogues and communities. Sometime in 1492, it is estimated that between 15,000 and 20,000 Jews from Spain, the so-called "Safaradim," settled in Thessaloniki, who were expelled from their homeland, following the decree of the "Catholic Kings" Ferdinand and Isabella. This fact played a decisive role both for the Jewish community and for the whole history. Thessaloniki then developed into an important center of theological studies and raised prominent personalities, leading Samuel Usqué, a Jewish poet from Ferrara, to call Thessaloniki "Mother of Israel" in 1537.

At the end of the nineteenth century, the Jews owned most and the most important trading houses, while they are dominating the industry. The population exceeded seventy thousand people, i.e., about half of the city's population. In the fire of 1917, 53,737 Jews were left homeless, while many synagogues and institutions were destroyed. As a result, many Jews were deported during the interwar period. Finally, in 1940, they numbered about fifty thousand.

On April 9, 1941, Thessaloniki was occupied by the Germans and strict restrictive measures were imposed on the Jews. In the last month of 1942, Jewish businesses were looted by the Germans, while the Old Jewish cemetery was completely destroyed.

On March 15, 1943, the first train departed from the old train station for the Auschwitz and Birkenau concentration camps. It is estimated that by August 2, 1943, a total of 46,061 Jewish residents of both Thessaloniki and other cities in Northern Greece were transported by 19 rail missions to their death destinations. Eventually, despite the efforts of top Christian clergy, senior civil servants, and some units of the National Resistance, and especially individual citizens, 96% of the Jewish community of Thessaloniki was eventually exterminated (Central Israeli Council, 2009).

After the war, those who survived the Holocaust gathered in Thessaloniki. However, bad memories and the difficult economic situation in post-war Greece led many Jews to emigrate to Israel and the United States. Characteristic of the multitude of Jews living in Thessaloniki is the Old Jewish Cemetery, which historians estimate should have numbered more than 300,000 graves in the early 1940s, making it perhaps the largest Jewish necropolis in all of Europe.

6   Israeli Monuments and Sites in Thessaloniki

Today, Thessaloniki has a number of cultural monuments, which can be grouped in synagogues, Schools and Institutions, modern historic buildings, markets, and places of remembrance. Thessaloniki had at the end of the nineteenth century over a hundred synagogues. Initially in 1492, there were three: the Romans, the Italians, and the Ashkenazis. Of the 100 synagogues, 32 could be considered parish centers, which, despite having a central community administration, from 1680 onwards, retained some responsibilities such as tax collection and birth registration, while being maintained by members' contributions (Alberto, 1985). The oldest of these 32 synagogues was the Ets ha Haïm (The Tree of Life) of the first century BC and the newer seventeenth century Mograbis. The largest synagogue was the Beit Shaul, built in 1898 by Fakima Modiano, wife of the great benefactor Saül Modiano (Veinstein, 1992).

Among the most important and historic synagogues are the synagogue of the Monasteriotes, whose building was inaugurated after its restoration on May 14, 2015, and the Yad Lezicaron "Synagogue," inaugurated in 1984 dedicated to the memory of the victims of the Holocaust. In the same place, since 1921, the synagogue of the market "Cal de la Plasa" was established (Christos, 2017).

An important shopping center even today is the Modiano market in the heart of the shopping center of Thessaloniki. At the place where the Emprar project located the city's bazaars in 1922, the construction of the first covered market of the city of European type begins in the plans of the architect Eli Modiano. The opening of the central food market took place on March 23, 1925, and the market operates uninterruptedly until today, going through periods of prosperity and decline. It consists of 144 small stores. The important historical shopping centers are the Saul Gallery, a shopping gallery that now occupies an entire building block and includes two T-shaped inner sidewalks. It was created in 1881 by Saul Modiano, one of the richest Jews in Thessaloniki (Nena, 2012).

A Holocaust memorial has been established in memory of the 50,000 Greek Jews of Thessaloniki who testified in the deadly Nazi concentration camps, which was inaugurated in 1997 by the President of the Greek Republic. It was designed by the brothers Glid and depicts the seven candled menorah and flames all entangled in a mesh of human bodies (Jewish community of Thessaloniki, 1997b).

The Museum of Jewish history is housed in the center of Thessaloniki in a building built in 1904. The museum is the most creative initiative of the Israeli Community of Thessaloniki and for which the Organization of the European Capital of Culture "Thessaloniki 1997" contributed, which restored the building (Fragoudi, 2015).

Today, there are also many modern Jewish monuments in Thessaloniki, which belonged to wealthy families of Thessaloniki, such as "Villa Allatini," perhaps the most imposing mansion, built in 1888 to designs by Vitaliano Pozelli. "Villa Modiano" was built in 1906 to house the family of banker Jacob (Yako) Modiano. Since 1970, it houses the Folklore and Ethnological Museum of Macedonia Thrace. "Villa Fernandez" (Casa Bianca) is one of the most impressing buildings. The construction was completed in 1913 for the family of the Jewish businessman Dino Fernandez Diaz Built in 1910 by the Italian architect Pierro Arigoni (Jewish community of Thessaloniki (1997c).

The Old Railway Station is a place of memory for the Jews of Thessaloniki. At this station, the Nazis gathered and transported the Jews to the concentration camps. In 2013, a memorandum of understanding was signed with the Municipality of Thessaloniki, the Railway Company, and the Israeli community of Thessaloniki, providing for the creation of both the Holocaust Museum at the Old Railway Station and the creation of the Conference Center within the Old Station (Tzimou, 2014).

7 Research Methodology

A primary survey was conducted using a structured questionnaire, which was addressed to tourists from Israel. The purpose of the research was to investigate:

A. The demographic characteristics and the motivations of the respondents, in order to outline the profile of the visitors.
B. The way of organizing the trip, the type of accommodation, as well as the length of stay.
C. The recording of the activities during their stay and the visitation of the most important cultural monuments.
D. The evaluation of Thessaloniki as a tourist and cultural destination and the formulation of proposals by visitors for the improvement of the tourist product of the city.

The survey was originally designed to be conducted in different seasons, in particular in Winter 2019–20, Spring 2020 and Summer 2020, in order to investigate any differences in the profile, motivation, and activities of tourists depending on the season. Eventually, the investigation was conducted only during the period November 2019–March 2020 and was stopped due to the restrictive measures taken for the pandemic caused by Covid-19. It should be noted the difficulty of finding tourists from Israel, as the period during which the survey was conducted is a low tourist season for them. Furthermore, the fact of the existence of the new regulation concerning personal data (GDPR) was an obstacle on the distribution of questionnaires in some hotels.

The questionnaires were collected mainly live, with the distribution in Jewish areas of Thessaloniki, and areas of cultural interest for tourists coming from Israel, but also in hotels in the city. Of the 93 questionnaires, 67 were answered in person. The remaining 26 questionnaires were answered through the "Trip advisor" platform by tourists who had recently visited Thessaloniki and had been sent the relevant questionnaire. This method was deemed necessary to complete a sufficient number of questionnaires. All the answers to the 93 questionnaires collected were recorded in the online questionnaire of "Google Forms,"

8   Results and Discussion

In terms of demographic characteristics, 40.9% of respondents were men, while 59.1% were women. The survey showed that the majority of Israeli tourists were mostly over 45 years old. In particular, 79.6% belong to the ages over 45 years. This may be due to the fact that these tourists had more cultural interests, in contrast to the younger tourists who visit Thessaloniki mainly in the summer more for the sea and entertainment (Table 1).

In terms of the profession, the research has shown that there was a very large dispersion. Most respondents were private employees (25.8%), self-employed (15.1%),

Table 1 Age groups of Israeli tourists participating in the survey

Age groups	%
18–24	2.2
25–34	8.6
35–44	9.7
45–54	30.1
55–64	34.4
65–74	12.9
75	2.2
	100.0

Source Research results

Table 2 Occupation of Israeli tourists participating in the research

Occupation	%
Private employee	25.8
Retired	15.1
Freelance	15.1
Educational	14.0
State employee	10.8
Business executive	8.6
Businessman	6.5
College student	2.2
Housekeeping	2.2
Unemployed	0.0
	100.0

Source Research results

retirees (15.1%), teachers (14%), and civil servants (10.8%), while none of them were unemployed (Table 2).

The highest level of education is considered very important for the profile of Israeli tourists, where according to their answers 74.2% had a University degree, Master or a Ph.D (Table 3).

The vast majority of the sample of tourists from Israel at a rate of 87.1% visited Thessaloniki for the first time, about 3.2% had come once in the past, 7.5% had come 2–3 times, and only 2.2% visited Thessaloniki four or five times in the past.

Regarding the length of stay, 82.8% answered that they spent 3–7 nights, proving that the city of Thessaloniki is not for the tourists from Israel a "city break" destination of one or two nights (Table 4).

During their stay in Thessaloniki, Israeli tourists prefer hotels, especially those of 3, 4, and 5 stars at a rate of 71%. Another 23.7% of travelers spent the night in "Airbnb" type accommodation, while a small percentage, of the order of 2.2%, stayed with relatives and friends in Thessaloniki (Table 5).

One of the most important findings of the survey was the majority of respondents at 95.7% answered that they organized their travel to Thessloniki themselves and only

Table 3 Level of education of Israeli tourists participating in the research

Level of education	%
Secondary school	8.6
College	17.2
University	39.8
Master degree	30.1
Ph.D.—postdoctoral	4.3
	100.0

Source Research results

Table 4 Nights of spend of tourist from Israel participating in the research

Nights of spent in Thessaloniki	%
1 night	6.5
2 nights	10.8
3 nights	49.5
4–7 nights	33.3
	100.0

Source Research results

Table 5 Type of accommodation of tourist from Israel participating in the research

Type of accommodation	%
1–2 star hotels	3.2
3 star hotels	34.4
4 star hotels	23.7
5 star hotels	12.9
Airbnb	23.7
Friends and relatives	2.2
	100.0

Source Research results

4.3% through a travel agency and only 7.5% travel in groups. This demonstrates the very low degree of mass organized tourism of tourists from Israel, excluding the summer season.

Regarding their main motivation, almost the half of the visitors (48.4%) answered that they came for entertainment and fun, while a large percentage (30.1%) answered that they came for cultural and 6.5% for religious motives. Also about 6.5% stated that the purpose of their trip was to visit friends and relatives, demonstrating family ties with the city of Thessaloniki (Table 6).

It was interesting to record the activities during the stay. Thus, the majority (81.7%) answered that they visited museums or monuments, despite the fact that the

Table 6 Motivation factors of Israeli tourists for visiting Thessaloniki

The main motives for visiting Thessaloniki	%
For leisure and fun	48.4
For cultural motives	30.1
For religious motives	6.5
For shopping, browsing and having fun	6.5
For visiting relatives and friends	6.5
For business motives	2.2
	100.0

Source Research results

Table 7 Activities of tourist from Israel during the visit in Thessaloniki

Visits to monuments—museums	81.7%
Tours—excursions	49.5%
Shopping	41.9%
Participation in cultural events	18.3%
Recreational activities (Fun)	11.8%
Educational activities	4.3%
Professional activities	2.2%

Source Research results

main motive for their visit to Thessaloniki was not culture. Also, a large percentage (49.5%) answered that they participated in tours and excursions; while 41.9% of travelers answered that during their stay, they did shopping or shopping. Thessaloniki is convenient for excursions to nearby areas, such as Chaklidiki, the Mount Athos, and the Archeological site of Vergina (the Tombs of the Macedonian Kings) (Table 7).

The Jewish monuments with the most visits by tourists from Israel are the museum of Jewish history, the synagogue of the Monasteriotes, the synagogue of Yad Lezikaron, the Holocaust Memorial, the Modiano market, and the Old Railway Station Thessaloniki. Note that based on the survey responses, tourists visited at least 3.3 points of Jewish interest on average (Table 8).

Also, 86% of respondents said that they visited the "White Tower," which is the landmark of the city, while more than half (59.1%) said that they visited the port of Thessaloniki. Tourists also visited an average of 3.3 places of cultural interest, in addition to Jewish cultural elements.

Table 8 Visited Jewish cultural point of interest in Thessaloniki by the tourist from Israel

Museum of Jewish history	69.9%
Monasteriotes' Synagogue	32.3%
Jad Lezicaron Synagogue	32.3%
Holocaust memorial	28.0%
Old railway station	26.9%
Modiano	26.9%
Joshua Abraham Salem Synagogue	19.4%
Old Jewish cemetery	18.3%
New Jewish cemetery	11.8%
Casa Bianca	10.8%
Villa Allatini	9.7%
Villa Modiano	8.6%
Saul's gallery	4.3%
Allatini Mills	4.3%

Source Research results

Table 9 Visited additional cultural sites in Thessaloniki by the tourist from Israel

White Tower	86.0%
Port of Thessaloniki	59.1%
Rotonda	35.5%
Castles	31.2%
Church of Hagia Sophia	24.7%
Church of St. Demetrios	22.6%
Byzantine Museum	22.6%
Archeological Museum	22.6%

Source Research results

Recording the visits of tourists, it was found that all Israelis (100%) and regardless of the purpose of the journey, they visited places of cultural interest in Thessaloniki during their stay (Table 9).

It should be noted that the tourists from Israel, for their local tours in Thessaloniki, used the bus only 9.7%, while 26.9% used the taxi. Also 22.6% rented a car, while the largest percentage 40.9% moved mainly by foot.

Finally, regarding the evaluation of Thessaloniki as a tourist cultural destination, some specific statements were made, presented in Table 10, for which the degree of agreement or disagreement was requested.

Table 10 Evaluation of Thessaloniki as a cultural destination by the tourist from Israel

	Strongly agree (%)	Strongly agree (%)	Neutral (%)	Disagree (%)	Strongly disagree (%)	Grad 1–5
It is one important destination for cultural tourism in Greece	35.5	40.9	19.4	4.3	0.0	4.07
There are plenty of important cultural resources	23.7	53.8	20.4	2.2	0.0	3.99
It is organized sufficiently (in services and information) so that is offered for cultural tourism	9.7	38.7	45.2	5.4	1.1	3.51

Source Research results

Table 11 Recommendation of Thessaloniki as a cultural destination by the tourist from Israel

	Yes (%)	Maybe (%)	No (%)
Would you visit Thessaloniki for cultural tourism in the future?	68.8	24.7	6.5
Would you recommend Thessaloniki for cultural tourism?	88.2	9.7	2.2

Source Research results

About 76.3% of Isreli tourists agree that Thessaloniki is an important destination for cultural tourism in Greece. Also 77.4% agree that Thessaloniki has many and important cultural resources. On the contrary, only 48.39% agree that it is sufficiently organized, in terms of services and information for cultural tourism. Thus, demonstrate the fact that Thessaloniki has further prospects for improvement as a cultural destination.

The majority of Israeli tourists (88.2%) would recommend Thessaloniki as a cultural destination. They themselves are highly satisfied with the destination Thessaloniki, and there is a high probability of repeating a journey in the future (Table 11).

9 Conclusions

Thessaloniki was a city in which the Israeli community had a significant presence, both in population and in the economic activity of the city. The result of this long-standing presence is that there are still many cultural and religious monuments today. It is not surprising that Thessaloniki is a popular tourist destination for Israelis, showing a significant increase after 2016, ranking Israelis in first place of the foreign traveler's overnights in the city.

The primary survey carried out between November 2019 and March 2020 in Thessaloniki revealed significant findings, such as the fact that about 80% of Israeli tourists are over 45–74 years old. It was also confirmed the fact that cultural tourists are highly educated, as 74.2% of the Israeli tourist in Thessaloniki had a university degree, a Master or a Ph.D.

It is also important that 95.7% of the city's visitors from Israel organized the trip themselves and that only 7.5% travel in groups. This fact gives great prospects for the development of organized tourism between Israel and Thessaloniki. Also important is the fact that according to the length of stay, Thessaloniki is not a short city break station, but 83% of visitors spend 3–7 days in Thessaloniki.

It is noteworthy that while only 36.3% stated that culture and religion were the main motivation for visiting Thessaloniki, all tourist (100%) made visits to monuments and museums. Consequently, tourists from Israel can be characterized as "purposeful cultural tourist" and "sightseeing cultural tourist," according to the classification of McKercher and Du Cros (2002).

It should also be noted that Israeli tourists are a combination of cultural heritage and religious heritage tourists, making it difficult to separate the two interest groups.

The intense interest stems from the fact that tourists from Israel, regardless of the travel motivation, visited on average at least 3.3 points of Jewish cultural or religious interest and at least another 3.3 sites of general cultural interest.

Although the visitors were very satisfied with their visit to Thessaloniki, the evaluation of cultural tourism services and information was rated 3.5 out of 5, which shows that there are opportunities for improvement.

In particular, 54% of the tourist suggested the creation of a network for the promotion and the information for the tourists who are interested in cultural monuments. About 44% suggested "Organizing cultural events with the active participation of tourists" and 43% suggested to "Creating a strategy plan to connect the cultural tourism to alternative forms of tourism."

In the recommendations of the tourists in a percentage of 50% was to "Creating lines of transportation that they allow moving to all cultural monuments." It should be noted that the public bus transportation in Thessaloniki offers two cultural and tourist lines (Line 50 and Line 22), which the visitors probably did not know, and therefore, more effective measures should be taken to inform the visitors of Thessaloniki.

The results of the research showed that there is great potential for the organization and promotion of tourist cultural packages for Thessaloniki. In this direction, the increase of direct flights between Thessaloniki and Israel is required. In the same context, the creation and promotion of joint cruise packages with stations in Israel, Cyprus, and Greece can contribute to the further increase of travelers from Israel.

Concluding, Jewish Thessaloniki is an invisible city in the context of a modern, developed urban environment, and it is important for it to be discovered, through its Jewish sites and monuments. From a tourism point of view, Thessaloniki has a lot to expect from the Israeli market and the Jewish diaspora, as the city, with the construction of the Holocaust Museum—a project that is under study and can be a global reference point for Jews everywhere.

References

Alberto, N. (1985). *The Synagogues of Thessaloniki*. Jewish community of Thessaloniki (in Greek).

Bonink, C. (1992). Cultural tourism development and government policy. MA dissertation, Rijksurniversiteit Utrecht.

Cater, E. A. (1987). Tourism in the least developed countries. *Annals of Tourism Research, 14*(2), 202–225.

Central Israeli Council. (2009). The Israeli community of Thessaloniki. Retrieved on August 4, 2020, from https://kis.gr/index.php?option=com_content&view=article&id=386&Itemid=49

Chevrier, F. G., & Clair-Saillant, M. (2006). Rejuvenating cultural tourism. What remains of the "tourist"? (Original title: Renouveau du tourisme culturel que reste-t-il du "touriste"?). Téoros, Revue de Recherche en Tourisme 2006 (Vol. 25, No. 2 pp. 72–74 ref. 5).

Christopher, L. M. (1993). *Urban Tourism: Attracting Visitors to Large Cities*, Mansell.

Christos, Z. (2017). *Thessaloniki of the Jews*. Thessaloniki: Publisher Epikentro.

City of Thessaloniki. (2017a). Monuments in History. Retrieved on August 4, 2020, from https://
    thessaloniki.gr/i-want-to-know-the-city/moments-in-history/?lang=en
City of Thessaloniki. (2017b). Monuments in History. The Roman period. Retrieved on August 4,
    2020, from https://thessaloniki.gr/i-want-to-know-the-city/moments-in-history/roman-period/?
    lang=en
D' Hauteserre, A.-M. (2000). Lessons in managed destination competitiveness: The case of
    Foxwoods Casino Resort. *Tourism Management, 21*, 23–32.
Devin, N. E. (2016). *Jewish Salonica.* . Between the Ottoman Empire and Modern Greece: Stanford
    University Press.
Dunne, G., Flanagan, S., & Buckley, J. (2010). Towards an understanding of international city break
    travel. *International Journal of Tourism Research, 12*(5), 409–417.
European Commission, Attitudes of Europeans Towards Tourism. (2013). Retrieved on August 4,
    2020, from https://ec.europa.eu/commfrontoffice/publicopinion/flash/fl_370_en.pdf
Fragoudi, C. (2015). Museum of Jewish History. Retrieved on August 4, 2020, from https://thessa
    rchitecture.wordpress.com/2015/10/23/mouseio-evraikis-istorias/
Fraport Greece. (2020). Air Traffic statistics. Retrieved on August 4, 2020, from https://www.skg-
    airport.gr/en/skg/air-traffic-statistics
Gilles, V. (1992). Salonique, 1850–1918. La "ville des Juifs" et le réveil des Balkans (Français).
    Translated in Greek (1994) Ekati Publishing, Athens.
Gould, G. (1994). *Evaluating tourism impacts: Study book.* Bournemouth University.
Hellenic Statistic Authority. (2011). Demographic characteristics of 2011. Retrieved on August 4,
    2020, from https://www.statistics.gr/en/statistics/-/publication/SAM03/2011
Herreman, Y. (2003). Museums and tourism: Culture and consumption. *Museum International,
    50*(3), 4–12.
Hughes, H. L. (2002). Culture and tourism: A framework for further analysis. *Managing Leisure,
    7*(3), 164–175.
Jewish community of Thessaloniki. (1997a). The History of the Thessaloniki Jews. Retrieved
    on August 4, 2020, from https://www.hri.org/culture97/eng/eidika_programmata/koinothtes/jew
    ish_community/#AA1
Jewish community of Thessaloniki. (1997b). *The holocaust memorial.* Retrieved on August 4, 2020,
    from https://www.jct.gr/HolocMemorial.php
Jewish community of Thessaloniki. (1997c). *Places of Jewish interest.* Retrieved on August 4, 2020,
    from https://www.jct.gr/villa_fernadez.php
Karathanasis, E. A. (2012). *History of Thessaloniki 323 BC-2012.* Kyriakidi Bros Publishing.
Kokkosis, Ch., Tsartas, P., & Grimba, E. (2011). *Special and alternative forms of tourism: Demand
    and supply of new tourism products.* Kritiki Publications.
McKercher, B., & du Cros, H. (2002). *Cultural tourism: the partnership between tourism and
    cultural heritage management.* The Haworth Hospitality Press.
Nena, K. (2012). The map of the city. In *The Saoul gallery.* Retrieved on August 1, 2020, from
    https://parallaximag.gr/thessaloniki/o-chartis-tis-polis-stoa-saoul.
Page, S. J., & Hall, M. (2003). *Managing urban tourism* (1st ed.). Prentice Hall.
Paul, H. (2013). *The history of the Jews of Salonika and the holocausta.*
Rinschede, G. (1992). Forms of religious tourism. *Annals of Tourism Research,* 54.
Silberberg, T. (1995). Cultural tourism and business opportunities for museums and heritage sites.
    Tourism Management, 16(5), 361–365.
Skg-airport. (2020). Air traffic statistics. Retrieved from https://www.skg-airport.gr/uploads/sys_
    nodeIng/2/2870/Thessaloniki_Traffic_2019vs2018.pdf. (Accessed the 4th of August 2020, at
    18:06).
Stylianou-Lambert, T. (2011). Gazing from home: Cultural tourism and art museums. *Annals of
    Tourism Research, 38*(2), 403–421.
Thessloniki Hotel Association. (2019). Profile & satisfaction of tourists.
Thessaloniki Hotels Association. (2020). Press releases. Retrieved on August 1, 2020, from https://
    www.tha.gr/default.aspx?lang=el-GR&page=52

Tighe, A. (1986). The arts/tourism partnership. *Journal of Travel Research, 24*(3), 2–5.
Tzimou, K. (2014). Old Thessaloniki: Old railway station. Retrieved on August 4, 2020, from https://parallaximag.gr/thessaloniki/chartis-tis-polis/i-thessaloniki-palia-paleos-sidirodr
Vakalopoulos, A. (1983). *History of Thessaloniki 316 BC–1983*. A. Stamouli Publishing.
Vukonic. (1996). *Tourism and religion*. Elsevier.

Place Attachment Genesis: The Case of Heritage Sites and the Role of Reenactment Performances

Simona Mălăescu

1 Introduction

The appearance and growth of the *Millennials* as a tourism market coincided for a long period of time with the experiential turn in tourism (*The Rise of Experiential Travel. Special report,* 2014; Mălăescu, 2017), the experiential tourism still remaining the most influential trend in 2019 (*Booking*, 2018; *Tripadvisor*, 2019). The previous "bread and butter" of the tourism industry—the *X Generation*—and also their predecessors (*The Boomers*), already proved themselves more experiential-inclined than the previous generations at their age. Two decades ago, Crouch (2002) already noticed that firstly, tourists are bodily engaged in the sense making of their encounters with tourism destinations and their experience are multisensorial. At the same time, senses also play the catalytic role in place attachment's genesis (Agapito et al., 2013). Trying to accommodate all this psychographic changes in tourists' motivation, in the case of heritage sites, in many respects, reenactment seems to be *the answer* for heritage site managers. "Perhaps because of this winning combination of imaginative play, self-improvement, intellectual enrichment, and sociality, reenactment is blooming" said Agnew (2004, p. 327) summing up its success. Another added-value resides in the fact that the (re-)presentation of cultural heritage in the forms of reenactment activities "creates a unique set of interactions between landscapes, local communities, tourists and heritage organisations" (Carnegie & Mccabe, 2008, p. 349) so there is no wonder that heritage managers fully promote it as a form of live heritage interpretation.

This sensorial turn, starting in the framework of the experiential paradigm—or, as Agapito et al. (2013) see it, as an actual embodiment paradigm—calls for revisiting or advancing the research literature on the mechanisms through visitors from their

S. Mălăescu (✉)
Babeş-Bolyai University, Cluj Napoca, Romania
e-mail: simona.malaescu@ubbcluj.ro

435

place attachment (PA), than leading to visitors' destination loyalty towards heritage sites. Dwyer et al. (2019) pointed out that, in order to progress in the understanding of consumer loyal behaviours, even marketing researchers shifted their focus on the attitudinal (or emotional) aspects, and paid attention to the construct of PA.

Based on his previous empiric evidence, Lee (2009) reached the conclusion that salient attachment antecedents vary depending on destinations with different physical features: antecedents like past experience, satisfaction, and tourists' age at their first visit along with destination attractiveness and family trip tradition were significant predictors in the case of a particular beach area but not in the case of attachment to the city were just the last two variable were relevant (Lee, 2001).

When we consider the role of destinations in tourist' self-identification and their functional attachment in the case of leisure aestival destinations compared to the heritage sites, places of memory tourism or, at the end of the continuum, the places for darktainment, we might realise their diversity. Park (2010) points out the fact that it is the socio-psychological dimensions of heritage sites, not the physical assets that make the implication of heritage crucial in a given society. Consequently, we still need to focus on a more contextual research of the antecedents, the eventual mediators or moderators of PA and especially, the mechanisms through which tourists form an attachment with a place in particular with heritage sites.

Previous literature found, in the context of natural environments (Lee et al., 2007; Gross & Brown, 2006) and also on beach tourism (Yuksel et al., 2010), that the measure of tourists' involvement in leisure activities could play a key role in developing emotional attachment. We embrace the same hypothesis that, the level of involvement in reenactment activities at heritage sites (with different levels of engaging from being in the audience, participating in mass-reenactment events, respectively, being a full status reenactor), could explain the variance in variables like satisfaction, PA, place dependence, place identity or destination loyalty registered.

The literature about the importance of PA's dimensions on tourists satisfaction, the probability to recommend and their loyalty grew considerably (Hosany et al., 2017; Chen & Segota, 2015, 2016; Whickam, 2000; Brocato, 2006; Lee et al., 2007; Kyle et al., 2004a; Yuksel et al., 2010; Chen et al., 2014). Some recent contributions in tourism loyalty literature still consider that less attention has been paid to how tourists experience and relate to destinations (Almeida-Santana & Moreno-Gil, 2018) although progress in researching how tourists live and experience their destinations has been made (Dwyer et al., 2019) considering that the rate of studies published from 2014 to 2018 on the destination/place attachment grew rapidly (12 articles in 2014, 25 in 2017 and 43 in 2018 indexed in Scopus only) (Dwyer et al., 2019).

The literature about the place attachment and its role as antecedent of destination loyalty, focussed in the last decade, on the *sunny side* of tourism (recreational and beach tourism), more consumer- or market(ing)-oriented, and overwhelmingly positive in respect of the researched emotionality tourists manifest in these destinations. In fact, not all the experiences at a site have to be *stricto-sensu* positive in order to be involved in the complex entanglement of individual's attachment to places, but the empiric support in literature is still unbalanced. As Manzo (2005) stressed out, attachment to the environment is not always positive. Biran and Buda (2018) pointed out the fact that in the case of sites like Gallipoli (Osbaldiston and Petray, 2011) the collective performance transforms the sense of horror, fear and isolation into feelings of patriotism and collective effervescence. There is more to be understood on the "deep links with the felt world, especially with the other-than-conscious, more-than-human, and hardly representable affective facets of experience" (Martini & Buda, 2018, p. 9) of the *dark side* of tourism, as the authors traces in their recommendations for future theoretical development. The concept of *place affect* that Halpenny (2010) derived from the environmental psychology was used in order to describe both positive and negative emotions of visitors towards a place. Although used in the previous literature as a sub-dimension of PA (Halpenny, 2010; Ramkissoon & Mavondo, 2015), Halpenny (2010)'s concept of *place affect* did not registered the expected echo on the more recent, more inclusive, models of PA as it shows the meta-analysis of the frameworks proposed in literature (Dwyer et al., 2019).

Considering the stated context, the issues this paper aim to address are: extending the nature of empirical contexts from which place attachment studies bring evidence to the heritage sites, balance the valence of emotionality, and finally, exploring and understanding the implications of reenactment activities in PA genesis in the case of tourists at heritage sites—considering the key role of their involvement in leisure activities in this genesis (Lee et al., 2007; Gross & Brown, 2006; Yuksel et al., 2010).

2 From the Bidimensional Sense of Place to the Six Dimensions Place Attachment Models

Visitors form on the long term, an emotional bond with some places they visit, no matter how short the visit, and sometimes even prior to their visit (Chen et al., 2014). The first concept pertaining to the individual–place relationship was *sense of place* (Low & Altman, 1992; Chen et al. 2014). Chen et al. (2014, pp. 324–325) conceptualised it as composed of two different aspects: *the relation to a place*— the typology of relationship people can have with a place (in terms of how this relationship is formed) and the *place attachment* seen as the degree of attachment (the depth) and types of attachments in terms of the psychological changes that the relationship with a place could determine on the individual.

Models on PA conceptualised it early as a twofold construct: physical attachment and social or based on interpersonal relationships developed within its environmental framework (Williams et al., 1992; Brocato, 2006). Soon, the affective component of attachment regained its undisputable place (Kyle et al., 2004b; Kyle, Graefe & Manning, 2005; Yuksel et al., 2010) under the concept of *affective attachment* (Kyle et al., 2004b). The recognition of the PA distinct cognitive component became more prominent. *Place identity* (PI) is described as a component of self-identity including beliefs about one's relationship with the place (Jorgensen & Stedman, 2006). Other works, building on the existent literature, underlines the symbolic importance of PI, as emotional attachment, the place saw as a "repository for emotions and relationships that give meaning and purpose to life" (Williams & Vaske, 2003, p.831). Whickam (2000) emphasised the emotional dynamic of the contribution of being at a particular place in self defining, and the fact that in return, the identification with a place, and belonging to that particular place could reinforce positive appraisal of the destination.

From the three dimensions of PA (Chen et al., 2014; Yuksel et al., 2010; Dwyer et al., 2019), PI is definitely the component more vulnerable to modifications during visits at a heritage site, considering its role in self-identity definition, especially in helping an individual define his/her social identity (Proshansky et al., 1983).

A destination also plays a functional role during tourists' stay based on its capacity to provide conditions and features to support specific activities or goals (Williams & Vaske, 2003) called *functional attachment* (Stokols & Shumaker, 1981; Williams & Roggenbuck, 1989) or *place dependence* (Yuksel et al., 2010; Chen et al., 2014). In this study, we embraced the previous conceptualisations of functional attachment (FA) as a setting for reenactment activities and its large spectrum of perceived instrumental roles in the personal motivation of visitors: from cultural and educational role (Howard, 2003; Cook, 2004) to the perceived patriotic feelings enhancement or, respectively, perceived role in national identity affirmation of reenactment activities (Park, 2010; Popa, 2016). However, our perspective on the case of heritage sites with a reputation of reenactment festivals is that a nuance imposes itself to their conceptualisation based on elements from the transactional perspective, suggesting that visitors evaluate places against alternatives, according to how well those places meet their functional needs (Brocato, 2006; Yuksel et al., 2010). If visitors at heritage sites are motivated by the historical significance for their social group or their nation, the process of evaluation of competitive sites could not take place.

Summing up the existent literature on the relation of individual with a place, Chen et al. (2014) defined the structure of the first component of sense of a place as biographical, spiritual, ideological, narrative, commodified and dependent type of relationship, and that PA could accommodate the type of *place identity, place dependence, affective attachment, social bonding, place memory and place expectation* (Chen et al., 2014, p. 325). However, as Dwyer et al. (2019) observed in their recent meta-analysis of the models on PA, most research embraced a bidimensional conceptualisation of PA (PI and PD).

A large body of literature brought evidence that PA proved to be also a predictor of destination loyalty (Brocato, 2006; Alexandris et al. 2006; Yuksel et al., 2010). The impact of PA on cognitive and affective loyalty was empirically tested (Yuksel et al., 2010).

When visiting heritage sites during reenactment festivals, tourists benefit from the collective-experienced recollection of crucial moments in history such as moment-by-moment reenactment of historic landmark events. In the light of previous literature, there is theoretically reasonable grounds to consider the role of reenactment activities at particular heritage sites as a key point in understanding how visitors engage with heritage sites and forge their PA through PI and PD. Consistent with previous perspectives on individual's social identification theories and group dynamics providing us with the subscales on measuring the in-group social identification strength (Stephan & Stephan, 1993), we interrogate about the possibility that the differences registered by visitors of sites relevant for their national history, in their level of involvement manifested during the trip and the variance of PA they develop, might be explained by personal pre-trip variables like *the national in-group identification strength* (NIIS) or, in other heritage sites, their *cultural* or *regional in-group identification strength* (RIIS). We argue that pre-trip individual variables should be explored in studying the mechanism of PA' genesis at heritage sites, people varying in their beliefs or level of affects towards their origins or, respectively, their in-group heritage social-defining landmarks.

3 Methodology

In this exploratory study, we surveyed 258 visitors and participants to reenact-ment events in 20 different tourism destinations in Romania, the majority of them presenting a historical and cultural heritage as a part of their tourism resources from different historical periods (from Antiquity, Middle Age to the beginning of Romania as a modern state). As a part of a broad research, we used a quantitative–qualita-tive approach based on questionnaires with closed and open-ended items, interviews and participant observation. In assessing NIIS and RIIS, we have adapted the in-group social identification subscale proposed by Stephan & Stephan (1993) based on *Integrated Threat Theory*. In regard of place attachment and its components, we used the same conceptualisation and instruments utilised by Yuksel et al. (2010) and their predecessors (Williams & Vaske, 2003 with modifications by Alexandris et al., 2006). In the correlational and multiple regression analysis, apart from surveying visitors and audience in several historical periods for statistic control, in the case of the same tourism destination like, for example, Alba Iulia—"The Romanians' Mecca" as the website of Union Museum (2020) labelled it, we were collecting data about visiting and assisting reenactment events reenacting antiquity period and other local events (like the celebration of town's days) and compared them with data collected on the mass-reenactment event held in 1 December 2018 (the National Day and the celebration of the first Centenary of Romania) (Fig. 1).

4 Results and Discussions

In our convenience sample, 57.3 % of the visitors considered the reenactment activity an attraction in itself, and 10.8% responded that their majority of visits in that loca-tion are motivated by the reenactment representations, but not exclusively. Another 8.9% of the respondents appreciated that in the absence of the reenactment repre-sentations, they would not visited that particular destination, despite the various tourism resources the 20 destinations surveyed presented. Data led us to reflect on the possibility that heritage sites holding reenactment festivals benefit from the presence of reenactment's enthusiasts that otherwise will not register as a target

Fig. 1 Reenactment of the Battle of Romans with Dacians and Sarmatians at *"Apullum"* Reenactment Festival (Alba Iulia, Romania)

tourists of heritage sites and calls for the necessity to explore the role of reenactment. Corroborating data with previous literature, we advanced the hypothesis that tourists' motivation to involve in exploring a heritage site through reenactment moments, their emotions and affects during their trip and their ulterior PA with destinations presenting heritage with particular significance for them, could be moderated or mediated, depending on their pre-trip personal characteristics, by their expectancies to feel a certain experience through reenactment activities, to fulfil some intrapersonal functions (Steiner & Reisinger, 2006) or just their personal quests for authentic existential experiences (Wang, 1999; Kim & Jamal, 2007).

Initially, at the level of the entire sample, the correlational analysis showed rather low indexes of correlation between the level of the identification of the heritage site as a reenactment destination and the perceived instrumental role of reenactment in enhancing subjects' patriotic feelings (PIRRPF)—0.49 or the perceived role of reenactment to the enhancement of national identity affirmation (PIRRNIA)—0.27). Also the indexes between PIRRNIA and the independent motivation to visit the heritage site (0.24) or place attachment (0.23), respectively, place identity (0.22) registered smaller values. However, by splitting the sample in subsamples, for a particular subsample, the indexes increased significantly. We have split the sample in three categories: respondents visiting Alba Iulia during the Celebration of National Day (S1), respondents visiting Alba Iulia during other events (S2), respectively, respondents visiting other destinations known for their reenactment activities (S3). In the S2, the correlation of PIRRNIA with place identity registered was 0.80, and the correlation with the identification of the destination with a reenactment activity destination was 0.74. The increased correlation index between PIRRNIA and place identity is consistent with the previous model of Chen et al. (2014) where *place expectation* "can either strengthen or weaken the individual–place bond, and it can change according to the information update of the expected activities or events" (Chen et al., 2014, p. 328). In both samples of Alba Iulia visitors, the correlation between PIRRPF and the degree of perception of this destination as a destination for reenactment was significant at 0.01 (0.43 in S1 and 0.49 in S2).

Although our aim was not to revisit the models on place attachment, we use regression analysis to explore to which extent, the variance of AA and PI could be explained by some predictor variables related to reenactment and its perceived instrumental role in the tourism experience at destination in general (Table 1).

With statistical indexes varying in amplitude between subsamples, variables like *reenactment motivation strength to visit a destination, the strength of the identification of the destination with reenactment activities,* PIRRPF and/or RIIS/PIRNIA were retained by the model due to their explanatory power and explained a sizeable part of the criteria variable's variance (Table 1). In the case of AA, these variables related to reenactment along with independent motivation to visit predicted almost half of its variance. In PI's case, the identification of the destination as a reenactment destination strength, reenactment motivation' strength, PIRRPF, and independent motivation explained up to 80% of the predicted variable.

Considering the results of the correlation and regression analysis, the implication of pre-trip personal factors like NIIS and RIIS in explaining PA's dimensions, we explored forward the role of reenactment in PA's component genesis and the possibility that the impact of personal factors, like NIIS, on affective attachment for that destination, could be actually partially mediated—as Baron and Kenny (1986) described the mediation model—by PIRRPF (Fig. 3). Similarly, the impact of the level of destination' identification as a reenactment destination on place identity (Fig. 2) could be partially mediated by the perceived role of reenactment in national identity affirmation. In a larger context, as relevant as it might seems the level of NIIS of visitors for the components of PA in the case of Alba Iulia destination, the model that could explain the differences in PI's genesis and emotions this destination elicits in visitors (participating or not participating in reenactment events) and furthermore, attachment formation might be its cognitive–emotional functionality in maintaining, on one hand individual's intrapersonal authenticity of the self, through PI, and on the other hand, its affective component could also be dependent on the emotional functionality attached to the experiences proposed in the destination at least in the case of heritage sites as opposed to the case of leisure-centred destinations.

The results in the fore-mentioned mediation relationship are consistent with the mediation model considering that: (1) predictors are associated with the potential mediator and also with the outcome variable; (2) after controlling the effect of the mediator, the impact of the predicting variable on the outcome variable decreased. Mediation's direction (e.g. if, for example, reenactment's relevant variable mediate the effect on place attachment) was determined by comparing the relative modification of the estimated parameters.

Obtaining different results on subsamples differing in their connection of the destination with the national history, hence their significance and relevance for visitors are

Table 1 Affective attachment and place identity' variability explained by relevant reenactment-related variables

Predictors in the model	Predicted	Sp	R	R²	Adj. R²	Std. Est. Err.	F	Sig.	Method
Reenactment motivation strength, independent motivation to visit, PIRRPF, identification of destination with reenactment strength, PIRRNIA	Affective attachment	2	0.83	0.69	0.48	104441	90003	0.01	Backwards
Reenactment motivation strength, NIIS, PIRRNIA, identification of destination with reenactment strength, independent motivation to visit, PIRRPF	Affective attachment	1	0.54	0.30	0.28	386482	15300	0.00	Backwards
PIRRPF, reenactment motivation strength, independent motivation to visit identification of destination with reenactment strength	Place identity	2	0.89	0.80	0.73	105976	7300	0.01	Backwards
Reenactment motivation strength, independent motivation to visit, identification of destination with reenactment strength, RIIS	Place identity	3	0.65	0.42	0.38	119391	9344	0.00	Backwards

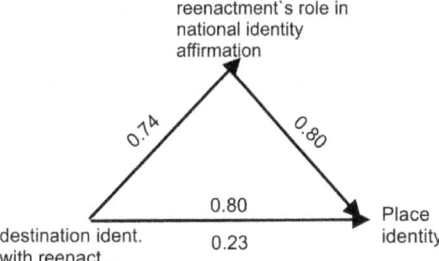

Fig. 2 Model of destination identification with reenactment destination's impact on place identity partially mediated by reenactment's perceived instrumental role in the national identity affirmation. (The values represent estimated standardised coefficients. The values above line reflect bivariate relationships; the values below reflect multivariate regression coefficients estimated for each predictor. All relationships are significant (at $p \leq 0.001$))

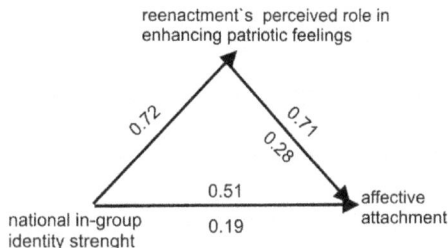

Fig. 3 Model of national in-group identity strength's impact on affective attachment mediated by reenactment perceived instrumental role in the enhancement of patriotic feelings. (The values represent estimated standardised coefficients. The values above line reflect bivariate relationships, the values below reflect multivariate regression coefficients estimated for each predictor. All relationships are significant (at $p \leq 0.001$))

consistent with the hypothesis that on the research context of heritage tourism sites, not all the destinations confirm the transactional perspective on functional attachment (Brocato, 2006; Yuksel et al., 2010) depending on functionality and personal significance of the destination. If visitors at heritage sites are motivated by the historical significance for their social group or their nation (as in the case of tourists visiting Alba Iulia), the process of evaluation of competitive sites is unlikely to take place. In the case of visitors declaring that they have not visited the destination in the absence of reenactment events (8.9%); hence the case of heritage sites with a reputation of reenactment festivals, the transactional perspective of considering the alternatives most likely is still undergoing. Other functionalities of the later destinations, like the reenactment events for *"The Booze and Bash Brigade"* where reenactors just enjoy a weekend where they get into fights and drink rather excessively (Howard, 2003) can be fulfilled by any reenactment destination which, again, put the likelihood of this evaluation process on the table.

5 Conclusion

Exploring the place attachment theories on the Romanian heritage sites research context—where reenactment activities are frequently the main attraction and ingredient in celebratory contexts—revealed the fact that the role of reenactment activities in this experiential era of tourists' motivation might play a more central role than tourism research literature gave it credit so far. First of all, it led us to reflect on the possibility that heritage sites holding reenactment festivals benefit from the presence of reenactment's enthusiasts that otherwise will not register as a target tourists of heritage sites. Also, tourists' place attachement towards destinations presenting heritage with particular significance for them, could be mediated or moderated by factors like their expectancies to feel a certain experience, a particular emotion (enhanced patriotic feelings, enhancement of national identity affirmation), and in that process, the reenactment activities could play nowadays a very important role opposite to the "dead heritage interpretation" era of tourism at heritage sites as Howard (2003, p. 260) so plastically labelled.

Our results revealing an increased correlation between tourists' expectation to enhance their national identity affirmation during reenactment activities and place identity are consistent with the previous model of Chen et al. (2014) where place expectation "can either strengthen or weaken the individual–place bond, and it can change according to the information update of the expected activities or events" (Chen et al., 2014, p. 328), in our case the absence/presence of reenactment performances. Findings also point in the direction that place expectation is a different dimension of place attachment than place identity (Chen et al., 2014).

Using different subsamples gave us also the opportunity to explore the possibility that the relation between personal factors and PA formation could not be as linear and as strong as it might appear in the case of destinations strongly connected with national history and personal NIIS (considering the fact that we obtained similar results on the subsample of other tourism destination hosting reenactment activities). Sometimes, the expectation to feel a certain patriotic feeling through reenactment activities is destination-independent and it might be more strongly connected with reenactment activities per se. In this direction, on the Romanian context, previous literature pointed out the fact that the (negative) extreme of this continuum—the nationalist exultation through reenactment is more connected with the Antiquity than with the modern history of Romania (Popa, 2016). Also, the difference in results obtained, depending on the sample destination's significance, sustains the hypothesis, that place dependence in the context of heritage sites hosting reenactment events, depends on the strength of the identification of the tourism heritage site as a reenactment destination and several personal pre-trip variables relevant for their choice to visit. Future studies would bring more evidence on this perspective.

References

Agapito, D., Mendes, J., & Valle, P. (2013). Conceptualizing the sensory dimension of tourist experiences. *Journal of Destination Marketing & Management, 2*(2), 62–73. https://doi.org/10. 1016/j.jdmm.2013.03.001.

Agnew, V. (2004). Introduction: What is Reenactment? *Criticism, 46*(3), 327–339. https://doi.org/ 10.1353/crt.2005.0001.

Alexandris, K., Kouthouris, C., & Meligdis, A. (2006). Increasing customers' loyalty in a skiing resort. *International Contemporary Hospitality Management, 18*(5), 414–425. https://doi.org/10. 1108/09596110610673547.

Almeida-Santana, A., & Moreno-Gil, S. (2018). Understanding tourism loyalty: Horizontal vs. destination loyalty. *Tourism Management, 65,* 245–255. https://doi.org/10.1016/j.tourman.2017. 10.011.

Baron, R. M., & Kenny, D. A. (1986). The moderator-mediator variable distinction in social psychological research: Conceptual, strategic, and statistical considerations. *Journal of Personality and Social Psychology, 6,* 1173–1182. https://doi.org/10.1037/0022-3514.51.6.1173.

Biran, A., & Buda, D. M. (2018). Unravelling fear of death motives in dark tourism. In P. R. Stone, R. Hartmann, A. V. Seaton, R. Sharpley, & L. White (Eds.). *The Palgrave handbook of dark tourism studies* (pp. 515–532). London: Palgrave Macmillan. https://doi.org/10.1057/978-1-137-47566-4_21.

Booking.com. (2018, October 18). Booking.com Reveals 8 Travel Predictions for 2019. Retrieved from https://globalnews.booking.com/bookingcom-reveals-8-travel-predictions-for-2019/.

Brocato, E. D. (2006). *Place attachment: an investigation of environments and outcomes in service context.* Doctoral Thesis, The University of Texas at Arlington.

Carnegie, E., & Mccabe, S. (2008). Re-enactment events and tourism: meaning, authenticity and identity. *Current Issues in Tourism, 11*(4), 349–368. https://doi.org/10.1080/136835008021 40380.

Chen, N. C., & Šegota, T. (2015). Resident attitudes, place attachment and destination branding: A research framework. *Tourism and Hospitality Management, 21*(2), 145–158. https://doi.org/10. 20867/thm.21.2.3.

Chen, N. C., & Šegota T. (2016). Conceptualization of place attachment, self-congruity, and their impacts on word-of-mouth behaviors. In *Travel and Tourism Research Association Conference.* Canada

Chen, N., Dwyer, L., & Firth, T. (2014). Conceptualization and measurement of dimensionality of place attachment. *Tourism Analysis, 19*(3), 323–338. https://doi.org/10.3727/108354214X14029 467968529.

Cook, A. (2004). The use and abuse of historical reenactment: Thoughts on recent trends in public history. *Criticism, 46*(3), 487–496. https://doi.org/10.1353/crt.2005.0002.

Crouch, D. (2002). Surrounded by place: embodied encounters. In: Coleman & Crang, M. (Eds.) *Tourism between place and performance* (pp. 207–218). New York: Berghahn Book.

Dwyer, L., Chen, C. N., & Lee, J. (2019). The role of place attachment in tourism research. *Journal of Travel & Tourism Marketing, 36*(5), 645–652 https://doi.org/10.1080/10548408.2019.1612824.

Gross, M., & Brown, G. (2006). Tourism experiences in a lifestyle destination setting: The roles of involvement and place attachment. *Journal of Business Research, 59*(6), 696–700. https://doi. org/10.1016/j.jbusres.2005.12.002.

Halpenny, E. A. (2010). Pro-environmental behaviours and park visitors: The effect of place attachment. *Journal of Environmental Psychology, 300*(4), 409–421. https://doi.org/10.1016/j.jenvp. 2010.04.006.

Hosany, S., Prayag, G., Veen, R. V. D., Huang, & S. Deesilatham, S. (2017). Mediating effects of place attachment and satisfaction on the relationship between tourists' emotions and intention to recommend. *Journal of Travel Research, 56*(8), 1079–1093. https://doi.org/10.1177/004728751 6678088.

Howard, P. (2003). *Heritage: management, interpretation, identity.* London; New York: Continuum.

Jorgensen, B., & Stedman, R. (2006). A comparative analysis of predictors of sense of place dimensions: attachment to, dependence on, and identification with lakeshore properties. *Journal of Environmental Management, 79*, 316–327.

Kim, H., & Jamal, T. (2007). Touristic quest for existential authenticity. *Annals of Tourism Research, 34*(1), 181–201. https://doi.org/10.1016/j.annals.2006.07.009.

Kyle, G., Graefe, A., Manning, R., & Bacon, J. (2004). Effect of activity involvement and place attachment on recreationists' perceptions of setting density. *Journal of Leisure Research, 36*(2), 209–231. https://doi.org/10.1080/00222216.2004.11950020.

Kyle, G., Mowen, A., & Tarrant, M. (2004). Linking place preferences with place meaning: An examination of the relationship between place motivation and place attachment. *Journal of Environmental Psychology, 24*(4), 439–454. https://doi.org/10.1016/j.jenvp.2004.11.001.

Kyle, G., Graefe, A., & Manning, R. (2005). Testing the dimensionality of place attachment in recreational settings. *Environment and Behavior, 37*(2), 153–177. https://doi.org/10.1177/001 3916504269654.

Lee, C. (2001). Predicting tourist attachment to destinations. *Annals of Tourism Research, 28*(1), 229–232. https://doi.org/10.1016/S0160-7383(00)00020-7.

Lee, J. (2009). *Investigating the effect of festival visitors' emotional experiences on satisfaction, psychological commitment, and loyalty (Unpublished doctoral dissertation).* College Station, Texas: Texas A&M University.

Lee, J., Graefe, A., & Burns, R. (2007). Examining the antecedents of destination loyalty in a forest setting. *Leisure Science, 29,* 463–481. https://doi.org/10.1080/01490400701544634.

Low S., & Altman, I. (1992). Place attachment: a conceptual enquiry. In I. Altman, & Low S. (Eds) *Place attachment.* New York: Plenum Press. https://doi.org/10.1007/978-1-4684-8753-4_1

Mălăescu, S. (2017). The experiential tourism approach in tourism product development and marketing: why not? In: N. Ciangă, Ș. Dezsi, V. Bodocan, F. Ipatiov, S. Mălăescu, C. Bolog, & M. Oprea (Eds.) *Sustainable Tourism Destinations Identity Image Innovation* (pp. 59–69). Cluj Napoca: Presa Universitara Clujeana.

Manzo, L. C. (2005). For Better or worse: Exploring multiple dimensions of place meaning. *Journal of Environmental Psychology, 25*(1), 67–86. https://doi.org/10.1016/j.jenvp.2005.01.002.

Martini, A. & Buda, M. D. (2018). Dark tourism and affect: Framing places of death and disaster. *Current Issues in Tourism.* Online I: https://doi.org/10.1080/13683500.2018.1518972

Osbaldiston, N., & Petray, T. (2011). The role of horror and dread in the sacred experience. *Tourist Studies, 1*(2), 175–190. https://doi.org/10.1177/1468797611424955.

Park, H. J. (2010). Heritage tourism emotional journeys into nationhood. *Annals of Tourism Research, 37*(1), 116–135. https://doi.org/10.1016/j.annals.2009.08.001.

Popa, C. N. (2016). The significant past and insignificant archaeologists. Who informs the public about their 'national' past? *The case of Romania, Archaeological Dialogues, 23*(1), 28–39. https://doi.org/10.1017/S1380203816000064

Proshansky, H. M., Fabian, A. K., & Kaminof, R. (1983). Place identity: Physical world socialization of the self. *Journal of Environmental Psychology, 3,* 57–83. https://doi.org/10.1016/S0272-494 4(83)80021-8.

Ramkissoon, H., & Mavondo, F. T. (2015). The satisfaction place attachment relationship: Potential mediators and moderators. *Journal of Business Research, 68*(12), 2593–2602. https://doi.org/10.1016/j.jbusres.2015.05.002.

Steiner, C. J., & Reisinger, Y. (2006). Understanding existential authenticity. *Annals of Tourism Research, 33,* 299–318. https://doi.org/10.1016/j.annals.2005.08.002.

Stephan W. G. & Stephan, C. W. (1993). Cognition and affect in Stereotyping: Parallel interactive networks. In Mackie & Hamilton (Eds.), *Affect, cognition and stereotiping: Interactive processes in group perception* (pp. 111–136), Orlando: Academic Press. https://doi.org/10.1016/B978-0-08-088579-7.50010-7

Stokols, D., & Shumaker, S. A. (1981). People in places: Transactional view of settings. In J. H. Harvey (Ed.), *Cognition, social behavior, and the environment* (pp. 441–488). NJ: Lawrence Erlbaum, Hillsdale.

Tripadvisor. (2019). 2019 experiential travel trends: new skills, health & wellness, family activities come into focus. Retrieved from https://www.tripadvisor.com/blog/experiential-travel-trends-hea lth-wellness-family/.

Union Museum Alba Iulia. (2020). Union Hall. Retrieved from https://mnuai.ro/sala-unirii/

Wang, N. (1999). Rethinking authenticity in tourism experience. *Annals of Tourism Research, 26,* 349–370. https://doi.org/10.1016/S0160-7383(98)00103-0.

Wickham, T. D. (2000). *Attachments to places and activities: The relationship of psychological constructs to customer satisfaction.* Unpublished doctoral dissertation. Pennsylvania: The Pennsylvania State University, University Park.

Williams, D., & Roggenbuck, J. (1989). Measuring place attachment: Some preliminary results. In L. H. McAvoy & D. Haward (Eds.), *Abstracts: 1989 leisure research symposium* (p. 32). Arlington, VA: National Recreation and Park Association.

Williams, D. R., & Vaske, J. J. (2003). The Measurement of place attachment: Validity and generalizability of a psychometric approach. *Forest Science, 49*(6), 830–40.

Williams, D., Patterson, M., Roggenbuck, J., & Watson, A. (1992). Beyond the commodity metaphor: Examining emotional and symbolic attachment to place. *Leisure Sciences, 14*(1), 29–46. https://doi.org/10.1080/01490409209513155.

Yuksel, A., Yuksel, F., & Bilim, Y. (2010). Destination attachment: Effects on customer satisfaction and cognitive, affective and conative loyalty. *Tourism Management, 31,* 274–284. https://doi.org/ 10.1016/j.tourman.2009.03.007.

Conditions for Creating Business Tourism Offers and the Regional Potential in Poland

Ewa Lipianin-Zontek and Zbigniew Zontek

1 Introduction

Business tourism is one of the dynamically developing forms of tourism in Poland in recent years. Lately, this type of tourism has been one of the most payable forms of tourism. Development of business tourism depends on many exogenous and endogenous factors. This paper is focused on some external factors, those related to a tourism destination.

Results shown herein refer to the tourism market in Poland. According to the world travel and tourism council (WTTC, 2020), contribution of travel and tourism in the gross domestic product has reached 4.7%, and it increased by 4% in comparison to 2018. Among all employed people in Poland, this sector constitutes 5% of total employment. The key factor affecting the tourism is the level of capital investments. Investments in traveling and tourism in 2016 accounted for 3.5% of total investments in Poland (WTTC, 2017, 5). Regarding the travel and tourism competitiveness index (TTCI) for 2019, the World Economic Forum (2019) ranked Poland 42nd of 140 countries. A significant improvement can be noticed when compared to the corresponding ranking from 2013, in which Poland was ranked 47th (World Economic Forum, 2013). When analyzed via the different sub-indexes, there are wide variations, Poland has a relatively high rank in "health and hygiene" (24th place) and 28th place in "cultural resources and business travel" (World Economic Forum, 2013, 31). Analyzing the dimensions which underlie this sub-index, Poland was rated high (19th place) in the dimension of the "number of international association meetings." However, it holds lower places in other sub-indexes, for example, "prioritization of travel and tourism" (58th place) and 81st place in "environmental sustainability" (World Economic Forum, 2019).

E. Lipianin-Zontek (✉) · Z. Zontek
University of Bielsko-Biala, Bielsko-Biala, Poland

© The Author(s), under exclusive license to Springer Nature Switzerland AG 2021   449
V. Katsoni and C. van Zyl (eds.), *Culture and Tourism in a Smart, Globalized, and Sustainable World*, Springer Proceedings in Business and Economics, https://doi.org/10.1007/978-3-030-72469-6_30

2 Literature Review

2.1 Destination Product as the Basis for Creating Business Tourism Products

The concept of destination is ambiguous and defined differently. Identification of its meaning contents includes some complex approach, expressed in the integration of different trends including geographical, economical, sociological and cultural approach (Buhalis, 2000; Gunn, 1994; Inskeep, 1994; Middleton & Hawkins, 1998; Page, 2005). The destination shall be then defined as a social-geographical system and managerial structure constituting a coherent, limited whole of a certain specification. According to UNWTO (2019), a local tourism destination is a physical space, in which a visitor spends at least one night. This space includes some tourism products, such as support services, attractions and tourist reserves. It has some physical and administrative borders that determine management, image and perception determining the market competitiveness. Local tourism destinations involve a variety of entities, including hosting community, which may form some slots and networks forming larger sites. By accepting such interpretation of a term "destination," not only its key values important for tourists are indicated (e.g., cultural, historical or natural heritage) but also tourist services, quality of environment and other aspects creating the entire offer and constituting the sum of experiences for recipients (Ashworth et al., 2012, p. 277).

The said destinations offer a combination of co-branded products and services. Harris and Leiper (1995, p. 87) explain that the destinations are places where people travel to and where they decide to stay for some time in order to experience some features or a certain type of attractions. Gomezelj and Michalič (2008) indicate a significant meaning of competitiveness in the scope of a tourism destination. Competitive dominance may be achieved when a general attractiveness of a tourist site exceeds the attractiveness of some alternative destinations that are open for potential tourists.

The regional product may be formed by a region, city or by each single element of the destination that determines the consumer's choice related to the direction of travel. Cooper et al. (1998) define destination as a concentration of objects and services designed for the purpose of meeting the tourists' needs. According to Saraniemi and Kylänen (2011), a tourism destination is one of the key conception of institutionalized tourism. Middleton and Hawkins (1998, p. 88) have shown that a tourism product related to the place of stay is the one which includes complex experience (sum of impressions and experiences) of a tourist from the moment he/she leaves their home to the moment it comes back. It is thus a combination of the following main elements: tourism development in a target place, accessibility, provided tourism services and some supporting products.

A destination tourism product is a combination of many factors which very often include a synergy effect. Buhalis (2000) captured this concept in the form of the 6A model, including attractions (natural, created by a man, artificial, purpose-forming heritage, special events); accessibility (system of transport which includes roads, terminals and vehicles); amenities (accommodation and catering facilities, retail sales, other tourism services); available packages (pre-arranged packages by intermediaries and principals); activities (all actions available within the destination and activities taken by tourists during their stay); ancillary services: services used by tourists (banks, post offices, hospitals, etc.).

The regional product may be formed by various elements which are adequately combined with each other and thus create a package of some material goods and services, meeting the expectations and needs of tourists who are ready to pay the determined price for them. A coherent image of a tourism destination forms a basis for creating some attractive business tourism products. The condition of business tourism in a destination depends on the tourism attractiveness of that destination. The said attractiveness shall be regarded as the force of attraction of a certain place, object or phenomenon. The degree of attractiveness of a tourism destination is influenced by many factors. A primary force for the formation and development of tourism in an area, is the quality and quantity of the natural environment including the levels of environmental pollution. Existing tourism development is also important, with accessibility, communication and appropriate mobility tools being key points for tourists, during their stay in the region.

2.2 Competitiveness of a Tourism Destination as a Premise for the Development of Business Tourism

Business tourism shall be defined as all travels of which purpose is closely related to work or interests of travelers. Davidson and Cope (2003) define the business tourism as an activity with a certain degree of freedom, most frequently as non-routine work activities, frequently carried out in a group. One may distinguish the following forms of business tourism: conferences, congresses, fairs, consumer parties, incentive events, integration trips, company meetings, trainings and various business trips. Business tourism is defined in the documents as the MICE (Meetings, Incentives, Conventions, Exhibitions) industry.

The significance of business tourism for economy is determined by its following, key features: focus on the high standard of tourist services (Bedradina & Nezdoyminov, 2019), lower sensitivity to the price level (Dominique-Ferreira et al., 2016) as well as definitely lower seasonality (Martin et al., 2019).

Competitiveness between tourism destinations and their products on domestic or foreign markets is a significant issue related to the creation of business tourism products (Fyall, 2019). As indicated by Brent et al. (2000, p. 6), competitiveness of destination has some significant consequences for the tourism industry and thus should be regarded as an important aspect for practitioners and decision makers. A competitive dominance of the destination may be achieved when a general attractiveness of a tourist site exceeds the attractiveness of some alternative destinations that are open for their potential guests (Gomezelj & Mihalič, 2008).

The main criteria for competitiveness of the destination is its potential formed by two groups of competitiveness factors (Enright & Newton, 2004). The first group includes so-called attractors, i.e., the key factors in building the attractiveness of a destination: dedicated tourist attractions, visual attractiveness, popular landmarks, interesting architecture, cultural variety, special events, festivals, climate, remarkable history, museums and galleries, gastronomy and safety. The second group of competitiveness factors includes general business elements: input elements (internal transport base and infrastructure, communication devices, staff qualifications, information accessibility, local management skills, financial system, geographical location, technological development, expenses for personnel, expenses for properties and others), industrial and consumer demand, competitiveness and cooperation between enterprises, trade and regional cooperation, internal organization and strategies, social structures, institutions and agenda as well as the market ties and tourism business structure (Enright & Newton, 2004).

Koo et al. (2016) have distinguished between comparative advantages and competitive advantages within the competitiveness analysis. Comparative advantages are the available reserves, their maintenance and development. Key reserves and attractors include form, culture, history and tourism development. Competitive advantages include the ability to allocate, use and manage resources. The authors include the following reserves in the key reserves for creating the competitiveness dominance: market relations, combination of activities and special events. Results of studies performed by Knežević et al. (2016) indicate the diversification of the determinant of the competitiveness of destinations due to the degree of economic development. The main determinants driving competitiveness in developing countries are as follows: tourism infrastructure and destination management, whereas in the developed countries, the competitiveness of the destination depends on the destination management and economic conditions, such as the general infrastructure, macro environment and business environment.

Due to complexity of the issue of competitiveness of tourism destinations and the high competitiveness of destinations attractive for business tourism, it is necessary to skillfully respond to the changing level of expectations on the market.

3 Methodology

The purpose of the conducted studies was the analysis of the business tourism potential in Poland. The analysis has been conducted on the basis of new statistics of the Polish Central Statistical Office concerning "tourist accommodation establishments equipped with conference facilities" (Central Statistical Office, 2017) and other data constituting stimulants for business tourism in Poland (Central Statistical Office, 2017). On the basis of the collected data, the components of the tourism potential have been specified, on the basis of which the business tourism products in Poland can be developed.

The structure of the hotel industry in Poland is quite diverse. According to official data, in the 1st quarter of 2019, the accommodation services were provided by 11,251 entitles in Poland. The aforementioned number included 4229 hotels what constitutes 39.4% of the whole classified accommodation base. The entities particularly designed for the provision of business tourism services are four- or five- stars hotels of which number amounts to 494, and it constitutes 11.7% in the hotel structure (Central Statistical Office, 2019).

The statistical analysis has been done on the basis of the Hellwig's (1968) method and the technique for order preference by similarity ideal solution method (TOPSIS) derived from it. In both cases, the following terminology has been adopted: m-number of objects (regions in Poland); n-number of features having the stimulant or destimulant character of business tourism. Both methods have been described by Hwang and Yoon (1981). Using the values of appropriate synthetic measures, these methods allow for a linear ordering of individual surveyed regions in terms of the potential of business tourism. Then, the results were compared with the corresponding values of the density and intensity of tourist traffic according to Schneider. The research procedure was created and used by Synówka-Bejenka (2017).

Using the Hellwig's method, a taxonomic measure of development d_i^H was determined, taking values from the range [0, 1], where 0 was the smallest and 1 was the most similar to the development pattern. The TOPSIS method was in turn used for indicating the synthetic factor d_i^T which determines the tourism potential in particular regions. The values accepted by the factor also fell within the range [0,1], where 0 meant the lowest and 1 the largest tourism potential.

The strength of interdependence of both rankings based on the above measures was analyzed by using the correlation factor of the Spearman's rank correlation coefficient r_s which adopts values from the range $[-1, 1]$ and is expressed by the following formula:

$$r_s = 1 - \frac{6 \sum_{i=1}^{m} o_i}{m(m^2 - 1)}$$

o_i-difference between the places in both rankings occupied by the ith region.

According to the assumption, the increasing consistency of orderings resulting from the methods used in the study was assigned to the correlation coefficient values that were closer to 1.

3.1 The Hellwig's Method

(a) Standardization of variables

$$Z_{ij} = \frac{x_{ij} - \bar{x}_j}{s_j}$$

$i = 1, 2, \ldots, m$
$j = 1, 2, \ldots, n$

(b) Determining the value of the development pattern

$$z_j^+ = \begin{cases} \max_i\{z_{ij}\} \\ \min_i\{z_{ij}\} \end{cases}$$

$\max = \text{variable} X_j$ is the stimulant.
$\min = \text{variable} X_j$ is the destimulant.

(c) Determining the distance of Euclidean objects from the development pattern

$$d_i^+ = \sqrt{\sum_{j=1}^{n} \left(z_{ij} - z_j^+\right)^2}$$

$i = 1, 2, \ldots, m$
$j = 1, 2, \ldots, n$

(d) Determining a taxonomic measure of development

$$d_i^H = -\frac{i_i^+}{\bar{d} + 2s_d}$$

where:

$$\bar{d} = \frac{1}{m} \sum_{i=1}^{m} d_i^+ \text{ and } s_d = \sqrt{\frac{1}{m} \sum_{i=1}^{m} (d_i^+ - \bar{d})^2}$$

3.2 The TOPSIS Method

(a) Normalization of variables

$$z_{ij} = \frac{x_{ij}}{\sqrt{\sum_{i=1}^{m} x_{ij}^2}}$$

$i = 1, 2, \ldots, m$

$j = 1, 2, \ldots, n$

(b) Determination of the value of the ideal object z^+ (from the Hellwig's method) and the anti-ideal object $\bar{z}$

$$z_j^- = \begin{cases} \max_i \{z_{ij}\} \\ \min_i \{z_{ij}\} \end{cases}$$

$\max_i \{z_{ij}\}$ when variable Xj is the stimulant.

$\min_i \{z_{ij}\}$ when variable Xj is the destimulant.

(c) Determining the distance of Euclidean objects from an ideal and non-ideal solution

$$d_i^+ = \sqrt{\sum_{j=1}^n (z_{ij} - z_j^+)^2} \quad d_i^- = \sqrt{\sum_{j=1}^n (z_{ij} - z_j^-)^2}$$

$i = 1, 2, \ldots, m$

$j = 1, 2, \ldots, n$

(d) Determining the synthetic index

$$d_i^T = \frac{i_i^-}{i_i^+ + i_i^-}$$

$i = 1, 2, \ldots, m$

The obtained values of both synthetic measures were used to divide the regions (voivodeships) into four typological groups, based on the arithmetic mean and standard deviation of these measures (Synówka-Bejenka, 2017). Scheme of assigning a given object to a particular class based on value d_i^H or d_i^T ($d_i^{H/T}$) was presented in Table 1.

Table 1 Typology by the value of the synthetic measure

Measure values $d_i^{H/T}$	Regional tourism potential	Class
$d_i^{H/T} \geq \bar{d} + s_d$	Very high	I
$\bar{d} + s_d > d_i^{H/T} \geq \bar{d}$	High	II
$\bar{d} > d_i^{H/T} \geq \bar{d} - s_d$	Medium	III
$d_i^{H/T} < \bar{d} - s_d$	Low	IV

Source Synówka-Bejenka (2017)

4 Results

4.1 Selection of Diagnostic Variables

11 diagnostic variables have been selected for the purpose of comparative statistical analysis, and ten of them were stimulants (variable x_4 was the destimulant):

X_1-number of accommodation objects with conference facilities per 100 km^2.
X_2-number of conference venues with a multimedia projector per 100 km^2.
X_3-number of conference venues with videoconferencing equipment per 100 km^2.
X_4-population per one cultural object (museums, galleries, art galleries).
X_5-number of conference venues with technical assistance per 100 km^2.
X_6-number of conference venues with a screen per 100 km^2.
X_7-number of conference venues with a flipchart per 100 km^2.
X_8-number of conference venues with a computer in the conference room per 100 km^2.
X_9-number of conference venues with Wi-Fi connection per 100 km^2.
X_{10}-density of municipal and poviat hard surface roads in one km per 100 km^2.
X_{11}-number of accommodation objects offering rehabilitation treatments (massages, physiotherapy, etc.) per 100 km^2.

Selection of variables has been made on the basis of analysis of significance of particular factors for the business tourism potential and adequately high variability coefficient. For the variables obtained in this way, the minima, maxima, arithmetic means $\bar{x}_j$, standard deviations S_j and classical coefficients of variation V_j. All values have been collectively presented in Table 2.

4.2 Regional Classification of the Polish Regions in Terms of the Business Tourism Potential

Table 3 shows values of taxonomic measures determined according to the Hellwig's and TOPSIS methods in the light of eleven selected diagnostic variables. Based on these values, a ranking of the studied regions was prepared.

Table 2 Characteristics of diagnostic variables

Specification	X_1	X_2	X_3	X_4	X_5	X_6	X_7	X_8	X_9	X_{10}	X_{11}
Minimum	0.49	0.95	77.53	0.03	0.42	0.46	0.16	12.46	0.91	0.13	0.93
Maximum	3.35	7.12	461.39	0.70	2.75	3.08	0.99	61.61	7.65	0.87	6.16
$\bar{x}_j$	1.25	2.69	183.32	0.13	1.07	1.11	0.33	26.98	2.72	0.31	2.33
S_j	0.70	1.55	99.62	0.15	0.59	0.64	0.21	14.70	1.76	0.18	1.29
V_j	0.56	0.58	0.54	1.15	0.55	0.58	0.63	0.54	0.65	0.59	0.55

Source Own calculations based on the Central Statistical Office (2017)

Table 3 Values of synthetic meters and ranges according to voivodships

Region	The Hallwig's method $d_i{}^H$	Ranking position R_H	Change of the ranking position	Ranking position R_T	TOPSIS $d_i{}^T$
Lower Silesia	0.4941041	3	5	5	0.1116238
Kuyavia-Pomerania	0.3280135	6	3	3	0.1119703
Lodzkie Voivodeship	0.2163861	14	4	10	0.1115762
Lubelskie Voivodeship	0.2343901	12	1	11	0.1115155
Lubuskie Voivodeship	0.2545597	11	6	5	0.1117660
Lesser Poland	0.8721739	1	1	2	0.1180125
Masovia	0.2931603	9	0	9	0.1115862
Opolskie Voivodeship	0.2256956	13	6	7	0.1117072
Subcarpathia	0.3247993	7	5	12	0.1113994
Podlaskie Voivodeship	0.1546119	16	3	13	0.1113972
Pomerania	0.4599607	4	12	16	0.1109299
Silesian Voivodeship	0.5785376	2	1	1	0.1401399
Holy Cross Voivodeship	0.2883316	10	6	4	0.1119165
Warmia-Masuria Voivodeship	0.2069735	15	1	14	0.1112724
Greater Poland	0.2951270	8	2	6	0.1117414
West Pomerania	0.3356007	5	10	15	0.1112518

Source Own calculations based on the Central Statistical Office (2017)

The results of conducted analyses have shown that in both rankings, the largest potential of business tourism in 2017 was found in Lesser Poland and Silesian Voivodeship. Those regions are characterized by the largest number of facilities equipped with conference rooms and equipment necessary for the preparation of business events. Additionally, within the territory of those regions, there are particularly attractive areas—large urban centers or mountain areas. In case of the Lesser Poland, the highest values of indicators for most variables were achieved (X_1, X_2, X_3, X_5, X_6, X_7, X_8, X_9 and X_{10}). In the Silesian Voivodeship, the highest level was achieved by the number of accommodation objects offering rehabilitation treatments (massages, physiotherapy, etc.) per 100 km^2. In this region, the lowest value was achieved by the only analyzed X_4 destimulant.

The determined value of the Spearman's rank correlation coefficient $r_s = 0.903$ has shown high consistency of the linear ordering of regions according to the values of taxonomic measures d_i^H and d_i^T. No change of the position has been noticed in case of one region (Masovia), one change of the position has been noticed in case of four regions (Lubuskie Voivodeship, Lesser Poland, Silesian Voivodeship and Warmia- Masuria Voivodeship. Differences in rankings have been distinguished after their division into four typological groups showing that both classifications in reference to nine voivodeships gave the same effect, whereas the other seven regions gave different results in both methodologies. Detailed classification of the studied regions was provided in Table 4.

In order to verify the manner, the voivodeships used their potential of attracting business tourists, a comparison of the obtained results and coefficients related to the number of people using the accommodation services in 2017 has been made. With consideration of data of the Polish Central Statistical Office (Central Statistical Office, 2017), two indicators have been used in the analysis-density of tourist movement (W_G-people using the accommodation services during the year in rela-

Table 4 Classification of the studied regions

Measure values $d_i^{H/T}$	Class	Tourism potential	Region
According to the value of Hellwig's development measure			
$d_i^H \geq 0.5215$	I	Very high	Lower Silesia, Lesser Poland, Silesian Voivodeship
$0.5212 > d_i^H \geq 0.3477$	II	High	Subcarpathian, Pomeranian, West Pomeranian
$0.3477 > d_i^H \geq 0.1738$	III	Medium	Kuyavia-Pomerania, Lubelskie Voivodeship, Lubuskie Voivodeship, Masovia, Opolskie Voivodeship, Holy Cross Voivodeship, Warmia-Masuria Voivodeship, Greater Poland
$d_i^H < 0.1738$	IV	Low	Podlaskie Voivodeship
According to the value of the synthetic TOPSIS meter			
$d_i^T \geq 0.1207$	I	Very high	Silesian Voivodeship
$0.1207 > d_i^T \geq 0.1137$	II	High	Lesser Poland,
$0.1137 > d_i^T \geq 0.1067$	III	Medium	Lower Silesia, Kuyavia-Pomerania, Lubelskie Voivodeship, Lodzkie Voivodeship, Masovia, Opolskie Voivodeship, Subcarpathia, Podlaskie Voivodeship, Pomerania, Holy Cross Voivodeship, Warmia-Masuria, Greater Poland, West Pomerania
$d_i^T < 0.1067$	IV	Low	-

Source Own calculations based on the Central Statistical Office (2017)

tion to the surface in square kilometers) and intensity of tourist movement according to Schneider (W_S-proportion of people using the accommodation services during the year to the number of inhabitants). Values of the indicators were provided in Table 5.

In order to be able to determine the strength of the degree of using the potential of business tourism, the values of the Spearman's rank correlation coefficient were determined again r_s. The analysis of the coefficients showed the high consistency of rankings as a result of using both taxonomic methods, with ordering on the basis of indicator W_G and W_S. Values of the indicators were provided in Table 6.

Due to the higher level of consistency of the results, the values obtained using the Hellwig's method were taken into account in the further analysis of the results. When starting the analysis in terms of density of tourist movement, one may observe high consistency between rankings R_H (Table 3) and R_G (Table 5). In both rankings, the positions of the first four regions were the same, and in case of other regions, the maximum discrepancy between the positions taken in individual rankings was

Table 5 Movement density indicator and Schneider indicator

Region	Movement density indicator W_G	Ranking position R_G	Change of the ranking position	Ranking position R_S	Schneider indicator W_S
Lower Silesia	166.60	3	1	4	114.50
Kuyavia-Pomerania	72.00	7	1	8	62.10
Lodzkie Voivodeship	40.90	15	1	14	48.20
Lubelskie Voivodeship	47.30	13	6	7	65.10
Lubuskie Voivodeship	71.80	8	5	13	52.70
Lesser Poland	322.80	1	1	2	144.70
Masovia	143.20	5	0	5	94.80
Opolskie Voivodeship	44.80	14	2	16	42.60
Subcarpathia	68.70	9	0	9	57.60
Podlaskie Voivodeship	33.00	16	4	12	56.30
Pomerania	156.40	4	1	3	123.40
Silesian Voivodeship	210.50	2	9	11	57.00
Holy Cross Voivodeship	49.60	12	3	15	46.40
Warmia-Masuria	52.20	11	5	6	88.00
Greater Poland	67.30	10	0	10	57.60
West Pomerania	120.60	6	5	1	161.70

Source Own calculations based on the Central Statistical Office (2017)

Coefficients	d_i^H	d_i^T
W_G	$r_s = 0.97$	$r_s = 0.89$
W_S	$r_s = 0.92$	$r_s = 0.85$

Source Own calculations based on the Central Statistical Office (2017)

4 positions. The results were affected mainly by the area of particular regions. The business tourism potential, adequately to the value of the earlier defined density indicator, was fully used by the following regions: Lesser Poland, Silesian Voivodeship, Lower Silesian Voivodeship, Pomerania.

Another analysis of the business tourism potential in terms of tourist movement density according to Schneider has shown much greater differences between the analyzed rankings R_H (Table 3) and R_S (Table 5). It was noticed that only the Lubelskie Voivodeship did not change its position, remaining on the first place in both rankings. Yet the regions with absolute consistency in the rankings R_H–R_G differ only by one position here. The exception is the Silesian Voivodeship—in this case, the difference amounts to 9 positions. This is evidently related to the high population density in this region. After the comparison of all three rankings: R_H, R_G, R_S, the Lesser Poland, the Lower Silesia, the Silesian Voivodeship and the Pomerania were characterized by the largest tourism potential, highly used. The West Pomerania also occupied relatively high positions in the rankings.

The regions subjected to the analysis are attractive because of their location, and they are situated in the mountains and by the sea and not far away from metropolitan area with large business centers. On the other hand, the example of the Masovia Voivodeship, with the capital city of Warsaw, shows that the analysis of the potential of business tourism requires deepening and analysis at the level of smaller administrative units.

5   Conclusion

Development of the business tourism products depends mainly on the potential of the destination where the products are offered. The greater the tourism potential, the greater the possibilities of meeting the needs of business tourists. The article compares the tourism potential of Polish regions using the Hellwig's development pattern method and the TOPSIS method. Eleven diagnostic features characterizing the conditions for the development of business tourism were selected for the analysis. The selection of factors for analysis was conditioned by the availability in the Polish statistical database and the expected usefulness of information about the resources of individual regions from the point of view of business tourists.

The obtained results showed the diversity of the analyzed regions in Poland, due to the possibility of meeting the needs of business tourism. The highest value of synthetic measures of the potential attractiveness of business tourism was characteristic for those regions which have a wide range of tourist attractions and are located in mountain areas (the Silesia region and Lesser Poland) and seaside destinations (the Pomerania and, to a lesser extent, the West Pomerania).

The analysis of movement density indicators and the Schneider analysis (related to the number of people using accommodation services) has shown that a part of the Polish regions (Lesser Poland, Silesian Voivodeship, Lower Silesia and West Pomerania) used their possibilities in an appropriate way. The following regions did not use much of their potential: Opolskie, Podlaskie and Lubelskie Voivodeships.

There is a real necessity to identify the key determinants of creating a tourist product in order to create tourist products of a destination, taking into account the requirements of tourists. The tourist product is a key factor in the economic development of the destination, and in order to ensure high quality and competitiveness of tourist services, it is necessary to create a network of enterprises and institutions in the region—implementing a common strategy for the development of a tourist product. Clusters are a particularly predestined form of network cooperation in order to develop a tourist product.

To sum up, it should be pointed out that the presented analysis may be useful both for economic entities creating business tourism products and for the authorities of individual territorial units. The presented analysis was limited by its regional scope, and therefore, future research should be refined to smaller territorial units.

References

Ashworth, G., & Goodall, B. (Eds.). (2012). *Marketing tourism places* (Vol. 2). Routledge.

Bedradina, G., & Nezdoyminov, S. (2019). Measuring the quality of the tourism product in the tour operator business. *Montenegrin Journal of Economics, 15*(2), 81–93.

Brent, J. R., Ritchie, J. R. B., & Crouch, G. I. (2000). The competitive destination: A sustainability perspective. *Tourism Management, 21*(1), 1–7.

Buhalis, D. (2000). Marketing the competitive destination of the future. *Tourism Management, 21*(1), 97–116.

Central Statistical Office. (2019). Local Data Bank. Retrieved from https://bdl.stat.gov.pl/BDL. Accessed the July 15, 2020.

Cooper, C. F., Gilbert, J., Wanhill, D., & Shepherd, S. R. (1998). *Tourism: Principles and Practice*. Harlow Addison-Wesley Longman.

Davidson, R., & Cope, B. (2003). *Business travel: Conferences, incentive travel, exhibitions, corporate hospitality and corporate travel*. Pearson Education.

Dominique-Ferreira, S., Vasconcelos, H., & Proença, J. F. (2016). Determinants of customer price sensitivity: An empirical analysis. *Journal of Services Marketing, 30*(3), 327–340.

Enright, M. J., & Newton, J. (2004). Tourism destination competitiveness: A quantitative approach. *Tourism Management, 25*(6), 777–788.

Fyall, A. (2019). *Tourism Destination Re-positioning and Strategies. The Future of Tourism* (pp. 271–283). Cham: Springer.

Gomezelj, D. O., & Mihalič, T. (2008). Destination competitiveness—applying different models, the case of Slovenia. *Tourism Management, 29*(2), 294–307.

Gunn, C. A. (1994). Emergence of effective tourism planning and development. *Tourism: The State Of The Art* 10–19.

Harris, R., & Leiper, N. (Eds.). (1995). *Sustainable tourism: An Australian perspective*. Routledge.

Hellwig, Z. (1968). Zastosowanie metody taksonomicznej do typologicznego podziału krajów ze względu na poziom ich rozwoju oraz zasoby i strukturę wykwalifikowanych kadr. *Przegląd Statystyczny, 4,* 307–326.

Hwang, C. L., & Yoon, K. (1981). *Multiple attribute decision making. Methods and applications*. Heidelberg: Springer Verlag.

Inskeep, E. (1994). *National and regional tourism planning: Methodologies and case studies*. Routledge.

Knežević Cvelbar, L., Dwyer, L., Koman, M., & Mihalič, T. (2016). Drivers of destination competitiveness in tourism: A global investigation. *Journal of Travel Research, 55*(8), 1041–1050.

Koo, C., Shin, S., Gretzel, U., Hunter, W. C., & Chung, N. (2016). Conceptualization of smart tourism destination competitiveness Asia Pacific. *Journal of Information Systems, 26*(4), 367–384.

Martín Martín, J. M., Salinas Fernández, J. A., & Rodríguez Martín, J. A. (2019). Comprehensive evaluation of the tourism seasonality using a synthetic DP2 indicator. *Tourism Geographies, 21*(2), 284–305.

Middleton, V. T., & Hawkins, R. (1998). *Sustainable tourism: A marketing perspective*. Routledge.

Page, S. (2005). *Transport and tourism: Global perspectives*. Pearson Education.

Saraniemi, S., & Kylänen, M. (2011). Problematizing the concept of tourism destination: An analysis of different theoretical approaches. *Journal of Travel Research, 50*(2), 133–143.

Synówka-Bejenka, E. (2017). Potencjał turystyczny województw Polski. *Wiadomości Statystyczne, 7,* 78–92.

UNWTO. (2019). Guidelines for Institutional Strengthening of Destination Management Organizations (DMOs)—Preparing DMOs for new challenges.

Central Statistical Office. (2017). *Tourism in 2017*, Warsaw 2018.

World Economic Forum. (2013). The Travel & Tourism Competitiveness Report 2013. Retrieved from https://reports.weforum.org/travel-and-tourism-competitiveness-report-2013/. Accessed the May 4, 2020.

World Economic Forum. (2019). The Travel & Tourism Competitiveness Report 2013. Retrieved from https://reports.weforum.org/travel-and-tourism-competitiveness-report-2019/. Accessed the July 15ᐧ 2020.

WTTC. (2017). Travel and Tourism Economic Impact 2017 Poland. Retrieved from https://www.wttc.org/-/media/files/reports/economic-impact-research/countries-2017/poland2017.pdf. Accessed the Decembers 14, 2018.

WTTC. (2020). Poland 2020 Annual Research: Key Highlights. Retrieved from https://wttc.org/Research/Economic-Impact. Accessed July 15, 2020.

The Role of Fashion Events in Tourism Destinations: DMOs Perspective

Dália Liberato, Benedita Barros e Mendes, Pedro Liberato, and Elisa Alén

1 Introduction

Tourist destinations promote the holding of events with the aim to ensure competitive advantage over its competitors (Ritchie & Beliveau, 1974; Hall, 1989,1992; Jago et al., 2010) expand its tourism potential (Grappi & Montanari, 2011; Getz & Page, 2016) and promote the development of the territory and quality of resident communities (Getz, 2007; Lamnot & Dowell, 2008; Lee, 2016). The events also allow to optimize limited resources (Stokes & Jago, 2007; Kellet et al., 2008; Hall, 2009) and distribute benefits to several stakeholders (Farley et al., 2016; Kelly & Fairley, 2018) and create opportunities for local business (Lamont & Dowell, 2008; Beesley & Chalip, 2011; Getz & Page, 2016; Lee, 2016).

In fact, the strategic development of events translates into an important way to increase tourist attractiveness and economic development of a destination (Getz, 2010), since event tourist, whose main motivation to visit the destination lies in participating in the event, in general, spend more time in the destination, spend more and travel in groups, which characterizes them as a segment of very lucrative market (Yoon et al., 2000; Tang & Turco, 2001; Gibson et al., 2003; Jones & Li, 2015).

Events allow several regions to take over as emerging tourism destinations, differing from competitors (Getz et al., 2012) through its unique characteristics, new market segments (Connel et al., 2015). In fact, events are symbolic elements of the brand image of several cities (Holt, 2004; Getz, 2008), for its strong visibility

D. Liberato (✉) · B. B. Mendes · P. Liberato
School of Hospitality and Tourism, Polytechnic Institute of Porto, Vila do Conde, Portugal
e-mail: dalialib@esht.ipp.pt

E. Alén
Faculty of Business Sciences and Tourism, University of Vigo, Ourense, Spain

D. Liberato · P. Liberato
CiTUR (Centre for Tourism Research, Development, and Innovation), Faro, Portugal

© The Author(s), under exclusive license to Springer Nature Switzerland AG 2021
V. Katsoni and C. van Zyl (eds.), *Culture and Tourism in a Smart, Globalized, and Sustainable World*, Springer Proceedings in Business and Economics,
https://doi.org/10.1007/978-3-030-72469-6_31

and transmitted confidence (Getz et al., 2012). Thus, events have the particularity of positively relating to the destination they occur in, as well as with their image (Todd et al., 2017).

Events break the routine of a city (Liu & Chen, 2007) and allow the desired development of a locality or region and its several components (Getz & Page, 2016). However, the range of benefits from event tourism, regarding tourism destinations, is wide. It is important that events are strategically designed and, with a view to outcomes, achieved in a strategic planning (short and long-term) and distributed across the several stakeholders of the destination (O'Brien & Chalip, 2007; Chalip, 2014; Smith, 2014; Kelly & Fairley, 2018). Planning should therefore consist of an integrated set of programmes and policies, including objectives desired by the destination (Getz, 2005).

In fact, although event tourism is often guided by the objectives associated with them, it is important to reflect on the social, cultural and environmental impacts that the organization of the event will have in the destination. In this sense, event tourism should be addressed as an open system which agglomerates the several components of a destination, and therefore, it is important to identify the results and impacts, positive or negative, of the several events to be held (Getz & Page, 2016).

With the motivation of knowing the impacts of the fashion events in the city of Oporto, the purpose of this article is to present and discuss the results of an empirical study with local and regional DMOs, presented by representatives' perceptions of how the organization of fashion events add value to the city of Oporto as a tourism destination.

2   Literature Review

Over time, local governments and institutions began to realize the role of fashion as a creator of identity and strategic advantage. Indeed, the shopping tourism is one of the main tools used by DMOs to promote a destination and increase demand flows (Kalabaska Ayala Ramírez & Cantoni, 2018), as tourists increasingly buy fashion products and crafts during their travels (Moscardo, 2004; Calderón et al., 2016; Liberato et al., 2020). Fashion events also attract many visitors, promoting aesthetic and creative characteristics of a city (Kalabaska et al., 2018) and are mostly held in recognized cities all over the world as fashionable cities. These destinations are mainly characterized by a wide range of business, financial, entertainment, cultural and leisure activities and have strong and unique identities (Capone & Lazzaretti, 2016), which distinguish them from competitors.

Tourists of fashion events are attracted by the image of fashionable cities, characterized by high status, elegance and dynamism. Thus, the events emerge as differentiating elements, capable of adding value to a destination in an increasingly competitive tourist market (Russo & van der Borg, 2002; Chilese & Russo, 2008).

In this sense, both fashion events and fashion weeks are crucial in the process of promoting a global tourism destination and attraction of a high number of visitors

(Hall, 1989), offering the city the opportunity to achieve direct and indirect economic and social benefits (Weller, 2008; Kalbaska, et al., 2018).

In fact, the organization of events allows to captivate tourists to a particular destination and offer them the opportunity to experience unique feeling and emotions in a unique place (Andersson, 2007; Morgan et al., 2010; Panoso et al., 2010; Ryan, 2010).

The diversity of events in a destination, including fashion events, is enhanced by building a successful cultural agenda and cultivating synergies between the different events and collaborations between several stakeholders (Merrilees et al., 2005; Stokes, 2008; Parent, 2010; Reid, 2011; Todd, Leask & Ensor, 2017).

The adoption of a cultural agenda contributes to economic, sociocultural and environmental development of a destination, once it is based on the principles of sustainability and is designed with the aim of achieving the objectives of the city and distribute the benefits to the several local stakeholders (Getz & Page, 2016; Ziakas, 2019).

In fact, for the continued success of a cultural agenda, it is relevant to be developed on the basis of local legislation, by a team of several stakeholders from the city/region, including members of the local administration, commercial enterprises and even elements of the national government (Ziakas, 2019).

The moments of networking between the several local stakeholders facilitate the management of the interests and needs of the several social groups and economic sectors, allowing outcomes and benefits to be distributed across all local actors (Werner et al., 2015; Kelly & Fairley, 2018) in a democratic, transparent and legitimate way (Dredge et al., 2011). Events should be organized and developed by the several local stakeholders, to ensure the destination development in accordance with its needs and priorities (Getz, 2005, 2013).

Adopting a stakeholder approach helps DMO's identify and enhance event tourism, positively develop relations between stakeholders and meet their expectations and needs (Todd et al., 2017). Relations between local actors in a tourism destination allow the sustainable development, increased competitiveness and the guarantee of sustainable success of rejuvenation strategies (Skinner, 2000; Tinsley & Lynch, 2001; Faulkner & Tideswell, 2006).

All tourist destinations must develop collaborative processes and effective partnerships between their stakeholders in order to share knowledge and acquire resources innovation (Brandão et al., 2019). Since tourism innovation requires the common effort of the several stakeholders, the territories play the crucial role of developing and enhancing their union (Sundbo et al., 2007).

The main role of a DMO is to promote tourism, whether it is business or a particular destination (Pike & Page, 2014). In this way, the DMOs promote and organize different types of events in order to achieve different types of tourists. The organization of several typologies of events promotes sustainable and balanced economic, social and environmental development of a destination (Ziakas, 2019). Local governments in many cities have already realized the importance of fashion events, since the higher the place that a destination ranks in the global hierarchy of fashion cities, the higher number of benefits and positive outcomes resulting from it (Jansson & Power, 2010).

Fashion cities are destinations recognized for their status (Gilbert, 1990) and therefore differentiate themselves from their competitors, are not easily replaceable and have unique, functional and symbolic attributes (Hankison, 2004) that make them known (Lewis et al., 2013).

2.1 Case Study

The city of Oporto is the second largest city of the country and the most representative of the northern region of Portugal. Over the last few years, there has been a positive development of the city of Oporto as a tourism destination, increasing its recognition internationally. The city of Oporto has been working in a remarkable way regarding its positioning on the European map. In fact, in addiction to winning international distinction, it has achieved the position of European destination. Oporto was considered the Best European Destination in 2012, 2014 and 2017—a distinction awarded annually by European consumers choice. This distinction offered attractiveness and notoriety to the city of Oporto as a tourism destination, also revealing itself as a potential for the region economic growth.

Indeed, the tourism sector has shown significant growth not only in the city of Oporto, but also throughout the northern region, supported by its historical, cultural, natural, architectural and gastronomic heritage, attracting, increasingly, a greater number of visitors (ERTPN, 2015). The northern region is also the core of the Portuguese fashion industry because most of the companies of the textile sector are located there (Table 1).

The city of Oporto is the stage of several fashion events, including Porto fashion week and Portugal fashion, which are effectively two of the main Portuguese fashion events, as well as in the Iberian Peninsula. The city of Oporto can offer fashion events, favourable conditions, including climate, infrastructure and equipment, the diversity of complementary leisure activities and good communications network.

Table 1 Guests (no.) and overnight stays (no.) in tourist accommodation establishments in the city of Oporto

	2014	2015	2016	2017	2018	2019	2020
Guests	1 144 376	1 287 725	1 426 863	1 536 798	1 958 645	2 223 458	278 150
Overnight stays	2 246 244	2 515 659	2 833 406	3 040 424	4 001 160	4 535 329	410 071

Source INE, 2020; ATP, 2020

3   Methodology

The objectives of the present research are to identify the results from fashion events to the destination Oporto; to understand if the fashion events are part of the tourism strategy of the city of Oporto and the north region; to understand how the DMOs cooperate in the organization and promotion of fashion events in the city of Oporto; to understand how fashion events contribute to the development of the tourism sector in the city of Oporto; and to understand how fashion events add value to the Oporto destination.

To fulfil, the objectives of this research were adopted the qualitative methodology, and the semi-structured interview was used as a data collection instrument. The interview is an advantageous data collection technique due to its efficiency and to the fact it promotes the collection of diversified data (Gil, 2008). Semi-structured interviews are characterized by the existence of a previously prepared script but offer the interviewer the freedom to include during the interview, several aspects considered as convenient. Interviews were conducted individually with three representatives of local and regional DMOs of the city of Oporto and the northern region of Portugal. (Table 2).

All interviews were audiotaped and transcribed verbatim. The interview schedule includes several open-ended questions:

- Are fashion events part of the tourism strategy of the city?
- What is the role of your entity in the organization, dynamization and promotion of events?
- Does the municipality promote fashion events? If so, at what level and how is the promotion of those events? What are the plans and campaigns developed with the aim to stimulate the demand for fashion events in the city of Oporto? How are they promoted, both nationally and internationally?
- Since fashion events are part of the municipal cultural agenda, which are the objectives that the municipality, as a partner, intends to achieve?
- The calendar of fashion events contributes to the decrease in seasonality at the destination?
- How is the local population affected by the fashion events? Do you think the local community is integrated and actively participates in the fashion events held in the city of Oporto?

Table 2 Interviews

No. interview	Interview date	Interview duration
1	04/29/2020	49:24
2	04/29/2020	49:17
3	05/04/2020	Written response

Source Authors

- There is concern about the appreciation of the city's tourism resources during the realization/organization of fashion events? They are carried out in tourism resources of the city? Heritage is used as resource to enhance fashion events?
- Is there any parallel programme to fashion events to ensure the extending tourists stay?
- Who are the existing partners in the organization of these events and who would you consider important to bring to the organization? How does it assess the importance of established partnerships for the success of fashion events in the city of Oporto?
- Do you believe that fashion events add value to the city as a destination, as well as their resources? Fashion events are related to the image and personality of the city of Oporto?
- From your perspective, what is the potential of fashion events that should be explored, viewing their increasing success?

Interview data were analysed using thematic analysis method—the comprehension of the phenomena must emerge from the data rather than from preconceived notions formulated by the researcher.

4 Results

To understand the relationship between the fashion events realized in the city of Oporto and the tourist development of the city, semi-structured interviews were conducted. The analysis of these interviews revealed the following integrative categories, presented in Table 3: fashion events; local community; tourism resources of the city; partnerships; and potential of fashion events. A deeper description of the results is presented in this section with participants' transcriptions to illustrate and facilitate understanding.

Table 3 Core categories and subcategories

Core categories	Subcategories
Fashion events	Tourism strategy Organization and promotion Objectives Results for the city
Local community	Integration and participation
Tourist resources of the city	Appreciation
Partnerships	Importance Parallel programmes
Potential of fashion events	Improvement suggestions Availability for support

Source Authors

Fashion events are part of the local and regional tourism strategy of the city of Oporto and north region *"since they have the capacity to internationalize, mobilize and disseminate the destination" (3)*.

About the organization of fashion events held in the city of Oporto, there is a balance between the three DMOs since none cooperates directly in this process. However, with regard to their promotion, DMOs representatives have different perspectives and positions: One of the regional DMOs does not promote or support these events, considering these as private events, which create competition between members; the other regional DMO cooperates in the processes of dynamization and promotion and may, when appropriate, finance communication and dissemination strategies; the local DMO, in turn, does not promote the fashion events held in the city of Oporto, but cooperates with their organization through the supply of spaces.

With regard to the objectives that DMOs intend to achieve with the organization of fashion events in the city of Oporto, they include, the tourism dynamization, the improvement of the city's image, the affirmation of the city as a destination for fashion events, the development of the local economy, the dynamism of regional infrastructures and the revitalization as conversion of existing structures.

All the respondents believe that the local community *"if there are conditions, participates actively" (3)* in the fashion events held in the city of Oporto, because residents *"welcome these events and understand the goals of these events" (1)*.

DMOs representatives believe that there is a concern for valuing of the city's tourism resources, by the organizers of fashion events. In addition to the reference to the building of the Alfândega of Oporto as the "mother house" of fashion events held in the city of Oporto, the representatives of DMOs recognize the bet made by event promoters in the choice of privileged places of the city, characterized by its beauty and symbolism.

The respondents believe that fashion events would be more advantageous for the tourism sector and the city itself if they were held in outdoor spaces, with open access, and not only to the professional public, as actually. *"Fashion is only of interest to the city when it passes the walls of business, exposes itself to the city or the region and brings with it other associated brands (...) it is necessary once again that fashion has the quality enough to communicate out. Otherwise, we were left with a very beautiful event, but for domestic consumption (...) what happens to events sometimes is a bit like this: they stay closed, they get stuck themselves and do not create this mission, this ambition to become vehicles of promotion" (2)*.

They understand, however, the difficulty of organizing an event in several spaces, mainly in public spaces. However, DMOs are available to support the issuance of spaces and several facilities.

Regarding the commitment to support parallel programmes to fashion events, capable to ensure the extension of tourists' stay, the municipal authority states that is unaware of their existence but believes that such programmes are developed by regional tourism entities. Thus, it reveals the need to create partnerships between the fashion sector, more specifically, the organizers if fashion events and the local and regional DMOs, so both sectors can interact, cooperate and receive benefits and positive outcomes of this union.

In fact, all respondents share the vison of creating partnerships between stake-holders as a factor of extreme importance for the success of the fashion events.

The creation of partner networks between various stakeholders would have the ability to ensure the positive development of several economic sectors of the city, such as the fashion industry, the tourism sector and the cultural sector, which covers music, theatre and architecture. *"We must work as a team and I think that at the level of economics should also be the same: If we unite we can create partnerships and learn more, because there is always someone who knows more about a certain area. And I believe fashion, by chance, is a sector that can cross with a series of activities" (1).*

The respondents believe that the holding of fashion events could add value to the city of Oporto, as a tourism destination, when associated with the image and personality of the city.

Regarding the potential of fashion events that must be improved so their success could be increased, the representatives of the DMOs believe that it is directly related to the objectives of the event promoters. However, revealed the aspects that, in their point of view, could be improved, such as the promotion and communication of the events, their realization in public spaces of the city, the openness to the general public and the creation of partnerships with the public sector.

However, respondents believe that the promotion of the city of Oporto as a fashion city should be primarily the responsibility of the fashion sector and the organizers of fashion events. *"Fashion must, in fact, have people behind it to give this dimension, this volume that otherwise does not adds value. (…) It needs, in fact, that actors are and have this international notoriety and create this dimension, being this an element of communication, that passes out"* (2) (Table 4).

5 Conclusion

Representatives of the DMOs reveal that fashion events have the ability to contribute to the achievement of several objectives of the local tourism sector, such as tourism promotion, the improvement of the image of the destination, the development of local economy and the affirmation of the city of Oporto as a city of fashion, among others. In fact, events have the ability to make destinations more attractive and active, as they break the routine of a city (Liu & Chen, 2007) and allow the desired development of a locality or region and its various components (Getz & Page, 2016). Fashion destinations are recognized by their status (Gilbert, 1990) and thus have the ability to differentiate themselves from their competitors, since they are not easily replaceable and have unique functional and symbolic attributes (Hankison, 2004) that make them known worldwide (Phillips & Back, 2011; Lewis et al., 2013).

They also believe that fashion events realization brings to the community which, in turn, is integrated and actively participates in this type of event. Fashion events are part of the local and regional tourism strategy provided, having the capacity to

Table 4 Summary of the results of interviews conducted with representatives of DMO's

Theme	Conclusions	Interviewed
Fashion events	They are part of the local and regional tourism strategy if they have the capacity to internationalize and promote the destination	E1/E2/E3
	None of the DMOs cooperate directly in the organization, but they support through promotion, actions and provision of facilities	E1/E3
	The objectives that the DMOs intend to achieve with the realization of fashion events are the promotion of tourism, improvement of the image of the destination, affirmation of the city as a destination for fashion events, development of the local economy, dynamization and revitalization of infrastructure, combating seasonality and improving the quality of life of the local population	E1/E2/E3
	They add value to the city of Porto and are associated with their image and personality	E1/E2/E3
Local community	It is integrated and actively participates in fashion events	E1/E2/E3
Tourism resources	There is a concern about the appreciation of the city's tourism resources when holding fashion events	E1/E2/E3
Partnerships	All DMOs recognize the importance of partnerships for the success of fashion events and understand the need to create partnerships between the fashion industry and the tourism industry	E1/E2
Parallel programmes	The DMOs recognize the importance of improving parallel programmes that enhance the enlargement of tourists' stay, after fashion events	E1/E3
Potential of fashion events	The DMOs believe that fashion events would be more successful if they were promoted on a wider scale and were held in open-air locations in the city, and access was not reserved only to professionals in the fashion industry	E1/E2/E3
	The DMOs understand the difficulty of organizing events in open spaces, aimed at a wider audience, and are available to support their realization	E1/E2

Source Authors

internationalize and promote the city of Oporto and the North region as tourism destinations.

None of the DMOs interviewed cooperate directly in the organization of fashion events. However, they support those events through promotion actions and the availability of several facilities and resources, highlighting the supply of spaces.

Respondents believe that the fashion events held in the city of Oporto would be more advantageous for the tourism sector if they were promoted on a larger scale, carried out in outdoor locations, allowing open access, and not only reserved to fashion professionals. According to Getz (2005, 2013), any event should be organized and developed based on joint strategies in order to facilitate the development of a destination according to its main needs and priorities. In fact, it is important that events are designed and held, from a strategic perspective, so that outcomes are maximized and distributed by the several stakeholders of the destination (O'Brien & Chalip, 2007; Chalip, 2014; Smith, 2014; Kelly & Fairley, 2018).

According to Ziakas (2019), the organization of several types of events allows a sustainable and balanced development, sustained by an economic, social and environmental valorisation of a destination. In this sense, the tourism destinations promote the holding of events in order to ensure competitive advantage over their competitors (Ritchie & Beliveau, 1974; Hall, 1989, 1992; Jago et al., 2010) and expand their tourism potential (Grappi & Montanari, 2011; Getz & Page, 2016) promoting the development of the territory and improving the quality of life of resident communities (Getz, 2007; Lamont & Dowell, 2008; Lee, 2016). However, DMO's understand the difficulties in organizing open spaces events and for a wider audience and are, therefore, available to support their achievement, recognizing the importance of partnerships between the public sector and the fashion industry.

In fact, over time, local governments and institutions have begun to realize the role of fashion as a creator of identity and strategic advantage, since fashion events attract a large number of visitors, promoting the aesthetic and creative characteristics of a city (Kalbaska et al., 2018), emerging as differentiating elements, capable of adding value to a destination, in an increasingly competitive tourist market (Russo & van der Borg, 2002; Chilese & Russo, 2008). Thus, fashion events become crucial in the process of promoting a world-class tourist destination and attracting a large number of visitors (Hall, 1989), offering the city the opportunity to achieve direct and indirect economic and social benefits (Weller, 2008; Kalbaska et al., 2018).

Acknowledgements The authors acknowledge the financial support of CiTUR, R&D unit funded by the FCT—Portuguese Foundation for the Development of Science and Technology, Ministry of Science, Technology and Higher Education, under the scope of the project UID/BP/04470/2020.

References

Andersson, T. D. (2007). The tourist in the experience economy. *Scandinavian Journal of Hospitality and Tourism, 7*(1), 46–58.

Beesley, L. G., & Chalip, L. (2011). Seeking (and not seeking) to leverage mega-sport events in non-host destinations: The case of Shanghai and the Beijing Olympics. *Journal of Sport & Tourism, 16*(4), 323–344.

Brandão, F., Breda, Z., & Costa, C. (2019). Innovation and internationalization as development strategies for coastal tourism destinations: The role of organizational networks. *Journal of Hospitality and Tourism Management, 41,* 219–230.

Calderón, H. G., Gonzalez, G. M., & Gardó, & T. F. (2016). Tourism and fashion: factors affecting trip length. *Universia Business Review,* 18–51.

Capone, F., & Lazzeretti, L. (2016). Fashion and city branding: An analysis of the perception of Florence as a fashion city. *Journal of Global Fashion Marketing, 7*(3), 166–180.

Chalip, L. (2014). From legacy to leverage. In J. Grix (Ed.), *Levaraging legacies from sports mega-events: Concepts and cases* (pp. 2–12). New York: Palgrave MacMillan.

Chilese, E., & Russo, A. P. (2008). Urban Fashion Policies: Lessons from the Barcelona Catwalks. Dipartamiento di Economia Universitá di Torino.

Connell, J., Page, S. J., & Meyer, D. (2015). Visitor attractions and events: Responding to seasonality. *Tourism Management, 46,* 283–298.

Dredge, D., Ford, F. J., & Wuthford, M. (2011). Managing local tourism: Building sustainable tourism management practices across local government divides. *Tourism and Hospitality Research, 11*(2), 101–116.

Farley, S., Cardillo, M., & Filo, K. (2016). Engaging volunteers from regional communities: Non-hot city perceptions towards a mega-event and the opportunity to volunteer. *Event Management, 20*(3), 443–447.

Faulkner, B., & Tideswell, C. (2006). Rejuvenating a maturing tourist destination: The case of Gold Coast, Australia. In: R. Butler (Ed.), *The tourism area life cycle: Applications and modifications* (Vol. 1). Clevedon: Channel View Publications.

Getz, D. (2005). *Event management and event tourism.* New York: Cognizant.

Getz, D. (2007). *Events studies: Theory, research and policy for planned events.* Oxford: Elsevier.

Getz, D. (2008). Event tourism: Definition, evolution, and research. *Tourism Management, 29,* 403–428.

Getz, D. (2010). The nature and scope of festival studies. *International Journal of Event Management Research, 5*(1), 1–47.

Getz, D. (2013). *Event tourism: Concepts, international case studies, and research.* New York: Cognizant.

Getz, D., & Page, S. J. (2016). Progress and prospects for event tourism research. *Tourism Management, 52,* 593–631.

Getz, D., Svensson, B., Petersson, R., & Gunnervall, A. (2012). Hallmark events: Definition, goals and planning process. *International Journal of Event Management Research, 7*(1/2), 47–67.

Gibson, H. J., Willming, C., & Holdnak, A. (2003). Small-scale event sport tourism: Fans as tourists. *Tourism Management, 24*(2), 181–190.

Gil. A. (2008) *Dados e técnicas de pesquisa social* 6a. Atlas São Paulo.

Gilbert, D. (1990). Strategic Marketing Planning for National Tourism. *Tourist Review, 10*(1), 9–33.

Grappi, S., & Montanari, F. (2011). The role of social identification and hedonism in affecting tourist re-patronizing behaviors: The case of an Italian festival. *Tourism Management, 32,* 1128–1140.

Hall, C. M. (1989). The definition and analysis of hallmark tourist events. *GeoJournal, 19*(3), 263–268.

Hall, C. M. (1992). *Hallmark Tourist Events: Impacts.* Managment and Planning: Belhaven Press.

Hall, C. M. (2009). Innovation and tourism policy in Australia and New Zealand: Never the twain shall meet? *Journal of Policy Research in Tourism, Leisure and Events, 1*(1), 2–18.

Hankison, G. (2004). The brand images of tourism destinations: A study of the saliency of organic images. *Journal of Product and Brand Management, 13*(1), 6–14.

Holt, D. (2004). How brands become icons: The principles of cultural branding. Harvard Business Press.

Jago, L., Dwyer, L., Lipman, G., van Lill, D., & Vorster, S. (2010). Optimizing the potential of mega-events: An overview. *International Journal of Event and Festival Management, 1*(3), 220–237.

Jansson, J., & Power, D. (2010). Fashioning a global city: Global city brand channels in the fashion and design industries. *Regional Studies, 44*(7), 889–904.

Jones, C., & Li, S. (2015). The economic importance of meetings and conferences: A satellite account approach. *Annals of Tourism Research, 52,* 117–133.

Kalbaska, N., Ayala Ramírez E., & Cantoni, L. (2018). The role of tourism destinations within the online presence of fashion weeks *Almatourism, 9,* 87–114.

Kellett, P., Hede, A. M., & Chalip, L. (2008). Social policy for sport Events: Leveraging (relationships with) teams from other nations for community benefit. *European Sport Management Quarterly, 8*(2), 101–121.

Kelly, D. M., & Fairley, S. (2018). What about the event? How do tourism leveraging strategies affect small-scale events? *Tourism Management, 64,* 335–345.

Lamont, M., & Dowell, R. (2008). A process model of small and medium enterprise sponsorship of regional sport tourism events. *Journal of Vacation Marketing, 14*(3), 253–266.

Lee, Y. K. (2016). Impact of government policy and environment quality on visitor loyalty to Taiwan music festivals: Moderating effects of revisit reason and occupation type. *Tourism Management, 52,* 187–196.

Lewis, C., Kerr, G. M., & Burgess, L. (2013). A critical assessment of the role of fashion in influencing the travel decision and destination choice. *International Journal of Tourism Policy, 5*(1/2), 4–18.

Liberato D., Liberato, P., & Silva, M. (2020). Shopping Tourism: Comparative analysis of the cities of Oporto and Lisbon as shopping destinations. In: V. Katsoni, & T. Spyriadis (Eds.), Cultural and tourism innovation in the digital era. Springer Proceedings in Business and Economics (pp. 365–379). Cham: Springer. https://doi.org/10.1007/978-3-030-36342-0_29

Liu, Y., & Chen, C. (2007). The effects of festivals and special events on city image design. *Frontiers of Architecture and Civil Engineering in China, 1,* 255–259.

Merrilees, B., Getz, D., & O'Brien, D. (2005). Marketing stakeholder analysis: Branding the Brisbane Goodwill games. *European Journal of Marketing, 39*(9/10), 1060–1077.

Morgan, M., Lugosi, P., & Ritchie, B. (2010). *The tourism and leisure experience.* Bristol: Channel View Publications.

Moscardo, G. (2004). Shopping as a destination attraction: An empirical examination of the role of shopping in tourists' destination choice and experience. *Journal of Vacation Marketing, 10*(4), 294–307.

O'Brien, D., & Chalip, L. (2007). Executive training exercise in sport event leverage. *International Journal of Culture, Tourism and Hospitality Research, 1*(4), 296–304.

Panosso Netto, A., & Gaeta, C. (2010) *Turismo de experiência Senac.* São Paulo.

Parent, M. M. (2010). Dacision-making in major sporting events over time: Parameters, drivers, and strategies. *Journal of Sport Management, 24*(3), 291–318.

Pike, S., & Page, S. J. (2014). Destination marketing organizations and destination marketing: A narrative analysis of the literature. *Tourism Management, 41,* 202–227.

Reid, S. (2011). Event stakeholder management: Developing sustainable rural event practices. *International Journal of Event and Festival Management, 2*(1), 20–36.

Ritchie, J. B., & Beliveau, D. (1974). Hallmark event: An evaluation of a strategic respond to seasonality in the travel market. *Journal of Travel Research, 13*(2), 14–20.

Russo, A. P., & van der Borg, J. (2002). Planning considerations for cultural tourism: A case study of four European cities. *Tourism Management, 6,* 631–637.

Ryan, C. (2010). Ways of conceptualization the tourist experience: A review of literature. *Tourism Recreation Research, 35*(1), 37–46.

Skinner, A. (2000). Napa Valley, California: A model of wine region development. In C. M. Hall, L. Sharples, B. Cambourne, N. Macionis, R. Mitchell, & G. Johnson (Eds.), *Wine tourism around the world: Development, management and markets* (pp. 283–296). Boston: Butterworth-Heinemann.

Smith, A. (2014). Leveraging sport mega-events: New model or convenient justification? *Journal of Policy Research in Tourism, Leisure and Events, 6*(1), 15–30.

Stokes, R. (2008). Tourism strategy making: Insights to the events tourism domain. *Tourism Management, 29*(2), 252–262.

Stokes, R., & Jago, L. (2007). Australia's public sector environment for shaping event tourism strategy. *International Journal of Event Manager Research, 3*(1), 42–53.

Sundbo, J., Orfila-Sintes, F., & Sørensen, F. (2007). The innovative behavior of tourism firms: Comparative studies of Denmark and Spain. *Research Policy, 36,* 88–106.

Tang, Q., & Turco, D. M. (2001). Spending behaviours of events tourists. *Journal of Convention & Exhibition Management, 3,* 33–40.

Tinsley, R., & Lynch, P. (2001). Small tourism business networks and destination development. *International Journal of Hospitality Management, 20,* 367–378.

Todd, L., Leask, A., & Ensor, J. (2017). Understanding primary stakeholders multiple roles in hallmark event tourism management. *Tourism Management, 59,* 494–509.

ERTPN. (2015). *Estratégia de Marketing Turístico do Porto e Norte de Portugal.* TPNP: TPNP.

Weller, S. (2008). Beyond global production networks: Australian fashion weeks's trans-sectoral synergies. *Growth and Change, 39*(1), 104–122.

Werner, K., Dickson, G., & Hyde, K. F. (2015). Learning and knowledge transfer processes in a mega-events context: The case of the 2011 Rugby World Cup. *Tourism Management, 48,* 174–187.

Yoon, S., Spencer, D. M., Holecek, D. F., & Kim, D. K. (2000). A profile of Michigan's festival and special event tourism market. *Event Management, 6*(1), 33–44.

Ziakas, V. (2019). Issues, patterns, and strategies in the development of event portfolios: Configuring models, design and policy. *Journal of Policy Research in Tourism, Leisure and Events, 11,* 121–158.

Wellness Tourism Resorts: A Case Study of an Emerging Segment of Tourism Sector in Greece

Marilena Skoumpi, Paris Tsartas, Efthymia Sarantakou, and Maria Pagoni

1 Introduction

The market of wellness tourism has shown great growth rates up to 6.5% annually, with revenues that reached from 563 to 639 billion dollars in the time period between 2015 and 2017 and the hopeful forecast to reach 919 billion dollars in 2022 (Global Wellness Institute, 2018). This trend is expected to bring a lot of profit to the countries that will take all necessary action to support and promote wellness tourism facilities and products. Greece can be the place to create wellness resorts, since it combines beautiful environment, natural local products, use of water for rejuvenation and relax and also offers a lifestyle which can reduce stress and anxiety. However, despite the steps taken to this direction, (Tsartas, 2010; Sarantakou, 2017; Tsartas et al., 2019; Coccossis et al., 2011) there is an important delay in developing specialized wellness tourism facilities, such as wellness resorts. This paper aims to study the profile and the viewpoints of the general public and the visitors of wellness resorts, as well as to draw conclusions that could be used in marketing plans by wellness tourism businesses and that would earn them more benefits and extend their customers' base. First of all, it should be noted that wellness tourism is not one of the well-known sectors in the tourism industry because of the lack of extended research—both in

M. Skoumpi
Hellenic Open University, Patra, Greece

P. Tsartas (✉)
Harokopio University, Kallithéa, Greece
e-mail: tsartas@hua.gr

E. Sarantakou
University of West Attica, Egaleo, Greece

M. Pagoni
Ministry of Tourism, Athens, Greece

© The Author(s), under exclusive license to Springer Nature Switzerland AG 2021
V. Katsoni and C. van Zyl (eds.), *Culture and Tourism in a Smart, Globalized, and Sustainable World*, Springer Proceedings in Business and Economics,
https://doi.org/10.1007/978-3-030-72469-6_32

theory and in case studies (Huang & Xu, 2014). Especially, in Greece, no studies have been conducted specifically on wellness resorts.

2 Literature Review

2.1 Wellness Tourism: Definition

Wellness tourism is the combination of relationships and phenomena resulting of the traveling and staying of people whose motivation is to rehabilitate, preserve and stimulate their physical, mental and spiritual health, as well as their social well-being (Smith & Puczkó, 2017). Wellness tourism is a subcategory of health tourism, distinctive from medical tourism, which constitutes another subcategory. This distinction also applies in the recent Greek law, following which health tourism includes three distinctive categories: medical tourism, spa-thermal tourism and wellness tourism (OGG Issue No. 208/ 11–12–2018).

2.2 Characteristics of the Wellness Tourism Market

The revenues of the global industry have been increased from 3.7 trillion dollars in 2015 to 4.2 trillion dollars in 2017. Specifically, 2017 revenues could be analyzed as follows:

- 1.083 billion dollars in personal care, beauty and anti-aging treatments
- 702 billion dollars in healthy eating and weight loss
- 639 billion dollars in wellness tourism
- 595 billion dollars in mind and body fitness (Global Wellness Tourism Economy, 2018)

Wellness tourism had an explosive increase from a $ 563 billion market in 2015 to a $ 639 billion market in 2017, which reaches a rate of 6.5% annually. Such increase is more than twice as fast as tourism overall (3.2%). It is forecast to grow even faster through 2022 (7.5% yearly), to reach $ 919 billion.

According to the table above, USA, China and Germany are the most successful national markets as to wellness trips in 2017, while Italy and Austria are the last two in the list.

In a CBI report (Center for the Promotion of Imports from Developing Countries), there has been an effort to categorize some European wellness tourists (cbi, n.d). The first conclusion is that all tourists could be potential wellness tourists, no matter their age or income. The first category which includes the larger number of wellness tourists is that of experienced wellness tourists who frequently chose wellness trips intentionally. Their average age is between 40 and 69 years old, and their aim is to enjoy massages and body treatments. There is also a large number of leisure tourists, about 25%, whose income is medium or low, the majority of whom are between 20 and 49 years old. The third category consists of people who are attracted by a social and luxurious lifestyle and reach 19% of wellness tourists with medium income. 14% of tourists are part of a fourth category which includes holidaymakers that love sports and energy. There is also another category in which 11% of tourists are included, who can be described as skeptics, and their main characteristic is prejudice and negative attitude toward wellness tourism. The last category represents 5% of tourists, who can be described as "beginners" and includes elderly people over 60 who do not have any experience of this type of vacation.

2.3   Wellness Resorts: Features, Services and Facilities

Wellness resorts are hotels specialized in services related to wellness, including personal health care and professional know-how in wellness facilities (Chen, 2012). The main departments of a wellness resort are the same as those of a hotel, while there is also the possibility to offer extra services, such as nutrition, exercise, self-improvement and motivation programs, as well as beauty, energy and relaxation treatments. Based on the services they provide, they may also provide visits to doctors, nutritionists, fitness trainers or physiotherapists. Most resorts offer spa, massages, hydrotherapy and thermotherapy, while some also offer electrotherapy, photo therapy and injury recovery exercises. Almost every wellness resort offers non-medical face and body beauty treatments.

In order for wellness resorts to survive and stand out in the competitive environment that has been formed in the wellness tourism sector, they are trying to show and promote their strongest features, whether these are the natural springs and the landscapes, or the experienced, well-known professionals that work for them (Smith & Puczko, 2017). FX Mayr wellness resort in Austria is addressed to people that wish to improve the condition of their digestive system and to detox (original-mayr, n.d.). In Japan, there is a successful development of curative spas and wellness resorts, because of the vast interest of elderly people over their good health and rejuvenation. In Germany, the trend is to combine wellness tourism and golf, as well as to use non-surgical anti-aging methods within the wellness resorts (wellnesshotels-resorts, n.d.). In Spain, the government has set as its 2020 goal to develop wellness tourism by advertising the mediterranean nutrition, healthy habits and quality of life of its people, good climate, good prices, quality services and well-trained professionals. (efesalud, 2016). In France, there are more than 9,000 institutes and curative spas, as well as a vast number of hotels and resorts that offer thalassotherapy and hydrotherapy (atout-france, 2016). The Russian Government aims at developing wellness tourism, taking into consideration the interest of modern tourists in alternative and traditional treatments, which are based on water and mud, especially since this type of tourism has many advantages, such as combating seasonality, a high profit margin and a wide range of ages that show their preference to it (globalwellnesssummit, 2015).

Wellness tourism is also a good case for Greece, thanks to the ideal natural environment and climate for vacation all year round, as well as thanks to the large number of curative springs. Spa-thermal tourism was the first systematic effort for tourism development of the Greek state. To serve that cause, a regulatory framework was created since the beginning of the twentieth century together with actions taken to create suitable spa facilities. Since the 1970s and due to socioeconomic changes and changes in tourist preferences, spa-thermal tourism was identified with medical social tourism for the elderly Greek tourists which led to the stagnation of thermal towns. Wellness reemerges as a way to enrich the Greek tourism product after 2000. L3498/2006 tried to regulate the complex characteristics of a diverse sector, such as health and wellness tourism. The law defines spas as a category of spa-thermal tourism facilities. Spas should have the adequate infrastructure and equipment to use curative natural resources or heated sea water, natural thermal water with added curative resources or thermal water with added mud, herbs, plants, aromas, volcano or quarz sand, light, heat, massage, different types of saunas in order to provide rejuvenation, wellness services and body beauty treatments. However, applying the law was very delayed. In 2018, the Ministry of Tourism of Greece issued the Ministerial Decision No. 2704/2018 (OGG Issue No B 603) of the delegated law 3498/2006, which defines the terms and conditions, as well as the process, the necessary supporting documents and all other details for businesses that want to acquire an operation license for spa–thermal tourism facilities (Tsartas et al., 2019). Gradually, the business world started to keep up with the market trend in wellness tourism, and many hotels enriched their product with wellness services. However, Greece still has not got a separate category for wellness resorts.

3 Field Research Methodology

The conducted research helps investigating the opportunities to develop and promote wellness resorts in Greece, which has not yet been in the scope of systematic research in the country. Specifically, the research aims at recording trends and findings related to demand characteristics for wellness resorts and to draw conclusions that could be used in marketing plans by wellness tourism businesses and that would earn them more benefits and extend their customers' base.

Field research has been conducted at two levels: the first level of research inquired the viewpoint of Greek tourists about wellness resorts, and the second level of research inquired wellness resorts business owners about the specific characteristics of their customers. Two types of questionnaires were used for conducting the research, via typeform.

The first questionnaire was sent to general public-Greek people that usually take vacation, regardless whether they have visited or plan to visit a wellness resort in the future. The questionnaire was distributed in the time period between April 11, 2019, and April 19, 2019 via Viber and Messenger, and at the same time, it was posted on Facebook. The questionnaire addressed to tourists had 27 closed-format questions, and it can be separated in three parts which included: general demographic data, personal details of the questioned and information on their visit to a wellness resort. 240 duly completed questionnaires were collected.

The second questionnaire was sent via email to wellness resort owners inn Greece in May 2019. In order to track down which hotels were appropriate to answer the questionnaires and fulfill the criteria of wellness resorts, it was necessary to search them on the website of the Hellenic Chamber of Hotels, where all hotels in Greece are registered, together with the type of services they provide. 4* and 5* hotels that offer spa services which have been chosen from the database. According to the data up until the end of December 2016, which can be provided by the Hellenic Chamber of Hotels, there are 9730 hotels in Greece, out of which only 100 are 4* and 5* hotels that offer wellness services. The questionnaire was sent to 50 hotel owners and managers of 4* and 5* hotels with wellness services (50% of the total number), and 23 duly completed questionnaires were finally collected (23% of the total number). The questionnaire aimed at inquiring the profile of wellness tourists and included 10 closed-format questions.

4   Results of the Study

4.1   Research on General Public-Greek Tourists

General Demographic Data: Out of 240 participants in the study, 146 were women (that is 60.8% of the participants) and 94 were men (that is 39.2% of the participants). 51.3% is a group of 123 people at the age of 26–40 years old, and the next big group represents 34.2% and includes 82 people at the age of 41–65 years old. 9.2%—that is 22 people—are in the age group between 18 and 25 years old, and 13 people (which represents 5.4%) are older than 65. The majority of the questioned tourists, which is 45% (108 people), are married, and they have children, while 39% (94 people) are single. 12% are married (28 people), 3% (6 people) have children but are either widowed or divorced, and 1% (4 people) are either widowed or divorced with no children.

Income, Education and Occupation: 45% of the questioned—that is 108 people— have a monthly family income between 1001 and 2000 euros. Moreover, 53 people have a monthly income of less than 1000 euros, and 49 people have a monthly income between 2001 and 3000 euros. 9.6% (23 people) have a monthly income of

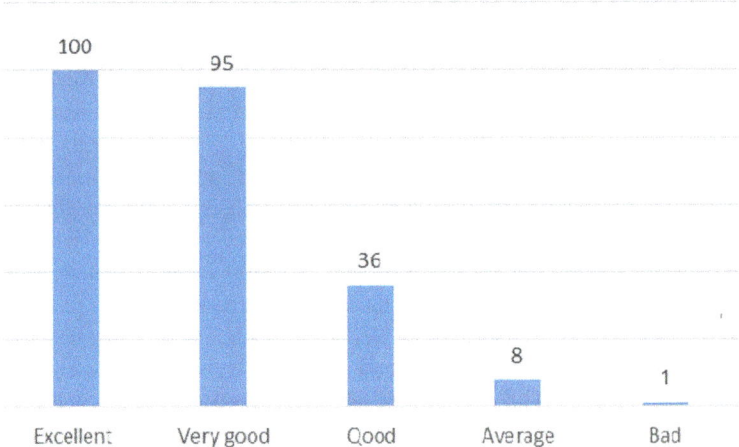

Chart 1 Health status

more that 4000 euros, and 7 people have a monthly income between 3001 and 4000 euros. 56.7% (136 people) of the questioned tourists have received a higher education (university degree). 27.5% (66 people) have a postgraduate degree, 14.2% (34 people) have a high school degree, and 1.7% have achieved secondary education. 35% of the sample (85 people) work in the private sector, and 25% (61 people) work in the public sector. 15% are self-employed (37 people), 8% (18 people) are students, 6% (16 people) are retired, 5% (11 people) are unemployed, 4% (8 people) are occupied in the household, and 2% (4 people) chose a different occupation.

Health State and Fitness Habits: It should be noted that over 2/3 of the people who responded to the questionnaire describe their health state as excellent. Specifically, as shown in Chart 1, 42% (100 people) think that their health is excellent, 40% (95 people) that it is very good, 15% (36 people) that it is good, 2% (8 people) that it is average, and only 1% (one person) thinks that their health is bad. On the question over, the frequency of fitness exercising (Chart 2), the 29% of people questioned, (69 people) answered that they do not exercise at all. 23% (55 people) exercise three times a week, 20% (48 people) two times a week, 15% (36 people) once a week, while 13% (32 people) exercise more than three times a week. When questioned which type of exercise they prefer, the majority answered that they prefer aerobics, since this is a type of exercise that has multiple advantages for the human body, and it can be found in most gyms. 11.2% prefer to use gym equipment, while 10.6% uses only weights. 9.4% prefers walking, while holistic fitness has become very popular and well-known in the recent years, with yoga being the first choice. Team sports are the choice of 6.1% of the people who answered the questionnaires, which shows how these sports are considered more suitable for younger people and children. Equally small is the percentage of those who prefer gymnastics, reaching

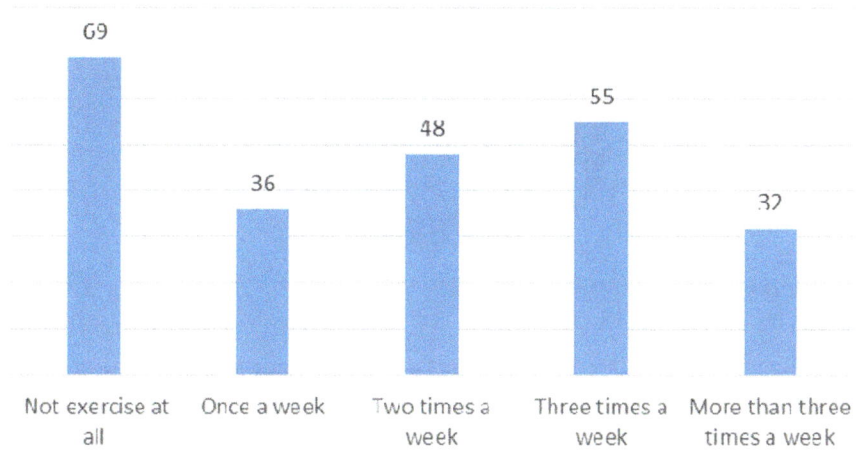

Chart 2 Frequency of exercise

only 4.5%, while some other type of exercise is preferred by a percentage of 3.6%. 75.4% of the people questioned are non-smokers, and 24.6% are smokers. As far as nutrition is concerned, 55% (132 people) adjust their habits to every season. 36.3% (87 people) have healthy nutrition habits, while only 7.5% believes that their habits are not healthy. 1.3% (3 people) cannot decide whether their nutrition is healthy or not. Only 7.5% of the questioned people do not have healthy nutrition habits. As to body weight, 172 people which represent 71.7% say they are of normal weight, while 24.6% (59 people) are overweight. There is nothing extraordinary to this result, since most people belong to one of these two groups. Only 8 people are underweight, and one is obese.

Touristic Choices: Most of the questioned people, i.e., 44% (105 people) enjoy their vacation once a year. 32% (76 people) are able to go on vacation twice a year, and 17% (42 people) travel more than twice a year. 7% (17 people) cannot afford to have a vacation. When questioned whether they have visited a wellness resort in the past (Chart 3) in Greece, the majority—83.8% (201 people) gave a negative answer, while only 16.3% (39 people) answered that they have already visited a wellness resort in the past. Those who gave a negative answer were then asked whether they

Chart 3 Visits to wellness resorts

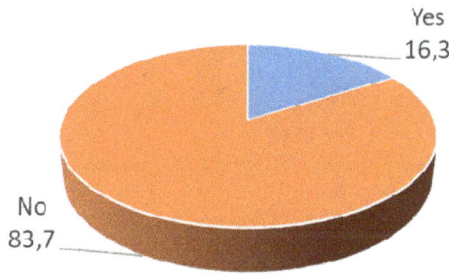

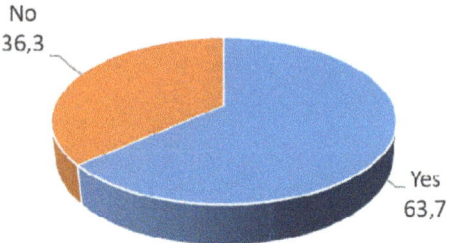

Chart 4 Intention to visit in the future

are willing to visit a wellness resort in the future. According to Chart 4, the majority (63.7%) gave a positive answer and 36.3% (73 people) said that they are not willing to visit a resort that offers wellness services. Then, the tourists that answered that they are not willing to visit a wellness resort were asked on the reasons they are negative toward this type of vacation. According to Chart 5, 31.5% (23 people) revealed that they are not interested in the services offered by a wellness resort. 30.1% (22 people) mentioned that they cannot visit a wellness resort due to economic reasons, and 22% (16 people) believe that the cost of such services is very high. 8.2% (6 people) mentioned family reasons, 2.7% (2 people) argued that they do not have the company that would follow them to a wellness resort, and 2.7% (2 people) said that they have not even thought of it and that they did not know the existence of such places. Finally, one person (1.4%) said that he/she does not have the time to visit a wellness resort, and another one (1.4%) said that he/she rather go to a destination simply for leisure.

At the end of this question, the tourists showed no interest in visiting a wellness resort receive a thank you message and their participation to the survey ends here. Thus, 167 out of 240 people continued to the next phase of the research, out of which 105 are women and 62 are men.

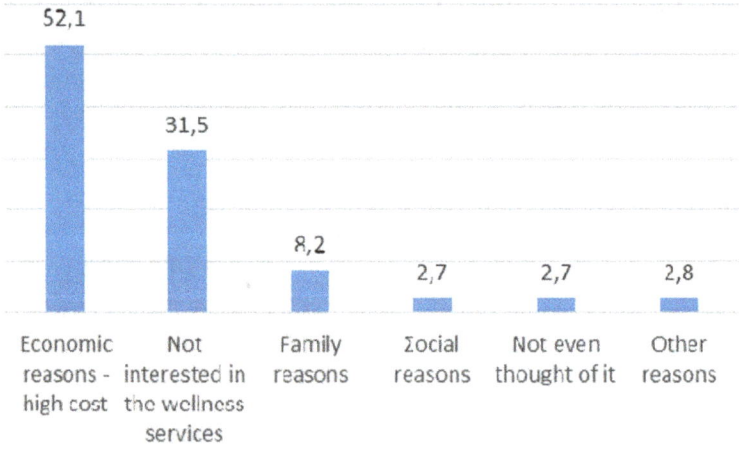

Chart 5 Reasons they do not wish to visit a wellness resort

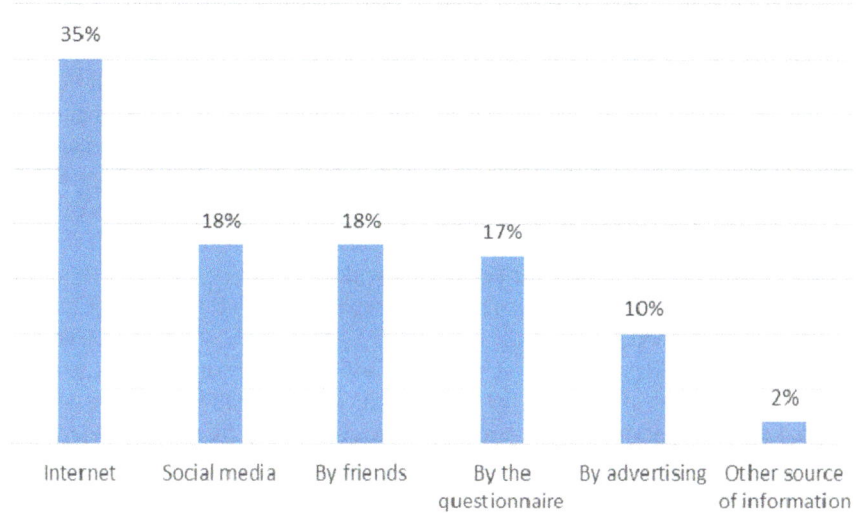

Chart 6 Information sources on wellness resorts

Tourist experience in wellness resorts: 35% of the sample having a previous experience in wellness tourism reported that they were informed through the Internet, 18% through the social media, 18% by friends, 17% by the questionnaire and 10% by advertising. 2% chose the option "other source of information," and more specifically, a person was informed because he/she works in tourism, one person was informed thanks to his/her studies, and one person knows about wellness resorts because of a TV show (Chart 6). The answers are indicative of the dynamics of the Internet and the social media in informing modern consumers on special touristic products. 75.4% (126 people) have never visited a wellness resort in Greece, 8.4% (14 people) have visited one only once, 7.8% (13 people) two times and 8.4% (14 people) have visited one more than twice. In total, 24.6%—which is not an insignificant percentage—have visited a wellness resort more than twice.

Motives, Options and Tourist Behavior in Wellness Resorts: When it comes to the motives of wellness tourists (Table 2 in the annex), most of the questioned answered relaxation (89.8%), seeking new experiences (88.6%), fighting stress (84.4%) and spa services (80.8%). Being in touch with nature also plays an important role for a percentage of 74.8%. Right after, the list of the most important criteria comes detoxing for a 62.3% of them, as well as detoxing from technology for 62.3%. More than half of the questioned tourists show interest in having self-awareness seminars (57.4%), in physical training (59.9%), but also in acquiring new good habits (58%). We can see that there is a wide range of tourists' needs that are covered by the services provided by a wellness resort. When recording the main services that the tourists seek in a wellness resort (Table 1 in the annex), most people wish to enjoy spas and massage treatments (80% and 88%, respectively), while a large number of people would like to participate in outdoor excursions. The use of sauna follows for a percentage of 60%, thalassotherapy for 59%, participation in group training for 59% and personal training in sports for 55%. Half of the questioned tourists would like to enjoy beauty treatments, while the same percentage would like to participate in activities such as skiing or kayaking. Anti-aging and cellularity treatments are of medium importance since half of the people questioned ask for them and half of them not so much. Less than half wish to attend cultural events (43%) or would dedicate time in visiting a museum or an archeological site (40%) or even take some cooking lessons (43%). These results show that in special touristic products there is significant interrelation among the services provided.

As to the amount of money they are willing to spend in wellness services (Chart 7), besides the accommodation expenses, 43.7% (73 people), are willing to spend 50–100 euros, 16.7% (28 people) 100–150 euros, 6.6% (11 people) more than 150 euros and 16.2% (27 people) less than 50 euros. Finally, 16.8% (28 people) are willing to spend any amount of money, as long as they are pleased by the quality and the level of services. Moreover, when it comes to the period, they would choose to spend their vacation in a wellness resort, most of the questioned chose spring, while—with a small difference—others said they would go any time during the year. 19% chose summer as the most suitable time to visit a wellness resort, and fall or winter

Table 1 Most successful national markets on wellness trips in 2017

Country	Number of wellness trips	Revenues from wellness trips
USA	176.5 MILLION	226 $BILLION
Germany	66.1 MILLION	65.7 $BILLION
China	70.2 MILLION	31.7 $BILLION
France	32.4 MILLION	30.7 $BILLION
Japan	40.5 MILLION	22.5 $BILLION
Austria	16.8 MILLION	16.5 $BILLION
India	56 MILLION	16.3 $BILLION
Canada	27.5 MILLION	15.7 $BILLION
United Kingdom	23.2 MILLION	13.5 $BILLION
Italy	13.1 MILLION	13.4 $BILLION

Source global wellness institute

Table 2 Age groups

Age	% of answers (%)
<30 y old	2.8
30–45 y old	44.4
46–65 y old	5
>65 y old	2.8

Chart 7 Money spent

are considered to be most appropriate by 13% of the questioned, respectively. 8% would visit a wellness resort during holidays. Most of the questioned people, who represent 53.3% (89 people) would stay in wellness resort for 3–7 days, while 35.3% (59 people) would rather stay for 2–3 days. 7–14 days are considered as the ideal vacation time in such a resort for 8.4% (14 people) and 3% wish to stay more than 14 days (4 people).

Development of Wellness Resorts in Greece: When asked if they believe that Greece has a potential for wellness resort development (Chart 8), the majority (77.10%) gave a positive answer. It should be noted that a significant percentage of 18% said that they could not give a clear answer. In the last question, tourists were asked on the reasons why wellness resorts are not so popular in Greece. The most popular answer given by 32.2% of the questioned tourists was that the cost of

Chart 8 Potential for wellness resort development in Greece

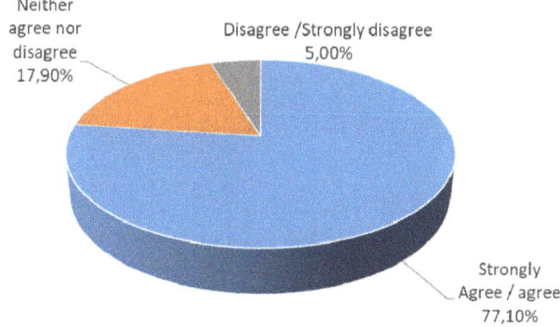

services provided is very high. 25.1% agrees to the opinion that there is lack of information regarding the services provided, and 13.1% thinks that there are other types of vacation that are more popular in Greece. 10.8% thinks that wellness tourism is not promoted enough by the state and 8.88% that people do not show much interest in this type of vacation. A small percentage of 8.36% believes that the reason that wellness resorts are not so popular in Greece is the lack of advertising and 1.56% gives other reasons.

4.2 Research on WELLNESS RESORTS Owners and Managers

Wellness Resorts Visitors Profile: According to the people questioned for the research, wellness resorts customers have a high level of income, education and social status. They are between 46–65 years old (Table 2) and are of both sexes, with women taking the lead. The majority of customers, which reaches 43.5%, spends 100–200 euros daily, 30.4% spend more than 300 euros, 21.7% less than 100 euros and 4.3%, 201–300 euros (Chart 9). According to the findings of the research, when it comes to nationalities of wellness tourists 23% are Greek, 20% Russian and British, respectively, followed by 12% German. At a percentage of 10% follow tourists from the USA and 8% come from France. The answer "other nationality" was filled with nationalities such as Balkans, Dutch and Polish and reached 5%. Last but not least Italians represent 2%.

Wellness Resorts Visitors' Choices: The research has shown that the most popular service (Chart 10), far from the rest, is massage, preferred by 32% of the customers. Right after that 19% of tourists prefer spa services and 16% beauty treatments. Anti-aging comes next with a percentage of 7%, as well as thalassotherapy and self-awareness with 5%, respectively. Personal fitness training is preferred by 3%

Chart 9 Average cost of staying

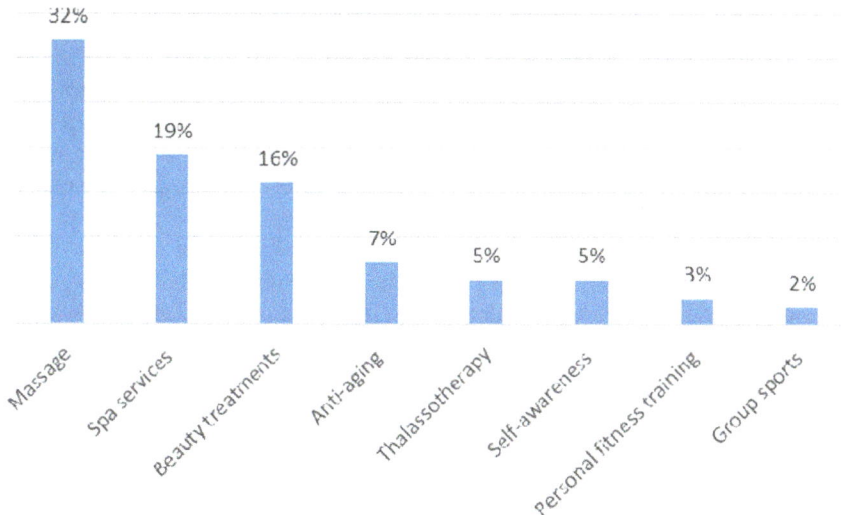

Chart 10 Services preferred

and group sports by 2% of the customers. For the next question, wellness resort owners and managers were asked whether visitors' choice of vacation that includes wellness services is intentional for the majority of them or random. The people questioned could choose among the three following answers: random, intentional and I do not know/no opinion. 60.9% answered that the choice of a wellness resort is intentional and only 13% answered that it is random. Finally, they were asked whether the customers are regular or not and 56.5% answered that the majority of wellness resorts visitors are in fact regular customers.

5 Conclusions, Policy Proposals

The statistical analysis so far has indicated that wellness and the form of tourism connected to it are very popular in the recent years. When taking into consideration the huge growth of trips related to wellness tourism, as well as the profits to be gained according to the Global Wellness Institute, it is evident that action should be taken in order to make Greece a world wellness destination. Moreover, in order to reinforce this form of tourism, both incoming and domestic tourists should be attracted.

According to the data of the research conducted to the general public, a large number knew neither wellness resorts nor any of the services they offer, which is rather strange considering the quick access to information one has nowadays. Therefore, this result could be attributed to insufficient information or lack thereof, as well as to a mistaken conception of the tourists that wellness resorts address to, or the type and cost of services provided therein. In any case, the number of people who knew and have used wellness resort services is also significant. The research also revealed the dynamics of the Internet and the social media in informing modern consumers on special touristic products. Moreover, the results also indicate a strong correlation of the provided services among special tourism products, such as wellness tourism, gastronomy and cultural tourism.

According to the research conducted in wellness resorts, the majority of the customers spend 100–200 euros or more than 300 euros on a daily basis. There is no significant difference between female and male visitors, while most of them are between 30 and 65 years old. Customers most likely have a higher education, income and social level; the most frequent visitors are Greek, British, Russian or German. The most popular services are massage, spa and beauty treatments, and the majority of tourists choose a wellness resort intentionally and visit it regularly.

The research showed that many Greek people ignored wellness resorts for many reasons. However, the majority believes that wellness resorts can be well developed in Greece. Therefore, since there is an appropriate background, some actions need to be taken to promote and advertise wellness resorts to the Greek public and to incoming tourists, such as

- Recording all hotels that fulfill the criteria for being wellness resorts and creating a special "wellness resorts" label.
- Promoting wellness resorts as a separate touristic product which should have a specialized marketing plan; special portals should be created to address to wellness tourists; participation in exhibitions related to wellness or tourism is also important.
- Promoting a multilevel destination marketing plan where wellness resort facilities and services is offered together with other specialized products (culture, gastronomy).
- Updating—and completing where necessary—the legal framework for the operation of the facilities where wellness services are provided.
- Promoting synergies among businesses with travel agents which manage, through mass bookings, to achieve better prices and better promotion of the product.

Annex

See Tables 3 and 4.

Table 3 Wellness resort visitors' motivation

Visitors' motivation	Number of people related to importance					% of people related to importance				
	No importance	Little importance	Medium importance	A lot of importance	Great importance	No importance (%)	Little importance (%)	Medium importance (%)	A lot of importance (%)	Great importance (%)
Relaxation	1	3	13	44	106	0.6	1.8	7.8	26.3	63.5
Self-awareness	6	16	49	51	45	3.6	9.6	29.4	30.5	26.9
Physical training	8	11	48	57	43	4.8	6.6	28.7	34.2	25.7
New experiences	1	5	13	42	106	0.6	3	7.8	25	63.6
Spa services	1	6	25	57	78	0.6	3.6	15	34.1	46.7
Weight loss	31	33	40	34	29	18.5	19.7	24	20.4	17.4
Detox	21	28	33	60	45	12.6	4.8	19.8	35.9	26.9
Adopt new habits	15	13	42	51	46	9	7.8	25.2	30.5	27.5
Solve sleeping problems	35	17	44	28	43	21	10.2	26.3	16.8	25.7%
Stop smoking	79	14	30	17	27	47.2	8.4	18	10.2	16.2
Fight stress	5	4	17	53	88	3	2.4	10.2	31.7	52.7

(continued)

Table 3 (continued)

Visitors' motivation	Number of people related to importance					% of people related to importance				
	No importance	Little importance	Medium importance	A lot of importance	Great importance	No importance (%)	Little importance (%)	Medium importance (%)	A lot of importance (%)	Great importance (%)
Rest	1	–	9	35	122	0.6	–	5.4	21	73
Contact with nature	3	6	33	52	73	1.8	3.6	19.8	31.1	43.7
Detox from new technology	9	13	41	39	65	5.4	7.8	24.5	23.4	38.9
Make new friends	22	27	51	42	25	13.2	16.2	30.5	25.1	15

Table 4 Services required by tourists in wellness resorts

Type of service	Number of people related to level of interest					% of people related to level of interest					
	Not interested	A little interest	Medium interest	Very interested	Great interest	Not interested (%)	A little interest (%)	Medium interest (%)	Very interested (%)	Great interest (%)
Spa	2	5	26	38	96	1.2	3	15.6	22.7	57.5
Massage	3	4	13	44	103	1.8	2.4	7.8	26.3	61.7
Sauna	14	9	42	46	56	8.4	5.4	25.2	27.5	33.5
Anti-aging therapies	22	25	39	32	49	13.2	15	23.4	19.2	29.2
Cellularity therapies	33	17	35	37	45	19.8	10.2	21	22	27
Injury recovery	32	20	45	34	36	19.2	12	27	20.3	21.5
Thalassotherapy	12	17	39	40	59	7.2	10.2	23.3	24	35.3
Beauty treatments	33	18	33	25	58	19.8	10.8	19.8	15%	34.6
Personal sports training	18	17	40	38	54	10.8	10.2	24	22.7	32.3
Group sports training	20	18	30	37	62	12	10.8	18	22	37.2
Sports activities such as skiing, kayaking	23	21	38	38	47	13.8	12.6	22.7	22.7	28.2

(continued)

Table 4 (continued)

Type of service	Number of people related to level of interest					% of people related to level of interest					
	Not interested	A little interest	Medium interest	Very interested	Great interest	Not interested (%)	A little interest (%)	Medium interest (%)	Very interested (%)	Great interest (%)
Cultural events	18	33	44	44	28	10.8	19.8	26.3	26.3	16.8
Visits to museums	26	27	46	38	30	15.6	16.2	27.5	22.7	18
Outdoor excursions	6	11	30	47	73	3.6	6.6	18	28.1	43.7
Cooking lessons	31	31	33	37	35	18.6	18.6	19.8	22	21

References

Chen, J. (2012). *Advances in hospitality and Leisure.* H.T: Bingley.

Coccossis, H., Tsartas, P., & Gkrimpa, E. (2011). In G. Zacharatos, & P. Tsartas (Eds.), S*pecial and alternative forms of Tourism: Demand and offer of new touristic products.* Kritiki Publications. ISBN: 9789602187241.

Global Wellness Tourism Economy, 2018 downloaded on 03.05.2019 Retrieved from https://global wellnessinstitute.org/industry-research/global-wellness-tourism-economy/

Hellenic Official Government Gazette, issue No. 208/ 11–12–2018, L 4582, "Thematic Tourism".

Huang, L., & Xu, H. (2014). A cultural perspective of health and wellness tourism in China. *Journal of China Tourism Research, 10*(4), 493–510.

Sarantakou, E. (2017). Approaches for the sustainable regeneration of mature destinations, chapter in the collective volume on Greek Tourism State-of-the-Art 'Tourism, Tourist Development. In P. Tsartas, & P. Lytras (Eds.), *Greek Scientists' Contributions* (pp. 201–213). Athens: Papazissis Publishers. (ISBN 978–960–02–3309–4).

Smith, M. K., & Puczkó, L. (2017). *The routledge handbook of health tourism.* New York: Routledge.

Tsartas, P. (2010). *Greek tourism development: Characteristics, research, proposals.* Kritiki Publishing. ISBN: 9789605860066.

Tsartas, P., Stavrinoudis, T., Sarantakou, E., Kontis, A., & Skoultsos, S. (2019). Spa and wellness tourism development: Cases from three Greek islands. In *3rd International Scientific Conference TOURMAN 2, Thessaloniki, Greece, Minutes in hard copy*

Digital Innovation

Policy Responses to Critical Issues for the Digital Transformation of Tourism SMEs: Evidence from Greece

Panagiota Dionysopoulou and Konstantina Tsakopoulou

1 Introduction

Digitilisation affects all stages of the tourism value chain and creates a new context for tourism SMEs to operate, as it transforms the way businesses organize their work and communicate with tourists, requires new skills by the tourism workforce, creates innovative tourism products, and enhances the visitor's experience (European Commission, 2017, Oberländer et al., 2020). It is estimated that $100 billion of industry value will migrate from traditional tourism businesses (travel booking intermediaries and traditional hotels) to Online Tour Operators (OTAs) and short-term rentals by 2025 (World Economic Forum, 2017: 11). These trends are accelerated amid the COVID-19 pandemic, stressing the need for coherent policies on the digital transformation of tourism SMEs.

Digitalization has therefore created a new business environment for tourism SMEs, one where they are struggling to remain competitive. The paper attempts to investigate the challenges facing tourism SMEs in the digital era and to review policy responses by illustrating the case of Greek tourism SMEs. Despite the fact that SMEs form the backbone of the Greek tourism industry, research on their adjustment to the changing landscape of the digital transformation era is scarce. This paper attempts to fill this gap. In the next section, a thorough review of the academic literature and recent reports at EU and OECD level identifies crucial issues concerning

P. Dionysopoulou (✉)
Director General of Tourism Policy, Ministry of Tourism, Athens, Greece
e-mail: dionysopoulou_p@mintour.gr

School of Social Science, Adjunct Professor, Post-graduate courses "Tourism Business Administration Msc", Hellenic Open University, Patras, Greece

K. Tsakopoulou
Head of Research Studies and Documentation, Ministry of Tourism, Athens, Greece

Department of Geography, Phd candidate, Harokopio University, Athens, Greece

the digital transformation of the tourism value chain. We then go on to present the main findings of research undertaken in autumn 2019, comprising a series of in-depth interviews held with key informants in Greek tourism industry associations and in the tourism departments of Regional Authorities. Interviews were supplemented with desk research of official policy documents and governmental and industry associations' sites. Research clarified Greek tourism SMEs' needs and revealed the internal and external barriers in their way to digital transformation, reviewed ongoing policies, and suggested new policy proposals. The paper concludes by relating results to the outcomes of the European Commission reports (2018, 2017) and the OECD report (2020) on the digital transformation of tourism SMEs at European and global level.

2 Literature Review: A Changing Landscape for Tourism SMEs

From the supply side perspective, digitalization has disrupted the distribution and marketing of the tourism product. The internet has permitted tourism businesses to dispense with powerful intermediaries, such as tour operators or Global Distribution Systems, and market through their own brand website (Hjalager, 2010: 3). At the same time, new online intermediaries have emerged (OECD 2020, Navío-Marco et al., 2018: 464, European Commission 2017). Artificial intelligence (AI) and big data analysis permit dynamic packaging and provide powerful insights on consumer behavior and business transactions, facilitating real-time decision making and helping businesses to structure the right offer (OECD, 2020, World Economic Forum, 2019: 17, Zsarnoczky, 2017: 87, 2018: 2; Boes et al., 2016). Web marketing has revolutionized communication with the client. Email, mobile-ready websites, social media, instant chat are used to improve the way businesses market their products and create loyalty to the brand (Navío-Marco et al., 2018: 465; European Commission, 2017; Law et al., 2014: 742). Internet banking, financial software, flexible home working by using the Cloud and digital collaboration tools, and interacting with governments online help small businesses save money and time and improve cost efficiency (European Commission, 2017). Automation helps tourism businesses to simplify their internal processes and increase productivity. Chatbots, for example, are a convenient form of communicating information to the customer (Zsarnoczky, 2017: 87, 89), whereas Property Management Systems (PMS) manage day-to-day hotel operations, from maintenance to booking (EPRS, 2018: 3). Digital and physical convergence applications, like the internet of things, where devices are connected to the internet and are capable of two-way communication, and augmented reality applications, where digital elements are projected into a real time space, create innovative tourism products (Kontogianni and Alepis, 2020: 6–7, Zsarnoczky, 2018). These include new tourism services, such as e-cars and e-scooters which require neither facilities nor personnel, as well as innovative experiences like real-time personalized tour guides (OECD, 2020; Happ and Ivancsó-Horváth, 2018: 13–14; EPRS, 2018:3).

From the demand-side perspective, digital transformation has revolutionized all stages of the touristic experience, from information search to the consumer's decision to purchase and the actual visit (Law et al., 2014). Online booking systems have shortened the whole process of ordering and paying for transportation, accommodation, transfer, and on-site activities. Prices are comparable and tourists make new bundles of the goods and services they need (OECD, 2020, Navío-Marco et al., 2018: 461, Zsarnoczky, 2017: 87). Decision support systems and context-aware systems create personalized and context-specific offers for travelers before, during, and after their stay (Zsarnoczky, 2018: 4–5, Buhalis and Amaranggana, 2015: 385, Gretzel et al., 2015a: 181). User-generated content is a great influence on consumers' decision-making, as tourists trust their peers for information rather than the company's marketing strategy (Navío-Marco et al., 2018: 463, Simeon and Martone, 2016: 682; Law et al., 2014: 741).

As their digital literacy increases, consumers have also become actively involved in the tourism business by offering tourism services to other consumers via digital platforms, giving rise to the concept of the prosumer, which refers to tourists co-creating value with the industry (Navío-Marco et al., 2018: 461–462; Richards, 2015). This is particularly evident in the sharing (or collaborative) economy model. An increasing number of people are willing to "share" what they own (their house or their car) or what they do (meals and excursions) and offer their spare capacities for collective use to maximize the exploitation of their resources and efficiency gains (WEF, 2019: 9, EPRS, 2018: 4, Zsarnoczky, 2018).

Nowadays, however, the collaborative economy includes tourism services provided by private people as well as businesses (EC, 2016: 3). The concept of the digital business ecosystem has been used to describe the new business model: new technologies help create loose networks of businesses, consumers, and organizations with global reach, that are not bound by hierarchy, but by symbiotic, yet at the same time antagonistic relationships (Kelly, 2015: 5; Gretzel et al., 2015b; WEF, 2019: 14). Digital platforms have come to dominate the tourism business ecosystem. Platforms shift production from inside the firm to the outside, as they coordinate and sell value created by users (OECD, 2020, WEF, 2019: 8–9, Zsarnoczky, 2018: 3). In TripAdvisor, Uber, and Airbnb, users create value for other users, attracting more users and increasing value, which in turn attracts even more users and so on. The modularity of the tourism product (a stay, a ticket, a tour, a meal) facilitates new entries, so that new service providers may easily join the business ecosystem to add their value (Gretzel et al., 2018). Such an approach however does not account for the numerous difficulties encountered by SMEs collaborating with platforms (Zsarnoczky, 2017: 87).

A major source of concern for policy making is the place of tourism SMEs in this changing landscape. SMEs, in general, are considered more flexible and adaptable to change than large organizations, so the speed, transparency and real-time exchange of knowledge offered by the electronic environment may act in their favor (Bocarando et al., 2017). In this context, policy reports and the academic literature both emphasize the opportunities offered to tourism SMEs in the digital era. Digitilisation allows small players a more equal exposition to the market together with industry leaders (Hjalager, 2010: 3) and enables SMEs to access distant markets and market niches at low cost, and, eventually, to attain scale without the negative effects of mass tourism (OECD, 2020; Dredge et al., 2018: 19). Electronic commerce entails benefits such as increased sales, cost reduction, time savings, better customer service, and productivity gains (Bocarando et al., 2017: 436). By analyzing online user-generated content, tourism operators understand the tourists' profile, adjust their offerings appropriately and manage visitors' satisfaction and business reputation by replying to critique and eliminating flows in the service (Simeon and Martone, 2016: 685).

Despite perceived opportunities, tourism SMEs are lagging behind in the digital era (OECD, 2020, Dredge et al., 2018: 18, European Commission, 2017 and 2019). While the digital gap between large and small tourism businesses has diminished in most countries when simple digital services such as web presence are concerned, it remains pertinent when it comes to adopting more advanced technologies like cloud computing. There is also a growing concern over the position of digital native tourism SMEs in the business ecosystem, as consolidation is taking place in innovative companies like e-scooters operators (Crowe, 2020). Barriers impeding SMEs in their way to digital transformation have to do with structural peculiarities of the tourism sector (such as fragmentation or demand fluctuations), as well as internal features of the tourism businesses themselves (like small size and ownership characteristics), both determinants being closely interconnected to each other (Najda-Janoska and Kopera, 2014).

Empowering tourism SMEs to adopt new technologies is increasingly becoming a strategic priority for governance. International bodies issue policy guidelines that have to do with connecting stakeholders to information and collaboration networks, training policies for employees and managers alike and enhancing digital awareness, enabling SMEs to access and process big data (OECD, 2020, Dredge et al., 2018).

3 Methodological Approach

Data on the digital transformation of tourism SMEs in Greece and on relevant policies are scarce. To gain insight into the industry and identify ongoing policy initiatives, we reviewed official policy documents as well as governmental and industry associations' sites. Relevant statistical data was provided by the Central State Aid Unit on EU funding in the framework of the Partnership Agreement 2014–2020. Desk research was coupled with in-depth interviews, conducted in autumn 2019, with key informants in the main Greek tourism industry associations at a national level and in the competent tourism departments of Regional Authorities to grasp eventual spatial differences (Table 1). Interview requests were made to all 13 tourism departments of Regional Authorities and seven interviews were eventually conducted. Research needs to be supplemented, at a later stage, by research in Greek tourism SMEs.

Table 1 List of key informants that have participated in primary research, autumn 2019

List of key informants		
Type	Key informants	Code
Industry stakeholders	Hellenic Chamber of Hotels	KI01
	Hellenic Association of Hotels	KI02
	Hellenic Association of Hotel Managers	KI03
	Institute of the Greek Tourism Confederation	KI04
Regional authorities	Region of Western Macedonia	KI05
	Region of Eastern Macedonia – Thrace	KI06
	Region of Attiki	KI07
	Region of Crete	KI08
	Region of Western Greece	KI09
	Region of Epirus	KI10
	Region of Peloponnese	KI11

A short list of five open-ended questions, regarding the main subject areas of the interview, was first sent by email to all participants, to encourage informants to share their experience on the subject. Interviews were then conducted over the phone, following an interview guide, varying from around 40 min to one hour. In some cases, input was received in writing. Written input was also received by the Ministry of Tourism of Greece, Directorate of Education, regarding governmental educational policies on the matter.

The main aim of the interviews was to explore the advance of digital transformation of Greek tourism SMEs and review related policies. Informants gave their opinion on the type of policy interventions most needed to effectively enhance the digital transformation of tourism SMEs in Greece and to describe policy initiatives already in place.

4　Results

4.1　Opportunities and Challenges to the Digitalization of Greek Tourism SMEs

Stakeholders acknowledge potential benefits from digital transformation, related primarily to access to distant markets and market niches. Sales through the brand website and social media marketing enhance visibility, integrate SMEs in global value chains (KI07), and provide an additional source of profit (KI03, KI04, KI05, KI06, KI08). However, it remains an issue whether small and medium-sized hotels' websites can reach first page results in search engines (KI01). Big data analysis offers insights to customers' profile (KI01, KI04), resulting in customer satisfaction and repeat customers (KI08). Data analysis is also helpful to overview and to assess competition (KI06, KI07). Practical benefits refer to better every day management and time saving through automation, so that employees may be occupied in customer service (KI01, KI02, KI07).

Despite the fact that the importance of digital tools is widely acknowledged, Greek tourism businesses present a varying degree of digitalisation depending on the type of technology and their size. Inputs show that while the gap between small and large firms has narrowed when it comes to web presence and connectivity, it remains prevalent when it comes to more advanced technologies like data analytics, channel management, PMS or virtual and augmented reality (KI01, KI03, KI04). Some enterprises feel that the use of more advanced tools is not required and "doesn't pay off" (KI04). Tourism SMEs that provide online booking and payment from their own website, substituting major aggregators, is a strong trend in Greece lately, as is marketing on social media (KI09). There is also a varying degree of digital awareness: small and medium-sized hotels may be able to successfully maintain their online presence in social media, while at the same time they may not understand the importance of using the cloud to manage their files (KI02).

Barriers have to do with internal and external limitations. Stakeholders stress that the small size of tourism businesses incurs a number of resource shortages, particularly lack of funds and competent staff, which are considered the main obstacles in their way to digital transformation (KI01, KI02, KI03, KI06, and KI07). Lack of digital skills and difficulties to recruit properly trained staff are widely mentioned. Stakeholders point out the lack of appropriate and continuous training programs and of digitally skilled labor supply (KI03, KI08, KI11). Insufficient know-how obscures the gains to be made by the digital upgrade of the tourism product (KI07, KI09).

Another issue is the high cost of bandwidth internet and of the relevant equipment (KI06), as well as poor infrastructure for fast and reliable internet connectivity, especially in remote areas (KI09, KI10). Lack of consistent cooperation between public authorities at all spatial levels that would effectively assess the needs and coordinate policies to adopt digitally enhanced business models by traditional tourism SMEs was also identified as a crucial barrier (KI05, KI09).

In the following, we provide an outline of policies undertaken by private and public bodies at different spatial levels that aim to address these issues.

4.2   Policies to Provide Financial Assistance

Despite the decrease of ICT costs (European Commission, 2017: 58), interviews have confirmed that technology solutions still involve a considerable cost for Greek tourism SMEs.

A number of sectoral and regional Operational Programs have focused on the digital transformation of tourism SMEs within the framework of the Partnership Agreement 2014—2020. At central government level, the Operational Program

Table 2 Number of tourism SMEs financed

Number of tourism SMEs financed, by business size			
Total expenditure on software and software services			

Operational Programme "Competitiveness, Entrepreneurship and Innovation" (EPAnEK), Partnership Agreement 2014–2020			
Project 1: Supporting the creation and operation of new tourism SMEs			
Size	Number of proposals	Total expenditure	(Of which) public spending
Micro	1,801	5,889,935.27	2,943,494.98
Small	70	235,579.00	117,614.50
Medium	6	23,450.00	11,725.00
Project 2: Supporting tourism SMEs to upgrade the quality of their services			
Size	Number of proposals	Total expenditure	(Of which) public spending
Micro	537	2,334,992.73	1,121,500.81
Small	145	865,444.58	415,492.65
Medium	38	317,304.08	147,079.26

Source Management Organization Unit of Development Programs SA, Central State Aid Unit

"Competitiveness, Entrepreneurship and Innovation" (EPAnEK), has funded two targeted projects to support new and existing tourism SMEs, supported by the ERDF (European Regional Development Fund) and the ESF (European Social Fund) (Blake et al., 2006). Both projects covered the entire tourism sector. Funding included the provision of software (including cloud computing and software as services). Results are presented below (Table 2). 90% of businesses financed for software and software services were microbusinesses (less than 10 employees) with a total expenditure of 8,224,928 euros. Most newly established tourism SMEs were situated in the South Aegean Region (326), the top tourism destination of the country, while a considerable number operate in the wider areas of Athens (Attiki, 253) and of Thessaloniki (Central Macedonia, 216).

Funding has also been available through Regional Operational Programs. In the region of Epirus, for example, financial support was provided for SMEs of all sectors to enhance e-business (KI10). A considerable number of these investment plans belong to tourism SMEs (108 out of 124). The majority deal with accommodation (67 in total), while a great part has to do with travel agencies, food service activities, and rent a car. Eligible costs include online trading and booking, digitalization of business organization, and production. In Western Macedonia funding aimed at enhancing innovation in existing and new micro, small and medium tourism businesses (KI05).

4.3   Policies to Build Digital Competence and Innovation

The need for continuous training to foster the digital skills of the tourism workforce is acknowledged by the majority of key informants. Discussion revealed the importance attributed to tourism education in general. Stakeholders stressed the role of state-managed Higher Schools of Tourism Education in Rhodes and Crete, that have proved successful in the past and could be self-financed (KI02). Post-secondary tourism education is considered pivotal as well, as it is a time when families and young people commit themselves to following a certain career (KI02). Programs should be upgraded, so that they will no longer be a forced choice for pupils who do not succeed in their exams to enter universities, which results in few graduates actually staying in the tourism workforce. Public–private partnerships with tech leaders in the sector, such as Google and online tour operators, are also put forward as a means to provide training in specialized areas like digital marketing (KI04). In this framework, the GNTO in cooperation with Google Hellas supports training in social networking of small-scale tourism entrepreneurs (Annual Report 2018: 17).

In response to market needs, Institutes of Vocational Training managed by the Ministry of Tourism, have adjusted their curriculum from 2019 onwards to comprise the specialty of "Management and Tourism Economy Executive", training students to the use of all essential travel market applications (from booking platforms to Content Management Systems). The Ministry of Tourism also organizes courses of Continuing Vocational Training addressed to employees in hotels, casinos, restaurants, and leisure centers that have been recently restructured in 2018 to comprise courses on hotel Property Management Systems, online reputation management, and channel management. E-learning has also been available for students during the COVID 19 pandemic lockdown (Bocarando et al., 2017).

Regions have also undertaken initiatives to support training. For example, the Strategic Action Plan for the Tourism Promotion of the Region of Attiki (KI07) promotes training programs addressed to tourism professionals. The region has also signed a Memorandum of Cooperation with the Hellenic Chamber of Hotels, to provide a roadmap for hosting scientific tourism conferences and promoting research synergies for the upgrade of tourism strategic planning.

A key challenge is the diffusion of knowledge. Some stakeholders put forward the need to connect businesses and form networks. Networks offer the chance to exchange know-how and information and group scarce resources to provide the much-needed technological tools and organize training. In the framework of Partnership Agreement 2014–2020, the Region of Attiki is planning to support the creation of innovation clusters within the region (KI07).

To support Greek tech start-ups offering products for the tourism sector and bring innovation to the market, the Hellenic Chamber of Hotels has recently taken the initiative to organize CapsuleT, the first travel and hospitality accelerator for Greek tourism startups (KI01). The aim is to provide hands-on learning opportunities and tools for the participating startups, as well as guidance on how to scale up their business. The project includes workshops, mentoring sessions, pitching opportunities, industry site visits, and networking activities (Boes et al., 2016).

4.4   Policy Initiatives for Market Intelligence and Promotion

Resource shortages hamper tourism SMEs efforts to unlock the possibilities of digital marketing. To tackle these issues local DMOs promote tourism SMEs operating in their area through their website (KI07).

Stakeholders have also stressed the need for timely and relevant data to assist tourism SMEs and destinations in understanding their customers and adjusting their services accordingly (KI01, KI06). The main barriers reported (KI06, KI07, KI08) concern the lack of know-how to collect and analyze such data. Insufficient understanding of the EU General Data Protection Regulation (GDPR) is also a problem, as well as safety issues concerning keeping and managing big data. However, GDPR is equally thought of as a challenge to train staff properly, design tourism products that respect private data, and prepare a plan to deal with violations of privacy. In the accommodation subsector, the Hellenic Chamber of Hotels cooperates with the private sector to offer Greek hotels and camping sites monthly data analysis of their social media reviews. The project also provides the Chamber with extended destination analysis to track long-term trends in the Greek tourism industry (KI01).

4.5   Policies to Enhance E-governance for the Tourism Sector

Stakeholders underline the importance of e-governance that would reduce costs incurred by administrative burdens. To address these issues, two projects are underway at central government level. The first project aims at upgrading the information services provided by the Ministry of Tourism to investors and has been underway during the last semester of 2018 in the framework of Operational Programme "Competitiveness, Entrepreneurship and Innovation" (EPAnEK). The project creates a comprehensive investor information system for spatial planning, that will inform investors on the urban, environmental, institutional, and regulatory framework in areas of tourism interest and on the existing investment environment in these areas. The project reaches a total budget of 985.854,00 €.

5   Conclusions

Although all key stakeholders embraced the opportunities offered by the digital transformation of Greek tourism SMEs, they identified a number of critical issues that require a prompt policy response, particularly the lack of adequate resources, traditional managerial practices, the scarcity of knowledgeable workforce, the absence of appropriate infrastructure and a need to enhance e-governance. Results are in line with the main barriers put forward in the academic literature and with the findings of the European Commission's reports (2018 and 2017) and the OECD report (2020) on the digital transformation of tourism SMEs at European and global level.

Greek tourism SMEs face similar problems with their European and global counterparts. Policies by central government, regional authorities, and industry stakeholders attempt to tackle these issues. They provide funding for investments in digital tools, training courses through state-managed schools or via partnerships with private sector tech experts, online promotion tools through DMOs' websites, e-governance tools for B2G interaction in the tourism sector, tools for big data analysis to improve market intelligence, incubator services to connect start-ups to the tourism market. On the whole, these policies adopt a realistic perspective and remain focused on the particular needs of extant Greek tourism SMEs with the aim to foster their digital competence.

An even greater emphasis must be put on producing relevant, coherent, and timely policies for digital transformation amid the COVID 19 pandemic. Policy planners can benefit from lessons learnt from global practice by keeping in mind the following principles (OECD, 2020; European Commission, 2017, 2019): the need to prioritize small and medium-sized businesses during policy making ("think small first") and to adjust policies to the structural limitations of tourism SMEs, the importance of integrating place and spatial parameters in policy making as neither digitalization nor the impact of the pandemic are felt the same in urban, rural and insular destinations, and, last but not least, the need to adopt the view of tourism businesses rather than tech companies when planning digital transformation for the tourism sector.

Acknowledgements Research has been undertaken within the framework of the Directorate General of Tourism Policy, Ministry of Tourism of Greece.

References

Ar, K., & Ef, A. (2020). Smart tourism: State of the art and literature review for the last six years. *Array, 6,* 1–12.

Blake, A., Sinclair, Th., & Campos Soria, J. A. (2006). Tourism productivity: Evidence from the United Kingdom. *Annals of Tourism Research, 33*(4), 1099–1120.

Bocarando, J. C., Perez, J. C., & Ovando, C. (2017). Tourism SMEs in a digital environment: Literature review. *European Scientific Journal, 13*(28), 429–451.

Boes, K., Buhalis, D., & Inversini, A. (2016). Smart tourism destinations: ecosystems for tourism destination competitiveness. *International Journal of Tourism Cities, 2*(2), 108–124. ISSN20565607, https://doi.org/10.1108/IJTC1220150032. Retrieved from https://centaur.rea ding.ac.uk/75357/. Accessed July 4, at 09:00.

Buhalis, D., & Amaranggana, A. (2015). Smart tourism destinations: Enhancing tourism experience through personalisation of services. In I. Tussyadiah & A. Inversini (Eds.), *Information and Communication Technologies in Tourism* (pp. 377–389). Switzerland: Springer International Publishing.

Calvino, F., Criscuolo, Ch., Marcolin, L., & Squicciarini, M. (2018). A taxonomy of digital intensive sectors. In *OECD Science, Technology and Industry Working Papers, No. 2018/14*. Paris: OECD Publishing. https://doi.org/10.1787/f404736a-en.

Chiappa, G., & Baggio, R. (2015). Knowledge transfer in smart tourism destinations: Analyzing the effects of a network structure. *Journal of Destination Marketing & Management, 4*(3), 145–150.

European Commission. (2016). *A European agenda for the collaborative economy. Communication from the Commission to the European Parliament, the Council, the European Economic and Social Committee and the Committee of the Regions.* COM (2016) 356 final, Brussels.

European Commission. (2017). *Management and Content Provision for ICT and Tourism Business Support Portal.* Retrieved from https://ec.europa.eu/growth/tools-databases/vto/. Accessed July 6, 2020, at 13:00.

European Commission. (2019). *18th European Tourism Forum: Digital transformation as the engine of sustainable growth for the EU tourism sector.* Concluding Document, 9–10 October 2019, Helsinki, Finland.

Crowe, C. (2020). *Seven trends that will define smart cities in 2020. Smart City Dives.* January 2 2020. Retrieved from https://www.smartcitiesdive.com/news/7-trends-that-will-define-smart-cit ies-in-2020/569471. Accessed January 10, 2020, at 08:00.

Dredge, D., Phi, G., Mahadevan, R., Meehan, E., & Popescu, E. S. (2018). *Digitalisation in Tourism: In-depth analysis of challenges and opportunities. Executive Agency for Small and Medium-sized Enterprises (EASME).* Virtual Tourism Observatory, Aalborg University: Copenhagen. https://ec.europa.eu/docsroom/documents/33163/attachments/1/translations/en/renditions/native.

European Parliamentary Research Service (EPRS) (2018). *Digital tourism in the European Union,* Briefing. Retrieved from https://www.europarl.europa.eu/RegData/etudes/BRIE/2018/628236/EPRS_BRI(2018)628236_EN.pdf. Accessed July 6, 2020, at 14:00.

Femenia-Serra, F., & Ivars-Baidal, J.A. (2019). DMOs: Surviving the Smart Tourism Ecosystem. In *Travel and Tourism Research Association Conference, European Chapter,* Bournemouth (United Kingdom), 8th-10th April.

Gretzel, U., & Scarpino Johns, M. (2018). Destination resilience and smart tourism destinations. *Tourism Review International, 22*(3), 263–276.

Gretzel, U., Reino, S., Kopera, S., & Koo, Ch. (2015a). Smart tourism challenges. *Journal of Tourism, XVI, 1,* 41–47.

Gretzel, U., Sigala, M., Xiang, Z., & Koo, C. (2015b). Smart tourism: Foundations and developments. *Electronic Markets, 25*(3), 179–188.

Hjalager, A. M. (2010). Progress in Tourism Management. A review of innovation research in tourism. *Tourism Management, 31*(1), 1–12.

Kelly, E. (2015). *Business ecosystems come of age.* Business Trends Series: Deloitte University Press.

Law, R., Buhalis, D., & Cobanoglu, C. (2014). Progress on information and communication technologies in hospitality and tourism. *International Journal of Contemporary Hospitality Management, 26*(5), 727–750.

Martínez-Román, J., Tamayo, J., Gamero, J., & Romero, J. (2015). Innovativeness and business performances in tourism SMEs. *Annals of Tourism Research, 54,* 118–135.

Milne, S., & Ir, A. (2001). Tourism, economic development and the global-local nexus: Theory embracing complexity. *Tourism Geographies, 3*(4), 369–393. https://doi.org/10.1080/146166800 110070478.

Ministry of Tourism of Greece. (2019). Annual Tourism Report—2018. Retrieved from https://ec.europa.eu/growth/tools-databases/vto/. Accessed July 4, 2020, at 10:00.

Najda-Janoszka, M., & Kopera, S. (2014). Exploring barriers to innovation in tourism industry: The case of the southern region of Poland. *Procedia—Social and Behavioral Sciences, 110,* 190–201.

Navio-Marco, J., Manuel Ruiz-Gómeza, L., & Sevilla-Sevilla, Cl. (2018). Progress in information technology and tourism management: 30 years on and 20 years after the internet—Revisiting Buhalis & Law's landmark study about eTourism. *Tourism Management, 69,* 460–470.

Oberländer, M., Beinicke, A., & Bipp, T. (2020). Digital competencies: A review of the literature and applications in the workplace. *Computers & Education, 146,* 1–13.

OECD. (2020). *OECD Tourism Trends and Policies 2020.* Paris: OECD Publishing.

Richards, G. (2015). Evolving gastronomic experiences: From food to foodies to foodscapes. *Journal of Gastronomy and Tourism, 1*(1), 5–17.

Ruiz-Gómez, L. M., Navío-Marco, J., & Rodríguez-Hevía, L. F. (2018). Dynamics of digital tourism's consumers in the EU. *Information Technology & Tourism.* https://doi.org/10.1007/s40 558-018-0124-9.

Shaw, G., & Wiliams, A. (2009). Knowledge transfer and management in tourism organisations: An emerging research agenda. *Tourism Management, 30*(3), 325–335.

Simeon, M., & Martone, A. (2016). Travel communities, innovative tools to support decisions for local tourism development. *Procedia—Social and Behavioral Sciences, 223,* 681–686.

World Economic Forum. (2019). *Platforms and Ecosystems: Enabling the Digital Economy. Briefing Paper.* Geneva. Retrieved from https://www3.weforum.org/docs/WEF_Digital_Platforms_and_ Ecosystems_2019.pdf.

Zsarnoczky, M. (2017). How does artificial intelligence affect the tourism industry? *Vadyba/Journal of Management, 31*(2), 85–90.

Zsarnoczky, M. (2018). *The digital future of the hospitality industry.* Boston Hospitality Review, Boston University School of Hospitality Administration, Spring.

URL

https://www.mintour.gov.gr/PressRoom/PressReleases/.
https://www.mintour.edu.gr/
https://www.mintour.gov.gr/el/
https://www.capsuletaccelerator.gr/, last accessed, 7 July 2020.

Security and Safety as a Key Factor for Smart Tourism Destinations: New Management Challenges in Relation to Health Risks

Salvador Ruiz-Sancho, Maria José Viñals, Lola Teruel, and Marival Segarra

1 Introduction and Objectives

Tourist destinations are basically configured, on the one hand, by natural and cultural attractions that make them unique and special and, on the other, by an infrastructure and facilities for tourists that allow the activities inherent in these attractions to be carried out. Upon these foundations, numerous tourist destinations have been set up all over the world, specialising in certain types of tourists according to the resources available in each place. In any case, it should be noted that not all tourist destinations have achieved similar market shares, since the competitiveness of destinations is also related to other factors, as suggested by the World Economic Forum (2019). It is clear that outstanding resources and quality services always help to position a place well, but without a tourism infrastructure it can never be considered a tourist destination. The quality of a tourist destination's services includes, among other things, the concepts of the safety and security of visitors. Over the years, although already underlined by the World Tourism Organization decades ago (UNWTO, 1996), security and safety are factors that have become an increasingly relevant key aspect of the tourism model (Istvan & Zimányi, 2011; Álvarez de la Torre & Rodríguez-Toubes, 2014) and also of the competitiveness of a destination, as highlighted by the OECD (2010) and the World Economic Forum (2019). As Jiménez García & Pérez Delgado (2018) recalled, the physical integrity of visitors must be a right of the people and, therefore, an indispensable requirement of a smart tourist destination.

Security is one of the most basic needs for human beings, as pointed out by Maslow (1954), who placed it second on his hierarchy behind physiological needs. Towards the end of the twentieth century, Sönmez (1998) pointed out that peace, calm and safety are basic prerequisites for attracting tourists to a destination, but one

S. Ruiz-Sancho (✉) · M. J. Viñals · L. Teruel · M. Segarra
Universitat Politècnica de València, Camino de Vera s/n, 46022 València, Spain
e-mail: salruisa@doctor.upv.es

of the most accurate definitions of the concept of tourist security was provided by Blanco (2006), who referred to the:

'set of measures, objective conditions and perceptions existing in the social, economic and political sphere of a tourist destination, which allow the tourist experience to take place in an environment of freedom, trust and tranquillity and with the greatest physical, legal or economic protection for tourists and their property and for those who contract tourist services in that destination'.

It should also be noted that security and safety have been key attributes of the image of many tourist destinations, although they have probably not made them sufficiently visible up to now in their marketing campaigns.

The main objective of this article is therefore to identify and analyse the elements that make up the concepts of security and safety in tourism, as well as to determine the factors that influence tourists' perception of security. In addition, it also emphasises how tourist destination security has become a factor of tourist quality and a determinant of the decision to purchase a tourist trip (OECD, 2010), based on the capacity of the destination to offer a safe and secure environment for its visitors.

Additionally, it studies how security issues have evolved in the global tourism context up until the current, hitherto unprecedented, situation when a virus has brought the tourism sector to a standstill worldwide and has highlighted the importance of addressing topics concerning the health and safety issues of a destination.

2 Methodology

The methodology consisted in performing a bibliographic review of previous research in order to clarify concepts and ideas on security and safety in tourism and its constituent elements. To this end, a qualitative analysis of scientific works by public bodies and studies conducted by prestigious publishing houses and cultural and scientific institutions was carried out. The key words (tourism security and safety, perceived image, safe destination) for the search were selected from vocabulary included in the Thesaurus, basically by means of Boolean, syntactic and comparative operators. The main search engine used was essentially Google Scholar, together with www.ask.com, www.alltheweb.com, www.excite.com, www.bing.com, www.nlsearch.com (Northern Light) and www.scirus.com (Elsevier).

At the time of writing, we are immersed in the COVID-19 pandemic, and so documents issued by international reference bodies and official government communications have also been analysed.

3   Conceptual Framework

Security is a concept that alludes to a state of physical and psychological comfort/discomfort in the individual linked to their perception of the real or potential risks from the environment (Viñals et al., 2014).

The risks can be said to be various, and the perception of them varies from one individual to another depending very much on factors relating to their personality traits and their actual knowledge of just how dangerous the risks are. It should also be noted that this is a concept that affects not only tourists, but all the individuals who make up a society, so it is usually an issue that governments assume as part of their political responsibilities. As pointed out by Durán & Bacigalupe (2018): '*There is no such thing as tourism risk, but rather natural risks and those resulting from human, cultural and social activity, which are projected onto it*'.

3.1   Risks Associated with the Territory

In order to understand the concept, we must start with the risks inherent in the territory. Thus, many places, including tourist destinations, may coincide with areas around the world where natural risks are common due to their geology and climate (volcanoes, earthquakes, tsunamis, hurricanes, etc.). To these, we must add the biological risks that exist in certain parts of the planet associated with the presence of microbial life that can interact with humans with fatal outcomes (bacteria, fungi, viruses and protozoa) or places where there is a high rate of endemic contagious diseases (Ebola, dengue, cholera, etc.). There may also be human-induced risks due to hazardous activities being carried out that can increase natural risks (landslides, fires, etc.) or can cause air, soil or water pollution (industrial, chemical, nuclear or radioactive activities, vehicular traffic, etc.).

The insalubrious conditions in certain areas (untreated wastewater with harmful bacteria, presence of pathogens on beaches, polluted atmosphere, food contaminated by pathogens, etc.) are also a major cause of health and hygiene risks for the population and for visitors. Serious health risks can be incurred through contact with or ingestion of contaminated water or food.

3.2   Socio-economic and Political Risks

Other risks that can cause insecurity for people have to do with economic stability and social behaviour. In relation to these elements, it should be noted that they are closely linked. For example, it has been shown that a country's gross domestic product (GDP) stops growing with high rates of homicide and violence (UNDP, 2012), since gatherings in public spaces, travel and recreational activities are very all limited under these conditions. In addition, insecurity generates significant costs ranging from public spending by institutions and private spending by citizens on security to the irreparable costs in terms of life and people's physical and mental integrity (UNDP, 2012).

Some of the most significant political risks are wars and terrorist threats (political violence), which give rise to violence and economic, social and political instability. They are therefore a great source of insecurity for the citizen that includes the risk of losing one's life. For example, according to data from Latin American Public Opinion Project—LAPOP (UNDP, 2012), the percentage of people who say they have limited their visits to recreational places in Latin America for fear of being a victim of crime ranges from 20.6% (Chile) to 59.1% (Dominican Republic). Some of the pioneering studies on crime and tourism development were those by Jud (1975), Pizam & Mansfeld (1996, 2011) and also Neumeyer (2004). Subsequently, a large body of the literature has been produced that addresses the decline of tourism in destinations subject to this type of problem, especially since the attacks on the Twin Towers in New York on 11 September 2001 (Hall et al., 2003).

In addition, it should be noted that there are forms of terrorism and extremism that specifically target tourism as a means of expanding their claims internationally (Pleterski, 2010).

To corroborate the theses on terrorism, we cannot fail to mention the work of Drakos & Kutan (2003) on the regional effects of terrorism in Mediterranean countries (Greece, Israel and Turkey between 1996 and 1999), in which they took the model for developed destinations of Enders et al. (1992) to analyse the regional effects of terrorism on the market shares of direct tourist competitors. Their results showed that an increase in instability in Greece contributed to an increase in the market share of Israel, while terrorism in Israel benefited the market share of Turkey.

3.3 Types of Risk Response

In view of the large number of potential risks to which the population and tourists travelling to a destination are exposed, the response of the public authorities is multi-dimensional and is approached, as proposed by Bonini (2010), from the perspective of protecting people, goods and services (tangible elements) and that of protecting intangibles (rights and guarantees, ethical principles, etc.).

Thus, and specifically with regard to tourism, it is deployed in a series of systems that, based on Grünewald's classification (2010), could be classified as:

- Public security: a system focused on controlling and minimising criminal acts (robbery and theft), acts of terrorism, public disorder, accidents, etc., and ensuring harmonious coexistence at the destination.
- Medical security: a system of emergency care and medical protection for visitors during their stay at the tourist destination.

- Hygiene and health protection: a system of measures adopted by the destination to ensure that tourist activities take place under hygienic conditions and in a healthy environment. The global pandemic caused by the COVID-19 has highlighted the importance of these measures on an international scale in today's globalised world.
- Food safety: a system designed to protect the health of consumers through a process that seeks to ensure food safety by controlling and eliminating unsafe food containing bacteria, viruses, parasites or chemicals that are harmful to health.
- Road and transport safety: a system that allows the visitor to move freely and safely on the roads and on the transportation at the destination.
- Environmental safety: a system that allows the protection of people against natural and human-induced risks.
- Information security: a communication system that allows true knowledge and transparency of information about the destination.
- Tourist safety: a system of protection for tourists that is put in place in some destinations or at specific times to safeguard the physical integrity of visitors in particularly complex situations (large events, areas with large crowds of people, etc.).

4 The Tourist's Perception of Security

Viñals et al. (2014) commented that tourists should be able to visit a place and carry out their activities in safety, and moreover, they should perceive it as such for their own peace of mind.

The safety and security of a destination can be viewed from two perspectives. The first is that of the manager responsible for the destination, who knows the risks and is responsible for providing the material and human resources necessary to prevent and manage them (*actual risk*). The second is the perspective of the visitors and tourists, who have their own perception of the safety and security of the destination (*perceived risk*).

This perception depends on various factors, namely:

- *The visibility and tangibility of how dangerous the risk is:* this factor is directly related to its perception, since the mechanisms of danger warning or detection in the most primitive human brain are linked to visible, tangible and sensory physical facts (Cao, 2011). Therefore, what creates a feeling of insecurity within us is basically produced by situations or scenarios involving physical threats (earthquakes, tsunamis, explosions, fires and so forth).
- *Information about security:* the level of information available, its ease of access and its accuracy are fundamental in the choice of a destination. According to Mansfeld (1992), information fulfils three basic functions in the choice of a destination: by minimising the risk of decision, by creating an image of the destination and by serving as a mechanism to justify that choice. Kim & Richardson (2003) pointed out that more familiarised tourists are more aware of the real opportunities of the destination, which creates a feeling of security and comfort that has

positive effects on their perception. Hence, as commented by Durán & Bacigalupe (2018), knowing how to manage information is essential to be able to cope with the uncertainty that conditions risk perception.

It should also be noted that the choice of destination made by many tourists is conditioned by information about safety during the trip and the stay there (Fernández Ávila, 2002; Xueqing Qi et al., 2009). Consequently, as pointed out by Echtner & Ritchie (2003), prior to the trip, tourists generate a 'projected image' from information related to historical, political, economic and social factors about the destination that they have stored throughout their lives, even without having visited it or having been exposed to sources of commercial advertising. This information may come from various sources, which Gartner (1994) called 'image formation agents' and Baloglu & McCleary (1999) called 'stimulus factors', consist of the official ones, the education received, the opinions of friends and family or other prescribers, and give rise to what Gunn (1972) and Martín Santana & Beerli (2002) called the 'organic image'. In addition, there are the commercial sources (providers of tourist services in the area) that give rise to the 'induced image', which is usually an 'interested' image of the place and, therefore, very often does not inspire sufficient confidence in the tourist, as it could be controlled by the promoters of the destination (Andreu et al., 2000). With regard to the first sources, it is necessary to highlight the importance of truthfulness and the existing emotional links with the traveller. Once at the destination, perception is produced, which is a process by which an individual receives information from the environment through the senses, selects it, organises it and interprets it so as to create a 'significant or perceived image' (Mayo & Jarvis, 1981). Therefore, it is necessary for tourists, once at their destination, to be well informed about the vicissitudes, risks and limitations of the place, as this will make their stay more peaceful, comfortable and relaxing because they will know what they can find there and how to behave in any contingency that might arise.

- *Capacity of the public authorities to respond to risks:* the provision of security is an important factor in the psychological comfort of tourists, as has been mentioned earlier. Security should be considered a public good. Grünewald (2010) noted that when security issues are not adequately addressed by service providers and authorities, and the tourist perceives insecurity, the attractiveness for tourists is depreciated and this has negative effects on the image of the destination. Elias & Dunning (1992) commented that the fear imagined by tourists during their trip vanishes when they find security devices at their disposal. Being perceived as a safe destination is an attribute that makes a destination even more important than it already is (Vargas-Sánchez, 2020).

Tourism security depends on the efficiency of the system to mitigate the hazards of the destination (Tarlow, 2011), with a distinction being drawn between risk management, disaster management and crisis management, this latter element having been previously noted by Sönmez et al. (1999) and Glaesser (2003). Tarlow also insisted on the need to correctly identify and correct the risks in order to be able to deal with them.

- *Personality traits of the visitor:* the perception of risk when choosing a destination is influenced by how dangerous the destination is, the information received or the public authorities' capacity to respond in the face of adversity. Yet, as several authors have suggested (Sönmez & Graefe, 1998; Pizam & Mansfiel, 1999; Shiffman & Lazar, 2000; Reisinger & Mavondo, 2006), there are also a series of personal factors such as the traveller's previous experiences and personality traits, their educational level, gender and country of origin, among others, that condition the perception of risk. Lepp & Gibson (2003) pointed out that there are tourists who are looking for new experiences and challenges—those known as 'allocentric travellers' according to Plog's classification (1974). There are also those seeking sensations, according to Zuckerman (2007), who have a low perception of risk and need strong stimuli to achieve a satisfactory recreational experience, and so they assume risky behaviour. Additionally, there are the 'psychocentric tourists' (Plog, 1974) who look for trips where everything is predictable and organised without leaving anything to improvisation; therefore, they want the security offered by a trip where everything is under control.

5 Tourism Security in the Midst of a Pandemic

Tourism worldwide has grown progressively in recent decades and with it the associated security problems. These problems differ depending on the moment in history. In the first decades of the development of tourism, between the 1950s and 1970s, in what was known as 'mass tourism', the security and safety of tourists were focused on tourism and health infrastructures and facilities (road safety, the state of roads at a tourist destination, etc.). Likewise, at the end of this period, aspects related to quality started to be considered, especially in customer service, and, as stated by Istvan & Zimányi (2011), health and hygiene began to be taken into account.

The next period, from the 1970s to the 1990s, was characterised by the important role played by political conflicts, mainly wars and terrorism, which caused intermittent security threats in various regions around the world, such as the Middle East, Northern Ireland or the Basque Country in Spain. In any case, it must be said that some destinations are highly resilient in the face of security problems, taking into account the harmful effects that these facts have caused, and that their predictability is sometimes difficult to manage (Sönmez et al., 1999). Such is the case of Spain, which is considered a safe country and neither the tourist boom nor the terrorist attacks have posed a serious threat to its image as a safe destination.

Since the 1990s, globalisation has brought with it other challenges associated with information and communication technologies, mobility and trade in goods and services. This period of 'global tourism' has security problems associated with it on a global scale. The most persistent have been of a political nature, followed by personal security problems, natural disasters and pandemics (Hall et al., 2003; Istvan & Zimányi, 2011). These security problems have not been as serious as the major tourism recessions due to economic causes, but since March 2020, the first concern in the security of destinations has been related to health, due to the spread of new viruses such as COVID-19.

Of the many types of risks that have threatened the development of tourism to date, never before has there been an event as unique as the health crisis caused by COVID-19, which has affected, to a greater or lesser extent, every country on the planet. Until then, terrorism was the main threat to tourism, which had replaced the wars that had taken place from 1945 to 2004 (Nigeria, Syria, Iraq, Afghanistan, etc.). Natural disasters, wars, terrorist attacks, etc., have been known to occur in specific places around the world, but no event has reached the virulence of this global pandemic to the same extent as the new coronaviruses (UNWTO, 2020).

It is true that there are precedents of contagion by viruses such as Ebola, influenza, Zika and SARS, but they have never before caused an effect such as the one we are currently witnessing. The international spread of the COVID-19 pandemic has been a turning point for tourism because of its global and fatal nature, which calls for reflection on the vulnerability of the different systems when faced with a virus that has travelled rapidly across continents in less than 36 h. In March 2020, the virus had spread across the planet, affecting millions of people and causing the deaths of hundreds of thousands. By June 2020, more than 9 million people had been affected and more than 470,000 had died. Tourism has been one of the biggest victims and practically all the hotels in the world have closed down; tight restrictions have brought travel almost to a standstill, with the general situation of tourism companies being particularly dramatic (Deloitte, 2020).

Reality has shown that no tourist destination had considered this possibility in its plans, and far less were they prepared to deal with the situation. It has become clear that many destinations have little resilience to tourism and that tourism is fragile and interdependent, as well as being crucial to the planet's economy. From now on, tourist destinations must consider health crises as a serious threat in the face of which they must take preventive management measures in order to minimise uncertainty and guarantee tourists' safety.

6   Discussion and Conclusions

In view of the above, the need for in-depth knowledge about all aspects related to the security and safety of a destination becomes clearly apparent, since they are indissolubly associated with the quality of the services provided and its competitiveness.

Knowledge about security problems and their causes helps to classify the type of risk and to provide the best response for crisis management.

In addition, the importance of the tourist's perception of security has been emphasised, bringing to the fore the need to draw attention to the safety and security of a destination and communicate in an adequate manner how they are addressed, since it forms part of the image of the destination.

Emphasis has also been placed on how safety and security condition the tourist in the process of choosing a tourist destination, and how they have become more important in this process over time. Tourists must visit a destination that can guarantee their physical integrity. It is therefore important that the destination manager knows the 'actual' risks and takes steps to provide the material and human resources necessary to minimise them and to manage them properly. However, it is also important that he or she considers how the tourist perceives it ('perceived risk'), since this factor will have a direct influence on satisfaction with the tourist experience. Thus, information and communication about safety provided by the destination must be truthful, accessible and available both before and during the trip, since this is a key issue in reducing visitors' uncertainty (Durán & Bacigalupe, 2018).

It has also been shown that the responsiveness and efficiency of the administration managing a destination in crisis situations increase the good image of the destination and offset the negative perception of the existence of risks in the place.

Moreover, it has become clear that the health crisis caused by the COVID-19 and its management has been the greatest concern so far known by tourism managers at all levels and in all countries around the world. The reasons for this concern are the great impact it has had not only on national economies but also at the international level, the effects it is having on the way people relate to each other and on tourism, and also the new perception of places and situations.

Beyond the health connotations of the crisis, it should be noted that no model of resilience has been identified so far for the tourism sector that would comprehensively take this risk into consideration in order to guarantee health safety and comfortable performance of tourism activities. It is clear that this pandemic can be considered a milestone in the history of tourism that will forever change the behaviour of both the competent authorities and the tourism service providers and tourists. Thus, if security was already a key element, it has now become the decisive factor in the competitiveness of destinations. Moreover, in the context of globalisation, it is clear that the response must be global, as Buus (2011) suggested when proposing a single security code applicable to all countries. On the other hand, and as a final reflection, it should be mentioned that no destination can be said to be 'safe' from this situation, and that public policies will be crucial to overcome this crisis. Furthermore, as regards recovery, it will be necessary to adapt and make extraordinary efforts to improve the image and reputation of destinations based on security arguments (Hernández-Martín, 2020), since the brand image has been seen not to be static, but to be capable of changing at any time and in the face of any event (Gutauskas & Valdez, 2019).

References

Álvarez de la Torre, J., & Rodríguez-Toubes, D. (2014). Safe and security as a component of quality in tourism: The case of Galicia. *Revista de Ocio y Turismo, 7*, 1–9.

Andreu, L., Bigné, J. E., & Cooper, C. (2000). Projected and perceived image of Spain as a tourist destination for British travellers. *Journal of Travel & Tourism Marketing, 9*(4), 47–67.

Baloglu, S., & McCleary, K. W. (1999). A model of destination image formation. *Annals of Tourism Research, 26*(4), 868–897.

Blanco, F. J. (2006). *Reflexiones sobre seguridad, poderes públicos y actividad turística (III): una aproximación al concepto de seguridad turística.* Retrieved from https://www.belt.es/expertos/HOME2_experto.asp?id=.2974. Accessed the May 20, 2020.

Bonini, J. (2010). Seguridad pública y social. Criterios y conceptos. In L.A. Grünewald (Ed.), *Municipio, Turismo, & Seguridad.* Universidad Nacional de Quilmes/Organización de Estados Americanos (OEA), 35–44.

Buus, S. (2011). The people's home goes Gulliver: Sweden and the 2004 Tsunami Crisis (Special issue on Narrives of Risk, Security & Disasters issues in Tourism and Hosp. Korstanje, M. (Ed.)). *International Journal of Tourism Antropology, 1*(3–4), 293–303.

Cao, J. (2011). Posibles causas que alteran las diferencias entre riesgo percibido y riesgo real. In *Instituto Nacional de Ciberseguridad.* Retrieved from https://www.incibe.es/. Accessed the May 12, 2020.

Deloitte (2020). *Impacto del COVID-19 en el sector 'Hospitality'. La experiencia del usuario en el mundo post-COVID.* Retrieved from https://www2.deloitte.com/es/es/pages/operations/articles/impacto-covid-19-sector-hospitality.html. Accessed the June 23, 2020.

Drakos, K., & Kutan, A. (2003). Regional effects of terrorism on tourism in Three Mediterranean countries. *Journal of Conflict Resolution, 47*(5), 621–641.

Durán, A. M., & Bacigalupe, M. A. (2018). El turista y la percepción del riesgo. El rol de la gestión de la información sobre la imagen de los destinos que han sufrido atentados terroristas. *Estudios y Perspectivas en Turismo, 27*(4).

Echtner, C. M., & Ritchie, B. J. R. (2003). The meaning and measurement of destination Image. *The Journal of Tourism Studies, 14*(1).

Elias, N., & Dunning, E. (1992). *Deporte y ocio en el proceso de civilización.* Madrid: F.C.E.

Enders, W., Sandler, T., & Parise, G. F. (1992). An econometric analysis of the impact of terrorism on tourism. *Kyklos, 45*(4), 531–554.

Fernández Ávila, V. (2002). El impacto del terrorismo en las llegadas de turismo internacional–algunos ejemplos. *Turismo y Sociedad, 1*, 70–79.

Gartner, W. C. (1994). Image formation process. *Journal of Travel & Tourism Marketing, 2*(2–3), 191–216.

Glaesser, D. (2003). *Crisis management in the tourism industry.* Burlington, MA: Oxford Butterworth-Heiinemann.

Grünewald, L. A. (2010). La seguridad en la actividad turística. La percepción desde la óptica de la demanda. In L.A. Grünewald (Ed.), *Municipio, Turismo, & Seguridad*, Universidad Nacional de Quilmes/Organización de Estados Americanos (OEA),19–34.

Gunn, C. (1972). *Vacationscape. Designing tourist regions.* Washington, DC: Taylor and Francis. University of Texas.

Gutauskas, A., & Valdez, R. (2019). Comunicación responsable de una marca turística en crisis: El caso de la Villa La Angostura en la erupción del Volcán Puyehue. *Revista de Estudios Latinoamericanos sobre Reducción del Riesgo de Desastres* (REDER), 53–68.

Hall, M., Timothy, D. J. & Thimoty-Duval, D. (2003). Safety and security in tourism: relationships management and marketing. *Journal of Travel & Tourism Marketing, 15* (2/3/4).

Hernández-Martín, R. (2020). Entender y afrontar la crisis turística. *El Día.* Retrieved from https://www.eldia.es/opinion/2020/04/10/entender-afrontar-crisis-turistica/1069164.html. Accessed the June 18, 2020.

Istvan, K., & Zimányi, K. (2011). Safety and Security in the Age of the global tourism (The changing role and conception of safety and security in tourism). *Applied Studies in Agribusinesss and Commerce,* 59–61.

Jiménez García, J., & Pérez Delgado, M. (2018). La seguridad como componente esencial del concepto de calidad turística. *Estudios y Perspectivas en Turismo, 27*, 921–943.

Jud, G. D. (1975). Tourism and crime in Mexico. *Social Science Quarterly, 56*, 324–330.

Kim, H. & Richardson, S. L. (2003). Motion picture impacts on destination images. *Annals of Tourism Research, 30*(1), 216–237.

Lepp, A., & Gibson, H. (2003). Tourist roles, perceived risk and international tourism. *Annals of Tourism Research, 30*(3), 606–624.

Mansfeld, Y. (1992). From motivation to actual travel. *Annals of Tourism Research, 19*, 399–419.

Martín Santana, J. D., & Beerli, A. (2002). El proceso de formación de la imagen de los destinos turísticos: Una revisión teórica. *Estudios Turísticos, 154*, 5–32.

Maslow, A. H. (1954). *Motivation and personality*. New York: Harper y Row.

Mayo, E.J., & Jarvis, L. P. (1981). *The psychology of leisure travel: effective marketing and selling of travel services*. Ed. CBI Publishing Company.

Neumeyer, E. (2004). El impacto de la violencia política en el turismo: Estimación dinámica transnacional. *El Diario de la resolución de conflictos, 48*(2), 259–281.

OECD (2010). Tourism 2020: Policies to promote competitive and sustainable tourism. In *OECD Tourism Trends and Policies 2010*. Paris: OECD Publishing.

Pizam, A., & Mansfeld, Y. (1996). *Tourism, crime and international security issues*. Chichester: Wiley.

Pizam, A., & Mansfeld, Y. (1999). *Consumer behavior in travel and tourism*. Haworth Hospitality Press.

Pizam, A. & Mansfeld, Y. (Eds.) (2011). *Tourism, security & safety: from theory to practice*. Routledge, Francis and Taylor, New York, USA

Pleterski, T. (2010). *El impacto del terrorismo sobre el turismo. Los efectos causados sobre la recepción de visitantes*. Trabajo Final de Grado. Universidad Politécnica de València.

Plog, S. C. (1974). Why destination areas rise and fall in popularity. *Cornell Hotel and Restaurant Administration Quarterly, 14*(4), 55–58.

Reisinger, Y. & Mavondo, F. (2006). Cultural differences in travel risk perception. *Journal of Travel & Tourism Marketing, 12*(1), 13–31.

Shiffman, L. C., & Lazar, L. (2000). *Consumer behavior*. Upper Saddle River, New Jersey: Prentice-Hall.

Sönmez, S. (1998). Tourism, terrorism, and political instability. *Annals of Tourism Research, 25*, 416–456.

Sönmez, S., & Graefe, A. R. (1998). Determining future travel behavior from past travel experience and perceptions of risk and safety. *Journal of Travel Research, 37*, 171–177.

Sönmez, S., Apostopoulos, Y., & Tarlow, P. (1999). Tourism in crisis: Managing the effects of terrorism. *Journal of Travel Research, 38*(1), 13–18.

Tarlow, P. (2011). Tourism disaster management in a age of terrorism (Special issue on Narrives of Risk, security & Disasters issues in Tourism and Hospitality). *International Journal of Tourism Anthropology, 1*(3–4), 254–272.

UNDP (2012). *Latin American Public Opinion Project (LAPOP), Vanderbilt University*. Retrieved from https://www.vanderbilt.edu/lapop/. Accessed the February 10, 2020.

UNWTO (1996). *Tourist safety and security: Practical measures for destinations*. Madrid: World Tourism Organization.

UNWTO (2020). *Evaluación del impacto del brote Covid-19 en el turismo internacional. Barómetro OMT del Turismo Mundial mayo 2020 / Con especial enfoque en el impacto de la COVID-19*. Retrieved from https://www.unwto.org. Accessed the July 9, 2020.

Vargas-Sánchez, A. (2020). El turismo post-cornavirus. Hosteltur. Retrieved from https://www.hosteltur.com/comunidad/003955_el-turismo-post-coronavirus-ii-html. Accessed the June 24, 2020.

Viñals, M. J., Morant, M. & Teruel, L. (2014). Confort psicológico y experiencia turística. Casos de estudio de espacios naturales protegidos de la Comunidad Valenciana (España). *Boletín de la AGE, 65*, 293–316.

World Economic Forum (2019). *The travel and tourism competitiveness report, 2019*. Ed. World Economic Forum.

Xueqing Qi, Ch., Gibson, H. J. & Zhang, J. J. (2009). Perceptions of risk and travel intentions: The case of China and the Beijing olympic games. *Journal of Sport Tourism, 14*(1), 43–67.

Zuckerman, M. (2007). *Sensation seeking and risky behavior*. Ed. American Psychological Association.

An Evaluation of Hotel Websites' Persuasive Characteristics: A Segmentation of Four–Star Hotels in Greece

Konstantinos Koronios, Lazaros Ntasis, Panagiotis Dimitropoulos, John Douvis, Genovefa Manousaridou, and Andreas Papadopoulos

1 Introduction

Information technology financing assumes a fundamental aspect in inspecting hotels consciously. Efficient and up-to-date distribution of the Internet and Web pages might propose opportunities for upgraded visitor services to reach increasing client presumptions, enhanced expense oversee, progressively compelling marketing strategies, and extended opportunities for hotels (Law & Jogaratnam, 2005; Piccoli, 2008). It is clear that Web pages funding will enhance hotels' profitability, diminish their expenses, and simultaneously increase the profit of the services and products offered to their clients. Thusly, investments into Web pages in hotels have expanded over the previous years (Piccoli, 2008).

The evaluation of Web pages has been the focal point of consideration for academics from the development of hotel Web pages in the later years of 90 s (Law, 2019). Amid this period, consumers had the option to save hotel rooms by conventional media (i.e., travel specialists, telephone, and fax). Since the mid-90s, academic analysts and industry professionals have underlined the importance of preserving travel significant services through online stages, to be specific hotel-booking Web pages, to enhance the nature of their services and income (Law, 2019). Subsequently, hotels began to incept Web pages to give global clients the information through the

K. Koronios (✉)
Department of Accounting and Finance, University of Peloponnese, Kalamata, Greece

L. Ntasis
Department of Economics, University of Peloponnese, Tripoli, Greece

P. Dimitropoulos · J. Douvis · A. Papadopoulos
Department of Sport Management, University of Peloponnese, Sparti, Greece

G. Manousaridou
Department of Physical Education and Sport Science, Aristotle University of Thessaloniki, Thessaloniki, Greece

V. Katsoni and C. van Zyl (eds.), *Culture and Tourism in a Smart, Globalized, and Sustainable World*, Springer Proceedings in Business and Economics, https://doi.org/10.1007/978-3-030-72469-6_35

Web. A few hotel sites have been accessible since the mid-90s through to the late 2000s. Hotel managers concentrated on site improvement as opposed to making an insignificant site by updating the Web innovations used to make their sites. Such enhancements were to display customized customer services and to establish the procedure of bookings (Baloglu & Pekcan, 2006). From 2010 onward, hotel directors have devoted to improve their Web pages by means of the adoption of present-day models and the advancement of different Web pages renditions (Fong et al., 2017).

In the present digital world, Web pages fill in as an immediate interface among hotels and clients. Giving a consistent and enjoyable Web page experience for clients is basic for hotels to enhance their room bookings (Li et al., 2017), just as for their maintainable competitiveness in the long haul (Geerts, 2014). Nonetheless, not every Web page is effective in accomplishing such purpose. Hence, a successful site evaluation is a noticeable issue for researchers and professionals (Chiou et al., 2010). To abridge, site evaluation is the focal point of industrial professionals and academic researchers (Law et al., 2010).

Nowadays, most of vacationers plan their days off, reserve requested services, and make their experiences public on the Web. Official Web pages are a necessary instrument in various viewpoints (Samad et al., 2019): (a) to inform possible or current travelers that are looking for places of interest, destination information, travel courses, opening hours and costs of museums, vacation spots and trails, accommodation and so forth (b) destination branding, for instance, imparting the fundamental characteristics and estimations of a brand for specific destinations, known as brand value in marketing and advertising terms (Sartori et al., 2012); (c) a method for assuring potential vacationers; (d) a channel for products and services applicable to advertising; and (e) a stage to share experiences and information between destination supervisors and voyagers, or among explorers themselves. It is basic information that the Web may fill in as a compelling marketing tool in the travel industry (Buhalis & Law, 2008). In view of Schmidt et al. (2008), plainly the Web is changing marketing techniques and hotels are capitalizing on the Web as a successful marketing tool. Company sites are significant since they speak to the company inside the Web condition. Henceforth, organizational Web pages must be effective and successful to have the option to equal their rivals. To survey the viability of such sites, it is imperative to decide assessment lists and assessment techniques (Schmidt et al., 2008).

In more detail, a requirement exists for broad research studies that completely analyze the way investments in sites can prompt making manageable antagonistic advantage for hotel organizations. The current literature on this subject in the accommodation field is not yet definitive, and there have been constrained theoretical and practical researches into this territory. Earlier specialists concentrated on the hospitality part have examined the different factors reliable for the accomplishment of hotel Web sites (Vrana et al., 2004). In any case, most of them have focused on singular perspectives and attributes rather than the general persuasiveness of Web pages. Having distinguished the gap, this examination plans to investigate Greek four-star hotels' procedure regarding their sites, assessing their general site persuasiveness and talk about how site investments can prompt antagonistic benefit in hotel organizations, and to suggest a hypothetical system that outlines the connections among site and competitive advantage.

2 Literature Review, Capitalize the First Letter of Every Word in the Title

The hotel business has constituted one of the quickest developing industry users of the Internet. The purposes behind this quick increment in Internet utilization are the various advantages that the Internet can give to both the hospitality sector and the visitor/tourist. Hotels can decrease their distribution expenses and therefore increment their benefits, just as reach out to a bigger prospective target market (Law & Hsu, 2005) and to upgrade service quality. Web page visitors would give advantages to organizations if the Web page can give important services to their visitors (Schmidt et al., 2008). A few articles have analyzed different elements adding to the achievement of a hotel site. Schmidt et al. (2008) assess hotels' site qualities and relate those attributes to site execution. Besides, studies on general Web design appear of being declining for better comprehending explicit site clients, for example, purchasers/programs and the visually impaired (Mills et al., 2008). Extra e-advertising research incorporated an examination of the effect of online client reviews on hotel room deals (Ye et al., 2009).

As indicated by Kim & Fesenmaier (2008), Web page visitors take quick decisions regarding a Web page dependent on their essential perception from a prompt interaction with the Web site. Such a condition is experienced in hotel sites, so it is basic to affect clients' essential perception. Because of this reality, hotel directors should be educated about explicit enticing strategies, their utilization and their effect on visitors, just as the manner in which audiences envision them. Almost no exploration has inspected the idea of persuasion and Web design (Zach et al., 2008). Persuasion constitutes the demonstration of persuading others to think, feel, or do what we need (Perloff, 1993). Within the framework of marketplace, consumers are convinced toward potential travel destinations through different methods incorporating relational discussions with friends and family and by advertising and publicizing in the broad media. Nonetheless, travelers today are progressively going to the Web and the sites as an information source. In setting of sites, persuasiveness is characterized as the capacity to bring out good impressions toward the Web page and to impact the attitudes on site users (Kim & Fesenmaier, 2008).

The powerful roles of computer technology have been investigated in the tourism setting (Law & Wong, 2003). As per Kim and Fesenmaier's (2008) characterization, the significance of six factors must be considered when assessing the convincingness of Web sites, which are: informativeness, ease of use, validity, inspiration, involvement, and correspondence. Their examination depended on the hypothetical setting proposed by Zhang and Von Dran (2001), who broadened Herzberg's (1987) motivation-hygiene hypothesis to the online setting, so as to recognize Web design aspects that help the data-seeking process (Cheung & Law, 2009). In more detail, the suggested system perceives and recognizes the sites' attributes which may be viewed as hygiene factors from those which may be viewed as motives in the Internet setting (Zhang &Von Dran, 2001). They recommend that hygiene factors are pivotal qualities, however deficient to guarantee Web site guest's fulfillment. Moreover, they recommend that inspirational variables conduce to guest fulfillment and to repeated site visits. In view of Kim and Fesenmaier's order (2008), informativeness and ease of use can be viewed as the hygiene factors. Accordingly, validity, motivation, involvement, and correspondence can be viewed as the inspiring variables. The demonstrated rousing variables assume an essential role in arranging the extraordinary segments of hotel Web pages. All the more explicitly, the categorization of hotel Web pages as per the convincing variables, advantage hotels in improving their openness and correspondence with customers. This order may sequentially increment online bookings and upgrade site attempts to improve their attractiveness to potential guests, by legitimately reassuring them. These components are critical to decide the distinctions among the current four-star hotels in the part.

Informativeness is characterized as the capacity to illuminate clients about product options that empower them to settle on decisions yielding the highest esteem (Lee & Hong, 2016). Informativeness is a perceptual construct that is estimated through self-revealed things (Pavlou et al., 2007). What is more, it envelops discerning intrigue because of its capacity to enable a buyer to make a knowledgeable judgment about the acknowledgment regarding the scope, and consequently is conceptually unmistakable from "emotional appeal". Researches have found informativeness to be significant in the arrangement of customer attitudes to electronic commerce Web pages (Gao & Koufaris, 2006). Hotel sites give a ton of data (e.g., services, costs, room photographs) so as to reinforce their service bundling and escalate their Web advertising. Generally, the related literature concurs that informative sites give a client the capacity to make an well-informed judgment for a future buy, expanding site's general persuasiveness (Lee & Hong, 2016).

Another significant attribute of sites is the level of their usability, which alludes to how well and how effectively a client, without formal preparing, can cooperate with an information framework or a site to achieve his/her assignment with the least time and cognitive efforts conceivable (Essawy, 2005). Scientists have proposed a few distinct methodologies for estimating site usability (Yeung & Law, 2006) one of which is functional comfort, that is an information system's usability. Usability is comprised of two particular highlights: (a) simplicity of comprehension, and (b) simplicity of navigation (Loiacono et al., 2002). The idea of simple of comprehension recommends that sites must be created with the end goal that visitors effectively comprehend who supports the Web page, what the objectives of the locales are and what they can accomplish on the site. It should likewise give convenient information to the user. Sites ought to be developed obviously so guests effectively locate the essential content and incorporate a Web guide and search instrument (Nielson, 2000). Along these lines, Web design permits client to effortlessly comprehend the information given and causes hoteliers to accomplish their goals.

Credibility is a fundamental segment of convincing correspondence generally. O'Keefe (2002, p. 67) characterizes validity as "decisions made by a perceiver (i.e., a message recipient) regarding the acceptability of a communicator". In general, scholars in general concur that two measurements are critical to source validity. One measurement is skill (e.g., capability, expertness, intelligence, definitiveness), and the other is reliability (e.g., character, sagacity, security, trustworthiness). Validity is particularly significant in the travel industry. Research shows that elusive, costly service purchases, for example, travel, include both economic and emotional risk. For the most part, customers secure data as a risk decrease system. The more solid or trustworthy the data, the lower the apparent risk (Loda et al., 2009). The significant signs of trust in a site are privacy and security, just as the recognizable proof of the Web site owner through the given information about the association and its history (Yang et al., 2003).

Inspiration is a strategy for influence intended to make a passionate response to a message by utilizing enthusiastic content. Motivation would a fortiori adequately convince a person who has little inspiration or little capacity to intelligently convert a message (Petty & Cacioppo, 1986). Motivation manages the manners by which emotional responses are stimulated as a function of the message's inspirational importance to people (Alhabash et al., 2013). At the point when an enticing message is received, affective reactions (i.e., emotions and moods evoked by the site) are joined with cognitive reactions (i.e., discerning assessment of the site) to develop the attitude toward the message (Lee & Hong, 2016). Accordingly, on account of hotel Web site, the extraordinary criticalness of aesthetics (e.g., utilization of visual features) outlines the yearning of the hotels to introduce a powerful and productive connection or perception and to incite engaging experiences so that the positive recognitions stimulate spiritually potential visitor to continue with the Web-based booking.

Involvement constitutes a focal idea that clarifies consumer conduct in marketing and retailing (Aldlaigan & Buttle, 2001). Involvement can be portrayed as "a progressing worry for an item class, that is, it is autonomous of procurement circumstances and is inspired by how much the item identifies with oneself and/or hedonic joy got from the item" (Richins & Bloch, 1986). The idea of involvement is portrayed in numerous structures. It is a person's continuous connection with the attitude or item. Blythe (1997) clarifies that association has both psychological and emotional components, which means inclusion of the brain and the feelings. In an online setting, Patwardhan et al. (2004, p. 418) expressed, "Cognitive involvement is the degree to which people attend, consider, center and apply mental effort while occupied with a specific online action. Involvement, in this research, alludes to the circumstance where a visitor feels emphatically about the Web page and invests energy scanning for information that can change the purchase choice".

The concept of reciprocity has been differently characterized from alternate points of view. The principal belief is that of qualities of the tool of a site (Sohn, 2011). Interpretations that attention on highlights try to distinguish either broad attributes like two-way communication or explicit qualities of sites, for example, search engines (McMillan & Hwang, 2002). Taking into account this point of view, one frequently referred to definition is that of Jensen (1998, p: 201) "a proportion of a media's potential capacity to let the user apply an impact on the substance as well as type of the mediated communication". The key components of reciprocity as indicated by McMillan (2000) where the key components are the highlights that encourage two-way communication and control. Ahren, Stromer-Galley and Neuman (2000) focused likewise on features that empower two-way communication just as on the interactive media features of the sites. The subsequent methodology characterizes reciprocity concentrating on process (Heeter, 2000). From the procedure point of view, definitions center around activities, for example, interchange and responsiveness (McMillan, 2000). As Bezjian-Avery et al. (1998 p.23) suggested, "In reciprocal frameworks, a client controls the substance of the connection mentioning or giving information". Most definitely, limits or specific citations are kinds of advantages normally allowed to online customers with the reason setting up a commonly valuable relationship.

The hotel business is customarily viewed as a critical piece of the service industry. Regarding the hotel business, there are numerous investigations that have focused on estimating service quality. Hotel services and consumer satisfaction of service change from culture to culture, and administration quality measurements contrast starting with one segment of hotel industry then onto the next help quality measurements vary starting with one area of hotel industry then onto the next. Then again, studies on service quality with respect to the hotel field in Greece are rare (Kamenidou et al., 2009).

Nonetheless, regardless of the significance of division in the hospitality area, there has been little study dependent on business segmentation. Past work has concentrated on hotel division dependent on consumer request qualities. Studies on the division of hotel types is generally rare, particularly those concentrating on hotel Web page design.

This present investigation distinguishes typologies of four-star hotels in Greece based on their own qualities, explicitly by assessing the convincing features of their Web pages. The decision of business division strategy depends on its commitment to the literature and its convenience for Web page hotel managers. This is on the grounds that utilizing these parameters can powerfully affect consumers and boost benefits for hotels.

Segmentation advances the viability of marketing efforts intended to draw in more straightforwardly and fitting item offering to a specific market segment advances consumer contentment. Profiling the structure enhances the comprehension of that section and permits an assessment of the qualities that segregate between fragments. Wedel and Kamakura (2000) note the significance of profiling with company descriptors to satisfy the accessibility necessity for successful market division. This information on the segments inside four-star hotels as indicated by the enticement of their sites adds to the hypothetical comprehension of Web page appropriation and use.

3   Methodology and Research Design

Content analysis was completed so as to investigate hotel Web pages. Every one of the sites dissected compares to one four-star hotel located in Greece. Deductive estimation necessitates the improvement of explicit coding categories prior to content investigation being done while preliminary estimation bolsters the act of new coding. The current study utilizes a mixture of the two since six principle coding categories were recognized before leading the examination: informativeness, usability, validity, motivation, involvement, and reciprocity. The complete populace of four-star hotels was assembled from the Hellenic Chamber of Hotels and the Greek Ministry of Tourism, and the sample of the investigation was 404 hotels.

As an initial step, an Internet search engine was utilized to discover the sites of four-star Greek hotels, beginning with the first in the positioning and continuing in alphabetical order. When the hotel site had been recognized it was entered in the examining list aimlessly and examined utilizing the factors suggested in this investigation. A few hotels were dismissed in light of the fact that their sites could not be found/does not existed.

Methodological proposals dependent on past examinations were carefully reviewed looked to build the unwavering dependability and legality of the discoveries (Krippendorff, 2004). To this end a codebook was made and utilized as rule in the investigation so as to "interview" the Webs. The codebook is like an organized poll made out of a continuation of things, each intended to identify the presence/absence of a specific property considered important to the examination (Rimano et al., 2015). The coding sheet was created based on the current literature on site content examination and incorporated the accompanying six measures to inspect the level of persuasiveness of hotels' sites:

i.   Informativeness: was calculated as the information abundance and quality demonstrated in the Web page.
ii. Usability: was characterized as the lack of difficulty of use of the Web site: understanding and navigation ease
iii. Credibility: is based on the security and trustworthiness communicated by the Web site
iv. Inspiration: refers to the Web site's visual features.
v.   Involvement: was characterized as the effort to enhance Web site interactivity
vi. Reciprocity: is conceptualized as the advantages offered to the users of hotel Web sites in the hope of building a reciprocal relationship

These measurements help sites to expand persuasiveness and hence, to impact site clients' attitudes. The examination was confined to the English variant of every site. The coders were prepared utilizing precise rules and operational definitions to decide if the variable was available or missing in each sample item. Additionally, a few sites were coded that were excluded from the example so as to distinguish and resolve any possible issues in coding and to advance intercoder understanding. In the wake of preparing, the coders freely investigated all hotels Web pages.

Assessing the continuous changes and updating of sites is one of the significant difficulties in site content investigation (Weare & Lin, 2000). Along these lines, the coders dissected similar sites around the same time. To additionally guarantee every one of the sites were seen under similar conditions, the coders utilized a similar program settings and a similar sort of computers. The coders made decisions autonomously of the sites for every one of the six measures. All estimates utilized an ostensible scale (dichotomous) to characterize the presence of every class with estimations of 1 or 0 doled out contingent upon the existence or lack thereof of the classification in the site (Neuendorf, 2016).

4 Results

Content analysis was accomplished in order to analyze four-star hotels Web sites in Greece. This study uses six main coding categories that were identified before the analysis based on the research of (Diaz & Koutra, 2013): informativeness, usability, credibility, inspiration, involvement, and reciprocity. The support technique employed to collect the data of 404 hotels in Greece. The hotels are around all the Greece and have Web site with at list English and Greek version. Furthermore, the analyzed hotels were located in different regions of Greece such as (Peloponnese, Macedonia, Central Greece, Aegean Island, and Crete). The Web sites under investigation were visited during July 2019. The coding framework was created in accordance with relevant literature concerning Internet page content analysis (Halpern & Regmi, 2013). Furthermore, the researches responsible for the content analysis visited each Web page independently and analyzed which one based on the six different factors. All the measures used a nominal scale (dichotomous) to define the presence of each item. More specifically, values of 1 or 0 were assigned depending on the appearance or non-appearance of each item among the various categories. Additionally, a number of rooms for each region are presented in (Table 1).

Hotels' Web pages must retain a user-friendly approach by assisting visitors to search and book online. Credibility concerns the security of and communication by the Web page. The content of a Web page ought to be credible, accurate, and

Table 1 Hotels rooms per region

Region	Frequency	Percentage	Cumulative percentage
Aegean Island	165	40.84	40.84
Central Greece	4	0.99	41.83
Crete	138	34.16	75.99
Macedonia	39	9.65	85.64
Peloponnese	58	14.36	100
Total	404	100	

comprehendible to assure that visitors will be convinced by the Web page. Inspiration pertains to the Web page's visual characteristics. The layout of a Web page has to be appealing, as a well-lit, spruce and properly ordered online booking indicates expertise, and must assist the visitor to locate desired services or information, or to single out services that are novel or unique and which strengthen the persuasiveness of the Web page. Involvement was delineated as the attempt to enhance Web page's interactivity. The outline of the Web page must include a range of social influences to motivate and influence visitors to buy a service online by offering them diverse options analogous to those that are possible when they purchase in the hotel reservation centers or travel agencies. Furthermore, Web pages should be collaborative and offer visitors relevant feedback to motivate them to book and purchase online. At last, reciprocity is conceptualized as the advantages provided to the visitors of hotel Web sites in the hope of creating a reciprocal relation. The aforementioned dimensions assist Web pages to enhance their persuasiveness and as a result, to increase Web page visitors' attitudes. This could led to an increase of actual bookings through the Web page.

Furthermore, Cronbach alpha reliability test coefficients were 0.78 for informativeness, 0.91 for usability, 0.76 for credibility, 0.87 for inspiration, 0.75 for involvement, and 0.81 for reciprocity. Even though that there is the standardized level for different reliability coefficients, it is suggested that those exceeding 0.70–0.95 indicate excellent agreement and 0.40–0.75 fair to good agreement (Neuendorf, 2016). Based on this criterion, all six variables had high levels of reliability, ranging from 0.75 to 0.91 ($p < 0.001$), with a confidence interval of 95%. In addition, mean difference test between five regions found that the total score of each region they do not differ. (Table 2).

For the aim of the present study, a latent class cluster analysis of four-star hotels' Web sites was performed. The analysis assessed the heterogeneity of four-star hotels in accordance with the degree of persuasiveness of their Web pages. In line with previous studies, Web page persuasiveness was calculated by the subsequent variables: informativeness, usability, credibility, inspiration, involvement, and reciprocity. Table 3 displays a summary of the estimation and adjustment indexes for each of the five models. Results indicated that there were two groups of four-star hotels, in accordance with the Bayesian information criterion (BIC). The lowest BIC value correlates with the best model based on (Vermunt & Magidson, 2002,2003). It was also feasible to take into consideration the various algorithms such as K-means,

Table 2 Mean difference between total score and mean score of each region

Region	Mean	p-value
Central Greece	31	0.397
Peloponnese	33.379	0.876
Crete	34.036	0.459
Aegean Island	36.133	0.086
Macedonia	34.948	0.498

Table 3 Latent class cluster of persuasiveness variable

Number of latent class	Log-likelihood	BIC	L_2	Classification error	Standard R-squared
Cluster 1	−5263.24	8932.42	5662.36	0.0001	1
Cluster 2	−5116.92	8859.95	5369.42	0.0048	0.9758
Cluster 3	−4970.6	8787.48	5076.48	0.0095	0.9516
Cluster 4	−4824.28	8715.01	4783.54	0.0142	0.9474

hierarchical clustering with the different exit parameter values. Indices showed that the model was well adjusted. In addition, the L_2 statistic, which shows the quantity of the relation noticed between the variables that cannot be explained by a model, was very high. Lastly, the R^2 were close to 1.

This analysis identifies that we have two main cluster based on the total score of each hotel and the number of rooms (see Table 4). The luxury hotels with over 100 rooms and the follower hotels with less than 100 rooms.

4.1 Group 1: Primary Hotels

This is the less various area and contains 48.76% of the hotels tests. This area is the most powerful and can create impact on clients so as to explore and make reservations through the site. This group has the most elevated potent with a high rate in the components examined. This group makes high qualities in the entirety of its segments. For instance, the sites join booking on the Web (90%), destinations (99%), travel data (91.4%), and helpful information (80.4%).

The ease of use measurement has a high rate right now, which are updated (98.3%), the sites are straightforward (98.6%), have a site map (86.3%), a search tool (95.7%), and have at any rate two languages (99.5%), which exhibits beneficiary universal profile and the significance of hotels that speak with target markets from a scope of nationalities.

The credibility measurement is ruled by three key things: data about the organization (84.7%), security policy (92.4%), and history (58.9%). The design and visual attributes of the sites show high figures of hotels. Under this framework, (98.2%) of the hotels in this category showcase a decent generally structure and design and the (89.9%) present great visual highlights.

This association has a high extent of inclusion things. In particular, (72.6%) of this group's Web pages contain feedback outcomes, so customers can contact the hotels, and have different instruments to impart among themselves and thusly share assessments about the organization and proposals, for example, social network (98.4%) or reliability programs (99.6%). Furthermore, RSS/WAP frameworks have a high presence in the sites (98.6%). RSS or Really Simple Syndication empowers repeated publication of refreshed site data to customers who need to get timely updates. WAP

Table 4 Size and profile of sections identified

Variable			Cluster 1: Primary hotels 48.76% (%)	Cluster 2: Follower hotels 51.24% (%)
Segmentation variables	Informativeness	Travel information	91.4	99.7
		During your residence	99.2	99.5
		Travel conditions	95.6	98.6
		Useful information	80.4	97.6
		Booking online	90.0	99.4
		Price	94.0	99.3
		Destinations	99.0	97.6
		Residence status	84.0	94.3
		Baggage	32.2	92.9
		Special assistance	35.0	91.6
		Career/Employment	45.6	89.2
		Hotel	65.3	86.3
		Car hire	52.3	88.0
		Hotel partners	35.9	86.9
		Refunds	56.0	85.7
		Innovation	81.3	84.6
		Sales area	47.6	73.4
	Usability	Updated	98.3	62.2
		Web map	86.3	81.1
		Search tool	95.7	79.9
		Easy of understanding	98.6	78.8
		Languages	99.5	77.6
	Credibility	About us	84.7	76.5
		History	58.9	75.3
		Privacy policy	92.4	74.2
		Financial information	56.7	85.2
	Inspiration	Layout	98.2	74.8
		Visual features	89.9	86.3
	Involvement	Feedback	96.8	72.6
		Social network	78.5	98.4
		Loyalty programs	95.9	99.6
		RSS/WAP	99.4	98.6
	Reciprocity	Customer service	99.9	92.3
		Promotion	97.2	79.6

(continued)

Table 4 (continued)

Variable			Cluster 1: Primary hotels 48.76% (%)	Cluster 2: Follower hotels 51.24% (%)
		Newsletter	89.1	99.7

or Wireless Application Protocol is a specialized standard for getting to data over a portable remote system during the residence. If the reciprocity variable is taken into account, there is a high extent of B2C correspondence with customers, for example, advancement (97.2%), news or newsletter (89.1%) or customer support (99.9%).

4.2 Group 2: Follower Hotels

Then again, this segment contains 51.24% of the hotel's examples. This segment is the less intense and conclude four-star hotels with under 100 rooms accessible for the clients. This group has the most elevated qualities on informativeness and association instead of validity and ease of use. This group makes high qualities in the accompanying parts. For instance, the sites fuse booking on the Web (99.4%), destinations (97.6%), price (99.3%), valuable data (97.6%), residence status (94.3%), and special help (91.6%).

All the more explicitly, the ease of use has a lower rate in this group's Web pages, which are updated (62.2%), the sites are straightforward (78.8%), have a site map (81.1%), a search tool (79.9%), and have at least two languages (77.6%), which exhibits their nearby profile.

The credibility measurement is commanded by three key things: data about the organization (76.5%), security policy (74.2%), and history (75.3%). The design and visual attributes of the sites show how hotels figures out. Under this framework, (74.8%) of the hotels within this group demonstrate an ordinary generally structure and format and the (86.3%) present typical visual highlights.

This association has a high extent of inclusion things. In particular, 96.8% of this group's Web pages contain feedback outcomes, so customers can contact the hotels, and have different instruments to impart among themselves and thusly share assessments about the organization and proposals, for example, social network (78.5%) or reliability programs (95.9%). Furthermore, RSS/WAP frameworks have a high presence in the sites (99.4%). RSS or Really Simple Syndication empowers repeated publication of refreshed site data to customers who need to get timely updates. WAP or Wireless Application Protocol is a specialized standard for getting to data over a portable remote system during the residence. If the reciprocity variable is taken into account, there is a high extent of B2C correspondence with customers, for example, advancement (79.6%), news or newsletter (99.7%) or customer support (92.3%).

5   Conclusion, Capitalize the First Letter of Every Word in the Title

This exploration adds to hypothesis, technique, and practice. The hypothetical contributions coordinate the modification and outline types that Kim and Fesenmaier (2008) have recognized in their components of persuasiveness to the marketing domain, especially to four-star hotels' Web design. In light of this investigation, the examination distinguishes the requirement for another point of view and portrays a creative way to deal with hotel segmentation.

Taking it into account from a manager's viewpoint, this research has underlined the significant groups of four-star hotels in regard to a series of indicators evaluating Web page persuasiveness. The outcomes showcase that there are two four-star hotels groups: (1) "Primary four-star hotel" and (2) "Follower four-star hotels". This division might help hotel directors to assess the level of persuasiveness of each group, estimate the attractiveness of each group's Web sites as well as perform the essential actions in order to enhance their Web sites.

Despite the fact that this examination gives a structure to proceeding with look into on this significant territory of the accommodation division, extra research is expected to propel the aftereffects of this investigation as certain impediments remain. One significant constraint is that the examination depended distinctly on Greek four-star hotels and along these lines it might lack generality. Hence, rehashing this examination with various examples would fortify the discoveries. In spite of this constraint, it is contended that the aftereffects of this examination contribute significantly to our comprehension of the persuasive architecture of hotel Web pages and give an establishment to future research exploring the Internet as a convincing tool.

References

Ahren, R. K., Stromer-Galley, J., & Neuman, W. R. (2000). Interactivity and structured issue comparisons on the political web: An experimental study of the 2000 New Hampshire presidential primary. *International Communication Association, June*, 1–5.

Aldlaigan, A. H., & Buttle, F. A. (2001). Consumer involvement in financial services: An empirical test of two measures. *International Journal of Bank Marketing, 19*(6), 232–245.

Alhabash, S., McAlister, A. R., Hagerstrom, A., Quilliam, E. T., Rifon, N. J., & Richards, J. I. (2013). Between likes and shares: Effects of emotional appeal and virility on the persuasiveness of anti cyberbullying messages on Facebook. *Cyberpsychology, Behavior, and Social Networking, 16*(3), 175–182.

Baloglu, S., & Pekcan, Y. A. (2006). The website design and Internet site marketing practices of upscale and luxury hotels in Turkey. *Tourism Management, 27*(1), 171–176.

Bezjian-Avery, A., Calder, B., & Iacobucci, D. (1998). New media interactive advertising vs. traditional advertising. *Journal of Advertising Research, 38*, 23–32.

Blythe, J. (1997). *The essence of consumer behaviour*. Pearson PTR.

Buhalis, D., & Law, R. (2008). Progress in information technology and tourism management: 20 years on and 10 years after the Internet—The state of eTourism research. *Tourism Management, 29*(4), 609–623.

Cheung, C., & Law, R. (2009). Have the perceptions of the successful factors for travel web sites changed over time? The case of consumers in Hong Kong. *Journal of Hospitality & Tourism Research, 33*(3), 438–446.

Chiou, W. C., Lin, C. C., & Perng, C. (2010). A strategic framework for website evaluation based on a review of the literature from 1995–2006. *Information & Management, 47*(5–6), 282–290.

Díaz, E., & Koutra, C. (2013). Evaluation of the persuasive features of hotel chains websites: A latent class segmentation analysis. *International Journal of Hospitality Management, 34,* 338–347.

Díaz, E., & Martín-Consuegra, D. (2016). A latent class segmentation analysis of airlines based on website evaluation. *Journal of Air Transport Management, 55,* 20–40.

Essawy, M. (2005). Testing the usability of hotel websites: The springboard for customer relationship building. *Information Technology & Tourism, 8*(1), 47–70.

Fong, L. H. N., Lam, L. W., & Law, R. (2017). How locus of control shapes intention to reuse mobile apps for making hotel reservations: Evidence from Chinese consumers. *Tourism Management, 61,* 331–342.

Gao, Y., & Koufaris, M. (2006). Perceptual antecedents of user attitude in electronic commerce. *ACM SIGMIS Database: The DATABASE for Advances in Information Systems, 37*(2–3), 42–50.

Geerts, W. (2014). Environmental certification schemes: Hotel managers' views and perceptions. *International Journal of Hospitality Management, 39,* 87–96.

Halpern, N., & Regmi, U. K. (2013). Content analysis of European airport websites. *Journal of Air Transport Management, 26,* 8–13.

Heeter, C. (2000). Interactivity in the context of designed experiences. *Journal of Interactive Advertising, 1*(1), 3–14.

Herzberg, F. (1987). One more time: how do you motivate employees? *Harvard business review, 65*(5).

Jensen, J. F. (1998). Interactivity. *Nordicom Review, Nordic research on media and comunication review, 19*(2).

Kamenidou, I., Balkoulis, N., & Priporas, C. V. (2009). Hotel business travellers satisfaction based on service quality: A segmentation approach in inner city five-star hotels. *International Journal of Leisure and Tourism Marketing, 1*(2), 152–172.

Kim, H., & Fesenmaier, D. R. (2008). Persuasive design of destination web sites: An analysis of first impression. *Journal of Travel Research, 47*(1), 3–13.

Krippendorff, K. (2004). Measuring the reliability of qualitative text analysis data. *Quality and Quantity, 38,* 787–800.

Law, R. (2019). Evaluation of hotel websites: Progress and future developments (invited paper for 'luminaries' special issue of International Journal of Hospitality Management). *International Journal of Hospitality Management, 76,* 2–9.

Law, R., & Hsu, C. H. (2005). Customers' perceptions on the importance of hotel web site dimensions and attributes. *International Journal of Contemporary Hospitality Management.*

Law, R., & Jogaratnam, G. (2005). A study of hotel information technology applications. *International Journal of Contemporary Hospitality Management, 17*(2), 170–180.

Law, R., & Wong, J. (2003). Successful factors for a travel web site: Perceptions of on-line purchasers in Hong Kong. *Journal of Hospitality & Tourism Research, 27*(1), 118–124.

Law, R., Qi, S., & Buhalis, D. (2010). Progress in tourism management: A review of website evaluation in tourism research. *Tourism Management, 31*(3), 297–313.

Lee, J., & Hong, I. B. (2016). Predicting positive user responses to social media advertising: The roles of emotional appeal, informativeness, and creativity. *International Journal of Information Management, 36*(3), 360–373.

Li, L., Peng, M., Jiang, N., & Law, R. (2017). An empirical study on the influence of economy hotel website quality on online booking intentions. *International Journal of Hospitality Management, 63,* 1–10.

Loda, M. D., Teichmann, K., & Zins, A. H. (2009). Destination websites' persuasiveness. *International journal of culture, tourism and hospitality research.*

Loiacono, E. T., Watson, R. T., & Goodhue, D. L. (2002). WebQual: A measure of website quality. *Marketing Theory and Applications, 13*(3), 432–438.

Magidson, J., & Vermunt, J. K. (2002). A nontechnical introduction to latent class models. *Statistical Innovations White Paper, 1,* 15.

McMillan, S. J. (2000). Interactivity is in the eye of the beholder: Function, perception, involvement, and attitude toward the web site. In *Proceedings of the conference-American academy of advertising* (pp. 71–78). Pullman, WA; American Academy of Advertising.

McMillan, S. J., & Hwang, J. S. (2002). Measures of perceived interactivity: An exploration of the role of direction of communication, user control, and time in shaping perceptions of interactivity. *Journal of Advertising, 31*(3), 29–42.

Mills, J. E., Han, J. H., & Clay, J. M. (2008). Accessibility of hospitality and tourism websites: A challenge for visually impaired persons. *Cornell Hospitality Quarterly, 49*(1), 28–41.

Neuendorf, K. A. (2016). *The content analysis guidebook.* Sage.

Nielson, J. (2000). *Designing web usability* (p. 105). Indianapolis, IN: New Riders.

O'Keefe, D. J. (2002). The persuasive effects of variation in standpoint articulation. *Advances in pragma-dialectics,* 65–82.

Patwardhan, B., Vaidya, A. D., & Chorghade, M. (2004). Ayurveda and natural products drug discovery. *Current Science,* 789–799.

Pavlou, P. A., Liang, H., & Xue, Y. (2007). Understanding and mitigating uncertainty in online exchange relationships: A principal-agent perspective. *MIS Quarterly,* 105–136.

Perloff, R. M. (1993). *The dynamics of persuasion: Communication and attitudes in the 21st century.* Routledge.

Petty, R. E., & Cacioppo, J. T. (1986). The elaboration likelihood model of persuasion. In *Communication and persuasion* (pp. 1–24). Springer, New York, NY.

Piccoli, G. (2008). Information technology in hotel management: A framework for evaluating the sustainability of IT-dependent competitive advantage. *Cornell Hospitality Quarterly, 49*(3), 282–296.

Richins, M. L., & Bloch, P. H. (1986). After the new wears off: The temporal context of product involvement. *Journal of Consumer Research, 13*(2), 280–285.

Rimano, A., Piccini, M. P., Passafaro, P., Metastasio, R., Chiarolanza, C., Boison, A., & Costa, F. (2015). The bicycle and the dream of a sustainable city: An explorative comparison of the image of bicycles in the mass-media and the general public. *Transportation Research Part F: Traffic Psychology and Behaviour, 30,* 30–44.

Samad, S., Nilashi, M., & Ibrahim, O. (2019). The impact of social networking sites on students' social wellbeing and academic performance. *Education and Information Technologies, 24*(3), 2081–2094.

Sartori, A., Mottironi, C., & Corigliano, M. A. (2012). Tourist destination brand equity and internal stakeholders: An empirical research. *Journal of Vacation Marketing, 18*(4), 327–340.

Schmidt, S., Cantallops, A. S., & dos Santos, C. P. (2008). The characteristics of hotel websites and their implications for website effectiveness. *International Journal of Hospitality Management, 27*(4), 504–516.

Sohn, D. (2011). Anatomy of interaction experience: Distinguishing sensory, semantic, and behavioral dimensions of interactivity. *New Media & Society, 13*(8), 1320–1335.

Vermunt, J. K., & Magidson, J. (2002). Latent class cluster analysis. *Applied Latent Class Analysis, 11,* 89–106.

Vermunt, J. K., & Magidson, J. (2003). Latent class models for classification. *Computational Statistics & Data Analysis, 41*(3–4), 531–537.

Vrana, V., Zafiropoulos, C., & Paschaloudis, D. (2004). Measuring the provision of information services in tourist hotel web sites: The case of Athens-Olympic city 2004. *Tourism and Hospitality Planning & Development, 1*(3), 255–272.

Weare, C., & Lin, W. Y. (2000). Content analysis of the World Wide Web: Opportunities and challenges. *Social Science Computer Review, 18*(3), 272–292.

Yang, X., Ahmed, Z. U., Ghingold, M., Sock Boon, G., Su Mei, T., & Lee Hwa, L. (2003). Consumer preferences for commercial web site design: an Asia-Pacific perspective. *Journal of Consumer Marketing, 20*(1), 10–27.

Yeung, T. A., & Law, R. (2006). Evaluation of usability: A study of hotel web sites in Hong Kong. *Journal of Hospitality & Tourism Research, 30*(4), 452–473.

Ye, Q., Law, R., & Gu, B. (2009). The impact of online user reviews on hotel room sales. *International Journal of Hospitality Management, 28*(1), 180–182.

Zach, F., Gretzel, U., & Fesenmaier, D. R. (2008). Tourist activated networks: Implications for dynamic packaging systems in tourism. In *Information and communication technologies in tourism 2008* (pp. 198–208). Springer, Vienna.

Zhang, P., & von Dran, G. (2001, January). Expectations and rankings of Web site quality features: Results of two studies on user perceptions. In *Proceedings of the 34th annual Hawaii international conference on system sciences* (p. 10). IEEE.

Zhang, P., Von Dran, G. M., Blake, P., & Pipithsuksunt, V. (2001). Important design features in different web site domains: An empirical study of user perceptions. *E-Service, 1*(1), 77–91.

Pilot Study for Two Questionnaires Assessing Intentions of Use and Quality of Service of Robots in the Hotel Industry

Dimitrios Belias and Labros Vasiliadis

1 Introduction

The use of robots for scientific and technological purposes is comprised of the design, development, and construction of robots, as well as the application of the use of robots in the industry (Belias & Varelas, 2018). The International Organization for Standardization defines a robot as an "automatically controlled, reprogrammable, multipurpose manipulator, and programmable in three or more axes, which can be either fixed in place or mobile for use in industrial automation applications" (ISO, 8373:2012). There are already many applications of robotics in automation and in manufacturing, like in the automobile industry (Chestler, 2016), while robots are also used as part of services provision during disasters (Kobres, 2018; Rauch, 2017).

In the tourist industry, where one of the most important challenges is the provision of high-quality services and the promotion of guest satisfaction (Barsky & Labagh, 1992), interest in robots has been rising, although the development of robotics hotel applications is still in its early stages (Ivanov & Webster, 2017). In the hotel and tourism industry, the concept of robot use is slowly but steadily gaining ground, particularly in high cost, high-quality tourist organizations (Belias, 2019a). Robots are being slowly introduced in the tourism industry and especially in hotels (Murphy et al., 2019; Pinillos et al., 2016). Some existing applications are primarily introduced through startup companies and include using robotics for bar automation and for welcoming the guests, as well as for some housekeeping tasks like floor cleaning (Ivanov et al., 2018; Kaivo-Oja et al., 2017).

D. Belias (✉)
University of Thessaly, 41500 Larissa, Geopolis, Greece
e-mail: dbelias@pe.uth.gr

L. Vasiliadis
National and Kapodistrian University of Athens, Athens, Greece

© The Author(s), under exclusive license to Springer Nature Switzerland AG 2021
V. Katsoni and C. van Zyl (eds.), *Culture and Tourism in a Smart, Globalized, and Sustainable World*, Springer Proceedings in Business and Economics,
https://doi.org/10.1007/978-3-030-72469-6_36

2   Literature Review

The use of robotics in the hotel industry is not without its critics, where concern has been voiced for the impact that the use of robots may have on labor markets (Thomas, 2017), as well as the potential impact on human jobs, including loss or shrinking of positions (Chan & Tung, 2019; Belias, 2019c; Belias & Varelas, 2018). In the past, innovations are sometimes initially met with skepticism by the public and by professionals. Examples of such innovations are online hotel booking and low-cost flights, which today comprise an integral part of the tourism industry, and innovator companies like Booking.com and EasyJet are global, well-known, and successful. In this context, robotics applications in the hotel and tourism industry may be a desirable investment for startup companies and entrepreneurship (Belias, 2019a). The use of robots in the hotel industry may have benefits like the reduction of labor costs, the improvement of guest experience, and the heightening of hotel's standards and operational efficiency (Ivanov & Webster, 2017; Pinillos et al., 2016). The use of robotics can have positive effects in the quality of service of hotels (Parasuraman et al., 1991), like performing simple routine tasks and reducing the workload of the reception (Belias, 2019a), in ways that are seamless and error-free.

There are still a limited number of robotic applications in the tourism industry (Belias & Varelas, 2018). Additionally, little research has been conducted regarding the use of robots in hotels (Belias, 2019b; Ivanov et al., 2019), and there is need for more research in this area (Belias & Varelas, 2018). Still, hotel customers may be becoming more open to the use of robots in the hotel and the individualized and novel experiences that they bring to the hotel service (Ivanov & Webster, 2019). Some studies have shown that hotel clients have positive attitudes toward using robots in the hotel and that there is a relationship between the use of robots and service quality (Ivanov & Webster, 2017; Ivanov et al., 2018). Another study showed that 214 participants had a higher intention to buy robotics hotel services, after they watched a video presentation regarding robot hotel services compared to a traditional video presentation of hotel services (Zhong et al., 2020). Another study with 240 participants found that using robots in the hotel led to positive brand evaluations for the sensory and intellectual aspects of the experience, but not for the affective aspect of the experience (Chan & Tung, 2019).

The present research is a pilot study of two questionnaires related to the use of robots in the hotel industry. The first assesses the intentions and the expectations of the use of robots in hotels, while the second is a version of the quality of services questionnaire (Servqual, Parasuraman et al., 1991), adapted for use with robots in the hotel industry.

3 Methodology

3.1 Sample

The sample of the pilot study was comprised of 157 adult participants, whose responses were collected as part of a convenience sample (Creswell, 2014). Participants were approached through the social and professional contacts of the researcher.

3.2 Materials

The materials that were used were two questionnaires. The first questionnaire examines intentions and expectations of robot use in the hotel industry. This questionnaire was created by the researcher after consulting the relevant literature, specifically to address participants' intentions about robotic applications in the hotel. It includes 28 items, which are rated on a five-point Likert-type scale from 1 ("completely disagree") to 5 ("completely agree"). The second questionnaire was an adaptation of the service quality questionnaire (Servqual, Parasuraman et al., 1991), for use with robots in the hotel industry. It is comprised of 22 items and uses the same five-point Likert scale (1 = "completely disagree" to 5 = "completely agree".

3.3 Data Analysis

Data were collected and analyzed using the software Spss Version 25. Two factor analyses were performed for each of the two questionnaires, in order to examine the factor structure of the scales. Principal component analyses were performed, with direct Oblimin rotation (delta = 0) and with Kaiser normalization, along with the tests of Kaiser–Meyer–Olkin's sampling adequacy (KMO) and Bartlett's data sphericity. Additionally, Cronbach reliability was performed for the two scales and for the factors that were extracted, along with Kolmogorov–Smirnov tests of data normality. Face, content and, concurrent validity was assessed for intention/expectations of robot use and service quality from robot use in the hotel industry (Creswell, 2014; Patten & Newhart, 2018). To examine concurrent validity between the factors of the two scales, non-parametric Spearman rho correlation tests were performed.

3.4   Results of the Pilot Study

This chapter presents the results of the pilot study regarding the intention and expectations from the use of robots in the hotel industry, as well as satisfaction from services provided through the use of robots in the hotel industry (Servqual). The sample of the study was comprised of 157 employees in hotel units in Greece. Initially, the demographic and other characteristics of the sample are presented. Reliability tests are then performed for the overall scales of intention and expectations from the use of robots and satisfaction with services provided using robots. Factor analyses were calculated for these two scales in order to study their factor structure analyses, given that the use of robots in general and in the hotel industry in particular is an innovative and emerging practice, and therefore, the factors that determine the intentions and expectations and the service quality of robots may be different than in other application and require study. The reliability of the factor structures is found to be high and acceptable. There was also evidence that the two questionnaires exhibit face, content, and concurrent validity. Finally, conclusions from the pilot study are provided, indicating that, overall, the two questionnaires and their factor structures are reliable and valid.

3.5   Characteristics of the Sample

The majority of employees were male (72%). The marginal majority of the sample were married (57%), with 33% being single and 10% being divorced. Almost all participants had the Greek nationality (97.5%). Almost one in three participants was graduates of technical schools (ATEI, 32.5%), 24% were university degree graduates (AEI), and 22% had a Master's degree. Most participants had a permanent job (76%), and the hotels in which participants worked were either five-star (51%) and four-star (49%) hotels. Table 1 presents these results.

Overall, 60% of the sample were between 31 and 50 years old. Fifteen percent were up to 30 years old (15%), 19% were between 51 and 60 years old, and 6% were over 60 years old. Almost one in two participants had more than 20 years of work experience (46%). Additionally, 11% had work experience of up to 5 years, 12% had work experience of 6 to 10 years, 20% had work experience of 11 to 15 years, and 11.5% had work experience of 16 to 20 years. One in five participants had worked in this hotel unit for 1 to 2 years (22%). Fifteen percent had worked in this hotel between 3 and 5 years (15%), 17% had worked at the hotel between 6 and 10 years, 21% had worked at the hotel between 11 and 20 years, and 25.5% had worked in that particular hotel for over 20 years. Years of employment in the hotel industry in general were 1–2 years for 8%, 3–5 years for 11%, 6–10 years for 18.5%, 11–20 years for 23%, and over 20 years for 39.5% of the employees (Table 2).

The main hotel departments that the employees worked at were the following. More than one in three participants were executive managers (36%). More than

Table 1 Gender, family status, nationality, educational level, type of work, and hotel class ($N = 157$)

		Frequency	Percent
Gender	Male	114	72.6
	Female	43	27.4
Family status	Married	90	57.3
	Single	52	33.1
	Divorced	15	9.6
Nationality	Greek	153	97.5
	Cypriot	2	1.3
	Georgian	2	1.3
Educational level	High school	17	10.8
	Vocational training (IEK)	12	7.6
	Technical school (ATEI)	51	32.5
	University degree (AEI)	37	23.6
	Master's degree	35	22.3
	Ph.D.	5	3.2
Type of work	Permanent position	120	76.4
	Seasonal position	37	23.6
Hotel class	5 stars	80	51.0
	4 stars	77	49.0

one in four participants were owners or board members (25.5%), and one in five participants worked in the reception (20%). Eight percent of the sample worked in the accounting department or the offices (8%). These and the remaining departments are presented in Table 3.

The locations of the hotels, where the participants of this study were employed, varied and are listed in the Appendix. Overall, 15% of the hotels were located in Thessaloniki, 21% were located in Kriti, 7.6% were located in Mykonos, 4.5% were located in Kefalonia, and 12.1% were located in Athens and the wider prefecture of Attica.

3.6 Reliability of Total Scales

Reliability tests were performed for the overall scales of intention and expectations of robot use in the hotel industry and quality of service of robots in the hotel industry (Servqual). The results, shown in Table 4, show that both scales had high and acceptable Cronbach reliability. Specifically, the intention and expectations of robot use in

Table 2 Age, work experience, and years of work at the hotel and in the hotel industry (N = 157)

		Frequency	Percent
Age	<30	23	14.6
	31–40	36	22.9
	41–50	58	36.9
	51–60	30	19.1
	>61	10	6.4
Work experience	0–5	17	10.8
	6–10	19	12.1
	11–15	31	19.7
	16–20	18	11.5
	>20	72	45.9
Years of work in this specific hotel unit	1–2	34	21.7
	3–5	24	15.3
	6–10	26	16.6
	11–20	33	21.0
	>20	40	25.5
Years of work in the hotel industry	1–2	13	8.3
	3–5	17	10.8
	6–10	29	18.5
	11–20	36	22.9
	>20	62	39.5

Table 3 Work departments

	Frequency	Percent
Managerial	57	36.3
Owner-Member of board	40	25.5
Reception	31	19.7
Accounting/office	13	8.3
Restaurant	6	3.8
Kitchen	4	2.5
Bar	2	1.3
Housekeeping	2	1.3
Spa	1	0.6
Public relations	1	0.6

Table 4 Reliability of total scales of intention and expectations of use of robots and service quality of robots in the hotel industry

	Cronbach reliability	No. of items
Intention and expectations of use of robots in the hotel industry	0.756	28
Service quality of robots in the hotel industry	0.962	22

hotels scale had a high reliability of $a = 0.76$ (28 items), and service quality of robot use in hotels had a very high reliability with $a = 0.96$ (22 items).

3.7 Factor Analyses of the Scales

3.7.1 Intentions and Expectations of the Use of Robots in the Hotel Industry

For the study of the factor structure of the scale of intention and expectations of robot use in the hotel industry, a principal component analysis was performed, with direct Oblimin rotation (delta = 0) and with Kaiser normalization, with Kaiser–Meyer–Olkin sampling adequacy (KMO) and Bartlett data sphericity tests. The Oblimin method was chosen, since this oblique, non-orthogonal rotation of the data led to an elimination of problems of item multicollinearity, or items that have a high loading (>0.4) on more factors than one. A simple factor analysis extraction without rotation provided a solution where three items showed multicollinearity, while a Varimax rotation led to five items exhibiting multicollinearity.

The results of tests of data suitability were positive, and the data were found to be suitable for use in the factor analysis (KMO $= 0.88 > 0.8$, Bartlett chi-square $=$ 2747.4, $p < 0.0005$). Table 5 presents these findings.

The factor analysis model extracted six factors, which explained 66.94% of the observed variance in the factor model of intention and expectations of robot use in the hotel industry (Table 6).

Table 7 presents the solution for the factor analysis that includes six factors. None of the items showed multicollinearity. It should be noted that two of the six factors consisted of only two items each. These were Items 27 and 28, with high loadings on Factor 3, and Items 14 and 15, with high loadings on Factor 6. While these two factors

Table 5 Kaiser–Meyer–Olkin and Bartlett tests (Intention and expectations of robot use)

Kaiser–Meyer–Olkin measure of sampling adequacy		0.882
Bartlett's test of sphericity	Approximate chi-square	2747.37
	df	378
	p	0.000

Table 6 Proportion of the variance that is explained by the factor analysis model (Intention and expectations of robot use)

Factor	Initial eigenvalues			Extraction sums of squared loadings			Rotation sums of squared loadings
	Total	% variance	Cum. %	Total	% Variance	Cum. %	Total
1	10.567	37.738	37.738	10.56	37.738	37.738	8.743
2	2.195	7.839	45.577	2.195	7.839	45.577	3.937
3	1.726	6.164	51.741	1.726	6.164	51.741	3.456
4	1.618	5.777	57.519	1.618	5.777	57.519	3.746
5	1.401	5.004	62.522	1.401	5.004	62.522	2.454
6	1.236	4.414	66.936	1.236	4.414	66.936	5.197
7	0.996	3.556	70.491				
8	0.876	3.129	73.620				

consisted of only two items each, and factors with less than three items are usually dismissed (Raubenheimer, 2004), in the present study it was decided to retain these two-item factors, which is acceptable especially in cases where there are important theoretical or practical reasons to do so (Gosling et al., 2003). Firstly, the two 2-item factors were retained because all items showed particularly high loadings on their respective factors. Secondly, the acceptance of these two 2-items factors led to all items of the questionnaire to be used, and there was no loss of data. Thirdly, the use of robots in the hotel industry is a relatively new endeavor in research and in application, and these two 2-item factors may be of interest for both the main study that will follow from this pilot study, but also perhaps for the future study of robot use in hotels, and specifically the intentions, expectations, and quality of service of robots in the hotel industry.

Factor 1 was named "Convenience and advantages of use of robots in the hotel industry" and consisted of 14 items (Items 1, 3, 4, 5, 6, 7, 10, 11, 12, 13, 17, 21, 22, 23). Items 3 to 7 had negative loadings and were reverse scored in order to be aligned with all other items that had positive loadings to the factor. Factor 2 was termed "Staff familiarization with new technologies and auxiliary work of robots" and consisted of three items (Items 18, 25, 26), which had high positive loadings on the factor (0.48, 0.95, 0.95, respectively).

Factor 3 was called "Anthropomorphic characteristics of robots" and consisted of two items (Items 27 and 28), which had high loadings on the factor (respectively, 0.89 and 0.95). Factor 4 was named "High cost of buying and using a robot" and consisted of three items (Items 8, 9, 19). Item 19 had a negative loading (−0.48) and was reversed in order to combine correctly with the other two items, which had a positive loading (0.90 and 0.73).

Factor 5 was called "Financial improvement in hotels and robot discretion" and consisted of three items (Items 2, 20, 24). Item 2 had a negative loading (−0.59) on the

Table 7 Final factor extraction for intention and expectations of robot use in the hotel industry

Factors	Items	Loadings
1. Convenience and advantages of use of robots in the hotel industry	1. The use of robots will facilitate financial transactions	0.427
	3. The use of robots will not complicate customer service	0.616
	4. The use of robots in the service market will not be difficult	0.734
	5. The use of robots in reception will not be difficult	0.629
	6. Robot use in cleaning services will not be difficult	0.615
	7. Robot use in alimentation services will not be difficult	0.778
	10. The use of robots will fit well with the character of hotel services offered by the hotel unit where I work	0.735
	11. The use of robots will improve the quality of services at the hotel unit where I work	0.694
	12. The use of robots will positively contribute to the work of the employees at the hotel unit	0.433
	13. The use of robots will make guest/tourist accommodation more comfortable at the hotel unit where I work	0.558
	17. I expect that the use of robots will be compatible with the hotel unit where I work	0.491
	21. Robots may come in contact with the personal items of the guest	0.753
	22. Robots may adequately replace the employee at the hotel unit where I work	0.689
	23. Robots will inspire a higher level of safety in the provision of services, compared to employees	0.597
2. Staff familiarization with new technologies and auxiliary work of robots	18. If the hotel unit where I work had robots, I would like their work to be ancillary to my work (routine tasks like cleaning the floor)	0.483
	25. I am familiar with the use of new technologies (educational technologies, information technology, icloud)	0.946

(continued)

Table 7 (continued)

Factors	Items	Loadings
	26. I am familiar with the use of smart devices (home automation, GPS, smart clocks)	0.949
3. Anthropomorphic characteristics of robots	27. The robots that will serve the clients of the hotel unit will need to have an external appearance that looks human	0.894
	28. The robots that will serve the clients of the hotel unit will need to exhibit human-like behavior (gestures, face expressions, voice)	0.952
4. High cost of buying and using robots	8. The use of robots will increase the operational cost of hotel units	0.902
	9. The purchase of robots may not be cost-effective for many hotel units	0.733
	19. The use of robots will not lead the hotel unit to an economy of scale	0.478
5. Financial improvement in hotels and robot discretion	2. The use of robots will not disrupt financial transactions	0.589
	20. The use of robots may improve the financial status of the hotel unit where I work	0.399
	24. Robots will be more discreet compared to the employees at the hotel unit where I work	0.416
6. Absence of concern for human position loss or role change	14. The use of robots does not trouble me in terms of loss of human positions in the work environment of the hotel unit	−0.934
	15. The use of robots does not trouble me in terms of role changes in the work environment of the hotel unit	−0.916

factor and was reverse scored in order to obtain a positive loading in alignment with Item 20 (0.40) and Item 24 (0.42). Finally, Factor 6 was named "Absence of concern for human position loss or role change" and consisted of two items (Items 14 and 15), which had high negative loadings on the factor (−0.93 and −0.92, respectively).

3.7.2 Quality of Service of Robots in the Hotel Industry (Servqual)

For the study of the factor structure of service quality of robot use in the hotel industry, the same methodology was applied as the case of intention and expectations of robot use in the hotel industry, since this practice showed the least cases of multicollinearity.

Principal components analysis with oblique Oblimin rotation and Kaiser normalization was used, with delta $= 0$, with Kaiser–Meyer–Olkin sampling adequacy (KMO) and Bartlett sphericity tests. The simple extraction of factors without rotation showed multicollinearity in 4 items, while orthogonal Varimax rotation was ineffective since it showed multicollinearity in 11 items. The use of the Oblimin method reduced the cases of multicollinearity to two. These cases were Item 12 ("Robots will be happy to help whenever needed"), with high loadings on Factors 1 (0.45) and 3 (0.54), and Item 17 ("Robots will help have the knowledge to answer customers' items"), which had high loadings on Factors 2 and 3 (0.42 and 0.47, respectively). These two items were excluded from the analysis.

The Kaiser–Meyer–Olkin and Bartlett tests (Table 8) were successful, and the data were deemed suitable for use in factor analysis (KMO $= 0.95$, Bartlett chi-square $= 2765.4$, $p < 0.0005$).

The second factor analysis model extracted three factors, which accounted for 68.09% of the observed variance in the data of the service quality of robot use (Table 9).

Table 10 presents the final factor model for the service quality of robot use in the hotel industry (Servqual). Factor 1 was named "Tangibles and reliability" and consisted of eleven items (Items 1–8, 11, 14–15). Overall, this factor is largely a combination or fusion of the first two factors in the classic Servqual model (Parasuraman et al., 1991), specifically service quality in terms of "tangibles" (Items 1–4) and "reliability" (Items 5–13). The loadings of the eleven items on this factor ranged from 0.54 to 0.85. Factor 2 was named "Individualized fulfillment of needs" and

Table 8 Kaiser–Meyer–Olkin and Bartlett tests (Service quality of robot use)

Kaiser–Meyer–Olkin measure of sampling adequacy		0.947
Bartlett's test of sphericity	Approximate chi-square	2765.36
	df	231
	p	0.000

Table 9 Proportion of the variance that is explained by the factor analysis model (service quality of robot use)

Factor	Initial eigenvalues			Extraction sums of squared loadings			Rotation sums of squared loadings
	Total	% Variance	Cum. %	Total	% Variance	Cum. %	Total
1	12.330	56.045	56.045	12.330	56.045	56.045	11.215
2	1.593	7.240	63.285	1.593	7.240	63.285	5.251
3	1.058	4.807	68.092	1.058	4.807	68.092	8.748
4	0.869	3.950	72.042				
5	0.744	3.383	75.425				

Table 10 Final factor extraction for service quality of robot use in the hotel industry

Factors	Items	Loadings
1. Tangibles and reliability	1. There will be better use of technological equipment	0.542
	2. Robots will contribute, so that the hotel facilities are clean and tidy	0.818
	3. The appearance of robots in the hotel facilities will be satisfactory	0.847
	4. The appearance of the hotel will improve	0.820
	5. Robots will live up to the promises of the hotel unit to its customers	0.771
	6. Robots will provide services at the right time	0.819
	7. There will be exceptional service performance by robots	0.791
	8. Robot services will be provided within a predetermined time schedule	0.806
	11. Robots will provide instantaneous service to clients	0.685
	14. Robots will inspire confidence to the clients	0.628
	15. Clients will feel safe in their interactions with robots	0.551
2. Individualized fulfillment of needs	18. Robots will provide individualized service to each client	0.705
	20. A robot will take a personal interest on each client	0.748
	22. Robots will understand the specific needs of each client	0.826
3. Exceptional service	9. Robots will keep flawless records of service	0.803
	10. Robots will inform as to the exact time that a service will be provided	0.749
	13. Robots will be able to respond at any time to client requests	0.472
	16. Robots will always be polite	0.610
	19. Robots ensure the security of the transactions	0.566
	21. The primary aim of a robot will be to provide exceptional service to clients	0.630

consisted of three items (Items 18, 20, 22). Loadings on Factor 2 were high and ranged from 0.71 to 0.83. Factor 3 was termed "Exceptional service" and consisted of six items (Items 9, 10, 13, 16, 19, 21). Loadings in Factor 3 ranged from 0.47 to 0.80.

Results show that Factor 1 consolidates the first two dimensions of the original Servqual scale, "tangibles" and "reliability" (Parasuraman et al., 1991), into a single factor. Factors 2 and 3 showed no similarity to the remaining four factors of the established Servqual scale, where "Responsiveness" involves Items 14, 16–18, "Assurance" consists of Items 15 and 19 and "Empathy" corresponds to Items 20–22 (Parasuraman et al., 1991). One explanation for this difference is that the nature of robot use in life and in the hotel industry in particular is an innovative and unprecedented practice, and therefore, the factors that determine satisfaction with the services provided by robots are likely to differ from the more traditional services provided in the hotel sector.

3.8   Reliability and Normality of the Extracted Factors

Following, Cronbach reliabilities and mean scores as well as normality tests were performed for the data for the six factors of intention and expectations of using robots in the hotel industry, as well as for the three factors of hotel service quality of robots. Table 11 summarizes the relevant results.

Factor 1 "Convenience and advantages of use of robots in the hotel industry" had a high and acceptable reliability with $a = 0.93$ (14 items). Factor 2, "Staff familiarization with new technologies and auxiliary work of robots," also had high reliability with $a = 0.78$ (3 items). Factor 3, "Anthropomorphic characteristics of robots," had very high reliability with $a = 0.92$ (2 items). Factor 4, "High cost of buying and using robots," had a moderate and marginally acceptable reliability with $a = 0.64$ (3 items). Factor 5, "Financial improvement in hotels and robot discretion," had an acceptable reliability with $a = 0.69$ (3 items). Finally, Factor 6, "Absence of concern for human position loss or role change," had a high reliability with $a = 0.88$ (2 items). Based on the results of the Kolmogorov–Smirnov normality test, the data for the six factors did not follow the normal distribution ($p < 0.200$, as per Lilliefors significance correction).

In the service quality of robot use in the hotel industry, the "Tangibles and reliability" factor had a very high Cronbach reliability with $a = 0.95$ (11 items). The "Individualized fulfillment of needs" factor had high reliability with $a = 0.86$ (3 items), as did the third factor "Exceptional service" with $a = 0.89$ (6 items). Kolmogorov–Smirnov's normality tests showed that the data for the three factors deviated from the normal distribution ($p < 0.200$).

Table 11 Reliability, normality, and means for intention/expectations of robot use and service quality from robot use in the hotel industry

		Cronbach reliability	No. of items	Mean	Std. deviation	Kolmogorov–Smirnov[a]		
						Statistic	df	p
Intention and expectations of use of robots	1. Convenience and advantages of use of robots in the hotel industry	0.934	14	2.67	0.802	0.078	157	0,020
	2. Staff familiarization with new technologies and auxiliary work of robots	0.779	3	3.75	0.743	0.216	157	0,000
	3. Anthropomorphic characteristics of robots	0.921	2	3.06	1.088	0.169	157	0,000
	4. High cost of buying and using robots	0.637	3	2.97	0.691	0.135	157	0,000
	5. Financial improvement in hotels and robot discretion	0.685	3	3.22	0.665	0.120	157	0,000
	6. Absence of concern for human position loss or role change	0.882	2	2.82	1.084	0.137	157	0,000
Service quality	1. Tangibles and reliability	0.947	11	3.08	0.796	0.117	157	0,000
	2. Individualized fulfillment of needs	0.861	3	2.48	0.987	0.107	157	0,000
	3. Exceptional service	0.893	6	3.43	0.832	0.111	157	0,000

[a]Lilliefors statistical significance correction

3.9 Validity of the Scales and Extracted Factors

Finally, the validity of the two scales, the scale of intention/expectations of robot use in the hotel industry and the service quality of robot use in the hotel industry, was examined, using measures of face validity, content validity, and concurrent validity (Creswell, 2014). The face validity of the questionnaires was confirmed through the study of the general appearance and the phrasing of the items of the two questionnaires by the researcher, who judged them to be satisfactory. Content validity (Creswell, 2014) was confirmed by five colleagues of the researcher, who received and studied the two questionnaires, the new intention/expectations of robot use in hotel questionnaire and the adapted service quality of robots in the hotel industry (Servqual), who found the content of the questionnaires to be clear and accurate, as well as that they contain sufficient items to address the issues of intentions and expectations of robot use in the hotel sector, as well as the quality of service from robot use in the hotel industry.

Concurrent validity (Creswell, 2014; Patten & Newhart, 2018) was assessed by performing a correlation analysis of the six factors of intention and expectations of robot use in hotels (Intention-Expectations of Robot Use) with the three factors of service quality of robot use in hotels (Servqual Robots). Following the normality tests, which showed that the data for these factors did not follow the normal distribution, non-parametric Spearman rho correlation tests were performed between the factors of the two scales.

The results, presented in Table 12, show that there is evidence for the concurrent validity of the intention and expectations of robot use in the hotel industry questionnaire and the quality of service through robot use in the hotel industry ($p < 0.01$). Factor 1 of the intention and expectations of robot use in hotels questionnaire had statistically significant and high correlations with all factors of service quality of robot use (Servqual Robots). In particular, Factor 1 "convenience and advantages of use of robots in the hotel industry" had positive high correlations with the Servqual robot factors of "tangibles and reliability" (rho = 0.86), "individualized fulfillment of needs" (rho = 0.73), and "exceptional service" (rho = 0.67). The fifth factor of the intention-expectations of robot use scale, "financial improvement in hotels and robot discretion," had positive medium to high correlations with Servqual robot factors "tangibles and reliability" (rho = 0.63), "individualized fulfillment of needs" (rho = 0, 41), and "exceptional service" (rho = 0.58).

Factors 2, 3, and 6 of intention-expectations of robot use had statistically significant and positive small to medium correlations with Servqual robots factors. Specifically, Factor 2 of "staff familiarization with new technologies and auxiliary work of robots" was associated with the service quality factors "tangibles and reliability" (rho = 0.35) and "exceptional service" (rho = 0.41), but not in this case with "individualized fulfillment of needs" (rho = 0.10, $p = 0.232$). Factor 3 of the intention-expectations of robot use scale, "anthropomorphic characteristics of robots," was associated with small positive correlations with all three service quality factors, "tangibles and reliability" (rho = 0.34), "individualized fulfillment of needs" (rho

Table 12 Correlation analysis between the dimensions of intention and expectations of robot use and service quality from robot use in the hotel industry

			Quality of service (Servqual)		
			1. Tangibles and reliability	2. Individualized fulfillment of needs	3. Exceptional service
Intention and expectations of use of robots	1. Convenience and advantages of use of robots in the hotel industry	Rho	0.857[a]	0.725[a]	0.669[a]
		p	0.000	0.000	0.000
		N	157	157	157
	2. Staff familiarization with new technologies and auxiliary work of robots	Rho	0.354[a]	0.096	0.409[a]
		p	0.000	0.232	0.000
		N	157	157	157
	3. Anthropomorphic characteristics of robots	Rho	0.337[a]	0.301[a]	0.206[a]
		p	0.000	0.000	0.010
		N	157	157	157
	4. High cost of buying and using robots	Rho	−0.490[a]	−0.298[a]	−0.483[a]
		p	0.000	0.000	0.000
		N	157	157	157
	5. Financial improvement in hotels and robot discretion	Rho	0.630[a]	0.408[a]	0.581[a]
		P	0.000	0.000	0.000
		N	157	157	157
	6. Absence of concern for human position loss or role change	Rho	0.457[a]	0.376[a]	0.473[a]
		p	0.000	0.000	0.000
		N	157	157	157

[a]Correlations are statistically significant at the $p > 0.01$ level

= 0.30), and "exceptional service" (rho = 0.21). Factor 6 of intention-expectations of robot use scale, "absence of concern for human position loss or role change," was associated with small to medium-sized relationships with all three Servqual robots factors "tangibles and reliability" (rho = 0.46), "individualized fulfillment of needs" (rho = 0.38), and "exceptional service" (rho = 0.47).

Finally, Factor 4 of the intention-expectations of robot use scale, "high cost of buying and using robots," had statistically significant and medium-sized negative correlations with the three factors of service quality scale, "tangibles and reliability" (rho) 0.49), "individualized fulfillment of needs" (rho = −0.30), and "exceptional service" (rho = −0.48).

4 Discussion—Conclusion

The present pilot study examined the factor structure, reliability, and validity of the intention-expectations of robot use in hotels scale and service quality of robots in the hotel industry Servqual robots, in research with 157 participants, primarily men (72%), owners, executives, and employees of hotels in various areas of Greece (including Kriti 21%, Thessaloniki 15%, Attica 12%, Mykonos 8%).

Two principal component analyses were performed, with oblique Oblimin rotation with Kaiser normalization. The first factor analysis showed the existence of six intention-expectations of robot use factors, which accounted for 67% of the observed variance in the data. The first factor was the "convenience and advantages of using robots in the hotel industry" (14 items), with very high reliability ($a = 0.93$). The second factor was the "staff familiarization with new technologies and auxiliary work of robots" (3 items), with high reliability ($a = 0.78$).

Factors 3 and 6 each consisted of two items, but it was decided to keep them in the study for three reasons (Gosling et al., 2003). First, the items showed high loadings to their respective factors. Secondly, because they have theoretical and practical importance for this type of research, specifically, thirdly, they are conceptually and practically interesting for this study subject in Greece, which has not been extensively researched. Thus, Factor 3 of the intention-expectations of robot use scale was called "anthropomorphic characteristics of robots" (2 items) and had a particularly high reliability ($a = 0.92$). Factor 4 was called "high cost of buying and using robots" (3 items) and was moderate but marginally acceptable with $a = 0.64$. Factor 5 was named "financial improvement in hotels and robot discretion" (3 items) and had acceptable reliability ($a = 0.69$). The reliability of the overall scale of intent/expectations of robot use in the hotel sector was high and acceptable with $a = 0.76$. Finally, the sixth factor was the "absence of concern for human position loss or role change" (2 items), which had high reliability ($a = 0.88$).

The second factor analysis, on the service quality of the use of robots in the hotel industry (Servqual Robots), extracted three factors, which accounted for 68.1% of the observed variance in the data of the Servqual scale. The first factor was termed "tangibles and reliability" (11 items), and it had a particularly high reliability with a

= 0.97. The second factor was "individualized fulfillment of needs" (3 items), with a high Cronbach reliability of $a = 0.86$. The third factor was "exceptional service" (6 items), and it too had high and acceptable reliability with $a = 0.89$. It is noted that the factorial structure of the adapted Servqual robots scale differed considerably from the established structure of the five Servqual factors (Parasuraman et al., 1991). Specifically, the first factor was an amalgamation of the two first factors of "tangibles" and "reliability," while the two remaining factors of the Servqual robots did not visibly correspond to the other three factors of the established five-factor Servqual model (Parasuraman et al., 1991). This finding can be explained by the fact that the use of robots in the hotel industry is a novel and innovative application, and the factors that determine the satisfaction with the services provided by robots may likely differ from the traditional structure of Servqual factors, namely tangibles, reliability, responsiveness, assurance, and empathy, in Parasuraman et al., 1991). However, it was noted that the first factor of the Servqual robots adaptation was very similar to the original structure in that it combined most of the items for the first two factors of the established Servqual scale ("tangibles" and "reliability").

In order to examine the validity of the intention-expectations of robot use in hotels and Servqual robots scales, measures of face validity, content validity, and concurrent validity were used. Face validity was established by the researcher, and content validity was established by five colleagues of the researcher, who examined of the appearance, the phrasing, and the adequacy of items to address the issues of intention/expectations and service quality of robot use in the hotel industry. To assess concurrent validity,

Kolmogorov–Smirnov normality tests showed that the data did not follow the normal distribution, and thus, non-parametric correlations were performed between the subscales of the intention-expectations of robot use in hotels scale and the service quality of robot use in hotels (Servqual Robots). Results showed that there was statistically significant support for the existence of concurrent validity between intention-expectations of robot use and Servqual robots in hotels ($p < 0.1$). specifically, the convenience and advantages of using robots in the hotel industry (Factor 1) had a high and positive relationship with the three factors of service quality, tangibles and reliability, and individualized fulfillment of needs and exceptional service (rho = 0.67 to rho = 0.86).

The factor of intention-expectations of robot use, "staff familiarization with new technologies and auxiliary work of robots" (Factor 2), was moderately correlated with tangibles and reliability (rho = 0.4) and with exceptional service (rho = 0.4), with the exception of individualized fulfillment of needs, which had no statistically significant relationship with this factor (rho = 0.1, $p > 0.05$). The anthropomorphic characteristics of robots (Factor 3) were associated to a lesser degree with all service quality factors, tangibles and reliability, individualized fulfillment of needs, and exceptional service (rho = 0.2 to rho = 0.3). The high cost of buying and using robots (Factor 4) had moderate negative correlations with tangibles and reliability, individualized fulfillment of needs, and exceptional service (rho = −0.3 to rho = −0.5). The financial improvement of the hotel and the discretion of robots (Factor 5) had medium to high positive correlations with tangibles and reliability, individualized fulfillment of needs, and exceptional service (rho = 0.4 to rho = 0.6). Finally, the absence of concern for possible job loss and role change due to the use of robots (Factor 6) also statistically significant, moderate and positive, and correlations with the three factors of service quality (rho = 0.4 to rho = 0.5).

The above suggests that the two scales may warrant further exploration in order for the factor structures to be further examined, as well as to study the issue of the use of robotics in the hotel and tourism industry and the related attitudes by hotel guests which seem to be growing favorable. The above pilot study will be followed by a full study with a larger sample size, in order to examine the views of the participants and to further assess the reliability and validity of these scales.

References

Barsky, J. D., & Labagh, R. (1992). A strategy for customer satisfaction. *Cornell Hotel and Restaurant Administration Quarterly, 33*(5), 32–40. https://doi.org/10.1177/001088049203 300524

Belias, D. (2019a). Entrepreneurship in the age of digital tourism: The future prospects from the use of robots. *Małopolska Journal* (Special Issue: Entrepreneurship theory and practice: current trends and future directions), *42*(2), 89–99.

Belias, D. (2019b). Examination of the current literature on how robots can contribute on hotel service quality. In A. Kavoura (Ed.), *International Conference on Strategic Innovative Marketing and Tourism. Springer Proceedings in Business and Economics* (pp. 835–841). New York: Springer.

Belias, D. (2019c). Research methods on the contribution of robots in the service quality of hotels. In A. Kavoura (Ed.), *International Conference on Strategic Innovative Marketing and Tourism. Springer Proceedings in Business and Economics* (pp. 939–946). New York: Springer.

Belias, D., & Varelas, S. (2018). To be or not to be? Which is the case with robots in the hotel industry? International Conference on Strategic Innovative Marketing and Tourism. In A. Kavoura, E. Kefallonitis, & A. Giovanis (Eds.), *Springer Proceedings in Business and Economics.* (pp. 935–941). New York: Springer.

Chan, A. P. H., & Tung, V. W. S. (2019). Examining the effects of robotic service on brand experience: The moderating role of hotel segment. *Journal of Travel & Tourism Marketing, 36*(4), 458–468. https://doi.org/10.1080/10548408.2019.1568953.

Chestler, D. (2016). *How robots are storming the travel industry* [online]. https://insights.ehotelier.com/insights/2016/11/15/robots-storming-travel-industry/

Creswell, J. W. (2014). *Research design: Qualitative, quantitative, and mixed methods approaches.* (4th ed.). Washington, DC: Sage.

Gosling, S. D., Rentfrow, P. J., & Swann, W. B., Jr. (2003). A very brief measure of the Big-Five personality domains. *Journal of Research in Personality, 37*(6), 504–528.

International Organization for Standardization. (2012). ISO 8373:2012: Robots and robotic devices—Vocabulary. https://www.iso.org/obp/ui/#iso:std:iso:8373:ed-2:v1:en

Ivanov, S. H., Gretzel, U., Berezina, K., Sigala, M., & Webster, C. (2019). Progress on robotics in hospitality and tourism: A review of the literature. *Journal of Hospitality and Tourism Technology, 10*(4), 489–521. https://doi.org/10.1108/JHTT-08-2018-0087

Ivanov, S. H., & Webster, C. (2017). Adoption of robots, artificial intelligence and service automation by travel, tourism and hospitality companies—A cost-benefit analysis. International Scientific

Conference 'Contemporary Tourism—Traditions and Innovations', Sofia University, October 19–21, 2017. https://ssrn.com/abstract=3007577

Ivanov, S., & Webster, C. (2019). Perceived appropriateness and intention to use service robots in tourism. In J. Pesonen & J. Neidhardt (Eds.), *Information and communication technologies in tourism 2019.* (pp. 237–248). Cham: Springer.

Ivanov, S. H., Webster, C., & Garenko, A. (2018). Young Russian adults' attitudes towards the potential use of robots in hotels. *Technology in Society, 55*, 24–32. https://ssrn.com/abstract=3193017

Kaivo-Oja, J., Roth, S., & Westerlund, L. (2017). Futures of robotics: Human work in digital transformation. *International Journal of Technology Management, 73*(4), 176–205.

Kobres, E. (2018). New technologies will revolutionize the hospitality industry. *Forbes*, June 28, 2018. https://www.forbes.com/sites/forbestechcouncil/2018/06/28/new-technologies-will-revolutionize-the-hospitality-industry

Murphy, J., Gretzel, U., & Pesonen, J. (2019). Marketing robot services in hospitality and tourism: The role of anthropomorphism. *Journal of Travel & Tourism Marketing, 36*(7), 784–795.

Parasuraman, A., Zeithaml, V., & Berry, L. L. (1991). Refinement and reassessment of the Servqual scale. *Journal of Retailing, 67*(4), 420–450.

Patten, M. L., & Newhart, M. (2018). *Understanding research methods: An overview of the essentials.* (10th ed.). New York: Routledge.

Pinillos, R., Marcos, S., Feliz, R., Zalama, E., & Gómez-García-Bermejo, J. (2016). Long-term assessment of a service robot in a hotel environment. *Robotics and Autonomous Systems, 79*, 40–57.

Raubenheimer, J. (2004). An item selection procedure to maximise scale reliability and validity. *SA Journal of Industrial Psychology, 30*(4), 59–64

Rauch, R. (2017). Hospitality trends 2018: *What's happening in hospitality* [online]. https://www.hospitalitynet.org/opinion/4085914.html

Thomas, M. K. (2017). The rise of technology and its influence on labor market outcomes. *Gettysburg Economic Review, 10*(1), Article 3. Retrieved from https://cupola.gettysburg.edu/ger/vol10/iss1/3

Zhong, L., Sun, S., Law, R., & Zhang, X. (2020). Impact of robot hotel service on consumers' purchase intention: a control experiment. *Asia Pacific Journal of Tourism Research, 25*(7), 780–798. https://doi.org/10.1080/10941665.2020.1726421

Spatial Patterns of Tourism Activity Through the Lens of TripAdvisor's Online Restaurant Reviews: A Case Study from Corfu

Christina Beneki and Thanassis Spiggos

1 Introduction

The aim of spatial data analysis (SDA) is to detect patterns and then trace and model how such patterns are interrelated to enhance understanding of the underlying mechanisms which produce them (Fisher, 2006). The key components of SDA are location, proximity, and distance. Hence, SDA findings logically arise from the locations of the subjects under analysis (Hao, 2019), while the data themselves are structured along lines of spatial dependence and heterogeneity, on the one hand, and scale and frame effects on the other (Wolf & Murray, 2017). Therefore, the decisions to be made before SDA is carried out include which scale of analysis is most appropriate; how distance is to be operationalized; and which neighbors are to be considered (Radil, 2016).

Spatial analysis is largely derived from Tobler's law (Sui, 2004; Tobler, 1970) and operated on the understanding that "pairs of features or observations taken nearby are more alike than those taken far apart." This relation underpins the notion of spatial autocorrelation, which assesses the extent of systematic spatial variation within a given mapped variable (Haining, 2001). Since spatial analysis was introduced, it has been modified and extended according to respective research interests, ranging, as Fisher (2000) points out, from "static data analysis and spatial data detection through neural networks and hierarchical models."

C. Beneki (✉) · T. Spiggos
Department of Tourism, Ionian University, Corfu, Greece
e-mail: benekic@ionio.gr

Regional Operational Programme (R.O.P.) Ionian Islands, Corfu, Greece

© The Author(s), under exclusive license to Springer Nature Switzerland AG 2021   559
V. Katsoni and C. van Zyl (eds.), *Culture and Tourism in a Smart, Globalized,
and Sustainable World*, Springer Proceedings in Business and Economics,
https://doi.org/10.1007/978-3-030-72469-6_37

Tourism is by no means a modern phenomenon; as Butler (2015) observes, evidence shows that people been engaged in tourism for hundreds of years, during which the practice has been largely defined in spatial rather than temporal terms (Romão et al., 2017). However, it is possible to see a distinct expansion of the geographical extent of tourism, as well as the time dedicated to the pursuit, from the mid-1950s. During the period 1950–2018, the number of international arrivals increased from 125.2 million to 1,404,000 million, or, in other words, there was a 55.5-fold increase over seven decades. Over the same period, the practice of tourism was taken up across more countries and regions. First, the Asia-Pacific (APAC) showed the same trend of increased tourist activity that had already been demonstrated in Europe and the Americas, and, most recently, an uptick has been seen in the importance of tourism to Middle East and Africa (MEA) nations. Although Europe's share of global tourist activity has fallen from 66.7% in 1950 to 50.8% in 2018, it retains its place at the top of the rankings in both spatial and temporal terms (Roser, 2017).

An analysis of tourism in any given area must take account of two basic dimensions, namely the spatial and the temporal (Batista e Silva et al., 2018). These dimensions are differently affected by various factors which constrain or promote international tourism, such as the resources devoted to the sector, natural limitations, fluctuations in tourist preferences, and so on, all of which affect the degree to which visitor numbers are concentrated in certain places and periods or dispersed more widely across a geographical region or the calendar (Williams, 1998).The temporal dimension of tourism has been of interest to researchers, particularly geographers, since the 1960s, while more recently, spatiotemporal analyses of tourism behaviors have been put on a more rigorous scientific footing (Xia & Arrowsmith, 2005).

Spatial patterns reveal how features are placed and arranged and the geographical relations between them. Such patterns are generated by spatial processes; hence, revealing and understanding these are fundamental to spatial analysis (Chou, 1995; Unwin, 1996).

The classic method of analyzing spatial patterns is through spatial autocorrelation, which is a means of identifying whether a global pattern exists within a given area and whether it is dispersed, clustered, or random, and of spotting anomalies to any emergent patterns. Moreover, where such a pattern is identified, the use of spatial autocorrelation also allows researchers to track it across locations and determine the relative contributions made to it by local features (O'Sullivan & Unwin, 2010). In other words, spatial autocorrelation methods yield data on where things are happening and suggest possible connections and relations between occurrences (Scott, 2015). Methods include cluster and outlier (COType) analysis and hot spot analysis, which are suitable for both global and local studies.

The current research aims to gain a more nuanced picture of patterns of tourist behavior on the Greek island of Corfu by applying spatial analysis to the "digital footprints" left on TripAdvisor. Analyzing the geographical distribution of the restaurants reviewed on the platform may yield a picture of the clusters, or hot spots, where several restaurants are reviewed, and, conversely, of the "cold spots" which are more rarely visited. These digital patterns can be leveraged to show clustering within the island as a whole, which, in turn, reveals a picture of how "tourismification" has differentially affected different parts of the island. Both hot spot and COType approaches are used.

The paper is organized as follows: Sect. 2 reviews the literature to lay out the theoretical background to the study. Thereafter, dataset and methods are described in Sect. 3, with findings discussed in Sect. 4. Finally, Sect. 5 offers conclusions and suggests the implications of the study for decision-makers.

2 Literature Review

2.1 Big Data

The literature clearly reveals that big data has assumed an ever more important place in the scientific analysis of spatial patterns of tourist behaviors. Three principal data sources are available for researchers in this field. The first, UGC data, consists of text and photos posted online. The second, device data, covers information yielded by GPS, mobile roaming, Bluetooth, etc., while the third, transaction data, refers to Web searches and online bookings, for example which Webpages are visited and on what dates. Given the differing nature of the data contained in these three repositories, they are of value to different areas of tourism research and can be leveraged to answer different queries, as outlined in the review of the most interesting recent papers below.

Pearce (1987) was one of the first researchers to use big data, in an analysis of the spatial range of charter flights in Europe. Connell and Page (2008) then studied patterns in automobile travel while Lynch et al. (2009) harnessed the power of big data to uncover links between the development of tourism in the Antarctic and the circulation of the ocean in the region. In 2011, Deng and Athanasopoulos analyzed spatiotemporal patterns of tourist behaviors in Australia and Guedes and Jiménez carried out a similar study in Portugal in 2015, with a focus on cultural tourism. Spatial analysis was used by Papageorgiou (2016) to demonstrate that Greece's National Tourism Plan had failed in its aim to change the geographical patterns of tourism in the country from a concentration along the shoreline to a multi-destination, country-wide business. Most recently, Gutiérrez et al. (2017) used big data analysis to demonstrate links between Airbnb locations and the spatial distribution of hotels in the city of Barcelona.

In a digital age, tourists are providing ever more online data, and an increasing number of ways to use it have been proposed. Two recent studies have leveraged digital data to explore tourism behaviors in the US state of Florida: Hasnat and Hasan (2018) examined tweets to trace the spatial patterns of both tourists and residents, while Kirilenko et al. (2019) focused on patterns of traffic circulation around tourist attractions in the state. Silva et al. (2018) used both big data and material from more traditional statistical sources to reveal the most important spatiotemporal patterns among tourists in Europe. Flickr has also proved a valuable source of primary data. Kim et al. (2019) collated Flickr geo-tagged photographs taken in an ASEAN Heritage Park to pinpoint hot spots, while Li et al. (2018) used the same data source to investigate how tourists and local interact, both spatially and socially, across ten US cities. Wood et al. (2013) also used Flickr data, specifically the locations in which posted photographs were taken, to estimate visitor numbers to 836 named global leisure destinations.

Several researchers have also harvested important data from TripAdvisor. Ganzaroli et al. (2017) used restaurant reviews on the platform to evaluate the relative popularity of restaurants in the Italian city of Venice, while Taecharungroj and Mathayomchan (2019) carried out a similar study for tourist attractions in Phuket, Thailand. Molinillo et al. (2016) used the same data source to analyze customer satisfaction ratings for 2211 hotels on the Spanish and Portuguese coasts. Ye et al. (2018) drew on online reviews of 314 hotels in Hong Kong to reveal patterns of tourist demand, and Vassakis et al. (2019) investigated tourist behaviors in the Greek island of Crete through an analysis of online visitor reviews of a variety of activities. Online data collected from platforms such as TripAdvisor are also of significant value to Destination Management Organizations (DMOs) seeking to make strategic decisions and formulate policy recommendations. A recent example of this is the analysis of spatial patterns in five Flemish cities principally known for their artistic heritage carried out by van der Zee et al. (2018).

2.2 Distance and Geographic Scale

The greater the distance from the matter under examination, the less able researchers are to discern associations between the features being observed (Getis, 1999). Hence, the role of distance is the key in investigating how spatial features are interrelated, whether through empirical research or insight, and the range of spatial statistics tools available to researchers take account of either distance between features or the radius of neighborhood influence (Wong, 1999).

Scale is another crucial parameter in investigating both the spatial and temporal dimensions of tourist behaviors. A hierarchical structuring of scale (Car et al. 2001; Freundschuh & Egenhofer, 1997) creates an analogy with macro and micro-levels of analysis: Researchers moving from one to another level are, in essence, moving from a planning-level examination to an enterprise-level one (Xia & Arrowsmith, 2005).

3 Methodology

The present research extracted data from TripAdvisor.com, believed to be the world's largest travel Web site. The main research question was whether spatial patterns in visitor behaviors in Corfu can be inferred from the number of customer reviews left about the island's restaurants on TripAdvisor. Spatially, the study operated on two scales: island-wide and in specific local areas taken as characteristic of Corfu's tourist development model.

3.1 Territory and Sector of Spatial Analysis

Corfu (pop. 102,052 in 2018) is the largest of the Ionian Islands. Located toward the north of the Ionian Sea, it has a total area of 592 km^2 and a 217-km shoreline. In terms of economy, the service sector is by far the most important (a 2016 figure suggests 90.7% of gross value added), largely driven by tourism and commerce. The island is one of the top tourist destinations within Greece and attracts a more cosmopolitan visitor cohort than many others in the region. It was named "Europe's leading Beach Destination" by an association representing travel agents in 2014 and presented with a European Film Location Award four years later. Tourism follows a marked seasonal pattern on the island: Most services are open between late March and late October, and there is a distinct bell-shaped curve in visitor numbers, with a sharp peak in the high season of mid-July to August.

Tourism is concentrated in specific areas within the island. The majority of resorts and tourist attractions lie along the coastline and are particularly dense toward the town of Corfu itself, which is located in the eastern part of the island. Figures gathered by the Institute of the Greek Tourism Confederation (InSETE) in 2019 reveal that the island has 408 hotels offering 47,262 beds, with a further 21,636 beds available in rooms for rent. Hotel stays appear to be increasingly popular, with 951,745 tourists choosing this accommodation type in 2018. Average duration of stay is 5.5 days, and non-Greeks are massively represented in visitor numbers, accounting for as many as 90% of tourists to the island. A further 735,832 cruise passengers also visit the island (2018).

Furthermore, approximately 1300 restaurants and related commercial enterprises operate on the island, according to figures published by the Corfu Chamber of Commerce. The present study is based on data collected from TripAdvisor in November 2019 and covers all the reviews posted to that date for each restaurant considered. The platform hosts data on 1,250 restaurants and related enterprises. A cursory look at the dates on which reviews were left corroborates the seasonality of tourism on the island, as 58% were posted within the June–August period.

3.2 The Choice of Spatial Features and Attribute

The aim of the study, namely to undertake spatial analyses of the tourist footprint in Corfu, required the scope to be limited in terms of geographical region, spatial features, population, and features considered, as outlined in further detail below.

As regards region, the whole of the island was considered as it constitutes a coherent and clearly defined spatial entity from geomorphological, administrative, and socioeconomic perspectives. The choice of spatial features to be examined was informed by the data available on TripAdvisor, which suggests three categories as key indicators of tourist behaviors: hotels, restaurants, and activities (things to do). Of these, restaurants were chosen for three reasons. Firstly, restaurant clients tend to be drawn from across international, domestic, and local groups; and secondly, visitors to restaurants offer customers a greater in situ choice than other categories, which allows an enhanced metric of tourist behaviors (Zee et al., 2018). Thirdly, restaurants were chosen for the Corfu-specific reason that this category generated more posts from tourists than hotels or "things to do." In terms of populating the sample for analysis, nearly all restaurants on the island were considered. Lastly, the number of reviews posted by clients was chosen as the attribute to be analyzed, because these not only offer a picture of long-term patterns of tourist behavior (and are thus of use in predicting future patterns), but also, on the basis of analyses carried out in other destinations, seem to be correlated with number of nights spent in a given location (Zee et al., 2018).

3.3   The Choice of Spatial Analysis Tools

Spatial statistics models add to classical ones by incorporating the concept of space and offering the necessary tools to determine the existence of spatial patterns. The first level of analysis undertaken for the present study sought to determine whether any correlation existed between feature locations and attribute values, that is, the location of restaurants under examination and the reviews left for them. For this purpose, spatial autocorrelation (Haining, 2015) was selected, using the Global Moran's I statistic (Getis & Ord, 2010). This model assesses whether a clustered, dispersed, or random pattern is expressed by a set of features (in this case, restaurants) in conjunction with a given attribute (here, customer reviews) (Scott, 2015). When statistical significance is conveyed by the p-value or z-score, a tendency toward clustering is indicated by a positive Moran's I index value, while a tendency toward dispersal by a negative one.

The next layer of analysis examined the locations of any statistically significant patterns determined in the first step. Local versions of global spatial autocorrelation statistics (Getis-Ord Gi * and the Anselin Local Moran's I) were used to analyze patterns by contextualizing restaurants within a locality and comparing their local statistical results to the global ones previously determined (Scott, 2015). A hot spot was indicated when the global mean was substantially lower than the local mean for a group of neighboring restaurants. Likewise, a local clustering of unexpectedly small values, that is, a low local mean, indicated a cold spot. Hence, the methods of analysis selected for this study considered a number of restaurants plus variation in review value to determine whether statistically significant differences existed on local and global levels. The ability to map the p-value and/or z-score generated for every feature considered allowed us to draw a picture of spatial patterns (Aldstadt, 2010).

A Gi* statistic (Getis & Ord, 2010) in the form of a z-score was generated for every feature under analysis. Hot spots were indicated by increasing z-scores, wherever these were statistically significant positive. Conversely, cold spots were indicated by diminishing z-scores, wherever these were statistically significant negative. Where the z-score was at or around zero, there appeared to be no spatial clustering.

Lastly, the hot spot/cold spot analysis described above was corroborated by a COType analysis, which detects spatial outliers through calculation of the "Anselin Local Moran's I" statistic (Anselin, 1995). This complementary layer of analysis allowed the researchers to verify whether any interesting findings had not been flagged up by the previous analysis. Under the Anselin Local Moran's I method, a cluster is indicated by a positive value for I, which in this case would indicate a restaurant with high attributes surrounded by others with similarly high attributes or, conversely, one with low attributes surrounded by similar others. An outlier, on the other hand, is identified by a negative value for I, indicating that the attributes for a given restaurant differ from those of neighboring restaurants. Hence, this type of analysis works by creating four categories of feature: a statistically significant cluster of high values (HH); a statistically significant cluster of low values (LL); a high-value outlier within a group of low-value features (HL); and a low-value outlier within a group of high-value features (LH).

4   Results

Case 1: Exploring spatial patterns in the scale of the Corfu Island based on the number of TripAdvisor restaurant customer reviews

The global Moran's I spatial autocorrelation index and ArcGIS Pro tool were used to carry out an incremental spatial autocorrelation analysis. First peak distance was 2840 m, and the first peak z-score was 7.21 (p-value < 0.001). The analysis revealed spatial clustering trends among the majority of restaurant reviews, as the Global Moran's I statistic was equal to 0.049 and statistically significant, as shown in Fig. 1.

This first level of analysis, seeking spatial patterns of restaurant reviews at global level, clearly indicates clustering. On that basis, we ran hot spot and COType analyses to pinpoint more precisely the locations where the statistically significant clusters appeared.

Findings of the hot spot analysis revealed that restaurants in locations along the western and northern coasts, frequently in the island's best-known tourist locations, garnered more reviews than expected, indicating the existence of hot spots. However, these patterns were not mirrored along the eastern coastline, home to the island's main town, and several of the principal points of entry. Moreover, cold spots were detected southwest of the town of Corfu, in the island's hinterland, as shown in Fig. 2.

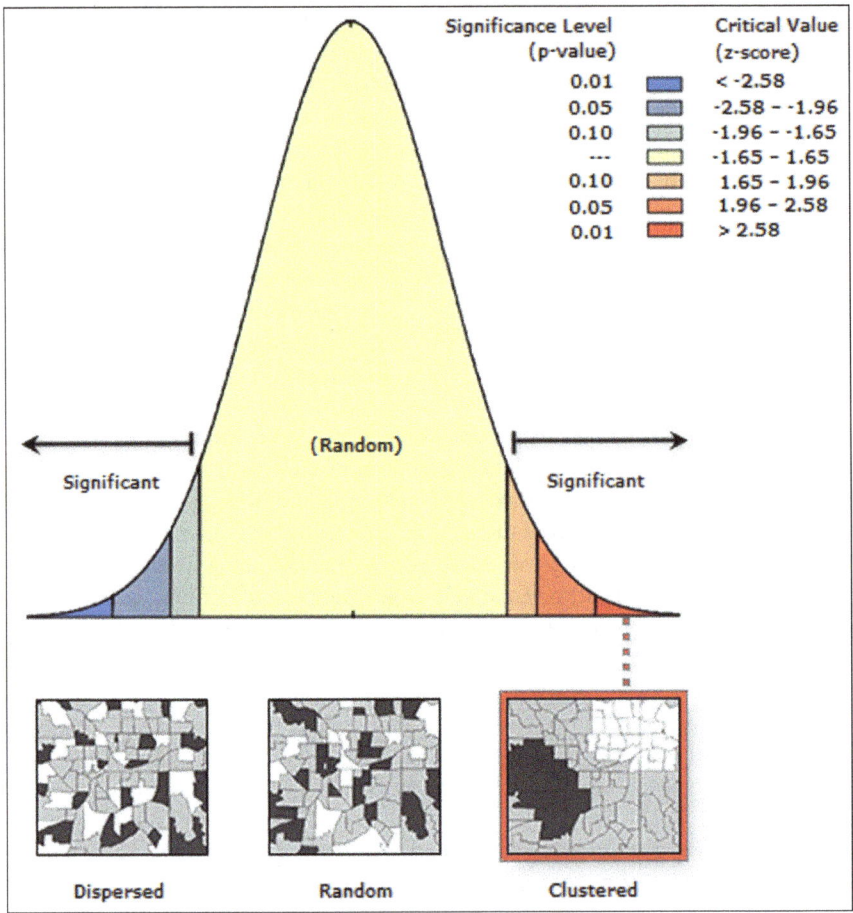

Fig. 1 .

The findings of the COType analysis, which revealed LH outliers, are depicted in Fig. 3. LH outliers were detected among the hot and cold spots identified by the previous analysis; in other words, restaurants with fewer reviews than expected existed among clusters of restaurants which generated large numbers of reviews and vice versa. Individual HL outliers were found along several stretches of the coastline. In general, the findings of the COType analysis tended to corroborate those of the hot spot analysis and also identified a further cold spot in the south of the island (see Fig. 3).

One noteworthy finding was that clusters could be formed on the basis of quality ratings, including customer satisfaction ratings, seasonality, and customer origin, as shown in Figs. 4, 5, and 6. Based on a prevalence of "excellent" and "very good" customer ratings, a hot spot can be identified along the southeast coast, encompassing the town of Corfu and reaching as far as the cold spot in the hinterland. When the

Fig. 2 Hot spot analysis results in Corfu Island

views of domestic Greek customers are taken alone, they undermine the hot spots identified on the western coast, causing a knock-on shrinkage in the cluster along the northern shore. Taking seasonality as the criterion also changes findings: Stronger clusters emerge on the northern shoreline during the September–November period, while the hinterland cold spot expands to touch the northernmost periphery of the town of Corfu.

Next, we tested the sensitivity of our analysis in terms of distance, repeating the hot spot analysis at island level but varying the distance band parameter across 2840 m (obtained from Euclidean distance), 3620 m (obtained from Manhattan distance), and 5000 m (obtained from empirical assessment), as shown in Figs. 7, 8, and 9.

The maps depicted in Figs. 7, 8, and 9 reveal that the spatial patterns of restaurant clustering are largely similar across all three distance bands. That said, some differences emerge in relation to the relative extent and intensity of each region's clusters, both among the northern hot spots and the central and southern cold spots.

Moreover, Figs. 10 and 11 reveal the importance of distance for visitors to Corfu in determining spatial patterns of restaurants and the generating customer reviews.

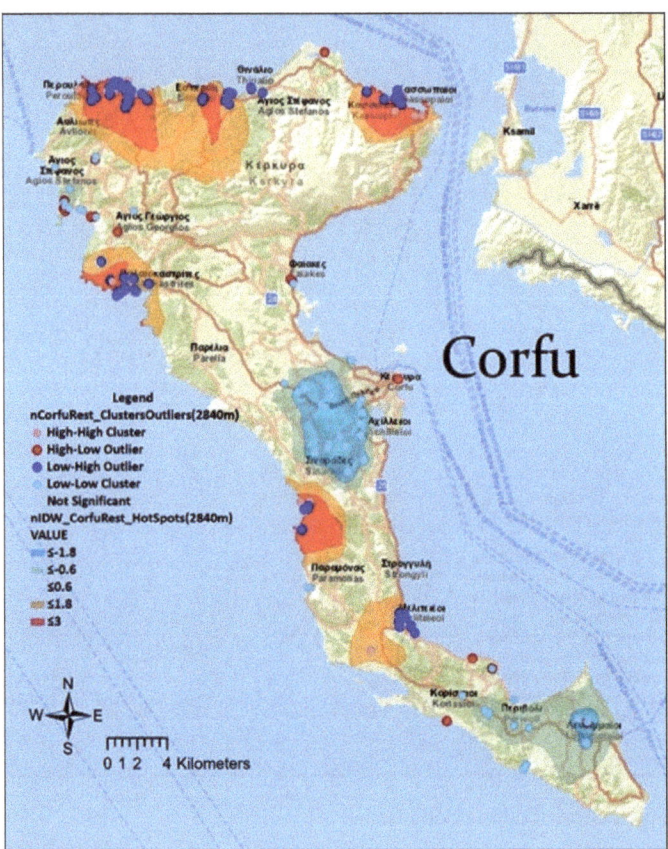

Fig. 3 Cluster and outlier analysis results in Corfu Island

Case 2: Exploring and location of spatial patterns in the scale of local destinations within the island of Corfu

We then changed the scale of the analysis in order to home in on three particular areas: the town of Corfu; the stretch of eastern coastline running from Gouvia to Pyrgi; and the stretch of (north)western coastline extending from Palaiokastritsa to Agios Stefanos. These three areas were selected as being representative of the model of tourist development which characterizes the island as a whole. The island-wide analysis showed no statistically significant spatial restaurant clusters for two of these selected areas. Applying spatial autocorrelation application indicated that the reviews of restaurants tended to be influenced by the spatial clustering trends across all three chosen regions.

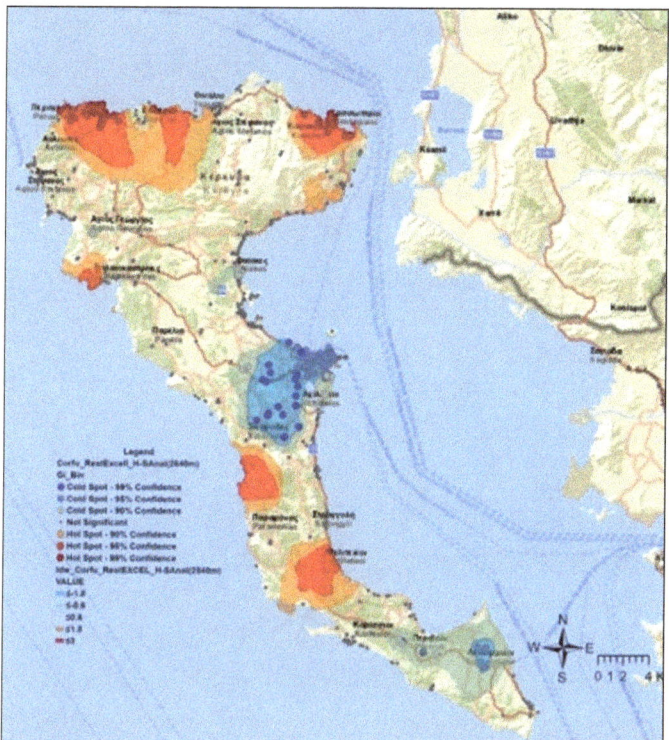

Fig. 4 Satisfaction-based clusters

Hot spot and COType analyses indicated the clusters depicted in the graphics below.

(a) *Corfu Town*

The town of Corfu has long attracted large numbers of travelers. Most visitors to the island arrive here, either by cruise ship into the port or by plane at the airport just outside town. The picturesque Old Town, classified as a World Heritage Site by UNESCO, is a particular draw. As depicted in Fig. 14, accommodation for tourists in the town is distributed across around 300 establishments, of different quality, type, and size. Visitors to the town encompass cruise passengers, university students, local people, and tourists undertaking brief trips from nearby islands, as well as the tourists who use the town and its facilities as a base throughout their stay.

Both types of spatial analysis, hot spot and COtype, clearly reveal the existence of statistically significant hot spots in the Old Town, which is to be expected given the high turnover of visitors it attracts. The COtype analysis also reveals LH outliers within or adjacent to these hot spots, such as restaurants garnering fewer reviews in close proximity to those garnering high numbers of reviews. Furthermore, clusters of LL outliers are found along the edge of the Old Town (Figs. 12 and 13).

Fig. 5 Greek customer-based clusters

(b) *Eastern Coast of Corfu: From Gouvia to Pyrgi*

The stretch of eastern coast which runs from Gouvia to Pyrgi, within easy reach of the airport, represents an extension of the town and its suburbs and encompasses residential areas as well as attracting significant numbers of tourists. Gouvia marina is a major draw, and the area is home to approximately 160 tourist accommodation establishments, including eight–ten large-scale complexes. This stretch of coast, taking in Gouvia, Dassia, and the Ipsos beaches, has been substantially affected by the development of tourist infrastructure, which has also extended a short distance inland. Patterns of mobility have developed accordingly among tourists moving from their accommodation to various attractions, from and to the marina, and making brief visits from neighboring or more distant destinations (Figs. 15, 16 and 17).

Both types of spatial analysis, hot spot and COtype, reveal the existence of a succession of statistically significant spots along this coastline, with hot spots in the Gouvia area, then cold spots appearing around Dassia, and weaker hot spots re-emerging toward the border with the Ipsos region.

The COType analysis also reveals LH outliers among the Gouvia hot spots. Interestingly, no significant spatial pattern was revealed in either the Komeno or the

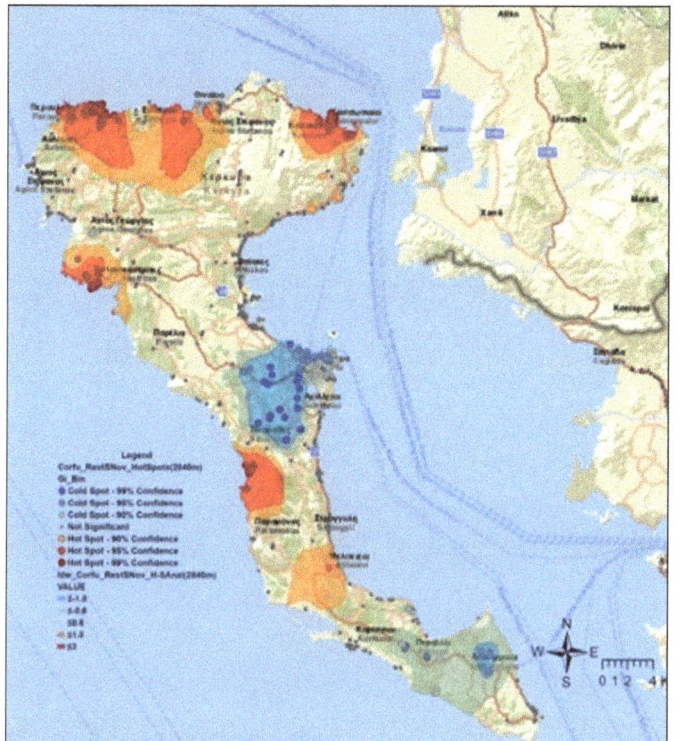

Fig. 6 Seasonality-based clusters (Sep–Nov)

Ipsos–Pyrgi region. Hence, although a strong "tourismification" footprint is revealed, high-quality hubs such as the major hotels remain.

(c)   *Northwestern Coast of Corfu: From Palaiokastritsa to Agios Stefanos*

The stretch of northwestern coast extending from the bay of Liapades to Agios Stephanos Avlioton has become a notable tourist destination. At some distance from the island's major entry points, the airport and the port of Corfu, the area is serviced by two smaller ports (Alipa and Agios Stefanos) and offers approximately 300 tourist accommodation establishments of a range of types, including a few small- and medium-sized hotels, as shown in Fig. 20. A distinctive morphology, with steep hills, an indented coastline, and wide beaches such as the famous sands of Pagon, Arila-Afiona, and Agios Stephanos, has largely influenced the development of tourism in this area. As a result, mobility patterns show intra-destination movement among visitors in local accommodation, as well as those coming in for shorter visits from neighboring or more distant destinations, and as part of domestic travel within the wider region.

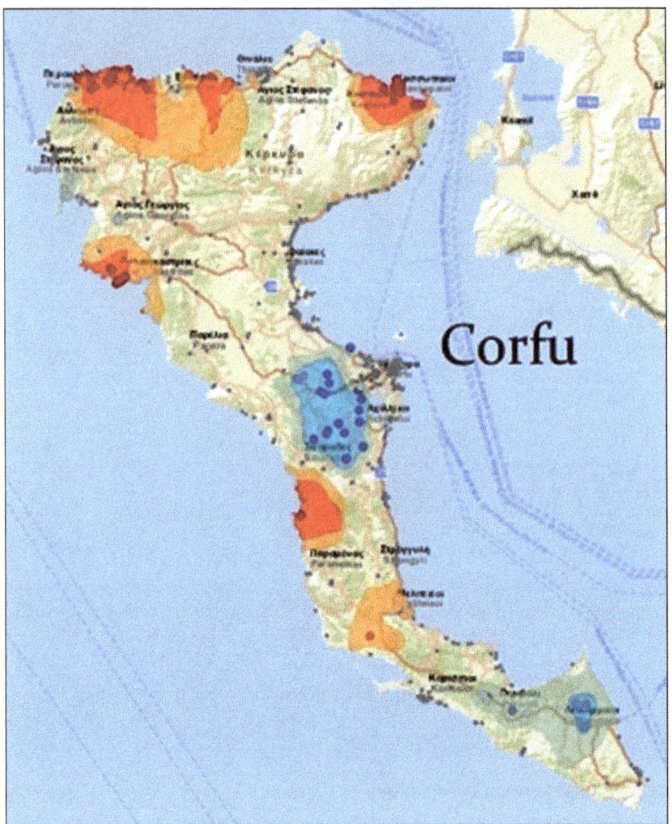

Fig. 7 "Hot spot analysis"—distance band: 2840 m

Both types of spatial analysis, hot spot and COtype, reveal the existence of a succession of statistically significant restaurant clusters on this stretch of coast. Their locations are morphologically distinct, with the first—hot spots—in Palaiokastritsa and the second—cold spots—in the Agios Georgios Pagon to the north.

The COType analysis also revealed several LH outliers among the hot spots of Palaiokastritsa. Notably, no statistically significant spatial pattern was revealed in Arila and Agios Stefanos, which are located even further northward (Figs. 18 and 19).

The overall pattern here varies from the "hubs" usually created by the "tourismification" footprint because of the region's distinct geomorphology (Fig. 20).

Findings from the COType analyses carried out also enable a comparison of restaurant clusters across the three regions chosen for study, as shown in Fig. 21, whereas the town of Corfu is characterized by hot spots with a prevalence of LH outliers, along the (north)western coast, and the pattern changes to fewer hot spots

Fig. 8 "Hot spot analysis"—distance band: 3620 m

and LH outliers, and significantly more cold spots. Finally, along the eastern coast, a pattern emerges of a greater prevalence of cold spots and a significant number of LH outliers.

When analyzed on the basis of the chosen attribute, however—customer reviews—hot spots appear more prevalent across all three selected regions, as shown in Fig. 22.

5   Conclusions

This paper has presented analyses of tourist behaviors in the Greek island of Corfu, a popular destination for local, domestic, and international visitors. Specifically, it has examined customer reviews of restaurants posted on TripAdvisor.com in order to better understand patterns of tourist movement. Two different analysis techniques were used, namely hot spot and COType. Running these analyses at different scales

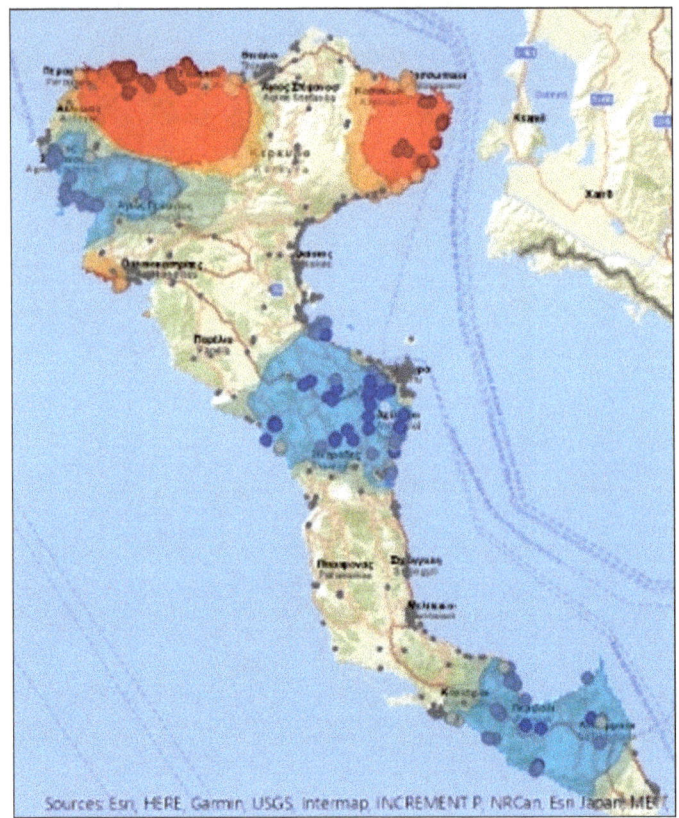

Fig. 9 "Hot spot analysis"—distance band: 5000 m

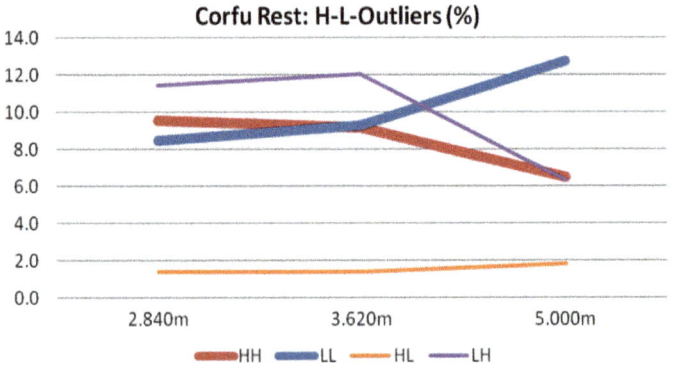

Fig. 10 Hot and cold spots and outliers of restaurants located in Corfu Island

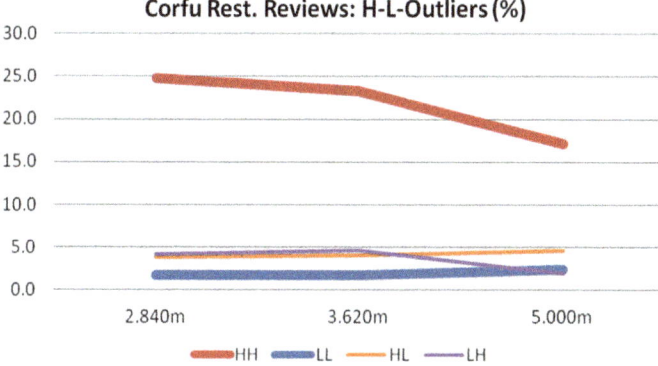

Fig. 11 Hot and cold spots and outliers of restaurants located in Corfu Island based on restaurant customer reviews from TripAdvisor

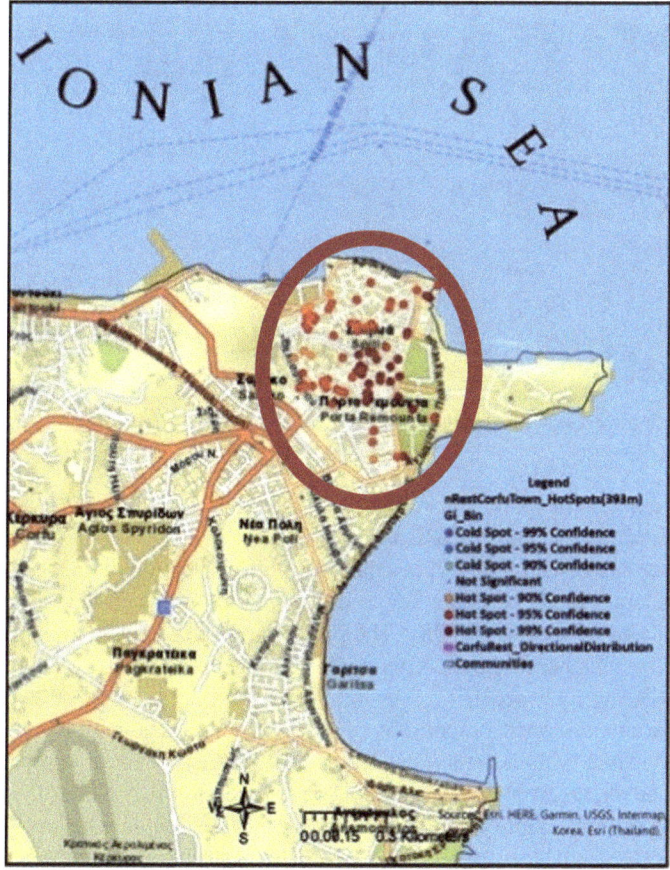

Fig. 12 Corfu town: hot spot analysis

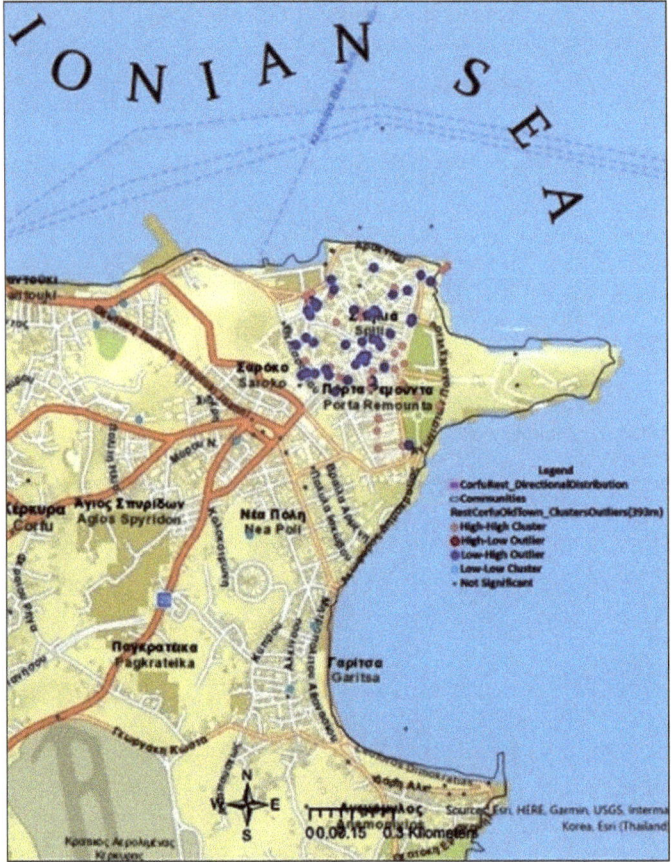

Fig. 13 Corfu town: cluster and outlier analysis

generated a picture of spatial clustering patterns which will add value to the study of how "tourismification" has impacted Corfu.

Moreover, our research underlines that the careful selection of distance and scale parameters, among others, is the key to differentiating between the various spatial patterns created by tourist mobility. The particular research question and approach will determine whether researchers choose to take an international, national, regional, local, or even sub-local scale.

More parameters must be considered when the intra-destination movement of restaurant visitors is under examination, as this produces a more complex pattern than distance alone. Among other factors, researchers must consider the influence on this type of movement of the tourist "product," for example whether visitors are more likely to remain in situ or be moving about to undertake sightseeing or adventure

Fig. 14 Corfu town: number
of accommodation
establishments

activities. They must also consider tourists' motives in choosing a certain destination, as these will significantly impact patterns of movement. Mobility is further constrained and molded by factors including the geomorphology of a region, which determines the distribution of tourist attractions, be they environmental or cultural, as well as the availability of accommodation, other elements of the hospitality industry, and transportation. Lastly, movement patterns are also determined to some extent by length of stay, both in terms of a tourist's total trip duration and in terms of how long they spend visiting a specific location during the trip.

Determining distance can, in some cases, be an even more complex matter for researchers. If a local or sub-local scale is used, researchers may be able to take an empirical approach; certainly, they will find it easier to trace real-life movements of particular visitors. When larger scales are used, however, more detailed analysis of tourist mobility is required.

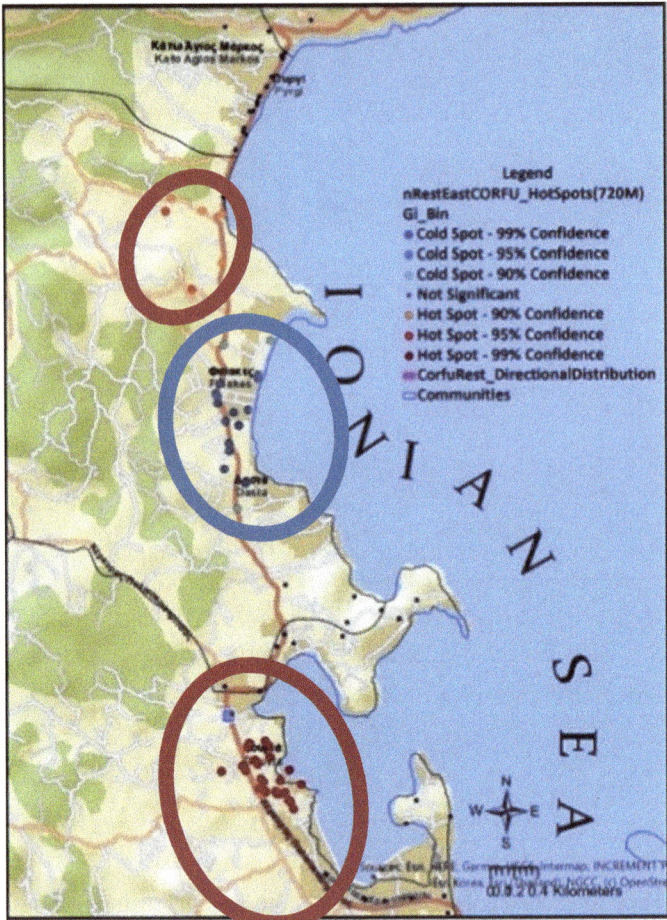

Fig. 15 Eastern coast: hot spot analysis

The findings of the present research in relation to the various types of clusters identified will be of varying use to local stakeholders. Where a picture of restaurant hot spots indicates intense tourist activity, it is suggested that local authorities study ways to mitigate the impact of saturation and that entrepreneurs consider carefully the viability or otherwise of enterprises. Where the picture is, instead, of restaurant cold spots, suggesting a comparatively low performance, local authorities should consider means to boost the attractiveness of the area, if necessary by "desaturating" other neighborhoods and redistributing the tourist presence more equally. Entrepreneurs, meanwhile, should consider how they can adjust their offerings to draw higher numbers of customers. Where HL and LH outliers are identified, it is recommended that the business community collaborates to raise the level of services offered to that of nearby competitors.

Fig. 16 Eastern coast: cluster and outlier analysis

Future researchers may fruitfully include an investigation of other elements of the tourist infrastructure, such as hotels and visitor attractions, in their analysis of tourist mobility to enhance the findings of the current study. A better understanding of spatial clustering may trigger collaboration among enterprises and localities and promote networking among the different elements of the tourist industry fabric, including higher education institutions offering tourist study programs as well as umbrella organizations which are instrumental in molding and promoting a destination profile.

Moreover, it would be interesting to also assess the temporal dimension of tourist mobility and the clustering patterns it creates. In this way, results in the future may be systematically differentiated according to both the geographical and temporal scales used for their analysis.

Fig. 17 Eastern coast:
number of accommodation
establishments

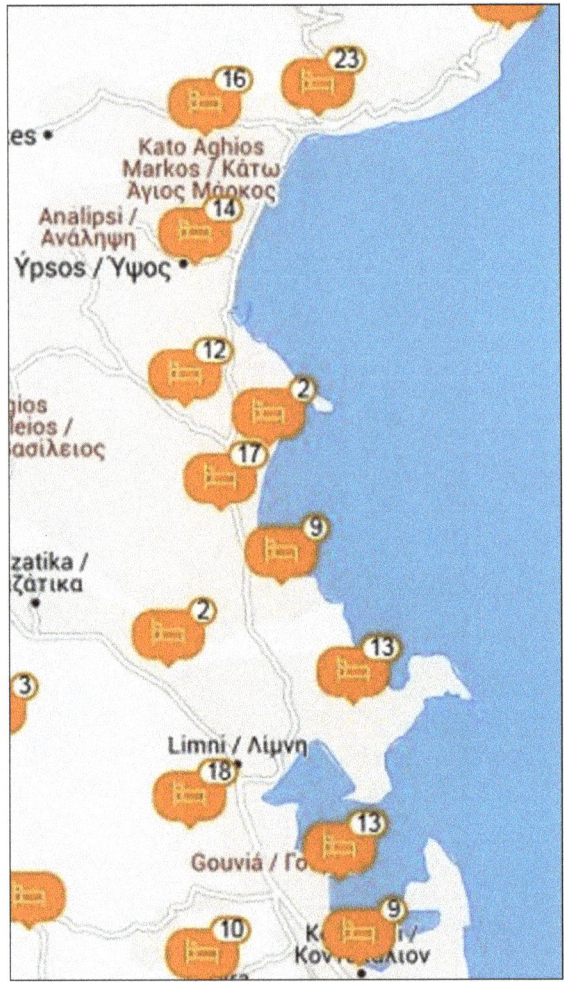

Lastly, we emphasize that analyses which leverage big data offer substantial improvements over those using field research or other traditional data sources, including saving on cost and time and access to a greater variety and amount of data. Nevertheless, the same caution must be exercised when using these new data sources as has always been used in using and interpreting the old ones. Indeed, combining old and new sources so that each complements and verifies the other might be the best way forward.

Fig. 18 Northwestern coast: hot spot analysis

Fig. 19 Northwestern coast: cluster and outlier analysis

Fig. 20 Northwestern coast: number of accommodation establishments

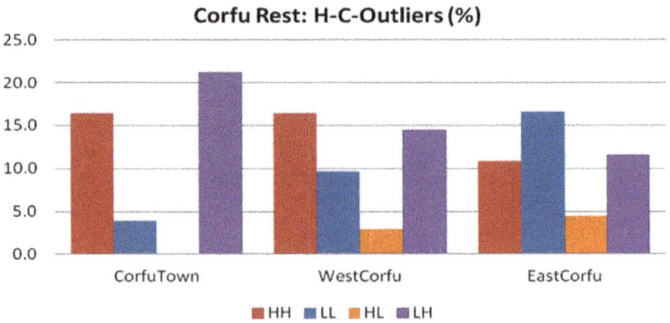

Fig. 21 Cluster and outlier analysis on the restaurants located in three regions of Corfu Island

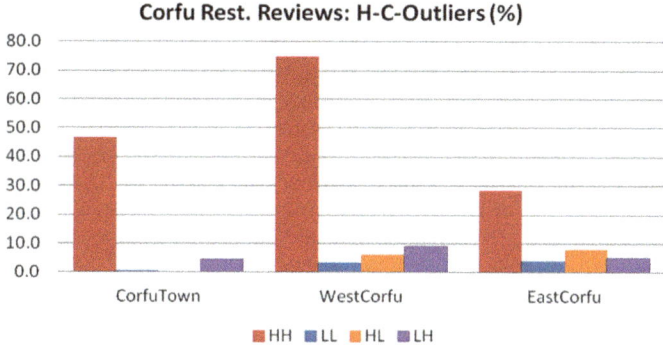

Fig. 22 Cluster and outlier analysis on the restaurants located in three regions of Corfu Island based on restaurant customer reviews from TripAdvisor

References

Aldstadt, J. (2010). Spatial clustering. In M. Fischer & A. Getis (Eds.), *Handbook of Applied Spatial Analysis*. Berlin, Heidelberg: Springer. https://doi.org/10.1007/978-3-642-03647-7_15

Anselin, L. (1995). Local indicators of spatial association-LISA. *Geographical Analysis, 27*(2), 93–115. https://doi.org/10.1111/j.1538-4632.1995.tb00338.x.

Batista e Silva, F., Marín Herrera, M. A., Rosina, K., Ribeiro Barranco, R., Freire, S., & Schiavina, M. (2018). Analysing spatiotemporal patterns of tourism in Europe at high-resolution with conventional and big data sources. *Tourism Management, 68*, 101–115. https://doi.org/10.1016/j.tourman.2018.02.020

Butler, R. (2015). The evolution of tourism and tourism research. *Tourism Recreation Research, 40*(1), 16–27. https://doi.org/10.1080/02508281.2015.1007632.

Car, A., Taylor, G., & Brunsdon, C. (2001). An analysis of the performance of a hierarchical wayfinding computational model using synthetic graphs. *Computers, Environment and Urban Systems, 25*(1), 69–88. https://doi.org/10.1016/s0198-9715(00)00036-3.

Chou, Y. H. (1995). Spatial pattern and spatial autocorrelation. In A. U. Frank & W. Kuhn (Eds.), *Spatial Information Theory A Theoretical Basis for GIS. COSIT 1995*. Lecture Notes in Computer Science (Vol. 988). Berlin, Heidelberg: Springer. https://doi.org/10.1007/3-540-60392-1_24

Connell, J., & Page, S. J. (2008). Exploring the spatial patterns of car-based tourist travel in Loch Lomond and Trossachs National Park, Scotland. *Tourism Management, 29*(3), 561–580. https://doi.org/10.1016/j.tourman.2007.03.019.

Deng, M., & Athanasopoulos, G. (2011). Modelling Australian domestic and international inbound travel: A spatial–temporal approach. *Tourism Management, 32*(5), 1075–1084. https://doi.org/10.1016/j.tourman.2010.09.006.

Fisher, M. M. (2006). Spatial analysis in geography. In *Spatial Analysis and GeoComputation*. Berlin, Heidelberg: Springer.

Fischer, M. M. (2000). Recent Advances in Spatial Data Analysis. Discussion Papers of the Institute for Economic Geography and GIScience, 70/00. WU Vienna University of Economics and Business, Vienna.

Freundschuh, S. M., & Egenhofer, M. J. (1997). Human conceptions of spaces: Implications for GIS. *Transactions in GIS, 2*(4), 361–375. https://doi.org/10.1111/j.1467-9671.1997.tb00063.x.

Ganzaroli, A., De Noni, I., & van Baalen, P. (2017). Vicious advice: Analyzing the impact of TripAdvisor on the quality of restaurants as part of the cultural heritage of Venice. *Tourism Management, 61*, 501–510. https://doi.org/10.1016/j.tourman.2017.03.019.

Getis, A. (1999). Spatial statistics. In *Geographical Information Systems* (2nd Ed., pp. 239–251). New York: Wiley.

Getis, A., & Ord, J. K. (2010). The analysis of spatial association by use of distance statistics. *Geographical Analysis, 24*(3), 189–206. https://doi.org/10.1111/j.1538-4632.1992.tb00261.x.

Guedes, A. S., & Jiménez, M. I. M. (2015). Spatial patterns of cultural tourism in Portugal. *Tourism Management Perspectives, 16*, 107–115. https://doi.org/10.1016/j.tmp.2015.07.010.

Gutiérrez, J., García-Palomares, J. C., Romanillos, G., & Salas-Olmedo, M. H. (2017). The eruption of Airbnb in tourist cities: Comparing spatial patterns of hotels and peer-to-peer accommodation in Barcelona. *Tourism Management, 62*, 278–291. https://doi.org/10.1016/j.tourman.2017.05.003.

Haining, R. (2015). Spatial Autocorrelation. International Encyclopedia of the Social & Behavioral Sciences, 105–110. https://doi.org/10.1016/b978-0-08-097086-8.72056-3.

Haining, R. P. (2001). Spatial autocorrelation. In N.J. Smelser & P.B. Baltes (Eds.) *International Encyclopedia of the Social & Behavioral Sciences*. Pergamon: Oxford, UK.

Hao, P. (2019). Spatial analysis. *The Wiley Blackwell Encyclopedia of Urban and Regional Studies* 1–7.

Hasnat, M. M., & Hasan, S. (2018). Identifying tourists and analyzing spatial patterns of their destinations from location-based social media data. *Transportation Research Part C: Emerging Technologies, 96*, 38–54. https://doi.org/10.1016/j.trc.2018.09.006.

Kim, Y., Kim, C., Lee, D. K., Lee, H., & Andrada, R-II. T. (2019). Quantifying nature-based tourism in protected areas in developing countries by using social big data. *Tourism Management, 72*, 249–256

Kirilenko, A. P., Stepchenkova, S. O., & Hernandez, J. M. (2019). Comparative clustering of destination attractions for different origin markets with network and spatial analyses of online reviews. *Tourism Management, 72*, 400–410. https://doi.org/10.1016/j.tourman.2019.01.001.

Li, D., Zhou, X., & Wang, M. (2018). Analyzing and visualizing the spatial interactions between tourists and locals: A Flickr study in ten US cities. *Cities, 74*, 249–258. https://doi.org/10.1016/j.cities.2017.12.012.

Lynch, H. J., Crosbie, K., Fagan, W. F., & Naveen, R. (2009). Spatial patterns of tour ship traffic in the Antarctic Peninsula region. *Antarctic Science, 22*(02), 123. https://doi.org/10.1017/s0954102009990654.

Molinillo, S., Ximénez-de-Sandoval, J. L., Fernández-Morales, A., & Coca-Stefaniak, A. (2016). Hotel assessment through social media: The case of TripAdvisor. *Tourism & Management Studies, 12*(1), 15–24.

O'Sullivan, D., & Unwin, D. J. (2010). *Geographic Information Analysis*. New York: Wiley.

Papageorgiou, M. (2016). Spatial planning theories, principles and paradigms for tourism activity: the Greek experience and practice, *Aeixoros, 26*, 37–66. ISSN: 1109-5008

Pearce, D. G. (1987). Spatial patterns of package tourism in Europe. *Annals of Tourism Research, 14*(2), 183–201. Pergamon Journals, Inc., Maxwell House, Fairview Park, Elmsford, New York 10523. *Journal of Travel Research, 26*(2), 50

Radil, S. M. (2016). Spatial analysis of crime. In *The Handbook of Measurement Issues in Criminology and Criminal Justice* (pp. 535–554).

Romão, J., Guerreiro, J., & Rodrigues, P. M. M. (2017). Territory and sustainable tourism development: A space-time analysis on European Regions. *The Region, 4*(3), 1–17. https://doi.org/10.18335/region.v4i3.142.

Roser, M. (2017). Tourism. Published online at OurWorldInData.org. Retrieved from. https://ourworldindata.org/tourism [Online Resource].

Scott, L. M. (2015). Spatial pattern, analysis of. *International Encyclopedia of the Social & Behavioral Sciences* 178–184.

Sui, D. Z. (2004). Tobler's first law of geography: A big idea for a small world? *Annals of the Association of American Geographers, 94*(2), 269–277. https://doi.org/10.1111/j.1467-8306.2004.09402003.x.

Taecharungroj, V., & Mathayomchan, B. (2019). Analysing TripAdvisor reviews of tourist attractions in Phuket, Thailand. *Tourism Management, 75*, 550–568. https://doi.org/10.1016/j.tourman.2019.06.020.

Tobler, W. (1970). A computer movie simulating urban growth in the Detroit Region. *Economic Geography, 46*, 234–240. https://doi.org/10.2307/143141.

Unwin, D. J. (1996). GIS, spatial analysis and spatial statistics. *Progress in Human Geography, 20*(4), 540–551. https://doi.org/10.1177/030913259602000408.

Van der Zee, E., Bertocchi, D., & Vanneste, D. (2018). Distribution of tourists within urban heritage destinations: A hot spot/cold spot analysis of TripAdvisor data as support for destination management. *Current Issues in Tourism* 1–22. https://doi.org/10.1080/13683500.2018.1491955

Vassakis, K., Petrakis, E, Kopanakis, I., Makridis, J., & Mastorakis, G. (2019). Location-based social network data for tourism destinations. In: M. Sigala, R. Rahimi, M. Thelwal (Eds.), *Big Data and Innovation in Tourism, Travel, and Hospitality*. Singapore: Springer

Williams, S. (1998). *Tourism geography. Routledge contemporary human geography.* . London: Routledge.

Wolf, L. J., & Murray, A. T. (2017). Spatial analysis. *International Encyclopedia of Geography: People, the Earth, Environment and Technology* 1–11

Wong, D. W. S. (1999). Several fundamentals in implementing spatial statistics in GIS: Using centrographic measures as examples. *Geographic Information Sciences, 5*(2), 163–174. https://doi.org/10.1080/10824009909480525.

Wood, S. A., Guerry, A. D., Silver, J. M., & Lacayo, M. (2013). Using social media to quantify nature-based tourism and recreation. *Scientific Reports, 3*(1).https://doi.org/10.1038/srep02976

Xia, J. & Arrowsmith, C. (2005). Managing scale issues in spatio-temporal movement of tourists modelling. In: A. Zerger & R. M. Argent (Eds.), *Proceedings of the MODSIM05: International Congress on Modelling and Simulation: Advances and Applications for Management and Decision Making* (pp. 162–169), Decmber 12–15, 2005. Melbourne, Victoria: Modelling & Simulation Society of Australia and New Zealand

Ye, B. H., Luo, J. M., & Vu, H. Q. (2018). Spatial and temporal analysis of accommodation preference based on online reviews. *Journal of Destination Marketing & Management, 9*, 288–299. https://doi.org/10.1016/j.jdmm.2018.03.001.

Online Platforms for Tourist Accommodation from Economic Policy Perspective in Greece: Case for Further Digitalization

Vesna Luković

1 Introduction

Tourism has the potential to significantly contribute to economic growth and development. In 2019, international tourist arrivals to Greece reached 31.3 million, which was a growth of 4.1% compared to 2018 (United Nations World Tourism Organization (hereafter: UNWTO), 2020). International tourism receipts in Greece reached 18.2 bn EUR in 2019, 13% more than in 2018 (UNWTO, 2020). In addition to those figures, the importance of tourism in Greece can be simply explained by the fact that one-quarter of all employment in Greece is in travel in tourism (World Travel Tourism Council (hereafter: WTTC), 2019).

Tourism today is considerably influenced by Internet. People can use Internet when they talk to friends, look for information or travel. Many economic activities have moved online. Access to Internet has proved necessary for a more competitive tourist industry in order to advance the accessibility of tourist services. With the development of digital technology and the Internet, online platforms have boomed and moved to important sectors of the economy including tourism, especially accommodation and rental services. Tourists and other travellers can now find accommodation and other products and services online. They can compare and review options from different accommodation providers and their search can be completed in seconds and without any costs. On the other side, owners of properties can list their spare accommodation on various online platforms such as Airbnb, Booking.com and others. Website algorithms at online platforms are made so as to provide information on apartments, houses, rooms, etc., offering various filters to navigate accommodation offers according to the tourists' needs. For people in Greece that has been hit hard by the latest global financial crisis and has still not fully recovered from it, offering

V. Luković (✉)
Thessaloniki, Greece

© The Author(s), under exclusive license to Springer Nature Switzerland AG 2021   587
V. Katsoni and C. van Zyl (eds.), *Culture and Tourism in a Smart, Globalized, and Sustainable World*, Springer Proceedings in Business and Economics, https://doi.org/10.1007/978-3-030-72469-6_38

houses and apartments and other tourist accommodation for short-term rental can be an important source of revenue.

This analysis was guided by concepts associated with digital technology and its impact on economic activity. The purpose was to explore the current level of digitalization in Greece, especially Internet connectivity and digital skills of the population and if there is room for further improvement. The analysis incorporates a general view that digital economy can help generate growth (European Central Bank, 2018) and that online platforms can bring additional tourist revenue (International Monetary Fund, 2018). Economic growth is crucial for Greece in order to get out of the debt spiral considering that public debt to GDP was 180% at the end of 2019. Looking at the context of numerous macroeconomic challenges Greece has to deal with, online platforms' role for accommodation is viewed as an opportunity to strengthen households' disposable income. In that respect, EU funds are an important source of financing and are part of an economic policy argument for further digitalization. The framework of this paper provides the window through which to view statistical analysis on the current level of digitalization in Greece, European Union (hereafter: EU) funding and macroeconomic challenges, all in light of Greece's commitment to international creditors. Before this analysis has been completed, the coronavirus (COVID-19) epidemic brought a new dimension to digitalization. Digitalization is especially relevant taking into account social-distancing strategies to tackle the coronavirus outbreak. In that respect, private accommodation has some advantages vis-a-vis standard hotel accommodation. That brings an additional argument for further digitalization in Greece.

2 Literature Review

Newer research has found that the potential benefits of digitalization are high (Charalabidis et al., 2015; Parviainen et al., 2017). One of the benefits is that by digitizing information-intensive processes, costs can be cut by up to 90% (Marcovitch and Willmott, 2014). On the other hand, digitalization has a proven impact on reducing unemployment (Parviainen et al., 2017) which is of high importance to Greece. Digitalization is typically understood as the adoption or increase in the use of digital or computer technology by an organization, industry, country, etc. IMFs definition of digitalization is that it "encompasses a wide range of new applications of information technology in business models and products that are transforming the economy and social interactions". (IMF, 2018: 1).

Academics have established that digitalization is transforming the world not only in terms of socializing, travelling and sharing information, but also in terms of working and new business models (Bouwman et al., 2017). Although "sharing economy" has no single definition (Goeroeg, 2018), common to all its definitions is that it is for those who need to make extra income and for those who want to save a bit of money. In this peer-to-peer system, companies use online platforms to connect people who have a spare dwelling with people who are searching for accommodation. IMF found that "platforms such as Airbnb have enabled rapid growth of short-term rentals in some economies, particularly those with a tourism industry, suitable housing stock, and a favourable legal environment" (IMF, 2018: 25).

Since "sharing economy" comes from companies that sell or rent services mostly via Internet, some reject the poetic label "sharing economy" by calling those companies "Internet-based service firms" (Bivens, 2019). Regardless of how they are labelled, they have several unique features. One is that online platforms have near zero marginal cost when adding or removing additional accommodation. Because of the near zero marginal cost, Internet-based service firms like Airbnb can easily increase supply to meet extra demand. That is what hotels cannot do easily. Bivens (2019) found that Airbnb increases the supply of short-term travel accommodations (and slightly lowers competitors' accommodation prices) in the area with high penetration of Airbnb. Also, in contrast to hotels, online platforms enable travellers to rent a wider range of accommodation types, from rooms to bungalows, villas, etc.

Online "sharing economy" platforms typically lead to reduced costs for travellers compared to hotels (Pierce et al., 2018), and on the other hand, generate revenue for owners of properties (Mao et al., 2018). It is no wonder then that online platforms for accommodation such as Booking.com, Airbnb, HomeAway and similar have been embraced by households as those platforms mean new opportunities for revenue (Clancy, 2018).

Digitalization has proven a positive impact on employment and has been included in policy strategies in many countries (Randall et al., 2017). In addition, digitalization does not only have a positive impact on economic growth and GDP per capita (Sabbagh et al., 2013; ECB, 2018), it improves the digital literacy of the population. To actively participate in the society and digital economy, access to technologies and services must be affordable, as well as the ability and skills to use them.

3 Methodology

Digitalization in this analysis is looked at from the viewpoint of Internet connectivity and digital skills to use the Internet. Internet, the system of interconnected computer networks, is in this paper analysed from the perspective of access to Internet, meaning that the term *access to Internet* includes all types of Internet connectivity (dial-up, fixed broadband, mobile broadband, etc.). Wired technologies most commonly used to access the Internet are divided between broadband and dial-up access over a normal or an ISDN telephone line. "Broadband lines are defined as having a capacity higher than ISDN, meaning equal to or higher than 144 kbit/s" (Eurostat, 2019:13).

Data for research have been derived from the Eurostat's statistical data warehouse and Eurostat's and European Commission's (hereafter: EC) special surveys. Particular surveys which are taken periodically may not provide the most recent data, and some Eurostat data are at the moment available only up to the end of 2018. Macroeconomic data for Greece were also taken from the International Monetary

Fund (hereafter: IMF) as well. Other reports and assessments from specialist organizations and EU bodies in regard to digitalization have been also analysed and
reviewed.

The research methods are descriptive and quantitative. Quantitative methods focus
on statistical data aiming to explore digitalization from the connectivity perspective
and looking at the digital skills and Internet use of people in Greece. I look at the
regional[1] level as well and some comparisons are done vis-à-vis other EU member
states. In terms of contributing to growth via digitalization, I look at short-term rental
opportunities for tourism[2] by analysing data on the growth of holiday and short-term
accommodations in the last few years. By exploring data on unemployment and
households' incomes from the beginning of the last financial crisis, I argue that
further digitalization has a significant economic appeal. Since disposable income
in Greece has been shrinking for almost a decade, especially in some regions that
are more disadvantaged, I argue that further digitalization would benefit household
incomes, domestic consumption and thus also aggregate demand. The economic
appeal is supported by the fact that the EU has committed itself to the digital single
market and that each EU member state, including Greece, can apply for relevant
funding.

4 Discussion

4.1 Access to Internet

The EU's digital single market aims to ensure access to Internet throughout the EU
so as to encourage participation in the digital economy and society. The EU strategy
has been based on improving access "for consumers and business, on creating the
right conditions for digital networks and on maximizing the growth potential of the
digital economy" (European Parliament, 2019: 11).

Part of the EU strategy was to aim to roll-out ultrafast connectivity to homes in
every EU region so that rural areas can also have better mobile coverage and fast
Internet (European Parliament, 2019: 43). In that respect, the aim of EU Telecoms
reform package was to "reduce the digital divide between urban and rural" (EC, 2007:
2). According to the Eurostat data warehouse, the average share of EU households
with access to Internet at home reached 90% in 2019 (Fig. 1). However, Greece is

---

[1] The first-level NUTS (statistical) regions of Greece are: EL3-Attiki (Attica),EL4-Nisia Aigaiou,
Kriti (North Aegean, South Aegean and Crete), EL5-Voreia Ellada(Eastern Macedonia and Thrace,
Central Macedonia, Western Macedonia, Epirus), EL6- Kentriki Ellada (Thessaly, Western Greece,
Central Greece, Ionian Islands, Peloponnese).

[2] According to Eurostat, tourism means the "activity of visitors taking a trip to a main destination
outside their usual environment, for less than a year, for any purpose, including business, leisure or
other personal purpose, other than to be employed by a resident entity in the place visited" (Eurostat,
2020a).

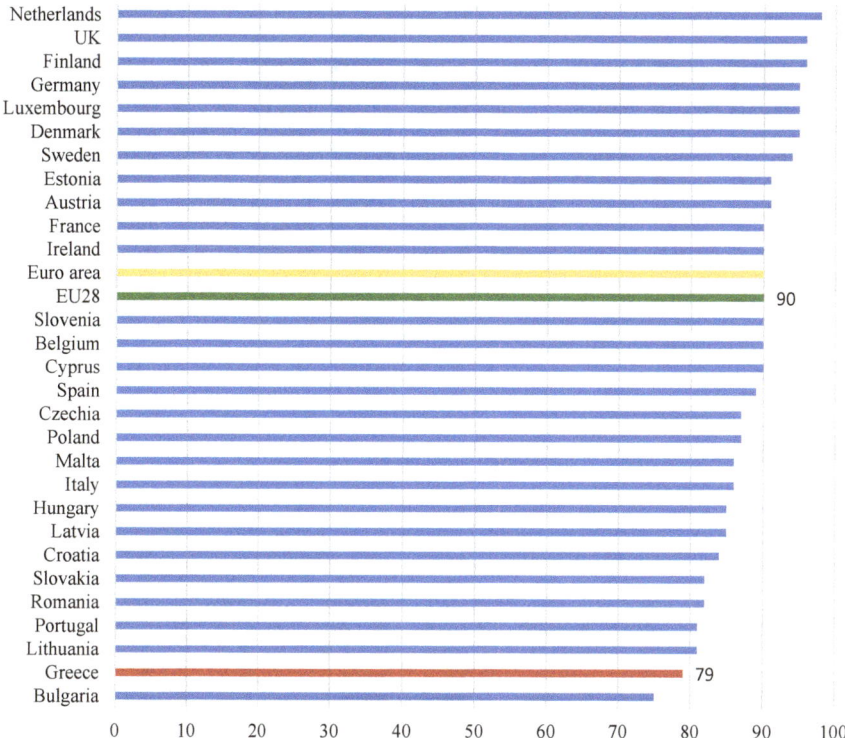

Fig. 1 Percentage of households with Internet access in 2019 in EU member states

below the EU average and considerably below some other countries in the EU. On the other hand, there has been an improvement since 2012, when only 54% of households in Greece had access to Internet at home compared to 76% in 2018 (Fig. 2) and 79% in 2019 (Fig. 1).

In terms of the type of Internet connection, the share of households in Greece with fixed broadband reached 76% in 2019 while mobile broadband reached 32% in the same year (Eurostat, 2020). As a comparison, the percentage in the EU is 78% and 56%, respectively.

Access to Internet has been steadily increasing in all regions across Greece, although urban areas are better off compared to rural areas. In some regions, especially in some islands and mountainous areas, there are still significant Internet infrastructure problems. In Attica (Attikki), the share of households with access to Internet has been the highest, over 80% in 2019 (Fig. 3). Many people are still unable to access digital platforms and lower income households are less likely to have Internet. The variation of households with Internet access is correlated with the population at risk of poverty or social exclusion, standing at 15.5% in Attiki (Attica), 22.3% for Voreia Ellada (Northern Greece), Kentriki Ellada (Central Greece) at 24.2% and Nisia Aigaiou and Kritti (the Aegean islands and Crete) at 21.6% in 2018.

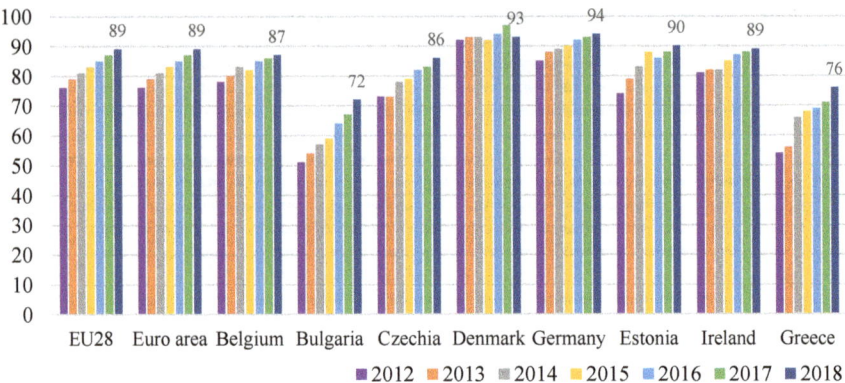

Fig. 2 Dynamics of Internet connectivity in Greece vis-à-vis some EU countries from 2012 to 2018, percentage (%) of households with access to Internet

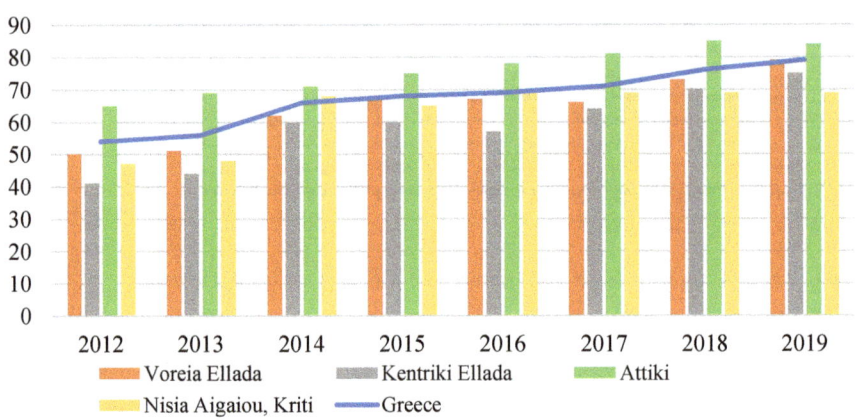

Fig. 3 Percentage (%) of households with access to Internet at home in Greece by region

4.2 Digitalization, Stylized Facts

Greece is a member of the EU and so also a member of the EU Digital Single Market. In line with that the country should keep up with the developments in the rest of the EU in regard to digitalization. Greece ranks low in the uptake of information technologies and high priority investment needs (EC, 2019b) have been identified to close the gap with respect to the Digital Agenda for Europe (EC, 2020). In 2016, Greece established the Ministry of Digital Policy, Telecommunications and Media which designed a National Digital Strategy (2016–2021). The strategy was expected to help closing the gap with other EU countries considering that, according to the data from the Eurostat, the EC and European Parliament, people in the north of Europe not only enjoy greater access to online services but they also have better digital skills than in southern Europe.

The lack of need or interest, insufficient skills and cost-related barriers are the most common reasons for not having Internet access at home. The share of households in the EU that do not have Internet at home because costs are too high, has been falling (Fig. 4). The same can be observed for Greece which has seen a significant drop in households that do not have Internet at home because costs are too high. In 2019, there were only 2% of such households (Fig. 4).

EC has introduced the Digital Economy and Society Index (DESI index) which measures digital performance with the following indicators: connectivity, human capital, use of Internet services, integration of digital technology and digital public services. Human capital is measured with two sub-dimensions covering *Internet user skills* and *advanced skills and development*. Internet user skills indicator is computed based on the number and complexity of activities involving the use of digital devices and/or Internet. Greece is not doing well in DESI index. According to DESI index in 2019, Greece is in the group (Fig. 5) of the least advanced digital economies.

In addition to having a reliable and fast Internet connection, digital skills are needed as well in order to benefit from digitalization. It is especially so considering that the new digital world is entering many aspects of society, including health, transport, security, energy, tourism, etc. The EU has established the digital single market strategy with an aim to "open up digital opportunities for people and business and enhance Europe's position as a world leader in the digital economy" (EC, 2020b). Advanced skills and development indicator is about information and communication

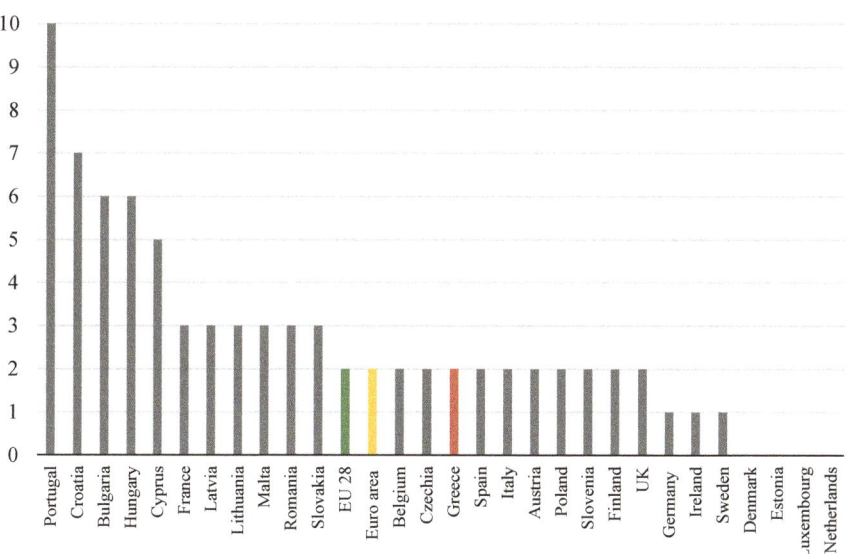

Fig. 4 Percentage (%) of households without Internet at home because access costs are too high in 2019 in EU member states

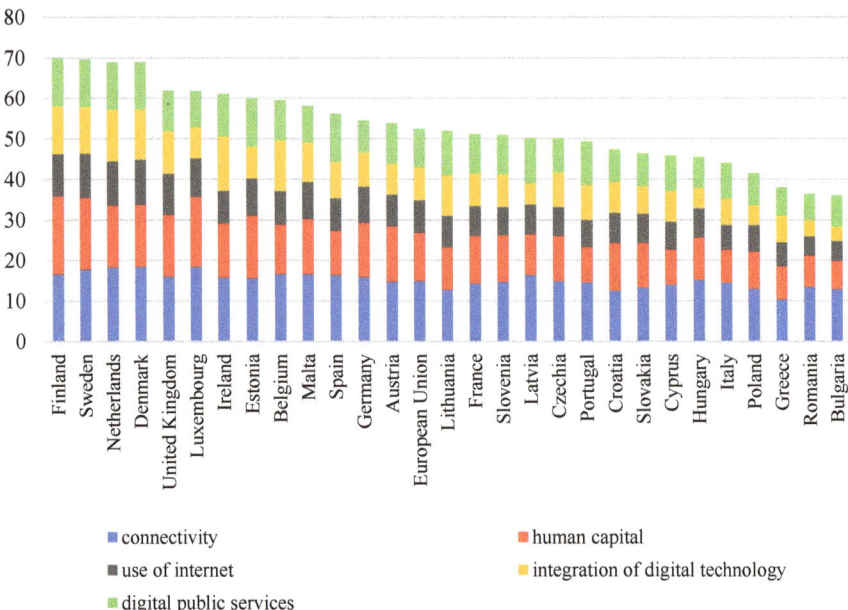

Fig. 5 EU: Digital Economy and Society Index (DESI)

experts and graduates. The number of information-communication experts in Greece is only a small fraction in total employment, standing at 1.6% compared to the EU average of 4% (Fig. 7). Although people in Greece are considered to be active users of Internet services, the country stands below EU average in terms of individuals who have above basic digital skills (Fig. 8). The share of people who have never gone online has been falling in Greece, although it still remains high, and reached on average 25% in 2018, with some regions in Greece even higher. The global financial crisis that started in 2008 hit all parts of Greece, but disparities between regions have widened significantly, not only in terms of social and labour market conditions but in terms of digitalization too and using Internet (Fig. 6).

4.3 EU Funds Available for Digitalization in Greece

Greece can close the gap vs other EU countries in regard to digitalization by setting adequate conditions for private investment and public financing resources for digital infrastructure, considering that "access to fast and ultrafast broadband-enabled services is a necessary condition for competitiveness" (EC, 2019).

The improvement of Internet connectivity in Greece over the last few years has been co-financed by the EU through the European funds. One such recent and important project, a public–private partnership project which aimed to connect 2260 remote

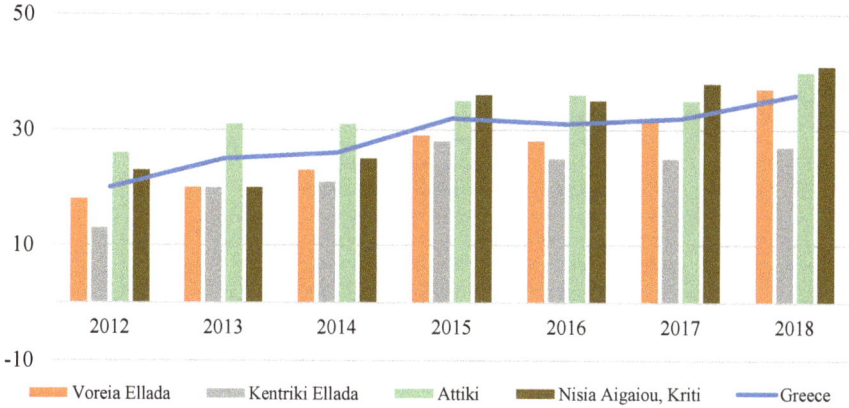

Fig. 6 Percentage (%) of individuals who ordered goods or services over Internet for private use, in Greece by region, 2012–2018

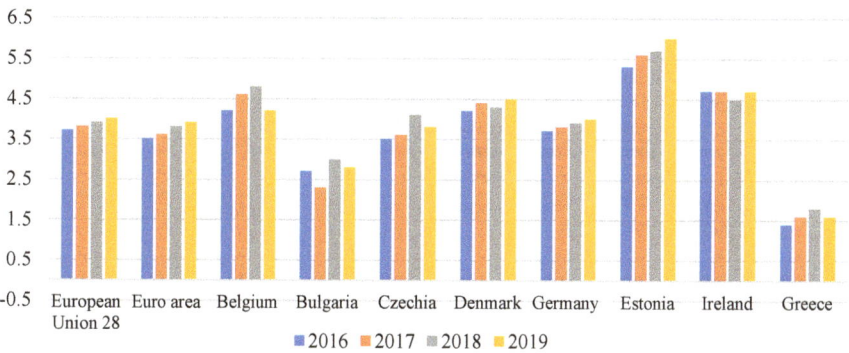

Fig. 7 Percentage (%) of employed information-communication specialists in total employment in Greece in comparison with some EU member states

villages, covering 45% of the Greek territory, was approved by the EU in 2017 and was called "Rural Broadband Greece". The project aimed, among other things, to bridge the digital divide by developing fibre-optic infrastructure in order to bring fast broadband to the most remote areas of Greece (EC, 2019c).

The national digital strategy (2016–2021) of Greece envisaged investments in all categories of the digital economy implying additional efforts to effectively implement it at all levels and extend its effects to individuals. Considering that Greece has limited capacity to invest, following the decade of contraction since the beginning of the crisis in 2008, the main source of public investments are European funds such as European Structural and Investment Funds (ESIF) (Fig. 9).

According to the EU, regional policy over 21 billion EUR of EU funds were available to Greece in the period 2014–2020 through ESIF. Total budget for Greece for 2014–2020 has been over 26 billion EUR. National contribution was determined

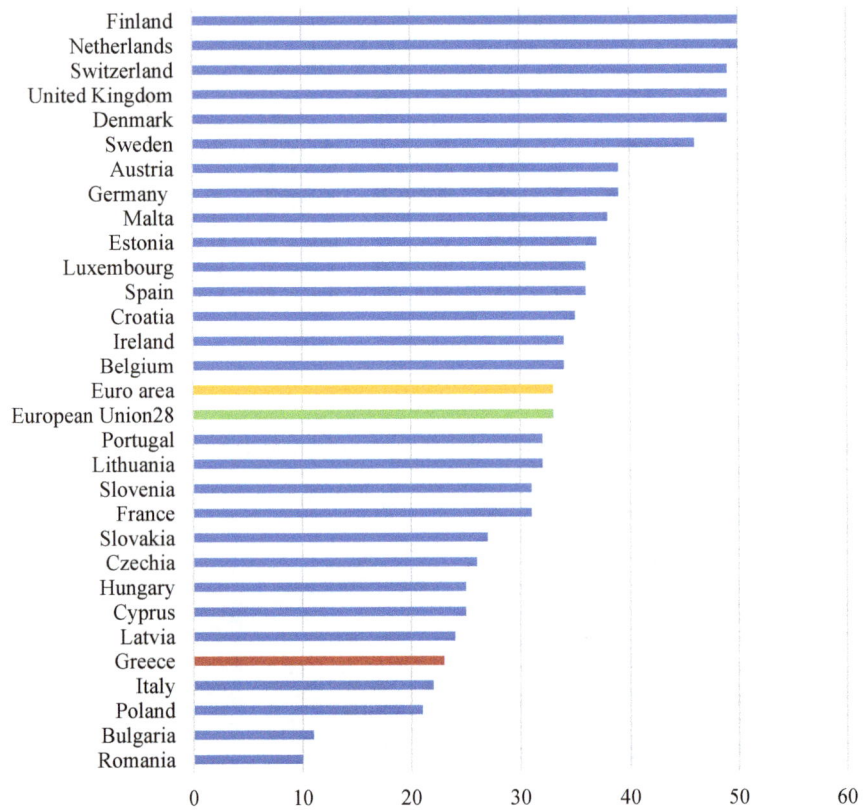

Fig. 8 EU: percentage (%) of individuals with above average digital skills, 2019

at 4.8 billion EUR, while the rest, that is, 21.4 billion EUR were to be available through the main structural and investment funds at the EU. According to the EC's website, national and regional programmes report financial data to the EC on their progress related to the total budget of the programme, decided financial resources and expenditures reported by the selection project and fund. Another important source of finance for investment in Greece is the European Fund for Strategic Investments from the "Juncker plan", making available 2.7 billion EUR. However, it seems that Greece has difficulties in implementing projects as it was only partially successful in implementing planned projects. An example was the project of broadband access in the programme information and communication technologies (broadband access information-communication infrastructure: additional households with broadband access of at least 30 Mbps). In 2016, it was planned to be delivered to 3,500.000 households, which was then in 2017 reduced to 63.185 households, and finally, in 2018, delivered to only 44.588 households. According to the same website, out of the planned information-communication technologies broadband projects, less than

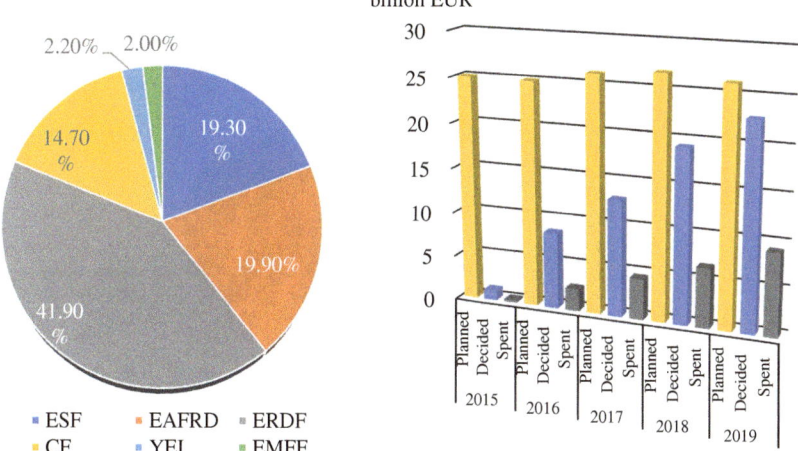

Fig. 9 EU Budget for 2014–2020 for Greece by funds and Implementation Progress as of 23rd April 2020 (Detailed information is available on the EU website https://cohesiondata.ec.europa. eu/countries/GR#. Abbreviations for funds stand for the European Social Fund (ESF), European Agricultural Fund for Rural Development (EAFRD), Youth Employment Initiative (YEI), European Regional Development Fund (ERDF), Cohesion Fund (CF), European Maritime and Fisheries Fund (EMFF). These are the main funds established to support economic development in the EU, in line with the objectives of the Europe 2020 strategy. The proportions of the funds used and implemented change daily, depending on the projects' implementation progress. Data are cumulative)

10% of it was implemented and only in 2017 and 2018. Plans were returned by the European Commission for corrections.

Greece registered more than 20 projects for the period 2014–2020 in regard to information and communication technologies. The approved budget for that has been more than 1 billion EUR, with the EU contributing over 842 million, and the rest is the national contribution. The aim of those projects was to enhance access to and use and quality of information and communication technologies.

Most of the EU financing in the period 2014–2020 has come via ERDF fund (785.8 million EUR), with an exception of the mentioned National Rural Development project (focusing on basic information and communication technologies' services) that was funded by EAFRD fund in the amount of 56.5 million EUR. In terms of the information-communication technologies, planned budget for the period 2014–2020 had over 60% of EU financial resources committed for less developed regions (Fig. 10). The average share of EU financing of the project is above 70%, and the rest is supposed to be national contribution.

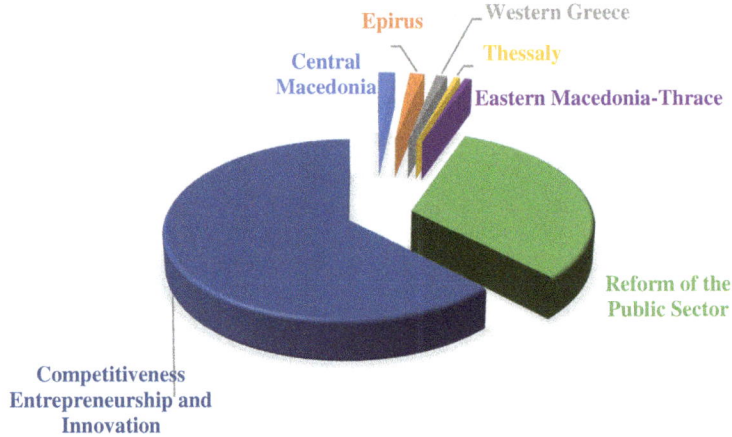

Fig. 10 Thematic objective: enhancing access to and use and quality of information and communication technologies (ICT) in less developed regions. Decided EU financial resources for information and communication technologies for less developed regions by programme titles

4.4 Tourism and Online Platforms for Accommodation

Greece is among the most specialized countries within the accommodation subsector in the accommodation and food service sectors in the EU (Eurostat, 2019). Online platforms such as Airbnb and others have been embraced by the landlords in Greece in the last few years. The result is that nights spent at the short-term accommodations have risen sharply (Fig. 13).

The rise has been in line with the change how tourists plan their holidays, access and buy their products and services influenced by information-communication technology. In the last decade, there has been an increase in the use of mobile devices (e.g. smartphones and tablets) to access information, products and services before and during tourism trips (European Parliament, 2015). The widespread use of information-communication technology produced a need for multi-skilled employees. That is one of the reasons why additional effort should be put into improving digital skills of the population.

The number of holiday and other short-stay accommodation has risen considerably in the last few years (Fig. 12), and many of them have been listed on online platforms. In 2018, the growth of holiday and other short-stay accommodation has been especially high in Attiki, by more than 12% while in the northern and central Greece it was about 10% (Fig. 14).

Online accommodation booking has been more common among middle-aged people (aged 25–54) than it was among either younger (aged 16–24) or older generations (aged 55–74). Most of these services were ordered through dedicated websites, that is, via online platforms such as Booking.com or Airbnb and similar platforms. In

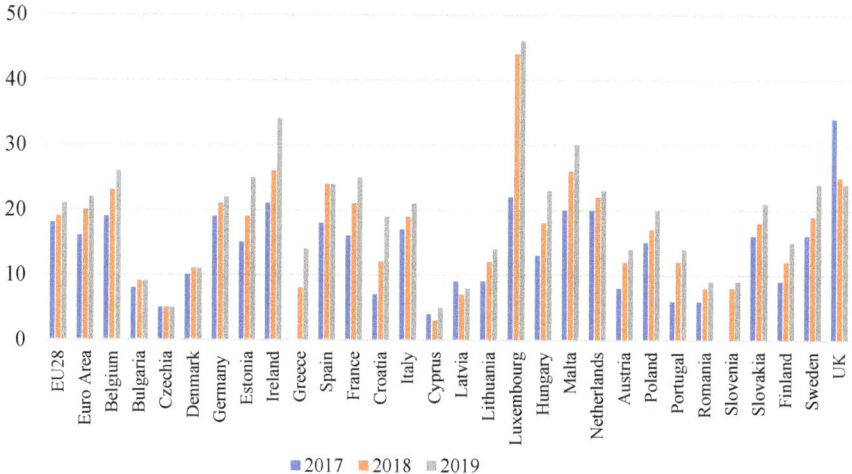

Fig. 11 Online accommodation booking, from 2017 to 2019: percentage (%) of individuals who used any website or application to arrange an accommodation from another individual

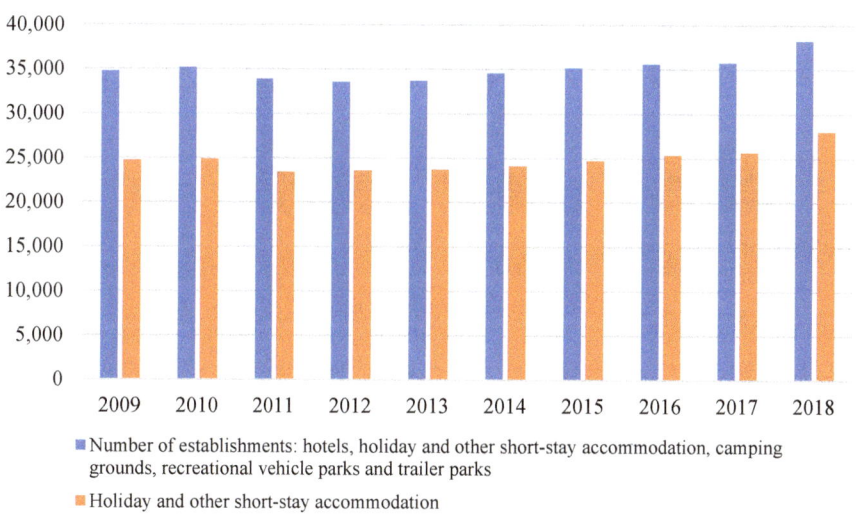

Fig. 12 Number of tourist establishments in Greece, 2009–2018

2017, about 18% of individuals (aged 16–74) in the EU used any website or application to arrange accommodation from another (private) individual during the preceding 12 months (Eurostat, 2019). From 2017 to 2019, the "collaborative" economy in the EU has grown further as the percentage of individuals using any website or application to arrange accommodation from another private individual grew to 21%. In Greece that share reached 14% in 2019, a 6 percentage points rise from 2018 (Fig. 11).

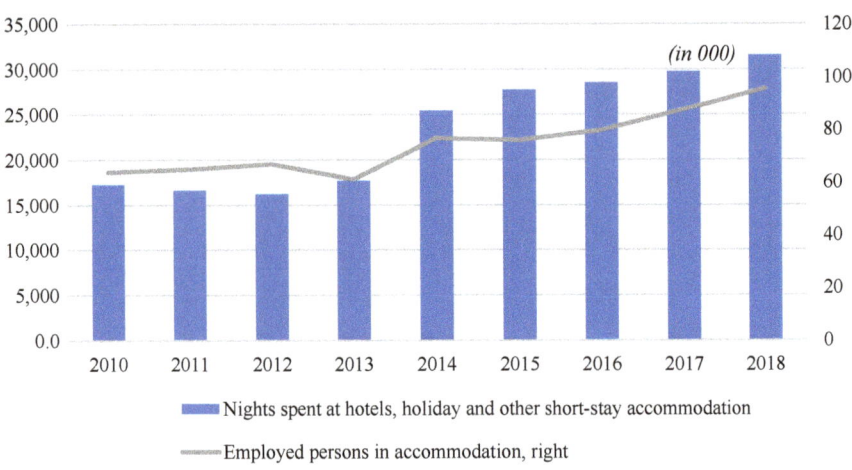

Fig. 13 Rise in short-stay accommodation and persons employed in accommodation

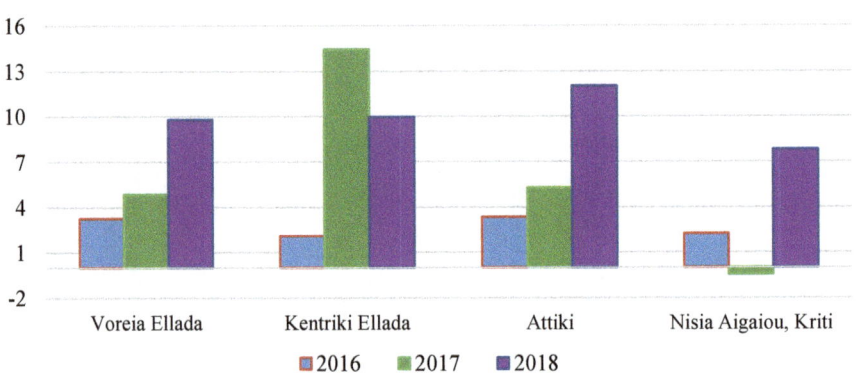

Fig. 14 Growth (%) of holiday and other short-term accommodation in Greece by region

Although data show that "sharing" or "collaborative" economy has risen in Greece too, it is still lagging considerably vis-à-vis the EU average and behind some other EU member states.

According to the research by Pierce et al. (2018), Airbnb had over 4 million listings in 191 countries in 2018. Booking.com, which also lists hotels in addition to all other types of accommodation, including accommodation offered by individuals and households, had as of the end of December 2019 over 29 million total listings, including more than 6.3 million listings of homes, apartments and other unique places to stay in 227 countries, according to the Booking.com's website. Responding to the evolving needs of tourists, Airbnb and Booking (as well as others such as HomeAway, VacationRentals), provide online platforms where travellers can access tourist services directly from other individuals. In line with that the number of tourist establishments in Greece has grown (Fig. 12) in almost all regions in recent years.

4.5 Macroeconomic Case for Digitalization

Greece is emerging from a decade-long debt crisis and its GDP per inhabitant has still not reached its pre-crisis level of 2008. After years of contraction and stagnation, the economy grew only modestly from 2017 to 2019. Growth was projected to exceed 2% in 2020 (EC, 2019) as the new government elected in July 2019 promised to follow pro-growth policies. However, the outbreak of coronavirus (COVID-19) and the renewed migrant crisis have negatively affected the growth projection. In response to that, Greece has to tackle many issues (Fig. 15).

After a decade of contraction, the unemployment rate remained high. In 2017, it was 21.5%, in 2018, it fell to 19.3%, while in 2019, it further dropped to 17.3% (EC, 2019). High unemployment has led to a significant reduction in households' disposable incomes and a sharp increase in poverty. In 2016, material deprivation in Greece was at 22.4% and has not improved much since then, in line with the fact that net wages fell considerably during the financial crisis and have not recovered much since then. The share of net wages in households' disposable income in Greece is the lowest among EU member states (Eurostat, 2019). The share of other income including income from owning a dwelling reflects high level of self-employment and high level of unemployment in Greece (Eurostat, 2019). Households are still relying on their savings which are well below the average in the euro area. Net saving of households has been falling since the onset of the crisis in 2008 and has been negative,

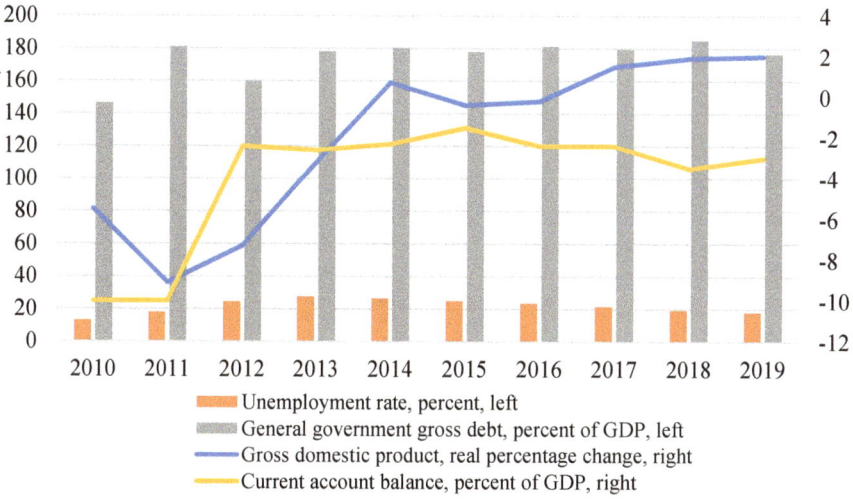

Fig. 15 Key macroeconomic indicators for Greece from 2010 to 2019

although stabilizing and slightly improving in recent years (Fig. 17). In 2017, GDP per inhabitant in Greece was still below 2008 level (Eurostat, 2020c). Besides Cyprus, the largest fall in household disposable income per inhabitant between 2008 and 2016 was in Greece. From 2008 to 2016, households' disposable income fell in 11 of 13 regions in Greece, by at least EUR 3400 per inhabitant. The largest fall of disposable income per inhabitant was in Attiki, by EUR 5800 per inhabitant (Eurostat, 2020c) (Fig. 16).

Economic recovery in Greece has been slower than expected, although tourism has been robust in recent years. In 2019, revenues from tourism grew by about 13% compared to 2018 and the rise in arrivals was about 4% (UNWTO, 2019). Tourism plays a significant role because jobs that are created or maintained because of tourism can help prevent economic contraction. Also, tourism has the potential to play a crucial segment in the economic development of remote and peripheral regions, such as the coastal, mountainous or distant regions. Tourism influences accommodation for

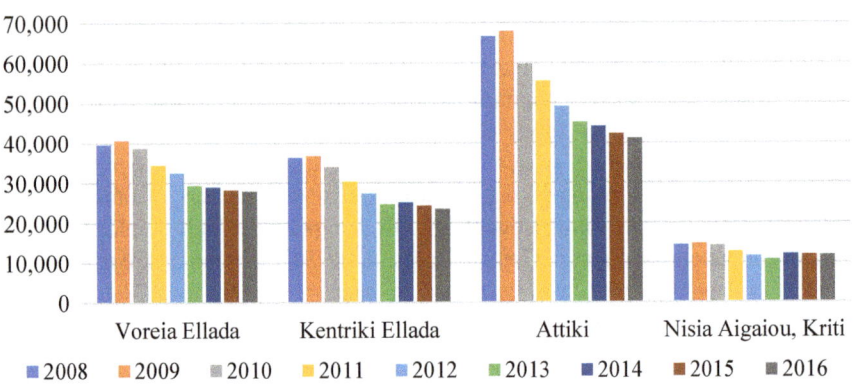

Fig. 16 Net disposable income by region, 2008 to 2016, million EUR

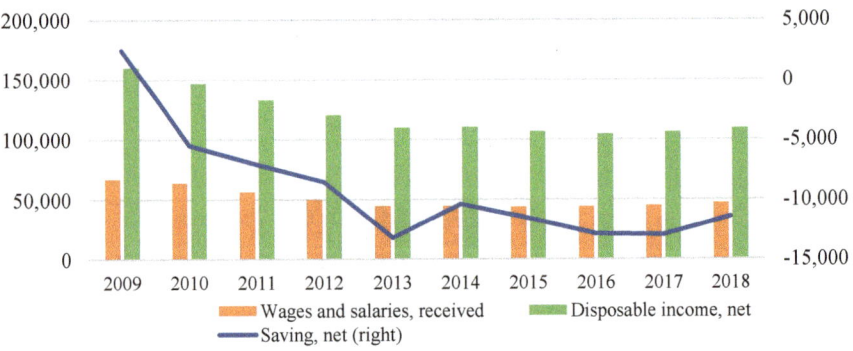

Fig. 17 Dynamics from 2009 to 2018 for some key households indicators: wages and salaries, disposable income and net saving (in million EUR)

visitors, food and beverage serving activities, railway, road, water and air passenger transport, and other activities.

International institutions found that much of the needed structural transformation of the Greek economy has not been achieved yet, and a "key question is whether the authorities can overcome long-standing vested interests that have traditionally blocked reforms" (IMF, 2019: 25). Macroeconomic challenges are many because, in addition to high unemployment and other problems in the labour market, Greece has high levels of public and private debt. Greece's public debt-to-GDP ratio is the highest in the euro area, standing at about 180% of GDP in 2019 (EC, 2019). Private debt is also high. Almost half of non-performing loans in banks has been due to households' loans and the low repayment capacity of public and private entities (IMF, 2019). As the income of households remains below Greece's pre-accession level, long-term projections suggest modest convergence (IMF, 2019).

EC found that "deep institutional and structural reforms initiated in recent years to modernize the economy and the State require many years of sustained implementation for their impact to fully unfold" (EC, 2019: 3). It concluded that Greece improved its budget balance and current account balance in recent years, but potential growth has fallen. In addition, high public debt, the negative net international investment position, the high non-performing loans on banks' balance sheets and high unemployment rate remain serious challenges.

In line with the above rests the case for further digitalization and participation on online platforms for holiday and other short-term stay. Arguments are many. First, if listing a property on online platforms does not require significant investments, except an access to Internet (to manage bookings) and a spare dwelling that can be offered for short-term stay, it can help raise households' income. Assuming that Internet's deployment and adoption of Internet-enabled digital services and products can be delivered to households under reasonable cost provisions, households have a possibility to use digital platforms by listing their property for short-term accommodation and other services. Hence, low search and communication costs over Internet are interesting for both, travellers and house owners as a possibility to bring revenue to property owners. Therefore, on a micro-level, from a household's perspective, participating in an online economy by listing a spare accommodation on platforms such as Airbnb brings extra income and raises household's disposable income.

Second, there is an argument related to taxation and state budget. In 2018, the government of Greece introduced mandatory registration for all property owners providing tourist accommodation online at the tax authority and reporting about revenue from short-term rental. As a result, the fiscal position of the government in regard to tax receipts from short-term rental has improved. Increased participation on online platforms improves fiscal receipts by the state, leading to the alleviation of the burden on the public budget. That strengthens government position and helps it fulfil its obligations vis-à-vis international creditors.

Third, there is a strong argument for further digitalization and households' participation in digital economy in Greece in terms of financial resources. Information-communication technologies' infrastructure and basic services have been so far co-financed by the EU. Within particular themes, there are many projects and financing them is available via EU structural and cohesion funds. Data on availability and implementation progress of EU co-financed projects show that the proportion of finances from the Greek state has on average been less than 30%. That is a considerable incentive to prepare projects that would bring Internet to remote and mountainous regions (for which, tourist demand is rising) while also providing faster and ultrafast Internet access elsewhere in Greece.

Fourth, the strongest argument for further digitalization is the fact that social consequences of the last crisis are still visible in all social indicators in Greece. By improving connectivity and access to Internet, households can list their accommodation to tourists via digital platforms. That would help many Greek households whose net wealth has fallen as a result of the crisis. In addition, reforms and pledges that the Greek government has made so far imply that many people have to deal with cuts to pensions and some of them are hardly surviving. Those platforms mean new opportunities for revenue.

Fifth, there is a strong macroeconomic argument for further digitalization. Participating in online economy would mean a low-cost impetus to GDP, which is particularly important taking into account numerous macroeconomic challenges in Greece, many of them relying on good GDP numbers. Additional revenue from short-term rental via online platforms and subsequent rise in domestic consumption contributing to GDP growth is in line with the conclusion that digitalization has proven a positive impact on employment and has been included in policy strategies in many countries.

Sixth, digitalization does not only have a positive impact on economic growth and GDP per capita, it improves the digital literacy of the population. Improving digital skills and increasing the proportion of information and communication technologies employment could help boost tourism (and other economic sectors), and even more so in the corona and post-coronavirus era. Improving digital skills among young adults should happen in line with digital transformation of the economy and society. In 2017, 54% of Greeks aged between 16 and 74 did not have at least basic digital skills, which is quite high relative to the EU average.

4.6   Coronavirus (COVID-19) Argument for Digitalization

The tourism sector has been hard-hit by the outbreak of COVID-19. In comparison with hotels, accommodation offered by households and other private entities on online platforms have certain advantages. First, tourists and other travellers do not have to meet their hosts or other people. Keys for the accommodation can be exchanged in a non-personal way (by leaving the keys in a designated place). The booking and all the other communication between guests and hosts can be done strictly online because travellers can book their stay online.

Second, hotels are typically large with a lot of guests, which is usually not the case with private accommodation provided in detached houses or buildings. Hence, the probability of interacting with people in person in accommodations offered by households on online platforms is significantly lower.

Third, hotels typically offer breakfasts and other services (fitness, pools, hair-dressers etc.) that imply the gathering of larger numbers of people. That is typically not the case in accommodation offered by households on online platforms for short-term tourist stay. Accommodation provided by households for short-term rental is a tourism accommodation establishment, defined as "the provision of accommodation, typically on a daily or weekly basis, principally for short stays by visitors, in self-contained space consisting of complete furnished rooms or areas for living/dining and sleeping, with cooking facilities or fully equipped kitchens. This may take the form of apartments or flats in small free-standing multi-storey buildings or clusters of buildings, or single storey bungalows, chalets, cottages and cabins. Very minimal complementary services, if any, are provided. This type of accommodation excludes daily cleaning, bed-making, food and beverage services" (Eurostat, 2020b).

Fourth, with slowly opening travel restrictions, people will be careful to travel. That might imply that travel will be mostly local or regional. In the case of Greece that means that, taking into account purchasing power of the neighbouring countries, that travellers would not seek luxury and expensive, but rather cheap accommodation. In that respect, it has been established that private accommodation listed on online platforms are, on average, cheaper than hotels (Bivens, 2019).

In sum, in the current and post-coronavirus era, an accommodation that can be booked online and is provided by households or other private parties listed on designated online platforms is in a better position to ensure social distancing and other response measures requested by the authorities. That is especially so if the accommodation is in coastal, rural and mountainous areas that are not hot-tourist spots and do not attract large crowds. Travelling to those areas can be by car or train, avoiding air travel which might also have problems with crowds. However, for rural and similar areas, Internet connectivity is crucial and that is one of the reasons why Greece should push for further digitalization. In the aftermath of the coronavirus outbreak, technology can be a key enabler because it is expected that travel will rebound rather quickly but will, for a certain amount of time, be different due to various pre-cautionary measures.

5 Conclusion

Greece needs high priority investments in order to close the gap with respect to the digital agenda in the EU. Not only that, Greece needs investments in all infrastructures, not only digital infrastructures in order to be able to overcome its economic slump following the last crisis. In that respect, Internet can be useful as it relates to many aspects of economic and tourism development. The country should increase higher-speed Internet connectivity and also invest in improving digital skills in order to be able to improve human capital in tourism. This is particularly important for households in Greece, considering that their disposable income has been shrinking following the last global financial crisis and bailout requirements.

On the other hand, changes in the way tourists and the tourism industry have used digital technology over the last decade have had significant effects on the industry and the way tourists travel. Digitalization is a challenge for traditional business models of tourism information and distribution. The sharing economy concept has been embraced by many as it means new forms of opportunity for local economies. Online platforms for short-term rental are enabled by digitalization and economic activity is increasingly moving online due to low cost of computation, storage and transmission of data. In regard to the online market for accommodation, travellers can have low search costs for any type of accommodation.

Given that the recovery of Greece is slow considering weak financial sector, combined with high unemployment, large external debt and an unfavourable net international investment position (now accompanied with coronavirus outbreak), digitalization, fast and ultrafast broadband, especially in less developed regions, would not only help Greek companies to operate better in a global business environment, but would also give a boost to households' income and thus improve overall macroeconomic picture in Greece as well.

References

Bivens, J. (2019). The Economic Costs and Benefits of Airbnb. *Economic Policy Institute*. https://www.epi.org/publication/the-economic-costs-and-benefits-of-airbnb-no-reason-for-local-policymakers-to-let-airbnb-bypass-tax-or-regulatory-obligations

Bouwman, H., et al. (2017).The impact of digitalization on business models: How IT artefacts, social media, and big data force firms to innovate their business model. *Econstor. Conference paper.* https://www.econstor.eu/bitstream/10419/168475/1/Bouwman-Reuver-Nikou.pdf

Charalabidis, Y., et al. (2015). Information systems in a changing economy and society in the Mediterranean region. In *Mediterranean Conference on Information Systems, 2015, Conference Proceedings, AIS Electronic Library* (pp. 1–6). Retrieved from https://aisel.aisnet.org/mcis2015/

Clancy, R. (2018). Landlords in London are embracing Airbnb, new research shows. *PropertyWire. (website).* https://www.propertywire.com/news/uk/landlords-london-embracing-airbnb-new-research-shows/

Eurostat. (2019). Digital Economy and Society Statistic, Households and Individuals, Statistics Explained. *(website).* Retrieved from https://ec.europa.eu/eurostat/statistics-explained/pdf scache/33472.pdf. Accessed on December 18, 2019.

Eurostat. (2020a). *Glossary: Tourism, statistics explained. (website).* Retrieved from https://ec.eur opa.eu/eurostat/statistics-explained/index.php?title=Glossary:Tourism. Accessed at February 21, 2020.

Eurostat. (2020b). *Glossary: Holiday and other short-term accommodation. Statistics Explained. (website).* https://ec.europa.eu/eurostat/statistics-explained/index.php?title=Glossary:Hol iday_and_other_short-stay_accommodation

Eurostat. (2020c). *Regional socio-economic developments-statistics. Statistics explained. (website).* https://ec.europa.eu/eurostat/statistics-explained/index.php/Regional_socioeconomic_develop ments_-_statistics

European Commission. (2007*). European initiative on an all-inclusive digital society: Frequently asked questions. (website).* https://ec.europa.eu/commission/presscorner/detail/en/MEMO_0 7_527

European Commission. (2019). *Digital Government Factsheet 2019, Greece. (website)* Retrieved from https://joinup.ec.europa.eu/sites/default/files/inline-files/Digital_Government_ Factsheets_Greece_2019.pdf

European Commission. (2020a). *Shaping Europe's digital future. (website).* https://ec.europa.eu/ info/strategy/priorities-2019-2024/europe-fit-digital-age/shaping-europe-digital-future_en

European Commission. (2020b). *the digital economy and society index (DESI). (website)* Retrieved from https://ec.europa.eu/digital-single-market/en/desi

EUR-Lex, Access to European Union Law. (2010). Communication from the Commission to the European Parliament, the Council, the European Economic and Social Committee and the Committee of the Regions. *A Digital Agenda for Europe (website).* Retrieved from https://eur-lex.europa.eu/legal-content/en/ALL/?uri=CELEX%3A52010DC0245.

European Parliament. (2019). *Contribution to growth—European digital single market, delivering improved rights for european citizens and businesses* (pp. 11–43). (website) Retrieved from https://www.europarl.europa.eu/RegData/etudes/STUD/2019/638395/IPOL_S TU(2019)638395_EN.pd

Foundation (2019). *Digital Transformation in Greece* (p. 19). *(website).* Retrieved from https://the foundation.gr/wp-content/uploads/2018/12/DigitalTransformation_2018_by_EIT-Digital_and_ Foundation.pdf

Goeroeg ,C. (2018). The definitions of sharing economy: A systematic literature review. *Management. University of Primorska, Faculty of Management Koper,* 13(2), 175–189

International Monetary Fund. (2018). Measuring The Digital Economy. *Policy paper.* (website) Retrieved from https://www.imf.org/en/Publications/Policy-Papers/Issues/2018/04/03/022818-measuring-the-digital-economy

International Telecommunications Unit. ITU-SG WTPF 13 Information Document 8. World Telecommunications ICT/Policy Forum (2013). Retrieved from https://www.itu.int/md/S13-WTPF13-INF-0008/en

Mao, Y., et al. (2018). *The Real Effects of Sharing Economy: Evidence from Airbn.* Cornell University School of Hotel Administration, The Scholarly Commons, Working Papers, 3-2018. https://sch olarship.sha.cornell.edu/cgi/viewcontent.cgi?article=1051&context=workingpapers

Marcovitch S., & Willmott P. (2014). Accelerating the digitization of business processes. *Mckinsey Digital (website).* https://www.mckinsey.com/business-functions/mckinsey-digital/our-insights/ accelerating-the-digitization-of-business-processes

Parviainen, P., et al. (2017). Tackling the digitalization chalenge: how to benefit from digitalization in practice. *International Journal of Information Systems and Project Management, 5* (2017), 63–77

Pierce, B. K., et al. (2018). Expanding economic opportunity through Airbnb in Thessaloniki, Greece. *Creativity Platform. Interactive Qualifying Projects at Worcester Polytechnic Institute Digital WPI. (website).* https://digitalcommons.wpi.edu/cgi/viewcontent.cgi?article=1550&con text=iqp-all

Randall, L. et al. (2017). *Digitalisation as a tool for sustainable Nordic regional development.* Discussion paper prepared for Nordic thematic group for innovative and resilient regions. Available at https://www.nordregio.org/research/digitalisation-as-a-tool-for-sustainable-nordic-regional-development

Sabbagh, K. et al. (2013). Digitization for economic growth and job creation: regional and industry perspectives. *The Global Technology Report 2013, World Economic Forum 2013* (pp. 35–42). Available at www3.weforum.org/docs/GITR/2013/GITR_Chapter1.2_2013.pdf

Tsakanikas, A. et al. (2014). *ICT adoption and digital growth in Greece* (pp. 53–62). Foundation of Economic and Industrial Research. Retrieved from https://iobe.gr/docs/research/RES_03_100 62015_REP_ENG.pdf

United Nations World Tourist Organization. (2019). *International Tourism Highlights*, 2019 Edition. *(website).* https://doi.org/10.18111/9789284421152.

United Nations World Tourist Organization. (2003). World tourism barrometer. *Statistical Annex,*
  18(2), 8 (2020). https://webunwto.s3.eu-west-1.amazonaws.com/s3fs-public/2020-05/UNWTO_
  Barom20_02_May_Statistical_Annex_en_pdf
World Travel Tourism Council. (2019). Greek tourism sector growing three times faster than wider
  economy. *WTTC Research (website)*. Retrieved from https://www.wttc.org/about/media-centre/
  press-releases/press-releases/2019/greek-tourism-sector-growing-over-three-times-faster-than-
  wider-economy-says-new-wttc-research/

The E-Tour Facilitator Platform Supporting an Innovative Health Tourism Marketing Strategy

Constantinos Halkiopoulos, Eleni Dimou, Aristotelis Kompothrekas, Giorgos Telonis, and Basilis Boutsinas

1 Introduction

Innovative Medical Tourism Strategy (In-MedTouR) is a project implemented within the frame of Interreg Greece–Italy V/A 2014–2020 program (Fig. 1). The project, which is co-funded by European Union, European Regional Development Funds (ERDF) and by National Funds of Greece and Italy, aims at creating a cluster of cooperation between Greece's and Italy's businesses engaged in Health Tourism (private or public hospitals, doctors, SPA, wellness centers, hotels, etc.) as well as research and development entities (universities, technological institutes, etc.), so that know-how and innovations can be exchanged between the two countries.

The overall goal of the project is to create an interregional network, which will enhance the cooperation, networking, and interaction between those involved in health tourism for the exchange of best practices. The network of cooperation will be fostered during the implementation of the project; nevertheless, the cooperation is going to be an ongoing procedure in the future, even after the completion of the project.

In-MedTouR has led to the introduction of a new touristic product, which is called health tourism and refers to the process of traveling outside the country of residence for the purpose of obtaining medical treatment and at the same time enjoying holidays in the destination country. For this purpose, an e-Tour Facilitator has been developed containing information about touristic and health providers in the cross-border area with a focus on the end-users of the health tourism product, namely the patients/tourists (Amodeo, 2010). The e-Tour Facilitator Platform offers matching services to the patients looking for health and hospitality units in order to receive the treatment they choose.

C. Halkiopoulos (✉) · E. Dimou · A. Kompothrekas · G. Telonis · B. Boutsinas
MISBILAB, University of Patras, Patras, Greece
e-mail: halkion@upatras.gr

© The Author(s), under exclusive license to Springer Nature Switzerland AG 2021
V. Katsoni and C. van Zyl (eds.), *Culture and Tourism in a Smart, Globalized, and Sustainable World*, Springer Proceedings in Business and Economics,
https://doi.org/10.1007/978-3-030-72469-6_39

Fig. 1 Web portal of In-MedTouR.eu project

The proposed e-Tour Facilitator is an intelligent information system that processes information related to providers, but also users (patients/tourists) and which will contribute to the automated overall proposal of medical tourism services covering individual needs. Being developed using state-of-the-art techniques for both the design and implementation phases of the project, e-Tour Facilitator's goal is to introduce innovative technological tools to enhance networking (e-Platform) and upgrade the overall end-user experience (e-Tour Facilitator).

Given the lack of cooperation between companies and research/educational institutions and limited access to innovation tools fact that made the development of health tourism difficult, In-MedTouR delivers a touristic product with high added value bringing together research industries with enterprises and improving their access to innovation. E-tour Facilitator supports sufficient functionality in order both to help end-users to select the proper health tourism product with respect to their profile as well as to provide end-users with an integrated view of available such health tourism products.

## 2	Literature Review

Targeted travel of people for their health is not a current phenomenon. Since ancient times, legends and myths have contained many travel stories made by people seeking filters and treatments for themselves or their fellow human beings (Heung et al., 2010). Travel for this purpose has created health tourism, or rather medical tourism. Medical tourism refers to biomedical procedures in combination with travel and tourism in the respective selected areas. Refers to the practice of traveling to different countries to find optimal healthcare facilities. The services that travelers are looking for include simple procedures, but also complex specialized surgeries such as heart surgery, dentistry, and cosmetics. At the moment, health tourism is gaining a large market share in the world with a volume of over 100 billion euros. Today, all types of health care are provided, including alternative therapies, psychiatric and health care, and even burial services. There are more than 50 countries worldwide that have declared medical tourism as a national industry (Hazarika, 2010).

In the twentieth century, the tendency of economically wealthy people from developing countries to travel frequently to developed countries in order to gain access to medical facilities that are considered more advanced and specialized compared to their country of origin (Hall and Williams, 2008). However, nowadays there is a shift in medical tourism as economically wealthy people from developed countries express a preference for medical care in developing countries for various reasons (Lunt et al., 2013). The rising cost of health care and medical services in general in developed countries is leading more and more people to choose to seek treatment in cheaper countries (Ciburiene et al., 2009; Mueller and Kaufmann, 2001; Sundbo et al., 2007). In addition, waiting times for emergency medical services are much shorter in developing countries than in developed ones. In summary, the basic principles of medical tourism that seem to apply in the twenty-first century relate to the following (Lunt and Carrera, 2010):

- Significant increase in the number of tourists traveling for medical purposes (health and medical issues).
- Shifting the trend of patients from developed countries to less expensive developing countries with the ultimate goal of gaining access to medical services.
- Availability of information regarding affordable medical facilities in various parts of the world where they are easily accessible via the Internet.

Medical tourism is defined as the sum of all relationships and phenomena resulting from a journey by individuals whose primary motivation is to cure a medical condition by utilizing medical services away from their usual place of residence, while usually combining this journey, with elements of holidays or tourism in the conventional sense (Voigt et al., 2010; Cook, 2010). Health tourism refers to cross-border healthcare, which is often determined by a variety of factors, such as competitive costs, avoiding long queues, or waiting times for treatment, or medical treatment. service that is not available in their country (Chuang et al., 2014; Szymańska, 2016; Pandey, 2018).

As in any industry, so in the medical tourism industry, success depends on the knowledge and proper execution of basic marketing strategies to focus on the needs of future consumers (Hjalager, 2010). Based on the above, attracting international patients is considered imperative as this target group is at the core of the medical tourism market (Katsoni, 2011). As the industry becomes increasingly competitive, the major players in this field will be distinguished by their ability to attract foreigners, medical visitors / travelers (Gilbert, 2014). Modern marketing tools and services should take into account the provision of potential medical tourists with valuable information about the services provided and the prevention of communication barriers through the use of specialized platforms that interact in many languages to ensure accessible communication with their patients (Halkiopoulos et al., 2019).

It is also important, by implementing modern marketing strategies, to strengthen contact with patients directly or through medical travel agents to provide medical transportation within and within the country of destination (Ejdys et al., 2015). It is now imperative through modern specialized digital tools and the provision of online services of immigration, nursing care, hotel reservations, leisure services, and tours during recovery and restaurant services in order to ensure the best experience of the patient/traveler (Cormany and Baloglu, 2010).

3   The Proposed E-Tour Facilitator Platform

3.1   The E-Tour Facilitator Architecture

The e-Tour Facilitator Platform focuses on the end-users of the health tourism product, namely the patients/tourists. This web platform (Fig. 2) offers matching services for the patients looking for health and touristic units to receive the appropriate treatment. Thus, the design of the platform is incredibly important since its usability is a vital aspect that can determine its future success. The proposed architecture of the e-Tour Facilitator Platform is shown in Fig. 3. The e-Tour Facilitator Platform is based on a recommender subsystem. Typically, the users of a recommender system insert product constraints and express needs and preferences by using the offered language, and then the recommender system matches the preferences inserted in attributes with relative items in a database.

3.2   The E-Tour Facilitator Web Portal

The e-Tour Facilitator Platform based on a database of user profiles and user preferences and a database storing all the availability information and schedules of those organizations/companies offers health tourism products. In order to promote the proper information to its users, the e-Tour Facilitator Platform uses an e-Shop

Fig. 2 Web portal of e-Tour Facilitator

module, which consists of a Commenting Subsystem and an e-mail subsystem. The commenting subsystem handles all comments for each different health tourism product uploaded on the e-Tour Facilitator Platform. Thus, end-users are able to reach those who have experienced a use/test of a particular health tourism product registered to the e-Tour Facilitator Platform. The automatic e-mail response subsystem manages the incoming messages after applying text analysis techniques distribute them to proper operators for further processing if needed. Also, the e-Tour Facilitator Platform based on a video database consisting of product videos, which is used in the promotion of health tourism products.

The e-Tour Facilitator Platform is focused on the final user who is the tourist-patient, and the user interface of the Platform is user-friendly and simple to navigate. The patient defines his kinetic state in order the platform to exclude destinations that do not meet the minimum requirements to support his disability with the appropriate facilities or services. Medical services are divided into categories and subcategories, and costs are differentiated per provider.

The functionality of the system is divided into three parts.The database contains the data to be stored, viewed, and edited, the front-end interface with users, and the back-end support system. The original specifications include data for the database. During the initial version of the system, the basic structure of the relational database is created based on specifications.

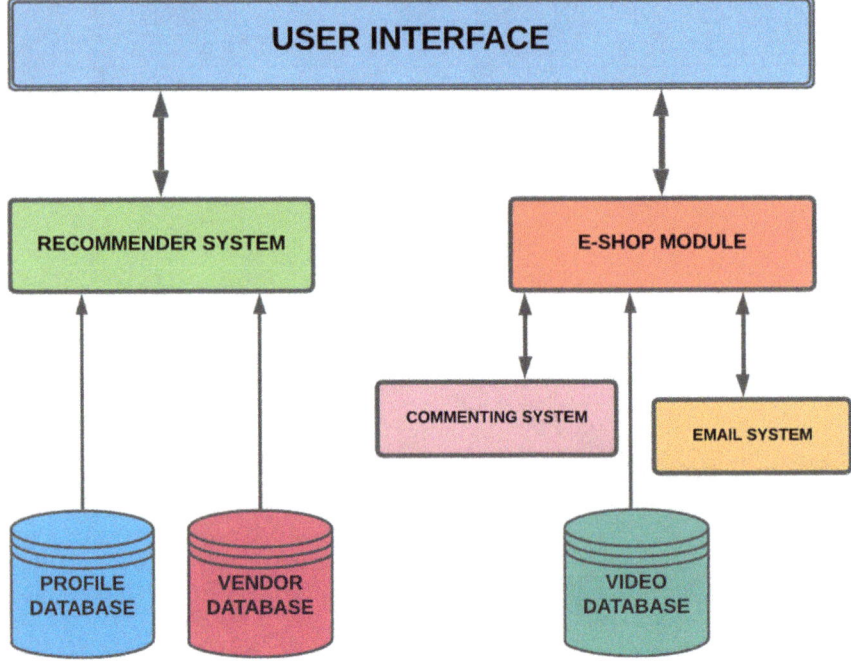

Fig. 3 Proposed architecture of the e-Tour Facilitator Platform

3.3 The E-Tour Facilitator Relationship Entity Diagram

The relational database (Fig. 4) is a data collection organized in correlated tables that simultaneously provide a mechanism for reading, writing, modifying, or even more complex data processes. The purpose of the database is the organized storage of information and the possibility of extracting this information, in more organized form, according to queries placed in the relational database.

According to the functional requirements of the experts, the workflow of the e-Tour Facilitator Platform was created, as shown in Fig. 5. The core entities of the system are the main building blocks of data collection and processing. These entities are:

- User
- The Medical Service
- The Medical Destination
- The Tourist Destination
- The Carrier
- The Middleman

Users will be able to declare their preferences for the tourist packages and medical services they want to access. This creates a more complete personal profile for each

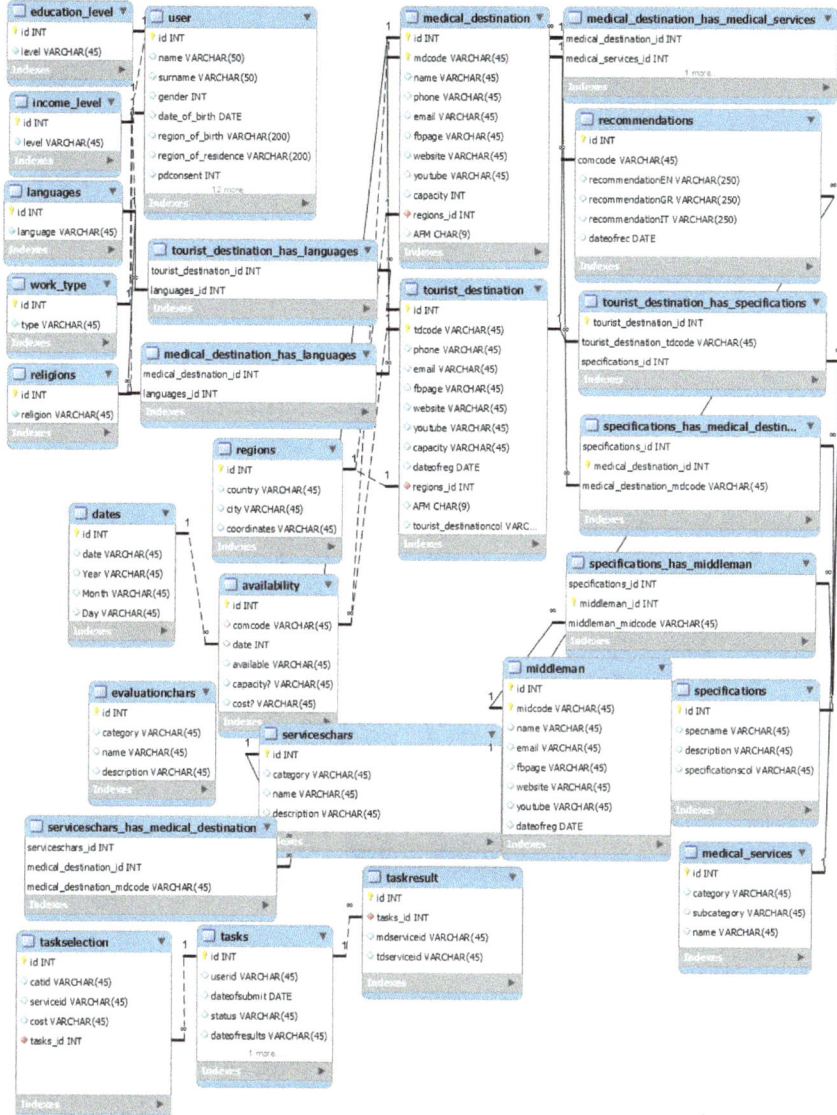

Fig. 4 E-Tour Facilitator Platform—part of the MySQL database schema

user. Health service providers or tour operators will be able to register with the system and add the service packages offered. The user chooses a medical service either on his own or through an intermediary who will give him a ready solution for a mixed package. The middleman combines transport options, tourist destinations, and medical destinations to support a medical service chosen by the user.

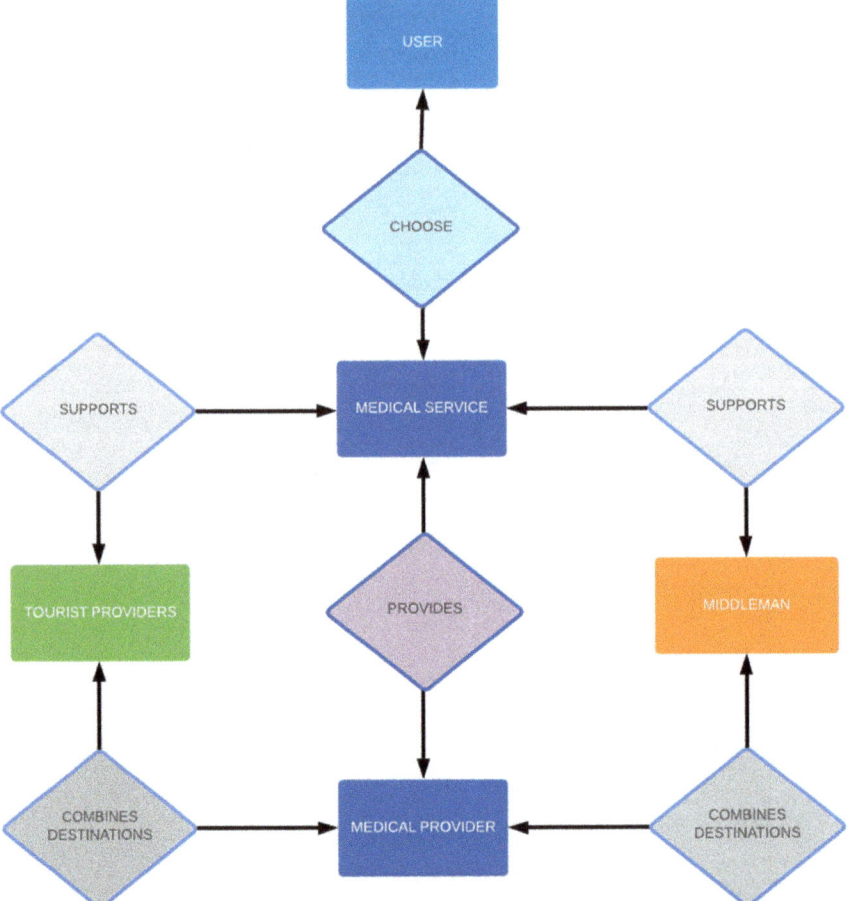

Fig. 5 Workflow for the e-Tour Facilitator Platform

Otherwise, when the user does not want the bundle to be solved by the middleman, he chooses the medical service he desires, and then chooses the medical destination provided by the medical service. The medical destination is then combined with a tourist destination (e.g., a hotel) and a mode of transport. The user has the option to first select the tourist destination that supports the medical service and then the medical destination (e.g., clinic, hospital).

In any case, the system will only offer, either through an intermediary or with a smart user choice, medical services, and medical or tourist destinations. The platform will not work as a reservation and availability management system, only suggestions, where registration of any reservation via the system will be ensured. Only service providers and intermediaries have the responsibility for the availability and management of the detention.

3.4   The E-Tour Facilitator Programming Framework

The e-Tour Facilitator Platform is based on the Laravel PHP Web App framework. Laravel is well known for its inbuilt lightweight templates that help the platform developers create amazing layouts using dynamic content seeding. In addition to this, it has multiple widgets incorporating CSS and JS code with solid structures. The templates of the.php framework are innovatively designed to create a simple layout with distinctive sections. It provides the platform with all the necessary tools to create well-structured and consistent user interface.

At the e-Tour Facilitator, the system model section consists of the core entities of the system. These are mainly the patients and the tour operators as well as medical service providers and intermediaries. The view section is the interface of the system users with the system services and functions that are implemented and run by the Back-End system. The controller is the link between model and view, where it implements and manages the business logic of e-Tour Facilitator.

Business logic is the part of the program that encodes the e-Tour Facilitator business rules that define how to create, store, and change data. It contrasts with the rest of the software that may involve lower level details about managing a database or displaying the user interface, system infrastructure, or generally linking different parts of the program. The programming languages used to install, configure, and operate the front-end system are html, php, JavaScript,, CSS and libraries such as jQuery. It is based on a smart search algorithm that is implemented, which is based on the user profile. This algorithm is used by the recommendation system, which will suggest users the packages that fit their profile before they even search. The algorithm evaluates the user's preferences when they register and identifies potential service packages of interest to the user.

Depending on the availability of health providers of each category and medical practice, only cities that have clinics offering the requested medical service are displayed and returned after the user's query. The search query can run by destination and/or by medical treatment. The system to find nearby medical or tourist providers based by the user's query, the platform contains pre-declared neighborhoods. For each custom neighborhood, it is declared the corresponding polygon in the map. A polygon is declared by its point coordinates. For each point of interest (city, town), we use its coordinates to clarify in which neighborhood it is contained.

We should underline that the calculation of cost is multifactorial. Firstly, it is linked to the cost of the medical treatment and the duration of stay at a hospital if required, while, secondly, it includes travel expenses to and from the country of permanent residence to the medical treatment destination, and an additional cost category is that of accommodation and tourist experience if required. Most of medical tourists travel with relatives.

3.5 The E-Tour Facilitator Commenting System

The advent of natural language processing (NLP) has armed each and every company with the means to analyze a plethora of data they have it empowered them to automate most of the processes involved in it, thereby enabling them to directly fetch actionable information and thus saving both time and human cost. Natural language processing (NLP) and text mining techniques are the main tools for information extraction. Text mining is the process of examining large collections of text and converting the unstructured text data into structured data for further analysis like visualization and model building. We will utilize the power of text mining to do an in-depth analysis of customer reviews on e-Tour Facilitator Platform. Customer reviews are a great source of "Voice of customer" and could offer tremendous insights into what customers like and dislike about a product or service. For the e-commerce business, customer reviews are overly critical, since existing reviews heavily influence buying decision of new customers in the absence of the actual look and feel of the product to be purchased.

Health tourism is no exception to the rule that prospective buyers of a service or product pay attention to evaluations of previous users. In many cases, ratings form a buyer's positive or negative attitude toward a health provider, thus affects his final decision. Evaluation and commenting of the offering services are a major part of the Platform. Every service can be evaluated and commented independently, and these evaluations are considered by the recommendation algorithm to propose the high rated providers first. People's opinion is building the electronic word of mouth (eWOM). When buying services, patients often look at information written by other patients on the Internet. In other words, they turn to electronic word of mouth (eWOM). Marketers can take steps to generate, support, and amplify eWOM and so influence consumers' decision-making process. For that reason, it is incorporated a commenting system on the e-Tour Facilitator Platform. Comments became an integral part of almost each website on the web—you cannot imagine there's a service out there that can't be commented. And it is only logical, because people want to talk about what they read, listen to, or see. And they want it right there—under what they have read or seen, not on their social networks. The commenting system of e-Tour Facilitator is developed from scratch to satisfy every future need of analysis.

It is developed an application that can analyze the content of user feedback in Greek, Italian, and English language. The application functions as a mining tool for the services offered and categorize them to continuously evaluate and improve the offered services. Thus, for the analysis of customer–patient reviews, a commenting system of the services is developed. For the user profile (customer–patient) recorded the name, surname, date of birth, sex, place of residence, education level, income level, type of work, religion the consensus in the management and processing of data, telephone, mobile, e-mail, the accessibility level (e.g., disabilities has), recording date, language, and medical data. A comment will consist of the following data fields (Fig. 6). The id of the comment, the id of the user who posted the comment, the message and its representation to a formal language, the id of the service for

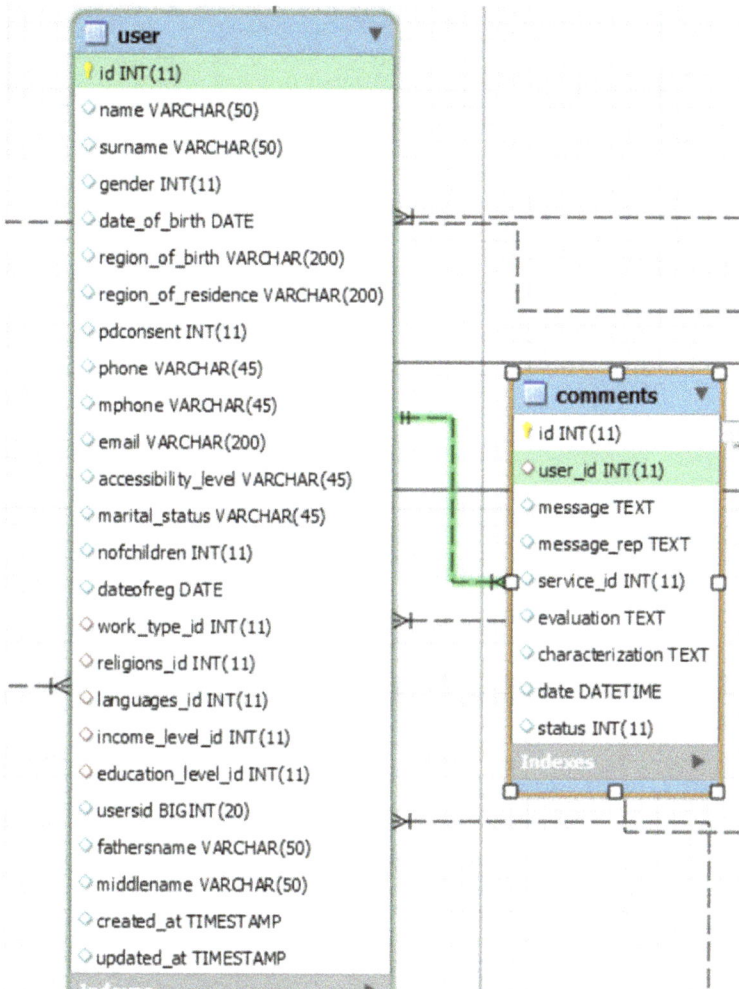

Fig. 6 E-R diagram of proposed commenting system for the e-Tour Facilitator Platform

which the comment is posted, the evaluation of the service, the sentiment analysis characterization of the comment, the date of comment, and the status of the comment.

3.6 Sentiment Analysis Process

For the evaluation and characterization of the comments, it is used sentiment analysis. Sentiment analysis—opinion mining, text analysis, emotion AI—determines the emotional tone behind words, to understand the attitudes and opinions being expressed (Fig. 7). The platform uses sentiment analysis to find and measure customer opinions and attitudes towards its providers of medical and tourist services. Analyzing sentiment on evaluation comments is an excellent source of data and will provide digital consumer insights that can (Fig. 8):

- Determine medical provider reputation
- Improve patient experience
- Stop issues becoming a crisis
- Determine future marketing strategies
- Improve marketing campaigns

Please select a characteristic.

| Medical Fees | × ▾ |

Please select minimum evaluation score.

| 50% | × ▾ |

Please select dates.

| 01/20/2020 → 01/27/2020 |

You have selected: Start Date: January 20, 2020 | End Date: January 27, 2020

Providers	Score
IVF Gynecology	100
Just4u	50
Marias Spa Center	50

Score for providers

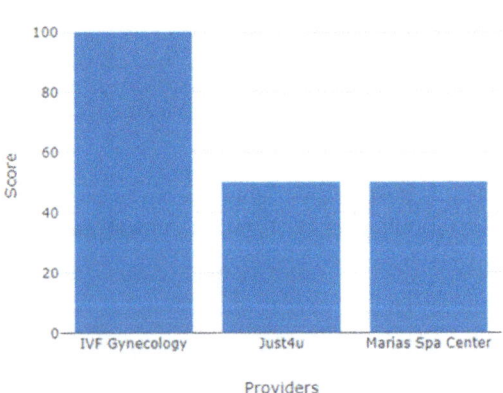

Fig. 7 Medical providers with minimum evaluation score

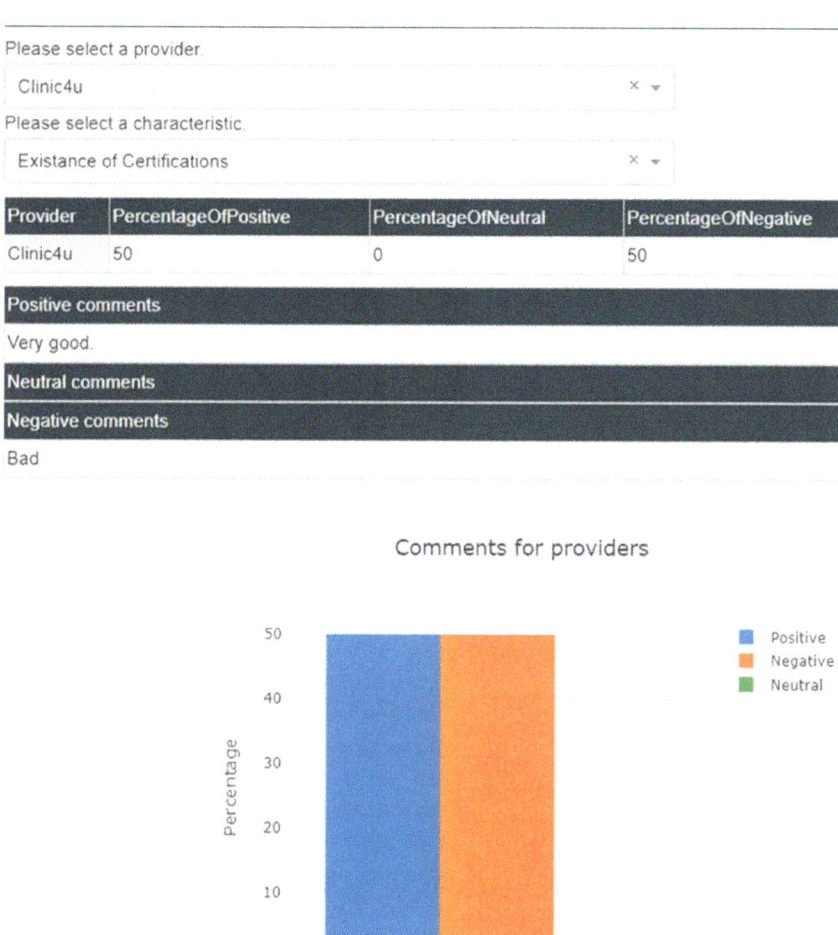

Fig. 8 Sentiment distribution of comments for specific medical provider/characteristic

For the sentiment analysis processes, a well-formed dictionary is used. Valence Aware Dictionary and Sentiment Reasoner (VADER) is a lexicon and rule-based sentiment analysis tool that is specifically attuned to sentiments expressed in social media. VADER uses a combination of a sentiment lexicon is a list of lexical features (e.g., words) which are generally labeled according to their semantic orientation as either positive or negative.

In order to store video clips for the offered services from medical and tourist providers, it is implemented a local video database that it is also used by the YouTube Data API, so as to upload the video to YouTube as well, using a back-end filter and upload operation. The YouTube Data API let us incorporate functions normally executed on the YouTube website into our own website and applications.

The ICT-Based Promotion Tools are implemented to help each simple user and service provider to have a better understanding about the quality of the services that are available through the In-MedTour Platform.

For the development of the various processing algorithms, it is used the Python programming language. Python is a programming language that let us work more quickly and is integrated into our applications more effectively. Python is powerful and fast and is adopted by all development members for marketing automation tasks (e.g., use of YouTube API, Mailing System, etc.).

For information retrieval and presentation, an open source reporting system is used. Dash is a productive Python framework for building web applications. It is written on top of Flask, Plotly.js, and React.js. It is ideal for building data visualization applications with highly custom user interfaces in pure Python where all important information about the offered services of the platform are presented in various ways. We deployed our dash applications to the main application server, and we share them through the Platform.

4   Conclusions

In summary, it is worth mentioning that encouraging results have emerged from the implementation of this e-Tour Facilitator Platform and in general of this whole project of In-MedTouR as the registration of health services and hospitality units in the cross-border area has been successfully carried out mainly for competition in medical tourism, event organization and publicity campaign to inform local communities about the prospects of medical tourism and networking opportunities between businesses and research and development institutions, the development of an institutional framework for the e-Cluster, to enhance interaction, collaboration and networking between businesses and research & development institutions as well as the matching of services—medical and tourism—according to the personalized needs of the end users.

References

Amodeo, J. (2010). Medical refugees and the future of health tourism. *World Medical & Health Policy, 2*(2), 65–81. https://doi.org/10.2202/1948-4682.1103.

Chuang, T. C., Liu, J. S., Lu, L. Y. Y., & Lee, Y. (2014). The main paths of medical tourism: From transplantation to beautification. *Tourism Management, 45,* 49–58. https://doi.org/10.1016/j.tou rman.2014.03.016.

Ciburiene, J. (2009). The Innovativeness of SME activity in the context of globalization in Lithuania. *Economics and Management, 14,* 723–730.

Cook, P. S. (2010). Constructions and experiences of authenticity in medical tourism: The performances of places, spaces, practices, objects and JBHOST. **02**(1), 250–265.

Cormany, D., & Baloglu, S. (2010). Medical travel facilitator websites: An exploratory study of web page contents and services offered to the prospective medical tourist. *Tourism Management, 32*, 709–716. https://doi.org/10.1016/j.tourman.2010.02.008.

Ejdys, J., Ustinovičius, L., & Stankevičienė, J. (2015). Innovative application of contemporary management methods in a knowledge-based economy—Interdisciplinarity in science. *Journal of Business Economics and Management, 16*, 261–274. https://doi.org/10.3846/16111699.2014.986192.

Gilbert, C. H. E. (2014). Vader: A parsimonious rule-based model for sentiment analysis of social media text. In *8th International AAAI*.

Halkiopoulos, C., Giotopoulos, K., Papadopoulos, D., Gkintoni, E., & Antonopoulou, H. (2019). Online reservation systems in e-Business: Analyzing decision making in e-Tourism. *Journal of Tourism, Heritage & Services Marketing (JTHMS)* Special Issue (ISSN: 2529–1947).

Hall, C. M., & Williams, A. M. (2008). Tourism and innovation. *Routledge.* https://doi.org/10.4324/9780203938430.

Hazarika, I. (2010). Medical tourism: Its potential impact on the health workforce and health systems in India. *Health Policy and Planning, 25*, 248–251. https://doi.org/10.1093/heapol/czp050.

Heung, V. C. S., Kucukusta, D., & Song, H. A. (2010). Conceptual model of medical tourism. *Journal of Travel and Tourism Marketing, 27*(3), 236–251. https://doi.org/10.1080/10548401003744677.

Hjalager, A. M. (2010). A review of innovation research in tourism. *Tourism Management, 31*, 1–12. https://doi.org/10.1016/j.tourman.2009.08.012.

Katsoni, V. (2011). The Role of ICTs in regional tourist development. *Regional Science Inquiry Journal, iii*(2).

Lunt, N., & Carrera, P. (2010). Medical tourism: Assessing the evidence on treatment abroad. *Maturitas, 66*, 27–32. https://doi.org/10.1016/j.maturitas.2010.01.017.

Lunt, N. et al. (2013). *Medical tourism: Treatments, markets and health system implications: A scoping review* (p. 6). OECD. Directorate for Employment, Labour and Social Affairs. https://www.oecd.org/els/healthsystems/48723982.pdf

Mueller, H., & Kaufmann, L. (2001). Wellness tourism: Market analysis of a special health tourism segment and implications for the hotel industry. *Journal of Vacation Marketing, 7*, 5. https://jvm.sagepub.com/content/7/1/5

Sundbo, J., Orfila-Sintes, F., & Sorensen, F. (2007). The innovative behavior of tourism firms—Denmark and Spain. *Reserch Policy, 36*, 88–106. https://doi.org/10.1016/j.respol.2006.08.004.

Szymańska, E. (2016, on-line 2017). Consumer participation in the health tourism innovation process. *Economics and Management, 8*(4), 28–38. https://doi.org/10.1515/emj-2016-0030

Pandey, P. (2018, September 23). Simplifying sentiment analysis using VADER in Python (on Social Media Text). *Medium.* https://medium.com/analytics-vidhya/simplifying-social-media-sentiment-analysis-using-vader-in-python-f9e6ec6fc52f

Voigt, C., Laing, J., Wray, M., Brown, G., Howat, G., Weiler B., & Trembath, R. (2010). Health tourism in Australia: Supply, demand and opportunities. In P. S. Cook (Ed.), *Gold Coast: CRC for Sustainable Tourism* (Op. Cit JBHOST, Vol 02 Issue 1, 2016, 250–265).

The Evolution of Online Travel Agencies in the Last Decade: E-Travel SA as an Exceptional Paradigm

Dimitra Psefti, Ioulia Poulaki, Alkistis Papaioannou, and Vicky Katsoni

1 Introduction

The rapid development of technology could not leave the tourism industry unaffected. In this context, some of the traditional travel agencies started to operate online, while at the same time, new offices were created with exclusive operation via the Internet. Online bookings have gradually begun to have a significant market share, consumers have become familiar with it, and online travel agencies are now growing significantly. In this development, it seems that the use of technology plays a decisive role and has reduced many of the obstacles that companies have encountered in the past. It is, in fact, terribly interesting that a company from any country now has the ability and opportunity of course to operate without additional fixed costs or strict legal restrictions in many other countries at the same time, to address many different consumer audiences and to grow independently from the economic situation of the country in which it is located. The dynamic of online travel agencies due to the utilization of technology is a key reason for choosing this topic. Consequently, this paper aims to present the way in which the tourism industry now operates utilizing technology and to highlight the dynamics of online travel agencies due to the evolution of technology. Therefore, what is attempted in the following sections is both

D. Psefti
Hellenic Open University, Patras, Greece

I. Poulaki (✉)
University of Patras, Patras, Greece
e-mail: ipoulaki@upatras.gr

A. Papaioannou
University of Peloponnese, Kalamata, Greece

V. Katsoni
University of West Attica, Athens, Greece

theoretical and empirical coverage of the topic through an e-Travel company which within 10 years has managed to be among the top online travel agencies in the world.

2　Literature Review

Admittedly, the tourism literature evidences that travelers usually rely on multiple information channels, both online and offline, for distribution and promotion, depending on their travel planning process (Bieger & Laesser, 2004; Katsoni, 2014, 2017) and on travelers' behavior to consume tourism products (Katsoni, 2015). According to Xiang (2018), the development of knowledge in the field of information technology and tourism is divided into two decades: the decade of "digitalization" and the decade of "acceleration". Regarding the era of "digitalization", this is characterized because initially a large part of electronic information was the digital version of information that existed in non-digital form. At this time, the Internet is developing and maturing as a commercial tool, while customers start and get acquainted with it by looking for information that would previously use other media such as television or book. During the age of "digitalization", technology is considered a simple mean, and its role focuses first on functionality and usability, then on communication and persuasion with the help of the Internet and finally on the understanding of personal needs and desires of travelers (Xiang, 2018).

Undoubtedly, company's digital environment reflects culture toward the customers and the quality of services that aims to offer in order to sustain its brand name in the market as well as its potentials from the new technologies and digital tools that include mainly distribution opportunities, in a strategic plan of travel purchase stimulation (Poulaki & Katsoni, 2020). As for the "era of acceleration", it is characterized because there is a huge increase in user-generated data on the Internet, but also because new technologies and devices spread very quickly not only in homes and workplaces, but also in many other natural environments. The "age of acceleration" is characterized by technologies such as Wi-Fi, search engines, Web 2.0, tablet, smartphone, laptops, open source, drones and the emergence of machine learning and artificial of intelligence (artificial intelligence). Also, users at this time use the Internet not only for simple navigation but also for more specific searches and subscriptions or shopping on sites. Social media such as Facebook, Twitter and Instagram and various other similar tools also help to redefine the role of the Internet which is changing from a posting platform to an engaging platform and social media. Currently, there is a tendency of exloiting user activity data as an ICT tool to support tourism management (Xiang, 2018). Another approach divides the evolution of technology specifically in the tourism industry into two periods: the first decade of the Internet and the years after 2001. During the first decade of the Internet (1991–2002), the tourism industry gained one of the leading positions in the use of the Internet as businesses communicated more effectively with their existing and potential customers through online channels. Internet and mobile technologies enabled consumers to acquire travel-related information and purchase core and ancillary tourism products from tourism suppliers directly (Buhalis, 2004; Morosan, 2014).

The tourism industry, after the advent of the Internet, supports the online distribution channels because they are the most appropriate means of selling tourism products and services as they are characterized by sensitivity to time and storage (Castillo-Manzano & López-Valpuesta, 2010). The Internet has become one of the most important platforms for professionals in the tourism industry providing services and transmitting information to their target customers, while the number of relevant Web sites has increased rapidly (Tsang et al., 2010).

This change undermined the role of traditional travel agencies in the distribution chain by forcing them to use new technologies in order to survive (Standing & Vasudavan, 2000). In addition, its adoption was the basis for the development of new systems that connect consumers to GDS, which significantly reduced barriers to entry for new players in the travel agent market. Consequently, during this period, almost every travel agency developed its own Web site, and many of them evolved from simple "e-booklets" to interactive systems as they could support bookings, searches and virtual tours. In fact, they made their Web sites the main (and in many cases, the only) source of contact with potential visitors (Xiang et al., 2015) in order to maintain their position in the tourism distribution market. This significant change was of course expected in the tourism sector as computers, and new technologies were already being used by companies in the industry, for example, CRSs and GDSs existed from the 60s and 80s, respectively. After 2001, there is an even bigger change especially for the tourism marketing, where the technology beyond the functions of the respective Web site and its usability began to focus on the persuasion and empowerment of the customers as well as the use mobile devices (Xiang et al., 2015). New technologies and Internet marketing have significantly influenced the development of the tourism industry, as evidenced by the fact that tourism products now hold the largest percentage of online sales compared to other product categories, while tourism in the last 5 years is growing rapidly in international level (Navío-Marco et al., 2018).

2.1 E-Travel as a Study Case

E-Travel is an online travel agency (OTA) distributing mainly airline tickets (98% of the total sales). At the same time, it provides its customers with the opportunity to complete reservations through its Web sites, such as hotel, car, travel insurance and ferry tickets. This travel agency was originally established as Pamediakopes and had as its main goal to become the leading online provider of Greek hotels, targeting both the Greek and international consumer public. The company very soon in an effort to increase its sales added a new product, airline tickets. This move proved to be extremely successful since—despite the minimal efforts of the company to promote airline tickets—it brought several new sales not only from Greece but also from countries outside it. As the public interest in airline tickets was much greater than that for hotels, the company changed direction and was re-established under the name "e-Travel". Now called "e-Travel", the company invests mainly in airlines, acquires additional domains for foreign countries and ends up having ten different Web sites which are translated into 40 languages and provide multiple currency options. In this way, it manages to meet the needs of customers from around 200 countries, which leads to a further increase in sales.

Reservations from countries outside Greece are slowly making up 95% of the total bookings with top markets in recent years Russia, the United Arab Emirates, Japan, China, Korea and some European markets. This diversity and continuous growth that characterize its markets are also evident in the people who work in e-Travel. The human resources consist of people of different culture and nationality who can and do support more than 15 different languages, and the company grows internally every year at an average rate of 30%, a percentage which reaches 50% in 2012. In order to better manage human resources and achieve growth in sales, e-Travel organizational structure changed to further divided departments with distinguished roles and responsibilities within the working teams. With all the above actions and options, e-Travel has managed in recent years to be among the top online travel agencies in the world, while from January 2018, it has merged with the Swedish company "Etraveli", which is the second largest online travel agency in Europe. It is understood, therefore, that the growth of the company in the decade we are studying is remarkable and the further examination and analysis of the factors that led to it will contribute to a better understanding of the general evolution of online travel agencies in the last decade.

2.2 Milestones

2004	The company is founded. Pamediakopes.gr initially starts as an online hotel provider with a group of 4 people and a very simple Web site
2005	Improves the appearance and design of the Web site as well as the user interface making it more user friendly. The airline tickets are now available to the users of the Web site. This move essentially determines the future of the company
2007	e-Travel SA is founded and obtains accreditation from International Air Transport Association (IATA) as well as cooperation with the first GDS. This year the company is starting to invest heavily in automating day-to-day operations such as ticketing, pricing and fraud prevention
2009	Expansion to international markets begins with Romania as the first country. Two new Web sites are being created—avion.ro and airtickets24.com—and the human resources reach 24 people
2010	The presence of e-Travel is established in Eastern and Southern Europe and especially in Russia, which is now its leading market. The human resources reach 38 people

(continued)

(continued)

2011	This year the corresponding mobile applications are available for the customers as well as trip.ru—the Web site addressed to the Russian market
2012	Russia becomes the first priority of the company, while the expansion in the Middle East continues. The number of employees is now 97, and the investment in new technologies continues with the creation of a new CRM that fully meets the needs of e-Travel
2013	Extends further in Europe by launching two new Web sites, trip.ua and trip.bg
2014	This year, the Web sites mytrip.com, trip.ae and trip.kz are available to the public as part of the expansion plan in the Middle East and Asia. At the same time, the language, the country and the currency are disconnected from the site, thus giving the customer the opportunity to choose the language he wants, the country of origin and the currency in which he wants to see the prices and finally to pay regardless of the site visit. The design of the Web sites provides the ability to adapt to any device, and the applications for the mobile devices are significantly improved. Another important change is the collaboration with another GDS. Investment is made in automation of some basic internal functions, and the possibility is given to smaller travel agents to buy the tickets of their customers through our Web sites, receiving cheaper prices. Customers can now rate their experience through the NPS and obtain their own "MY" account to track their bookings and store useful information for next purchase. More than 1,000,000 passengers "fly" with e-Travel this year, while employees reach the number of 139
2015	It is the year that focuses on mytrip.com sales and sales from many new countries. In 2015, the company starts a new venture, that of business-to-business sales, i.e., sales from other travel agencies that use the e-Travel platform with some natural benefit. The application of pamediakopes.gr is voted one of the top 20 in Europe, while the first customer support packages are being developed which give customers the opportunity to enjoy some privileges in the service by simply paying a small price. This year the number of employees reaches 162
2016	Sales through partnerships with other—usually smaller—travel agencies continue to grow, and a new sales channel is created. This year is also an investment in low-cost carriers which until now did not actually appear on e-Travel Web sites. In addition, new services are added that relate either to customer service or to additional travel-related services such as the purchase of extra luggage while at the same time developing technologies that contribute to customer self-service
2017	In 2017, new additional services are added for the customer, and the employees have now reached 200. Sales continue to rise course and come from many different markets
2018	This year is also a milestone year for e-Travel as it merges with the Swedish company Etraveli and now forms the Etraveli Group. This merger creates various synergies in the technologies and products that the two companies have at which point they begin to exploit them with the aim of further developing sales

The purpose of the company is to offer travelers the best possible price combined with excellent customer service and valuable products for customers, utilizing cutting-edge technology. Their strategy is adjusted according to the prevailing market conditions and developments in technology but always includes the following:

- new markets and branding
- low prices, high level of service and services related to each market/country
- new sales channels (e.g., B2B)

Fig. 1 E-Travel Business
Model attributes. *Source*
Authors' production

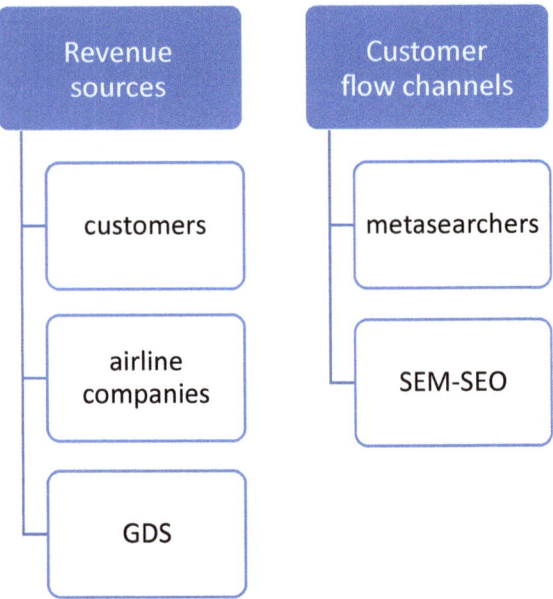

- unique content on the Web site, additional services & personalized experience for each customer.

  Further to the above, Fig. 1 presents the business model attributes of e-Travel SA.

3 Methodology

The methodology followed for the empirical part of this study concerns secondary data collection and analysis along with on-site primary research that concerns with a qualitative on-site primary research that includes interviews. Evidently, due to the author's working relationship with the company under study, it was possible to collect data, information and procedures through on-site research by discussing with the heads of the respective departments of e-Travel, in order to further study the above and draw conclusions (e-Travel as a study case literature review section and empirical part of the paper). The process followed to collect all the necessary qualitative data is the following:

1. Meeting with heads of e-Travel departments in which a discussion was held from which information on technological development of the company during the period under consideration and the corresponding projects.
2. Study of corporate material which included the description, the particular characteristics and objectives of each project.

3. Monitoring specific tasks in the various departments of the company which were affected by these projects.
4. Discussion with the heads of departments in order to answer some clarifying questions after monitoring the work.
5. Recording the basic elements of each project.
6. A summary of these in the specific chapter of the paper.

It should be emphasized that the data collected is mainly related to airline tickets as they account for more than 95% of sales and bookings for e-Travel.

4 Results

The data collected concern the period 2009–2017 (Fig. 2). Besides, 2008 is essentially the first year that e-Travel manages airline tickets and records important sales for its size. The basic sizes of the company from year to year are improving with a great upward course, making the company one of the most important in the field both in Europe and globally. The end of 2017 finds e-Travel 7th in Europe and 11th in the world in terms of online airline ticket sales. In terms of sales, growth rates vary from year to year; however, the sign remains consistently positive throughout the period under review. Indicatively, the increase of sales from 2009 to 2010 is of the order of 200%, then continues at a decreasing rate reaching 20% from 2015 to 2016 and finally records an upward trend from 2016 to 2017 with a 50% increase. Reservations completed through e-Travel Web sites follow a similar pace, and the growth rate of which for the above years is 210%, 20% and 35%, respectively. At this point, it is worth mentioning that the drop in the pace recorded in the period 2012–2014 which reflects the economic recession years.

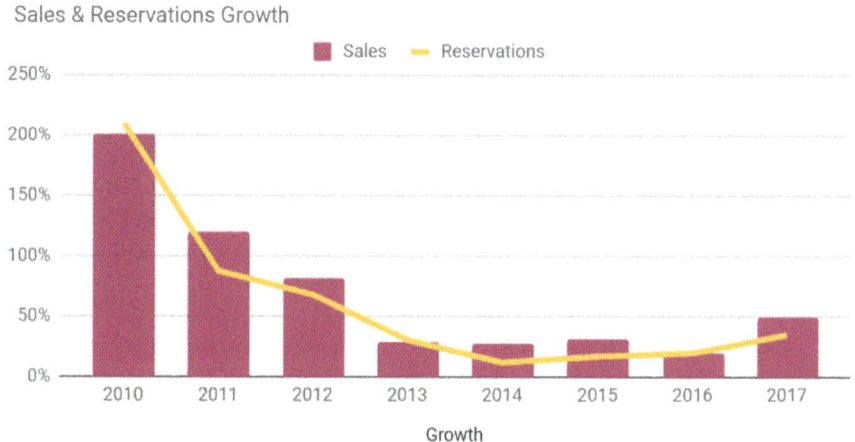

Fig. 2 Sales and reservation growth 2010–2017. *Source* e-Travel SA

5 Conclusions

This study includes the main changes in technology which in turn has affected the tourism industry and in particular travel agencies. An attempt was made to capture these changes both in theory and in practice by examining in depth a successful company in the field of travel agencies, e-Travel SA. Initially, technologies such as machine learning, artificial intelligence and the Internet of things have contributed significantly to the automation of many functions of a travel agency as well as to the more intelligent and direct management of data received by the company either from its own systems or from third party systems. This is evident by observing the improved services that e-Travel now provides by using the above technologies to automatically produce various reports generated by bid data analyses towards more efficient pricing strategies and to send related emails to the customers. In addition, it is clear that "smart" mobile devices are now used by the majority of consumers, as well as social media. The access to information 24 h a day and from anywhere is given and certainly creates the desire for self-service on the part of customers. In this context, e-Travel has taken important decisions by implementing in mobile applications (and not only) projects such as the automatic resolution of simple cases, the automation of a large percentage of changes and cancelations, the page of frequent questions and answers but also the automatic sending of invoices.

Having many challenges to face, an online travel agency needs to be flexible in terms of its sources of revenue, and up-selling can definitely help. Facing similar challenges, e-Travel also began to add gradually many services that complement the basic product, such as adding luggage or lunch, booking a specific seat on the aircraft or compensating for a flight delay. What has been found is that too many customers eventually needed such services but also revenue increase was achieved. Another finding is that there are many ways to advertise on the Internet with and without cost. Social media, search engines, email marketing and some paid advertising are tools that every company should use—as well as e-Travel—if wants to continue to gain sales and customers. Finally, it is important that any business does not rest and constantly strives to create the ground for new sales and profits since the environment of this industry is volatile. Observing, for example, e-Travel, it was found that on the one hand it is constantly finding new ways to gain sales and on the other hand it maintains a very good control of the situation so that it can make quick decisions and identify internal and external threats and opportunities. It has achieved this through its detailed and automatic reports on all the operations and departments of the company, through the introduction of new payment methods and additional currencies, through the effort to offer the lowest possible price and the most unique combination of flights but also trying to listen carefully the needs of its customers. Closing what is definitely perceived through this study is that for a company in the industry the use of technology is one way to its profitability but also its survival.

Acknowledgements The authors would like to thank all the departments of the company e-Travel SA for the provision of the data used in order to conduct this study.

References

Bieger, T., & Laesser, C. (2004). Information sources for travel decisions: Toward a source process model. *Journal of Travel Research, 42*(4), 357–371.

Buhalis, D. (2004). eAirlines: strategic and tactical use of ICTs in the airline industry. *Journal of Information and Management, 41*(7), 805–825. https://doi.org/10.1016/j.im.2003.08.015

Castillo-Manzano, J., & Lopez-Valpuesta, L. (2010). (2010) The decline of the traditional travel agent model. *Transportation Research Part E: Logistics and Transportation Review, 46*(5), 639–649.

Katsoni, V. (2014). The strategic role of virtual communities and social network sites on tourism destination marketing. *e-Journal of Science & Technology, 9*(5), 107–117

Katsoni, V. (2015). ICT applications in the hotel industry through an e-CRM systems theory approach. Academica turistica (Spletna izd.), letnik year 8, številka 1, str. 15–23, 67. Available at: https://www.dlib.si/details/URN:NBN:SI:doc-VES5GDHR

Katsoni, V. (2017). The effects of ICTs on tourism distribution channels and DMOs marketing strategies. In Z. Andreopoulou, N. Leandros, G. Quaranta, & R. Salvia (Eds.), *New media, entrepreneurship and sustainable tourism development* (pp. 58–66). Italy: Francoangeli. EAN: 9788891751058, ISBN: 8891751057.

Navío-Marco, J., Ruiz-Gómez, L.M., Sevilla-Sevilla, C. (2018). Progress in information technology and tourism management: 30 years on and 20 years after the internet—Revisiting Buhalis & Law's landmark study about eTourism. *Tourism Management, 69*, 460–470. Retrieved from: https://www-sciencedirect-com.proxy.eap.gr/science/article/pii/S0261517718301134

Morosan, C. (2014). Toward an integrated model of adoption of mobile phones for purchasing ancillary services in air travel. *International Journal of Contemporary HospitalityManagement, 27*(2), 246–271.

Poulaki, I. & Katsoni, V. (2020) Current trends in air services distribution channel strategy: Evolution through digital transformation. In V. Katsoni & T.H. Spyriadis (Eds), *Cultural & Tourism Innovation in the Digital Era. 6th International Conference of IACUDIT, Athens, Greece*, 12–15 June 2019. Springer Proceedings in Business and Economics. Available at: https://doi.org/10.1007/978-3-030-36342-0_21

Standing, C., Vasudavan, T. (2000.) Diffusion of internet technologies in travel agencies in Australia. In *ECIS 2000 Proceedings*. Retrieved from: https://www.researchgate.net/publication/221407345_Diffusion_of_Internet_Technologies_in_Travel_Agencies_in_Australia_Craig

Tsang, N., Lai, M., & Law, R. (2010). Measuring E-Service quality for online travel agencies. *Journal of Travel & Tourism Marketing, 27*(3), 306–323. https://doi.org/10.1080/1054840100374 4743.

Xiang, Z. (2018), From digitalization to the age of acceleration: On information technology and tourism. *Tourism Management Perspectives, 25*, 147–150. Retrieved from: https://www-sciencedirect-com.proxy.eap.gr/science/article/pii/S221197361730137X

Xiang, Z., Magnini, V.P., Fesenmaier, D.R. (2015) Information technology and consumer behavior in travel and tourism: Insights from travel planning using the internet. *Journal of Retailing and Consumer Services, 22*, 244–249.

Do Hotels Care? A Proposed Smart Framework for the Effectiveness of an Environmental Management Accounting System Based on Business Intelligence Technologies

Christos Sarigiannidis, Constantinos Halkiopoulos,
Konstantinos Giannopoulos, Fay Giannopoulou, Anastasios E. Politis,
Basilis Boutsinas, and Konstantinos Kollias

1 Introduction

Hotels are customer-oriented organizations and operate within a complex structure that aims to provide services to people traveling for a variety of reasons such as business and leisure. They have the difficult task to satisfy their customers in a manner that they are profitable businesses; however, in doing so, they consume a large amount of resources such as energy and water. As such, their operation results in the generation of various pollutants and waste that affect the environment in many ways and at the same time to the depletion of non-renewable resources.

Hotel businesses perhaps more than any other form of business are facing increasing demands for better environmental performance. These requirements traditionally originate from governmental and international authorities as well as by customers and local communities. In addition, there is an increasing awareness among investors, as well as banking and insurance companies on environmental issues,

C. Sarigiannidis (✉) · C. Halkiopoulos · K. Giannopoulos · B. Boutsinas
MISBILAB, University of Patras, Patras, Greece

C. Halkiopoulos
e-mail: halkion@upatras.gr

F. Giannopoulou
Iris Consulting, Athens, Greece

A. E. Politis
University of Western Attica, Attica, Greece

K. Kollias
Department of Economics, Democritus University of Thrace, Thrace, Greece

© The Author(s), under exclusive license to Springer Nature Switzerland AG 2021
V. Katsoni and C. van Zyl (eds.), *Culture and Tourism in a Smart, Globalized, and Sustainable World*, Springer Proceedings in Business and Economics,
https://doi.org/10.1007/978-3-030-72469-6_41

which inevitably makes environmental performance an integral part of business policies.

Within this framework, environmental performance requirements become quite important factors in the tourism industry. Following this, hotel leadership and management may face difficulty finding the most appropriate means for analyzing, evaluating, and documenting their environmental performance. Therefore, it is imperative for hotels to incorporate tools that will contribute to improvement of their environmental performance, along with the collection and visualization of objective and comparable data. At the same time, the incorporation of such tools will enable the proper targeting of public investments toward sustainable development.

2 Preliminaries

2.1 Environmental Management Accounting System (EMA) in Hotels and Tourism

Sustainable development and environmental protection are a major challenge for mankind as it requires a complex balance between the development of all aspects of society and the economy as well as the preservation of the environment. Issues related to environmental activities and actions are of high priority and importance to all types of organizations and at all levels, i.e., global, continental, regional, governmental, and business.

Tourism, which has grown rapidly worldwide over the past decades, is fully interconnected with the quality of the environment. In particular, many activities in the hospitality sector have adverse effects on the environment. On the other hand, it has been proven that "Tourism is one of the largest industries in the world and one of the most important income producers in the world."

In this context, environmental performance requirements become important decision-making factors in the tourism industry. However, tour operators may have difficulty finding the most appropriate means for analyzing, evaluating, and documenting their environmental performance.

Environmental accounting is the interdisciplinary field, where the concepts of sustainability, the protection of the environment, and the conservation of natural resources are met with financial management. As a scientific field, environmental accounting contributes to the integration of environmental parameters as measurable values into the accounting and tax system by systematically monitoring and analyzing the impact of economic activity on the environment.

Environmental Management Accounting EMA aims to identify, collect, and analyze two types of information for internal decision-making:

- Physical information on use and flows, energy, water and materials as well as pollutants and waste generated and

- Monetary information on costs, earnings, and savings associated with the environment

EMA is further defined as the environmental accounting which focuses mainly on providing information for internal decision-making purpose and as the accounting process, which aims to achieve sustainable development, seeks to identify the costs and benefits of preserving the natural environment in the performance of the economic or other activity of the organizations, and to display them in monetary values in their Reports (UN DSD, 2001).

2.2 Waste Structures in the Tourism Industry and the Hotels Sector

Hotels are one of the most important sources of solid waste generation, as well as water consuming.

Solid waste from the hotel industry consists of kitchen waste, garden garbage, paper, books, cardboard, plastic, synthetic fibers, glass, bulb, rubber, leather, metal containers, clothes—clothing, and electronic waste (among others). Much of the waste in hotels consists of food waste, which can account for more than 50% of the hospitality waste and up to one-third of the total food used in hospitality (Marthinsen et al., 2012).

For water, this precious fluid is the basis of life, science, and reality sound the alarm. Stocks in many parts of the planet are depleted, and its quality is getting worse. Lack of water clearly sets out existential challenges for local communities but has also become an increasingly important strategic vision in corporate planning, including tourism businesses. Strategic questions include, for example, how tourism operators can contribute to responsible water supply in local management as they are responsible for water management in the tourist destination and if the tourism industry has to really deal with water management planning. Such wider assessments of the use of tourism resources for local communities, however, are largely lacking in the academic literature (Becken, 2014).

2.3 Industry 4.0—Concept and Elements

The first Industrial Revolution took place with hydraulic power and a locomotive, giving people access to a wide range of products for the first time. The second revolution involved electricity, which led to mass production. The third revolution caused computerization, which led to huge advances in product quality, but also to networks that required the foundations for further change. The fourth revolution, Industry 4.0, will now facilitate extensive personalization, further changing the way products are produced and sold. Industry 4.0 wants to take data collection to new levels, monitoring the entire manufacturing process, collecting, and storing information from individual measurements by production machines. A lot of data must be collected, the collection process must be easy and transparent for the operator without the need for additional work to delay the manufacturing / production process.

The concept of Industry 4.0 appears as a global trend regarding the evolution in industrial manufacturing in the years to come. As such, it seems as a necessity for all industrial and manufacturing sectors to take under consideration the evolution that this concept brings. Industry 4.0 represents the coming fourth industrial revolution on the way to an Internet of Things, Data and Services. Decentralized intelligence helps create intelligent object networking and independent process management, with the interaction of the real and virtual worlds representing a crucial new aspect of the manufacturing and production process.

The fourth industrial revolution offers big data analytics techniques utilizing machine learning techniques from the field of artificial intelligence and integrating them into business intelligence tools for use by businesses. A large volume of data is now also emerging from the Internet of Things (IoT), which is also adopted by the fourth industrial revolution.

Regarding Environmental Accounting and despite the progress that has been made in it, there is still a lack of commitment for integration in the accounting systems by the business entities. One reason for these constrictions may be the lack of appropriate data or the appropriate data collection technology, combined with the inherent complexity of corporate viability (Searcy & Elkawas, 2012).

The utilization of Industry 4 in EMA could contribute:

- Higher data accuracy, which will be collected free from the risk of estimation error
- Improving data immediacy
- Convenience of data collection
- Elimination of intervention opportunities by managers on the type and results of measurements.
- Higher results reliability
- Higher comparability of results
- Ability to share data internally and externally, etc.

3   The Proposed Waste Management Accounting System

This paper aims to investigate the requirements for the design of a smart framework of an environmental management system in hotels, based on Industry 4.0 and business intelligence technologies. The main objective is to contribute to the transparency of environmental performance, using a system of accounts in accordance with the Statistical Framework for Measuring the Sustainability of Tourism (SF-MST), facilitating the implementation and use of Environmental Management Accounting Systems. This will be implemented by exploring the innovations provided by the fourth

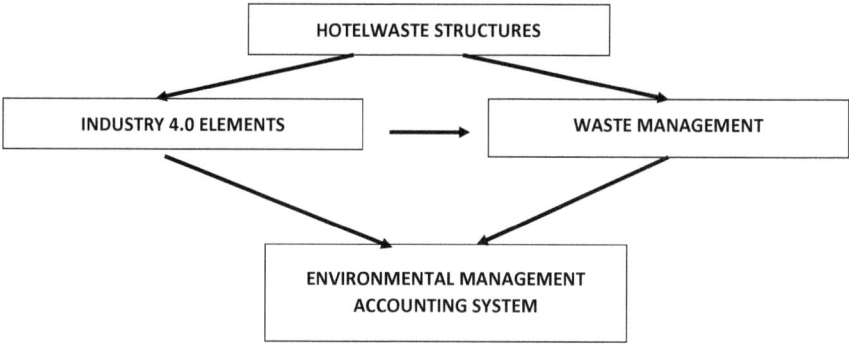

Fig. 1 Developed framework

industrial revolution—namely the Industry 4.0 concept and the selection of specific elements to be applied for data collection, analysis, and quantification of waste in physical—measurable units.

The appropriate methodology for the structure of the proposed framework is based on a hierarchical analysis of the required factors, as mentioned previously. Following this, the paper investigates the environmental issues concerning hotel sector, and the types and structures of waste. The framework to be developed is shown in Fig. 1.

As it can be seen in the diagram, the system consists of the convergence of the principal identified pillars, namely

- The identification of the waste structures originating from hotels operation.
- The application of Industry 4.0 elements and tools for the identification, analysis and processing and visualization of data on waste, based on IoT and business intelligence (BI).
- The construction of an environmental management accounting system—EMA, specifically for the sustainability and the protection of the environment.
- The three pillars mentioned above can lead to a precise analysis of waste, their quantification, and the allocation of actual cost in financial—monetary units, enabling these to be inserted into the accounting management systems.

3.1 The Proposed IoT Analytics Platform

IoT analytics is a platform that helps to analyze the huge volume of data generated by connected IoT devices. In other words, the Internet of Things is just an ecosystem and the physical things are connected to the Internet, while we analyze results through the IoT analytics platform. These physical things include smartphones, tablets, wearable device, sensors, etc.

The layers in this IoT architecture (Fig. 2) that we proposed are, (a) **Hardware/Sensing Layer** is the first layer of the IoT architecture. It consists of the hardware device which is connected to the network layer. Wireless sensor networks

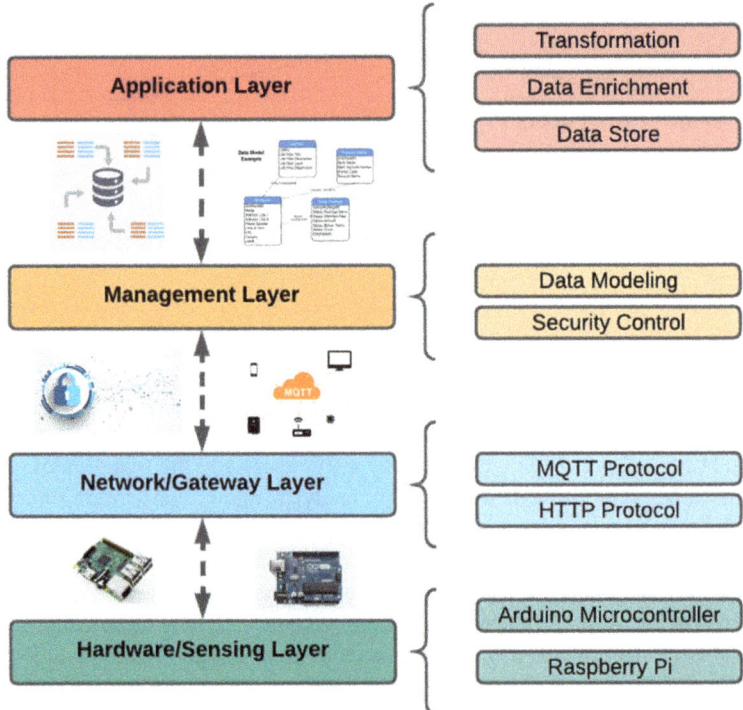

Fig. 2 Proposed IoT architecture

(WSN) and radio-frequency identification (RFID) are considered as the two main building blocks of the IoT. The Arduino microcontrollers are directly connected with the sensors. The Arduino microcontrollers are also connected to the Raspberry Pi which is also connected to the Internet using Ethernet or Wi-Fi. It transmits the data collected from the sensor in real time to the server, (b) **Network/Gateway Layer**, the second layer acts as the bridge in between hardware/sensing layer and management layer. This layer receives the digitized data and routes it over wired LANs, Wi-Fi, or the Internet for further processing. There is a different protocol which can be used to communicate between IoT gateway and servers like MQTT, AMQP, COAP, and HTTP, (c) **Management Layer**, the third layer which is responsible for data modeling and security control. At this layer, the necessary data is extracted from the data transferred by the sensor and (d) **Application Layer**, the final layer of the IoT analytics architecture. This layer uses the data processed by the management layer.

The application layer supports the real-time stream processing for IoT. Real-time stream processing refers to the data processing with the data stream collected from the IoT device in real time. The tasks which can be included in this process are, (a)**Transformation**, that includes the conversion of the data which is collected from the IoT device, (b) **Data Enrichment**, the operation in which the sensor collected raw data is combined with the other dataset to get the results, and (c) **Data Store** which includes storing the data at the required storage location. The protocols we used for the IoT platform was the MQTT and HTTP. MQTT protocol uses a publish/subscribe architecture.

Finally, the application layer supports the real-time streaming data processing platform in order to process data from ingestion platforms. Streaming is defined as that we can instantly process the data as they arrive and then does processing and analyzing them at ingestion time.

3.2 Arduino/Raspberry Pi Electronic Platforms

Regarding collection of data, the use of IoT devices is suggested implementing Arduino Technology (Arduino.cc, 2015). Arduino technology can be directly interconnected with sensors (voltage, photoelectric, temperature, magnetic, digital tilt, gas, gravity sensors, etc.). More, specifically, the sensors that have been chosen for this research project are for the calculation of the carbon footprint. The footprint considers all six of the Kyoto Protocol's greenhouse gases: carbon dioxide (CO_2), methane (CH_4), nitrogen monoxide (N_2O), hydrofluorocarbons (HFCs), perfluorocarbons (PFCs), and sulfur hexafluoride (SF_6). The Arduino microcontrollers are also connected to the Internet via a Raspberry Pi hardware platform and Ethernet or Wi-Fi protocols (Raspberrypi.org., 2015).

An effective solution for data management and remote control of the Hotel Unit is the creation of a wireless sensor network (WSN), which can be a useful part of the architecture of an automation and control information system in modern hotel units. In recent years, the design of wireless sensor networks has gained immense importance due to the ever-growing number of commercial and military applications. The sensors are usually scattered in a space called sensor field. Each of these sensors (or nodes) has the ability to collect data and then route them to the sink via a multi-hop route (Fig. 3). The base station can communicate with the task manager node via the Internet or via a satellite.

Wireless sensor networks have been proposed, instead of other implementations due to their specialized features, some of which are the flexibility provided in remote observation to control the physical environment with great precision, due to the use of optimal hardware and software as well as the use of protocols directly related not only to the type of application but also to the wireless network that will be implemented with optimal management of energy consumption.

The data collected by the sensors is transmitted in real time to a server. Apache MiNiFi (Apache MiNiFi, 2018) on the Raspberry Pi is used to collect data from the sensors. Subsequently, from Apache NiFi, data will be routed to multiple destinations as required. Apache Spark is used at the data processing stage, which is a platform used to as a data analytics platform for the data collected from the sensors (Apache Spark, 2019). Apache Spark is a unified analytics engine for large scale data processing. Basically, Apache Spark is a computing technology which is specially

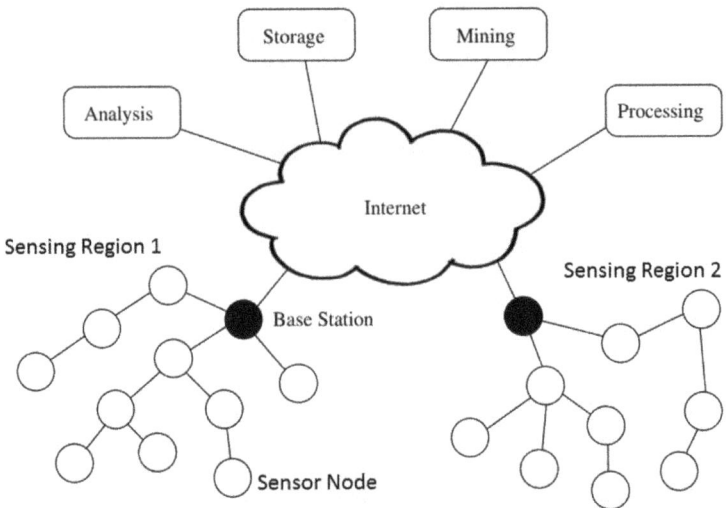

Fig. 3 Architecture of a wireless sensor network (WSN)

designed for faster computation. Spark is designed in order to cover batch applications, interactive queries, algorithms, and streaming. The main feature of spark is that it is in-memory cluster computing which means that this will increase the processing speed of an application.

3.3 Integrating Apache Spark and NiFi for Data Lake

Apache Spark is used widely for large data processing. Spark can process the data in both, i.e., batch processing mode and streaming mode. Apache NiFi to Apache Spark data transmission use site to site communication. And output port is used for publishing data from the source.

Then, a Data Lake is used as a storage repository (Fig. 4) that we can store large amount of structured, semi-structured, and unstructured data. It offers high data quantity to increase analytic performance and native integration.

Finally, the last step is the visualization of the data. In the paper, the design of an OLAP model is proposed, implemented in the Apache Kylin which is an open source distributed analytics engine designed to provide SQL interface and multidimensional analysis (OLAP). Apache Kylin (Fig. 5) allow query massive dataset and build Cubes from the identified data tables via Queries with an ODBC API (Giannopoulos and Boutsinas, 2014; Apache Kylin, 2015).

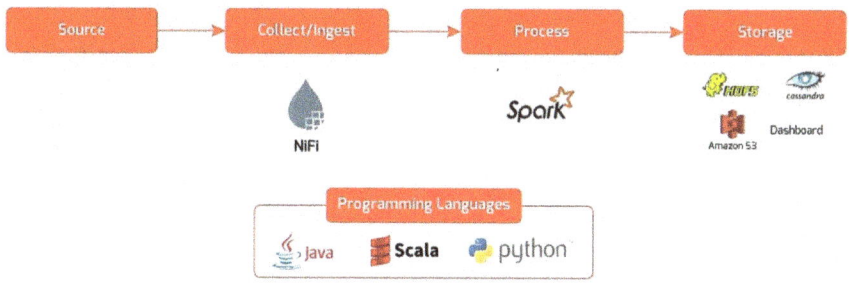

Fig. 4 Integrating Apache Spark

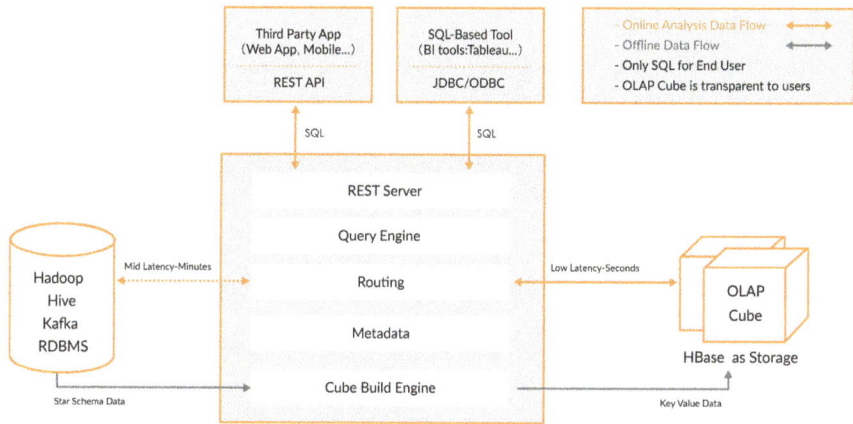

Fig. 5 Apache Kylin architecture

The use of IoT devices of the appropriate technology is further identified enabling to interface directly with sensors (electrical consumption, photoelectric, temperature, magnetic, gas, weight, etc.). These are used for the design and implementation of an environmental accounting system for hotels using the SF-MST UNWTO account system, which will monitor the environmental impact of tourism and hotels in particular, on key environmental issues such as

- Greenhouse gas emissions,
- solid waste,
- sewage,
- ecosystems and biodiversity disorders,
- water,
- energy efficiency,
- beaches,
- ecosystem protection,
- environmental taxes.

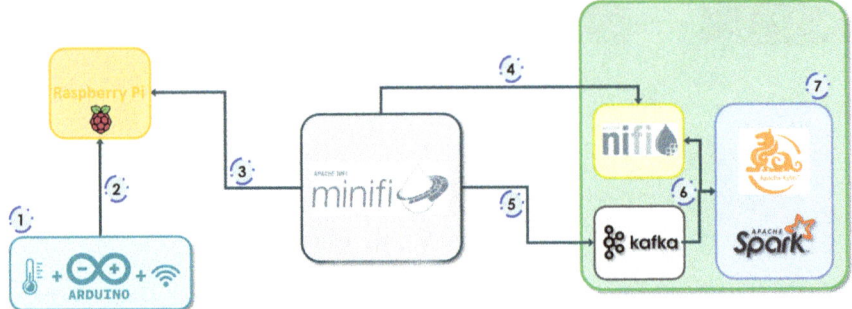

Fig. 6 Proposed smart framework of waste management accounting system (WMAS)

Data collected by the sensors are to be streamed in real time on a server. The data are routed appropriately for processing from a dedicated data analysis platform and the Data Lake technology, which is applied to store a large volume of structured, semi-structured, and unstructured data. The analysis and processing of such data are facilitated with the implementation and use of Environmental Management Accounting Systems using Business Intelligence techniques. The convergence of EMA and BI under SF-MST accounting system leads to the thorough analysis, evaluation of documentation, and the transparency of the environmental performance for hotels.

The proposed model for IoT architecture is illustrated in Fig. 6:

4   The Proposed EMA System

Research conducted has led to an initial identification of waste in the hotel sector, and the Industry 4.0 elements for data collection, analysis, and processing. Hence, this is not enough, since environmental management procedures, need to be further developed by their transformation into specific actions for a better environment. The identification of the actual—precise cost of waste is a fundamental procedure and its allocation into the accounting system, is an absolute necessity for business operation. As such, the present study revealed the necessity of combination of various tools principles, applications, and tools to a new—innovative system for waste identification and an efficient environmental management system. An integral part of the system is the transformation of data into monetary units by the principles of an Environmental Management Accounting system.

The system identifies the structure and characteristics of the waste, in the context of the operation of a hotel. It is focused mainly on water, energy, greenhouse gas emissions, and solid waste. Based on these initial data, the financial dimension of waste is defined. The application of business intelligence techniques and the use of Environmental Management Accounting knowledge and methodologies in the management of greenhouse gas emissions, water, and soil pollution from harmful substances discharge and water consumption in the hotel industry. The system will record the environmental impact on physical units, with the help of special sensors, of suitable technology (electrical consumption, photoelectric, temperature, magnetic, gas, weight, water quality, etc., so that they can be interconnected) directly with computers. The data collected by the IoT analytics platform will then feed the Environmental Accounting Cost System which will convert the input information into monetary units.

In terms of design, the system will include in addition to the information system with integrated technologies Industry 4.0, and an integrated accounting and cost organization, components of which will be establishment of:

- Physical capital accounts, within the framework of SEEA (System of Environmental-Economic Accounts)
- Environmental cost and environmental benefit accounts
- Cost centers based on the activities carried out.
- System for monitoring the flow of materials, energy, pollutants, and waste
- Methods of valuation of environmental goods
- Modern costing methods, such as activity-based costing, material flow cost analysis, full costing, life cycle costing.

Based on the above, the operation of the system will include the following stages:

- Collection of physical EMA information
- Calculation of monetary EMA, through valuation, or budget tables
- Distribution of costs in activities
- Production of reports that will include information on changes in physical capital accounts, as well as environmental and cost–benefit accounts.

5 Conclusion and Further Research

The proposed structure comprises a system of measuring environmental costs in both physical units and in monetary terms, based on an EMA for the hotels sector, implementing Industry 4.0 elements. It can be used as the innovative tool for waste management, and the enhancement of sustainability and environmental protection.

The implementation of an Environmental Management Accounting system in the context of environmental management in the hotel sector is expected to offer a number of advantages not only for the hotel itself and its customers but also for the society as a whole. Further, based on the study, a pilot system, in the form of an application, will be developed. It is expected that the financial dimension of waste and its contribution in creating costs need to be defined.

Following this, the pilot application needs to be applied as a case study on the waste management practices. In particular, the application of the specific Industry 4.0 elements, concerning the collection of data, needs to be tested with case study research at 5* and 4* hotels in Greece. This part of the research to be further conducted will explore and evaluate the current situation of the hotels' operation, which will be compared with the situation of the environmental performance and its financial dimension, after the implementation of the system that will be developed in this project.

References

Apache Kylin. (2015). Apache Kylin (online). Retrieved from https://kylin.apache.org. Accessed the June 4, 2019.

Apache MiNifi. (2018). Retrieved from https://nifi.apache.org/minifi. Accessed the June 4, 2019.

Apache Spark. (2019). Retrieved from https://spark.apache.org. Accessed the June 4, 2019.

Arduino.cc. (2015). *Arduino –Switch Case* (online). Retrieved from https://arduino.cc. Accessed the June 4, 2019.

Becken. (2014). Water equity—Contrasting tourism water use with that of the local community. *Water Resources and Industry, 7–8*, 9–22.

Giannopoulos, K., & Boutsinas, B. (2014). Tourism satellite account support using OLAP. *Journal of Travel Research, 55*(1), 95–112.

Ghulam Rabbany, Md., Afrin, S., Rahman, A., Islam, F., & Hoque, F. (2013). Environmental effects of tourism. *American Journal of Environment, Energy and Power Research, 1*(7), 117–130.

IFAC. (2005). IFAC's Statement Management Accounting Concepts.

JAPAN. (2015). Environmental Accounting Guidelines 2005, Ministry of Environment. https://www.env.go.jp/en/policy/ssee/eag05.pdf

Katsoni, V. (2011). The Role of ICTs in regional tourist development. *Regional Science Inquiry Journal, iii*(2)

Lanquar, R. (2012). Tourism in the mediterranean: Scenarios up to 2030. MEDPRO Report No. 1, July 2011. Retrieved from https://www.medpro-foresight.eu. 5 Sept 13.

Marthinsen, J., Sundt, P., Kaysen, O., & Kirkevaag, K. (2012). How to increase prevention of food waste in restaurants, hotels, canteens and catering: Report prepared for the Nordic Council of Ministers. https://norden.divaportal.org/smash/record.jsf?pid=diva2%3A701203&dswid=-7274. Accessed the May 5, 2016.

MQTT. (2019). MQTT (M2M)/Internet of Things Protocol(online). Retrieved from https://mqtt.org. Accessed the June 4, 2019.

Ostdick. (2017). Ostdick proposes 5 key elements of Industry 4.0. https://blog.flexis.com/author/Nick-Ostdick

Raspberrypi.org. (2015). What is a Raspberry Pi?|Raspberry Pi (online). Retrieved from https://www.raspberrypi.org. Accessed the June 4, 2019.

Ruettimann, B. G., & Stoeckli, M. T. (2015). From lean to Industry 4.0: An evolution? From a visionary idea to realistic understanding. In Held at Fertigungs- technischesKolloquium (Industrie 4.0—Industrie 2025), Institute for Machine Tools and Manufacturing (IWF) of ETH Zürich, November 26, 2015. Retrieved from https://www.scirp.org/journal/PaperInformation.aspx?PaperID=72522

Ryu, K., Bordelon, B. M., & Pearlman, D. M. (2013). Destination-image recovery process and visit intentions: Lessons learned from Hurricane Katrina. *Journal of Hospitality Marketing and Management, 22*(2), 183–203.

Sarigiannidis, Ch., Vutsinas, V., & Politis, A. (2018). MFCA, ABC and EFCM systems: Comparative analysis and combination for the improvement of environmental management accounting (EMA). In *VIII GECAMB 2018 Conference on Environmental Management and Accounting, Setubal Portugal, Conference Proceedings.*

Sarigiannidis, Ch., & Politis, A. (2018). Packaging and sustainability—Application of MFCA, ABC και full costing methods in environmental management accounting for packaging production. In

1st Panhellenic Scientific Conference on Graphic Arts and Printing, University of Ioannina, July 1–4, 2018. Accepted for publication in the conference proceedings (in Greek).

Searcy, C., & Elkhawas, D. (2012). Corporate sustainability ratings: An investigation into how corporations use the Dow Jones sustainability index. *Journal of Cleaner Production, 35,* 79–92.

Shen, H., & Zheng, L. (2010). Environmental management and sustainable development in the hotel industry: A case study from China. *International Journal of Environment and Sustainable Development* 1–12.

UN DSD. (2001). United Nations Department of sustainability and Development.

Vahatiitto, J. (2010). Environmental quality management in hospitality industry—Case Hotel K5 levi. Retrieved from https://epub.lib.aalto.fi/en/ethesis/pdf/12433/hsee-thesis12433.pdf

An Innovative Recommender System for Health Tourism

Antiopi Panteli, Aristotelis Kompothrekas, Constantinos Halkiopoulos, and Basilis Boutsinas

1 Introduction

Nowadays, the introduction of the Internet and the new opportunities that arise regarding e-business services frequently overwhelms users. Huge amounts of data are available to them, and thus, the processing procedure and the extraction of useful knowledge are difficult. Also, mass customization tends to replace mass marketing, and personalized services are offered to users. Firms need a better understanding of customers' needs and preferences in order to offer personalized products or services.

Recommender systems (RS) can facilitate all these procedures. They are tools that can process the recorded data, and based on users' preferences and needs that are explicitly acquired or predicted by the system, they can keep only the part of the data that is relevant to the user and suggest customized services (Lorenzi & Ricci, 2003). The firms with the use of RS can increase the volume of their sales, since they can offer products and services that satisfy customers' needs or they can suggest products and services that the user might not be able to find without this recommendation or might not imagine that he is in need of them. The current trend in RS is the diversity of the recommendation list, which means that the recommended products or services are similar to the query inserted by the user and diverse to each other. This leads to a variety of choices for the user.

In a common recommender system, users insert queries by selecting some attributes out of the offered ones and by giving values to them. In this way, they express their needs and preferences. Then, the RS try to match the inserted query to the relevant items of the electronic database. In other cases, user profiling is exploited, where users are asked to classify themselves in one of the profiles provided by the

A. Panteli (✉) · A. Kompothrekas · C. Halkiopoulos · B. Boutsinas
MISBILAB, University of Patras, Patras, Greece
e-mail: halkion@upatras.gr

system. In this case, implicit needs that have not been provided by the user may be guessed by the system (Ricci et al., 2002).

The main purpose of this article is to present an innovative recommender system in the field of health tourism. This system aims at matching health tourist preferences and needs to health/tourism providers. It focuses on providing complete health tourism products, by matching user profile characteristics of both health and tourism service providers in order that users receive the treatment they choose, in the right location, the right period and at the right cost. It incorporates a database of cases, i.e., medical, wellness and tourism service providers. Usually, the contemporary databases contain huge amounts of data that have to be processed, and thus, recommender systems need fast and effective techniques since exact methods might be ineffective. In our case, we use a facility location problem—solved by a fast heuristic technique presented in Panteli et al. (2019)—in the recommendation process, which employs a parameter that controls the diversity of the recommendation list and thus the variety of the proposed results. Our general objective anywise is to broaden the perspective of the techniques that can be used in the recommendation process.

2   Literature Review

2.1   Recommender Systems and Health Tourism

The recommender systems (RS) are software tools that collect data related to preferences and needs through the interaction of a user with a Web site or a Web application and then organize and process them in order to keep only the relevant part of data and make personalized suggestions (Fesenmaier et al., 2003).

The motivation behind these systems is to eliminate or reduce the information overload. In fact, users are overwhelmed by the massive volume of data since the Internet became a part of every aspect of their everyday life. The RS are effective tools that can help them filter the abundant part of data and extract only the part that is relevant to their needs and preferences. Recommender systems organize the data in order to be processed by effective heuristic algorithms that operate within these systems, and they provide precise and appropriate results based on users' queries.

The recommendation process is divided in the information collection phase, where data about the user are collected, in the learning phase where a learning algorithm is used in order to extract the useful knowledge from the previous phase and in the recommendation/forecast phase where the suggestion of relevant objects to the user's preferences takes place (Núñez-Valdéz et al., 2012).

RS are present in many domains such as the entertainment where recommendations for movies, music, theatrical plays are frequent, the tourism, the health care, the education, the e-commerce where product recommendations for consumers take place and so on.

The algorithm that will be chosen depends on the application domain of the RS (Montaner et al., 2003) and the type of the available data that have to be processed. Clustering techniques are usually incorporated in RS because they are thought to reduce dimensionality—i.e., reduce the number of operations—and thus, increase efficiency. However, clustering has to be carefully adapted in order not to compromise accuracy to improve efficiency.

The basic components of RS are the items, the users and the transactions. The items are the recommended objects that are characterized by a series of attributes. The users interact with the system, and they insert queries by giving values to the attributes. The system usually asks the users to provide basic information about preferences and needs and tries to offer a more customized experience. The interaction between the user and the RS is usually called a transaction. It contains log-like data that can help the recommendation algorithm to operate properly (Ricci et al., 2011).

In our study, an innovative recommender system is used in health tourism. Many definitions exist about health tourism. At first, we refer to health tourism when a tourism facility tries to attract tourists by offering at the same time health care and tourism services. The International Union of Tourist Organizations defines health tourism as "the provision of health facilities utilizing the natural resources of the country and more specifically mineral water and climate (IUTO, 1973, Health Tourism, Geneva: United Nations)". This definition has been further enriched in the following years and has been stated that it refers to the process of receiving medical services in a leisure setting away from home (Goeldner, 1989) or else the process of traveling outside the country of residence in order to receive medical treatment. Travelers visit health tourism resorts in order not only to receive a medical treatment of an ailment but to relax and rest as well. Typical examples of this concept are the cruise lines that offer health tourism services.

Usually, an important factor that encourages people to health tourism is the poor health care system of their countries. Long waiting lists in operations cause people to go abroad to seek medical care. This fact in combination with the high medical cost in some countries motivates people to seek abroad for lower-cost options and thus contributes to the development of health tourism.

In the past, related analysis of health tourism data proved that the corresponding resorts in that field did not advertise properly the offered services. According to the specific motives of the health tourists, the health tourism market is divided into some market segments. For example, if the primary reason for travel abroad is health in a different climate or context or wellness activities combined with health activities or medical treatment along with leisure activities and so on.

It was stated in the literature that there are at least two possible segmentation aspects of the health tourism market, health and income. As for the health segment, marketing campaigns could target people with minor ailments such as high cholesterol, anorexia, dental, dermatological, infertility problems and so on. Furthermore, marketing campaigns could target people who would like to maintain their youth and their external appearance through cosmetic surgeries, spa treatments, holistic medicine approaches, leisure activities for an active everyday life and so on. As for the income segment, hotels or spa resorts might target high-income people that are able to spend much money for the luxurious offered health services or middle-class people who can afford medium-level prices (Goodrich, 1994).

Nowadays, user's feedback data combined with log data from Web sites leads to a more efficient segmentation of users based on similar preferences, needs and ways of acting. Hospitality marketers diversify their marketing plans by targeting health tourists since health tourism seems a profitable sector. In fact, there are facilities that integrate wellness or alternative medical therapies with the conventional medicine. Therefore, health tourists may have an operation, but at the same time, they may receive treatment or recuperation in a more pampering, pleasant and refreshing form. For example, these hotel-like or spa-like facilities might offer not only general surgeries, but also massages, aromatherapy, facial services and so on.

Searching for tourism-related information and services is a quite attractive Web activity, and many Web sites incorporate contemporary technology and tools in order to support the traveler in the decision-making process of the selection of a tourism destination or a tourism-related service (Katsoni, 2011). Users frequently search in Web sites for information such as mostly visited places, sightseeing, weather conditions, flights, hotels, transport, currency, restaurants and so on.

Travel and tourism-related services contain much information, and thus, the corresponding industry is the primary sector for the application of electronic commerce (Werthner, 2003). Smart applications like the recommender systems assist users in the process of making a decision about tourism-related activities. Moreover, when the tourism industry tries to analyze consumer-related data, it is necessary to perform data pre-processing in order then to conduct efficient data mining. Recommender systems assist in this process as well (Cooley et al., 1997).

In the tourism domain, usually case-based recommender systems are used since their operational efficiency gets improved over time by learning from past interactions and by updating their case bases with new experience. In fact, these systems focus on the reuse of experience, which is modeled as a case. They also exploit the explicit feedback that users give about their experience regarding the services that they received. New problems are solved by transferring and adapting solutions that were used for similar problems in the past. Actually, in the retrieval step, the system receives a problem specification (in the form of a query), searches through the case base, scores each case for similarity to the query and selects the highest-scoring case(s), which are the subject of subsequent processing steps (Bridge et al., 2005). The similarity measure that is used in order to score the matching of the case to the query is extremely important and depends on the type of data and other factors.

The health tourism recommender system of our study focuses on the patients and tourists and tries to offer them health and touristic services according to their preferences and needs. In that way, it matches patients and tourists to health and touristic units in order that they receive the treatment they choose.

3 Health Tourism Recommender System

The recommender system of our study consists of a case base which contains information about the health (HS_Provider) and tourism (TS_Provider) service providers such as location, medical service category and subcategory, tourism service category and subcategory, dates that the facilities are available for services and so on. Moreover, it contains ratings from users (HT_Evaluation), who have already acquired a health tourism package. In fact, the users evaluate the provided services according to their experience. For example, the medical cost, the quality of medical services, the infrastructure quality, the accommodation quality, the medical staff's responsiveness and so on. Figure 1 represents the case base of the health tourism recommender system.

Figure 2 describes the data flow in the system. A user inserts a query which specifies her needs and preferences in the form of values in specific attributes. The system processes this type of information and applies a matching procedure. It selects those *health providers* that (1) offer services which belong to the category and subcategory that the user has selected in her query, (2) are located in the "neighborhood" of the location that the user has stated and (3) are available during the time period that the user has mentioned. It should be mentioned that the "neighborhood" expands the specific location that the user selects to a greater radius, i.e., to a larger region around the location stated by the user. Then, the system searches for *tourism providers* in

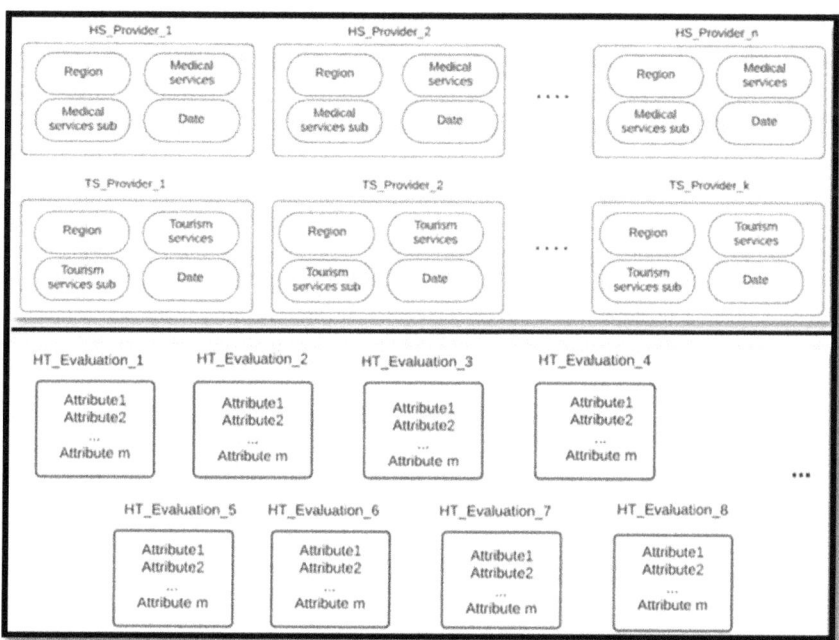

Fig. 1 Representation of the case base of the health tourism recommender system

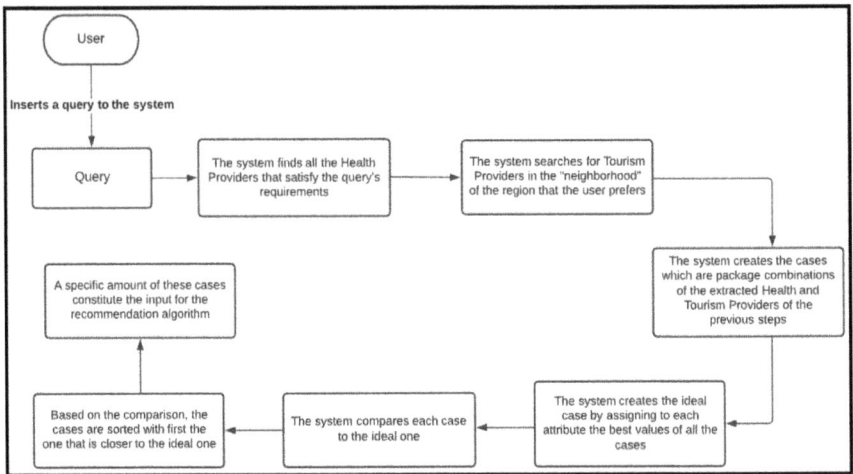

Fig. 2 Representation of the data flow in the health tourism recommender system

this "neighborhood" in order to create all possible package combinations. In fact, a package combination offers health services by a health provider based on the requirements of the user's query in combination with tourism services offered by a tourism provider in the same "neighborhood" of the health provider. In this way, it creates the available "cases", the package combinations. Moreover, by taking into consideration the ratings that exist in the case base and the requirements of the user's query, the system creates an ideal case that contains the best values for all the attributes that exist in the case base. Each case is compared to the ideal one, and a sorting procedure is the next step. After these steps, the system employs a recommendation method that is analyzed in Panteli and Boutsinas (2019) utilizing association rule mining (Fayyad et al., 1996; Boutsinas et al., 2008).

This method focuses on the diversity of the recommendation list which is a major challenge in contemporary recommender systems since it provides important value and leads to the user's satisfaction. Bradley and Smyth (2001) defined the diversity in the process of recommendation as the opposite of similarity. The items that are suggested to the target user have to be similar to the inserted query in order to be close to the requirements of the user but at the same time diverse to each other since users may get bored or unsatisfied after receiving many recommended objects that present a high level of similarity to each other (Hurley & Zhang, 2011). Usually, users cannot express their needs and preferences in a detailed way or they might not imagine exactly what they want or what options do exist but might be open to explore new and diverse directions. As a result, a list with cases that constitute the same exact match with a more implicit query might be ineffective, whereas a more diverse recommendation list that provides a variety of cases might be more useful to the user because of the existence of alternative options. However, the most common problem when the diversity and the similarity of the recommendation list are studied is the trade-off between these two measures since they have an almost reverse relationship. When the overall diversity between the objects of the recommendation list is high, this might lead to a lower similarity ratio of each recommended object to the query.

In our application, we incorporate a specific facility location problem to the recommendation process, the multiple p-median problem (MPMP) which contains a parameter that has been proven in our previous work (Panteli & Boutsinas, 2019) that plays the role of the regulator of the similarity–diversity trade-off. The value of this parameter in combination with the size of the recommendation list can affect the similarity to diversity ratio. This parameter refers to the alternatives that are recommended to the user by a recommender system and how these alternatives satisfy the requirements of the user.

Therefore, the health tourism system recommends some possible health and tourism providers that satisfy the requirements of the user and have the best ratings. At the same time, the recommended cases except from satisfying the user's requirements, i.e., being similar to the query inserted by the user, are diverse to each other in order that alternative choices are offered. The facility location problem can be solved by a heuristic technique since the health tourism database contains large volumes of data and exact techniques might be ineffective.

Consider the following example that describes the operation of the health tourism recommender system in a simple way. Given that the case base consists of only 100 available health and tourism providers, a user inserts a query and defines her preferences.

Attributes	Query inserted by the user
Preferred region	Patras
Preferred medical service category	Wellness tourism
Preferred medical service subcategory	Cosmetic surgery
Preferred dates	9/8–12/8

The system tries to match the inserted by the user query to the health providers that exist in the Case Base. As a result, it tries to find the health providers in the neighborhood of the specified by the user region (Patras) that offer cosmetic surgery services and are available for the dates specified by the user. The neighborhood of the specified region can be calculated by a common distance measure.

At the same time, the system tries to find the tourism providers that exist in this neighborhood in order to create all possible package combinations. As a result, the cases are package combinations of medical and tourism providers that are located in the neighborhood of Patras and offer the services requested by the user in the specific dates.

After the creation of all the possible cases (package combinations) that contain the corresponding ratings, the system creates the ideal case. The ideal case consists of the best ratings for each of the attributes that are included in the package combinations.

Given the ideal case, the system compares each of the available package combinations to the ideal case. In a real case base with many available providers, this step would end up with a large number of comparisons. Then, the system sorts these cases starting from the ones who are the most close to the ideal case and selects a specific number of them that are used as the input of the recommendation algorithm that is based on the MPMP, as stated before. This problem takes an input data matrix that contains the distances between the ratings of the cases and these ideal instances. Given a specific threshold that depends on the nature of the input data, a binary matrix is created that shows if a package combination is appropriate for the recommendation list. Then, the system, by the solution procedure of the multiple p-median problem (MPMP), tries to end up with p recommended cases (package combinations). The value of p is defined initially according to the system's requirements. This problem also incorporates a parameter mc that states how many cases out of p have to match the ideal case's values. According to the size of the recommendation set, i.e., the value of p and the value of the mc parameter, as it has been already proven in Panteli and Boutsinas (2019), the similarity to diversity ratio takes different values. As a result, the mc parameter in combination with the size of the recommendation set constitutes a regulator for the trade-off between similarity and diversity measures. The metrics that are used in order to calculate the diversity and the similarity ratios can be found in Smyth and McClave (2001).

In the end, the system, by taking into consideration the fact that the diversity of the recommendation list is a current need, creates the final set of recommender cases to the user. The user after her experience provides her personal evaluation, and in this way, the case base is updated with more cases.

4 Conclusion

In this paper, we presented the operational procedure of a recommender system that aims at matching the preferences of the health tourists to health and tourism providers. Actually, this system's innovative character derives from the fact that provides package combinations of medical and tourism providers. The user asks for medical services, and the system tries to satisfy these needs in combination with tourism services from providers in the neighborhood of the region that the user prefers. These package combinations consist of ratings regarding some attributes. The ratings are recorded by users that have already received the corresponding medical and tourism services from these providers. The package combinations constitute the cases of the case base that are then processed by a facility location-based method. In fact, the multiple p-median problem from the facility location field is employed in the recommendation procedure. This problem incorporates a parameter that plays the role of the regulator of the similarity and diversity trade-off among the objects of the recommendation list. As a result, the proposed health tourism recommender system suggests health tourism package combinations according to the needs and preferences of the user and based on the experiences of other users that have provided the corresponding ratings regarding the received services. These package combinations, since the multiple p-median problem is employed, present satisfactory levels of diversity between them in order that the user selects from a variety of alternative choices.

References

Antiopi, P., & Basilis, B. (2019). Improvement of similarity-diversity trade-off in recommender systems based on a facility location model, 2019 10th International Conference on Information, Intelligence, Systems and Applications (IISA), Patras, Greece, 2019, pp. 1–7. https://doi.org/10.1109/IISA.2019.8900661.

Boutsinas, B., Siotos, C., & Gerolymatos, A. (2008). Distributed mining of association rules based on reducing the support threshold. *International Journal on Artificial Intelligence Tools, World Scientific Publishing Company, 17*(6), 1109–1129.

Bradley, K., & Smyth, B. (2001). Improving recommendation diversity. In *Proceedings of the Twelfth Irish Conference on Artificial Intelligence and Cognitive Science*, Maynooth, Ireland (pp. 85–94).

Bridge, D., Göker, M. H., McGinty, L., & Smyth, B. (2005). Case-based recommender systems. *The Knowledge Engineering Review, 20*(3), 315–320.

Cooley, R., Mobasher, B., & Srivastava, J. (1997, November). Web mining: Information and pattern discovery on the world wide web. In *Proceedings Ninth IEEE International Conference on Tools with Artificial Intelligence* (pp. 558–567). IEEE.

Fayyad, U. M., Piatetsky-Shapiro, G., & Smyth, P. (1996). *Advances in knowledge discovery and data mining*. Cambridge: AAAI Press/MIT Press.

Fesenmaier, D. R., Ricci, F., Schaumlechner, E., Wöber, K., & Zanella, C. (2003, January). DIETORECS: Travel advisory for multiple decision styles. In *ENTER* (pp. 232–241).

Goeldner, C. (1989). 39th congress AIEST: English workshop summary. *Revue De Tourisme, 44*(4), 6–7.

Goodrich, J. N. (1994). Health tourism: A new positioning strategy for tourist destinations. *Journal of International Consumer Marketing, 6*(3–4), 227–238.

Hurley, N., & Zhang, M. (2011). Novelty and diversity in top-n recommendation—analysis and evaluation. *ACM Transactions on Internet Technology (TOIT), 10*(4), 1–30.

International Union of Tourism Organisations (IUTO). (1973). *Health tourism*. Geneva: United Nations.

Katsoni V. (2011). The role of ICTs in regional tourist development. *Regional Science Inquiry Journal, 3*(2).

Lorenzi, F., & Ricci, F. (2003, August). Case-based recommender systems: A unifying view. In *IJCAI Workshop on Intelligent Techniques for Web Personalization* (pp. 89–113). Berlin, Heidelberg: Springer.

Montaner, M., López, B., & De La Rosa, J. L. (2003). A taxonomy of recommender agents on the internet. *Artificial Intelligence Review, 19*(4), 285–330.

Núñez-Valdéz, E. R., Lovelle, J. M. C., Martínez, O. S., García-Díaz, V., De Pablos, P. O., & Marín, C. E. M. (2012). Implicit feedback techniques on recommender systems applied to electronic books. *Computers in Human Behavior, 28*(4), 1186–1193.

Panteli, A., & Boutsinas, B. (2019, July). Improvement of similarity-diversity trade-off in recommender systems based on a facility location model. In *2019 10th International Conference on Information, Intelligence, Systems and Applications (IISA)* (pp. 1–7). IEEE.

Panteli, A., Boutsinas, B., & Giannikos, I. (2019). On solving the multiple p-median problem based on biclustering. *Operational Research 1–25.*

Ricci, F., Arslan, B., Mirzadeh, N., & Venturini, A. (2002, September). ITR: A case-based travel advisory system. In *European Conference on Case-Based Reasoning* (pp. 613–627). Berlin, Heidelberg: Springer.

Ricci, F., Rokach, L., & Shapira, B. (2011). Introduction to recommender systems handbook. In *Recommender systems handbook* (pp. 1–35). Boston, MA: Springer.

Smyth, B., & McClave, P. (2001, July). Similarity vs. diversity. In *International Conference on Case-Based Reasoning* (pp. 347–361). Berlin, Heidelberg: Springer.

Werthner, H. (2003, August). Intelligent systems in travel and tourism. In *IJCAI* (Vol. 3, pp. 1620–1625).

Cultural Innovation

Enhancing Revisit Intention Through Emotions and Place Identity: A Case of the Local Theme Restaurant

Alexander M. Pakhalov and Liliya M. Dosaykina

1 Introduction

In recent years, food tourism has changed its role from a niche segment to a popular form of tourism (Tsai & Wang, 2017) that gained interest from both researchers and practitioners (Ellis et al., 2018).

Empirical evidence from several countries shows that local food experience can positively affect overall trip satisfaction and revisit intention (Stone et al., 2019). Local food can also be used as an element of a marketing strategy for a destination or even for interdestination networks such as the Slow Food movement (Berg & Sevón, 2014). Thus, food tourism can be used as a working tool for regional tourism development (Boyne & Hall, 2004) and a factor of destination's attractiveness (Bukharov & Berezka, 2018).

Nevertheless, the contribution of food tourism to the destination's development is possible only in case of availability of events (such as local food festivals) and attractions (such as theme restaurants). These events and attractions are key touchpoints that create an authentic emotional experience for gastronomic tourists.

Our study aims to examine the relationships among tourist experience, emotions, and behavioral intentions based on a case of Zaboi theme restaurant located in Kemerovo, Russia.

Zaboi (the name of the restaurant translates as "Coalface") is an upscale restaurant and the most popular food tourist attraction (TripAdvisor, 2020) in Kemerovo region (also known as Kuzbass region). Kemerovo region is the largest coal-mining region in Russia (Selyukov, 2015) and it actively uses the coal theme in its place branding activities (Vasyutin et al., 2018). The value proposition of Zaboi restaurant combines traditional Siberian food with the authentic atmosphere of a coal mine. The theme of the coal-mining region is used as a central point of the restaurant's experience

A. M. Pakhalov (✉) · L. M. Dosaykina
Lomonosov Moscow State University, Moscow, Russia

design (Chen & Guan, 2008). Zaboi restaurant is a good example of a regional theme restaurant (Kim & Moon, 2009) that reflects regional identity in several ways including branding, atmosphere, cuisine, and entertainment.

Based on the empirical evidence from this attraction we will try to answer two research questions (RQ):

RQ1. Is there a relationship between emotional experience and visitors' behavioral intentions in a case of the local-specific theme restaurant?

RQ2. What role does place identity (and its perception) play in shaping the visitors' emotional experience?

2 Literature Review

Our research lies at the intersection of three large areas of tourism research (Fig. 1): revisit intention studies that focus on determining factors of tourist loyalty, food tourism studies that cover various issues associated with this form of tourism, and theme restaurant studies that discuss marketing strategies and consumer behavior features in the case of these attractions.

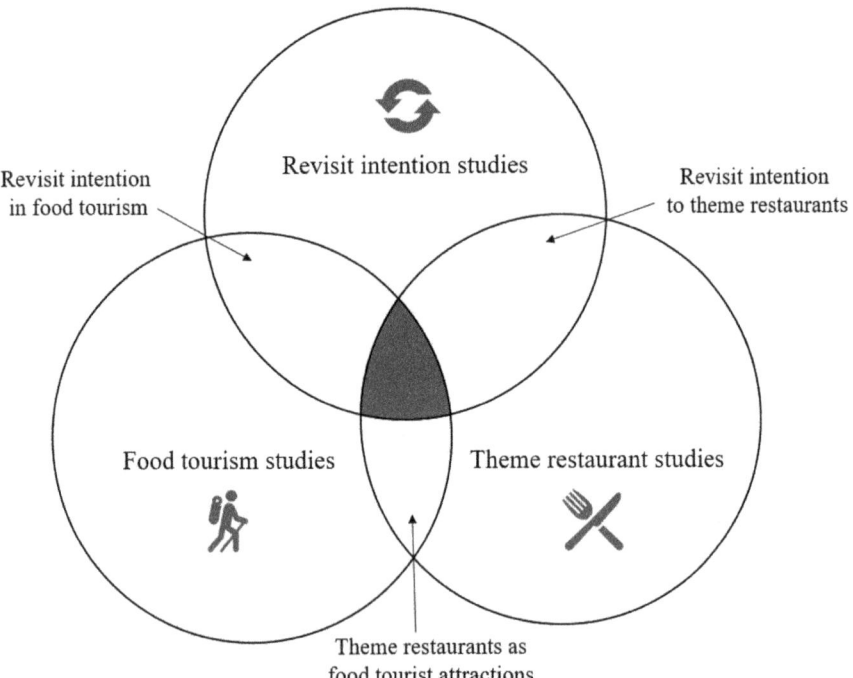

Fig. 1 Intersections between three key research fields that contribute to the present study

Revisit intention is an important topic for researchers from various domains such as destination branding, tourism management, and experience marketing. Early empirical studies on this topic were published in the 1990s. These studies were focused both on an intention to revisit destinations (Sampol, 1996; Kozak & Rimmington, 1999), and on an intention to revisit specific tourist attractions such as theme parks (Kim & Choi, 2000) or heritage places (Davies & Prentice, 1995). In the first half of the 2000s, academics propose some conceptual approaches (Oppermann, 2000; Yoon & Uysal, 2005) to study revisit intention and some other intentions that reflect tourist loyalty such as an intention to recommend (also known as "word-of-mouth intention"). At the same time, the number of empirical studies on loyalty features in special forms of tourism and tourist attractions continued to grow.

Revisit intention studies in food tourism reveal that gastronomy plays an important role in the tourist experience and indicate that some travelers would return to the destination to savor its unique gastronomy (Kivela & Crotts, 2006). However, these studies usually focus on emotions and intentions of food festivals' visitors (Lee & Arcodia, 2011; Blichfeldt & Halkier, 2014). In one of the recent studies (Chang et al., 2018), the authors found the effects of novelty-seeking and food involvement on the intention to revisit the Jeonju Bibimbab Food Festival in Korea. Pieces of evidence on the correlation between satisfaction, perceived value, and intention to revisit were also found based on the visitors' survey at one of the food festivals in the USA (Kim et al., 2015). Local-specific theme restaurants that are a part of larger attractions also have a potential for value creation in food tourism (Sørensen et al., 2020).

The most extensive evidence on drivers of revisit intention came from theme restaurants' studies. Many studies highlight a special role of unique emotional experiences as a source of return intention (Han et al., 2009; Kim & Moon, 2009, Chen et al., 2015). Factors that indirectly or directly influence the intent to return to the theme and upscale restaurants also include atmosphere (Weiss et al., 2005; Nawawi et al., 2018), authenticity of experience (Meng & Choi, 2018), nostalgia about past experiences (Hwang & Hyun, 2013), and food quality (Weiss et al., 2005; Tsaur et al., 2015). However, there is still a lack of evidence on the relationship between visitors' experiences of regional (Kim & Moon, 2009) or ethnic (Song et al., 2019) theme restaurants and place identity of a country, a region, or a particular destination. In this paper, we want to take the first step toward closing this gap.

3 Research Design and Methodology

Our empirical research includes two stages. At the first stage, we get preliminary insights based on the analysis of online reviews posted by Zaboi's visitors at TripAdvisor, the most popular global tourist experience resource (Onorati & Giardullo, 2020). At the second stage, we formulate hypotheses and test them based on quantitative research, including an online survey followed by regression analysis. Online reviews and visitors' surveys are typical data sources for restaurant customer experience. However, researchers relatively rarely combine these sources within one study.

We tried to do this to get a comprehensive picture of the emotions and experiences of the theme restaurant's visitors.

3.1 Analysis of Online Reviews

For this part of the study, we have selected and analyzed visitors' online reviews from TripAdvisor Web service for the period from August 2013 to December 2019 using an approach proposed in previous studies (Yan et al., 2015; Pakhalov & Rozhkova, 2020). We excluded from the main sample six negative reviews, which should be studied separately (but this analysis is impossible due to their very small number). Thus, our final set of reviews included 198 neutral and positive reviews.

Based on the literature review we have identified five main factors that form visitors' experience in the case of theme restaurants:

- Atmosphere and environment (Atmosphere);
- Food (Food);
- Service and staff (Service);
- Location (Location);
- Price (Price).

We used these factors as codes assigned to reviews. We also used "revisit" and "recommend" as two additional codes for the reviews which explicitly reflected two basic loyalty behavioral intentions. The reviews were encoded manually, and then the frequency of each code was calculated using QDA Miner software.

3.2 Online Survey

The online survey was based on the questionnaire that included three main parts: (1) the respondent's attitudes to restaurants in general; (2) emotions and experiences from visiting the Zaboi restaurant; (3) social and demographic characteristics.

The key task of this study's phase was to test three hypotheses about the relationship between emotions and behavioral intentions of the theme restaurant's visitors:

H1. Visitors' emotions are significantly associated with revisit intention.
H2. Visitors' emotions are significantly associated with intention to recommend.
H3. Visitors' emotions are significantly associated with intention to write an online review.

We measure seven types of visitors' emotions by the 7-point Likert-type scales proposed for upscale restaurants (Ryu & Jang, 2007). This approach is shown in Fig. 2.

Unhappy	1	2	3	4	5	6	7	Happy
Annoyed	1	2	3	4	5	6	7	Pleased
Bored	1	2	3	4	5	6	7	Entertained
Disappointed	1	2	3	4	5	6	7	Delighted
Depressed	1	2	3	4	5	6	7	Cheerful
Calm	1	2	3	4	5	6	7	Excited
Indifferent	1	2	3	4	5	6	7	Surprised

Fig. 2 Emotion scale for assessing theme restaurant customers' emotional experiences (Ryu & Jang, 2007)

The set of behavioral intentions in our study included three intentions associated with restaurant customer loyalty: intention to revisit, intention to recommend, and intention to write an online review. The first two intentions were used in many previous studies in this field (Tsaur et al., 2015; Stone et al., 2019; Rousta & Jamshidi, 2020), while the third intention was included to the survey due to the growing importance of online reviews as tourist experiences' reflections (Dixit et al., 2019). All three intentions were assessed using the 7-point Likert-type scales.

Social and demographic parameters were included in the questionnaire to look for possible differences between the emotions of visitors of different demographic and social groups and regions since some studies claim that emotions of the authentic theme restaurant's visitors can vary depending upon geographic location (Muñoz & Wood, 2009) or demographic features (Nawawi et al., 2018).

We invite to the survey only the respondents who had visited the Zaboi restaurant at least once during the last year. The link to the survey was posted on social media. We also used snowball sampling to expand the sample. Data collection was carried out in February–March 2020.

The data collected in the online survey were used in regression analysis in the Gretl software. We built three ordinary least squares (OLS) regressions, each using one of three loyal behavioral intentions (to revisit, to recommend, and to write an online review) as the dependent variable.

4 Results of the Online Reviews' Analysis

This section presents the results of the online reviews' analysis. Table 1 contains the frequency of different thematic codes (percentage of cases) in the final sample of reviews from TripAdvisor (2020).

Users in their reviews paid attention to the unusual atmosphere, the authentic interior that resemble a real mine, and interesting serving of local cuisine (stylized as a mine, in a mining theme, original interior, etc.) The taste and quality of the dishes, taking into account the specifics of the region, are in second place (from the Siberian

Table 1 Frequency of codes in online reviews (own work using QDA software)

Topic (code)	Frequency ($N = 198$) (%)
Experience factors	
Theme and atmosphere	35.86
Menu and food	32.32
Service quality	18.18
Location	2.53
Price	2.02
Intentions	
Revisit	3.03
Recommend	8.08

menu, everything is very tasty, etc.). Users also appreciated the high level of service and staff training (good service, friendly staff, etc.)

Our analysis of online reviews shows that visitors' experiences are closely related to the image of Kemerovo region. For example, one of the reviewers writes: "*I was pleasantly surprised by the visit to the restaurant. The stylish atmosphere immerses you in the exoticism of Kuzbass from the very first steps. The walls are made in coal, giving the impression of a mine. You can try on a helmet, touch the mine telephone, sit in the director's office (and even dine there).*" Another visitor shares a similar experience by pointing out that the restaurant's atmosphere is "*the quintessence of Kuzbass.*"

Most of the users who expressed their intentions to revisit and to recommend the restaurant to their friends focus in their reviews on food, service, and the restaurant's atmosphere. However, the extremely small number of reviews with an explicit mention of these behavioral intentions does not allow us to gain reliable conclusions about the significance of these factors' influence on visitors' loyalty.

5 Results of Online Survey

This section summarizes the main findings and observations from an online survey of restaurant diners. Our final sample included 188 respondents. Table 2 shows the structure of the sample.

Table 3 summarizes the results of three regressions. In each of these regressions, we use one of the behavioral intention indicators as a dependent variable (intention to revisit, intention to recommend, and intention to write an online review). The set of explanatory variables in each regression is the same: seven "emotional" indicators based on Ryu & Jang's scales and four control variables: "gender" (1 for female 0 for male), "age", "income", and "tourist" (1 for respondents from other regions, 0 for local residents).

Table 2 Survey sample structure (own work)

Characteristics	% of sample ($N = 188$) (%)
Gender	
Female	49.47
Male	50.53
Age	
18-25	36.70
26-35	21.28
36-45	24.47
46-55	14.36
Over 55	3.19
Place of permanent residence	
Kemerovo region	68.09
Other regions	31.91
Income	
High	31.91
Medium	54.26
Low	13.83

The results of our regression analysis provide evidence for partial confirmation of all three hypotheses of the study. We obtain statistically significant coefficients for at least two "emotional" variables in each of the three regression models. Thus, emotions are significantly associated with three behavioral intentions that characterize visitors' loyalty.

Both revisit and recommend intentions are significantly positively associated with visitors' pleasure and delight. The set of emotions significantly associated with an intention to write an online review is different. Visitors that feel themselves cheerful and surprised are more willing to share their experience online. This regression also gives us an unexpected result: the intention to write an online review is significantly higher for respondents who described their emotional experience as boring. This result may be partly explained by the fact that some customers expect more variety of entertainment from the theme restaurant. They find the real Zaboi experience more boring than expected, and they tell about this mismatch in their online reviews. For example, one of the reviewers tells: "*the place is cute and cool but boring… not allowed to play cards.*"

The regression analysis does not reveal any significant socio-demographic differences for revisit and recommend intentions. At the same time, there is a gender difference in the case of intention to write an online review: male respondents are more willing to share their experience online. This can be explained by the restaurant's specificity: mining theme and the "brutal" atmosphere of Zaboi restaurant probably better meets the interests of the male audience.

Table 3 Regression analysis (own work using Gretl software)

	Intention to revisit	Intention to recommend	Intention to write an online review
Happy	−0.15 (0.18)	−0.15 (0.17)	0.20 (0.22)
Pleased	0.48*** (0.18)	0.46*** (0.14)	−0.01 (0.19)
Entertained	−0.16 (0.13)	0.01 (0.11)	−0.31** (0.14)
Delighted	0.41** (0.16)	0.32* (0.17)	0.31 (0.21)
Cheerful	−0.03 (0.15)	0.05 (0.16)	0.39* (0.20)
Excited	−0.06 (0.08)	−0.08 (0.07)	0.01 (0.10)
Surprised	0.22 (0.14)	0.18 (0.11)	0.25* (0.15)
Gender	−0.06 (0.20)	0.13 (0.18)	−0.61*** (0.22)
Age	0.00 (0.01)	0.01 (0.01)	0.01 (0.01)
Income	−0.21 (0.16)	−0.22 (0.14)	−0.34** (0.16)
Tourist	−0.07 (0.17)	0.09 (0.17)	−0.03 (0.25)
Constant	2.62*** (0.75)	1.84*** (0.63)	1.57** (0.79)
R^2	0.29	0.35	0.30
R^2 (adjusted)	0.25	0.30	0.26
p-value(F)	<0.0001	<0.0001	<0.0001
N	188	188	188

$* = p < 0.10$, $** = p<0.05$, $*** = p<0.01$
Standard errors are shown (in brackets)

The most interesting research question is what is behind the positive emotions experienced by Zaboi restaurant customers. We asked respondents an open-ended question about the main drivers of their positive emotional experience. Free-form respondents' answers were subsequently coded using subject codes based on an approach similar to online reviews analysis. Table 4 shows the frequency of occurrence of topic codes in respondents' answers.

More than 30 respondents mentioned in their answers two types of emotional drivers. In addition to "food + atmosphere" combination (that is quite typical for theme restaurants), many respondents associated their positive emotions with a combination of authentic atmosphere and regional identity. The synergy of these

Table 4 Analysis of emotional drivers (own work using QDA software)

Type of association	Frequency (N = 188) (%)
Theme and atmosphere	49.4
Menu and food	22.34
Regional identity	16.49
Purpose of visit	7.45
Service quality	3.19
Price	1.60
Other	14.36
No answer (left blank)	2.66

factors works for both tourists and locals. Tourist get strong emotions by revealing the main theme of the destination in the atmosphere of the restaurant (*"Must-visit place with the atmosphere of a mining region"*). Locals gain their positive emotional experience through a sense of belonging to the regional community (*"Coal mine. Kemerovo. Homeland"*). An interesting insight appears in the answers of several respondents who were born in Kemerovo and later moved to Moscow. Their emotional responses relate to nostalgic feelings and hometown memories (*"Place that I visit to recall something lovely from my youth"*). This insight is in line with one of the previous researches that confirmed positive influence of personal nostalgia on revisit intention in the case of luxury restaurants (Hwang & Hyun, 2013).

6 Conclusions

In line with previous studies (Han et al., 2009; Chen et al., 2015; Meng & Choi, 2018), our results confirm a significant positive relationship between emotions and revisit intention. Pleased and delighted visitors express a greater intention to revisit and to recommend the theme restaurant. We did not find any significant differences in the emotions and intentions of different age and gender groups of respondents, except for the intention to share an online review.

Our results provide new evidence on the role of place identity in shaping visitors' emotions. The authentic coal-mining atmosphere provokes a positive emotional response for both tourists (by enhancing the destination's main theme) and locals (by using a strong sense of regional community). This response, in turn, is positively associated with intentions to revisit and to recommend.

Our results allow us to suggest a set of recommendations for Zaboi restaurant, which are also partially applicable to other theme restaurants. Recommendations can be structured depending on the behavioral intentions and the stages of loyalty of the current client (from potential customer to adherent). Table 5 contains the entire set of recommendations.

Table 5 Guidance to increase loyalty intentions for "Zaboi" restaurant

	Intention to revisit	Intention to recommend	Intention to write an online review
Potential customer	Social networks mass media Web site interests targeting + geo-targeting	Social networks mass media Web site interests targeting + geo-targeting	Creating interactive entertainment content based on the restaurant's theme local brand promotion + geo-targeting
Casual customer	Basic needs must be satisfied to avoid the customer's refocus of attention	Creating unique experience for customers and permanent for the restaurant	Local brand promotion + geo-targeting
Customer	Developing the thematic courier delivery network	Creating the "unique cues"; themed gifts developing the thematic courier delivery network	Tracking the theme in everything creating the "unique cues"; themed gifts
Regular customer/adherent	Client survey + email-marketing + look-alike personalization of the offer developing a thematic courier delivery	Creating the "unique cues"; themed gifts developing the thematic courier delivery network the theme nights the dish of week	Creating a strong association (with a place brand) creating the "unique cues"; themed gifts

7 Limitations and Further Research

Our study has some limitations that provide some suggestions for future research in the field. First, our study is based on evidence from the one theme restaurant. Additional research based on the experience of restaurants in other regions can lead to more reliable conclusions and recommendations. Second, most of our respondents and online reviewers are residents of Russia (and speak Russian). An important direction of further research is a cross-cultural study with the participation of foreign tourists, whose participation in this study was impossible due to travel restrictions in time of the COVID-19 outbreak. The coronavirus pandemic also caused the third limitation of our study: with restaurants temporarily closed, we had to abandon our original on-site survey idea in favor of an online survey. Unfortunately, memories of a visitors' experience might have been less exact than data collected on-site (Meng & Choi, 2018). Finally, our evidence on the relationship between place identity and visitors' emotions is rather tentative. Future research may reveal the role of different mediators (such as novelty or nostalgia) in the influence of place identity on emotions and experiences of local theme restaurants' visitors.

Acknowledgements We are grateful to Elena Sharko (Lomonosov MSU), Daria Saks (PepsiCo Russia), and Natalia Rozhkova (Lomonosov MSU) for their helpful comments on the study draft.

References

Berg, P. O., & Sevón, G. (2014). Food-branding places–A sensory perspective. *Place Branding and Public Diplomacy, 10*(4), 289–304.

Blichfeldt, B. S., & Halkier, H. (2014). Mussels, tourism and community development: A case study of place branding through food festivals in rural North Jutland Denmark. *European Planning Studies, 22*(8), 1587–1603.

Boyne, S., & Hall, D. (2004). Place promotion through food and tourism: Rural branding and the role of websites. *Place Branding, 1*(1), 80–92.

Bukharov, I., & Berezka, S. (2018). The role of tourist gastronomy experiences in regional tourism in Russia. Worldwide Hospitality and Tourism Themes.

Chang, M., Kim, J. H., & Kim, D. (2018). The effect of food tourism behavior on food festival visitor's revisit intention. *Sustainability, 10*(10), 3534.

Chen, A., Peng, N. S., & Hung, K. (2015). The effects of luxury restaurant environments on diners' emotions and loyalty: incorporating diner expectations into an extended Mehrabian-Russell model. *International Journal of Contemporary Hospitality Management, 27*(2), 236–260.

Chen, C., & Guan, Y. (2008). Experience design of the theme restaurant make the dining be a memorable experience. In 9th International conference on computer-aided industrial design and conceptual design (pp. 982–985). IEEE.

Davies, A., & Prentice, R. (1995). Conceptualizing the latent visitor to heritage attractions. *Tourism Management, 16*(7), 491–500.

Dixit, S., Badgaiyan, A. J., & Khare, A. (2019). An integrated model for predicting consumer's intention to write online reviews. *Journal of Retailing and Consumer Services, 46,* 112–120.

Ellis, A., Park, E., Kim, S., & Yeoman, I. (2018). What is food tourism? *Tourism Management, 68,* 250–263.

Han, H., Back, K. J., & Barrett, B. (2009). Influencing factors on restaurant customers' revisit intention: The roles of emotions and switching barriers. *International Journal of Hospitality Management, 28*(4), 563–572.

Hwang, J., & Hyun, S. S. (2013). The impact of nostalgia triggers on emotional responses and revisit intentions in luxury restaurants: The moderating role of hiatus. *International Journal of Hospitality Management, 33,* 250–262.

Kim, W. G., & Moon, Y. J. (2009). Customers' cognitive, emotional, and actionable response to the servicescape: A test of the moderating effect of the restaurant type. *International Journal of Hospitality Management, 28*(1), 144–156.

Kim, S. H., & Choi, Y. M. (2000). Evaluating customer's satisfaction level as an influential factor to revisit theme park. *International Journal of Tourism Sciences, 1*(1), 111–126.

Kim, Y. H., Duncan, J., & Chung, B. W. (2015). Involvement, satisfaction, perceived value, and revisit intention: A case study of a food festival. *Journal of Culinary Science and Technology, 13*(2), 133–158.

Kivela, J., & Crotts, J. C. (2006). Tourism and gastronomy: Gastronomy's influence on how tourists experience a destination. *Journal of Hospitality and Tourism Research, 30*(3), 354–377.

Kozak, M., & Rimmington, M. (1999). Measuring tourist destination competitiveness: conceptual considerations and empirical findings. *International Journal of Hospitality Management, 18*(3), 273–283.

Lee, I., & Arcodia, C. (2011). The role of regional food festivals for destination branding. *International Journal of Tourism Research, 13*(4), 355–367.

Meng, B., & Choi, K. (2018). An investigation on customer revisit intention to theme restaurants. *International Journal of Contemporary Hospitality Management, 30*(3), 1646–1662.

Muñoz, C. L., & Wood, N. T. (2009). A recipe for success: Understanding regional perceptions of authenticity in themed restaurants. *International Journal of Culture, Tourism and Hospitality Research, 3*(3), 269–280.

Nawawi, W. N. W., Kamarudin, W. N. B. W., Ghani, A. M., & Adnan, A. M. (2018). Theme restaurant: influence of atmospheric factors towards the customers' revisit intention. *Environment-Behaviour Proceedings Journal, 3*(7), 35–41.

Onorati, M. G., & Giardullo, P. (2020). Social media as taste re-mediators: Emerging patterns of food taste on TripAdvisor. *Food, Culture and Society,* pp. 1–19.

Oppermann, M. (2000). Tourism destination loyalty. *Journal of Travel Research, 39*(1), 78–84.

Pakhalov, A., & Rozhkova, N. (2020). Escape rooms as tourist attractions: Enhancing visitors' experience through new technologies. *Journal of Tourism, Heritage and Services Marketing, 6*(2), 45–50.

Rousta, A., & Jamshidi, D. (2020). Food tourism value: Investigating the factors that influence tourists to revisit. *Journal of Vacation Marketing, 26*(1), 73–95.

Ryu, K., & Jang, S. S. (2007). The effect of environmental perceptions on behavioral intentions through emotions: The case of upscale restaurants. *Journal of Hospitality and Tourism Research, 31*(1), 56–72.

Sampol, C. J. (1996). Estimating the probability of return visits using a survey of tourist expenditure in the Balearic Islands. *Tourism Economics, 2*(4), 339–351.

Selyukov, A. V. (2015). Technological significance of internal dumping in open pit coal mining in the Kemerovo region. *Journal of Mining Science, 51*(5), 879–887.

Song, H., Van Phan, B., & Kim, J. H. (2019). The congruity between social factors and theme of ethnic restaurant: Its impact on customer's perceived authenticity and behavioural intentions. *Journal of Hospitality and Tourism Management, 40,* 11–20.

Sørensen, F., Fuglsang, L., Sundbo, J., & Jensen, J. F. (2020). Tourism practices and experience value creation: The case of a themed attraction restaurant. Tourist Studies (OnlineFirst).

Stone, M. J., Migacz, S., & Wolf, E. (2019). Beyond the journey: The lasting impact of culinary tourism activities. *Current Issues in Tourism, 22*(2), 147–152.

TripAdvisor. Zaboi. Retrieved from https://www.tripadvisor.com/Restaurant_Review-g811326-d2652489-Reviews-Zaboi-Kemerovo_Kemerovo_Oblast_Siberian_District.html (Last Accessed 01.07.2020 23:55)

Tsai, C. T. S., & Wang, Y. C. (2017). Experiential value in branding food tourism. *Journal of Destination Marketing and Management, 6*(1), 56–65.

Tsaur, S. H., Luoh, H. F., & Syue, S. S. (2015). Positive emotions and behavioral intentions of customers in full-service restaurants: Does aesthetic labor matter? *International Journal of Hospitality Management, 51,* 115–126.

Vasyutin, S. A., Deniskevich, E. N., Kim, O. V., Selezenev, R. V., Yumatov, K. V., Yakimova, N. S., & Gavrilov, M. I. (2018). Kemerovo Region's experience of implementing and promoting cluster initiatives in tourism industry. In International conference" Economy in the modern world"(ICEMW 2018) (pp. 270–276) Atlantis Press.

Weiss, R., Feinstein, A. H., & Dalbor, M. (2005). Customer satisfaction of theme restaurant attributes and their influence on return intent. *Journal of Foodservice Business Research, 7*(1), 23–41.

Yan, X., Wang, J., & Chau, M. (2015). Customer revisit intention to restaurants: Evidence from online reviews. *Information Systems Frontiers, 17*(3), 645–657.

Yoon, Y., & Uysal, M. (2005). An examination of the effects of motivation and satisfaction on destination loyalty: A structural model. *Tourism Management, 26*(1), 45–56.

Anticipated Booking on Touristic Attractions: Flamenco Show in Seville

Fernando Toro Sánchez

1 Introduction

Flamenco is an expression of the people that you have transcended to the consideration of good within the Intangible Cultural Heritage of UNESCO. Its origin is very indeterminate and, in its consolidation, converge the union of different influences from different cultures, Arabic, Castilian romanzas and is even associated with traditions of distant India and its nomadic character. The flamenco sits in the south of the Iberian Peninsula in the middle of the fourteenth century, a period of great historical confluence in Spain with the taking of Granada and the discovering of America, where Seville becomes a metropolis of first order for its hegemony in the "Carrera de Indias",[1] being an area where for its commercial activity come together different towns with their distant cultural influences being, the flamenco, a form of expression of the plain people that manifests through singing and dancing, their way of interpreting the feelings that have occurred whether of joy, protest, sorrow, or pain. Then the musicality is incorporated with the introduction of various instruments in the "coplas", such as the guitar of clearly Arabic influence, the castanets coming from Castile and coming from America.

It is not until the nineteenth century with the late Spanish Illustration Period and after the time of Napoleonic overtook, flamenco begins to confer itself as a

---

[1]La Carrera de Indias reference the traffic of ships from the America's Overseas Territories of the Spanish Empire from the sixteenth century to the end of the nineteenth century, which in principle reached exclusively to the port of Seville, to later move to the port of Cadiz and already in the middle of the nineteenth century to other Spaniards harbours.

Tablao—etymologically from the Turkish Tavla "wood" refers to the material of the stage on which the flamenco show takes place and by extension gives name to the local.

F. Toro Sánchez (✉)
University of Seville, Seville, Spain
e-mail: fennantoro@gmail.com

© The Author(s), under exclusive license to Springer Nature Switzerland AG 2021   673
V. Katsoni and C. van Zyl (eds.), *Culture and Tourism in a Smart, Globalized, and Sustainable World*, Springer Proceedings in Business and Economics,
https://doi.org/10.1007/978-3-030-72469-6_44

more regulated cultural expression, appearing the transcription of traditional letters to manuals, with writings and with the appearance of shows containing flamenco as the place of the performance, first in Seville and later in other places in Andalusia in its different manifestations, finally settling in the capital, Madrid.

Currently, flamenco is strongly rooted in Andalusian culture, being a vehicle of cultural transmission between generations and among peoples, so that it has transcended beyond the borders of Spain in order to find a flamenco space anywhere in the world being an artistic expression that allows the fusion with other types of musical expressions, which has led to its universal consideration. Anticipation of bookings is a fundamental objective that any tourism manager should prioritize in their resource management. Specifically, the different tools of "revenue" have been applied in the different elements of the "tourist value chain" (Porter, 1985/Gereffi, Humphrey and Sturgeon 2005/Zuñiga Collazos et al. 2012), such as transport, especially air, hotel reservations and progressively in the tourist attractions, following the logical scheme of booking "Mobility, Residence and Attraction", which can be altered by the traveller under own consideration or by the suggestion of the marketing elements.

Local operators of tourist attractions present two disadvantages at their objective of anticipating the reservation, regardless of local internal competition with other operators: on the one hand, sequential time consideration in the decision of the traveller/tourist, which is supposed that relegates it to a non-priority element; on the other hand, its secondary role in user trust part of the prominence in the global digital operators (OTAS) market, such as Trip Advisor, Get Your Guide (GYG) or Booking that after taking a primary role in the reserve in the plot of the stand, they are progressively doing so in the tourism experience industry. To the latter factor, we must add the appearance of the booking platforms that bring together the ease of the management of the reservation with exposure of the offer of the experience of the tourist attraction in many channels until reaching the end user, which combines in different proportions the advantages offered by each channel with the satisfaction of the experience, in its final decision: the medium is the message (McLuhan, 1967). Among other considerations, maintaining the own marketing channel gives a competitive advantage to any marketer of goods and services (Michael, 1990) both for the control of the value proposal and by winning in the contribution margin, given with the commission percentage with which the different OTAS and booking platforms work.

The objective is to observe whether by observing the empowering of the network of own digital marketing channels in a local tourist activity such as a Tablao flamenco in Seville you can get some clear conclusion about the anticipation in sales and what conclusions we can draw on the consumer behaviour of these activities, traveller or tourist that can be used for marketing and sales purposes by local operators of tourist activities. To do this, the study uses the machine learning tool, BIG ML [2], with analysis monitored with logarithmic regressions where the incidence of use of the digital channels themselves is perceived in each time slot.

---

[2]https://bigml.com/.

2 Literature Review

The consumer's behaviour of tourist experiences, whether traveller or tourist, has not been discussed so far in depth. However, there is a extend literature on works in the anticipation of the reservation in tourist products or services, specially flight tickets and in the reservations of hotel accommodation, fitting the behaviour of the consumer to models of behaviour in decision-making and sequentially in the search for information and collection of different products (Kazmi & Rahman 2007); so that the more information the individual has, the better decision he usually makes (Kulviwat et al., 2004). Mostly in these studies, planning is considered as a majority factor in the hiring of any tourist product or service, and for the key understanding of the choice of a tourist destination (Song et al., 2012). More specifically in the level of the tourist experience we can find behaviours further away from a planned scheme, the smaller the time difference between the reservation and the date of completion of the tourist activity or experience (Toro-Sánchez et al., 2021). The individual will perform two types of information searches, one pre-purchase and one when in the "ongoing" travel process, both different in the framework that are performed (Kulviwat et al., 2004) and two different behaviour schemes can be differentiated; the first insane to the Theories of Planned Behaviour (Azjen, 1991) and the second most concerning to the Prospect Theory (Kahneman & Tversky, 1980, 2013) and decision based on emotional intelligence (Goleman, 1995)

(a) Pre-purchase decision

The planned behaviour theory (TPB) (Azjen, 1991) which develops the theory of reasoned action (TRA) by (Azjen and Fishbein, 1975)(Madden et al., 1992) is justified in the use of behaviour in tourist activities in works and is developed in various studies by the model of goal directed behaviour (MGDB) (Perugini and Bagozzi, 2001) including two factors, desire and previous attitude, both positive and negative, as factors before that link first to intent and subsequent to the behaviour, in the addition of the influence of subjective norms (SN) and the contemplation of a series of descriptive rules as constructor model of behaviour (Esposito et al., 2016). Subjective norms, and always following the same scheme, are developed within the model of goal directed behaviour (MGDB), which influence both desire and cause close to intent (Malle, 1999) and the subsequent state of behaviour action (Malle & Knobe, 1997), as illustrated by the Diagram 1.

Desire as predictor of intent (Bagozzi, 1992) is given among other factors by the existence of anticipated emotions (Meng & Han, 2016) establishing a bond of affectivity that are fed in part by anticipated behaviours that tend to automate decision processes (Van Aart et al., 2010). Desire as a mood functions as a motivational agent towards action in the decision-making towards the target (Perugi & Bagozzi, 2001), in this case in the planning actions of tourist activities (Meng & Han, 2016). Desire does not fully influence behaviour; subjective rules and perceived control over behaviour also include. Subjective rules, intention, which may or may not favour desire, in a positive or negative way (Marcus & Nurius, 1987) in the case of the tourist experience

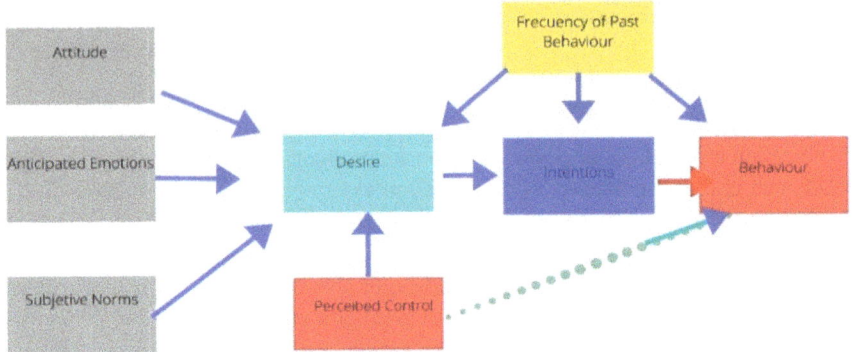

Diagram 1 Model of goal directed behaviour

and given the affective predisposition of the activity generate a positive stimulus to desire and will that is based on reasonable facts (Bagozzi, 1992).

The user, by increasingly using the technological means in their purchasing behaviour, dramatically affects channels and operators (Drucker, 2002). Although the information received by the technological means is not the only source and cannot be viewed in isolation (Rowley, 2000), technological means enhance perceived control over the decision (Athyaman & Adee, 2002) and it reinforces the nature of desire (Bandura 1997 and Rejeski et al., 2005). Following this line, the reviews (reviews of Trip Advisor or Google) also function as a tool of persuasion by which, once the channel is selected by the user, in this case the OTAS, and online according the social influence (Guadagno et al., 2013), vary according to the nature of this. Social norms, personified in valuations, take on a role of authority, as the individual is strongly influenced by his behaviour by the opinions and advice, he/she identifies as an authority (French & Raven 1959/Hinkin et al. 1989). Thus, once the channel is selected, the user in his decision that he perceives not influenced—or perceives dominated by the one who identifies as an authority—exercises a domain over it, regardless of whether he/she notifies that there is this influence of the channel, even if he does not know how to distinguish the distributor from the tourist experience (OTAS) of the supplier of the same activities, or confusing the roles of each one. It prioritizes control over its decision, which it interprets rational and not influenced, for self-interest and calculated on terms of benefits, costs and risks (Cialdinni, 1984). When the subject is influenced, the model becomes more complex and the less predictable behaviour (Cialdinni, 1984), so the channelling of behaviour and perception of self-control of the subject over the digital media (Guadagno et al., 2002) play in favour of the direction of behaviour.

Behaviour can also be influenced by previous behaviours, which complements the MGDB Meng & Han, 2016) and in the repeated purchase, of different activities or by the same channel or type of channel as is the case of OTAS, adapting in each case to the inclusion of different variables to enhance their predictive power (Perugini & Bagozzi, 2001). In this sense, reviews function as rules of acceptance which direct behaviour based on past experiences that the subject adopts as his own, where past behaviour is defined as a kind of methodological control that shows a more accurate form of assessment of the effects on theoretical variables (Azjen, 1991) and is shown, past behaviour, as an independent predictor of behaviour (Armitage & Corner, 2001). This past behaviour is also determined by its influence on the decision based on how often similar behaviour is performed the time lapse between similar behaviour action (Triandis et al., 1971). In the same way they directly influence the behaviour, the memory about the last behaviour "Recency" that may or may not be different from the current behaviour (Leone et al., 2004). When the behaviour is repeated and becomes usual, as in the model of travel, over time it is guided through an automatic cognitive process, rather than by an elaborate and precise decision (Van Aarts et al., 1998). TPB ignores repetitive base behaviours, even though habits successfully perfect stocks (Triandis, 1980/Hull 1947) and instrumentally or asymmetrically if it is unsatisfactory (Savignon, 1972).

Behaviour is strongly driven by stimuli as long as they have a specific objective and it functions as such (Bargh 1984); e.g., Lau and Hsu validates the MGB for tourist behaviour at the Las Vegas destination and is based on expectations of making money as a goal. So, when the behaviour of the traveller/tourist is common, the usual mode of search and choice of the channel is strongly associated with the specific tourist destination (Van Aart, 1997). We can say that the more common the behaviour is, the less reasoned is within the planning process (Greenwood et al., 1989). Closer to the action being facilitated by the better presence of communication channels, the user understands as optimal and adopts as its own trusting them, this being a tool of high persuasion used generally by the OTAS; knowing that the habit as it is repeated es more strength than the intention in the decision (Triadis, 1980). Therefore, the attitude, in this context, is strong, as positive, as the purest influential agent (Athiyaman, 2002), being the form of purchase (the chosen channel) an instrument more associated with the search for expected value through the comparison of different alternatives (Dabholher, 1996)

(b)   Ongoing purchase decision scheme

The user's behaviour of tourist activities can be affected by a feeling of scarcity. Especially when it influences the time between the reserve and that of the activity, and it is reduced generating a feeling of pressure. In this case, since the opportunity cost given the time constraint is evident, in most cases, behaviour planning loses weight against other types of models in decision-making closer to prospects models (Kahnemman & Tversky, 1979) and risk adversion. The tourist product can be seen by the user as an investment, in the time most precious by himself which is leisure time and vacation: when the individual perceives that he is investing, the buying behaviour is facilitated (Mano & Oliver, 1993), increases expectations and decreases the risk adversion (Lynn & Borget, 1996). The scarcity of time is not sufficient condition of the apprehending value of a service or product as indicated by the classical theory (Smith,1776) and that is applicable to commodities (Brock, 1968), but if the increase of this sense of scarcity since the preferences of the individual are independent of the restrictions at first and are oriented towards the target (Lynn, 1991). In the face of scarcity, the individual's attitude strengths to the target to the detriment of planning and, aspects such as emotional intelligence, where the value perceived by the individual is imbued by emotions.

The decisions in this point are patronized by different taxonomies, which will lead to the behaviour of individuals, more relaxed in spending, as a component of the decision as the pattern is more consistent in values such as education (Lynn, 1991), professional experience and monetary habits (Furham & Ayrsile, 1998). Spending on goods or services that the individual belonging to his desirable objectives is most favoured in his behaviour and corresponds to multi-organizational models where empathy, emotional stability, self-realization and resiliency come into play (Goleman, 1995). When the decision of spending is conditioned by emotional intelligence factors, there is no clear relationship between them (Engelber & Sjoberg, 2006)

The following model of buying behaviour in airline tickets (Kazni & Rhama, 2007) is a development that combines planning with the domain of information by the user and the decision with the purpose of purchase. This diagram is extended with the Kulviwaat et al. (2004), two different periods of tourist search and purchase behaviour (Diagram 2).

In the model, online search can be perfected in the case of tourist experiences with two different sub-models either the model of "pre-purchase", which is perfected with direct and direct action directed to the purchase or, the model of "ongoing", which would reduce the influence of previous factors although they will exist, with the marked influence of pressure on the shortage of time.

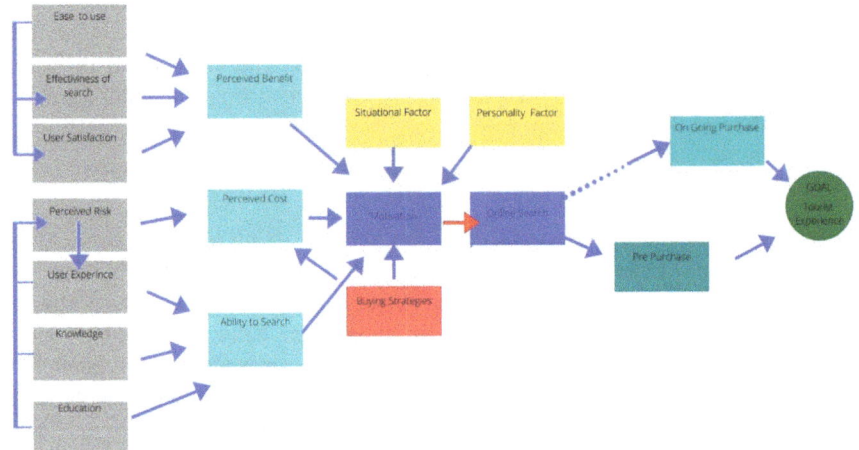

Diagram 2 Two different periods of online tourist purchase model

3 Methodology

For the study, we have taken two types of samples about tickets sold in the same show of tourist experiences, specifically a flamenco Tablao in Seville, in two time periods differentiated by a milestone in the change in more intensive employment in own digital marketing channels to the detriment of the marketing of the product by OTAS, with the aim of observing the impact on the anticipation in the reserve and the behaviour of the traveller/tourist in both phases and globally and finding indicators that are applicable to the marketing actions of local operators.

- Period I: The first sample takes 1752 values from 1 May 2018 to 31 May 2019, where marketing takes place through external, digital and offline channels.
- Period II: The second sample takes 1007 entries from 1 October 2019 to 29 February 2020, where the effort in the digital marketing of the website itself is also implemented.

In both samples we carry out the study with the Big ML tool for mass data processing (machine learning) and in its functionality of logarithmic scales which will offer us the percentage weight of contracting tickets through the different channels used and with the objective variable that is the anticipation of the reservation evaluated in the difference in days between the date of booking and that of the realization of the tourist experience. The tool allows us to cross data, so we can add to the study nuances such as the language in which the reservation is made or the combination in the reservation of different experiences, which deepens the study in the behaviour of the traveller/tourist depending on the anticipation in the reservation..

Work hypotheses

H1: traveller/tourist when in the pre-purchase stage of tourist activities or experiences uses the means of intermediation of reference, OTAS as a priority channel online information search.

H2: traveller/tourist when in the ongoing stage of activities tends to look for information from the operator itself as long as the operator has an outstanding online presence (Fig. 3).

4 Results

Explanation of variables:

- OTAS: GYG/gyg/TripAdvisor
- Own Online Channel: WEB
- Local Offline Channels: Hotels/PIT (Tourist Information Point)/Door

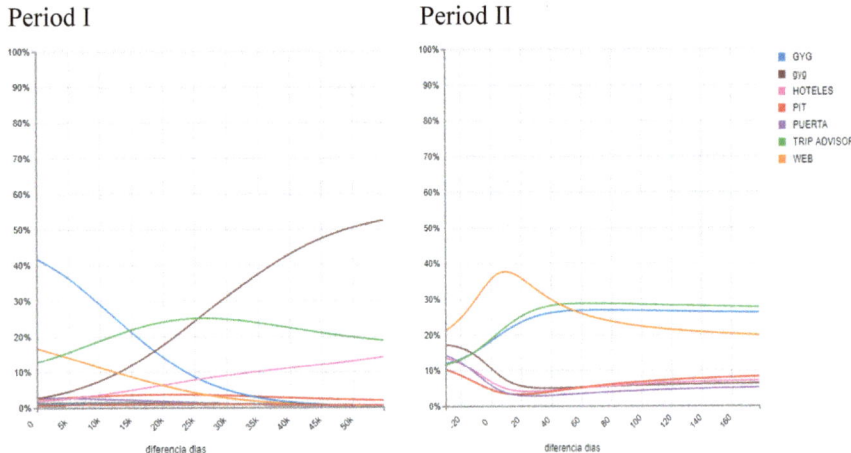

Fig. 3 a Advance bookings per channel (logarithmic regression)

We see that the first period the marketing is strongly dominated by the action of the OTAS, especially in the advance sale and in any of the three companies contracted to market these services, although different behaviours are distinguished in different OTAS, taking into account the different marketing actions used. The online channel itself takes on more weight in marketing as the anticipation of the reservation is lower. The actions carried out to enhance the own online channel (Web) lead to the direct action of increasing the weight of sales, as was its objective. However, it has three interesting consequential effects for our study:

(a)   Anticipation of the reservation is encouraged on the channel itself
(b)   The behaviours of the external OTAS channels are levelled
(c)   At small levels of anticipation time, ongoing decision, there is a behaviour more tending to the choice of the own channel (Fig. 4).

In addition to the chart above, the total reserve scatters are shown in sorted by number of days in advance in abscissa, where you can clearly identify the two reservations trends:

- OTAS, in blue-GYG and green-Trip Advisor, with only presence in pre-purchase and (approximately up to 45 days before the event)
- Own web channel, which increases its activity increasingly in an individual behaviour ongoing purchase

The intense red zone corresponds to the marketing in OTA (GYG) to which the same impact actions have been applied as to the website itself so that we can compare the behaviours of the individual when there is practically no anticipation. In this sense it is observed that at the time of the event, the behaviour clearly varies from the contracting by the OTA, to the direct contracting of the website itself. The basic difference in the exposure of the two OTAS studied lies basically in the way

Period II

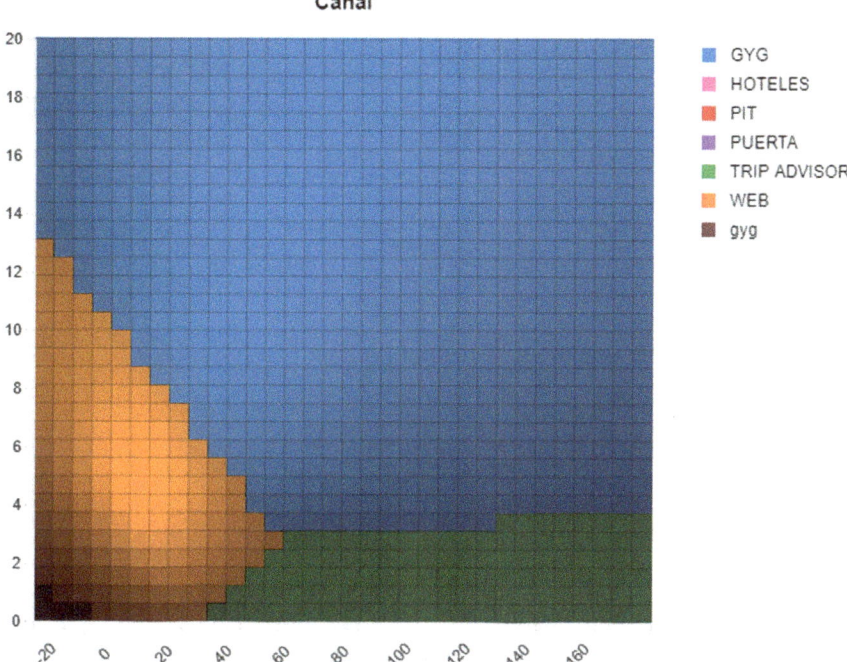

Fig. 4 b Advance bookings per channel (Logarithmic Scale)

of appearing in online positioning, strong in promoted advertising or SEM, in the case of GYG (Blue) and less intense in the case of Trip Advisor/Viator (Green) who enhances its positioning especially in the reputation of its references (Reviews). As for the website itself, it follows a positioning strategy consistent with the activity of the OTAS, operating point SEM positioning and come the close moment of the events, on a daily basis and a few hours after its celebration (Fig. 5).

This graphic analysis shows in clear behaviour ongoing on the purchase by the website itself, where the anticipation of the minimum reservation, up to 48 h before the event, with a repetition rate only of 0.8%, which means that the website is viewed only for the purchase with a rebound rate of 55.71%: The user accesses the website only with the intention of buying and if not the terms do not convince them, abandon the purchase: at this point it should be added that the average duration of the session is 1 min 40 s, an estimated period insufficient to perform a planning task.

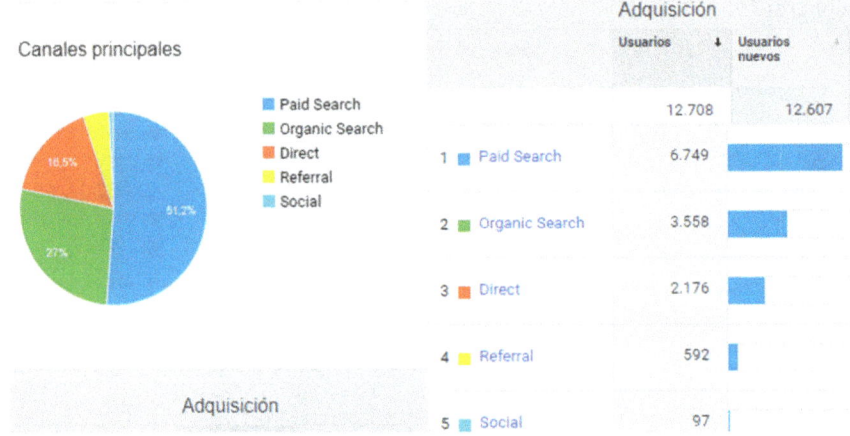

Fig. 5 c Own web channel acquisition behaviour during period II

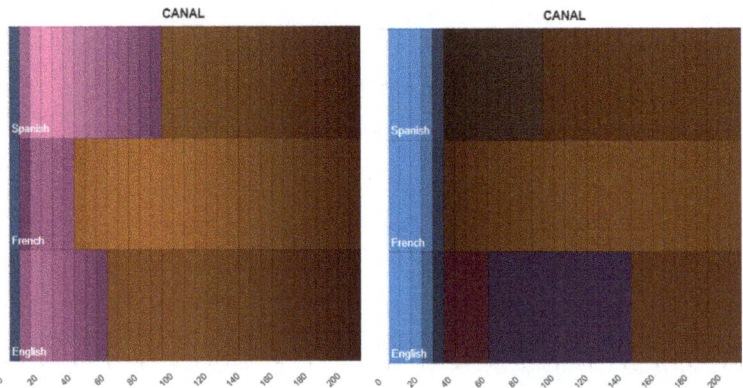

The brown areas, in different gradients corresponding to the weight they take over the total percentage, correspond to the GYG channel that is an OTA, just like the purple areas of the first chart, such as the blue of the second, that identify the weight of TripAdvisor (OTA).

In the second period, the intensification of the sale on the web is observed in the red zone of the client that book the experience on English language that is the majority, and at the same time, the geographically furthest from the locality where the experience is made. The blue areas correspond to offline sales "on site" and with a truly short period between the booking and the celebration of the event (Fig. 6).

Looking at both graphs, we see that little exposure of activity in online marketing in the period favours the boosting of the sale of the offline and "in situ" (APP) channel, clearly in an ongoing behaviour and in this case induced by a prescriber who is the sales agent in the same locality. By intensifying online sales, both through external and own channels, offline sales (Gate/PIT/Hotels) are practically testimonial. Online

Period I   Period II

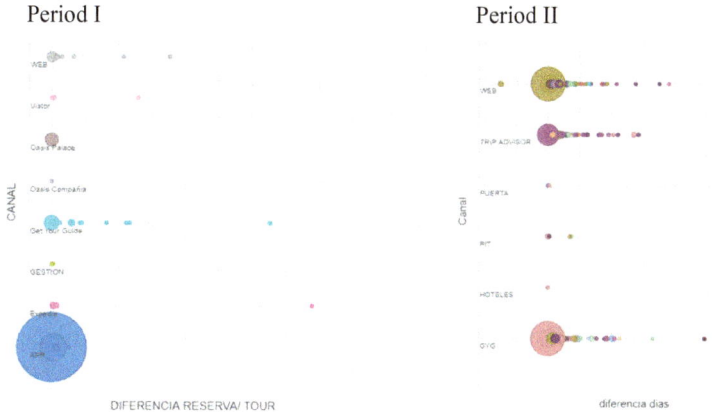

Fig. 6 a Advance bookings by channel (Absolute Scale)

sales also exhibit ongoing behaviour, similar across all channels used and to a much lesser extent, anticipation behaviour in the purchase.

As a result of the observation of the list of graphs, we can conclude that the establishment of the starting hypotheses is as follows:

H1: traveller/tourist when in the pre-purchase stage of tourist activities or experiences uses the means of intermediation of reference, OTAS as a priority channel online information search. ACCEPTED, it is already observed that it is the main and practically only source of information in this search stadium.

H2: traveller/tourist when in the ongoing stage of activities tends to look for information from the operator itself as long as the operator has an outstanding online presence. PARTIALLY ACCEPTED. The user shuffles several sources of information in a distributed way, being more influenced by external purchasing determinants in addition to their own control over the decision and mainly on the device screen listed presence from all the operators.

5 Conclusion

The logarithmic analysis of this study offers us data that relates the specific weight of each channel to the total samples observed, although the fact of intensifying marketing actions on online sales, both in own and external channel, generates an intensification of searches for information in general that are subsequently transformed into sales, as seen in graph 2b) that handles absolute values. The user values the online presence when searching for information and is limited to obtaining the information of the main positions that he displays on the screen, maximum if the search is carried out through a mobile phone device and at an ongoing stage, which is the one that appears mostly in this study. The fact that the three online channels used Web (own) and GYG and Trip Advisor (external) are distributed to almost the same proportions the weight of sales makes us think about the importance of online positioning in consumer behaviour and in this sector of tourist activities, and for this particular activity, and as a strong indicator in the purchase process, relegating the search for information to a much less important plane.

Moreover, taking into account the factor of the recommendations, which in the attraction observed are the highest scores in each of the channels, taking the Google score the score of the own channel leads us to think that the recommendation replaces the research in the information search phase, taking into account that the behaviour of tourist/traveller lasts this decision on the tourist attraction until the last moment and in the ongoing stage. Therefore, the decision is based on the confidence of the review provided that the activity offer has maximum online presence, either externally or by own channel. The study leads us to weigh the weight of OTAS marketing actions against the real possibility of local operators of tourist activities, always carrying out advertising promotion actions (SEM) located in investment and time and segmented towards the target and geolocated advertising. In this case, SEM's actions have not been automated, so it was intended to use a communication of greater suggestion. The analysis of the traffic of the website concludes that the stage of searching for information is scarce, as well as repetition in visits.

In the limitations of the study in only evaluating the final reservations, we do not have data on previous searches for information by the user. A survey to observe the importance that the user gives to the search for information in this regard could open paths of study and above all to give more meaning to online positioning practices based on SEO and not promoted as a proposal of own and unreferenced value.

Acknowledgements We would greatly like to thank Tablao Pura Esencia Flamenco in Sevilla (www.puraesenciaflamenco.com) and his entire team for the provision of the data on which the basis of the study is obtained.

Also thank the Profesores D. Jesus López Bonilla and D.Luis Miguel López Bonilla from the Department of Market Research and Marketing of the University of Seville, for their continued support in the direction of this and other published articles that in complete form the PhD work thesis.

References

Ajzen, I. (1991). The theory of planned behavior. *Organizational Behavior and Human Decision Processes, 50*(2), 179–211.

Armitage, C. J., & Conner, M. (2001). Efficacy of the theory of planned behaviour: A meta-analytic review. *British Journal of Social Psychology, 40*(4), 471–499.

Aarts, H., Verplanken, B., & Knippenberg, A. V. (1998). Predicting behavior from actions in thepast:Repeated decision making or a matter of habit?*Journal of Applied Social Psychology, 28*(15), 1355–1374.

Athiyaman, A. (2002). Internet users' intention to purchase air travel online: An empirical investigation. *Marketing Intelligence and Planning, 20*(4), 234–242.

Bargh, J. A. (1984). Automatic and conscious processing of social information. In American Psychological Association convention, 1982, Washington, DC, US; Portions of the research discussed in this chapter were presented at the aforementioned conference, and at the 1982 meetings of the Society for Experimental Social Psychology in Nashville, Indiana. Lawrence Erlbaum Associates Publishers.

Bandura, A. (1997). The anatomy of stages of change. *American journal of health promotion: AJHP, 12*(1), 8–10.

Bagozzi, R. P. (1992). The self-regulation of attitudes, intentions, and behavior. *Social Psychology Quarterly*, 178–204.

Brock, T. C. (1968). Implications of commodity theory for value change. In *Psychological foundations of attitudes* (pp. 243–275). Academic Press.

Savignon, S. J. (1972). Communicative Competence: An Experiment in Foreign-Language Teaching. *Language and the Teacher: A Series in Applied Linguistics, 12*

Dabholkar, P. A. (1996). Consumer evaluations of new technology-based self-service options: an investigation of alternative models of service quality. *International Journal of Research in Marketing, 13*(1), 29–51.

Cialdini, R. B. (1984). Influence: The new psychology of modern persuasion. Morrow.

Drucker, P. (2002). *Administração na próxima sociedade*. NBL Editora: A-Exame.

Engelberg, E., & Sjöberg, L. (2006). Money attitudes and emotional intelligence. *Journal of Applied Social Psychology, 36*(8), 2027–2047.

Esposito, G., van Bavel, R., Baranowski, T., & Duch-Brown, N. (2016). Applying the model of goal-directed behavior, including descriptive norms, to physical activity intentions: A contribution to improving the theory of planned behavior. *Psychological Reports, 119*(1), 5–26.

Fishbein, M., & Ajzen, I. (1975). Belief. Attitude, Intention and Behavior: *An Introduction to Theory and Research, 578.*

Furnham, A., & Argyle, M. (1998). The psychology of money. Psychology Press.

Gereffi, G., Humphrey, J., & Sturgeon, T. (2005). The governance of global value chains. *Review of International Political Economy, 12*(1), 78–104.

Goleman, D. (1995). *EI: Why it can matter more than IQ*. London: Bloomsbury.

Guadagno, R. E., Muscanell, N. L., Rice, L. M., & Roberts, N. (2013). Social influence online: The impact of social validation and likability on compliance. *Psychology of Popular Media Culture, 2*(1), 51.

Guadagno, R. E., & Cialdini, R. B. (2002). Online persuasion: An examination of gender differences in computer-mediated interpersonal influence. *Group Dynamics: Theory, Research, and Practice, 6*(1), 38.

Hinkin, T. R., & Schriesheim, C. A. (1989). Development and application of new scales to measure the French and Raven (1959) bases of social power. *Journal of Applied Psychology, 74*(4), 561.

Kahneman, D., & Tversky, A. (1979). On the interpretation of intuitive probability: A reply to Jonathan Cohen.

Kahneman, D., & Tversky, A. (1980). Prospect theory. *Econometrica, 12.*

Kahneman, D., & Tversky, A. (2013). Prospect theory: An analysis of decision under risk. In *Handbook of the fundamentals of financial decision making*: Part I (pp. 99–127).

Kazmi, Y., & Abdul Rahman, M. (2007). Information search and evaluation of alternatives in online airline ticket purchase: Two case studies investigating consumers' online purchase decision-making process and influence of education.

Kulviwat, S., Guo, C., & Engchanil, N. (2004). Determinants of online information search: A critical review and assessment. *Internet Research.*

Leone, L., Perugini, M., & Ercolani, A. P. (2004). Studying, practicing, and mastering: A test of the model of goal-directed behavior (MGB) in the software learning domain. *Journal of Applied Social Psychology, 34*(9), 1945–1973.

Lynn, M. (1991). Scarcity effects on value: A quantitative review of the commodity theory literature. *Psychology and Marketing, 8*(1), 43–57.

Lynn, M., & Bogert, P. (1996). The effect of scarcity on anticipated price appreciation. *Journal of Applied Social Psychology, 26*(22), 1978–1984.

Madden, T. J., Ellen, P. S., & Ajzen, I. (1992). A comparison of the theory of planned behavior and the theory of reasoned action. *Personality and Social Psychology Bulletin, 18*(1), 3–9.

Malle, B. F. (1999). How people explain behavior: A new theoretical framework. *Personality and Social Psychology Review, 3*(1), 23–48.

Malle, B. F., & Knobe, J. (1997). The folk concept of intentionality. *Journal of experimental social psychology, 33*(2), 101–121.

Mano, H., & Oliver, R. L. (1993). Assessing the dimensionality and structure of the consumption experience: Evaluation feeling and satisfaction. *Journal of Consumer Research, 20*(3), 451–466.

Markus, H., & Nurius, P. (1987). Possible selves: The interface between motivation and the self-concept.

McLuhan, M., & Fiore, Q. (1967). *The medium is the message.* New York, *123*, 126-128.

Meng, B., & Han, H. (2016). Effect of environmental perceptions on bicycle travelers' decision-making process: developing an extended model of goal-directed behavior. *Asia Pacific Journal of Tourism Research, 21*(11), 1184–1197.

Perugini, M., & Bagozzi, R. P. (2001). The role of desires and anticipated emotions in goal-directed behaviours: Broadening and deepening the theory of planned behaviour. *British Journal of Social Psychology, 40*(1), 79–98.

Michael, E. (1990). Porter. The competitive Advantage of Nations.

Porter, M. E. (1985). Value chain. The Value Chain and Competitive advantage: Creating and sustaining superior performance

Rejeski, W. J., Katula, J., Rejeski, A., Rowley, J., & Sipe, M. (2005). Strength training in older adults: Does desire determine confidence? *The Journals of Gerontology Series B: Psychological Sciences and Social Sciences, 60*(6), P335–P337.

Rowley, J. (2000). Product search in e-shopping: A review and research propositions. *Journal of Consumer Marketing.*

Greenwood, D. J. (1989). Culture by the pound: An anthropological perspective on tourism as cultural commoditization. *Culture by the pound: an anthropological perspective on tourism as cultural commoditization,* (Ed. 2), 171–185.

Smith, A. (1776). La riqueza de las naciones.

Song, H. J., Lee, C. K., Norman, W. C., & Han, H. (2012). The role of responsible gambling strategy in forming behavioral intention: An application of a model of goal-directed behavior. *Journal of Travel Research, 51*(4), 512–523.

Triandis, H. C., Malpass, R. S., & Davidson, A. R. (1971). Cross-cultural psychology. *Biennial Review of Anthropology, 7,* 1–84.

Triandis, H. C. (1980). Reflections on trends in cross-cultural research. *Journal of Cross-Cultural Psychology, 11*(1), 35–58.

Toro-Sánchez, F., López-Bonilla, J. M., & López-Bonilla, L. M. (2021). Early purchase in tourist activities: Evidence from a UNESCO world heritage site. *Revista TURISMO: Estudos e Práticas, 10*(1).

Van Aart, C., Wielinga, B., & Van Hage, W. R. (2010, October). Mobile cultural heritage guide: Location-aware semantic search. In *International conference on knowledge engineering and knowledge management* (pp. 257–271). Berlin, Heidelberg:Springer.

Zúñiga Collazos, A., & Castillo Palacio, M. (2012). Turismo en Colombia: Resultados del sector (2007–2010).

The Fisheries Local Action Groups (Flags) and the Opportunity to Generate Synergies Between Tourism, Fisheries and Culture

Luis Miret-Pastor, Ángel Peiro-Signes, Marival Segarra-Oña, and Paloma Herrera-Racionero

1 Introduction

The *European Maritime and Fisheries Fund* (EMFF) is one of the five *European Structural and Investment Funds* (ESI Funds) for the period 2014–2020 that contributes to the achievement of the objectives from the *Common Fisheries Policy* (CFP), the *Integrated Maritime Policy* and the *Europe 2020 Strategy*. This fund, like the previous *European Fisheries Fund* (EFF), incorporates the aid to the territorial development of fishing areas as one of its priorities and does it by seeking to implement *community-led local development* strategies (CLLD) through the *Fisheries Local Action Groups* (FLAGs).

The creation of FLAGs opens up the possibility to finance projects of a very diverse nature that help territorial development and job creation in fishing areas. This possibility has led to an intense debate about the need to seek diversification alternatives to the traditional fishery (Gallizioli, 2014). Morgan et al., (2014) argue that diversification approaches are needed which complement and maintain a direct or indirect link to fishing, so that fishers can exploit their professional skills, knowledge and social networks gained through fishing (Symes, 2007).

In countries such as Spain where fishing shares territory with a very powerful tourism industry, the opportunity that these funds can represent when designing projects that seek to create complementarities and synergies between tourism and fishing becomes evident. Traditional fisheries are the repository of an enormous cultural heritage, both tangible and especially intangible. Many of the major Mediterranean tourist destinations are located in former fishing villages where, in just a few years, fishing was pushed aside in favor of tourism. However, in many cases, fishing and fishermen are still there, contributing with their products to the local gastronomy

L. Miret-Pastor (✉) · Á. Peiro-Signes · M. Segarra-Oña · P. Herrera-Racionero
Universitat Politècnica de València, Valencia, Spain
e-mail: luimipas@esp.upv.es

687

and providing prints that travelers admire and photograph because they recognize in them part of the lost essence of these places.

The fishing diversification projects linked to tourism and financed from the FLAGs are numerous and have been a subject of analysis (Miret et al., 2018, Herrera et al., 2018). However, another line of financing that explicitly appears in the fishing funds is linked to cultural projects. These projects are not easy to identify and analyze because many times they are confused with touristic projects and in their epigraph, they are mixed with social projects that have nothing to do.

This paper hypotheses is that in recent years have emerged numerous projects seeking to recover and to value the richness, both tangible and intangible, linked to fishing cultural heritage. These projects have made flourished numerous cultural routes, museums, fairs, etc. However, there is no work quantifying and analyzing these projects and showing the relationship between the European fishing funds, the FLAGs and the emergence of an entire cultural infrastructure linked to the fishing and marine world. For all these reasons, this work aims to identify, quantify and characterize the cultural projects financed in Spain through the European fishing funds.

To achieve this objective set out in this first section, we will analyze the operation and expenses made at European and Spanish level by the European fishing funds in the second section; the methodology and database used will be explained in the third section, and in the fourth section, we will present and discuss the results. Finally, we will present the conclusions, the limitations and the related future research opportunities linked to this study.

2   The European Fisheries Founds

The last two European fisheries funds, both *the European Fisheries Fund* [EFF] (2007–2013) and the *European Maritime and Fisheries Fund* [EMFF] (2014–2020), incorporate an axis (4th axis in the EFF or UP4 in the EMFF), which highlights the specific objective of increasing employment and territorial cohesion in fishing areas. This supposes the novelty of incorporating a territorial vision in a sectorial fund.

For the management of these funds, numerous *Fisheries Local Action Groups* (FLAGs) have emerged in recent years. The FLAGs are public–private partnerships formed by representatives of the fisheries and aquaculture sectors, as well as by other members of the local community, working to implement a *community-led local development strategy* (CLLD) for their territory.

There are currently 367 active FLAGs in Europe, which therefore act as facilitators of the territory, assuming direct responsibility in managing and applying of the *European Maritime and Fisheries Funds* (EMFF). While in the EFF period, there were 30 FLAGs in Spain. They covered six regions: Galicia, the Canary Islands, Andalusia, Catalonia, Asturias and Cantabria. In the period of the EMFF 10, Spanish coastal regions joined the founds. In addition to the above-mentioned areas, Murcia, the Valencian Community, the Basque Country and the Balearic Islands, currently totalizing 41 FLAGs.

The main objective of the FLAGs is achieving the UP4 objective, ensuring the sustainable development of fisheries communities in social, economic and environmental terms (Marciano & Romeo, 2016), with decision making coming from a bottom-up approach which brings together representatives from the public, private and civil sectors (Budzich-Tabor, 2014).

For the whole of the planned period (2014–2020), the EMFF has a total budget of € 6,4 billion (of which € 1,161 billion are managed by Spain). The *Union Priority* 4 (UP4), which is the one we are studying here, has assigned a total of 547 million euros. In the specific case of Spain, 107,6 millions are dedicated to the 4th axis.

Each country has assigned a portion of the fishing fund depending on the size of its fishing sector. Each country develops its own national *Operational Program* (OP) where the funds are distributed among the different union priorities (UPs). The funds in the UP4 are distributed among the different FLAGs. FLAGs select the projects that they are going to finance based on their own criteria, although the national authorities are in charge of reviewing and finally approving the projects.

The normative of the EMFF foresees that UP4 funding can be used to achieve the following objectives (FARNET, 2019):

1. *Adding value,* which includes adding value to local fisheries products.
2. *Diversification* of fisheries activities into other sectors.
3. *Sociocultural,* promoting social well-being and cultural heritage.
4. *Environmental,* including operations to mitigate the climate change.
5. *Governance,* reinforcing the role of fishing communities.

The five objectives set out in the regulation Article 63 (EU, 2014) are mandatory and used for the categorization and reporting of projects once they are approved. The EMFF projects in Spain therefore follow this classification that it is common to the rest of the countries. The EFF projects followed another similar classification but based on four sections: (1) diversification; (2) governance and management, (3) environment, culture and society; and (4) added value from fishery products.

According Miret et al. (2020) work, based on the latest data on the FLAG projects reported by countries to the European Commission, in Spain, a total of 461 projects had been financed, with an amount of € 9,998,202. The projects were distributed as follows: 165 added value projects, 168 diversification projects, 39 environmental projects, 75 sociocultural projects and 14 governance projects (Fig. 1).

The sociocultural projects, from which we will extract to be analyzed in this study, are a total of 75 projects with a total amount of €1.616.870. These projects are not all those that appear in our analysis, since the database used here has been updated more recently, so the number of projects considered is greater. Remember, on the other hand, that these are data from the EMFF, to these projects we added those from the previous period (the EFF).

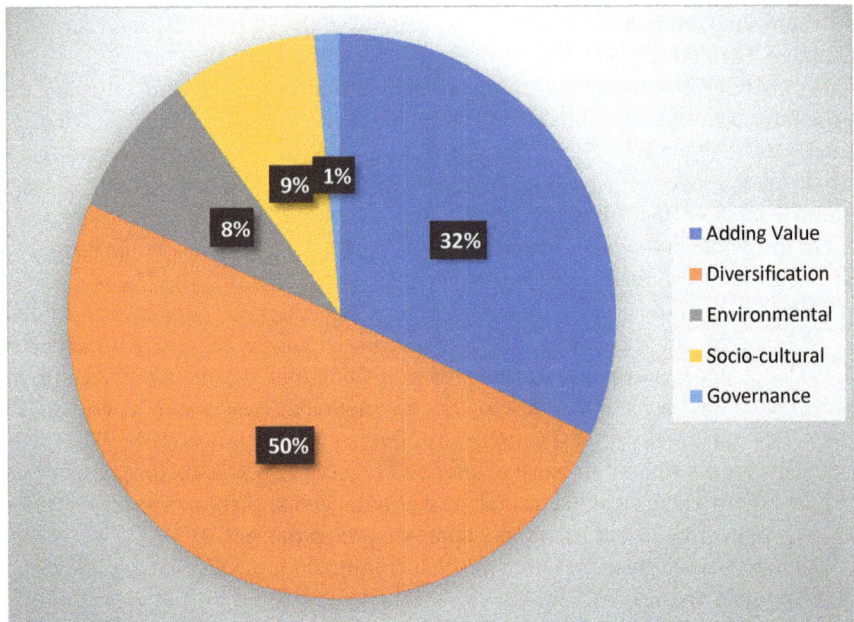

Fig. 1 Average expenditure in Spain for each objective in UP4

3 Methodology

We collected the data used in this analysis from *Spanish Network of Fishing Groups* website database. The website includes information on the different projects financed by both the EFF and the EMFF. The database is constantly being updated. We used data from July 2020 to this study. Therefore, the data covers until the first half of the last year's EMFF. However, we should consider that, in practice, European projects work with the $n + 2$ rule. That is, EFF projects do not end in 2013, but go all the way through 2015. Similarly, EMFF projects they do not end in 2020 but span two more years.

For the EFF period, we can filter projects based on the fourth axis theme. As mentioned: diversification; governance and management; environment, culture and society and added value for fishery products. This classification into four large groups can be broken down into subgroups. Thus, within "Diversification," we found three sections: "use of by-products," "new activities and products" or "tourism." Similarly, the "Environment, Culture and Society" section is divided in "Social area" subgroups into "Culture and Fishing Heritage," "Infrastructures," and "Environment." In this case, it is easy to identify the target projects within the "Culture and Fishing Heritage" section. Within this section, we identified 90 projects. The project initial screening suggested that some projects should be removed to be able to compare with those in the EMFF. We finally remove 21 projects to maintain coherence between the two programs. For example, we decided to remove those actions carried out in schools, those aimed at promoting the consumption of fish, etc.

Unlike, EMF, cultural projects were more difficult to select in the EMFF program since subsections have been eliminated and there are only the five thematic axes: added value, diversification, governance, environment and sociocultural. That is, the classification is too broad, heterogeneous and difficult to shape.

For example, tourism projects have usually been classified within diversification projects, but since there is a section on sociocultural projects, this epigraph has been also used, creating some confusion. Cultural projects (museums, routes, exhibitions, etc.) can be classified either as tourism projects (diversification) or as sociocultural projects. In practice, tourism projects are usually projects related to hospitality, accommodation, fishing tourism, recreational fishing, etc. While most cultural tourism projects, appear in the section on sociocultural projects (as they were already for EFF).

So for the EMFF, we have focused on sociocultural projects, but bearing in mind that a large part of the projects that appear are social projects (gender, youth support, emigrants, etc.) and also considering that this section often acts as a junk box where many projects difficult to catalog are incorporated.

4 Results

Firstly, we proceeded to identify cultural projects within the database, and later, we classified them into five major groups: routes, tangible heritage, intangible heritage, museums and festivals. Within the routes section, we have the adaptation, enhancement and signaling of tourist routes with very diverse characteristics, both nature and heritage.

The tangible heritage section usually refers to the recovery, rehabilitation or enhancement of fishing- or sea-related buildings. They are usually historical buildings or monuments, and sometimes, they are the buildings that house the Fishermen's Guilds. However, we can also find projects for the recovery of traditional boats or the recovery and cataloging of documentary collections.

In the intangible heritage section, projects for the recovery, dissemination and enhancement of culture, history or seafaring traditions are incorporated. On many occasions, this work ends up being reflected in a book or in some type of audiovisual material.

Festivals, fairs, markets, days, etc., are included in the festivals section. These are specific actions, where the marine culture and its products are exposed and promoted.

Finally, the museum section incorporates the permanent exhibitions for which buildings have been adapted. However, in many cases, they go beyond simple exhibitions and become interpretation or research centers of seafaring culture. In most cases, they are sea or fishing museums.

Table 1 Number of cultural projects financed with fishing funds

	EFF period (2007–2013)	EMFF period (2014–2020)	Total
Routes	12	8	20
Tangible heritage	15	14	29
Intangible heritage	17	14	31
Museums	17	19	36
Festivals	6	14	20
TOTAL Cultural projects	67	69	**136**

We have identified a total of 136 cultural projects for the two periods analyzed. We classified the projects in Table 1 in groups five groups: routes, tangible heritage, intangible heritage, museums and festivals.

Over the EFF period, there were a total of 710 projects in the database, while over the EMFF period, we found 834 projects. Therefore, this implies that cultural projects were just under 10% of the total projects in the EFF period and around 8% of the total in the EMFF period.

Data indicated that the number of cultural projects is similar between in the two periods analyzed. However, two aspects must be taken into account. First, the EFF period is already completed while the EMFF period, due to the $n + 2$ rule will still be running for at least two and a half more years. On the other hand, the EFF only covered six of the ten Spanish coastal autonomous communities, while EMFF covers a larger territory. The FLAGs over the two programs have increased from 30 to 41. Figures in each section indicate that similar number of projects related to tangible and intangible heritage and museums are presented. On the contrary, we observed a decrease in projects linked to routes and an increase in the number of festivals.

5   Conclusion

This work has made it possible to identify and quantify the cultural projects linked to the fishing and maritime world that have been financed through the FLAGs and from the 4thaxis of the last two fishing funds.

The results show a total of 136 cultural projects (67 in the EFF period and 69, so far, in EMFF). Cultural projects, once identified, have been classified into five different categories: routes; tangible heritage; intangible heritage; museums and festivals. The distribution of projects among these five categories is quite uniform. It can be said that the European fishing funds are helping to create a network of cultural and environmental routes, as well as a network of sea and fishing museums. They are also contributing to the enhancement and recovery of the numerous and valuable tangible and intangible heritage linked to fishing, as well as to value and spread it through different kinds of materials and events.

All these efforts are important to highlight, but we must not lose sight of the fact that the objective of the 4th axis of the fishing funds is the development and job creation in the fishing-dependent territories, as well as the job creation. Therefore, this financial effort must be rewarding the fishing communities. Valuing the marine culture is a laudable goal, but it is necessary to combine this with the job creation and an emphasis on the territory development, which involves creating synergies between tourism, culture and fishing.

The creation of all these cultural networks linked to fishing must also be linked to tourism policy. Tourism is in a time of a profound transformation where a new tourism model is been shaped (Segarra et al., 2012). Sun and beach tourism is no stranger to these changes and seeks to move away from a massive, Fordist and impersonal model. For this, the destinations are betting on more sustainable and original models (Aguiló et al., 2005), where destinations position themselves offering their own heritage, cultural, gastronomic products, etc. In this context, adding value to cultural products linked to fishing can be increase the touristic offer of many beach destinations.

This work is a first step to study the emergence of tourism linked to the marine culture from the fishing funds. However, the work is very limited due to the scarcity of existing quantitative data. Undoubtedly, important questions remain to be analyzed. Among others, to analyze whether these projects favor fishing communities: Do they increase their income? Do they create jobs? Do they increase the value of the fish they capture? For many people, valuing their culture is a necessary step to value their product and their work, but others understand that it is a way to divert funds from fishing to tourism or, even worse, one more step in the attempt to end fishing as an extractive activity and transform it into a folkloric and tourist product. In any case, these issues do need other methodologies and a much deeper discussion.

Acknowledgements This work was supported by National Plan for Scientific and Technological Research and Innovation (Spanish Economy and Competitiveness Ministry). Research Project PID2019-105497 GB-I00, the European Fisheries Funds, opportunities for the fisheries sector through diversification and the FLAGs management (DivPesc).

References

Aguiló, E., Alegre, J., & Sard, M. (2005). The persistence of the sun and sand tourism model. *Tourism Management, 26*(2), 219–231.

Budzich-Tabor, U. (2014). Area-based local development—A new opportunity for European fisheries areas. In *Social issues in sustainable fisheries management* (pp. 183–197). Springer:Dordrecht.

FARNET FLAG Factsheet. Retrieved from https://webgate.ec.europa.eu/fpfis/cms/farnet2/on-the-ground/flag-factsheets-list_en (Accessed the 6th of March 2020).

Gallizioli, G. (2014). The social dimensions of the Common Fisheries Policy: A review of current measures. In *Social issues in sustainable fisheries management* (pp. 65–78). Springer:Dordrecht.

Herrera-Racionero, P., Miret-Pastor, L., & Lizcano, E. (2018). Traveling with tradition: Artisanal fishermen and fishing tourism in valencian region (spain). *Cuadernos de Turismo, 41*, 679–681.

Marcianò, C., & Romeo, G. (2016). Integrated local development in coastal areas: The case of the "Stretto" Coast FLAG in Southern Italy. *Procedia-Social and Behavioral Sciences, 223,* 379–385.

Miret-Pastor, L., Molina-García, A., García-Aranda, C., & Herrera-Racionero, P. (2018). Analysis of the fisheries diversification funds in Spain during the period 2007–2014. *Marine Policy, 93,* 150–158.

Miret-Pastor, L., Svels, K., & Freeman, R. (2020). Towards territorial development in fisheries areas: A typology of projects funded by Fisheries Local Action Groups. *Marine Policy, 119.*

Morgan, R., Lesueur, M., & Henichart, L. M. (2014). Fisheries diversification: a case study of French and English fishers in the Channel. In *Social Issues in Sustainable Fisheries Management* (pp. 165–182). Springer:Dordrecht.

Segarra-Oña, M. D. V., Peiró-Signes, Á., Verma, R., & Miret-Pastor, L. (2012). Does environmental certification help the economic performance of hotels? Evidence from the Spanish hotel industry. *Cornell Hospitality Quarterly, 53*(3), 242–256.

Symes, D. (2007). The future for fishing dependent areas. *Prospects for the Development of the Fisheries Sector, 2013,* pp. 15–16.

Cultural and Tourism Promotion Through Digital Marketing Approaches. A Case Study of Social Media Campaigns in Greece

Constantinos Halkiopoulos, Maria Katsouda, Eleni Dimou, and Antiopi Panteli

1 Introduction

The focus on digital marketing for designing a successful marketing strategy is the only way an organization can effectively reach its audience. With the technological progress and the increasing use of social networks, especially by young people, traditional marketing techniques are no longer able to reach such a large audience. Television, radio, newspapers and magazines have been replaced by YouTube, Facebook and other social media apps and sites. That is why advertisement has become digital. There are many digital marketing techniques ranging from search to social and email marketing that can improve the digital experience of an organization (Van de Zee and Vanneste, 2015). Digital marketing allows businesses to see accurate results in real time as there is more interaction with users (Aziz et al, 2012). According to (Kannan, 2017) an important characteristic that sets the digital environment apart from the traditional marketing environment is the ease with which customers can share word-of-mouth information, not only with a few close friends but also with strangers on an extended social network (Van der Borg et al, 1997).

The vast majority of businesses and organizations participate in social networks and invest in their social media presence to rapidly disseminate their brand through viral content, social media contests and other consumer engagement efforts (Kumar & Mirchandani, 2012). Social media as a marketing communication channel differ from traditional marketing channels, and even other digital marketing channels given the fact they provide a two-way conversation or exchange (Keegan & Rowley, 2017). The new communication channel challenges marketers because they appreciate the need to engage in social media in order to protect their brand reputation (Lee & Youn, 2009), enhance customer engagement (Gummerus et al., 2012) or augment

C. Halkiopoulos (✉) · M. Katsouda · E. Dimou · A. Panteli
MISBILAB, University of Patras, Patras, Greece
e-mail: halkion@upatras.gr

© The Author(s), under exclusive license to Springer Nature Switzerland AG 2021
V. Katsoni and C. van Zyl (eds.), *Culture and Tourism in a Smart, Globalized, and Sustainable World*, Springer Proceedings in Business and Economics,
https://doi.org/10.1007/978-3-030-72469-6_46

online sales (Tsimonis and Dimitriadis, 2014). They acknowledge the contribution of social media as a powerful mean to generate sustainable and positive word-of-mouth (Katsoni, 2011). Furthermore, marketers comprehend the crucial role of social media in the context of marketing, as they influence consumer preferences and purchase decisions (Halkiopoulos et al, 2019). In addition, consumer engagement became one of the key objectives of many marketing professionals (Dessart et al., 2015). Interesting enough that despite the growth of their social media presence, the plethora of businesses and organizations lacks formal marketing strategies or policies regarding social media activity (Ansah et al, 2012).

The project "Regio-Gnosis", funded by the European Commission, Directorate General Regional and Urban Policy and by own contribution, aims to provide support for the production and dissemination of information and content linked to the EU Cohesion policy. The planned communication strategy has 3 dimensions: (1) to increase the visibility of the implemented projects funded by the EU in four selected regions of Greece (Attica, Central Macedonia, Epirus and Western Greece), (2) to highlight the impact/benefits (social, economic, cultural etc.) of these projects to the local communities and (3) to encourage civilian participation, engagement and social dialogue in relation to the projects.

An important part of the promotion of the Regio-Gnosis program is digital and social media marketing. The internet is now a major part of informing the public, which is why it is essential to pay particular attention to that. So far, more emphasis has been placed on Facebook and Instagram, because of the high popularity of these social networks among the young. Specifically, through marketing campaigns on Facebook and Instagram, the Regio-Gnosis' Facebook page has reached over 4.5 million impressions (total count), which is very encouraging for the project.

The aim of the Regio-Gnosis program is to inform people about the EU-funded projects in Greece and highlight their effects on citizens' lives. Road, cultural, social and economic projects that facilitate tourism in Greece and promote cultural heritage are promoted by Regio-Gnosis' site and its pages on social media. In Regio-Gnosis' Facebook page have been conducted four full succeeded photography contests with over 300 participants. The goal of these contests was the public's acquaintance with the various EU-funded projects in Greece and the encouragement to visit them for photographing in order to take part in the contest. Except from online actions, Regio-Gnosis program has implemented cultural events in the four selected regions which are also promoted in social media. According to (Giudici et al., 2013) given the greater importance for tourists of a deeper understanding of holiday destinations, and the search of unique, involving and authentic experiences, and of culture as a driver to obtain sustainability, leads on a sustainable tourism perspective. Regio-Gnosis program not only promoted the EU funded projects, but through that, encouraged cultural tourism through cultural events and photography contests.

2 Methodology

As social media presence growths it is vital to clearly define the marketing objectives, select the right social media platform, design the suitable content, engage the right users to spread the message and be able to evaluate the impact of social media marketing activity. The key objective is to maintain and expand the brand's positioning in social channels while minimizing negative impacts (Killian & McManus, 2015). A crucial factor that could augment popularity of an enterprise or an organization is through social media campaigns which have a critical role in promotion of social networks or the organization's website. Advertising is the main way businesses interact with consumers, so building attractive campaigns could raise both interaction and popularity. The most important part for a successful marketing strategy is the creation of campaigns that serve the purpose of the organization either with the purchase of products or the customer growth or in our case, the raise of information about cohesion policy in Greece through various actions. There are many social networks with a really large number of users nowadays, but Facebook is the most popular.

Facebook has almost 5 million users in Greece which population is about 10 million people. Traffic and engagement are high, and studies show that consumer behavior is influenced by Social Media. Facebook is a platform designed around the user experience, and advertisements do not differ from this feature. Facebook advertising appears in a variety of formats such as text, banners (images–photos) and videos. International studies show that Facebook is a great tool for increasing a brand's recognition, awareness for a new service–product or theme, the increase of users, strengthening loyalty to the public and increasing interactions. An important feature of Facebook advertising is the possibility of targeting the appropriate target group, not only based on language and country but also based on gender, age, educational level or even hobbies. Another important factor is the ability of campaigns to run across both Facebook and Instagram. All these features contribute to the best results of an advertisement.

Two main types of Regio-Gnosis' campaigns were promoted in social media, especially on Facebook. The first type was the promotion of the photography contests that took place through social media in four selected regions (Western Greece, Central Macedonia, Attica, Epirus). The strategy followed for the promotion of the above contests was the same. Initially, the campaign was promoted to the concerned region and in the next stage the promotion was expanded to neighbor regions aiming for audience expansion. The second type of campaigns that the Regio-Gnosis project created was about the promotion of workshops and cultural events that would take place in the four selected regions. Promotion of these events was aiming to increase event attendance. Both campaigns have been targeted to increase the visibility of the implemented projects funded by the EU, to highlight the impact/benefits (social, economic, cultural etc.) of these projects to the local communities and to encourage civilian participation, engagement and social dialogue in relation to the projects.

Regio-Gnosis project had to meet a target of attracting 2.5 million users on Facebook and Instagram, target which has been achieved. In this paper, we will focus on Facebook campaigns and the way the project achieved its target of informing people about EU funded projects in Greece through social media. Photography contests and promotion of program's cultural events would encourage both the information and the raise of touristic and cultural concern. Campaigns were based on the promotion of the program's actions to a specific audience, mainly young people and in the area in which each action took place. Specifically, the program implemented 4 photography contests, one for each region, promoting in this way the EU funded projects. Targeting the audience who are interested in photography and travel is a factor that increased the effectiveness of advertising.

Apart from promoting a campaign on Facebook and choosing the right audience, an important element is the type of the campaign. The content of a campaign needs to be relevant and interactive, like, for example, a Facebook contest. The right planning and execution of a contest, especially with interesting gifts can clearly attract much more audience than promoting a workshop or a cultural event. For this reason, special emphasis was placed on finding valuable gifts for the photography contests that would manage to attract the audience to participate. Through this, the audience would be encouraged to travel for taking a picture and participating to the photo contest discovering both the EU funded projects and the visiting area. Contests related to the page's subject with attractive gifts are also being helpful to raise the page's followers.

3 Results

An exploratory research was conducted using Facebook Insights as the main data collection tool. In the Fig. 1, we demonstrate the steps followed to evaluate the presence of Regio Gnosis project on Facebook.

Initially, the social media manager of Regio Gnosis program creates written or visual engaging and attractive content in order to promote our program's actions (Fig. 2). Afterwards, the Page administrators check Facebook Audience Insights which provide aggregated information about people connected to the page and people

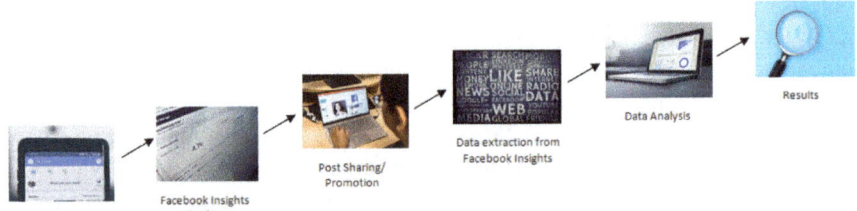

Fig. 1 Schematic representation of the evaluation process

Fig. 2 Post creation

on Facebook so they can share the content in order to find people matching the current audience.

In Fig. 3, Facebook Insights show the day and the time frame in which the majority of the followers are online and therefore the post will have a greater impact. Based on this information, the Regio-Gnosis page's administrators make sure to check this graph before sharing each post for greater performance. After post sharing, according to our project's goals, the social media manager decides whether the post needs to be promoted.

A critical step in our analysis is the data extraction using Facebook Insights. Figure 4 shows the growth of the page likes from the page's creation until today. Specifically, for the Regio-Gnosis' page managed to reach 6000 likes within a period of almost 1 year (Figs. 5, 6, 7 and 8).

The main posts of the Regio-Gnosis' page mostly concern photo contests, cultural events, workshops, live streaming videos and photos of these events. After the data analysis of Facebook Insights, the posts with the greatest impact were the photo contests. Indicatively, the first photography contest (Fig. 5) with subject "The Inter-action of the Rio-Antirio Charilaos Trikoupis Bridge with Air, Water, Land and Light" conducting from 30/06/2019 until 30/09/2019 contributed to the growth of the newly formed page of Regio-Gnosis. In particular, at the beginning of the 1st

Fig. 3 When the fans are online

Fig. 4 Total page likes

contest the Page "Regio_Gnosis" had 628 likes while at the end of all contests almost tripled the total likes (Fig. 4).

During the second photography contest (Fig. 6) which was about the New Waterfront of Thessaloniki and held from 07/09/2020 until 31/10/2020, page's likes were increased by around 3200, an enough satisfying number. It's important to mention that the photo campaigns (Fig. 9) as well as the video campaigns (Fig. 10) contributed not only to the increase of page likes but also to the raise of page engagements.

Subsequently, we used data mining techniques to discover and extract useful patterns and correlations in our data. In particular, we applied decision trees which is a widely used method in machine learning and association rule mining. Data Mining is an emerging knowledge discovery process of extracting previously unknown, actionable information from very large scientific and commercial databases and drawing useful conclusions, especially in the case of marketing and advertising. Usually, a

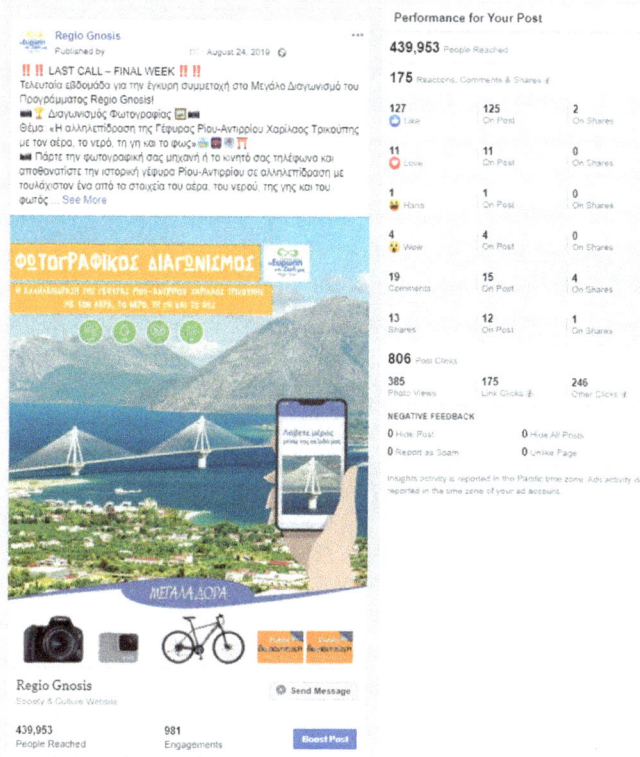

Fig. 5 The performance of the announcement of the 1st photography contest

data mining process extracts rules by processing high dimensional categorical and/or numerical data. The steps of the knowledge extraction process consist of (a) Selection, (b) Preprocessing, where in this phase the purpose is to clean them, i.e. to correct incorrect, problematic or missing data, (c) Data Transformation, where in this phase the data is smoothed and noise is removed, for the aggregation of the data, i.e. for the production of their summary, for their normalization, i.e. for the scaling of the characteristics of the data set in a specific and limited price range or to create new features from existing ones, (d) Data Mining, using algorithm (s) to produce a model and (e) Interpretation and Evaluation, to evaluate the results, which were produced throughout the process (Fig. 11).

In addition, in the data analysis we performed, we also checked the case of outliers. Outlier detection is the identification of objects, events, or observations that do not conform to an expected pattern in relation to the data set. In the rapidly changing business environment, it is imperative for companies to analyze their multiple and complex data and look for outliers that can reveal important information about the operation of a business, leading to saving money or creating new business opportunities. In our dataset, outliers mostly indicate the posts that have really large reach to

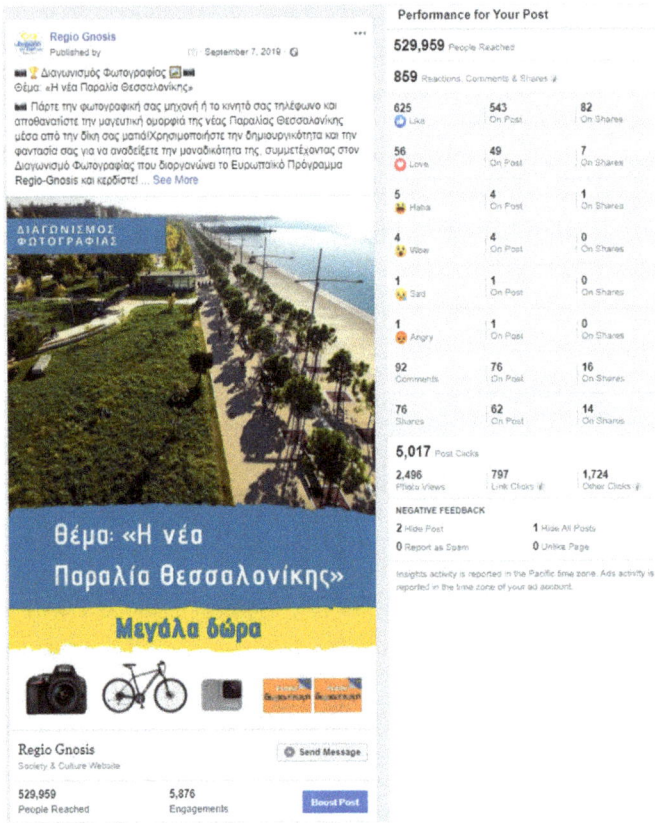

Fig. 6 The performance of the announcement of the 2nd photography contest

the public because of the paid advertisement. These outliers affect the assumptions and interfere the analysis, so we removed them from the dataset.

In the next stage we proceeded to the application of classification analysis. Classification is one of the most important data mining tasks and it is a supervised learning task. Displays data in predefined groups or classes. One of the most well-known classification methods is Decision Trees: These are tree structures that graphically represent the relationships of data. Decision trees are used to categorize and predict data. A decision tree is constructed according to a set of pre-categorized data training. Each of its internal nodes identifies the control of the attributes. Each branch that connects the internals with the offspring corresponds to a possible value for the trait. The technique of decision trees is quite common in the field of data mining. In the present work we applied the most common algorithm, J48. Below we present a decision tree created based on the WEKA Software (Waikato Environment for Knowledge Analysis, data mining software is one of the most popular machine learning software, it is open source software and is written in Java programming language) and captures

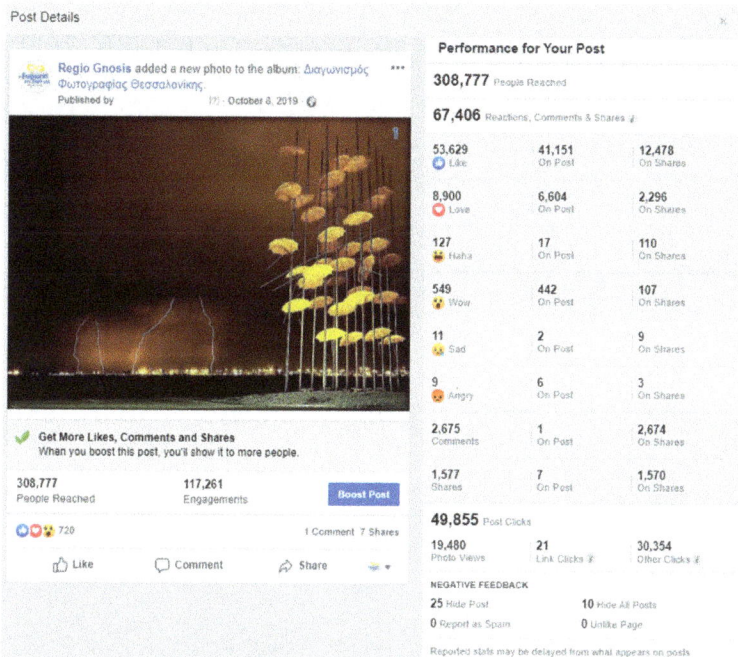

Fig. 7 Performance of the 2nd photography contest's call for vote

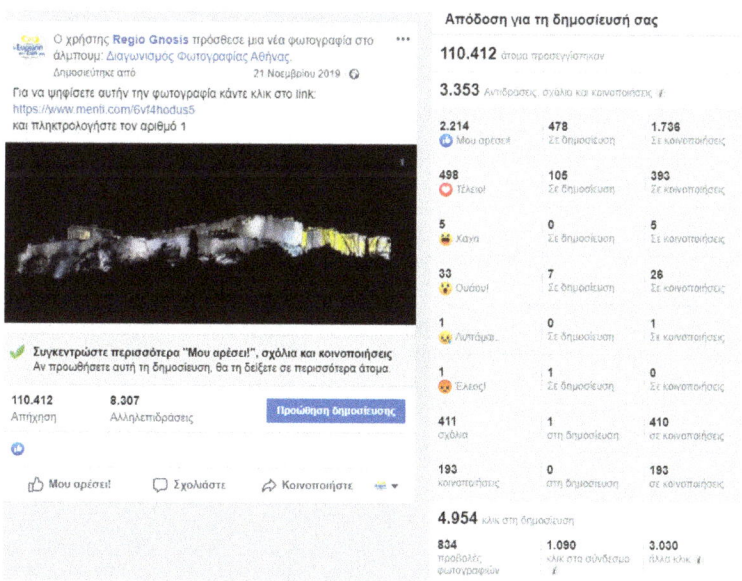

Fig. 8 Performance of the 3rd photography contest's call for vote

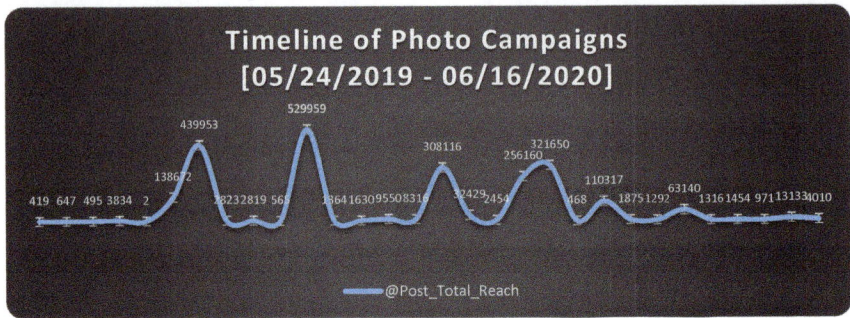

Fig. 9 Timeline of photo campaigns

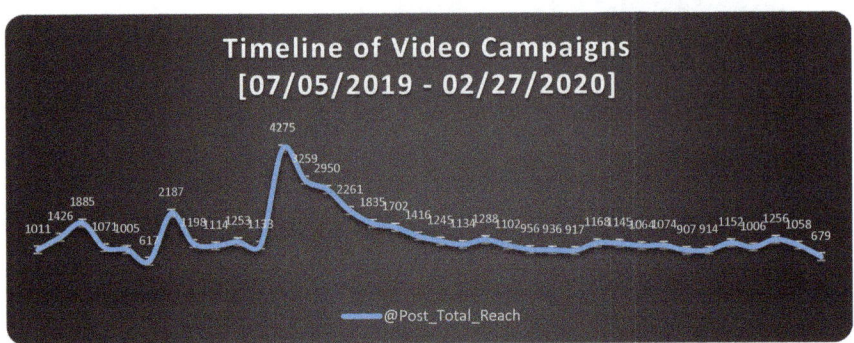

Fig. 10 Timeline of video campaigns

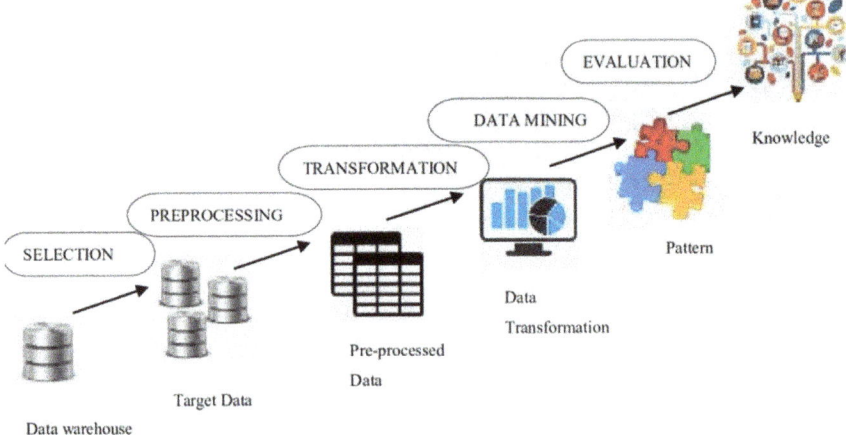

Fig. 11 Data mining typical procedure

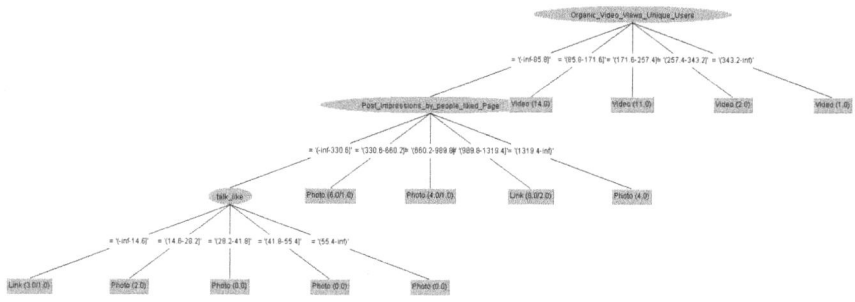

Fig. 12 Decision tree (J48)

by category of advertising campaign all the important relationships extracted based on the J48 algorithm (Fig. 12).

Detailed Accuracy By Class ===

	TP Rate	FP Rate	Precision	Recall	F − Measure	MCC	ROC Area	PRC Area	Class
	0.824	0.184	0.667	0.824	0.737	0.608	0.845	0.611	Photo
	0.333	0.065	0.500	0.333	0.400	0.318	0.801	0.375	Link
	0.966	0.068	0.815	0.818	0.811	0.748	0.905	0.769	Video
Weighted Avg	0.818	0.068	0.815	0.818	0.811	0.748	0.905	0.769	

In the final stage of data analysis, we proceeded with application of association algorithms. Association Rule Mining is a common technique used to find associations between many variables. In Data Mining, Apriori is a classic algorithm for learning association rules (Fayyad et al., 1996, Boutsinas et al., 2008). Apriori is designed to operate on databases containing transactions (for example data collected from surveys in this case) (R Development Core Team 2008). As is common in association rule mining, given a set of item sets, the algorithm attempts to find subsets which are common to at least a minimum number C of the item sets.

Apriori uses a "bottom up" approach, where frequent subsets are extended one item at a time, and groups of candidates are tested against the data. The algorithm terminates when no further successful extensions are found. Apriori uses breadth-first search and a tree structure to count candidate item sets efficiently. It generates candidate item sets of length k from item sets of length k-1. Then it prunes the candidates which have an infrequent sub pattern. According to the downward closure lemma, the candidate set contains all frequent k-length item sets. After that, it scans the transaction database to determine frequent item sets among the candidates.

Association rules present association or correlation between item sets. An association rule has the form of A "B, where A and B are two disjoint item sets.

The Goal: studies whether the occurrence of one feature is related to the occurrence of others. Three most widely used measures for selecting interesting rules are:

- Support is the percentage of cases in the data that contains both A and B,
- Confidence is the percentage of cases containing A that also contain B, and

- Lift is the ratio of confidence to the percentage of cases containing B.

Apriori
========
Minimum support: 0.85 (47 instances)
Minimum metric < confidence > : 0.9
Best rules found:

post_share = '(-inf-1.8]' 50 ==> talk_share = '(-inf-1.2]' 50 < conf:(1) > lift:(1.1)
lev:(0.08) [4] conv:(4.55)
talk_share = '(-inf-1.2]' 50 ==> post_share = '(-inf-1.8]' 50 < conf:(1) > lift:(1.1)
lev:(0.08) [4] conv:(4.55)
post_comment = '(-inf-1.2]' 50 ==> talk_comment = '(-inf-1]' 50 < conf:(1) >
lift:(1.08) lev:(0.07) [3] conv:(3.64)
post_share = '(-inf-1.8]' post_consuptions_link_clicks = '(-inf-7.8]' 49 ==>
talk_share = '(-inf-1.2]' 49 < conf:(1) > lift:(1.1) lev:(0.08) [4] conv:(4.45)
talk_share = '(-inf-1.2]' post_consuptions_link_clicks = '(-inf-7.8]' 49 ==>
post_share = '(-inf-1.8]' 49 < conf:(1) > lift:(1.1) lev:(0.08) [4] conv:(4.45)
talk_comment = '(-inf-1]' post_share = '(-inf-1.8]' 48 ==> talk_share = '(-inf-
1.2]' 48 < conf:(1) > lift:(1.1) lev:(0.08) [4] conv:(4.36)
talk_share = '(-inf-1.2]' talk_comment = '(-inf-1]' 48 ==> post_share = '(-inf-
1.8]' 48 < conf:(1) > lift:(1.1) lev:(0.08) [4] conv:(4.36)
post_comment = '(-inf-1.2]' post_consuptions_link_clicks = '(-inf-7.8]' 48 =
=> talk_comment = '(-inf-1]' 48 < conf:(1) > lift:(1.08) lev:(0.06) [3] conv:(3.49)
talk_share = '(-inf-1.2]' post_comment = '(-inf-1.2]' 47 ==> talk_comment =
'(-inf-1]' 47 < conf:(1) > lift:(1.08) lev:(0.06) [3] conv:(3.42)
post_share = '(-inf-1.8]' post_comment = '(-inf-1.2]' 47 ==> talk_share = '(-
inf-1.2]' 47 < conf:(1) > lift:(1.1) lev:(0.08) [4] conv:(4.27)

We also have generated the Correlation Matrix (Fig. 13) to exploit how the variables are related to one another. The most important finding is that the number of unique users who created a story about a page post by interacting with it (variables: @talk_share, @talk_link, @talk_comment) is highly correlated with both the number of unique users who engaged the page post and the number of unique users who matched the audience targeting and clicked anywhere in the page post on News Feed (variables: @engaged_users, @match_audience_targeting_consumers_on_post). Proportionally, the number of stories created about a page post (variables: @post_share, @post_link, @post_comment) is highly correlated with both the number of unique users who engaged the page post and the number of unique users who matched the audience targeting that clicked anywhere in the page post on News Feed. In addition, we observe a strong correlation between the number of unique users who engaged the page post and the number of unique users who matched the audience targeting that clicked anywhere in the page post on News Feed. Another interesting finding is that

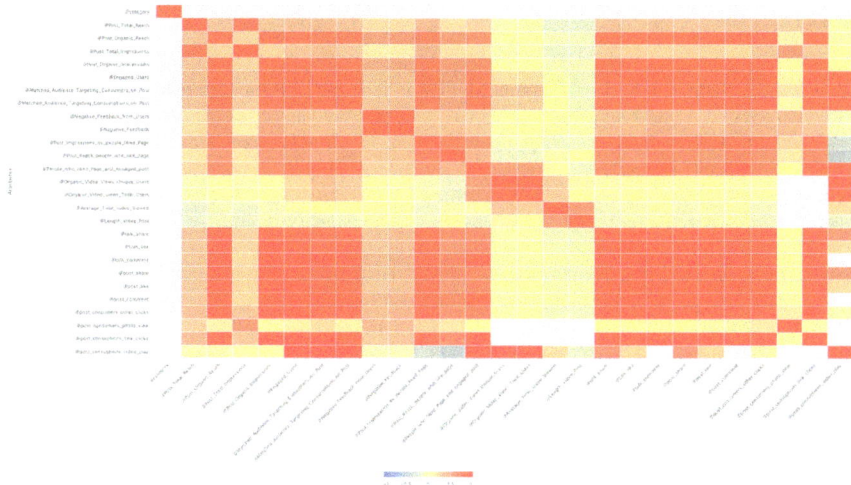

Fig. 13 Correlation Matrix

the number of times people have given negative feedback to a page post (variable: @negative_feedback) is not highly correlated with the number of unique users who engaged the page post (variable: @engaged_users).

4 Conclusion

The study and analysis of data extracted from Facebook Insights could assist a business or an organization to establish a brand or improve the brand positioning and enhance the engagement of its users through the launch of a targeted marketing social media campaign. In our case study, posts related to photography contests contributed the most not only to the increase of page likes but also to the raise of page engagements. From the data analysis derives that Facebook campaigns related to photography contests, matched the audience targeting and led to the raise of the number of unique users who engaged the page posts.

Successful Digital Marketing is a crucial factor for a business or an organization to reach its audience. Through Digital Marketing, Regio-Gnosis program not only promoted the EU funded projects, but also encouraged cultural tourism through cultural events and photography contests. As a new kind of tourism is characterized by continuously searching for new places and new experiences, it's important for a social media page to motivate the public to visit places that may not be widely known by enhancing their interest in exploring and discovering. Facebook advertising can be very useful to that adjusting the target audience according to their interests or hobbies. There is a general need for continuing research into social media marketing strategies and their impact on reaching large audiences.

References

Ansah, A. K., Blankson, V. S., & Kontoh, M. (2012). The use of information and communication technologies (ICT) in front office operations of chain hotels in Ghana. *International Journal of Advanced Computer Science and Applications, 3*(3), 72–77. Retrieved from http://dx.doi.org/10.14569/IJACSA.2012.030313

Aziz, A. A., Bakhtiar, M., Syaquif, M., Kamaruddin, Y., & Ahmad, N. (2012). Information and communication technology application's usage in hotel industry. *Journal of Tourism, Hospitality, and Culinary Arts, 4*(2), 34–48.

Boutsinas, B., Siotos, C., & Gerolymatos, A. (2008). Distributed mining of association rules based on reducing the support threshold. *International Journal on Artificial Intelligence Tools, World Scientific Publishing Company, 17*(6), 1109–1129. Retrieved from https://doi.org/10.1142/S0218213008004321

Dessart, L., Veloutsou, C., & Thomas, A. M. (2015). Consumer engagement in online brand communities: a social media perspective. *Journal of Product and Brand Management, 24*(1), 28–42. Retrieved from https://doi.org/10.1108/JPBM-06-2014-0635.

Fayyad, U. M., Piatetsky-Shapiro, G., & Smyth, P. (1996). In *Advances in knowledge discovery and data mining*. AAAI Press/MIT Press.

Giudici, E., Melis, C., Dessı, S., & Ramos, B. F. (2013). Is intangible cultural heritage able to promote sustainability in tourism. *International Journal of Quality and Service Sciences*. 101–114. Retrieved from https://doi.org/10.1108/17566691311316275.

Gummerus, J., Liljander, V., Weman, E., & Pihlström, M. (2012). Customer engagement in a Facebook brand community. *Management Research Review, 35*(9), 857–877. Retrieved from https://doi.org/10.1108/01409171211256578.

Halkiopoulos, C., Giotopoulos, K., Papadopoulos, D., Gkintoni, E., & Antonopoulou, H. (2019). Online reservation systems in e-business: Analyzing decision making in e-Tourism. *Journal of Tourism, Heritage and Services Marketing (JTHMS)* Special Issue (ISSN: 2529–1947).

Kannan, P. K. & Li, H. (2017). Digital marketing: A framework, review and research agenda. *International Journal of Research in Marketing*. 22–45. Retrieved from https://doi.org/10.1016/j.ijresmar.2016.11.006.

Katsoni, V. (2011, December). The role of ICTs in regional tourist development. *Regional Science Inquiry Journal, 3*(2)

Keegan, B. J., & Rowley, J. (2017). Evaluation and decision-making in social media marketing. *Management Decision, 55*(1), 15–31. Retrieved from https://doi.org/10.1108/MD-10-2015-0450

Killian, G., & McManus, K. (2015). A marketing communications approach for the digital era: Managerial guidelines for social media integration. *Business Horizons, 58*(5), 539–549. Retrieved from https://doi.org/10.1016/j.bushor.2015.05.006

Kumar, V., & Mirchandani, R. (2012). Winning with data: Social media-increasing the ROI of social media marketing. *MIT Sloan Management Review, 54*(1), 55–61.

Lee, M. & Youn, S. (2009). Electronic word of mouth (eWOM) how eWOM platforms influence consumer product judgement. *International Journal of Advertising, 28*(3), 473–499. Retrieved from https://doi.org/10.2501/S0265048709200709

Tsimonis, G., & Dimitriadis, S. (2014). Brand strategies in social media. *Marketing Intelligence and Planning, 32*(3), 328–344. Retrieved from https://doi.org/10.1108/MIP-04-2013-0056.

Van der Borg, J., Minghetti, V., & Riganti, L. (1997). The attitude of small and medium sized tourism enterprises towards information and telecommunication technologies: The case of Italy. pp. 286–294. New York, NY: Springer. Retrieved from https://doi.org/10.1007/978-3-7091-6848-6_30

Van der Zee, E., Vanneste, D. (2015). Tourism networks unraveled; A review of the literature on networks in tourism management studies. *Tourism Management Perspectives 15*, 46–56. Retrieved from https://doi.org/10.1016/j.tmp.2015.03.006

Challenges and Opportunities for the Use of Indoor Drones in the Cultural and Creative Industries Sector

Virginia Santamarina-Campos, María de-Miguel-Molina, Blanca de-Miguel-Molina, and Marival Segarra-Oña

1 Context and Overall Objectives

Nowadays, Aerial filming and photography are indispensable resources for the Creative Industries (CIs) since they expand their creative possibilities. The options for filming special shots from different angles and heights arise as a competitive advantage for them. For example, film directors like tilting or panning in movies when filming actors stepping out of a car or leaving a building. Until now, for shots like these, helicopters needed to be rented which resulted in high costs and only financially sound companies could afford them. Or, if the scene didn't require high angle shots but just a shot from a moderate height, scaffolding, jibs or lifting platforms were usually hired or bought. With the rise of the drone industry and the resulting drop of acquisition costs for those "flying robots", the Creative Industries are employing more and more drones for aerial filming. This is quite easy to understand, since there are several advantages as cost reduction, flexibility, less risky and also less invasive. However, professional drone use by CCIs is limited to outdoor applications as existing drones lack a precise, robust and affordable indoor positioning system (IPS), as well as advanced safety features. Positioning systems are key for drone flights. When filming outdoors, drones use a GPS, which not only helps the operator to locate the drone and to program flight paths, but also to stabilise the drone while avoiding drifting issues. This is an important factor for camera operators, but GPS localisation does not work indoors. Firstly, the signal is too subdued, and secondly, even if it were not so, the GPS's precision is far too inaccurate with a meter range margin of error.

V. Santamarina-Campos (✉)
Department of Conservation and Restoration of Cultural Heritage, Universitat Politècnica de València, València, Spain
e-mail: virsanca@upv.es

M. de-Miguel-Molina · B. de-Miguel-Molina · M. Segarra-Oña
Universitat Politècnica de València, Management Department, València, Spain

© The Author(s), under exclusive license to Springer Nature Switzerland AG 2021 709
V. Katsoni and C. van Zyl (eds.), *Culture and Tourism in a Smart, Globalized, and Sustainable World*, Springer Proceedings in Business and Economics,
https://doi.org/10.1007/978-3-030-72469-6_47

Be that as it may, CCIs would also like to use drones indoors. Unfortunately, there is not safe and affordable technology available for the CCIs to do so. The technologies available to work in confined spaces are the following:

(1) The motion capture system, that has a very reliable and precise (mm-range) IPS. However, this system is very expensive (> 200,000 euros) and it needs several auxiliary devices, which as a result makes setting up time consuming.
(2) The vision position system, which is affordable. However, how it functions depends on its surroundings. For instance, on transparent or reflective surfaces the device might malfunction. Moreover, localisation is just in z-axis and with a maximum altitude of 2.5 m.

To overcome these issues, within the AiRT project, we have developed a drone which includes a new indoor positioning system with cm range precision and an intelligent flight control system. Additionally, the integration of active and passive safety measures guarantees safe and user-friendly experiences for the drone operator. Also, an important consideration should be made to the time needed for assembling and disassembling the filming set, which can be dramatically reduced. Data obtained from the performed demonstrations tests showed results of "ready to fly settings" of 20 min and just 10 min for disassembling. Another interesting feature of this innovative project is that self-calibration is included, which also reduces time needed for setting-up the filming set.

The professional camera control will make the AiRT drone an attractive tool (Fig. 1), especially for the Creative Industries.

This product, with its innovative features, will allow most of small and medium-sized companies to provide new services which, in turn, will increase their chances to grow within the European and international market.

Fig. 1 Detail of the use of AiRT software by experts *Source* AiRT project

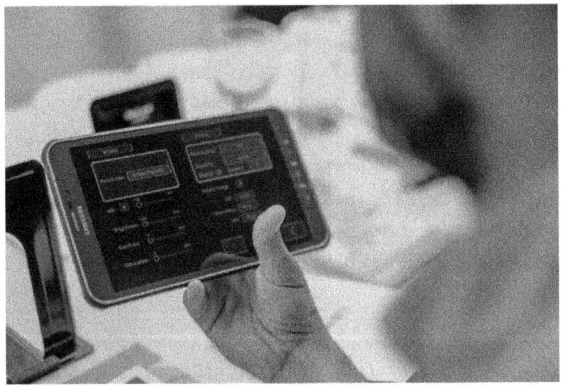

2 Work Performed and Results Achieved

Four interconnected phases have been developed in our approach to transfer the AiRT system to the creative industries in an efficient way (Fig. 2):

(a) Identification of needs
(b) Adaptation and optimisation of different technologies
(c) Integration and validation of the system
(d) Evaluation and demonstration of the system

At the same time, our methodology is end-user centred with the goal of using technological innovation to provide feasible solutions to the Creative Industries. To do it, this technique relies on three fundamental pillars (Fig. 3):

Fig. 2 Approach of AiRT in four phases *Source* Own elaboration

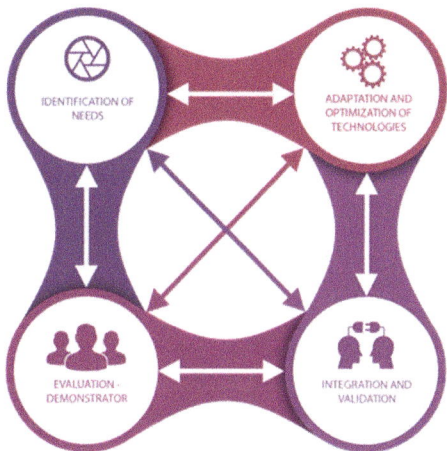

Fig. 3 The three pillars. Requirements to be satisfied in AiRT project methodology *Source* own elaboration

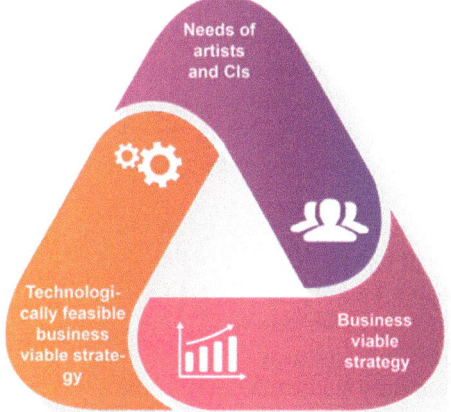

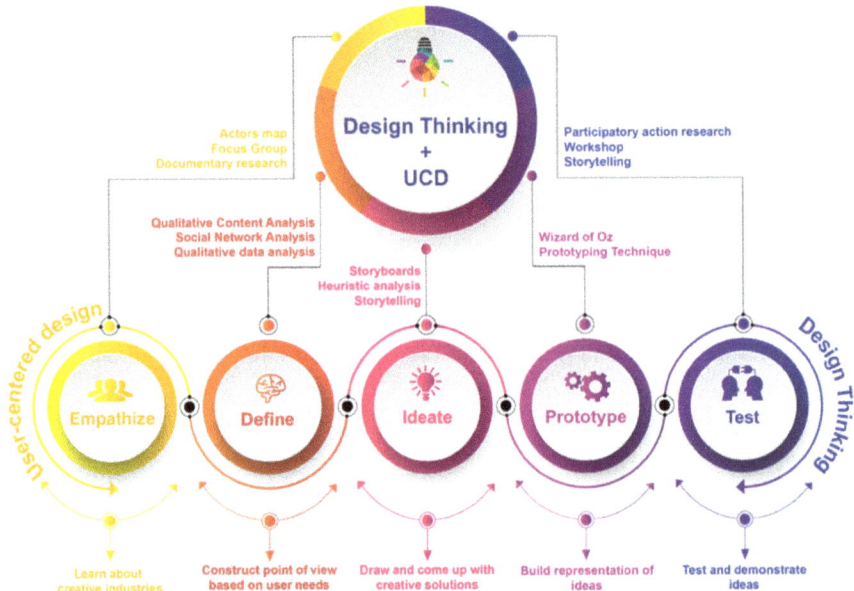

Fig. 4 Design thinking method applied to AiRT project *Source* Own elaboration based on Both (2009)

- Empathise with creative industries to understand their problems and requirements.
- Generate collaborative activities with end-users. Feedback from all the different stakeholders will be considered for incorporation.
- Prototype development validated by end-users. The evaluation may lead to improvements before producing the final product.

After identifying these three pillars, we applied the design thinking methodology (Rowe, 1991, Both, 2009, Dorst, 2011), where the participation of the CCIs is present throughout the process (Fig. 4), to AiRT's project.

This resulted in five interrelated phases in which the end user is always involved: empathise, define, ideate, prototype and test.

In the first phase three focus groups (Liamputtong, 2011) were held in three different countries-Spain, the UK and Belgium—to identify customers' needs. From these meetings, we were able to obtain information from thirteen different sectors of the CCIs.

Informants (experts) were chosen strategically from 13 different sectors to guarantee that all the Creative Industry's needs are detected, including experienced drone pilots working in different CCI sectors (40% of the informants). The Photography sector was the most numerous (21% of the total), followed by Movie and Advertising (19%), being these three sectors the most directly related to the industries that are currently leading the use of drones.

From the analysis of the information gathered in these sessions, we obtained the needs, the ethical issues and the risk analysis. Results from focus groups were obtained by using a Qualitative Content Analysis method. Qualitative data was obtained through the analysis of interviews of participants in focus groups, which were recorded and transcribed. The software QDA Miner (Provalis Research) was used to undertake the analysis and all the information in the transcriptions was codified. Codes were organized in relation to the predefined questions, trying to guide the focus group participants' answers to these questions.

Additional codes were defined to cover information that was also considered to be important and, moreover, strategies for the commercialisation phase. That is why results from focus groups have been incorporated in the basic design of the drone, the development of the European Policy book and the redefinition of the exploitation plan.

In the Third phase, from the synthesis of the information obtained in the focus groups, we wrote a script and a breakdown of that script which were used to apply the storyboard method (Van der Lelie, 2006), representing the use of the AiRT system in different creative settings, which allowed us to convey our main ideas more clearly by applying scientific knowledge to a real environment (Maguire, 2001).

On the other hand, by using the storyboard, we were able to extract the requirements to define the functionalities of the AiRT system, which will include the Ground Control Station (GCS) software, the intelligent flight control system and the final design of the drone.

The main focus of the fourth phase was on the integration process. All of the components of the AiRT system have been tested and efforts were devoted to the system integration and technical validation.

In the last phase, we did the demonstration of the AiRT system with the collaboration of the CCIs. Through the use of PAR, Participation action research (Whyte, 1991), we carried out the demonstration in real environments chosen by the users.

Finally, after the demonstration, we presented the AiRT system at the workshop celebrated in Valencia in June 2018, and at the ICT2018 event in Vienna in December 2018, not only to the CCIs but also to other sectors of the industry. The different research methodologies, problems and solutions, future benefits and applications, main features of the innovative product were explained and also a demonstration was developed to show the objectives achieved by the AiRT team.

The methodology used has allowed to carry out a responsible innovation process, in which the creative industries, the business sector and the research community have participated together throughout the project process. Open access to the results obtained, gender equality tried to be achieved in all the activities, scientific education through the dissemination of results, the definition of ethics in the area of drones and participatory governance have been pursued.

3 Progress Beyond the State of the Art

Four major technical innovations have to be highlighted in the AiRT Project (Fig. 5):

(1) The RPAS itself, especially designed for professional indoor use by CCIs with passive and active safety measures that integrates (1) a user-friendly fully automated flight control system, (2) an IPS based on Ultra-wideband technology (UWB), (3) an on board RGB-D camera which allows 3D reconstruction of the space prior to the flights for filming, (4) a new flight mode system (from totally manual flight to totally autonomous flight) as well as (5) internal protocol's communication among components were developed considering secure emissions as the main variables (instead of interoperability or speed, that are the current mainly used protocols).

(2) The IPS, based on novel UWB wireless radio technology, especially suitable for indoor spaces.

(3) The 3D mapping software for indoor environments, that allows 3D reconstruction of the indoor space (3D model), achieved in real-time from data that it is collected from the RPAS prior to the filming process.

(4) Multiplexers approach to integrate a number of sensors and systems for managing a complex data flow and providing an adequate decision-making process on operation.

Apart from technological innovations, the AiRT system also addresses socially relevant topics, such as helping to suppress risky auxiliary means and thus improving the safety of the workers. Users will not need auxiliary devices which present unnecessary risks to their safety or the safety of the workers on set.

Fig. 5 Main technical innovations of the AiRT project *Source* Own elaboration

Fig. 6 Lean management of the airt project *Source* Own elaboration

To conclude, AiRT contributes, not only to the economic growth of CCIs, but also to other sectors and thus to European economic prosperity. Therefore, industries will be able to offer new services to their clients, helping them expand within the European market and as a result, employment within CCIs and other sectors will be secured and new jobs created.

This subdivision led to the AiRT workplan, which consisted of 7 work packages (WP). In the following sections we describe in more detail the progress and achievements we have made during the project—in relation to each work package.

Work Package 1 Project management and coordination

This work package develops activities which have ensured the smooth workflow of all the WPs and their compliance with the overall work plan.

As communication channel between partners the consortium agreed to use the web-based project management tool Basecamp. All relevant documents (deliverables, drafts, minutes, photos, etc.) have been uploaded and made available to each member of the consortium. Moreover, on 31 March 2018 a Quality Plan (QP) was elaborated which played a big part in the success of this project. In this work package, the biggest achievement we have had is to accomplish lean management (Fig. 6) by establishing open communication even if the consortium is multi-disciplined.

Work package 2 Analysis of CCIs needs, ethical/security issues and risk analysis

In workpackage two, through three focus groups from Spain, the UK and Belgium, we were able to obtain information from thirteen different sectors of the creative industries of which forty percent were drone pilots (Fig. 7).

As a result, we obtained the needs analysis, the ethical issues and the risk analysis (Fig. 8). Furthermore, on one hand, the design of the drone, the GUI interface and the development of advanced functionalities were based on the results. On the other, it was useful for the development of the European policy book and to enrich the Exploitation plan (Fig. 9).

Fig. 7 Participants focus group in luton *Source* AiRT project

Fig. 8 Analysis and results
of the focus groups *Source*
Own elaboration

Fig. 9 Analysis and results
of the focus groups *Source*
Own elaboration

Fig. 10 Detail of the automatic calibration of the IPS in AiRT software *Source* AiRT project

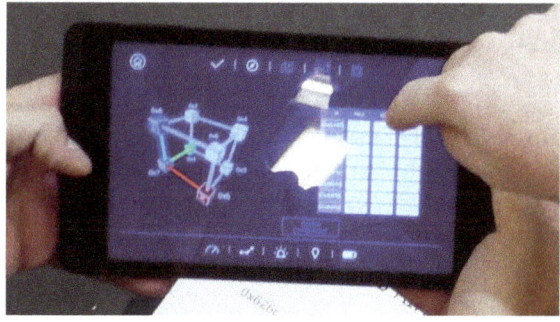

Work package 3 Analysis of CCIs needs, ethical/security issues and risk analysis

This WP is mainly dedicated to adapt the indoor positioning system (IPS) from Pozyx to the drone. The work done can be divided into three parts:

(a) **The Indoor Positioning System**

The Pozyx positioning system was adapted for use on drones. The adapted drone positioning system consists of a central processing unit connected to four ultra-wideband (UWB) units to be mounted on the corners of the drone. The four UWB units consist of a small wireless transceiver and an antenna. To make optimal use of the novel hardware, a specialized positioning algorithm was developed and an improved system for automatic calibration of the IPS was developed to automatically determine the positions of the IPS infrastructure (Fig. 10).

(b) **Communication IPS and drone**

For the development of WP3, the main task assigned to AeroTools has been to provide technical support to Pozyx (e.g.communication between IPS and RPAS, providing support and performing functioning tests of the IPS in RPAS environment, mechanical requirements of the tags to a better adaptation to the RPAS, etc.).

The support has been provided in the form of both technical knowledge and material resources and spaces dedicated to perform required tests.

(c) **On-board Control System and Mapping**

During the development of the project, the main results achieved by the Universitat Politècnica de València development team has been achieved, from the hardware design of an On-board Control System (OCS) suitable to be installed on a RPAS, to the implementation, troubleshooting and refining of the (1) communication protocol to transfer the commands from the OCS to the Gimbal controller, (2) the communication protocols (defined by the recording camera manufacturer) to transfer the commands from the OCS to the Camera controller or the troubleshooting and refining of the communication protocol (defined by Pozyx) to transfer the IPS information from IPS to the OCS, and to calibrate the IPS from the GCS. Also the secure the transmissions between the OCS and the GCS has been achieved.

Work package 4 RPAS Design and optimization

The main goal of the WP4 was the definition and manufacturing of a flexible flying platform that meets the needs of the Creative Industries (end-users), set by the focus groups. The final design considered parameters like weight, easy to deploy, flexible platform capable of allocating a complex electronic architecture and outstanding performer drone. Considering the RPAS design, the AIRT drone is a solid and reliable flying platform, very stable and gentle when moving, as it has demonstrated during the test flights.

Work package 5 Integration, validation and demonstration

This WP5 can be divided into 4 main tasks:

(a) **The GUI (graphical user interface)**

Figure 11 shows the five non-linear phases that have been applied to develop the AiRT software: empathise, define, ideate, prototype and test.

- **1st phase. Empathise:** This phase started with the Focus Groups activities in the three participating countries (Spain, UK and Belgium). There, the topic of "software" was included in the discussions.
- **2nd phase. Define:** The evaluation of the 1st phase was carried out, using the Qualitative Content Analysis, Social Network Analysis (SNA). This process is known as the define phase.

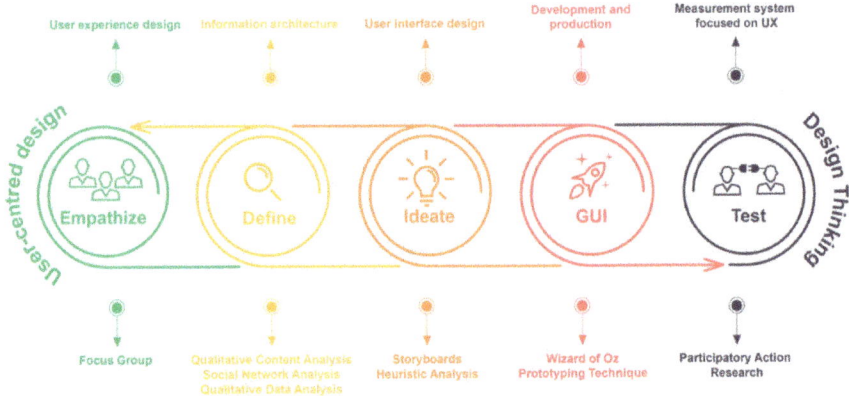

Graphical User Interface's methodology

Fig. 11 Methodologies used in the design of the GCS software and GUI *Source* Own elaboration, adapted from (Both 2010)

Fig. 12 Participation, action, and research interoperation generates PAR methodology *Source* own elaboration

- **3rd phase. Ideate:** In the third phase, based on the synthesis of the information obtained in the Focus groups, written scripts were prepared and then transferred to Storyboards representing the possible functionalities of the GCS software in different creative scenarios. This helped to communicate the main ideas and needs more clearly. This step included also the analysis of available software on the market. In this phase the requirements of the software were defined more precisely, based on simulated real case scenarios (the storyboards).
- **4th phase. Prototype:** Once the heuristic evaluation of the flight plan software available in the market was completed, and the design elements extracted from the Storyboards were defined (user interface of the client application), an iterative design process was initiated using the Wizard of Oz Prototyping Technique (Both, 2010). To do this, the models for the GUI were designed, first using paper prototypes as visual language and then the online tool NinjaMok© (Figs. 13, 14a and 15b). All these results have been integrated in the final user interface.

Fig. 13 Detail of the lollipop icons (obverse) *Source* AiRT project

Fig. 14 Detail of the use of AiRT software by experts *Source* AiRT project

Fig. 15 'Recall' procedure, applied for both communication and dissemination monitoring and evaluation *Source* Own elaboration based on Kusek and Rist (2004)

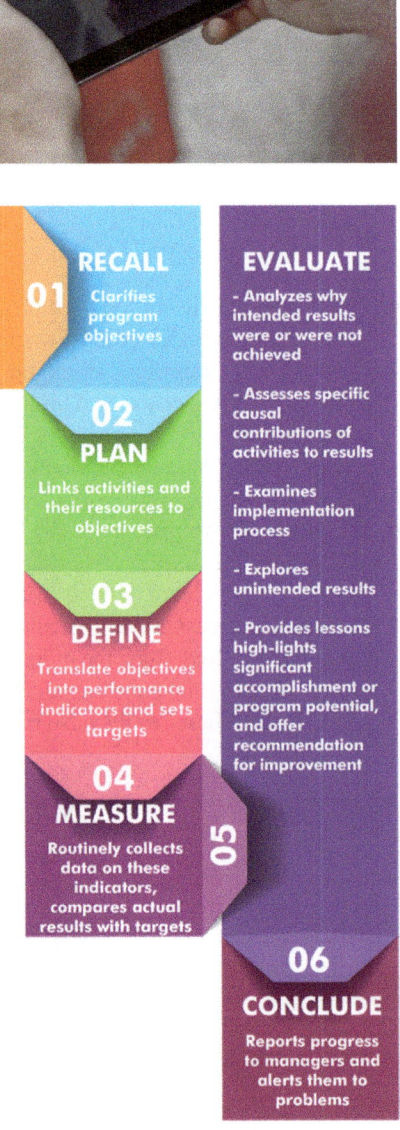

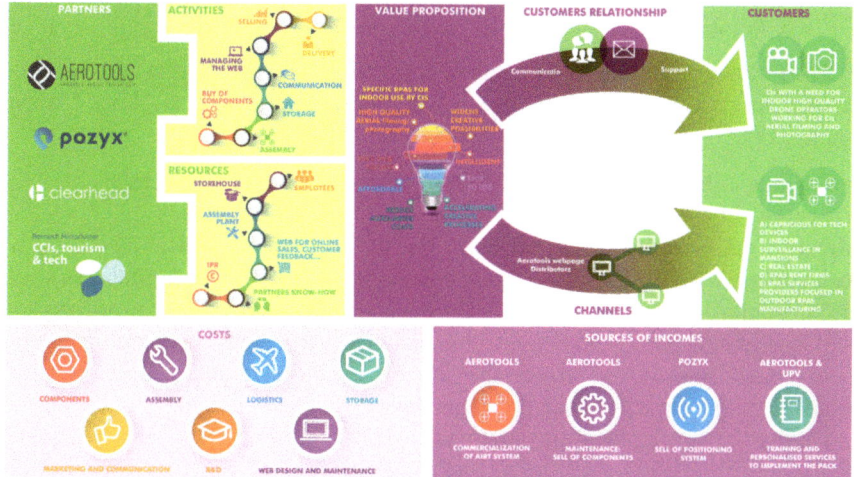

Fig. 16 canvas business model of the AiRT project *Source* Own elaboration

- **5th phase. Tests:** In this last phase, end users made use of the prototypes, based on the selection of relevant scenarios for the creative industries, in the three participating countries. The objective of this stage was to identify failures or to provide new improvements, through the Participation Action Research tool (PAR). The evaluation of this step helped to identify potential usability issues, and to check compliance with previously established usable design principles.
- The results of the UCD are presented in the following sections.

(b) **Design and implementation of a multiplatform Ground Control System (GCS)**

The objective of this task, led by UPV, was to develop an application for allowing users to command remotely the RPAS via the OCS and to receive and display telemetry information from the drone. We designed a new workflow for helping users of the creative industry to design their desired shots within a friendly, easy to use and complete interface. The application has been developed using a framework that allows us to build it for different platforms, mainly Android and Windows.

The AiRT Software GUI is feature complete. It contains all the planned functionality and some new functionality that has arisen during the development. The interactive and autonomous control of both the recording camera and the RPAS itself has been integrated. The GUI allows the end-users to control all parameters of the AiRT system in a simple and intuitive way. The application is divided into four sections, for allowing the user to perform all the tasks available in AiRT:

1. Configuration
2. Mapping
3. Planning

4. Recording

The following sections show some screenshots of each of the four sections. The detailed instructions for using the application are available in the User Manual (zenodo.org/communities/airt_project).

(c) **Integration process of the different components and flight tests**

Although many of the Advanced Functionalities were already working correctly, such as the distance sensors or the safety elements, or the camera & gimbal control, the integration stage affects the whole drone and systems. Therefore, the process and workload in integration and optimization had to be carried out very carefully.

In the last stage of the project, the IPS was tightly integrated with the onboard autopilot in order to achieve autonomous flight.

During the development of the drone several innovations must be highlighted, such as:

- A protocol oriented to a safety automatic flight that allows system to change between manual and automatic flight.
- A modular design and integration of a variety of systems and sensors.
- A Control Multiplexor as a central system that coordinates the OCS with the rest of the electronic areas, like the FCS, sensors, cameras, IPS...
- A camera and gimbal with outstanding performance:
- Free 360° turn
- Interchangeable lenses
- Remote control of both, camera parameters and function, and gimbal.

Demonstration

In this task, end-users tested the prototype, based on the selection of a relevant scenario for the creative industries, the Science Museum designed by the Spanish architect Santiago Calatrava, which is located in the City of Arts and Sciences (CAC).

The objective of this stage was to identify failures and to provide new improvements through the PAR tool (Whyte, 1991) (Fig. 12). The PAR was divided into two phases. In the first phase, two user tests were developed, one for the user guide (Fig. 13) and another for the prototype. Also, the heuristic evaluation of the AiRT system was carried out with the aim of identifying potential usability problems, checking for compliance with previously established good design principles (Von Hippel & Tyre, 1995; Kemmis et al., 2013). The PAR methodology allowed us to evaluate the software (Fig. 14) and the system.

Results for the software indicated that some elements needed to be improved, for example, the support offered in the help link and the information given about the remaining time until a task is completed.

The aim of this tool was to integrate three basic aspects in the dynamics: participation (democratic participation of the CCIs), action (engagement with experience), and research (soundness in thought and the growth of knowledge).

After the usability test, twelve basic tasks were carried out for the control and management of the AiRT system in the process of filming in an interior space. In these tasks, experts and developers worked collaboratively. This second stage was very enriching, as the developers were able to identify improvements and even to discover some new development opportunities. Once this stage was finished, a second test was carried out in which the experts provided many situations in which the drone could work indoors. However, they specified some improvements related to helping the user to understand more quickly how to use the system. After considering all our analyses, we can conclude that some of the improvements said by the experts were considered when the business model was defined.

Work package 6 Dissemination of project results

A Project communication and dissemination plan was designed in order to provide practical advice for all participating parties on how to communicate, disseminate and engage with stakeholders committed with the community. Three main topics were elaborated: Website (http://airt.eu), social media and internal/external communication. Moreover, different types of communication material and its purpose/target audience were produced. Simultaneously, a report on monitoring and evaluation of communication and dissemination activities was elaborated, clarifying the programme objectives and analysing whether the intended results were achieved or not. To do so, AiRT followed for both dissemination and communication strategies the 'recall' approach which relies on a 6-step model (Fig. 15).

The goal of the dissemination activities was to draw people's attention to the AiRT project and Horizon 2020 funding. This was achieved on one hand by a PCPD especially tailored for the different stakeholders, dissemination material and templates, and a communication material report. At the same time, the results were presented to the various stakeholders throughout the project. The overall results of the dissemination are very positive:

(1) Scientific publications. 20 publications.
(2) Social networks. The result obtained after two years of work, has been of great impact, exceeding in most of the channels the objectives set out in the communication plan. Among the objectives exceeded we can highlight, e.g., 1175 Facebook Likes, 113,800 Twitter Impressions or 3995 Instagram Likes.
(3) Press. We published 36 articles, which is three times the number expected. We have also participated in other events such as fairs (Belgium, Germany and Spain), exhibitions such as NEM (Madrid) and ICT 2018 (Vienna), and other workshops by invitation, that have been disseminated through their own channels.
(4) Website. We have used our website to disseminate as much information of the project as possible, such as the newsletters or the workshop, among others. In these twenty-four months, we have got around 13,800 visits.

(5) Videos. We have produced eight videos. Three of them were about the focus groups. Other two were videos showed the project, being one focused on the consortium and the other three on demos, workshop and Vienna exhibition.

Work package 7 Exploitation strategy

This WP had two main goals:

1. To give a tool to the CCI SMEs which might help them to grow within the European and international market.
2. To make research data generated by selected Horizon 2020 projects accessible with as few restrictions as possible, while at the same time protecting sensitive data from inappropriate access.

Here, too, the design thinking method was applied. As a consequence, AiRT developed a Business Strategy Plan (BSP), including IPR strategy and agreement and Data Management Plan (DMP). The BSP was based on the Canvas Business Model (Fig. 16) approach by Osterwalder & Pigneur (2010), analysing:

Customer segments; Value proposition; Channels; Customers relationship; Key resources and activities; Key partners and suppliers; Sources of incomes; Costs.

4 Conclusions

The first conclusion we can draw is that the application of the Design thinking method for the AiRT project can be considered as a successful approach. This methodology has allowed us to carry out an innovation process in which the creative industries, the business sector, the research community, the education sector and government institutions have participated jointly throughout the project.

We have sought open access to the results obtained, gender equality in all the activities carried out, scientific education through the dissemination of results and the involvement of primary and secondary school students, the definition of ethics in the area of drones and participatory governance (Fig. 17).

The AiRT project has made possible to create a drone with the cultural and creative industry and for the benefit of it, facing the enormous challenges of this sector, and generating responsible practices in innovation. The participation of end users has allowed innovation policies to be more acceptable and responsible, improving competitiveness and creativity within the framework of the project. Ethics has been considered in every step as a drive instead of as a restriction for innovation.

The AiRT project has placed special emphasis on openness, transparency, diversity, inclusion and adaptation to changes in European creative industries. It has helped us to redefine our role in the small business sector, and boost the interactions between science and the creative industry. All actors were involved in the innovation process with the aim of obtaining results more appropriate to the needs of this sector (Fig. 18).

Overall, it can be stated that the AiRT project was a success. The goals were largely achieved and innovative technologies integrated. In some areas, expectations were

Fig. 17 Actors participating
in the process of research
and responsible innovation
Source Own elaboration

Fig. 18 Target audience of
the Airt project *Source* Own
elaboration

even exceeded. For example, in the field of dissemination, it is worth mentioning the impact of scientific publications. In the case of the book "Ethics and Civil Drones, European Policies and Proposals for the Industry", it is among the top downloaded in Law in the Springer publishing house, with 15.000 downloads. On the other hand, the book "Drones and the Creative Industry. Innovative Strategies for European SMEs" also published by Springer, reached around 10,000 downloads six months after its publication. This publisher is in the fourth position in the Scholarly Publisher Indicators Ranking (RSPI).

We have also checked that our paper "Ethics for civil indoor drones: A qualitative analysis", published in the International Journal of Micro Air Vehicles (in August 2018, as an open access format), is the third most read of the journal. This journal is indexed in the Journal Citation Report.

In the first step of the project, the identification of the needs of the CCIs (focus groups activities) were held successfully and made up the basis for next steps of the project (RPAS design, GUI, future market uptake etc.). In order to enable indoor operation, a novel indoor positioning system was designed and tailored to integrate tightly with the RPAS. More specifically, a multi-antenna system resulted in improved accuracy, reliability and orientation estimation. In the last stage of the project, the integration with the autopilot was completed, enabling the autonomous flight. The innovative integration of a new flight mode system, consisting of different flight modes with different degrees of freedom (from totally manual flight to totally autonomous flight, and a number of intermediate flight modes with configurable restrictions) must be highlighted. Also, the multiplexor approach to improve the safety of the drone during the flight is worth mentioning. Finally, we want to point out that our system was tested first in a relevant space (Science Museum at the City of Arts and Sciences, Valencia). To do so, we applied the PAR method, based on the critical analysis of the AiRT RPAS & software, with the active participation of some CCIs stakeholders. This method combines two processes, to know and to act, involving the CCIs in both of them. The extracted information from the analysis of this activity was used for cycles of product improvements/optimizations.

It is important to note that the three main developments of the project (drone, IPS and software) can be acquired separately or together, facilitating the diversity of applications. The IPS can be purchased together with the drone, or independently and adapt it to other equipment.

On the other hand, the on-board control of unmanned aerial systems for indoor flight, and the user interface for the design of flight plans (with unmanned aerial vehicles for indoor flight), are free software (FOSS) and open source (FLOSS) licensed, in a way that users can research, modify and improve their design through the availability of their source code. For this reason, both products are available on GitHub, with the aim of expanding the possibilities of adaptability to other equipment or needs, enhancing collaborative work and making accessible to small European companies a tool without cost. In conclusion, the potential market of the software was demonstrated by its impact in Instagram. The number of "likes" that the post received in 24 h was close to 19,000.

Another important fact to be highlighted is that the AiRT Project has resulted in growth and development for the small companies involved in it. For example, the number of employees in these companies has grown between 44 and 275% in two years. On the other hand, all the companies involved in the project have started new lines of research and innovation linked to the results of the AiRT, with the prospect of presenting new proposals in the EU Research and Innovation programme of Horizon 2020 (Santamarina et al., 2019).

Finally, the AiRT team is aware of the essential role of scientific education in responsible research and innovation. On December 2018, they developed and implementated the permanent workshop "Drones" in the Science Museum of the City of Art and Sciences from Valencia. This activity was developed with the support of the Science Museum, the Valencian Innovation Agency and The Generalitat Valenciana, the main governmental institution in the Valencian Region The focus of this workshop was primary and secondary school children. In this way, through the collaborative construction of drones and piloting them in an indoor circuit, students learn in a playful and co-creative way the challenges that the world of indoor drone presents, and also the ethical and legal implications that they will entail. Involving students in the innovation process and providing them with fundamental competences, educators can implement scientific education.

Acknowledgements This project has received funding from the European Union's Horizon 2020 research and innovation programme under grant agreement n° 732433.

We want to thank the European Commission for their support through the concession of this grant and the experts that have participated since the beginning in the joint development of the final product offering their insights and sharing their knowledge.

References

Both, T. (2009). *Design thinking bootleg*. Stanford: Institute of Design Thinking at Stanford. Retrieved from https://dschool.stanford.edu/resources/the-bootcamp-bootleg.

Both, T. (2010). Bootcamp Bootleg. Stanford d.school. Stanford: Institute of Design Thinking at Stanford. Retrieved from http://dschool.stanford.edu/wp-content/uploads/2011/03/BootcampBootleg2010v2SLIM.pdf

Dorst, K. (2011). The core of 'design thinking' and its application. *Design Studies, 32*(6), 521–532.

Kemmis, S., McTaggart, R., & Nixon, R. (2013). *The action research planner: Doing critical participatory action research*. Springer Science & Business Media.

Kusek, J. Z., & Rist, R. C. (2004). *Ten steps to a results-based monitoring and evaluation system: A handbook for development practitioners*. World Bank Publications.

Liamputtong, P. (2011). In *Focus group methodology: Principle and practice*. Sage Publications.

Maguire, M. (2001). Methods to support human-centered design. *International Journal of Human-Computer Studies, 55*(4), 587–634.

Osterwalder, A., & Pigneur, Y. (2010). Business model generation: A handbook for visionaries, game changers, and challengers. Wiley

Rowe, P. G. (1991). In *Design thinking*. MIT press.

Santamarina, V., Miguel, M. de, & Miguel, B. de. (2019). Deliverable 1.3. Final public report.pdf. Retrieved from https://doi.org/10.5281/zenodo.2556536

Van der Lelie, C. (2006). The value of storyboards in the product design process. *Personal and Ubiquitous Computing, 10*(2–3), 159–162.

Von Hippel, E., & Tyre, M. J. (1995). How learning by doing is done: Problem identification in novel process equipment. *Research Policy, 24*(1), 1–12.

Whyte, W. F. E. (1991). *Participatory action research*. Sage Publications, Inc.

Prof. Dr. Virginia Santamarina-Campos Virginia Santamarina-Campos is a Professor at the Department of Conservation and Restoration of Cultural Property of the Universitat Politècnica de València, and expert in the mural art and creative spaces. Coordinator of the Research Microcluster "Cultural and Creative Industries, Tourism & Tech" of the International Campus of Excellence. In the past 5 years, she has led 15 international R+D projects in the area of cultural heritage.

Prof. Dr. María de Miguel Molina Law Degree Universitat de València, PhD on Business Administration, Universitat Politècnica de València. Associate Professor at the Management Department UPV. Head of Studies at the Management School UPV. Main research on the public sector, public policies, public management, inclusive policies. Visiting Professor at the Lakehead University in Canada and the UC Berkeley in the Unites States, as well as some European institutions.

Prof. Dr. Blanca de Miguel-Molina Economics Degree Universitat de València, International MBA Ford-Anglia Ruskin University- Universitat Politècnica de València, Master of Business Administration Anglia Ruskin University-UK, PhD on Business Administration UPV. Associate Professor at the Management Department UPV. Main research on innovation in creative and cultural industries, corporate community involvement, scientometrics.

Prof Dr. Marival Segarra Oña Industrial Engineer Degree and PhD on Management UPV. Professor at the Management Department, Universitat Politècnica de València. Director of the MBA at the Management School UPV. Main research on the eco-innovation services performance and social innovation. Visiting Professor at the Cornell Hospitality Centre in the United States and the University of Bath in the UK.

Intangible Cultural Heritage in Spata Greece: From Mythology to Gastronomic Folklore and from Tradition to Contemporary Culture

Dionysia Fragkou, Loukia Martha, and Maria Vrasida

1 Introduction

The importance and role of cultural heritage in the shaping of modern life has been greatly appreciated and studied by many scholars (Jurkovic et al., 2019). Still it is important to distinguish between tangible and intangible cultural heritage and provide a clear distinction between them. Tangible heritage is related with monuments, objects, buildings, landscapes, books and more and it has been protected and appreciated. Intangible cultural heritage is a relatively new term and is mainly based on the definition of culture as "the way of life of a community". It is strongly related to the stories, habits and traditions of a community that are passed down verbally through generations. According to the classification of folklore, traditional songs and poems are considered part of the folk tradition of a locality. Stories, values, heroes, local identity and even in some cases potential foes can all be traced in those tales and stories (Harun and Jamaludin, 2018). A Conceptual Model of Folktale Classification as a Visual Guide to a Malaysian Folktale Classification System Development. MATEC Web of Conferences.). Some of the folk tales, poems and songs have been documented and written and have managed to be salvaged but traditional fairs have been a different approach to the historic continuum such as the case of Spata.

The aim of this paper is to create a connection between tradition, culture and folk tales as they are manifested in the case of Spata. Through the description and historic trace of this gastronomic tradition a link between mythology, culture and tradition is attempted. Finally the overall goal is to provide a viable sustainable future for such events by connecting them to the modern day technology, acknowledging the

D. Fragkou (✉) · L. Martha
University West Attica, Athens, Greece
e-mail: dfrangou@uniwa.gr

M. Vrasida
EGTC Amphictyony, Athens, Greece

© The Author(s), under exclusive license to Springer Nature Switzerland AG 2021
V. Katsoni and C. van Zyl (eds.), *Culture and Tourism in a Smart, Globalized, and Sustainable World*, Springer Proceedings in Business and Economics,
https://doi.org/10.1007/978-3-030-72469-6_48

event as part of the intangible heritage of the area and listing it in UNESCO and finally adding an economic viability axis through tourism. Recent studies, globally, emphasize that culture in the broadest sense and according to the anthropological approach—values, symbols, beliefs, traditions—is the driving force for the social, political and economic structure of modern societies. This will act as the rationale for this paper emphasizing the importance of including St. Peters Stew in the UNESCO catalogue of intangible heritage.

2   Event Description

Intangible cultural heritage will be discussed and analyzed through the description and analysis of the traditional fair in Spata that has been taking place on the 29th of June. The day is related in the orthodox calendar with St. Peter and St. Paul and in folk tales it is related to the summer equinox. The fair is better known as St. Peters Stew. It is a Christian tradition that can be traced back to the Othman occupancy times. The tale describes a Christian being chased by Turks. The Turks stopped under a tree where the Christian had climbed to escape. While he was holding his breath on the tree in order not to be found, he was praying to St Peter to help him and spare him and in return he would bring a sacrifice to the Saint. In a few moments the Turks left, the Christian was saved and he kept his promise to the Saint by sacrificing a lamb in the chapel of St. Peter and St. Paul on the 29th of June and then they made the first stew. Every year this tradition was kept, and the family kept growing with an increasing number of people participating in the sacrificial stew cooking (Toundas, 2005).

Later the lamb was replaced with bull which is a clear reference to the Greek mythology, and the sacrifice of Iphigenia, the daughter of King Agamemnon and Queen Clytemnestra. In the story, Agamemnon offends the goddess Artemis on his way to the Trojan War by accidentally killing one of her sacred deers. She retaliates by preventing the Greek troops from reaching Troy unless Agamemnon kills his eldest daughter, Iphigenia, at Avlis as a human sacrifice. A sacrifice (thusia in Greek) to the gods was the most important activity in Greek religion. According to the Greek philosopher Theophrastus, the Greeks sacrificed to the gods for three reasons: to honor them, to thank them, or to request a favor from them. Sheep, goats, pigs, and cattle, as well as fish and birds, were offered to the gods. The sacrifice of an animal was carried out according to strict guidelines and following a specific ritual (Sorum, 1992).

One of the main rituals of both Greek and Roman religion was animal sacrifice. Sacrifices established the appropriate relationships among gods, humans, and animals. The gods were superior and immortal, whereas humans were mortal and ought to be pious and submissive to the deities. Animals existed to be used by humans in their worship of their gods. Sheep and goats were the most common sacrificial animals, although some special sacrifices involved bulls. In recent times animal sacrifice is not an acceptable practice and has been replaced or is symbolized by cooking and gastronomic rituals that do not involve animal cruelty and sacrifice. More theatrical and symbolic expression of the traditional folk tale have been used to keep the memory alive but still obey to the constantly changing needs and principles of a modern society (Fischer, 2005). Theater, sacrifice and ritual present a radical re-definition of ritual through the analysis and the performance.

Moving on to Cristian times, as a religious symbol St Peter is a characteristic patron saint of the residents of the city of Spata with St. Peters chapel being located in a historic area where older relics were found (Zafeiris, 1977). The festivity for St Peter is a central element that connects the people of Stata with their devotion and promise to the saint and with each other, thus acting as a timeless bonding agent for the locality. The ancient times the orthodox tradition, the present days and the future of Spata are all connected on a continuous timeline through their initial sacrifice and the folklore, thus creating a contemporary cultural event (Markou, 1981). Despite the fact that there is a chapel in the area, this tradition and cultural event can be categorized under intangible cultural heritage because the emphasis is not on the location and the chapel but on the ritual and the communal cooking as an activity. Activity based cultural events are mainly characterized as intangible and this offers a flexibility and an opportunity for community bonding that is not dependant on physical locations or physical participation (Meraklis, 1984).

The tradition of the St Peters Stew is kept active throughout the years in the area and every family had to bring part of the materials. In very poor years during 2nd WW the communal sharing of food which was scarce, created a strong bond between the locals. The years that followed and as the tradition was growing even more, the need for better quality raw material and the growing number of participants, made the tradition less spontaneous and more organized (Bertolis, 2004). A citizens association was created and took over the organization of the event. Still the procedure was formalized but the bonding and sense of community remained unchanged.

St Peters Stew tradition is strongly related to the orthodox religious faith of the locals. The preparation of the stew starts after the evening mass and the stew is cooking all night. After next morning's mass people gather at the chapel and every family takes their pre-determined quantity of Stew to eat at home. The process of preparing this food, with the "cleaning of the onions" and the preparation of all the necessary materials belongs to the collective responsibility of the women of the city. Cooking in huge cauldrons (now numbering around seventy) on wood fires is the sole responsibility of men. During the preparation of this meal, the festival unfolds. In traditional tales and folk tales, culture has been greatly linked to gastronomy and cooking, clearly establishing the link between gastronomy and culture. Furthermore, the social dimensions of communal cooking with food that is later distributed to the community with emphasis on the poorest families create a culture of participation, social inclusion and sharing.

A series of unexplained events relate St Peters Stew with religious spirituality, folk tales and local legends offering a mystic dimension and connecting it with the meaning of ritual sacrifice as it is passed down. In ancient Greek and Roman religion, performing a ritual according to specific tradition and custom was crucial. Failure to do so rendered the act meaningless. Thus, preserving rituals and passing them from one generation to the next became an important social function. Dundes (1965) also provides another definition of folklore consisting of an itemised list of the forms of folklore and I think this could help us understand what is really referred to when one talks of folklore. According to him, folklore includes myths, legends, folktales, jokes, proverbs, riddles, chants, charms, blessings, curses, oaths, insults, retorts, taunts, teases, toasts, tongue-twisters, and greeting and leave-taking formulas…It also includes folk costume, folk dance, folk drama, (and mime), folk art, folk belief (or superstition), folk medicine, folk instrumental music (e.g., fiddle tunes), folksongs (e.g., lullabies, ballads), folk speech (e.g., slang), folk similes (e.g., as blind as a bat), folk metaphors (e.g., to paint the town red), and names (e.g., nicknames and place names)… oral epics, autograph-book verse, epitaphs, latrinalia (writings on the walls of public bathrooms), limericks, ball-bouncing rhymes, jump-rope rhymes, finger and toe rhymes, dandling rhymes (to bounce the children on the knee), counting-out rhymes (to determine who will be « it » in games), and nursery rhymes games; gestures; symbols; prayers (e.g. graces); practical jokes; folk etymologies; food recipes; quilt and embroidery designs; house, barn and fence types; street vendor's cries; and even traditional conventional sounds used to summon animals to give them commands;… mnemonic devices (e.g. the name Roy G. Biv to remember the colors of the spectrum in order), envelope sealers (e.g. SWAK−Sealed With A Kiss), and the traditional comments made after body emissions (e.g., after burps and sneezes),… festivals and special day (or holiday) customs (e.g., Christmas, Halloween, and birthday). The past is omnipresent in the daily life. It comes up in many ways, and it is something that people seek for relentlessly. Memory, history, and relics continually furbish our awareness of the past (Lowenthal, 1985). In fact for what is said to be the past we openly relay on what is written in the history books and the archives, all the data that has been collected by the historians through the centuries. In this concern the often-neglected difference between the notions of memory and history becomes relevant (Nora, 1989).

It is important to realize that folk traditions as part of the intangible cultural heritage of an area are not reviving history which is based mainly on events. They are reviving collective memories that are part of the common shared experiences of a community, thus enhancing and preserving the bond of a community regardless of time and distance. Folklore is the generic term to designate the customs, beliefs, traditions, tales, magical practices, proverbs, songs, etc.; in short the accumulated knowledge of a homogeneous unsophisticated people. According to Leach, (1996), folklore is that art form, comprising various types of stories, proverbs, sayings, spells, songs, incantations, and other formulas, which employs spoken language as its medium. In Aurelio N. Espinosa's terms, folklore, or popular knowledge, is the accumulated store of what mankind has experienced, learned, and practiced across the ages as popular and traditional knowledge, as distinguished from so-called scientific knowledge. In the case of Spata it is an tale that has never been officially recorded and is passed down from generation to generation and more related to family traditions.

As Katan (1999) puts it, all people instinctively know what culture is and the culture they belong to, but it does not follow that they can define it with ease. People of Spata when asked in open ended questions, they could not easily Identify the elements that defined their culture and their particularities in relation to the specific locality. Yet when the Stew was mentioned and the questions were specific, they could easily define themselves as Spatans and they could identify their cultural differences in relation to other citizens of Attica. Their sense of local identity and pride seemed to increase when they associated themselves with the tradition of the Stew. However, it seems that, according to the same author (1999: 16), defining culture is imperative, particularly for anthropologists, because it delimits how it is perceived and taught. Still, although many anthropologists have attempted to define culture, they have not reached any agreement regarding its nature. Following this lack of agreement as to the nature of culture, different anthropologists have come up with different definitions. Edward Burnett Tylor (in Katan, 1999) defines culture in the following terms: Culture is that complex whole which includes knowledge, belief, art, morals, law, customs and any other capabilities and habits acquired by man as a member of society. For the American anthropologists Alfred Louis Kroeber and Clyde Kluckholm (in Katan, 1999), culture can be defined as follows: culture consists of patterns, explicit and implicit of and of behaviour acquired and transmitted by symbols, constituting the distinctive achievement of human groups, including their embodiment in artefacts; the essential core of culture consists of traditional (i.e. historically derived and selected) ideas and especially their attached values. Folklore, Culture, Language and Translation even if these models consider culture from different points of view, they have much in common and constitute a fund of information for the understanding and the definition of culture.

Culture systems may, on the one hand, be considered as products of action, on the other hand, as conditioning elements of future action. In her definition of the term 'culture', Gail Robinson (in Katan, 1999) argues that culture can be defined as a system consisting of two levels. The first is the external level which consists of behaviours (language, gestures, customs and habits) and products (literature, folklore, art, music, artefacts). The second is the internal level which is related to ideas (beliefs, values and institutions). Other definitions of culture can be drawn from the models of culture known as Trompenaars' layers, Hofstede's onion, the Iceberg theory and Hall's Triad of culture.

Folklore, language, culture and translation. These four elements are closely related to one another. Firstly, folklore is related to culture in the sense that it is, as previously mentioned, a mirror of culture. Folklore reflects culture because it relates to the way of life of the people who produce it: their ceremonies, their institutions, their crafts and so on. It also expresses their beliefs, customs, attitudes and their way of thinking. Folklore actually gives a penetrating picture of the way of life of the people who produce it (Dundes, 1965: 284). For that reason, it is, as pointed out by Malinowski (in Dundes 1965: 281), important to understand the setting of folklore in its actual life if one wants to understand it. The relationship between culture and folklore can also be shown in the definition of folktales, as part of folklore, by Lester (1969: vii) (found in Held and Moore, 2008), who says that Folktales are stories that give people a way of communicating with each other about each other- their fears, their hopes, their dreams, their fantasies, giving their explanations of why the world is the way it is. It is in stories like these that a child learns who his parents are and who he will become. Arbuthnot (1964: 255) corroborates this idea by saying that '…folktales have been the cement of society. They not only expressed but codified and reinforced the way people thought, felt, believed and behaved.'

3   The Future of St Peters Stew

As it has been established above, customs, rituals and various events associated with popular culture, play a central role in local communities. They preserve the collective memory alive and demonstrate in the most eloquent way the connection with the cultural heritage the cultural identity and the social and cultural development (Sheenagh, 2009). Cultural Research and Intangible Heritage. Culture Unbound: Journal of Current Cultural Research). Popular culture archives have traditionally existed in the form of friable paper, brittle sound tapes, or fragile photographs. However, during the past decades the challenges of the digital era have been met and digital archives have been created. Traditionally, the main role of folklore archives has been to protect the collected items and making them accessible. As technology evolves, it must not be viewed simply as an information storage tool. Through technology traditions and culture can become accessible to everyone, offering a link and a connection to the locality and a local community regardless of time and distance. Participants in the events can be informed and learn, using the internet as an educational tool but with the very recent turn of global events, they can also find new ways of participating. Physical presence is no longer implied and the experiences have been transformed to serve the new digital reality. Proximity is not always a prerequisite for creating a sense of community sharing and the links that bind people to their history and locality become intangible as well. Common memories, shared experiences, and real time participation enhanced by digital representations, augmented reality and many more, can act as the new social binding tools. The importance is not in the medium (technology or physical access) but in the narrative and the collective memory (folklore, culture) (Balling and Christensen, 2013).

4   Conclusion

All the customs and events offer the opportunity for the members of the community to meet, celebrate together and show their respect to the traditions, the history and the commitment to the common bonds that bind them to their locality. In this way the various events contribute to the preservation of a collective memory, mutual understanding and the strengthening of social identity. Very often they are an occasion for people who have moved away from the locality to return to their origin to reconnect with other members of the community. At the same time, through technology, they create a channel of wider communication with other groups that are attracted by the events and arrive in the area as tourists or as visitors or as researchers and observers (Portalés et al., 2018).

Such events, like St Peters Stew, prove in the most apparent way that beyond differences and diversity, all events have common elements that shape the common European heritage as they reflect not only the rural cultural wealth but also the similarities that connect the rural populations (Satterthwaite and Tacoli 2003), "The urban part of rural development: the role of small and intermediate urban centres in rural and regional development and poverty reduction", Rural-Urban Working Paper 9, IIED, London; this can be downloaded from IIED's website (http://www.iied.org/rural_urban/ downloads.html#UPWPS). They offer a common European historic timeline emphasizing the agents that bond communities and create a common collective memory. The various symbols that have always been used have been the means for cultures to express their relationship to various phenomena and to structure their survival. Customs and popular cults were not only elements of dealing with nature and its consolation but also a means of strengthening ties and collective identity. Customs and traditions emphasize the similarities in the tiles that form the common European mosaic of communities. At the same time, new forms are developed that utilize the cultural resources of quality tourism services, which highlight the folk heritage as an important element of economic development. Local development with a European perspective and is enhanced and nurtured under the preservation of intangible cultural heritage. Traditions and customs are the everlasting memories developing the common collective memories of present and future citizens. In many cases, culture and heritage tourism have been associated with the immovable heritage and the protection of various monuments. In recent years, however, along with the emergence of monuments and historic buildings, attention has also turned to the intangible heritage, local traditions, rituals and customs.

Thus, the various events offer an opportunity for the rural population to benefit themselves both in their 'internal environment'—strengthening ties, strengthening relationships, strengthening identity—and in the 'external environment' by developing exchanges and communication and consequently tourism and economic development, giving the opportunity to young people to remain and/or return to the countryside, as they will be able to experience an environment in which tradition and modernity coexist (Keat, 2000), a new model of development, which will not be based exclusively on material production and consumption, but on a system of social values and an economy of knowledge and relations.

References

Arbuthnot, M. H. (1964). Children and books, 3d ed. Chicago, IL: Scott, Foresnlan and Company.

Balling, G., Christensen, N. K. (2013) What is a non-user? An analysis of Danish surveys on cultural habits and participation. *Cultural Trends, 22*(2), 67–76

Bertolis, P. A. (2004). *The stew of St. Peter*. Spata Attica, (found in Greek).

Dundes. (1965). In Simon J. Bronner (Ed.), *Meaning of folklore: The analytical essays of alan dundes* Introduced by Copyright: 2007.

Fischer Lichte, E. (2005). In *Theatre sacrifice ritual: Exploring forms of political*, Routlrdge.

Harun, H., & Jamaludin, Z. (2018). A conceptual model of folktale classification as a visual guide to a malaysian folktale classification system development. MATEC Web of Conferences.

Held, D., & L. Moore, H. (eds). (2008). *Cultural politics in a global age, uncertainty, solidarity and innovation*. Oneworld Publications.

Jurkovic, et al. (2019). The perception and social role of Heritage Buildings in modern society. In *Innovation in intelligent management of heritage buildings* Pub. IRCLAMA, University of Zagreb.

Katan, D. (1999). *Translating cultures, An introduction for translators*. Interpreters and Mediators: Manchester, St. Jerome Publishing.

Keat, R. (2000). *Cultural goods and the limits of the market*. Basingstoke and New York: Macmillan Press/St. Martin's Press.

Leach, M. (1996). Definitions of folklore. In Funk & Wagnall (Ed.), Standard dictionary of folklore, mythology and legend. *Journal of Folklore Research, 33*(3).

Lowenthal, D. (1985). *The past is a foreign Country*. Cambridge University Press.

Markou, S. (1981). *The offering"*, Spata Attica (found in Greek).

Meraklis, G. M. (1984). *Greek folklore social cohesion*. Odysseas pub, Attica. (found in Greek).

Nora, P. (1989). Between memory and history: Les lieux de mémoire. *Representations, 26*: 7–24, The Regents of the University of California.

Portalés, C., Rodrigues, J., Gonçalves, A., Alba, E., & Sebastián Lozano, J. (2018). Digital cultural heritage. *Multimodal Technologies and Interaction, 2*(3), 58.

Satterthwaite, D., & Tacoli, C. (2003). The urban part of rural development: the role of small and intermediate urban centres in rural and regional development and poverty reduction. Rural-Urban Working Paper 9, IIED. Retrieved from http://www.iied.org/rural_urban/downloads.html#UPWPS.

Sheenagh, P. (2009). Cultural research and intangible heritage. Culture Unbound: *Journal of Current Cultural Research, 1*, 10.

Sorum, Elliott C. (1992). Myth, Choice, and meaning in euripides' iphigenia at aulis. *The American Journal of Philology, 113*(4), 527–542.

Toundas, N. (2005). *Spata attica: Myth*. History and Folklore: AO Publisher, Keratea Attica (found in Greek).

Whitaker, F. (2017). *UNESCO moving forward the 2030 agenda for sustainable development*. Scienti-c and Cultural Organization: United Nations Educational.

Zafeiris, N. (1977). *St Peters Chapel in Spata Attica, Scientific Association for Byzantine Civilizations (pub)*, (found in Greek).

Silk Road Regionalism and Polycentric Tourism Development

Stella Kostopoulou, Evina Sofianou, and Dimitrios Kyriakou

1 Introduction

Over the postwar period, tourism has become one of the largest and fastest-growing global industries, characterized by continuous growth, spatial expansion and product diversification, a multidimensional crucial development parameter of the world economy, gaining ground as an essential source of revenue for many countries (Batala et al., 2017; Dimadama & Chantzi, 2014). Tourism, widely interconnected to several other sectors of the economy, has a major impact on the economic and social development of countries and regions opening up for business and trade, creating jobs, motivating entrepreneurship and protecting heritage and cultural values.

Over the last few decades, evolving tourism trends have led to a shift away from the standardized mass tourism model popularized since the early postwar period, to more individualistic creative tourism patterns, where visitors try to attain qualitative experiences and gradually tend to become more concerned about brands (Gilbert, 1989; Manhas et al., 2014). Many scholars emphasize that creating brand can help to insert a positive identity and illustration of a destination, making the destination better recognized and easier to be advertised as a tourism product (Cai, 2002; Krajnović et al., 2013).

In such a perspective the Silk Road regions may provide a contemporary branding approach as an array of emerging destinations over the historic route that served as a bridge between East and West. Nowadays, the legendary Silk Road revives again as a wide tapestry of rising tourism destinations and products, engaging numerous countries with rich natural and cultural heritage. Silk Road can thus be used as a unique branding tool to further develop a broad variety of tourism policy goals (Kostopoulou, 2019). Silk Road tangible and intangible heritage, such as monuments, archeological

S. Kostopoulou (✉) · E. Sofianou · D. Kyriakou
Aristotle University of Thessaloniki, Thessaloniki, Greece
e-mail: kostos@econ.auth.gr

© The Author(s), under exclusive license to Springer Nature Switzerland AG 2021   737
V. Katsoni and C. van Zyl (eds.), *Culture and Tourism in a Smart, Globalized, and Sustainable World*, Springer Proceedings in Business and Economics,
https://doi.org/10.1007/978-3-030-72469-6_49

sites and architectural complexes, arts, music, customs, traditions, clothing, lifestyles, etc., if developed as an international brand connecting West and East, may attract tourists to visit destinations, attaining a lot more excitement than on an ordinary trip. Therefore, determining the concept of Silk Road destination, may offer great opportunities for the expansion of the tourism industry, allowing to better conquer new tourism markets (Kour, 2016; Douglas et al., 2001; Manhas et al., 2014).

This requires the efficient cooperation of chains of tourism resources, forming networks of destinations located over many regions and countries. In terms of tourism regionalization, developing a transnational inter-regional network approach of destinations could open up new tourism markets and economic opportunities within the Silk Road areas. Regionalism can promote tourist cooperation among destinations in order to create inter-regional tourism supply, that would eventually integrate the sub-regional tourism resources of destinations. Furthermore, incorporating tourism activities and resources into integrated polycentric networks can stimulate the collaboration and partnership among destinations, while also create opportunities for the tourism industry to set the standard at the international level (Manhas et al., 2014).

In this paper we seek to discuss the concepts of cultural heritage tourism regionalism and polycentric networking along Silk Road areas, to highlight the need for the systematic organization of tourism intelligence and a common methodological Silk Road approach, enabling country-specific, yet comparable tourism analysis of emerging Silk Road destinations. Introducing a regionalism and polycentricity networking approach for Silk Road tourism development may help to improve inter-regional and international cultural heritage tourism synergies to support sustainable destination management and offer a dynamic potential to local tourism business. Undertaking the term of tourism networks that exhibit the characteristics of linkages within the Black Sea area among destinations, we take into consideration for our research the traditional bazaars as Silk Road cultural heritage assets, considered as nodes for the regionalization of tourism destinations through cultural heritage tourism polycentric development. The research methodology of this conceptual study is based upon secondary data on Silk Road heritage as an international tourism prospect, collected from different sources i.e. articles, reports, books, journals, magazines, internet documents, websites and other online sources. The case of a Black Sea cultural heritage tourism network based on a specific category of Silk Road cultural assets, the urban bazaars, is used to highlight the emergence of the proposed approach.

2 Silk Road Overview

Silk Road is widely recognized as a historical vast trade network, the longest and oldest trading route in the world, consisting of land routes, many of them in the form of caravans, and maritime routes of sea trade, that linked the great civilisations of the East and the West (Map 1).

Early activities along the Silk Road are generally believed to have started during the Chinese Han dynasty (206 B.C.–A.D. 220), stimulated by the mission to the

Map 1 The old silk road *Source* China discovery, silk road maps collections https://www.chinad iscovery.com/china-silk-road-tours/maps.html

West of the Chinese ambassador Zhang Qian (138–139 B.C.). However, the route was already several thousand years old by then, when Alexander the Great from the ancient Greek kingdom of Macedonia, followed much of it during his eastern conquests (Green,1991; Hammond, 1989; Kostopoulou et al., 2016). For many centuries valuable Chinese silk, spices, handicrafts, precious stones and minerals, porcelain, and other goods were transported to Europe, while China received gold and other precious metals, glass and many other products from the west (Stanojević, 2016). Hence, the Silk Road has not only been a trade route, but also a bridge for contact between people and civilizations promoting cultural, religious, lingual and information exchanges among the flourishing economies of China, Persia, India, Arabia, Greece and Rome. The historic Silk Road meant connections and challenges, but also safe long distance travelling (Wojciech, 2006). Many cities were built along the Silk Routes, originally founded as small transportation gateways, then gradually turned into major trade and exchange centers (Vasiliev & Shmigelskaia, 2016).

The term Silk Road 'Die Seidenstraße' was first introduced in 1877 by the German geologist and geographer Ferdinand von Richthofen (1833–1905) (Chin, 2013), deriving its name from the silk first exported from China, considered to be an especially expensive and highly valuable product at the time. Richthofen laid the foundation for a whole direction of historical research, which later became interdisciplinary, since the subject of analysis concerns the processes of cultural, economic and political interaction of states along the Silk Roads (Kylasov, 2019). It should be noted that Richthofen also used the term "Silk Roads" in the plural for routes both east and west (Waugh, 2007). Other scholars and explorers added further detail by gathering manuscripts, artefacts and artworks, providing unique insights into some of the multivalent, multi-directional flows and networks of a pre-modern world system (Winter, 2020).

Today the Silk Road covers a vast geographical area and a large number of countries, heterogeneous in their economy, politics, society and culture. The areas along the Silk Road have a multidimensional character, as they include a multitude of interconnected aspects, such as natural and artificial barriers, as well as cultural, social, economic and political interactions. Given this reality, a significant number of regional tourism development goals can only be effectively addressed through successful collaborative processes, which require continuous networking, collaboration and integration between different entities (EC, 1999, 2011), based also on the regionalism concept.

3   Literature Review

3.1   Overview of the Concept of Regionalism

Regionalism as a cognitive example and a new type of scientific thinking, utilises the term region and its derivatives explaining the essence and principles of cooperation between national and global (supranational) levels of world organization (Tolpakova & Kuchinskaya, 2015). The term "regionalism" refers to a set of ideas, values and policies aimed at forming a region, or it can mean a type of world order, while in the first sense is usually associated with a regional project or organization (Soderbaum, 2016).

According to the academic literature, regionalism may be distinguished in two eras. First, the early 1950s and 1960s that were largely inspired by the early successes of European integration and the growing popularity of dependency theories, with an emphasis on collective autonomy in the greatest part of the Third World. The second, also known as New Regionalism, can be traced to the mid-1980s and is usually closely linked to the evolving forces of economic liberalization and globalization (Ruland, 2001). New Regionalism approaches underlined the importance of non-state actors in a more multidimensional and pluralistic way (Söderbaum, 2015), claiming that there are no 'natural' regions, but these are made, remade and unmade, intentionally or non-intentionally, in the process of global transformation, by collective human action and identity formation.

The scientific concept of regionalism in its contemporary dimension thus elaborates the objective processes of formation of large multidimensional areas and other territorial communities (e.g. supranational groups, regional and intra-country communities) that have a common cultural code, seeking the most effective use of internal and external development factors (Dergachev, 2008). Modern ideas about the region as a form of spatial organization created by integration processes in international relations, are limited to the problem of forming a new (global) comparative regionalism, developing as a modern version of multipolarity. Having been formed "from bottom up", it contains elements of spontaneity and autonomy of its actors and seeks to implement the ideas of open regionalism that are compatible with economic interdependence. Being multidimensional, the modern region includes economic, environmental, social, political and other dimensions, as well as non-state and sub-national bodies (Hettne et al., 1999).

As a result of the above, the outcome of regionalism is regionalization, where regionalization of a geographical area is considered as a consequence or an objective process resulting from the historical, political, socio-economic, administrative and cultural development of a particular territory (Shopova, 2017). Moreover, regionalization is based on the notion that regions are important geographical entities that should foster endogenous growth and thus create stronger regions and stimulate development and cohesion (Camagni & Capello, 2013).

As Shopova (2017) claims, tourism regionalization or zoning plays an important role in tourism management, forming regional structures as a result of various territorial typologies. According to this latter, regionalization can also be considered as a territorial classification according to certain criteria (e.g. production, demography) and indicators (e.g. labor, population, urbanization) that are similar to a given territory. Therefore, tourism regionalization refers to a *modus operandi* of inter-regional cooperation, integration and cohesion that creates a regional space. In the most basic sense, it may mean no more than a deepening or widening of activity, trade, people, ideas at the regional level (Fawcett, 2005:25).

The emergence of Silk Road tourism is a typical example for inter-regional cooperation among neighboring countries, aiming at transforming the advantages of geographical proximity and tourism complementarity into practical cooperation and joint development, as well as reference to a heritage of common cultural values, mixed with a highlight of principles of pragmatism and flexibility (Grimmel & Li, 2018, Long & Xu, 2017). Therefore, the Silk Road vast geographical area, vaguely defined based on historical knowledge, could be regionalised into tourism regions with specific distinct characteristics, e.g. Central Asia, Black Sea, Mediterranean, for better analysis and research. Traditionally, knowledge of the Silk Road and its tourism assets have focused on an area ranging from China to Istanbul in Turkey, the so-called 'Classic Silk Road'. In 2016, the United Nations World Tourism Organization (UNWTO) in collaboration with the European Commission (EC) launched the Western Silk Road (WSR) Tourism Development Initiative, a joint tourism project focusing on the impact of the Silk Road also on European destinations (UNWTO, 2017). The WSR Initiative aims at revitalizing the Silk Road heritage located in the European region, from the Caspian Sea, around the Black Sea and along parts of the Mediterranean basin, where sub-regions may be further identified.

For Silk Road tourism to be successful, it may incorporate a comprehensive regionalization logic and encourage the voluntary involvement of targeted states and/or regions by providing a comprehensive framework for cooperation (Qoraboyev & Moldashev, 2018). Getting in depth, paraphrasing Teye (1988), tourism development at the regional level means that regions must work together and integrate attractions, capital, infrastructure, natural and human resources to serve the domestic and international (inter and intra-regional) tourism sector. Since this typically refers to extensive geographical areas, the polycentricity concept is introduced as a new methodological approach in the development of cultural tourism networks.

3.2 Polycentric Tourism Networks

In spatial terms, a network is defined as a set of items, called vertices or nodes, with connections between them, called edges or links describing whether the nodes are interconnected or not (Boccaletti et al., 2006).Tourism as a socio-economic phenomenon has spatial dimensions characterised by three elements, the place of origin, the destination (attractions) and the links between them (Tang & Li, 2016) that form spatial connections creating interactions among the nodes. Tourism networks are viewed to function as systems which can organise and integrate tourism destinations, creating benefits for the stakeholders involved, enhancing destination performance and quality (Żemła, 2016; Zach & Racherla, 2011). Tourism networking may be expressed as the cooperation among destinations, their local authorities, tourism stakeholders, local societies e.tc. that favors the competitiveness of the destination, as the tourist area is expanding boundaries within the network.

According to Gaman et al., (2017), the lack of effective tourism planning led to polarisation phenomena, where tourism flows were mainly oriented towards popular destinations and thus, the rest of territory did not receive the adequate amount of attention. However, tackling down territory disparities, brought to surface the definition of polycentricity or polycentric development, aiming at territorial cohesion, economic performance and tourism development through establishment of functional spatial cooperation, taking into account the socio-economic relations between settlements.

Polycentric spatial development involves numerous and varying definitions (Lambregts, 2009). In a narrow, literal sense, the term refers to a spatial unit that consists of more than one centre (Brezzi & Veneri, 2014; Schmitt et al., 2015). According to Riguelle et al. (2007), an area is polycentric if its population or employment is not concentrated to a substantial extent in one single centre. Polycentricity, is a concept that encourages regions and cities working with neighboring territories, to explore common strengths and reveal potential complementarities (Cowell, 2010; Meijers, 2008; Nordregio, 2005) bringing added value that cannot be achieved by the individual regions and cities in isolation (ESPON, 2016).

Polycentricity as a multi-scalar concept (Rauhut, 2017) has been put to a wide variety of different uses at different spatial scales (You, 2017), urban, regional, interregional, national and international, also with regard to the cross-border issue (ESPON, 2013). On a regional scale, polycentricity often refers to the development of functional relationships within a network of cities, reflecting the fact that cities establish some forms of cooperation or interaction with their neighbors (usually labelled as 'city region') (Dembski, 2015; Evers and de Vries, 2013; Hall & Pain, 2006), with common characteristics.

In the growing body of literature on polycentric spatial configurations, polycentricity has two dimensions, focusing on two aspects, the size and distribution of cities in polycentric regions (morphological dimension) and their interrelationships (functional dimension) (Brezzi & Veneri, 2014; Burgalassi, 2010; Burger & Meijers, 2012; Finka & Kluvánková, 2015), which may reflect "multidirectional flows", such as economic, cultural, trade (Burger et al., 2014; ESPON, 2013; Davoudi, 2008), as well as travel, social visits and leisure trips (Hewings & Parr, 2007; Parr, 2005). The morphological dimension focuses on the characteristics of nodes, basically addressing the rank-size classification of the urban centres in the network, the territorial distribution or location of the urban centres within the network, and their connectivity (Burger & Meijers, 2012; Lambregts & Kloosterman, 2006; Meijers & Burger, 2010; Parr, 2004). Within the functional dimension the focus lies on the level and spatial distribution of inter-dependencies between the urban centres (Schmitt et al., 2015). Functional specialisation is an important dimension of polycentricity as it is these functions that make cities or areas different from each other produc the flows necessary for economic and political integration (Nordregio, 2005).

Cultural heritage assets can be represented as a variety of nodes on a network, providing cultural interaction spaces and may focus on a certain theme (Radosavljević et al., 2019; Chun & Bin, 2008). Cultural heritage networks can therefore be defined as mechanisms for the conservation and protection of cultural heritage, encouraging tourism cooperation of all actors and authorities involved in the tourism sector. The promotion of interactions between areas can lead to functional synergies, offering enriched tourist products to visitors. Developing cultural heritage tourism networks along the Silk Road stands on the idea to put collectively a range of destinations, activities and cultural assets under an integrated theme, and thus encourage industrial prospect with the growth of auxiliary products and related services (Greffe, 1994 in Kour (2016)).

This paper focuses on the morphological polycentricity approach, meaning the analysis of the nodes (specific Silk Road heritage assets), their attributes and interactions. According to Limtanakool et al. (2007), the node-attribute approach focuses on the concentration of activities or functions in a node, and it is considered that the largest concentration of functions e.g. products, facilities, and services one node concentrates, the most important is its position within a network. The paper considers a specific category of Silk Road cultural heritage assets, the traditional urban bazaars, as nodes of the morphological tourism polycentric network. The Silk Road bazaars with common typological and morphological characteristics, e.g. traded products linked with the Silk Road cultural heritage, are represented as nodes on interregional tourism networks that denote a topological structure taking under consideration spatial functional linkages between the nodes.

4 Silk Road Bazaars and Old Traditional Markets

Bazaars are considered among the most typical Silk Road cultural heritage assets. The word "bazaar" is derived from the ancient Persian word "Waazaar" (Porushani, 1995 cited in Assari et al., 2011:21). Historically, agora, forum and bazaar, basilica and bedesten, kaisariya and tcharshi, souq, khan and stoa, indicate architectural and urban typologies which were developed in order to accommodate production, exchange and distribution of goods in different civilizations that succeeded one another (Yerolympos, 2007). Silk Road bazaars were formed along the caravan routes, where caravans stopped for rest and overnights in the caravanserai, khans, etc. way stations, built with the aim of providing a secure and comfortable journey for travellers. Bazaars are known as the marketplaces or assemblage of shops where a variety of goods, domestic animals and services were displayed to buy and sell. Out of the spontaneous trade places that appeared, later forming constant bazaars, often suburbs developed as settlements that gradually evolved into villages and cities.

Usually bazaars had specialized sections of various land uses and functions, like silver and gold sellers, leather, shoes, bags, carpet sellers, wooden works, glass and mirror sellers, silk, cotton, copper sellers and many others (Pourjafar et al., 2012), while the central area was also hosting the gathering of other public activities such as social services, administration, trade, arts and crafts and baths (Assari et al., 2011). Therefore, bazaars were more than local markets for the barter of traditional goods and handicrafts, they were also marketplaces where national and international trade was conducted (Pourjafar et al., 2012).

The urban bazaar was part of the public street network as a unique place in the city to function as a market, and it was organized according to professions, guilds and corporations (Geist, 1989 cited in Yerolympos, 2007). In architectural terms, the urban bazaar was a covered public passageway surrounded by shops and stores in two sides (Assari et al., 2011). Comparison of the bazaars along the Silk Road shows many similarities in terms of types and elements of spatial and architectural configuration, social network, proportion parameters and function in their various traditional urban spaces (Pourjafar et al., 2012). However, the typologies of these trade places varied between East and West (Yerolympos, 2007).

Nowadays, traditional bazaars, are part of the contemporary urban fabric, endowed with cultural and functional uniqueness and authenticity. Compared to typical shopping centres, traditional bazaars in modern cities are more than just a retail environment, including activities such as baking, sewing, shoe making, metal smithing and so on, since most of the goods are often produced within the bazaar. These activities create an active, enlivened environment designed to strike the senses, attracting consumers and visitors that wish to learn how handmade goods come about in small, private retail stores, possessing quality and local identity (Najdjavadipour, 2011). Another distinctive morphological element of traditional bazaars is the mix of indoor and outdoor space, by the use of courtyards, often beautifully landscaped. Moreover, bazaars are pedestrian-oriented, offering an exclusive shopping promenade experience, that keep visitors occupied and entertained in the dense web of shops and corridors. Traditional bazaars, as spatial, architectural and functional ensembles, could indicate new tourism destinations, combining entrepreneurial opportunities through polycentric networking and synergies.

5 Methodological Approach

The main aim of this paper is to identify morphological elements and functional values incorporated in the form of bazaars and traditional markets polycentric networks. The methodological approach used for the research is a descriptive-analytical method, based on qualitative and quantitative data. Qualitative data were collected on citing historical and objective observations, using a plethora of documents and indexing, i.e. articles, reports, books, internet documents and other online sources, while quantitative data were based upon a classification system of bazaar trading activities.

Based on secondary research, the identification and analysis of the main bazaars is performed on regional and inter-regional level. The data was organized in an inventory according to the bazaars' attributes. All bazaars/traditional markets under study with their traded products are recorded and classified following a regionalization pattern, on the basis of the following territorial units: Country, NUTS I II (Regions) (Nomenclature of territorial units for statistics), and LAU I (Local Administrative Units) (Municipalities). Then, the bazaars with products and attributes connected to the Silk Road culture are selected.

In order to filter the bazaars under investigation, the methodological step implies the identification of the products connected to the Silk Road cultural heritage to be included in the final inventory. For this research the following products or trading activities in traditional bazaars related to Silk Road culture are considered for analysis: spices, tobacco; fabric and textile products, silk, cotton; clothing and accessories; carpets; gold/silver jewellery and goldsmiths/silversmiths articles; leather works, shoe, purse; wooden works and carpenters, furniture; glass and mirror; antiques; copper makers and sellers, metal works, iron works. The bazaars with an increased abundancy of traded products linked to the Silk Road heritage are classified as main nodes within the potential polycentric functional networks. These nodes are considered as Silk Road bazaar destinations of high-level hierarchy, with major interregional and regional impact in the network. The bazaars with less traded products connected to the Silk Road are classified as nodes with local impact in the network.

The Black Sea region is the area selected to highlight the proposed methodology, as a sub region of the Western Silk Road (UNWTO, 2017). The Black Sea region, at the crossroad between East and West, is a developing tourist destination that presents slow, yet steady growth within the global tourism market (UNWTO, 2019). The Black Sea countries Silk Road are endowed with significant cultural and ethnic diversity, rich historical and architectural heritage that shape a unique tapestry of diversified communities, languages, cultures and religions, as potential exclusive poles of attraction for tourists (Dimadama & Chantzi, 2014).

For the present analysis, ENI CBC Black Sea Basin Programme eligible areas[1] in Greece, Bulgaria and Turkey have been selected as study areas, based on the criterion of cross-border tourism development potential. For the application of the methodology introduced several criteria are prescribed:

- The study areas selected consist of the NUTS II Regions of Central Macedonia and of Eastern Macedonia and Thrace in Greece, the NUTS II Regions of Severoiztochen and Yugoiztochen in Bulgaria and the NUTS II equivalent regions of TR10 (İstanbul) and TR21 (Tekirdağ, Edirne, Kırklareli) in Turkey.
- The selected Regions are neighbouring and accessible by road and/or rail transport for further cross-border cooperation in tourism development.
- The LAU I equivalent spatial entities are considered for Municipalities and as sphere of influence the urban and rural areas within their boundaries.
- The bazaars/traditional markets are considered as nodes of the networks.

The research adopts a twofold perspective, addressing both the regional level (sub-regions defined in each country) and the inter-regional level (sub-regions networking across Greece, Bulgaria and Turkey). The main aim of the research is to support building polycentric tourism networks based on the bazaars as Silk Road cultural heritage assets. The ultimate goal is to propose various polycentric tourism networks with potential for functional specialization in order to further establish a unified 'Silk Road' bazaar brand name and improve the performance of cross-border regional cooperation in cultural heritage tourism.

6   Silk Road Cultural Heritage Polycentric Tourism Networks in the Study Area

Within the study area, there is a plethora of urban and rural bazaars and traditional markets, with similar attributes concerning the products sold, as well as typological and morphological characteristics. In this research, urban bazaars and traditional markets with linkages to the Silk Road cultural heritage are identified, located over the selected study regions within the three countries. The bazaars identified are then mapped and classified in tables, and finally represented in graphs as nodes on the proposed polycentric bazaar functional networks with interregional and local impact.

---

[1] https://blacksea-cbc.net/programme/eligible-areas/.

6.1 Identification of Urban Bazaars Over the Study Areas

Some indicative examples of bazaars and traditional markets within the study area, with the above mentioned characteristics are presented below.

Bedesten of Thessaloniki, Region of Central Macedonia, Greece

Historically, the city of Thessaloniki has been an important hub on the major land and sea routes between East and West. The Bedesten covered market of Thessaloniki, located in the central urban core, was traditionally a point of reference for the commercial life of the city, the shopping center that defined the urban planning of the market area (Cezar, 1983; Falmpos, 1961), where imported and non-imported goods were safely stored and sold, mainly luxurious textiles and fabrics, precious stones, metals, gold, silver and jewelery (Kuran, 1968). Thessaloniki Bedesten, dates back to the "classical" era of Ottoman architecture and is a listed monument. Located in the heart of the old Ottoman market, which is also a listed historical area, Bedesten is a lively, popular market with small shops with jewellery, textiles, fabrics and local products, attracting locals and visitors.

Old Market of Komotini, Region of Eastern Macedonia and Thrace, Greece

The city of Komotini, an important regional hub, is seat of the Region of Eastern Macedonia and Thrace and a node of geopolitical importance between Greece, Bulgaria and Turkey. The Old Market of Komotini. located in the city centre, dates back to the Ottoman era, and retains the morphological and typological characteristics of oriental spatial patterns, designated as a historic place under special state protection, Among the main characteristics of the traditional market are the alleys and narrow streets, the one or two storey buildings and the small shops with local products, such as spices and herbs, nuts, textiles, jewellery, coffee, syrup sweets, traditional delicacies, antique stores, clothes and shoes, while also tin and copper smiths, tailors, knives and textile merchants.

Kapaliçarsi, Istanbul, Istanbul Subregion, Turkey

The Kapaliçarsi or Covered Bazaar or Grand Bazaar, is one of the main markets and tourism attractions of Istanbul. It is a labyrinth complex of two bedestens built in the fifteenth century, the Cevahir and Sandal Bedestens, that became the hub of Istanbul's commerce, where numerous stalls and shops were built and gradually roofed.[2] The streets of the covered bazaar are named after its artisans, i.e. silk-thread makers (kazazcilar).[3] The building complex, hit by many fires and earthquakes, consists of over 3000 shops and 61 streets including wells, fountains, mosques, cafeterias and restaurants. In the bazaar there are many traditional items and products, e.g. antiques, copperware, icons, mirrors, water pipes, clocks, old coins and jewellery, traditional sweets, spices and herbs.

---

[2]http://www.turkishculture.org/architecture/bazaar/the-covered-bazaar-98.htm.
[3]https://archnet.org/sites/3472.

Bedesten of Edirne, Tekirdağ Subregion, Turkey

The Bedesten is the oldest of Edirne historical bazaars dating back to 1418, built after the completion of the Old Mosque. In the Bedesten a variety of valuable goods were sold including gold, jewellery, weapons, and carpets. Caravans of traders from Persia arrived to Edirne to sell their merchandise and buy local products, and then continue their journey to the Balkans. European merchants who visited the Bedesten, returned home with beeswax and leather products, silk from Bursa and wool from Ereğli. The Edirne Bedesten now comprises 36 rooms for merchants located along two streets inside the building and is roofed by 14 domes and a row of shops around the external facade.

Central outdoor market, Varna, Severoiztochen (Northeastern) Region, Bulgaria

The outdoor market is located in the central part of Varna, between Dr. Piskyuliev, Drin and Angel Kanchev streets. It is the largest outdoor market in Varna, where the main local products traded are seafood products, meat, agricultural products, spices e.tc. There are also jewellery shops, accessories, flower shops, e.tc.

Sozopol Old Street Market, Yugoiztochen (Southeastern) Region, Bulgaria

Sozopol is the oldest town in Bulgaria, founded in 610 BC as a Hellenic colony.[4] The Old Town of Sozopol is an architectural reserve and the most popular tourist attraction on the Bulgarian Black Sea coast. It features narrow stone laid streets, with tiny sidewalks, churches, antique houses, of small hotels, galleries, restaurants, souvenir shops and street vendors and performers.[5] Among the main products sold in the Old Street market area are food products, paintings, candles, household items, pottery, metal and copper items. The majority of the shops are low-rise, wooden small shops, following the architectural typology and morphology of the Old Town.

6.2 Mapping of Urban Bazaars Over the Study Areas

See Map 2.

6.3 Classification of Urban Bazaars Over the Study Areas

See Table 1.

---

[4] https://www.europeanbestdestinations.com/destinations/sozopol/.

[5] https://www.myguidebulgaria.com/attractions/the-old-town-of-sozopol.

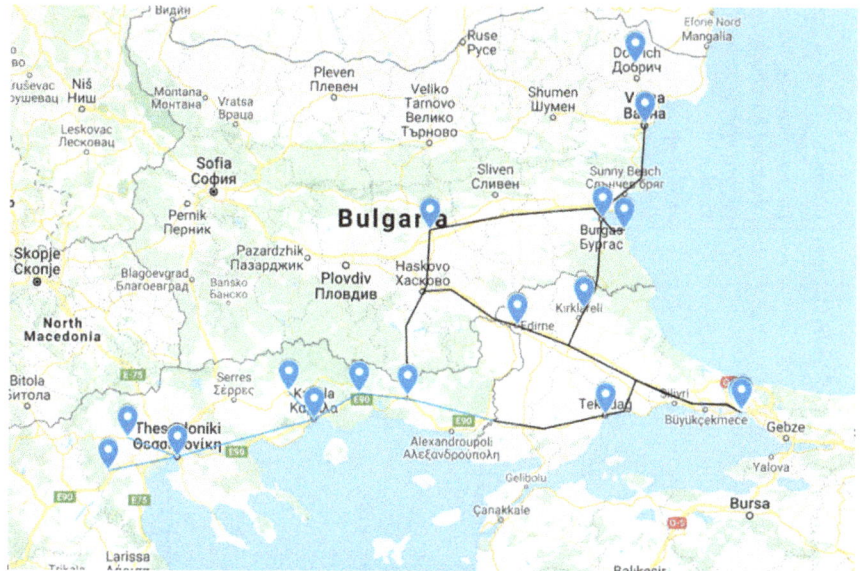

Map 2 Case study areas with bazaars/traditional markets connected to the silk road cultural heritage. Map background: Google Maps, authors' elaboration

6.4 Visualisation of Bazaars Polycentric Tourism Networking in the Study Area

In the following graphs, examples of potential polycentric bazaars' networks are presented. The bazaars identified are considered as the nodes of the potential polycentric tourism networks. Functional links between nodes are based on the existence of a main Silk Road product, as an example the silk textiles, featuring on regional level (Graphs 1, 2, 3) and copper items on interregional level (Graph 4).

6.5 Discussion

The case study research over the selected Black Sea areas identified a complex system of urban destinations that have the potential of connectivity through their bazaars as Silk Road cultural assets, for the development of functional relationships for cultural heritage tourism development. According to the theoretical analysis, the contribution of traditional local products connected to the Silk Road heritage could act as a lever to local development, since through their link to the culture, history, customs and traditions of the region, they signify a local identity. In conjunction with the preference of consumers for quality tourism products, they can boost the region's popularity and attract visitors.

Table 1 Classification of bazaars/traditional markets in the study areas. Authors' elaboration

Country	NUTS II	NUTS III	Municipality	Bazaar	Info/Site/Promotion	Products
GREECE	Central Macedonia (EL52)	Thessaloniki (EL.522)	Thessaloniki	Bedesten	X	fabrics, metals, gold, silver and jewellery, copper, spices, herbs, textiles
				Kapani market	X	workshops, coffee shops, herbs, spices, dry cardamom bread, household items, textiles, clothes, meat, fish, ecclesiastical objects, copper items
				Vatikioti-Athonos sq.	X	spices, tea, nuts, legumes, songbirds, baskets, packaging and shoe-making
				Modiano market	X	agricultural products, spices, nuts, herbs, meat, fish
				Nea agora		agricultural products, spices, nuts, herbs, meat, fish
				Vasileos Herakleiou-Fragkon str.		agricultural products, spices, nuts, herbs, meat, fish, household items, clothing, carpets, textiles, perfumes, baskets, metal or copper items

(continued)

Table 1 (continued)

Country	NUTS II	NUTS III	Municipality	Bazaar	Info/Site/Promotion	Products
				Bit bazaar	X	antiques, coins, jewellery, books, metal or copper items, household items
		Imathia (EL521)	Veria	Traditional market		Agricultural products, spices, herbs, sweets, household items
		Pella (EL524)	Yannitsa	Old (Ottoman) market		Agricultural products, spices, sweets
	Eastern Macedonia & Thrace (EL51)	Rhodope (EL513)	Komotini	Traditional old market	X	workshops, coffee, sweets, spices, herbs, nuts, legumes, jewellery, metal or copper items, household items
		Xanthi (EL512)	Xanthi	Traditional market		traditional sweets, spices, herbs, textiles
		Kavala (EL515)	Kavala	Traditional market		nuts, jewellery, traditional syrup sweets

(continued)

Table 1 (continued)

Country	NUTS II	NUTS III	Municipality	Bazaar	Info/Site/Promotion	Products
		Drama (EL514)	Drama	Traditional market	X	spices, herbs, cold cut meats (i.e. pastrami, sausages the recipes of which have roots from Cappadocia), legumes, beans, potatoes, tins and cooper utensils and objects, baskets e.tc., workshops
BULGARIA	Severoiztochen (Northeastern) (BG33)	Varna Province (BG331)	Varna	Central outdoor market	X	Food products, spices, herbs, nuts, textiles
		Dobrich Province (BG332)	Dobrich	Central market		Food products, spices, herbs, nuts
		Shumen Province (BG333)	Shumen	Shumen market		Food products nuts, spices
	Yugoiztochen (Southeastern) (BG34	Burgas Province (BG341)	Sozopol	Sozopol old street	X	Food products, spices, household items, copper items
		Sliven Province (BG342)	Silven	Silven market	X	Food products, spices

(continued)

Table 1 (continued)

Country	NUTS II	NUTS III	Municipality	Bazaar Marketplace	Info/Site/Promotion	Products
		Stara Zagora Province (BG344)	Stara Zagora		X	Food products, spices
TURKEY	Istanbul Subregion (TR10)	Istanbul Province (TR100)	Istanbul city centre	KapaliCarsi	X	textiles, jewellery, spices, herbs, clothing, shoes, carpets, metals, traditional sweets, spices, mirrors, copper, gold, silver
			Aminoo area	Egyptian Bazaar, spice bazaar or Mısır Çarşısı	X	Spices, herbs, nuts, textiles, traditional sweets
			Sultanahmet district	Arasta market		pottery, spices, textiles, carpets, jewellery
			Amenio district	Mahmud Pasha market	X	food, clothing, textiles
			Beyazit area	Journalists Market	X	Books, tobacco products
			Beyazit area	Tentmakers' market	X	Food products, spices, clothing

(continued)

Table 1 (continued)

Country	NUTS II	NUTS III	Municipality	Bazaar	Info/Site/Promotion	Products
			Beyazit area	Sahaflar Carsisi old book bazaar		Stationery, calligraphy materials, textbooks, novels and holy literature, tobacco-chewing gents peddle watches, badges, old coins, copper items
			Fatih area	Kadınlar Pazarı		local products, nuts, spices, herbs, food products
			Fatih area	Amasya Pazari		textiles, fabrics, jewellery
	Tekirdağ Subregion (TR21)	Edirne Province (TR212)	Saraçlar Street	Ali Pasha Bazaar		Textiles, jewellery, traditional sweets
			Edirne city centre	Bedesten	X	Jewellery, food products, spices, textiles, traditional sweets
			Edirne city centre	Selimiye Arasta		Food products, traditional sweets, textiles, clothing
		Kırklareli Province (TR213)	Hizirbey complex	Kirklaeri bazaar	X	Textiles, clothing
		Tekirdağ Province (TR211)	Çerkezköy district	Covered market place	X	Food products, clothing, nuts, textiles

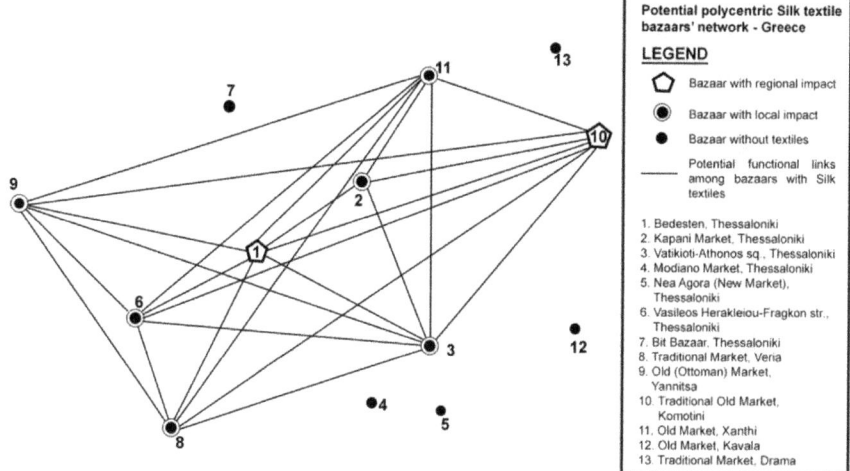

Graph 1 Potential polycentric silk textile bazaars' network in the NUTS II Region of Central Macedonia and Region of Eastern Macedonia and Thrace, Greece

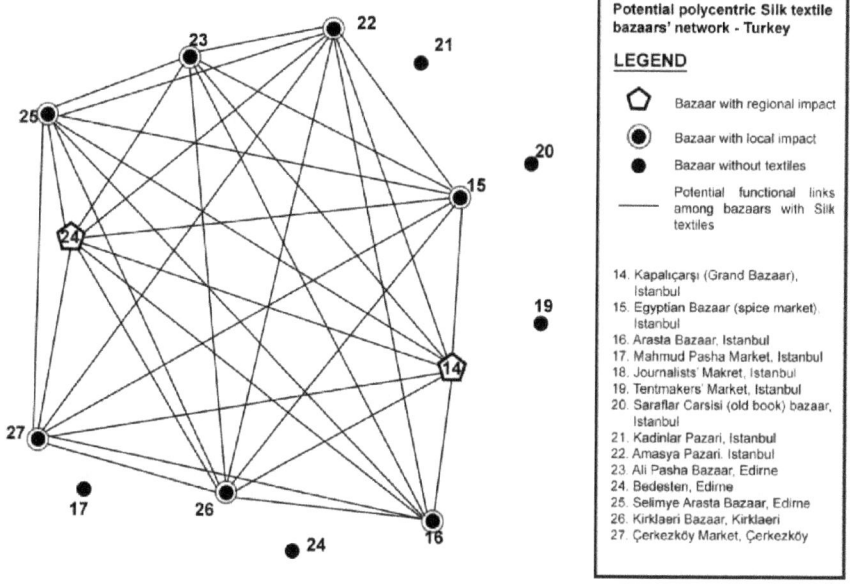

Graph 2 Potential polycentric silk textile bazaars' network in the NUTSII equivalent Istanbul and Tekirdağ sub-regions, Turkey

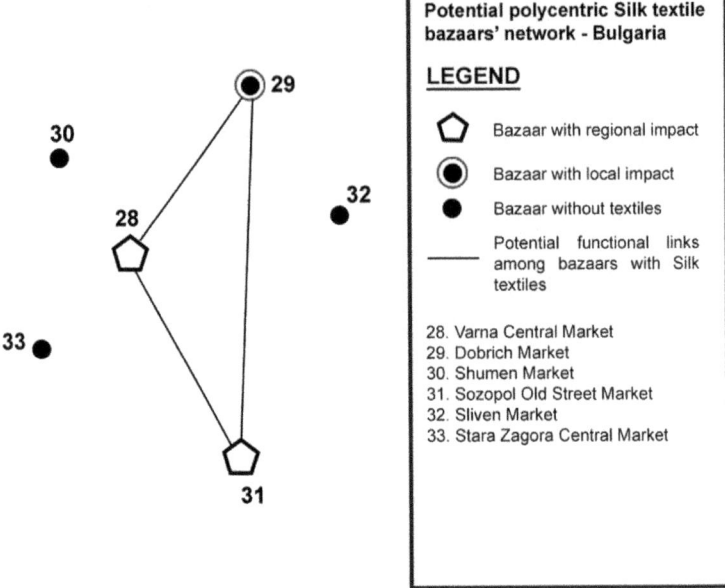

Graph 3 Potential polycentric silk textile bazaars' network in the NUTS II regions of Severoiztochen (Northeastern) and Yugoiztochen (Southeastern), Bulgaria

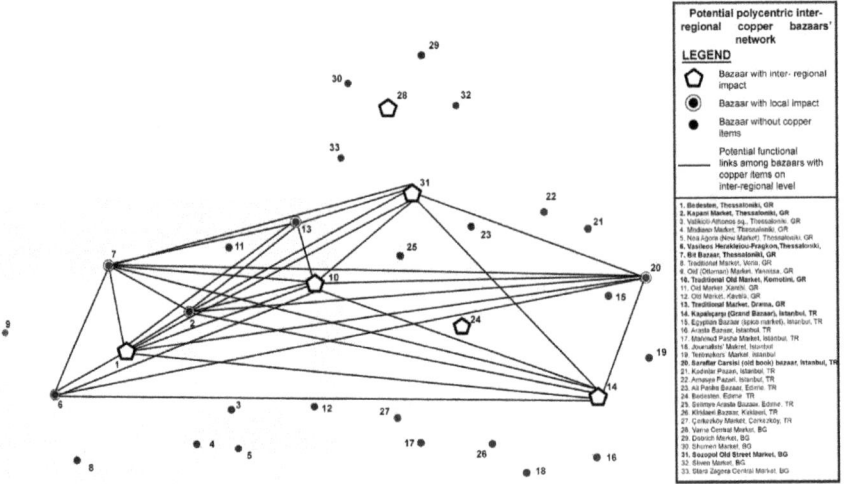

Graph 4 Potential polycentric inter-regional copper bazaars' network within the study areas

Research results demonstrate that the selected regions under study in Turkey and Greece are endowed with a significant list of Silk Road bazaars, while the list of Bulgaria is rather limited. However, polycentric networking perspectives emphasize the potential for cross-border cooperation on Silk Road cultural tourism development. More specifically, on intra-regional and regional level, the analysis has proven that there are many complementarities among the bazaars in terms of morphological Silk Road characteristics. Bazaars and traditional markets within each region/country present a significant plethora of traded products linked to Silk Road culture within the study regions namely, Thessaloniki Bedesten and Komotini Traditional Old Market in Greece, Istanbul KapaliCarsi and Edirne Bedesten in Turkey, Varna Central Market and Sozopol Old Town market in Bulgaria. These bazaars and traditional markets are classified as nodes of regional and inter-regional dominance due to their strong morphological connection to the Silk Road cultural footprint. They are thus considered as main nodes of the potential polycentric networks although they do not share the same list of traded products, and because of their impact in the wider area. The rest of the bazaars are classified as nodes with local impact and they can be integrated in various functional polycentric networks based on the traded product categories. In this aspect, the potential of local and regional polycentric networks is relatively high, increasing the opportunities for enhancing local bazaars as Silk Road tourism destinations and for entrepreneurial synergies in tourism and commerce sector.

On the inter-regional scale, research results provided evidence that there is strong differentiation among the bazaar characteristics. The bazaars in Greece and Turkey are highly related because of the relevance of Silk Road traded products and the common bazaar architecture typology. However, the degree of interdependence among the bazaars throughout the study area, appears to be rather limited in the case of Bulgarian bazaars and traditional markets, nevertheless functional relationships exist. The research displays bazaars with traded products linked with the Silk Road cultural heritage that can stimulate networking with other regions. Although these relations are not numerous, they are significant in terms of tourism development potential. Further research that would include including more regions and countries is expected to support the research's findings with regard to the polycentric inter-regional functional networking towards cross-border tourism development. Cross-border, inter-regional and/or transnational Silk Road bazaar networks over the study areas may prove to be typical examples to highlight the benefits of polycentric networking on the productivity and efficiency of tourism business activities within the networks. Furthermore, Silk Road bazaars' inter-regional networks could provide additional support especially for peripheral destinations, e.g. Silk Road bazaar circuit tourism packages designed also to reduce spatial tourism inequalities.

7 Conclusions

The present paper focuses on the cultural and tourism value of the Silk Road bazaars and traditional markets as Silk Road cultural assets in the Black Sea area. The main aim is to identify the bazaars connected to the Silk Road cultural heritage in the study area, and highlight their significance as tourism destinations of local, regional and inter-regional scale. The ultimate goal is to propose polycentric Silk Road bazaar morphological and functional networks and promote cross-border collaboration opportunities, through featuring new Silk Road tourism destinations and products.

Although bazaars are included in the tourism attractions lists as places to visit, there is limited research on their potential as Silk Road tourism products, offering an intriguing and rather unexplored field of academic study. However, Silk Road is a complex tourism system, on various spatial scales (international, national, inter-regional, regional, local) and thus, it is necessary to be systematically organized. In order to develop a common methodological Silk Road approach enabling country-specific, yet comparable Silk Road tourism analysis, the concepts of regionalism at the inter-regional level and polycentricity at the intra-regional and regional levels are implied, as a *modus operandi* for regional cooperation, integration and cohesion.

To achieve the formulated objectives, this paper assigns the following tasks: (i) delimitation of the study area; (ii) identification of the urban bazaars connected to the Silk Road cultural heritage within the study area; (iii) collection of information and data on their typological characteristics and traded products with connection to the Silk Road; (iv) classification of urban bazaars and creation of a Silk Road bazaar morphological inventory; (v) visualization of the potential polycentric bazaar networks based on the Silk Road cultural heritage.

One of the main difficulties of the research was the demarcation of the bazaars connected to the Silk Road, due to the differences on typological and morphological bazaar characteristics among countries and regions. Furthermore, the lack of available information in English about bazaars facilities and functions, as in most cases all the information was in the country's/region's native language, was a significant hindrance for the analysis.

The research analysis findings showcase the plethora of bazaars in the study area as Silk Road cultural tourism resources. These resources feature the potential of enhancing alternative tourism destinations with a unified 'Silk Road' brand name, promoting an authentic experience for tourists and share the way of life of local communities. Furthermore, results clearly show the polycentric networking tourism potential based upon the Silk Road bazaars through numerous opportunities for entrepreneurial synergies across the study area.

The development of collaboration synergies between bazaar and tourism entrepreneurs involved within the study areas is expected to have a major impact on reinforcing interregional cultural and economic relations. Furthermore, research has clearly spotted the opportunity for further developing tourism by using Silk Road bazaars as key tourist attractions for promoting cross-border synergies among local communities over the Black Sea Basin region.

References

Assari, A., Mahesh, T., Emtehani, M. R. & Assari, E. (2011). Comparative sustainability of bazaar in Iranian traditional cities: Case studies in Isfahan and Tabriz. *International Journal on Technical and Physical Problems of Engineering,3*(9), No. 4, 18–24.

Batala, L. K., Regmi, K., & Sharma, G. (2017). Cross border co-operation through tourism promotion and cultural exchange: A case study along Nepal and China (TAR) OBOR—prospective. *Open Journal of Business and Management, 5*(01), 105–118.

Brezzi, M., & Veneri, P. (2014). Assessing polycentric urban systems in the OECD: Country, regional and metropolitan perspectives. *European Planning Studies, 23*(6).

Burger, M. J., van der Knaap, B., & Wall, R. S. (2014). Polycentricity and the multiplexity of urban networks. *European Planning Studies, 22*(4), 816–840.

Burger, M., & Meijers, E. (2012). Form follows function? Linking morphological and functional polycentricity. *Urban Studies, 49*(5), 1127–1149.

Burgalassi, D. (2010). Defining and measuring polycentric regions. The Case of Tuscany, Discussion Papers del Dipartimento di Scienze Economiche - Università di Pisa, n. 101. Retrieved from http://wwwdse.ec.unipi.it/ricerca/discussion-papers.htm.

Cai, L. A. (2002). Cooperative branding for rural destinations. *Annals of Tourism Research, 29*(3), 720–742.

Camagni, R., & Capello, R. (2013). Toward smart innovation policies. *Growth and Change, 44,* 355–389. https://doi.org/10.1111/grow.12012.

Cezar, M. (1983). Typical commercial building of the Ottoman classical period and the Ottoman construction system, tr. by A. E. Uysal, İş Bankası Yayınları, İstanbul.

Chin, T. (2013, Autumn). The invention of the silk road, 1877. *Critical Inquiry, 40*(1), 194–219. Published by: The University of Chicago Press.

Chun, Z., & Bin, L. (2008). Cultural approach to planning of inner city regeneration. In *Cultural approach/Inner city regeneration, 44th ISOCARP Congress*, Dalian. Retrieved from http://www.isocarp.net/Data/case_studies/1245.pdf. (Accessed the 4th of August 2020).

Cowell, M. (2010). Polycentric regions: Comparing complementarity and institutional governance in the San Francisco Bay Area, the Randstad and Emilia-Romagna. *Urban Studies, 47*(5), 945–965.

Dai, X. (2016). Assessing the development of the silk road tourism in jiangsu using content analysis of network text. *International Journal of Simulation–Systems Science and Technology, 17*(40), 1–5.

Davoudi, S. (2008). Conceptions of the city-region: A critical review. *Proceedings of the Institution of Civil Engineers-Urban Design and Planning, 161*(2), 51–60. https://doi.org/10.1680/udap.2008.161.2.51.

Dembski, S. (2015). Structure and imagination of changing cities: Manchester, Liverpool and the spatial in-between. *Urban Studies, 52*(9), 1647–1664. https://doi.org/10.1177/0042098014539021.

Dergachev, V.A. (2008). Regional studies. Available from: http://www.dergachev.ru/book-7; Tolpakova, T., V., & Kuchinskaya, T., N. (2015). China's "New Regionalism" as a mechanism to strengthen the influence of china in the global integration processes: An example of EAEU cooperation. *International Journal of Economics and Financial Issues, 5*(Special Issue) 109–115, Eurojournals.

Dimadama, Z., & Chantzi, G. (2014). A new era for tourism in the black sea area. *International Journal of Cultural and Digital Tourism, 1*(2), Autumn 2014 Copyright © IACUDIT ISSN (Online): 2241–9705 ISSN (Print): 2241–973X, 23-29.

Douglas, S., Craig, S., & Nijssen, E. (2001). Integrating branding strategy across markets: Building international brand architecture. *Journal of International Marketing, 9*(2), 97–114.

EC. (1999). European spatial development perspective: Towards a balanced and sustainable development of the territory of the European Union. Luxembourg: Office for Official Publications of the European Communities

EC. (2011). European territorial cooperation–Building bridges between people. September 2011. Brussels: European Commission.

ESPON EGTC. (2016). Policy brief: Polycentric territorial structures and territorial cooperation. Retrieved from https://www.espon.eu/topics-policy/publications/policy-briefs/polycentric-territorial-structures-and-territorial.

ESPON. (2013). *GEOSPECS-European Perspective on Specific Types of Territories.*

Evers, D., & de Vries, J. (2013). Explaining governance in five mega-city regions: Rethinking the role of hierarchy and government. *European Planning Studies, 21*(4), 536–555. https://doi.org/10.1080/09654313.2012.722944.

Falmpos, F. K. (1961), Bedestens and hans in Smyrna (in Greek). *Mikrasiatika Chronika, 8,* 131–156.

Fawcett, L. (2005). Regionalism from a historical perspective. In M. Farrell Mary, B. Hettne & L. van Langenhove (Eds.), *Global politics of regionalism:Theory and practice* . London: Pluto Press.

Finka, M., & Kluvánková, T. (2015). Managing complexity of urban systems: A polycentric approach. *Land Use Policy, 42,* 602–608.

Gaman, G., Racasan, S. B., Potra, A. C. (2017, May). Tourism and polycentric development based on health and recreational tourism supply. *GeoJournal of Tourism and Geosites, 19*(1), 35–49.

Geist, J.F. (1989). Le Passage. Un type architectural du XIXe siècle. Ed. Mardaga, Liège 1989, 52–96. Original title: Johann F. Geist, Passagen, ein Bautyp des 19. Jahrhunderts, 1969, translated into English in 1982 and French in 1989.

Gilbert, D. (1989). Rural tourism and marketing: Synthesis and new ways of working. *Tourism Management, 10*(1), 39–50.

Greffe, X. (1994). Is rural tourism a lever for economic and social development? *Journal of Sustainable Tourism, 2,* 23–40.

Grimmel, A., Li, Y. (2018). The belt and road initiative: A hybrid model of regionalism, Working Papers on East Asian Studies, No. 122/2018, University of Duisburg-Essen, Institute of East Asian Studies (IN-EAST), Duisburg.

Hall, P., & Pain, K. (2006). *The polycentric metropolis: Learning from mega-city regions in Europe.* London: Earthscan.

Hettne, B., Inotai, A., & Sunkel, O. (Eds.). (1999). *Globalism and the new regionalism.* Basingstoke: Macmillan Press.

Hewings, G. J. D., & Parr, J. B. (2007). Spatial interdependence in a metropolitan setting. *Spatial Economic Analysis, 2*(1), 7–22.

Kolpakova, T. V., & Kuchinskaya, T. (2015). China's "new regionalism" as a mechanism to strengthen the influence of china in the global integration processes: An example of eurasian economic union. *International Journal of Economics and Financial Issues, 5,* 109–115.

Kostopoulou, S. (2019). Silk road cultural heritage tourism network. In *Proceedings 5th international conference of international association for silk road studies IASS SUN "Silk road: Connecting cultures, Languages, and Ideas,* September 26–28, 2019, pp. 232–260. Moscow, Russia.

Kostopoulou, S., Toufengopoulou, A., Kyriakou, D., Malisiova, S., Sofianou, E., & Xanthopoulou–Tsitsoni, V. (2016) The Western silk road in Greece. Aristotle University of Thessaloniki, Silk Road Programme 2016, Western Silk Road Tourism Initiative, A UNWTO-EU Initiative. Retrieved from https://webunwto.s3.eu-west-1.amazonaws.com/2019–09/aristotleuniversity-nationalswotanalysis-westernsilkroadgreece.pdf

Kour, P. (2016). Strategic destination branding of silk route for maximizing its tourism potential in India. PhD Thesis, University of Jammu, Faculty of Business Studies. Retrieved from https://shodhganga.inflibnet.ac.in/handle/10603/158170 (Accessed the 27th of July 2019).

Krajnović, A., Bosna, J., & Jašić, D. (2013). Umbrella branding in tourism–Model regions of istria and dalmatia. *Tourism and Hospitality Management, 19*(2), 201–215.

Kuran, A. (1968). The mosque in early Ottoman architecture. Publications of the Center for Middle Eastern Studies.

Kylasov, A. V. (2019). Traditional sports and games along the Silk Roads. *International Journal of Ethnosport and Traditional Games, 1*, 1–10. https://www.doi.org/10.34685/HI.2019.1.1.006.

Lambregts, B, & Kloosterman, R. (2006). Randstad Holland: Multiple faces of a polycentric role model. In P. Hall, & K. Pain (Ed.), *The polycentric metropolis, learning from mega-city regions in Europe* (pp. 137–145). London: Earthscan.

Lambregts, B. (2009). The polycentric metropolis unpacked: concepts, trends and policy in the Randstad Holland. Amsterdam: Amsterdam institute for Metropolitan and International Development Studies (AMIDSt). Retrieved from https://pure.uva.nl/ws/files/1066533/144701_06.pdf. (Accessed the 4th of August 2020).

Limtanakool, N., Dijst, M., & Schwanen, T. (2007). A theoretical framework and methodology for characterising national urban systems on the basis of flows of people: Empirical evidence for France and Germany. *Urban Studies, 44*(11), 2123–214. https://journals.sagepub.com/doi/pdf/10.1080/00420980701518990.

Long, M., & Xu, Q. (2017). The research on enhancing the competitiveness of tourism industry in Hubei Province under the background of one belt and one road, advances in social science. Education and Humanities Research (ASSEHR). In *2017 International conference on social science (ICoSS 2017)*, Vol. 117. Atlantis Press.

Manhas, P. S., Kour, P., & Bhagata, A. (2014). Silk route in the light of circuit tourism: An avenue of tourism internationalization. *Procedia—Social and Behavioral Sciences, 144*, 143–150. https://doi.org/10.1016/j.sbspro.2014.07.283.

Meijers, E. J., & Burger, M. J. (2010). Spatial structure and productivity in US metropolitan areas. *Environment and Planning A, 42*(6), 1383–1402.

Meijers, E. (2008). Measuring polycentricity and its promises. *European Planning Studies, 16*(9), 1313–1323.

Najdjavadipour, S. (2011). Using the concept and architectural components of a Bazaar as a means of creating architectural spaces that stimulate and awaken the senses. A Research Project submitted in partial fulfillment of the requirements for the degree of Master of Architecture Professional, Unitec Institute of Technology, ID: 1261626.

Nordregio. (2005). ESPON 1.1.1 Potentials for polycentric development in Europe. Retrieved from https://www.espon.eu/sites/default/files/attachments/fr-1.1.1_revised-full_0.pdf

Parr, J. (2005). Perspectives on the city-region. *Regional Studies, 39*(5), 555–566.

Parr, J. (2004). The polycentric urban region: A closer inspection. *Regional Studies, 38*(3), 231–240.

Porushani, I. (1995), Bazaar, Encyclopedia of the Islamic world.

Pourjafar, M., Samani, G., Pourjafar, A., & Hoorshenas, R. (2012). Archi-cultural parallel of persian and turkish bazaar along the silk road case studies: Rey, tabriz and istanbul bazaar. Archi-Cultural translations through the silk road. In *2nd International conference, Mukogawa Women's Univ.*, Nishinomiya, Japan, July 14–16, Proceedings.

Qoraboyev, I., & Moldashev, K. (2018). The belt and road initiative and comprehensive regionalism in central Asia, Chapter 7, 115–130. In M. Mayer (Ed.), *Rethinking the silk road*.

Rauhut, D. (2017). Polycentricity - one concept or many? *European Planning Studies, 4313*(January), 1–17. https://doi.org/10.1080/09654313.2016.1276157.

Riguelle, F., Thomas, I., & Verhetsel, A. (2007). Measuring urban polycentrism: A European case study and its implications. *Journal of Economic Geography, 7*(2), 193–215.

Ruland, J. (2001). ASEAN and the European union: A bumpy interregional relationship, Zentrum für Europäische Integrationsforschung Center for European Integration Studies, Rheinische Friedrich Wilhelms-Universität Bonn

Schmitt, P., Volgmann, K., Münter, A., & Reardon, M. (2015). Unpacking polycentricity at the city-regional scale: Insights from Dusseldorf and Stockholm. *European Journal of Spatial Development, 59*, 1–26.

Shopova, I. (2017). Regionalisation of tourism in Bulgaria–outcomes and implications. In *Jubilee international scientific conference Bulgaria of regions, proceedings*, 27–28 October 2017, pp. 368-376. Plovdiv, University of Agribusiness and Rural Development

Söderbaum, F. (2015). Early, old, new and comparative regionalism: The scholarly development of the field. KFG Working Paper Series, No. 64, October2015, Kolleg-Forschergruppe (KFG) "The Transformative Power of Europe", Freie Universität Berlin.

Söderbaum, F. (2016). In *Rethinking regionalism*, Palgrave Macmillan. ISBN: 9780230272415

Stanojević, N. (2016, January–March). The new silk road and Russian interests in Central Asia. *The Review of International Affairs RIA, LXVII*(1161), pp. 142–161. Belgrade

Tang, J., & Li, J. (2016). Spatial network of urban tourist flow in Xi'an based on microblog big data. *Journal of China Tourism Research, 12*(1), 5–23. https://doi.org/10.1080/19388160.2016.1165780.

Teye, V. B. (1988). Prospects for regional tourism cooperation in Africa, *Tourism Management*, September, pp. 221–234. Butterworth & Co (Publishers) Ltd.

Tolpakova, T. V., & Kuchinskaya, T. N. (2015). China's "New Regionalism" as a mechanism to strengthen the influence of china in the global integration processes: An example of EAEU cooperation, *International Journal of Economics and Financial Issues, 5*(Special Issue) 109–115, Eurojournals.

Vasiliev, I. A., & Shmigelskaia, N. A. (2016). The revival of the silk road: Brief overview of the 4th China-Eurasia Legal Forum, Vestnik of Saint Petersburg University. *Russian law survey, (2)*, 94–101, UDC 340.155.2(47). https://doi.org/10.21638/11701/spbu14.2016.209.

Waugh, D. C. (2007). Richthofen's "Silk Roads": Toward the archaeology of a concept. *The Silk Road, 5*(1), 1–10.

Winter, T. (2020). The geocultural heritage of the Silk Roads. *International Journal of Heritage Studies, 00*(00), 1–20. https://doi.org/10.1080/13527258.2020.1852296.

Wojciech H. (2006). Why is the silk road so important? *Silk Road Initiative Newsletter*, April. Retrieved from www.silkroad.undp.org.cn (Accessed the 16th of June 2019).

World Tourism Organization (UNWTO). (2017, June). Research on the potential of the western silk road highlights WP3. In *Western silk road tourism development. Grant agreement, enhancing the understanding of European tourism*. Co-funded by the COSME programme of the European Union.

World Tourism Organization (UNWTO). (2019). Tourism Barometer.

Yerolympos, A. (2007). Typologie et mutations des quartiers de commerce dans les villes de la Méditerranée Orientale. In M. Cerasi, A. Petruccioli, A. Sarro, & S. Weber (Eds.), *Multicultural urban fabric and types in the south and eastern Mediterranean* (pp. 241–264). Würzburg, Beirut: Egon Verlag.

You, Y. (2017). The classification of urban systems: A review from monocentric to polycentric, *42*(Isbcd), 1–4. https://doi.org/10.2991/isbcd-17.2017.

Zach, F., & Racherla, P. (2011). Assessing the value of collaborations in tourism networks: A case study of Elkhart County, Indiana. *Journal of Travel and Tourism Marketing, 28*(1), 97–110.

Żemła, M. (2016). Tourism destination: The networking approach. *Moravian Geographical Reports, 24*(4), 2–14.

https://www.chinadiscovery.com/china-silk-road-tours/maps.html

https://webunwto.s3.eu-west-1.amazonaws.com/s3fs-public/2019–12/tourism-in-the-bsec-region_0.pdf

https://blacksea-cbc.net/programme/eligible-areas

https://www.kapalicarsi.com.tr/

https://visit.varna.bg/en/objects.html/shop/261

http://thessaloniki4all.gr/places/ottoman-monument/bezesteni

https://archiscapes.wordpress.com/2015/02/23/typology-covered-market/

https://archnet.org/sites/3472

http://www.turkishculture.org/architecture/bazaar/the-covered-bazaar-98.htm

https://turkisharchaeonews.net/object/bedesten-edirne

https://www.myguidebulgaria.com/attractions/the-old-town-of-sozopol

https://www.europeanbestdestinations.com/destinations/sozopol/).

Lightning Source UK Ltd.
Milton Keynes UK
UKHW022210270622
405042UK00002B/27